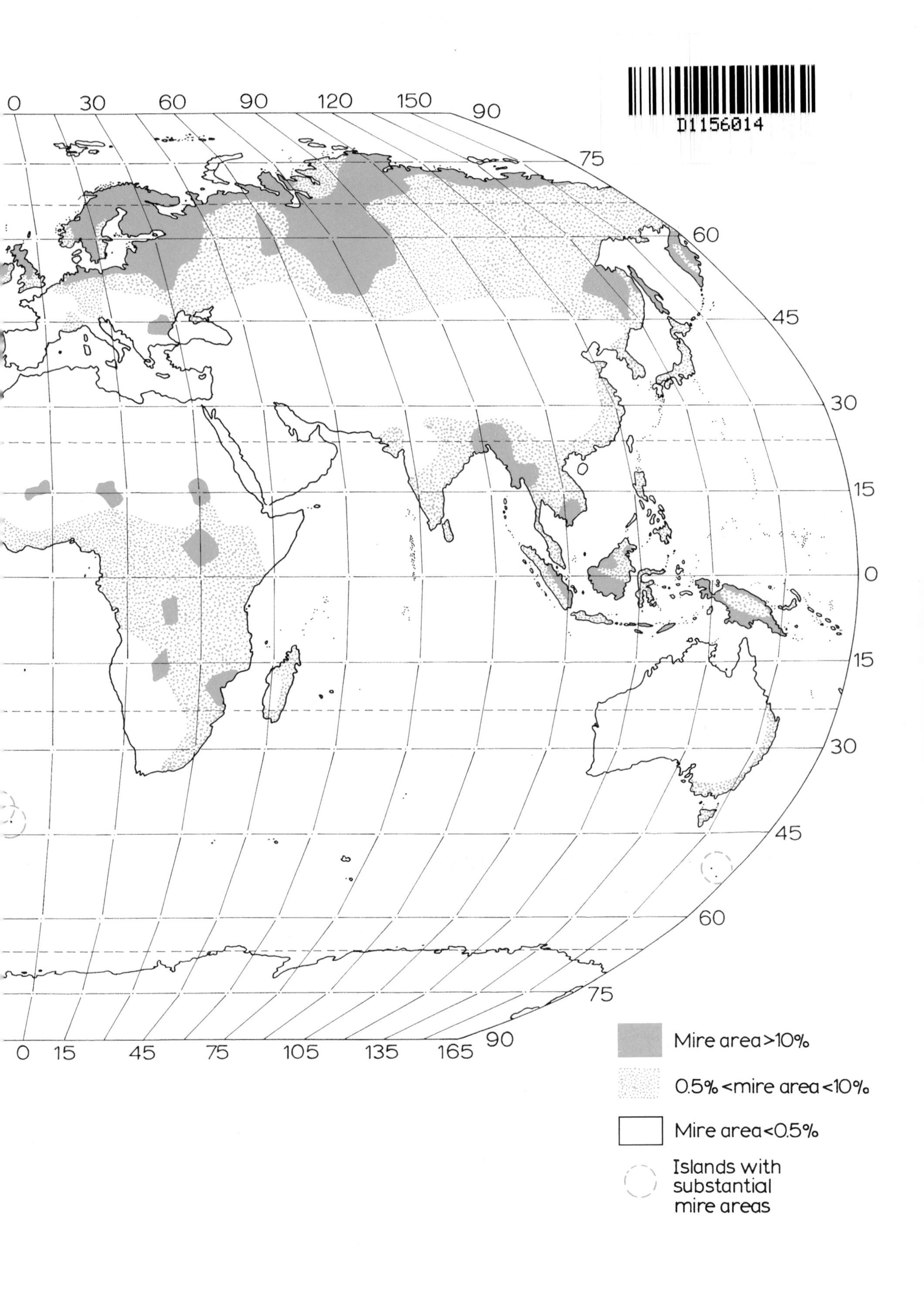
0
30
60
90
120
150
90
75
60
45
30
15
0
15
30
45
60
75
90
0
15
45
75
105
135
165
Mire area>10%
0.5% <mire area<10%
Mire area<0.5%
Islands with substantial mire areas

ECOSYSTEMS OF THE WORLD 4A

MIRES: SWAMP, BOG, FEN AND MOOR

GENERAL STUDIES

ECOSYSTEMS OF THE WORLD

Editor in Chief:

David W. Goodall

CSIRO, Midland, W.A. (Australia)

I. TERRESTRIAL ECOSYSTEMS

A. Natural Terrestrial Ecosystems

1. Wet Coastal Ecosystems
2. Dry Coastal Ecosystems
3. Polar and Alpine Tundra
4. Mires: Swamp, Bog, Fen and Moor
5. Temperate Deserts and Semi-Deserts
6. Coniferous Forests
7. Temperate Deciduous Forests
8. Natural Grasslands
9. Heathlands and Related Shrublands
10. Temperate Broad-Leaved Evergreen Forests
11. Mediterranean-Type Shrublands
12. Hot Deserts and Arid Shrublands
13. Tropical Savannas
14. Tropical Rain Forest Ecosystems
15. Wetland Forests
16. Ecosystems of Disturbed Ground

B. Managed Terrestrial Ecosystems

17. Managed Grasslands
18. Field Crop Ecosystems
19. Tree Crop Ecosystems
20. Greenhouse Ecosystems
21. Bioindustrial Ecosystems

II. AQUATIC ECOSYSTEMS

A. Inland Aquatic Ecosystems

22. River and Stream Ecosystems
23. Lakes and Reservoirs

B. Marine Ecosystems

24. Intertidal and Littoral Ecosystems
25. Coral Reefs
26. Estuaries and Enclosed Seas
27. Ecosystems of the Continental Shelves
28. Ecosystems of the Deep Ocean

C. Managed Aquatic Ecosystems

29. Managed Aquatic Ecosystems

III. UNDERGROUND ECOSYSTEMS

30. Cave Ecosystems

ECOSYSTEMS OF THE WORLD 4A

MIRES: SWAMP, BOG, FEN AND MOOR

GENERAL STUDIES

Edited by

A.J.P. Gore†

Institute of Terrestrial Ecology
Monks Wood Experimental Station
Abbots Ripton, Huntingdon (Great Britain)

ELSEVIER

Amsterdam – Lausanne – New York – Oxford – Shannon – Tokyo

ELSEVIER SCIENCE B.V.
Molenwerf 1
P.O. Box 211, 1000 AE Amsterdam, The Netherlands

First edition 1983
Second printing 1996

Library of Congress Cataloging in Publication Data
Main entry under title:

Mires--swamp, bog, fen, and moor.

(Ecosystems of the world ; 4A-)
Includes bibliographies and indexes.
Contents: A. General studies.
1. Wetland ecology. I. Gore, A. J. P. (Anthony John Poynter) II. Series: Ecosystems of the world ; 4A, etc.
QM541.5.M3M57 574.5'26325 81-15162
ISBN 0-444-42003-7 (vol. 4A) AACR2

ISBN 0-444-42003-7 (Vol 4A)
ISBN 0-444-41702-8 (Series)
ISBN 0-444-42005-3 (Set)

Printed in the Netherlands on acid-free paper.

PREFACE

This volume covers most types of ecosystems developing in non-saline wetlands, for which a variety of names have been used, the most general one being "mires". Most of these ecosystems are characterized by the accumulation of organic matter, which is produced and deposited at a greater rate than it is decomposed, leading to the formation of peat. Saline wetlands have largely been covered in the first volume of this series, dealing with salt marshes and mangrove swamps. Some of the ecosystems dealt with in this volume include trees, but a volume (No. 15) devoted specifically to forested wetlands will appear later.

After an introduction, the physical factors involved in mire formation are first discussed. The second chapter deals with climatic aspects, and the following one considers at length the hydrological aspects of mire formation and maintenance. The fourth chapter is concerned with peat — its structure, chemistry and formation.

The next seven chapters discuss some general aspects of mire ecosystems — micro-organisms and nutrient transformations in the substrate, root growth, primary production, succession, animals and secondary production. Part B of this volume will contain eleven chapters describing mire ecosystems in different geographical regions. The volume ends with a chapter describing human impact on one type of wetland ecosystem — the fens of England — and a glance into the future.

It was a tragedy that Mr. A.J.P. Gore (Tony Gore to his friends), the editor of this volume, died on November 2, 1979. He was at that time at the peak of his career; the heart attack to which he succumbed occurred without forewarning; and his death came as a real shock to his colleagues and friends throughout the world.

At the time of his death, Tony Gore had already prepared for press the majority of chapters in this volume. As it happened, I had visited him in England, as Editor in Chief of the series, during the previous week, and we had discussed a number of points together. This made it easier, when the tragic news reached me, to decide that the best course of action at that stage would be for me to take personal responsibility for the volume. We had, for instance, agreed that Tony would write a concluding chapter looking at the world-wide future of mire ecosystems; Tony had not put pen to paper on the subject, but (despite my lack of the necessary expertise) I decided to "stand-in" for Tony, and contribute a brief chapter to include the points Tony had specifically mentioned.

The volume as it appears, then, is conceptually Tony Gore's, and some three-quarters of it was actually prepared by him as editor. I have completed the task, and seen the volume through the press, as he would have done had he lived.

I would like to thank the authors for their help and patience in the difficult circumstances surrounding the change in editorship, and apologize for the delays in publication that many of them have suffered. I would also like to express my appreciation to Tony's widow, Mrs. Anne Gore, for her help in various ways.

In conclusion, I know that Tony would have wished to express his thanks to colleagues at Monks Wood Experimental Station, and in particular to his secretary Valerie Burton, who collaborated closely with him in the preparation of the manuscripts.

DAVID W. GOODALL
CSIRO Division of Land Resources Management
Wembley, W.A

LIST OF CONTRIBUTORS TO VOLUME 4

J.A.R. ANDERSON
Room 408, 4th Floor
Katong Shopping Centre
Singapore 18

ROBERT E. BLACKITH
Department of Zoology
Trinity College
Dublin 2 (Ireland)

MARINA S. BOTCH
Komarov Botanical Institute
2 Popova Street
197 022 Leningrad (U.S.S.R.)

I.K. BRADBURY
Department of Geography
University of Liverpool
P.O. Box 147
Liverpool L69 3BX (Great Britain)

ELLA O. CAMPBELL
Department of Botany and Zoology
Massey University
Palmerston North (New Zealand)

R.S. CLYMO
Department of Botany and Biochemistry
Westfield College
London NW3 7ST (Great Britain)

R.M.M. CRAWFORD
Department of Botany
University of St Andrews
St Andrews KY16 9AL (Great Britain)

COLIN H. DICKINSON
Department of Plant Biology
University of Newcastle upon Tyne
Newcastle upon Tyne NE1 7RU (Great Britain)

BURKHARD FRENZEL
Universität Hohenheim (02100)
Institut für Botanik
Postfach 106
7000 Stuttgart 70 (Germany G.F.R.)

DAVID W. GOODALL
CSIRO Division of Land Resources Management
Private Bag, P.O.
Wembley, W.A. 6014 (Australia)

A.J.P. GORE†
3 Kettering Road
Stamford, Lincs PE9 2LR (Great Britain)

J. GRACE
Department of Forestry and Natural Resources
University of Edinburgh
Mayfield Road
Edinburgh EH9 3JU (Great Britain)

ALAN C. HAMILTON
Department of Environmental Sciences
New University of Ulster
Coleraine, Co. Londonderry (Northern Ireland, U.K.)

RONALD H. HOFSTETTER
Department of Biology
P.O. Box 249118
University of Miami
Coral Gables, Fla. 33124 (U.S.A.)

H.A.P. INGRAM
Department of Biological Sciences
University of Dundee
Dundee DD1 4HN (Great Britain)

WOLFGANG J. JUNK
I.N.P.A.
C.P. 478
BR 69 000 Manaus AM (Brazil)

D.R. KEENEY
Department of Soil Sciences
University of Wisconsin
Madison, Wisc. (U.S.A.)

VIKTOR V. MASING
Plant Taxonomy and Ecology
Tartu State University
40 Michurin Street
202 400 Tartu (U.S.S.R.)

C.F. MASON
Department of Biology
University of Essex
Wivenhoe Park
Colchester CO4 3SQ (Great Britain)

EDMUNDO PISANO
Instituto de la Patagonia
Casilla 102-D
Punta Arenas, Magallanes (Chile)

F.C. POLLETT
Newfoundland Forest Research Centre
P.O. Box 6028
St Johns, Nfld. A1C 5X8 (Canada)

RAUNO RUUHIJÄRVI
Department of Botany
University of Helsinki
Unioninkatu 44
Helsinki SF 0010 (Finland)

JOHN SHEAIL
Institute of Terrestrial Ecology
Monks Wood Experimental Station
Abbots Ripton
Huntingdon, Cambs PE17 2LS (Great Britain)

L.J. SIKORA
Biological Waste Management and Organic Resources Laboratory
A.E.Q.I.
U.S.D.A.–S.E.A.–A.R
Beltsville, Md. 20705 (U.S.A.)

HUGO SJÖRS
Växtbiologiska Institutionen
Uppsala Universitet
Box 559
751 22 Uppsala 1 (Sweden)

MARTIN C.P. SPEIGHT
Forest and Wildlife Service, Research Branch
Sidmonton Place
Bray, Co. Wicklow (Ireland)

VALERIE STANDEN
Department of Zoology
University of Durham
South Road
Durham DH1 3LE (Great Britain)

J.H. TALLIS
Department of Botany
University of Manchester
Manchester M13 9PL (Great Britain)

J.A. TAYLOR
Department of Geography
University College of Wales
Llandinam Building
Penglais
Aberystwyth, Dyfed SY23 3DB (Great Britain)

KEITH THOMPSON
Department of Biological Sciences
University of Waikato
Private Bag
Hamilton (New Zealand)

CLAUDIO VENEGAS C.
Instituto de la Patagonia
Casilla 102-D
Punta Arenas, Magallanes (Chile)

T.C.E. WELLS
Institute of Terrestrial Ecology
Monks Wood Experimental Station
Abbots Ripton
Huntingdon, Cambs PE17 2LS (Great Britain)

S.C. ZOLTAI
Canadian Forestry Service
Northern Forest Research Centre
Edmonton, Alta. T6H 3S5 (Canada)

CONTENTS OF VOLUME 4A[1]

PREFACE V

LIST OF CONTRIBUTORS TO VOLUME 4 . . . VII

Chapter 1. INTRODUCTION
by A.J.P. Gore† 1

The volume title. 1
Purpose and scope 1
Mire classification 2
Mechanisms and models 12
Conclusions 18
Appendix I: Brief notes and references to other countries with important mire areas 18
Appendix II: Definitions of terms commonly used in mire studies 26
References. 30

Chapter 2. MIRES — REPOSITORIES OF CLIMATIC INFORMATION OR SELF-PERPETUATING ECOSYSTEMS?
by B. Frenzel 35

Introduction 35
Main requirements for valid interpretation. 35
A short survey of postglacial climate history 36
Palaeoecological sources of information. 40
The developing mire — self-perpetuating mechanisms versus climate-controlled processes. 44
Conclusion 56
References. 59

Chapter 3. HYDROLOGY
by H.A.P. Ingram 67

Introduction 67
The hydrological cycle and the water balance 69
The water input to mires: recharge 70
Evaporation 73
Water storage 99
Liquid water discharge 118
Mire water relations and theoretical telmatology. . . 145
Acknowledgements. 150
References. 150

Chapter 4. PEAT
by R.S. Clymo 159

Peat structure 159
Peat chemistry 172
Accumulation of peat 196
Acknowledgements. 217
Appendix I: Outline of a commercial classification of peats 217
Appendix II: Treatment of organic soils in the U.S.D.A. 7th Approximation soil classification 218
Appendix III: Outline of a geological/ecological classification of biogenic sediments 218
References. 218

Chapter 5. MICRO-ORGANISMS IN PEATLANDS
by C.H. Dickinson. 225

Introduction 225
Nutrient input to mire systems. 226
Nutrient uptake by higher plants 238
Nutrient transformations within peat. 241
General summary 243
References. 243

Chapter 6. FURTHER ASPECTS OF SOIL CHEMISTRY UNDER ANAEROBIC CONDITIONS
by L.J. Sikora and D.R. Keeney 247

Introduction 247
Redox potential. 247
Hydrogen ion concentration. 248
Ion exchange and complex formation 249
Microbial ecology 250
Carbon transformations 250
Nitrogen transformations 251
Phosphorus 252
Sulfur 253
Toxic substances produced in waterlogged soils . . . 254
Iron and manganese 254
Summary 254
References. 255

Chapter 7. ROOT SURVIVAL IN FLOODED SOILS
by R.M.M. Crawford. 257

Introduction 257
Swiftness of death 260
Plant morphology in relation to flooding tolerance . . 262
Metabolic adaptations to flooding. 271
The evolution of flooding tolerance 277
References. 280

[1]For short contents of Volume 4B, see p. XI.

Chapter 8. PRIMARY PRODUCTION IN WETLANDS
by I.K. Bradbury and J. Grace 285

Introduction 285
Methods for estimating productivity 285
Productivity estimates in wetlands. 287
Maximum productivity 292
Photosynthetic performance. 295
Water relations 298
Characteristics of stands 301
Conclusions 305
References. 306

Chapter 9. CHANGES IN WETLAND COMMUNITIES
by J.H. Tallis 311

Introduction 311
The parameters of wetland communities 311
Factors bringing about change in wetland communities 313
Documented patterns of change in wetland communities 322
Rates of change deduced from stratigraphic data . . 331
The ecological requirements of hydroseral communities 332
Discussion. 341
References. 344

Chapter 10. THE ANIMALS
by M.C.D. Speight and R.E. Blackith. . 349

Introduction 349
Survival in wetlands 349
Trophic characteristics of wetland faunas 353
The cost of dispersion from wetlands. 358
Zoogeographical considerations 359
Is there a generalized animal assemblage of accumulative wetlands? 362
References. 363

Chapter 11. ASPECTS OF SECONDARY PRODUCTION
by C.F. Mason and V. Standen 367

Introduction 367
Swamp and fen 367
Bogs and oligotrophic mires. 374
Discussion. 378
References. 380

SYSTEMATIC LIST OF GENERA. 383

AUTHOR INDEX. 387

SYSTEMATIC INDEX 397

GENERAL INDEX 409

CONTENTS OF VOLUME 4B

Chapter 1. THE PEATLANDS OF GREAT BRITAIN AND IRELAND
by J.A. Taylor

Chapter 2. THE FINNISH MIRE TYPES AND THEIR REGIONAL DISTRIBUTION
by R. Ruuhijärvi

Chapter 3. MIRES OF SWEDEN
by H. Sjörs

Chapter 4. MIRE ECOSYSTEMS IN THE U.S.S.R.
by M.S. Botch and V.V. Masing

Chapter 5. MIRES OF AUSTRALASIA
by E.O. Campbell

Chapter 6. THE TROPICAL PEAT SWAMPS OF WESTERN MALESIA
by J.A.R. Anderson

Chapter 7. WETLANDS IN THE UNITED STATES
by R.H. Hofstetter

Chapter 8. WETLANDS IN CANADA: THEIR CLASSIFICATION, DISTRIBUTION, AND USE
by S.C. Zoltai and F.C. Pollett

Chapter 9. ECOLOGY OF SWAMPS ON THE MIDDLE AMAZON
by W.J. Junk

Chapter 10. THE MAGELLANIC TUNDRA COMPLEX
by E. Pisano

Chapter 11. PEATLANDS AND SWAMPS OF THE AFRICAN CONTINENT
by K. Thompson and A.C. Hamilton

Chapter 12. THE FENLANDS OF HUNTINGDONSHIRE, ENGLAND: A CASE STUDY IN CATASTROPHIC CHANGE
by J. Sheail and T.C.E. Wells

Chapter 13. CONCLUSIONS — THE FUTURE OF MIRES
by D.W. Goodall

SYSTEMATIC LIST OF GENERA
AUTHOR INDEX
SYSTEMATIC INDEX
GENERAL INDEX

Chapter 1

INTRODUCTION[1]

A.J.P. GORE†

THE VOLUME TITLE

The term **mire** includes all those ecosystems described in English usage by such words as **swamp**, **bog**, **fen**, **moor**, **muskeg** and **peatland**. These more specific words are used in a variety of contexts throughout the volume but, in setting the stage, the use of the generic word **mire** should help to avoid confusion in readers to whom swamp, etc., may mean different things.

PURPOSE AND SCOPE

Apart from some limited excursions, as for example those referred to by Gorham (1957, p. 146), into concepts implying that mires are living organisms in themselves, the primary functional influences governing the initiation and formation of mires have been known, or guessed at correctly, for a long time. The relationships between these primary influences are much less well understood. It is perhaps the particular feature of mires that, in their natural state, they have an autoregulatory capacity that makes them especially interesting among all terrestrial ecosystems. The purpose of the first part of this volume is to describe the primary functional factors of mires and to provide as full an account of their interactions as is now possible. The other major aim is to draw on examples of mire types from different parts of the world to illustrate as much as possible of the actual manifestation of the influencing factors and their interactions.

The primary influences and their functional relationships

Part A of this volume examines the main primary variables such as climate, water, soils, plants and animals. Included here are the related topics of primary (plant) productivity and secondary (animal) productivity together with aspects of decomposition and recycling, microbiological and chemical, and related processes under aerobic and anaerobic conditions. As with all ecological writing there is an acute need to present issues interactively, and it is this need which emphasizes the potential value of dynamic models. It is fortunate that, unlike many ecological systems, mires lend themselves especially well to the development of such models, and some of these are discussed later in this chapter. Among terrestrial ecosystems, mires have the advantage of being recognizably discrete [see, for instance, Gorham (1957, p. 163) and Ingram (1978, p. 1053)] and therefore amenable to effective comparisons involving related formations in all parts of the world.

The world's examples

Part B of this volume consists of contributions from authors describing a wide range of mires in many parts of the world. National boundaries

[1]This chapter was left unfinished by Tony Gore at the time of his death. He had not, for instance, compiled the list of references, and some of the papers he mentioned could not afterwards be identified. However, I have had the invaluable assistance of Professor Hugo Sjörs, who both offered a careful and detailed editing of the text with regard to subject matter, and contributed a substantial proportion of the items missing from the list of references. The opportunity was also taken to add references to some recent papers and others which Tony Gore would doubtless have wished to mention had he known of them.
David W. Goodall.

rather than natural ones have often determined the individual areas described, but as far as possible these national fragments have been selected to represent the world distribution of mires. In Appendix I to this introductory chapter, some material has been included to describe mires from those parts of the world not covered by individual chapters. Fig. 1.1 shows the general distribution of mires on a world scale.

Seven chapters deal with different parts of the Arctic, Boreal and North Temperate Zones, with their many opportunities for mire formation. By contrast, the Southern Hemisphere has so little land at latitudes inherently suitable for the formation of mires that two chapters, on southern Chile and on Australasia, provide a fair representation. One chapter is devoted to the description of truly tropical mires (west Malesia), and another to Africa. A third chapter on the tropics deals with riverine swamps in the central Amazon. This latter is not a mire ecosystem within the ordinary definition of "peat-forming" or "accumulative", but it demonstrates the effect of seasonally wet and dry conditions commonly found in swamps of tropical areas. Swamps having this meaning are otherwise not representatively included in this volume, although some similar conditions are described in the chapters on Africa and Australasia.

There are places where the subject matter overlaps with material contributed in one or even several other chapters, but where this has occurred an alternative aspect of the common material is often described. For example, Sjörs and Ruuhijärvi in their respective accounts of the mires of Sweden and Finland each deal with the detailed aspects quite differently. The Swedish chapter contains much floristic information about the micropatterns in mires, while the Finnish chapter examines the broader issues of the mire landscape. Furthermore, the very wide range of coverage in different chapters presents a number of problems for consistency of presentation. Chapters such as those on Finland and Sweden deal with the subject matter in considerable detail. Others which have to include enormous regions of the earth, such as Canada and the U.S.S.R., can only provide equivalent detail for small areas. The problem of classification and communication experienced by authors of such extensive chapters reflect those of the book as a whole. At a lower level of communication are the terms in common usage, and a glossary of these is provided as Appendix II to this introductory chapter. By use of this glossary, possibly via the General Index, and by the system of cross-referencing, it is the aim to improve the relating of ideas between different parts of the book.

Influence of man and conservation

The penultimate chapter of the volume (Vol. B, Ch. 12) deals with a single case study of drastic change. It is not necessarily an account widely representative of human intervention in mires, but is concerned with part of a large area of eastern England which has been transformed over the past 200 years from a complex of mire systems into a region of intensive agriculture. Only a few small fragments of wet peatland are precariously preserved. Examples of a similar kind are common among the regional descriptions, and the quest for extensions of agriculture and forestry, for energy (peat fuel), for use in horticulture, for soil conditioning, and recently for the manufacture of activated carbon, raises important questions about the continued existence of many types of this particular type of ecosystem (see Goodwillie, 1980). It is notable that there is now increasing interest, in Europe at least (e.g. Tüxen, 1976), in restoring exploited areas to mire-forming conditions where possible.

MIRE CLASSIFICATION

The fundamental subdivision of mires adopted in this volume is into **fen** and **bog**. Even at the widest scale of enquiry it appears that these two words can adequately subdivide all **mires. Swamp** has a very wide range of meanings in international usage, hence its pride of place in the volume title; it can include not only mires (peat-forming) but **marsh** and **wetland** as well. The use of **swamp** here is generally but not exclusively restricted to the sense of "mires". In North America, for example, **swamps** are usually wooded or forested mires (Heinselman, 1963; Zoltai and Pollett, Vol. B, Ch. 8). **Marshes** (non-peat forming; seasonally waterlogged areas) are not systematically included in this volume. **Wetland** is another general term not very different from **swamp** in its coverage. Further discussion of these terms will be found in the Glossary (Appendix II, pp. 26–30).

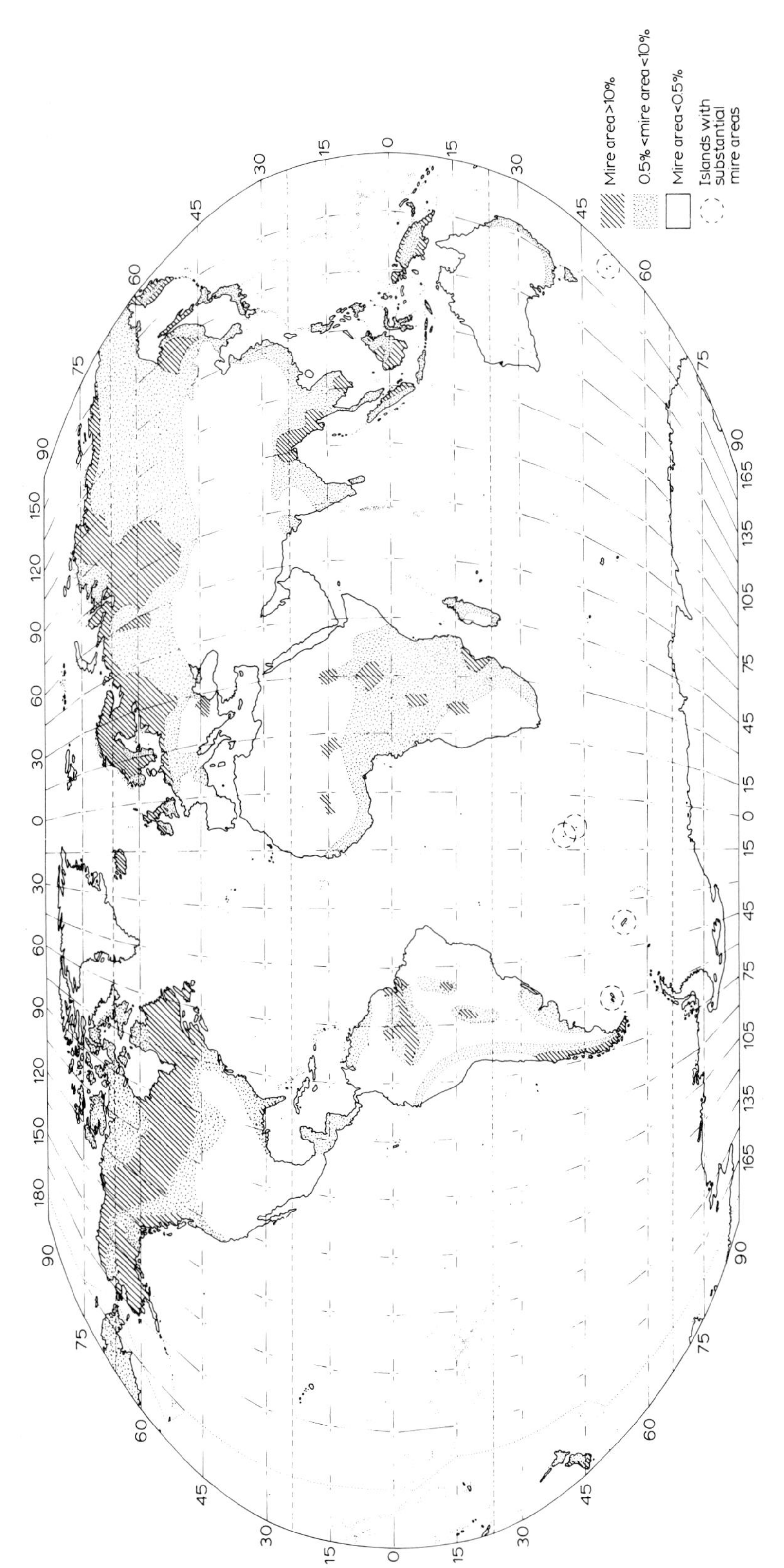

Fig. 1.1. World-wide distribution of mires (from various sources, and approximate only; detailed information is available only for rather limited areas).

Thus, while **fen** and **bog** distinguish the essentials, more elaborate classifications have been constructed and these are reviewed briefly below. The meaning of words in languages other than English is not easily resolved, although excellent technical dictionaries dealing specifically with mires and peat have been prepared, notably by Masing (1960) (German, Estonian, Russian, English, Swedish and Finnish), Bick et al. (1976) (German, Polish, English and Russian) and Heikurainen (1977) (English, German, Russian, Swedish and Finnish — mainly terms referring to drainage). The glossary in Appendix II also explains some of the words more important in mire ecology, and Clymo (Ch. 4) has reviewed ten classification systems for peat, again mainly from an ecological point of view. However, the main contributions to meaningful understanding in this sense come from authors in this volume whose native language is not English.

Classifications group like with like on the basis of one or more criteria. In such large-scale features as mires the criteria have to be fairly easily accessible, and shape is an obvious one. Chemistry, especially pH, is suitable at a more detailed level of classification, and plant community associations are of value at a similar scale. Examples of each of these are summarized below to provide a basis of reference in the chapters which follow. For example, in Chapter 3 of Part B Sjörs gives a diagram showing the distribution of mire types in northwest Europe. There are many others in the literature (see Fig. 1.4 below; Kats, 1971, fig. 14; and Overbeck, 1975, fig. 51).

Classifications based on shape

The **bog** and **fen** terminology is broadly equivalent to *Hochmoor* and *Niedermoor* respectively, of the German and Fennoscandian schools. The wide usage of these German words, or equivalents, in mire classifications throughout the world reflects the historical lead of northwest continental Europe in peatland investigations. Elsewhere in Europe similar ideas developed; Gorham (1953), for example, quotes Gerard Boate as having classified several types of Irish bog in 1652, and there is much evidence that the patterns of vegetation change which accompany peat development have been appreciated for a long time, certainly not exclusively by German-writing people. It is nevertheless to them that the main stream of ideas on mire classification is due, although, at the turn of the century and later, Fennoscandian authors (R. Sernander, A.K. Cajander, L. von Post, H. Osvald and many others) made important contributions to the development of mire science.

Osvald (1925b) classified European raised bogs as:

(1) Continental raised bogs (*Waldhochmoore*).

(2) Baltic raised bogs (*eigentliche Hochmoore* or *echte Hochmoore*).

(3) Atlantic raised bogs (*Flachhochmoore* or *Planhochmoore*).

(4) Upland raised bogs (*terrainbedeckende Moore*).

The raised-bog type which is often quoted as the ideal (cf. Weber, 1902) is the Baltic raised bog type 2, having the form shown in Fig. 1.2. Continental raised bogs or *Waldhochmoore* are more wooded and tree-covered versions of type 2, and the Atlantic raised bogs were supposed to have a plateau-like surface instead of being domed. Ruuhijärvi (Part B, Figs. 2.4A and B) gives an illustration of the theoretical distinction in shape, but there are not many adequately surveyed mires reported in the literature to provide a clear distinction between types 2 and 3. In any case Kulczyński's example (Fig. 1.2) lacks a definite **rand** (steep marginal bank), and it is not clear whether this is due to the choice of a particular mire section for survey or to a genuine difference of profile shape between the Baltic raised bogs of Poles'ye (U.S.S.R.) and these elsewhere. Fig. 1.3, taken from Granlund (1932), gives a more detailed example of a raised-bog shape, in this case the Fastebo mosse in southeastern Sweden. Upland raised bogs, type 4, are misnamed since they are not really raised bogs at all, but were included by Osvald (1925b) in his classification in an attempt to account for their evident vegetational similarity to raised bogs. The essential difference is that a raised bog proper is usually limited in extent and definable, for most if not all its perimeter, by fairly easily recognizable boundaries (the **rand**). Blanket bog is rarely confined in such a way. Taylor (Part B, Ch. 1) describes many aspects of blanket bog, literally *terrainbedeckende Moore*.

Walter (1968, p. 497) gave a map showing broadly the distinctions of different types of *Hochmoore* in Europe. Some of his zones have been subdivided

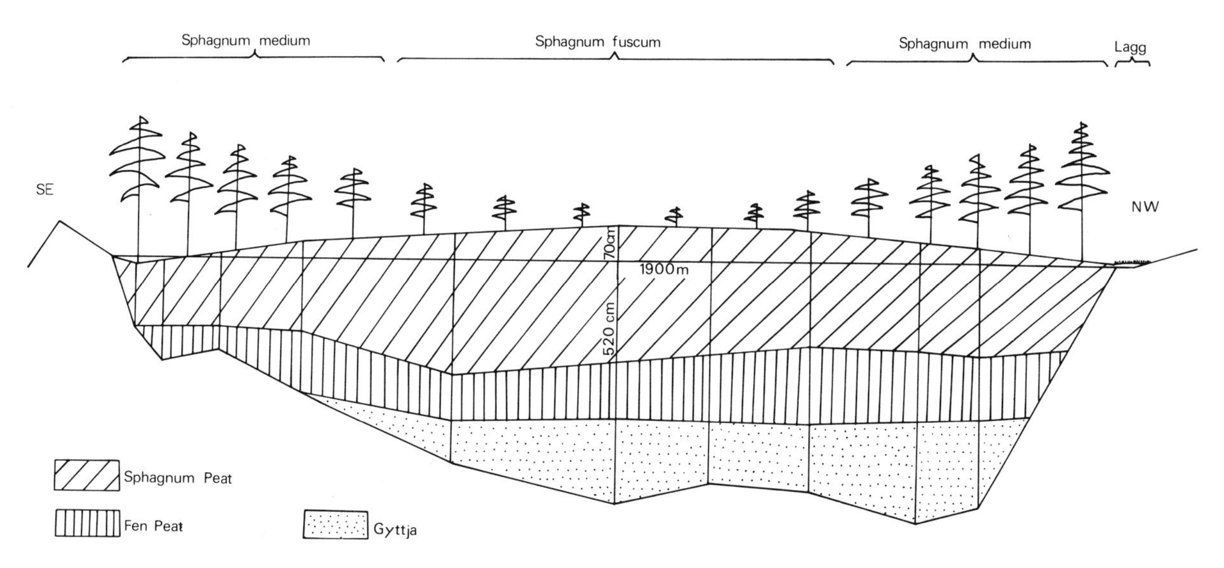

Fig. 1.2. Form of Baltic raised bogs (after Kulczyński, 1949, fig. 26).

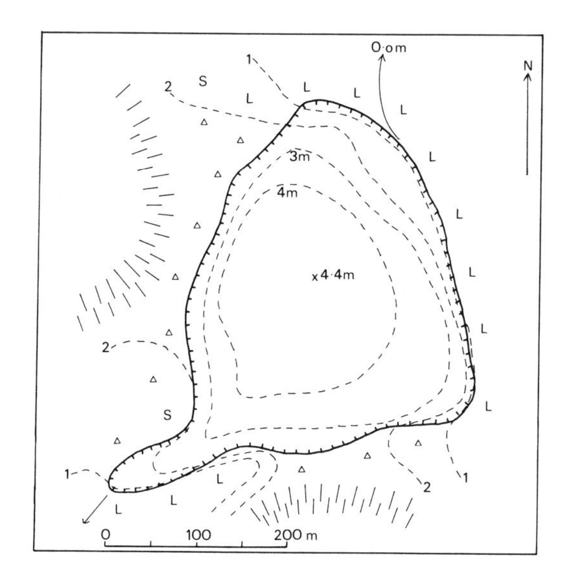

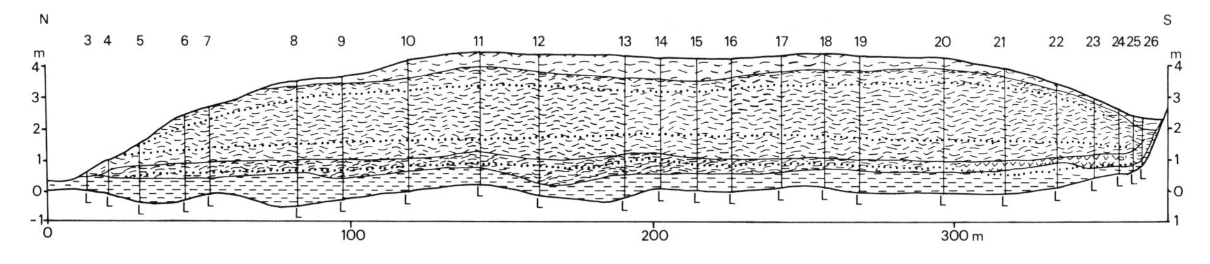

Fig. 1.3. Plan and section of the Fastebo mosse, southern Sweden (after Granlund, 1932).

by Moore and Bellamy (1974, pp. 13 and 14); the results are shown in Fig. 1.4. These authors had the advantage of later work, especially in Finland and Sweden, which in turn benefited from aerial photography. The resulting improved appreciation of mire shape helped to refine Osvald's (1925b) classification. For example, the terms "concentric" and "eccentric" (Sjörs, 1948) refer to general topography, often revealed in the disposition of pools or hollows on the bog surface which can most easily be appreciated from the air. Some examples are illustrated in Part B (e.g., pp. 18 and 40). The terms "primary", "secondary" and "tertiary" used by Moore and Bellamy (1974) will not be developed here since they are essentially references to peat quality, not mire shape. However, their ideas are a reminder that classifications of mires based on shape must include a reference to fens. ***Aapa*** mires are the most easily distinguished "fen type" in terms of shape, but in many so-called "string mires", sections through the long axes of such mires show that they are essentially **geogenous** (see next section). Ruuhijärvi (Part B, Figs. 2.4C, G and H) and Sjörs (Part B, Figs. 3.4B, C and D) illustrate the point.

Table 1.1 indicates some correspondences between the zones recognized by Moore and Bellamy (Fig. 1.4) and those of Walter (1968). Zones *1*, *2*, *7b* and *9* in Fig. 1.4 (excluded from Table 1.1) are not so easily identified or related to other classifications. Zone *1* consists essentially of areas with only lowland swamps and fens of very localized distribution. These types occur throughout the other zones. A better idea of the distribution and local occurrence of mires in Zones *1* and *2* can be gained from Part B, Chapter 4 by Masing and Botch (see Part B, pp. 106–111, especially).

Zone *7b* is of special interest to the British Isles since it refers to a type of raised bog which occurs mainly in lowland areas of western Britain. In his Appendix of examples (Part B, pp. 24–41), Taylor shows vertical sections of a few such bogs, which illustrate the idea of "ridge raised bogs" of Zone *7b*. Two entirely different kinds of ridge-shaped raised bogs occur in North America: in Minnesota (forest-covered), and in the Hudson Bay Lowlands (in interfluves) (Sjörs, 1963). Kulczyński (1949, pp. 75 and 131) appreciated the importance of vast flat interfluves — very low watersheds and river partings — in providing the basis for "raising" the bog surface which often have all the appearance

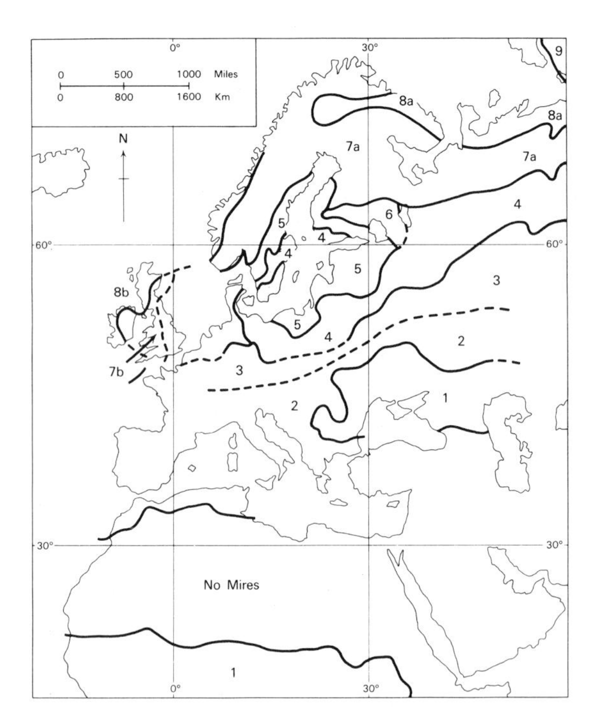

Fig. 1.4. Spatial zonation of European mire complexes (after Moore and Bellamy, 1974): *1* = valley bogs and flooded fens; *2* = tertiary valley mires; *3* = continental raised bogs; *4* = plateau raised bogs; *5* = concentric raised bogs; *6* = eccentric raised bogs; *7a* = aapa fens; *7b* = ridge raised bogs; *8a* = palsa bogs; *8b* = blanket bogs; *9* = Arctic bogs.

[Professor H. Sjörs writes: "There are faults in this map to which attention should be drawn. It is, for example, not possible to make out what is No. *4* and No. *5* in southern Scandinavia, and in the Baltic region as a whole. Probably, for Sweden, the *4* should have been *5* and the *5* definitely *6*. In reality, both *4* and *5* and intermediates are represented intermingled in south Sweden, south of the area marked *5* (which should be *6*). Moreover, it seems quite unlikely that zone *4*, which in Poland is more continental than *5*, would be the opposite in Germany (in the southern Polish parts of *4*, real *Hochmoore* are however practically lacking except in the Sudety Mountains and the Carpathian foreland). The North Sea area lacks a number on the map, but it should be *5* (despite the floristic difference from the Baltic type). As is quite clear now, an area with many raised bogs (of types *4* or *5*) occurs in northeast U.S.S.R. east of the White Sea mainly within the area marked *7a*, although aapa mires (*7a*) do prevail again closer to the Ural Mountains. The location of *7b* seems queer as in the text the examples are from Poles'ye and Bavaria. The *8a* (palsa mires) are placed a little too much south on the map (cf. the map taken from Ruuhijärvi, Moore and Bellamy 1974, fig. 2.8 p. 24, which however, gives a too broad (not too long but too wide) extension of the palsas towards W–SW)."]

of true raised bogs (*echte Hochmoore*) of Zone *5*, although they may lack marginal slope or **rand**. Tansley (1939, p. 686) referred to the raised **moss** or bog type principally found in the central plain of

TABLE 1.1

Correspondence between the mire zones of Moore and Bellamy (1974) (Fig. 1.4) and those of Walter (1968, fig. 343)

Zone 1	valley bogs and raised bogs
Zone 2	tertiary valley mires
Zone 3	continental raised bogs: largely equivalent to *Waldhochmoore*, which however occur in the following zones as well
Zone 4	plateau raised bogs
Zone 5	concentric raised bogs: *typische Hochmoore* (Atlantic, Baltic and north Russian subtypes)
Zone 6	eccentric raised bogs
Zone 7a	aapa fens: *Aapamoore*
Zone 8a	palsa mires: *Palsenmoore*
Zone 8b	blanket bogs: *terrainbedeckende Moore, Deckenmoore*

N.B.: In Zones 3 to 6, the classification takes only the ombrotrophic bogs into account; but minerotrophic mires are also widespread, and locally dominant.

Ireland and the midland valley of Scotland. These do appear to be nearer in form to the Zone *5* (*echte Hochmoore*) type, but detailed surveys of these large mires, before exploitation, have rarely been reported.

Zone *8a* is an expression of discontinuous permafrost, located in the ***palsas***. Zone *9* consists of Arctic peat formations much influenced by low temperatures and continuous permafrost. The zone falls towards the periphery of European mires, but forms a link between the Arctic U.S.S.R. and Canada and Alaska. In this huge area, mire shape is much more related to the dominant effects of frost and ice than to the effects of liquid water drainage so predominant in the North Temperate and Boreal Zones.

Radforth and Bellamy (1973) have attempted to relate the Canadian **muskeg** to the European classification. Their task was made especially difficult, not only by the vast proportions and inaccessibility of these peatlands, but also by the special dependence of the patterning of the terrain on frost. Radforth (1977, p. 132) wrote that "variability arising from cryogenic factors applies to probably more than 40 per cent of Canada's land area". He pointed out that when cryogenic phenomena persist in muskeg, as they do in the north of Canada, it makes a marked difference to the topographic characteristics of organic lands. There are many features of muskeg having the forms described for northern Europe, but they are often scattered within a continuously peat-covered landscape. Sjörs (1959, 1963) provided a European mire ecologist's view of the shapes and patterns in the Hudson Bay region of continuous mire. He commented that one difficulty in classifying the variety of patterns easily observable from the air was the lack of figures for gradients, and only few vertical mire sections could be made. However, Zoltai and Pollett (Part B, Ch. 8) provide an integrated account of Canadian mires, relating the roles of ice and liquid water in different regions.

One of the most thorough contributions to all the essential aspects of mire shape relevant to a more southerly group of North American mires is by Heinselman (1963). The region he studied in Minnesota (U.S.A.) provided a basis of classification by shape and, more importantly, a basis for understanding the mechanisms of mire formation. This latter theme is taken up again both in this and in other chapters of this volume.

It is sufficient to conclude that simple classifications by shape and surface pattern are generally the most useful type of classification in mire ecology. This is because of the immediate relevance of shape, both past and present, to hydrology, a topic which is developed in relation to theoretical concepts in a later part of this chapter. However, shape, surface patterns and hydrology are dependent on local and general land-forms, a fact which is often neglected or underestimated when the "regionality" of peatlands is discussed.

Classifications based on chemistry

The general equivalence of **bog** and **fen** to *Hochmoor* and *Niedermoor* respectively has been referred to already. Used in context, the English terms like the German ones are usually unambiguous; but they can refer to mire shape (see above), to the chemical properties of the peat (see Clymo, Ch. 4), or to the floristic or ecological characteristics of the present surface (Sjörs, Part B, Ch. 3).

Tansley (1911, pp. 208–213) gave definitions of **moor** and **fen** which long persisted in the English ecological literature, though later (see below) he gave different definitions. He emphasized the chemical definition and distinguished the German terms as being primarily concerned with shape. It is doubtful however whether the German usage has been any more discriminating in practice with regard to shape (see, for example, Weber, 1908, p. 28). Tansley (1939, pp. 673–674) proposed that **moor** in the mire sense should be replaced by **moss** or **bog**, because **moor** in English common usage generally may refer equally to heaths or to wet bogs (e.g. the Yorkshire Moors, Dartmoor). **Moss**, although used occasionally in some of the ensuing chapters, is not favoured because of the confusion with moss plants themselves. This is mainly a matter of semantics however, and the main distinction between acid, mineral-poor, present-day soil conditions for **bogs**, and less acid or even alkaline mineral-rich conditions for **fens**, remains.

The main separation between bogs and fens thus lies in the origin and chemistry of their respective water supplies. Ombrotrophic (rain-fed) conditions are unambiguous in their meaning for **bogs**, even though it is not always easy in practice to decide in detail where such conditions apply (see the reference to Ingram, 1967 below). Fens, as far as Tansley (1911) was concerned, meant the East Anglian Fens of England; these are usually distinctly alkaline and calcium-rich. The idea of **oligotrophic fens** was not contemplated in British literature until later. **Oligotrophy** and **eutrophy** are terms commonly used in limnology — originally in relation to plankton productivity rather than chemistry itself. Ombrotrophy has a clear meaning in relation to bogs, and invariably implies oligotrophy, whereas several terms have been proposed in the fen context, some of which have been less helpful to understanding and clear communication. The terms **minerotrophic**, **geogenous** and **rheophilous** all convey the idea of water supplies which have been influenced by soils or rocks before they reached the mire surface. Like the contrasting terms **topogenous** and **soligenous**, they may all refer to almost any trophic status from eutrophic to oligotrophic. Sjörs (1948, p. 279) starting from ideas of Von Post and Granlund (1926), proposed the terms **geogenous**, covering **limnogenous**, **topogenous** and **soligenous** origins. **Geogenous** is self-evident in its meaning and seems logically acceptable as having a contrasting meaning to ombrogenous. **Limnogenous** indicates that water is derived from lakes or rivers, while **topogenous** and **soligenous** refer to soil-derived water, respectively static and flowing. In the special circumstances of extensive low relief of the Poles'ye "marshes", Kulcyński (1949, p. 12) found it necessary to introduce the term **rheophilous**; but its sense appears indistinguishable from soligenous. The word has, however, also been used in the wider sense of geogenous or minerotrophic — that is, including topogenous, though this is incompatible with a term derived from the Greek ῥέος, meaning a stream or flow. Kulczyński's **bogs** were not ombrogenous, and therefore were misnamed as bogs. The terms topogenous, soligenous and rheophilous mix physical concepts with chemical ones, and this has led to confusion in their use. For example, Tansley (1939, p. 719) refers to "topogenous" raised bogs. The fen which was assumed to precede the present-day bog could have been described as topogenous, but the existing bog surface is probably ombrogenous. The term rheophilous is widely used at present but causes other difficulties. Water may be channelled in running off ombrogenous bogs, and this may result in local changes in the vegetation. Ingram (1967) referred to several authors who had noted this type of effect. Crawford (Ch. 7) points out that the very survival of certain higher plants growing in such waterlogged conditions may depend on effects of flowing water in aeration and removal of toxins. Such an effect can only be regarded very indirectly as one of chemistry in the sense used here.

The glossary (Appendix II, pp. 26–30) should help to make clearer references to these terms where they have been used in the text. In general one is inclined to agree with Moore and Bellamy (1974, p. 80) that many of these terms are cumbersome to use and a more broadly based use of the simpler **fen** and **bog** is to be preferred. These authors conclude with what seems to be a satisfactory definition of fens and bogs: **fens** are mires influenced by water derived predominantly from outside their own immediate limits; **bogs** are influenced solely by water that falls directly on to them as rain or snow. Fens (and also to some extent bogs) vary widely in the actual chemistry of water and peat. To provide some quantitative data, Table 1.2 illustrates the range of ionic concentrations in run-off and rain water from a number of widely scattered localities before the widespread sulphur pollution of today.

Classification based on plant species composition

Despite the value of shape and of chemical properties for classification purposes, in practice it

TABLE 1.2

Ionic composition (p.p.m.) of rain water and of run-off from rocks of different geological origin (after Gorham, 1961)

Site/rock type	Ca^{2+}	Mg^{2+}	Na^{+}	K^{+}	HCO_3^{-}	SO_4^{2-}	Cl^{-}
Rain water							
Newfoundland	0.8	–	5.2	0.3	–	2.2	8.9
Wisconsin	1.2	–	0.5	0.2	–	2.9	0.2
Minnesota	1.0	–	0.2	0.2	–	1.4	0.1
Northern Sweden	1.2	0.2	0.4	0.3	–	2.5	0.7
Central Sweden	0.6	0.1	0.3	0.2	–	2.6	0.5
Guyana	0.8	0.3	1.5	0.2	–	1.3	2.9
Run-off water							
Nova Scotia							
Granite	1.0	0.5	5.2	0.4	*	5.9	7.7
Quarzite and slate	2.1	0.4	3.0	0.6	1.8	5.2	4.9
Carboniferous strata	3.0	0.6	3.6	0.5	6.1	5.3	5.4
Bohemia							
Phyllite	5.7	2.4	5.4	2.1	35.1	3.1	4.9
Granite	7.7	2.3	6.9	3.7	40.3	9.2	4.2
Mica schist	9.3	3.8	8.0	3.1	48.3	9.5	5.4
Basalt	68.8	19.8	21.3	11.0	326.7	27.2	5.7
Cretaceous rocks	133.4	31.9	20.7	16.4	404.8	167.0	17.3

*Not detected.

has often been plants which have been used to distinguish mire types and their extent. It was, for example, the Central Committee for the Survey and Study of British Vegetation (Tansley, 1911) that first identified the mire (and other major) vegetation types in Britain. The difficulty of using vegetation for purposes of ecology is that circular arguments can easily develop. Such pitfalls are increased in mire studies by the fact that the vegetation largely contributes to the formation of the soil. Nevertheless, used with care, effective grouping for mapping purposes is achievable. An important value of classifications based on vegetation criteria has been the recognition of differences, and, as importantly, regular similarities, in what appear to be similar mires judged on the basis of their shape. This application of phytosociologic classifications are especially suited to those regions of the earth where mires have physiognomically short vegetation. Where forested mires are concerned, such approaches are more complicated, but have been successfully used in Poland and Finland, for instance. In areas where the species or their indicative values are less well known, classifications can be based on vegetation structure and "life forms". Such classifications are briefly reviewed below.

Apart from differences due to degree of oceanic influence, the subdivision of ombrotrophic or bog vegetation has been easier than the subdivision of fens. This is because bogs often have extensive, uniform or regularly recurring, vegetation and by definition a limited variation in habitat conditions. By contrast, fens vary enormously in size and have unlimited possibilities of combinations of mineral influences (see also McVean and Ratcliffe, 1962, p. 127). Swamps (fens with trees) combine this variability with the effects of the trees themselves, which may or may not reduce the discriminating usefulness of vegetation classifications. This may help to explain the wide range of approaches to classification used in different regional chapters of Part B of this volume.

To illustrate the extent to which classifications based on vegetation types can improve the appreciation of finer differences between mires, three classifications (Tansley, 1939; McVean and Ratcliffe, 1962; and Daniels, 1978) of British mire types are presented in Table 1.3. Tables 1.4 and 1.5 compare two of these types in detail.

TABLE 1.3

Bog (mainly blanket bog) vegetation types in Britain and Ireland with probable equivalents of different authors

Tansley (1939)	McVean and Ratcliffe (1962)	Daniels (1978)
Sphagnetum (wide range of sites in British Isles)	–	–
Rhynchosporetum (western Ireland)	–	–
Schoenetum (western Ireland)	*Trichophoreto-Eriophoretum typicum*[1] (see McVean and Ratcliffe, 1962, p. 102)	–
Eriophoretum (southern Pennines, England)	*Calluneto-Eriophoretum* (eastern Highlands, Scotland)	*Calluna–Eriophorum vaginatum*; upland *Calluna–Eriophorum vaginatum*
Scirpetum[1] (western and northern Scotland)	*Trichophoreto-Eriophoretum typicum*[1] (western Highlands, Scotland)	*Erica tetralix–Scirpus*[1]; *Calluna–Narthecium*
Molinetum (Scotland and western Ireland)	*Molineto-Callunetum* (western Highlands, Scotland)	*Molinia–Erica tetralix*
Callunetum ("not to be regarded as part of the bog or moss formation")	–	mixed *Calluna*

[1] *Scirpus cespitosus* has been called *Trichophorum caespitosum* and more recently *Baeothryon caespitosum*.

The results obtained by the different authors working in the same region are in fairly good agreement, but the discrimination is less satisfactory when a larger region is studied, even though the original is included within the larger region. Clearly, sampling methods are of considerable importance. When different regions are surveyed comparisons of classifications become still more indefinite. A more extreme example for this type of comparison is found in the classifications of Du Rietz (1949) for bogs in Sweden. Table 1.6 possibly suggests that the flora of the bog expanse in eastern, south-central (II) and southwestern (III) Sweden is similar to that of the "red bogs" found in central Ireland, but Tansley (1939, p. 696) emphasized the **physiognomic** dominance of *Scirpus cespitosus* on a typical Irish raised bog at Athlone.

The relevance of physiognomic status remains an unresolved question in almost all vegetation classification. The methods used by Tansley (1939) and by McVean and Ratcliffe (1962) only made use of quantitative (cover) information in a general way. The method used by Daniels (1978) made explicit use of cover values. "Pseudo-species" are a technique devised by M.O. Hill (see Hill et al., 1975) for including some of the effects of physiognomic status in a classification otherwise based on "incidence" (presence and absence).

Classifications based solely on vegetative structure

Some classifications depend solely on physiognomy, but their value for mire ecology is limited. The advantage of such methods is to facilitate large-scale mapping exercises not specifically concerned with any given habitat type such as mires. On a smaller scale, structural sketch maps of individual mires are common in the Fennoscandian literature, and are also used in the U.S.S.R. Radforth (1977) described a physiognomic system of classification as follows:

TABLE 1.4

Species composition (frequency %) of *Scirpetum* (Tansley, 1939) and its equivalents in Table 1.3; an oligotrophic fen (soligenous) variant, *Trichophoreto-Eriophoretum caricetosum*[1], is included for comparison with the ombrogenous bog types

	Scirpetum[1,2]		*Trichophoreto-Eriophoretum*[1,3]		*Calluna–Narthecium*[4]	*Erica tetralix–Scirpus*[1,4]
	Scottish Highlands	SW England	*typicum* W Scotland	*caricetosum* W Scotland	Scotland (oceanic mires)	Scotland (oceanic mires)
	10 or 11 sites	2 sites	12 sites	9 sites	109 quadrats	112 quadrats
Calluna vulgaris	100	100	100	89	97	96
Empetrum nigrum	0	50	0	0	0	0
Erica tetralix	82	100	100	100	98	99
Myrica gale	64	0	83	11	0	0
Carex echinata	0	0	33	89	0	0
C. nigra	0	0	8	44	0	0
C. panicea	18	0	42	89	0	0
Eriophorum angustifolium	82	100	100	100	95	86
E. vaginatum	64	50	100	22	0	0
Scirpus cespitosus[1]	91	100	100	56	81	89
Molinia caerulea	73	50	100	89	82	99
Narthecium ossifragum	91	100	100	89	94	76
Drosera rotundifolia	36	0	83	89	88	0
Potentilla erecta	55	100	58	89	0	0
Hypnum cupressiforme	40	50	83	44	0	0
Sphagnum capillifolium[5]	0	0	0	0	82	0
S. papillosum	60	0	100	67	81	0
S. rubellum[5]	60	0	92	44	0	0

[1]See note, Table 1.3.
[2]Tansley (1939); [3]MacVean and Ratcliffe (1962); [4]Daniels (1978, noda 3 and 4 respectively).
[5]*Sphagnum capillifolium* may not be distinct from *S. rubellum*.

Cover class	Description
A	trees >15 ft [4.5 m]
B	trees <15 ft [4.5 m]
C	non-woody grass-like
D	woody, tall shrubs or dwarf trees 2–5 ft [60–150 cm] high
E	woody shrubs <2 ft [60 cm] high
F	sedges and grasses <2 ft [60 cm] high
G	non-woody broad-leaf plants <2 ft [60 cm] high
H	leathery to crisp mats of lichens <4 in. [10 cm] high
I	soft mats of mosses <4 in. [10 cm] high

A map of Canada based on this scheme (Radforth, 1955, 1958) would, however, have to be used in conjunction with a map of mires derived in another way to be useful ecologically. Radforth himself paid great attention to patterns as seen from the air.

Phytocenoses (based on life forms) are widely utilized by Russian vegetation surveyors. Botch and Masing (Part B, Ch. 4) tabulate the main forms and their constituent species used in the regional descriptions of the mire zones of the U.S.S.R. In their examples and in Part B, Chapter 10 by Pisano, who uses the system published by UNESCO (1973), vegetation structural properties are supplemented by species lists.

There are thus limits to the usefulness of classifications based on vegetation alone. In the Boreal and North Temperate Zones plant communities are often distinctly related to mire formations so that descriptions of the vegetation can give a reliable measure of the mire ecosystem and its variants. Unfortunately this is not true in many other parts

TABLE 1.5

Species composition of *Eriophoretum* (Tansley, 1939) and its equivalents in Table 1.3 (% frequency, except for first column); the *Erica tetralix–Scirpus*[1] nodum is included for comparison

	Eriophoretum[2]	*Calluneto-Eriophoretum*[3]		*Calluna–Eriophorum vaginatum*[4]	Upland *Calluna–Eriophorum vaginatum*[4]	*Erica tetralix–Scirpus*[1,4]
		lichen-rich facies	shrub-rich facies			
	S Pennines (England)	mainly E Scottish Highlands		Upland areas in England and Scotland		
	several sites	13 sites	8 sites	70 quadrats	106 quadrats	112 quadrats
Andromeda polifolia	r[5]	0	0	0	0	0
Calluna vulgaris	lsd	100	100	99	92	96
Empetrum nigrum	lsd	15	63	0	55	0
E. nigrum subsp. *hermaphroditum*	–	85	75	0	0	0
Erica tetralix	r-la	15	50	74	0	99
Vaccinium microcarpum	–	15	13	0	0	0
V. oxycoccos	r-la	92	88	0	0	0
V. myrtillus	r	0	0	0	0	0
Eriophorum angustifolium	l	31	50	71	70	86
E. vaginatum	d	100	75	93	91	0
Scirpus cespitosus[4]	r-la	31	25	67	0	89
Molinia caerulea	r	0	0	0	0	88
Narthecium ossifragum	r	8	0	0	0	76
Potentilla erecta	–	39	25	0	0	0
Rubus chamaemorus	lsd	85	88	0	0	0
Hypnum cupressiforme	–	31	38	67	56	0
Sphagmum capillifolium[6]	–	0	0	63	0	0
S. papillosum	–	8	12	0	0	0
S. rubellum[6]	–	8	0	0	0	0

[1]See note, Table 1.3.
[2]Tansley (1939, p. 703); [3]MacVean and Ratcliffe (1962, table 50); [4]Daniels (1978, noda 11, 14 and 4, respectively).
[5]r = rare; l = local; lsd = locally subdominant; la = locally abundant; d = dominant.
[6]*Sphagnum capillifolium* may not be distinct from *S. rubellum*.

of the world. The chapters by Pisano (Part B, Ch. 10), Campbell (Part B, Ch. 5), Anderson (Part B, Ch. 6) and Thompson and Hamilton (Part B, Ch. 11) all show to the contrary that many communities of plants *may* or *may not* form peat. Identification by vegetational characters, though easy, is insufficient for ecosystem characterization in such circumstances. In Europe this applies to alder swamps, reed swamps, and other wetlands with a variable substrate. Even though vegetation is often characteristic of soil and other natural environmental conditions, the effects of human interference can confuse the interpretations.

All the methods described in this section lead to arbitrary classifications which may be transient, *ad hoc* stages, *en route* towards a clearer, truly ecological understanding of the relative importance of each of the underlying influences. In the following section, some of the progress in this understanding of relationships and presumed interactions of shape (hydrology), development, chemistry and vegetation is considered.

MECHANISMS AND MODELS

Qualitative hypotheses — large-scale

The concept of **terrestrialization**, the so-called ***Verlandung*** well described by Weber (1908) and at least foreshadowed by Dau (1823), has had a great influence in mire ecology. Weber's stratigraphical evidence was from the lower horizons of raised bogs in northwestern Germany, but much evidence

TABLE 1.6

Bog (raised bog) vegetation types in southern Sweden (Du Rietz, 1949)

I. Eufuscion subformation (*Ryggmosse* type — eastern Sweden)
  1. Ledo-Parvifolion alliance
     on wooded bogs and pinewood slope (rand) with *Sphagnum recurvum* var. *tenue* dominant
  2. Eufuscion alliance
     on central bog expanse with *Sphagnum fuscum* hummock dominant, also *S. balticum* and *S. cuspidatum* in wet hollows

II. Rubello-Fuscion subformation (*Skagershultmosse* type — south-central Sweden)
  1. As in I
  2. Rubello-Fuscion alliance
     on central bog expanse with *Sphagnum capillifolium* and *S. magellanicum*; in addition to species of the Eufuscion alliance

III. Rubellion subformation (*Komosse* type — southwestern Sweden)
  1. Parvifolion alliance
     without *Ledum palustre*, usually with *Vaccinium uliginosum* or *Erica tetralix* dominant
  2. Rubellion alliance
     on central "regeneration complexes" (see below, p. 29), with *Erica tetralix*, *Sphagnum capillifolium* and *S. magellanicum* but also *S. imbricatum* and *S. papillosum*
     N.B.: Sphagnum imbricatum and *S. papillosum* grow only in fens in eastern Sweden, where the former species is rare

of a similar kind exists elsewhere. The frequent occurrence of *Equisetum fluviatile*, *Phragmites australis* and *Carex* spp. among the earliest identifiable plant remains strongly suggested that such bogs originated from shallow lakes which subsequently became overgrown (terrestrialized). They were mainly influenced by local water flowing into the lake and hence frequently topogenous, or rather "limnogenous" (Sjörs, 1948). Peat accumulated later, in many cases, progressively raised the surface above the catchment water and thus removed it from the mineral influence of mineral soil and rock. Given sufficiently moist climate, rain water thus eventually formed the sole source of moisture and mineral nutrients. This process has been inferred both from the shape of the cupola which gives such raised bogs their name, from chemical analyses, and from the presence of extremely oligotrophic species, such as *Sphagnum balticum*, *S. capillifolium* (*S. rubellum*) and *S. fuscum* in the upper layers of the peat.

Attractive to the ecologist as this mechanism for progressive succession is, early stratigraphic studies on other peatlands had shown that the above sequence is far from general (Tallis in Chapter 9 describes many of the sequences of deposition of organic residues which have been observed). Even in the classical area of the mainland of north-western Europe, an alternative mechanism of formation (**paludification**) was documented, and was known to be widespread farther west and north. For example, Cajander (1913) remarked that there were three main ways of peat formation in Finland: (1) by infilling of shallow lakes — that is, terrestrialization; (2) by swamping (***Versumpfung***) of flood-plains of rivers; and (3) by swamping of forest soils. Cajander admitted that the first process was better known than the others; and even today swamping or paludification is a less easy process to visualize. Tansley (1939) for example, remarked on the profile of a raised bog at Edenderry (Ireland), obtained from a peat core made by H. Osvald, "Here the fen underlying the bog started with carr on a wet clay, with no preceding lake as at Athlone [a core from which Tansley described previously], the sedges and the trees probably colonising a wet depression in the ground-moraine left on the disappearance of the last ice sheets." If this was so, it would be an example of a common method of peat formation described from Finland and Sweden (see Part B, Ch. 3, p. 72 etc.).

For blanket bogs and other distinctly or even strongly sloping mires, paludification is in general the obvious origin. The term paludification has come to include the formation of peat directly, or via terrestrial humus, on podzolized soils and even bare rock surfaces. Frenzel (Ch. 2), Taylor (Part B, Ch. 1) and Sjörs (Part B, Ch. 3), among others, all discuss aspects of this process which has probably been responsible for the initiation of the great majority of the world's peatlands. Sjörs also remarks that several Fennoscandian authors regard "primary peat formation" on wet surfaces (formed, for instance, by land uplift) as a process distinct from paludification proper — for instance upon formerly forested land. Granlund (1932), in the raised bog zone of southern Sweden, found that paludification had probably been an active process only at two periods of increased climatic wetness —

that is, after 4300 B.P. and after 2600 B.P. (datings by pollen analysis). Farther north, more extended periods may have occurred (Malmström, 1930).

Two notable contributions to this topic have been made independently, namely by Kulczyński (1939/40, 1949) in Poles'ye, the area once known as the Pripet Marshes and now part of the western U.S.S.R. (Belorussiya and Ukraine) (see Botch and Masing, Part B, Ch. 4), and by Heinselman (1963) in northern Minnesota, U.S.A. (see also Hofstetter, Part B, Ch. 7).

Kulczyński (1939/40, 1949) found terrestrialization inadequate to explain either the topographic distribution or the stratigraphy of the extensive mires in Poles'ye. He proposed (1949, p. 128) an upslope development of mires in an area of very low relief, in which peat formations in shallow valleys progressively impeded the drainage of land of slightly higher elevation and thereby *induced* conditions suitable for accumulation in places which had previously been free-draining. Because the solid geology consisted of permeable strata the ground-water level was the primary factor. On the other hand Heinselman (1963), working in an area with essentially impermeable soils, also visualized a somewhat similar process of upslope progress of peat formation. "Upslope growth of the peatland seems to be achieved by damming up the incoming waters from mineral soils. This creates a wet area into which the swamp forests can advance and prepare the way for the bog flora. It is noteworthy that the initial invasion on mineral soils does occur in a wet area but the communities are still forests not aquatics." Heinselman considered that the terrestrialization concept had had too dominant an influence in the development of mire ecology in North America. Processes similar to that which he described were believed to have caused extended paludification on slightly sloping till soils in Sweden (Malmström, 1923), and even on stronger slopes (Lundqvist, 1951).

Thus it is not easy to distinguish mature ombrogenous bog types on the basis of present-day vegetation characteristics alone, as it is obvious that apparently similar hydrologic and chemical conditions can apply at the surface of peat which has been initiated by either terrestrialization or paludification, although a distinctly unidirectional or fan-like slope of the bog usually rules out large-scale terrestrialization.

Climatic rhythms. Conway (1948) elaborated on an idea formulated by Von Post (1944) in an attempt to account for the ***Grenzhorizont*** of Weber (1908) and the "recurrence surfaces" of Granlund (1932). A series of three harmonic variations (Conway, 1948, fig. 6) was thought to represent fluctuations in climate. "The (vertical) ordinates represent the hypothetical magnitude of whatever climatic factor it is that favours more rapid and abundant growth of *Sphagnum* species on the bog surface". Conway suggested that what Wickman (1951) later termed "effective rainfall" was the nearest equivalent to this climatic factor. Her proposal was that there are thresholds of climate which can only be crossed at certain stages of the overall cyclic trend. The ideas are very speculative and while not wholly compatible with other fragments of theory discussed in this volume, they cannot be accepted uncritically in the light of the actual long-term climatic trends as described by Frenzel (Ch. 2). Moreover, the "recurrence surfaces" are neither as regular nor as simultaneous as was postulated by Granlund.

Qualitative hypotheses — small-scale

The hummock-and-hollow regeneration hypothesis of Von Post and Sernander (1910) and of Osvald (1923) is described in detail by Tallis in Chapter 9. Probably more exceptions to this idea have been discovered than cases which are clearly in support of a regular cyclic mode of succession. Other cases of variation over mire surfaces, such as those on aapa or palsa mires, or on bogs with real pools, are clearly irreversible, but in most cases secondary — that is, formed later than the mire itself.

Quantitative models — large-scale

A model of raised-bog development which has been given more exact form is that of Granlund (1932). The maximum convexity of domed raised bogs in parts of southern Sweden was shown by Granlund to be distinctly related to rainfall distribution (see Tallis, Ch. 9). Werenskjold (1943) pointed out that there is a very high correlation ($r = 0.998$) for a linear relation between the square of the height of the cupola of raised bogs surveyed by Granlund and the annual rainfall, for bogs with a diameter of up to 1000 m. Wickman (1951) thereafter, recognizing the essential physical characteristics of raised bogs in a remarkably succinct and prescient way, went on to propose a simple theoretical model based on Darcy's Law (see Rycroft et al., 1975, p. 536):

$$v = k \cdot \frac{dH}{dl} \qquad (1)$$

where v is the discharge across unit cross-sectional area of a porous medium, k is a proportionality constant (the hydraulic conductivity), H is the hydrostatic head and l is the length of flow line. The form of a raised bog was assumed to be a very slightly domed but essentially flat disc, having most of its height sharply gained at the perimeter — the **rand**. Ruuhijärvi shows typical sections in Chapter 2 of Part B, but such diagrams can be misleading because of the great exaggeration of the vertical scale. Rain water falling on the bog plane is assumed to follow a series of radiating flow lines towards the perimeter. The discharge from any given flow line will depend on the height difference $(h-z)$ where h is the height of the bog plane and z is the height of the margin. Wickman pointed out that because the height of a raised bog is small in relation to its diameter, the influence of the relation between height variation and the length of the flow lines (dH/dl in eq. 1) on their discharge can be neglected.

Integrating from the surrounding ground level to the bog plane for a small sector of the bog, the discharge is proportional to:

$$\int_0^h (h-z)\,dz = \frac{h^2}{2} \qquad (2)$$

The discharge Q from the whole margin will, for constant size, be:

$$Q = Ah^2 \qquad (3)$$

where, to quote Wickman "A is a constant depending on the size and form of the bog, the permeability of the peat and eventually other local factors such as vegetation etc.".

Thus, Wickman formulated the following: "The height of an ombrogenous bog in equilibrium is proportional to the square root of the effective annual rainfall, all other factors being constant". The "effective annual rainfall" is the difference between the potential evaporation and the observed rainfall.

Fig. 1.5 illustrates the test of this relationship using Granlund's (1932) data. As can be seen the fitting is exceptionally good, particularly for bogs having diameters of up to 500 m. Size is obviously likely to be an important variable in the hydrology of raised bogs, and Wickman illustrated this by comparing two bogs of different size but of similar shape. If the linear dimensions are proportional to k, then the amounts of water to be transported from the bogs are proportional, to a first approximation, to the areas of their planes — that is, to k^2. In a region of constant effective rainfall, the discharge per unit length of margin of a bog must consequently be proportional to $k^2/k = k$ (assuming reasonably regular convexity). This suggests that the response to this increased flow per unit of margin would be a greater waterlogging on the bog

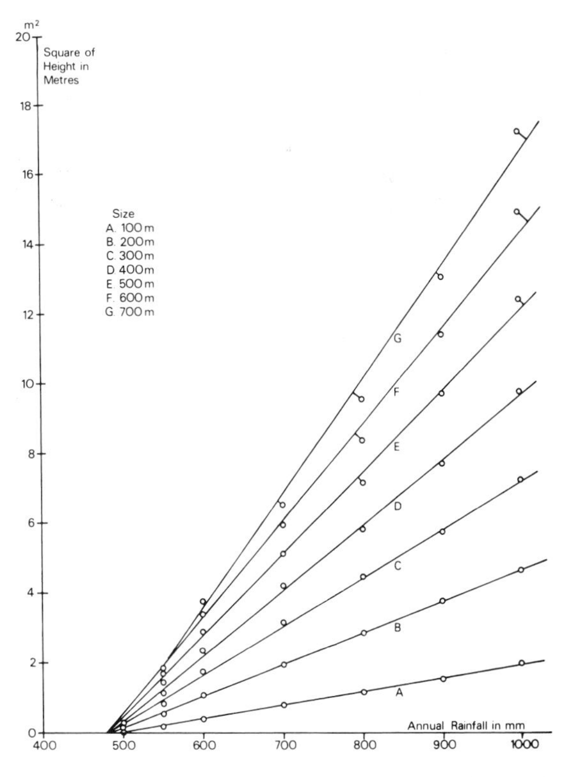

Fig. 1.5 Relation of the height of an ombrogenous bog to the annual precipitation and the bog size (data from Granlund, 1932; after Wickman, 1951).

plane. Under certain conditions of mild climate and non-excessive rainfall, a corresponding response in terms of increased height would result.

Thus Wickman went on to state that "The heights of similar ombrogenous bogs in equilibrium are proportional to the square root of their linear enlargement factors" [i.e. their dimensions]. Again using Granlund's data, Wickman illustrated this as shown in Fig. 1.6. Wickman considered that the increasing poorness of fit above 500 m in Fig. 1.6 was due, firstly, to the paucity of primary values in Granlund's data, among which there were many fewer measurements for bogs of diameter greater than 500 m; secondly, to the probability that the selected transects Granlund used for peat inventory would miss the point of maximum height on large bogs; and thirdly, no attempt was made to distinguish between bogs of various shapes. While this last reason could be particularly important in departures from the theoretical constraints already outlined, the question of maximum height seems dependent on the regular disc form assumed for raised bogs; in reality the shape is quite variable, even when only convex bogs are considered.

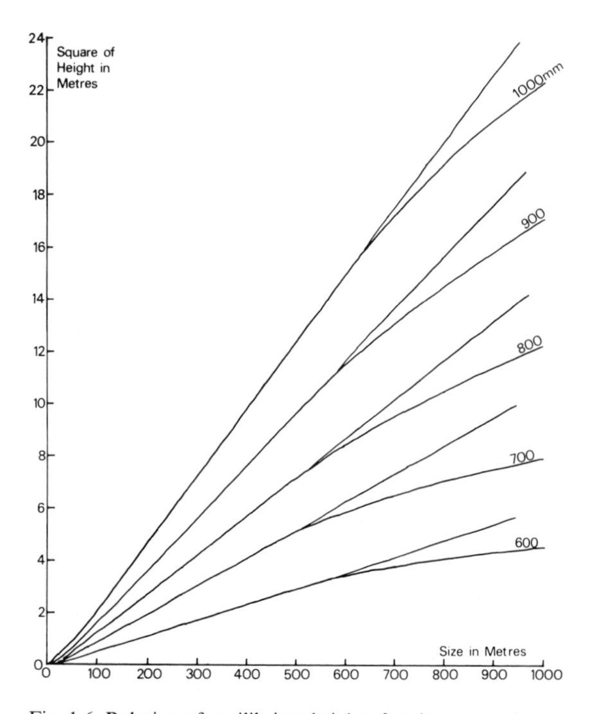

Fig. 1.6. Relation of equilibrium height of ombrogenous bogs to linear dimensions under different precipitation (based on data of Granlund, 1932; after Wickman, 1951).

Of the remaining elements in the proportionality constant A in eq. 3, permeability of the peat is important. Wickman argued that the good agreement in the simple model itself suggests that permeability of the surface layers of the peat bogs studied by Granlund are virtually constant over southern Sweden. Subsequent measurements of horizontal hydraulic conductivity of peats of low humification in a number of circumstances have shown rather similar values. For example, Rycroft et al. (1975) have given values in the range from 1×10^{-3} to 1×10^{-2} for peats with humification values on the Von Post scale (Von Post and Granlund, 1926) of 3, 2 and 1 and thus appear to support Wickman's contention. On the other hand, Malmström (1923) found widely varying values for peats in northern Sweden, and M. Pearson (pers. comm., 1976) found that the velocity of water movement diminished greatly with depth below the water table. In fact, run-off from peatland increases greatly with rises in water level, making effective permeability dependent on the discharge itself.

The term "equilibrium" is particularly significant in connexion with Granlund's observations on **recurrence surfaces**. Godwin (1946, and particularly 1952) assumed that the hypothesis of Granlund, that climate controls and limits the height of raised bogs, must imply continuously increasing climatic wetness, but Granlund himself made no such assumption, not really discussing whether bogs with maximum height had reached their mature convexity. This point is discussed at greater length elsewhere in this volume, notably by Frenzel (Ch. 2) and by Tallis (Ch. 9). Quoting Godwin (1952) and Walker and Walker (1961) as accounting for Granlund's results in terms of fluctuating climate, Dahl (1969) progressed with this explanation which takes both the facts of Granlund and the theory of Wickman into account. According to this, fluctuations in the **effective rainfall** could cause adjustments to the height of raised bogs in relation to their areal dimensions as outlined above, these dimensions being presumably controlled by local topography and catchment drainage properties. A decline in the effective rainfall should cause reduced plant (especially *Sphagnum*) production, increased decomposition at the bog surface (and thus formation of peat of a darker more humified texture) and a corresponding *loss* of height. Conversely an increase in effective rainfall should encourage

Sphagnum growth and decrease decomposition with a corresponding *increase* in height of the bog plane. Small bogs would be more sensitive than large ones, because of their smaller run-off per unit length of margin. Variation in the effective rainfall could lead to "hunting" for an equilibrium height with no finite end-point ever being reached. Such a simple model ignores the fact that excessive moisture could also cause *reduced* growth in bog hollows (algal hollows ultimately developing into bog pools). The actual complexity resulting either from the underlying topography and geology or from variations, both general and local, in surface relief, both known to be very much more variable than the classical terrestrialization hypothesis would suggest, is also ignored. Under conditions of such variation it is perhaps hardly surprising that the dating of recurrence surfaces even on the same bog is rarely synchronous. Frenzel discusses this point further on p. 44. The presence of a regeneration complex on one part of Tregaron Bog in Wales (Godwin and Conway, 1939; see also Taylor, Part B, Ch. 1) can be cited as one of many present-day examples of such spatial heterogeneity. Deposits of volcanic ash, "tephra layers", can provide more reliable date markers for whole bogs as shown by Yoshioka (1963) (see Fig. 1.10 below), but surveys of all the relevant sections of such bogs are required because many so-called "raised bogs" have shapes far from the ideal, flattened discs of Wickman's hypothesis.

The foregoing ideas lead naturally to the modelling of bog development at the detailed level, and such models are now described.

Quantitative models — small-scale

Computer simulations. Jenny et al. (1949) formulated the production and decay of forest litter using the equation:

$$\frac{dW}{dt} = I - kW \qquad (4)$$

where W is the dry mass of litter (with dimensions[1] ML^{-2}) on the forest floor, I is the annual litter fall ($ML^{-2}T^{-1}$), k is a fractional decomposition factor (T^{-1}) and t is time. It was assumed that annual steps were small in relation to the total periods involved. Gore and Olson (1967) proposed a compartment model, and applied this equation to successive compartments using data from a blanket bog site at Moor House in northern England (see Taylor, Part B, Ch. 1). Most of these data were obtained from a long-term clipping experiment (Gore, 1975) in which the accumulation of standing-dead litter, mainly of *Eriophorum vaginatum* tussocks, provided curves in which the application of eq. 4 could be easily visualized. The approach was essentially a process of curve-fitting, by trial and error using a computer, to simulate form and relative behaviour of two linked compartments: (1) above-ground live matter and (2) standing-dead litter. Gore and Olson thereafter went on to simulate the development of a whole peat profile by extending the compartment model to include two further compartments: the aerobic peat (surface layers or acrotelm) and anaerobic peat (subsurface layers or catotelm). Estimates of the decay constants and the periods for the formation of each of these layers were again estimated by curve fitting, where I in eq. 4 was an input, either from standing-dead litter or directly from live material formed below the surface, and the equilibrium levels of the curves produced by simulation were made to agree with the known mass [depth (L) × bulk density (ML^{-3})] of each peat layer.

Clymo (1978) avoided the additional complexity of modelling bog vegetation consisting of a mixture of higher plants (with organs above and below ground) and mosses by adopting for study a carpet of pure *Sphagnum*. In this case it is easier to visualize a height increment with time. The opportunity for checking model predictions for moss growth against profile stratigraphy is obvious. Clymo's (1978) model included compaction effects; and, unlike the model of Jones and Gore (1978) which simply adapted eq. 4 by replacing W (mass) by H (depth), Clymo's model allowed depth and dry matter to vary independently, thus providing for the prediction of profiles with varying bulk density. Jones and Gore (1978) found however that, in spite of the limitations of their simple modification of eq. 4, the selective decay of different species components, principally *Calluna vulgaris*, *Eriophorum vaginatum* and *Sphagnum capillifolium*, itself produced a profile of changing bulk density.

---
[1] The usual symbols for dimensions employed are: M = mass, L = length, T = time.

However, from neither of these applications has it been possible to predict profiles satisfactorily beyond a few decimetres depth. As Clymo pointed out, this is hardly surprising since so many of the known variables have been omitted from the models. Apart from compaction and selective decay, the effects of past climatic and vegetational variation were not considered. The effects of direct penetration by deep roots of higher plants into the anaerobic layers were simulated later by Gore (unpublished), and, while the effects were not large enough to modify substantially the earlier conclusions, the phenomenon is one that may have to be allowed for in realistic models of many bog types, as well as in age determinations by radiocarbon dating. Studies which provide further ideas of the essential processes and their relative magnitudes have been conducted on reclaimed fen peats in the western Netherlands by Schothorst (1977) (see p. 20). Thus, in summary, the models can be used in two ways: either, when given the bulk density profiles (or the surface vegetation mass; Gore and Olson, 1967), to provide best estimates of parameters such as I and K in eq. 4, and such as the nine parameters used by Clymo (1978, p. 218, eq. 8); or, when given the measurements (from independent observations) of the parameters themselves, to predict age and bulk density profiles (Clymo, 1978; Jones and Gore, 1978).

A regression "model". Relations between temperature and redox potential have provided a simpler type of small-scale quantitative model. Urquhart and Gore (1973) found that linear regression coefficients of redox potential against seasonal temperature in the upper layers of two wet bogs, one at sea level, the other at 560 m in northern England, became more negative with increasing depth. The results were in accord with the hypothesis that increasing temperature in the spring caused: (1) increasing microbial activity and therefore a decrease in redox potential, which was more marked deeper in the profile; and (2) lowering of the water table and increased effects of atmospheric oxygen with greater effect near the surface. Further results of the study on two less waterlogged peat sites were not free from anomalies; but from other evidence (Bohn, 1971) redox measurements at very low concentrations of free oxygen appear to give consistent results (see also Clymo, Ch. 4).

CONCLUSIONS

The models, hypotheses and concepts outlined above each attempt to integrate some dynamic aspects of mires. The chapters which follow on climate, water, chemistry, plants and animals, and on the world's mires as they are (Part B), mainly approach the study of the ecology of mires in more detail than is assumed by the models described, and it is not surprising that there are often large gaps between the complexity of the observed detail and the simplicity of the models. The gap between experimental and observational hydrology on the one hand, and the suspected effects of "climate" and physiography on the other, is perhaps the widest and most relevant in mire ecology. The scope here for interactive dynamic models seems large indeed.

APPENDIX I: BRIEF NOTES AND REFERENCES TO OTHER COUNTRIES WITH IMPORTANT MIRE AREAS

Norway. With some 3 million ha of peatlands Norway is one of the most serious omissions from the regional accounts. According to Løddesøl (1948, 1968), about 2 million ha of mires are below the timberline, in lowland and upland areas which comprise some 50% of the total area of the country. Another 1 million ha of mires are in mountainous areas. Nordhagen (1943) and Dahl (1956) have described the vegetation of the mire types found in the mountains of southern Norway. Næss (1969) has surveyed mires in southeastern Norway, and Vorren (1967) has covered the distribution of palsas in the north. The oceanic mires on an island north of Narvik and one west of Trondheim Fjord have been described in broad vegetational terms by Osvald (1925a).

More recently, Dierssen and Dierssen (1978) have studied mire vegetation, and Hornburg (1972) and Moen (1973, 1975) have suggested typical mires worthy of preservation. Such designation of nature reserve areas is urgent because exploitation for agriculture and forestry is already large, and the latter is increasing. The peat industry is still modest; but even in 1966–1967, the famous Jiffy-Pot enterprise produced 400 million pots. Moen (1973) has emphasized the important fact that Norway, probably unlike any other single country with the exception of Canada, has almost every known type of mire; from raised bogs to blanket bogs, and from flat fens, sloping fens and aapa mires to palsas in the far north.

The reason for this rich diversity lies in the extremes of latitude, altitude and oceanicity, combined with a wide variation in geological conditions, encompassed by such a relatively small country.

Iceland. Iceland is geographically situated in the peat-forming zone (Fig. 1.1), and has been reported by Einarsson (1968) as having 10% of its total land area of 103 000 km^2 covered with peat bogs. There are no raised bogs and all mires are geogenous, but they are unusual in that the mineral influence is due to wind-blown dust and the deposition of volcanic ash. Two kinds of mire are distinguished, *floi* and *halamyri*. *Floi* mires are more topogenous and occur in lowland areas, *halamyri* are found in sloping situations and in mountains and are the more widespread of the two types. Palsas, known locally as *rustir*, occur in the highlands. Steindórsson (1975) has described the mire vegetation. Bjarnasson (1954) stated that most of the larger peat deposits were in the southwestern extremity of the country not far from Reykjavik. Bjarnasson's interest was primarily with peat for fuel, and the high ash content of Icelandic peats (see also Bjarnasson, 1968) raises special problems infrequently met with in the utilization of peat elsewhere. However, the intrinsically high fertility of Icelandic peats is an advantage to agriculture, which is exclusively for hay or pasture. The layers of volcanic ash (tephra) are of exceptional value as synchronous reference data over great areas (Einarsson, 1968). The age of the older layers can be determined from pollen analyses and ^{14}C studies, and that of layers formed within the past 1200 years from written records of volcanic eruptions. The scope for estimating peat deposition rates seems large.

The Netherlands, Belgium, Luxembourg, Denmark, France and Italy. In The Netherlands most of the peat deposits occur in two main areas — namely, to the southwest of Utrecht and in the northeast, near Emmen, almost on the border with Federal Germany. The former are mainly eutrophic, essentially fen types, and the latter are mainly bog types. The Netherlands Soil Survey (1961) has published a soil map of The Netherlands on a scale of 1:200 000 and this not only shows a more exact distribution of peat types in Holland, but illustrates how much peatland has been exploited and also the forms which exploitation has taken. Only a very small area of mire with natural vegetation still remains, near Schoonebeek, southeast of Emmen. This is a remnant of once much larger deposits extending southeastwards in a wide band from Groningen. Today there is a keen interest in the regeneration of peatlands now under agriculture or already extensively exploited for fuel and horticultural products, and latterly for activated carbon. The mires of the eastern Netherlands and of neighbouring Germany are classed as true raised bogs (*echte Hochmoore*), but as pointed out by Overbeck (1975) they appear only infrequently to have followed a classical *Verlandungs*- or terrestrialization sequence. A typical profile at Schoonebeek, shown to the author by Mr. E. Ensing (Fig. 1.7) of The Netherlands Office of Nature Conservation, has minerotrophic peat (but including *Scheuchzeria palustris* remains) at the base, overlying a shallow sandy podzol. A dark deposit of mesotrophic *Sphagnum recurvum* peat interspersed with layers of *Sphagnum cuspidatum* higher in the profile was followed by a lighter deposit of relatively undecomposed *Sphagnum papillosum* and *S. magellanicum*. The *Grenzhorizont* of Weber, discussed for example by Frenzel (Ch. 2) in this volume, can be very clearly seen in this locality (see Fig. 1.7). The

Fig. 1.7. Peat profile at Schoonebeek, The Netherlands.

peat deposits near Utrecht in the western Netherlands are utilized exclusively for dairy grassland now. It has been thought that this form of utilization is in the best interests of conserving the peat, but a recent study by Schothorst (1977) indicates that oxidation (and subsidence) is proceeding quite rapidly in the aerated layers above the water table. This work is also referred to on p. 18 above.

The peat deposits in Belgium are referred to by Vanden Berghen (1951), Eurola (1962) and Goodwillie (1980), and are confined to small, widely scattered, raised bogs in the Ardennes/Eifel cross-border region with Germany, to the north of Luxembourg (see also Fig. 1.8, Federal Republic of Germany) and to the Hautes Fagnes a little to the west. Froment (1975) has described the effects of draining for forestry in the latter area. The experience is similar to those in Britain; such plants as *Scirpus cespitosus*, *Erica tetralix*, and *Sphagnum* spp. are replaced by monospecific stands of *Molinia caerulea*.

Denmark is reported (Kroigaard, 1968) as having had some 60 000 ha of peatland, and Eurola (1962) briefly mentioned the vegetation of a plateau-shaped raised bog in northern Jylland as described by Mentz (1912). In a personal communication Mr. J.K. Øvig of the Danish Land Development Service stated that, while most fen areas in Denmark are small and widely scattered, there were notably large raised bogs at Store Vildmose and Halsmose in northern Jylland, at Lille Vildmose south of Aalborg, Knudsmore near Herving, Skastmose northwest of Tønder, and Kongsmose south of Løgumkloster. Only a small proportion, however, remains in a reasonably natural state.

Peatland ecosystems in France are sparsely scattered. Eurola (1962) refers to Lemée (1938) who described ombrogenous bog vegetation at Perche in the Pyrénées. There are several other accounts of bog-forming vegetation from elsewhere in France, notably Bretagne (Brittany) and the Vosges (Kaule, 1974). Although fairly numerous, French peatlands are small and heavily exploited (Goodwillie, 1980).

Most of the peat deposits of Italy are located in the Piemonte, Lombardia and Venezia regions in the north. They occur in morainic basins and depressions in the Po Valley, in river deltas, and along the northern coast of the Adriatic. In central Italy, and even in parts of the south, peat deposits are small but are widespread. The review by Moretti and Balboni (1968) gives details of locations and types, from an essentially utilitarian viewpoint. Most Italian wetlands, however, have been intensively drained.

Germany. The broad distribution of mire types in the Federal Republic has been given by Kuntze (1971), whose map is reproduced as Fig. 1.8. Kuntze distinguished raised bogs from fens, the former occurring mainly in the north and in some upland areas, such as the Harz (Jensen, 1961), the Hohe Rhön, the Schwarzwald (Hölzer, 1977) and the

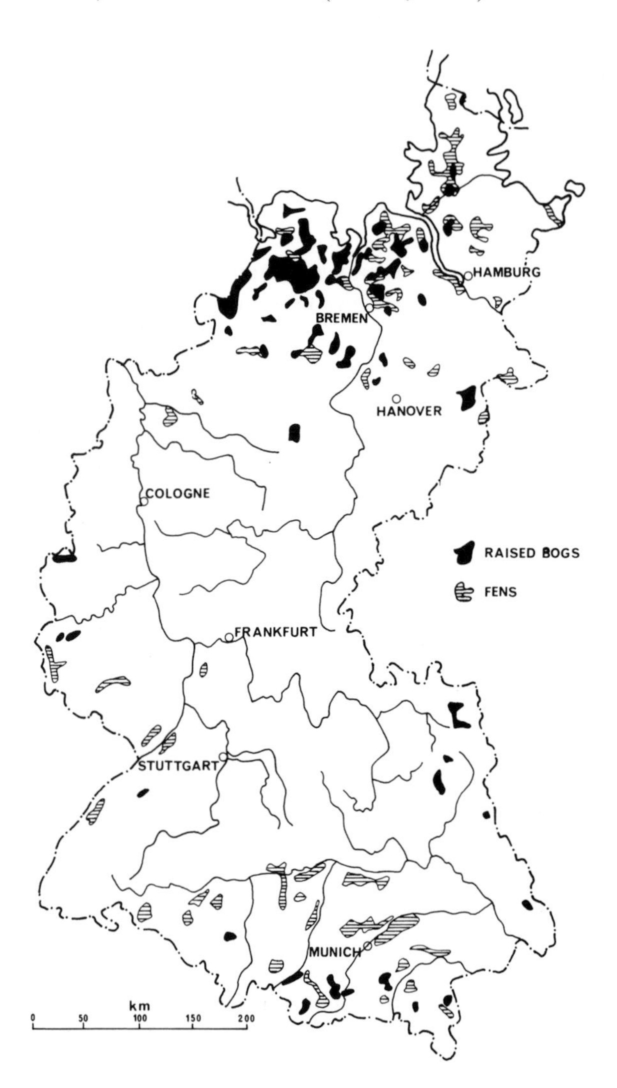

Fig. 1.8. Distribution of mires in the Federal Republic of Germany (from Kuntze, 1971).

alpine foothills (Kaule, 1974), while fens occur predominantly in the south. Today, few of these mires remain unexploited. Of the 1.1 million ha of peatland, almost equally divided between fens and bogs, all the former and about 80% of the latter are now cultivated or otherwise exploited. Schneekloth (1968a) has reviewed the range of maps which have been produced showing the distribution of German peat deposits. According to Overbeck (1975), a mere 400 ha of original mire vegetation remains in northern Germany. Overbeck has provided an extensive description of thr peatlands of Nieder-Sachsen and Schleswig-Holstein in northern Germany, although because of their extensive modification his account is essentially a historical one. A different type of review was edited by Göttlich (1976). This book covers many aspects of mire ecology with special reference to German work. Unlike conditions in many other countries, the few remaining natural mires in the Federal Republic are either protected or being considered for protection (Goodwillie, 1980).

In the German Democratic Republic there are very few bogs, and almost all the 0.5 million ha of peatlands are fens. As in the Federal Republic, virtually the whole of these mire deposits have been exploited.

Poland. The inclusion of this section on Poland has been almost wholly due to Dr. A. Szczepański of the Polish Academy of Sciences, Institute of Ecology, Mikołajki, Poland, whose contribution is gratefully acknowledged.

There are about 1.5 million ha of peatlands (90% or more exploited) in Poland, equivalent to about 5% of its territory. Of these peatlands, 95% are geogenous (fens), while the remaining much smaller proportion are ombrogenous (raised bogs). Fig. 1.9 (from Jasnowski, 1975) gives a general idea of the distribution of mires, showing their concentration in Pomerania and among the Mazurian lakes. The lakes vary in size, and overgrowth of the smaller water bodies is a common means of origin of the mires in this area. The largest mire complexes however are found in the valleys of the rivers in the northern half of Poland; for example the valley mires of the Biebrza River (Oświt, 1973; Pałczyński, 1975) are 100 000 ha in extent and those of the river Notec some 50 000 ha. Valley mires are also extensive along the rivers Krzna, Łeba, Narew, Obra,

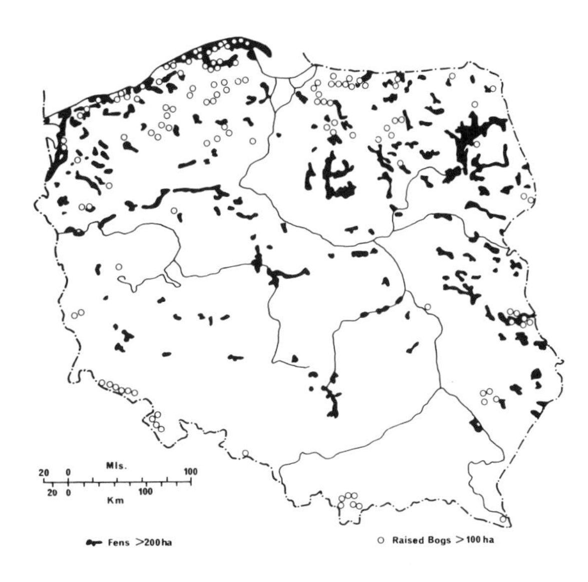

Fig. 1.9. Distribution of mires in Poland (from Jasnowski, 1975).

Odra, Pisa and Warta. Raised bogs are few and are confined to the western parts of the lake district of Mazuria, the Baltic coast, the Sudety Mountains and the Carpathian foothills. More detailed accounts of the distribution of fens and raised bogs in Poland are given by Jasnowski and Pałczyński (1976) and by Ilnicki (1977).

Much attention has been directed to the plant associations of Poland's wetland areas, with emphasis on the more eutrophic types. Szafer and Zarzycki (1972) referred to twenty alliances of plants associated with different kinds of wetland:

Water-bodies
1. Litorellion
2. Lemnion
3. Potamion eurosibiricum
4. Charion fragilis
5. Phragmition
6. Glycerio-Sparganion

Edges of water, flooded areas
7. Magnocaricion elatae

Eutrophic and mesotrophic non-wooded wetlands
8. Rhynchosporion albae
9. Caricion canescentis-fuscae
10. Caricion davallianae
11. Filipendulo-Petasition
12. Calthion
13. Molinion coeruleae

Eutrophic swamps
14. Alnion glutinosae
15. Alno-Padion

Raised bogs with and without tree cover
16. Sphagnion fusci
17. Ericion tetralicis
18. Dicrano-Pinion

Springs and flushes
19. Cardamino-Montion
20. Cratoneurion commutati

A zone of contact between the Atlantic flora and the more continental flora occurs over Polish territory (Herbichowa, 1979), and palaeoecological studies (Marek, 1965; Pacowski, 1967; Jasnowski, 1972; Pałczyński, 1972, 1975; Oświt, 1973) are assessing the long-term changes in plant associations. The high proportion of lacustrine habitats has encouraged great interest in the phytosociology and succession in littoral vegetation, for example by Kępczyński (1960, 1966), Dąmbska (1961, 1965), Olkowski (1966), Bernatowicz and Zachwieja (1966), Polakowski and Dziedzic (1972), Rusińska (1974) and Michna (1976). Similar studies on valley marshes and mires have been made by Olaczek (1967), Pacowski (1967), Oświt (1973), Polakowski (1969) and Pałczyński (1972). Succession in the rather uncommon dystrophic bog lakes has been described by Sobotka (1967). There has been some work done on the productivity of the littoral vegetation of lakes in Poland — for instance, in several papers by Bernatowicz and coworkers (1965, 1966, 1968) — and the subject was further studied under the International Biological Programme; but these studies are still few in number. The problems of mire conservation in Poland were reviewed by Jasnowski (1977).

Phragmites australis plays an important role in many of Poland's wetlands, and this species has formed the object of intensive study at the Wetland Research Laboratory of the Institute of Ecology in Mikołajki. A summary of this work is being published by Szczepanski (1978).

Austria, Switzerland, Czechoslovakia, Romania and Hungary. The broad zonation of European mires shown in Fig. 1.4 is placed in better perspective by quoting the existence of a small raised bog in the Austrian Alps described by Ullmann and Stehlik (1972). This bog is situated 700 m above sea level about 105 km southwest of Vienna in the Steiermark. It appears to owe its similarity to Fennoscandian mires, with a marked concentric pool and hummock system, to local climatic conditions much influenced by complete shading during two winter months by high mountains immediately to the south. This small bog has a well-defined raised plane, at least in its northeast–southwest section. The height of the plane is some 3 m above the local base datum, and the length of the section is 300 m. Studies on some of the relatively few other bogs in Austria and in neighbouring regions of southern Germany, Switzerland and the hilly uplands of Czechoslovakia are referred to by Ullmann and Stehlik (1972). Imhof and Burian (1972) have worked on the productivity of reed (*Phragmites australis*) communities in the eastern region of Austria near the Hungarian border. The Neusiedler See is a very shallow lake (average depth 1 m) of some 230 km^2, almost half covered by a reed belt some 1 to 3 km in width. A recent inventory of Austrian mires has been summarized by Wendelberger (1973).

The classical Swiss work by Früh and Schröter (1904) should be mentioned here, though today only some 125 km^2 of Swiss mires are in a reasonably natural state (Goodwillie, 1980).

Czechoslovakia was reported by Mejstřík (1968) as having 30 750 ha of peatlands, with oligotrophic (raised bog), transitional mire, and eutrophic fen types in almost equal proportions; 87% of these peat deposits were in mountain or upland areas in Czech lands, the remainder in Slovakia. The mire vegetation has been described, for instance, by Neuhäusl (1972), and Ferda (1968) referred to mires in the Šumava (Böhmer Wald), Krušné Hory (Erzgebirge) and Krkonoše (Riesengebirge). Few of these mountain mires exceed 100 ha. Exploitation is common, but about a quarter of the peatlands are under protection in their natural state. Rybníček (1974) has described mires in the uplands between Bohemia and Moravia, and Ingram elsewhere in this volume (Ch. 3) mentions hydrological studies in various habitats of modified mires in the same area.

As in Austria, research has been done on reed-swamp areas in Czechoslovakia in connexion with the International Biological Programme. The littoral ecosystems of two old fishponds, namely the

Nesyt fishpond (320 ha) in southern Moravia and the Opatvicky fishpond (165 ha) near Třeboň in southern Bohemia, were selected for intensive study (Hejný, 1973). Emphasis was placed on reed-swamp communities dominated by *Phragmites australis*, but including *Acorus calamus*, *Glyceria maxima*, *Scirpus lacustris*, *Scirpus maritimus*, *Sparganium erectum*, *Typha angustifolia* and *T. latifolia*. Temporary flooding between 1960 and 1966 brought about the re-establishment of reed-swamp communities after over 100 years of drainage (Kvét et al., 1969). The fishponds have been developed from the thirteenth century onwards from previously swampy lowlands. They appear from general accounts to have much in common with the English Norfolk Broads (see Taylor, Part B, Ch. 1).

Apart from reed swamps, peat deposits in Romania are relatively scarce, but (as elsewhere) include bogs, mostly in the Carpathian Mountain area, and fens and reed swamps, which occur in the valleys of the same region. In an extensive account, Pop (1960) has placed the mires of Romania in their ecological order, and provided details of many individual mires. In his figs. 21 and 22 Pop (1960) shows the location of eutrophic mires and raised bogs respectively. Most of these mires occur in the Carpathian Mountains and in the Transylvanian Alps (Carpaţii Meridionali). In the delta of the Danube there are huge areas of reed swamp. These have been described by Pallis (1916) and Rodewald-Rudescu (1974).

Unlike her immediate neighbours, Hungary has very few upland regions, so that peat deposits are almost all of lacustrine or riverine types (Soó, 1954; Kovács, 1962). Belak (1968) mentions relatively large areas of reed swamps and fens along the lakes of Balaton and Fyerte and the rivers Danube, Tisza and Körös.

Japan. Mires in their natural state now occur most commonly in the northern island of Japan, Hokkaido, and in the north of the main island Honshu. Sakaguchi in his general text on peatland geology (1978; see also 1979), has provided a broad picture of peatland distribution in Japan, pointing out that most of the lowland peat areas which occur widely in all parts of the country have been changed by human exploitation. Botch and Masing (Part B, Ch. 4) set the wider climatic context of Japanese mires in their consideration of the mires of the extreme eastern U.S.S.R.

Details of the wide range of mire types found in Japan have been given by Suzuki and his coworkers (1973) and Yoshioka (1974). These include swamp forest with *Alnus japonica* and *Fraxinus mandshurica*, and there are true reed swamps of *Phragmites australis*. Among the other wetland areas a few marshes remain in their natural state, but many have been converted to rice paddies and irrigation ponds. There are a wide range of types including sedge fen, carr, raised and blanket bogs. Notable among the fen types is the occurrence of *Molinopsis japonica* with quite similar ecological properties to *Molinia caerulea*. Yoshioka reports on two "moorland" areas in northern Honshu, namely Akaiyachi Moor and Ozegahara Moor. Akaiyachi Moor with a peat depth of 3 to 5 m is thought to have originated from a swamp forest of *Fraxinus mandshurica*, not from lacustrine conditions. There is no eutrophic fen in the area, but a transition community occurs of *Osmunda cinnamomea* including *Moliniopsis japonica* and *Phragmites australis*, and the peat mosses *Sphagnum magellanicum*, *S. palustre* and *S. papillosum* are also present locally.

Ozegahara Moor in northeastern Honshu, which began its development 8000 to 10 000 years ago, has a peat depth of 5 to 6 m. Gallery forests flourish on the banks of creeks which wind through the mire and these forests include a wide range of tree species such as *Aesculus turbinata*, *Betula ermanii*, *Tilia japonica* and *Ulmus davidiana*. Swamp occupies less than 10% of the area, and eutrophic fens are also limited to areas near the foot of gentle slopes bordering the mire. In the central more oligotrophic parts there are some 400 bog ponds which, unlike most pools of true bogs described elsewhere in the world, include islets of floating peat. The pool areas form the highest parts of the raised bog habitats of Ozegahara Moor; common species here include *Eriophorum vaginatum*, *Narthecium asiaticum* and *Rhynchospora alba*, together with *Sphagnum compactum* and *S. papillosum*. On areas of slight slope there is a hummock-and-hollow complex, with the hummocks containing *Carex middendorfii*, *Molinopsis japonica* and *Vaccinium oxycoccos*, together with *Sphagnum fuscum* and *S. papillosum*. In the hollows are *Drosera anglica*, *D. rotundifolia*, *Lepidotis inundata*, *Rhynchospora alba*, *Scheuchzeria palustris* and *Sphagnum pulchrum* with *S. papillosum*.

On the marginal more steeply sloping areas of the bogs, *Molinopsis japonica* is dominant in association with *Carex middendorfii*, *Eriophorum vaginatum*, *Gentiana thunbergii* var. *minor*, *Myrica gale* var. *tomentosa* and *Sanguisorba officinalis*.

As the slopes steepen, a fern community of *Osmunda cinnamomea* develops, and this in turn gives way to scrub dominated by *Prunus grayana* and *Sorbus commixta* in association with *Hydrangea paniculata* and species of *Malus*, *Prunus*, *Rhododendron* and *Rhus*. The undergrowth of this scrub includes *Osmunda cinnamomea* and *Pteridium aquilinum* var. *latiusculum*.

The botanical interest and the mire characteristics of Ozegahara Moor are therefore striking. From the map given by Yoshioka it is notable that the first three of the above communities occur in several distinct areas within a very small zone having a length of about 600 m between the creeks Yoppi and Nushiri.

Blanket bog in Japan usually develops on the slopes of volcanoes where the precipitation, especially snowfall, is high. The effect of the gentle slopes, combined with the fine ash, cause impeded drainage and lead to bog formation directly on to the dry land surface. Blanket bogs are found on Mount Taisetsu in Hokkaido, and on several mountains including Mount Hakkōda, Mount Hachimanti and Mount Gassan on Honshu. They have an almost identical flora to that of the raised bogs, but lack hummock-and-hollow microtopography. There are frequent islets of woodland or scrub occupying hillocks within the bog.

A detailed study of Takadayachi Moor in the Hakkōda Mountains was made by Yoshioka and his associates, (Yamanaka, 1963; Yoshioka, 1963). The stratigraphy through sections of this small raised bog (Fig. 1.10) show a remarkable similarity to sections of raised bogs in Sweden described by Granlund (1932) (see Fig. 1.3). Instead of relying on "recurrence surfaces" as measures of bog development, however, Yoshioka had the advantage of well-defined layers of volcanic ash (see notes on Iceland, p. 19 above). Although the dates of the corresponding eruptions are not known, the rates of peat accumulation are clearly shown to be least at the mire margin and greater over the bog plane, thus generally supporting the ideas of Granlund (1932) and Wickman (1951). From Yoshioka's data, however, it appears that Takadayachi Moor fits rather poorly into the curves Wickman derived from Granlund's observations on Swedish raised bogs.

China. On recent vegetation maps of China, mire areas are indicated mainly in the extreme northeast, but also in the central mountainous parts. The peatland area is said to be some 30 000 km^2 (world survey by the International Peat Society: see Kivinen and Pakarinen, 1980).

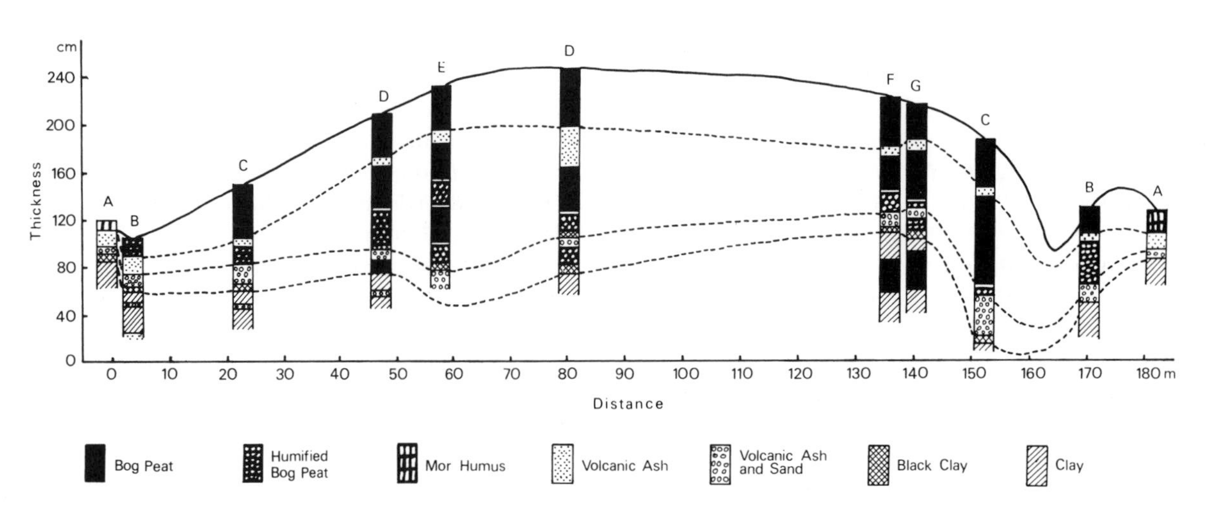

Fig. 1.10. Section through Takadayachi Moor in the Hakkōda Mountains, as an example of a Japanese raised bog (after Yoshioka, 1963).

India and Pakistan. Gopal (1973) has briefly reviewed the literature on the wetlands and marshlands of India in the context of productivity, within the International Biological Programme. He has emphasized the role of the monsoon rains in the seasonal development of marshland vegetation. Rapid growth occurs from perennating underground parts with the onset of the monsoon. After the monsoon this vegetation flourishes throughout the winter and later, as summer approaches, the vegetation reverts to meadowland and the marsh species perennate as below-ground parts. The main species are: *Cyperus exaltatus*, *C. rotundus*, *Eichhornia crassipes* (introduced), *Eleocharis plantaginea*, *Hygrophila auriculata*, *Ipomoea aquatica*, *Monochoria vaginalis*, *Oryza sativa* (an escape from cultivation), *Paspalum scrobiculatum*, *Phragmites* spp., *Spirodela polyrhiza* and *Typha angustata*.

Misra (1946) has described the dynamics of Indian marshland vegetation in relation to habitat factors; but in general the majority of published work lays emphasis on floristics and zonation. Water chemistry is often notable for its extremely high pH values, ranging between pH 7.5 and pH 11.0. Calcium carbonate is frequently precipitated, forming encrustations over submerged macrophytes such as *Hydrilla verticillata*.

As far as utilization of Indian marshlands is concerned, Gopal is more forthright than many authors on tropical wetlands. For example it has been suggested that *Eichhornia crassipes*, introduced from South America, might be cultivated and exploited as a protein food source. Harvesting would help to control this noxious water weed.

Such seasonally dependent conditions of soil waterlogging are rarely if ever conducive to peat formation, but substantial deposits of peat have been reported (Anonymous, 1968) underlying silts of the delta of the Ganges and Brahmaputra rivers in Bangladesh. The origin of these is not mentioned, but one speculates that they may have been formed under swamp forest as in Central Africa.

Southern latitudes. Besides the extensive mires on Tierra del Fuego and adjacent islands dealt with by Pisano (Part B, Ch. 11), and those in Tasmania and New Zealand (Part B, Ch. 5; see also Dobson, 1979), peat deposits are known to exist in most southern temperate and subantarctic islands (Schwaar, 1979), such as Tristan da Cunha, Gough Island (see below), Marion, Macquarie and Easter Islands, the Falkland Islands, South Georgia (Smith, 1979b), the South Orkney and South Shetland Islands (including Elephant Island), and even on the west coast and offshore islands of the Antarctic Peninsula itself (Smith, 1979a). Here, however, peats are exclusively terrestrial and are formed from mosses other than *Sphagnum*, decay being hampered by low temperatures and permafrost close to the surface rather than by waterlogging.

Gough Island lies at a latitude of 40°S, almost midway between the southernmost tips of South Africa and South America. It has an average temperature of 11.7°C, and an annual rainfall of 3250 mm. Schwaar (1977), reporting on the properties of the blanket peat of the island, describes a profile of some 105 cm depth. The existing vegetation and the peat consists of remains of ferns and the shrub *Phylica arborea* (Rhamnaceae). The profile contained a marked lighter upper peat of some 70 cm overlying a more decomposed, darker one below. There was, however, no change in botanical composition at this level. Chemical analyses of the upper 20 cm showed an ash content of 2.2%, pH 3.1, nitrogen 2.1%, phosphorus 0.09%, potassium 0.14%, calcium 0.3% and magnesium 0.19%. Although these values differ in certain respects from those obtained from raised bogs in Germany and Argentina, having generally higher nutrient contents than those for *Hochmoor* types, the results are similar to analyses for blanket bog of similar depth at Moor House in northern England (see Taylor, Part B, Ch. 1). In reporting these values Gore (1961) compared them with data from a raised bog at sea level some 70 km from Moor House, and also noted particularly higher phosphorus and calcium contents in the Moor House peat; but the exceptionally high values of nitrogen and phosphorus on Gough Island must, as Schwaar records, be due to the effects of sea-bird droppings.

Hawaiian Islands. Selling (1948) has described the montane mires of the four islands Kauai, Molokai, Maui and Hawaii. He also reported on vegetation coming close to that of true bogs on Oahu. These mires occur at altitudes of 1200 m to 1765 m, under rainfall conditions of 5000 mm annually. Temperature is thought to be less important and, although few figures were available to Selling, one "true open bog" (Wahiawa bog) reported by Fosberg (1936) is found at 600 m on Kauai. The disposition and orientation of the mires is related to the direction of the trade winds.

Selling had a clear appreciation of the words "mire" and "bog" as used in this volume and reported the presence of raised bogs and soligenous mires. The former are illustrated by an example at Kilohaua, Kauai (1220 m elevation). The section of this "bog hillock" shows a maximum peat depth of about 3 m beneath the "cupola" and a total diameter of 50–60 m. The other sections illustrated in Selling's paper all reveal a ridged substrate, and strongly suggest that these peat deposits are blanket bogs. However, it is doubtful whether more than a small proportion are truly ombrogenous. Selling emphasized the role of soligenous influences, but used the word "bog", clearly intending it to mean ombrogenous.

The vegetation of the soligenous mires consists of densely tufted species of *Rhynchospora*. The surface is fairly even and includes *Oreobolus furcatus*, the principal peat-forming species, and species of *Aira* and *Panicum*. These associations merge into "wet grass heath" in which occur scattered low shrubs (*Coprosma*, *Metrosideros polymorpha*, *Styphelia*, *Vaccinium*), herbs (*Acaena exigua*, *Drosera longifolia*, *Schizaea robusta* and *Selaginella deflexa*; *D. longifolia* in Kauai only), and mosses, including *Racomitrium lanuginosum* (*Sphagnum* spp. are rare, and in any case do not form peat in these localities), and liverworts and lichens. Bog hillocks are built up from associations of such cushion-forming plants. There is no **lagg**, but the boundary between the hillock and the rest of the swampy area can be quite sharp.

Cultivation is quite impracticable so that draining, which was tried in western Maui, is merely destructive of an important resource of late Quaternary historical information.

Conclusion. There are further omissions known to the author, but for which it has not been possible to obtain more than a hint of the extent and nature. Cuba is reported by Moore and Bellamy (1974) as having 200 000 ha of peatland, and there are also peatlands in Jamaica, in danger of being exploited for fuel. Israel, having but 5000 ha of peat, is nevertheless active on the subject of reclamation (Schallinger, 1968) and agricultural conservation (Shoham and Levin, 1968). Very little has been reported from the tropical swamps of Mexico and Central America. As a final caveat furthermore, individual contributors of the national or regional accounts in this volume often stress the lack of information available to them in respect of parts of the areas they discuss.

Up-to-date figures for peatland areas in most peat-containing countries have been given by Kivinen and Pakarinen (1980), totalling to 420 million ha for the whole world. However, there is still probably an underestimation for certain areas difficult of access, and a total lack of figures for many tropical and southern lands, so Kivinen and Pakarinen estimated the true figure to be about 500 million ha. At least 6% of this area is exploited, and the total exploited area may well be greater, since some countries exclude the more heavily exploited or cultivated peatlands from the areas they report. In some instance, there might have been some inclusion of mineral wetlands.

It would be inappropriate to close this Appendix without reference to the work of Von Bülow (1925, 1929) and Kats (1971), both of whom addressed the subject of mires on the world scale. Kats's paper, in particular, contains a very substantial bibliography of mire literature from the Soviet Union and the world as a whole.

APPENDIX II: DEFINITIONS OF TERMS COMMONLY USED IN MIRE STUDIES

In different countries, a wide range of different terms have been applied to the mires that are the subject matter of this volume, and their exact connotations often present difficulties. The more important and relevant words in common usage are given below with sufficient explanation to link them to the ecological conditions under which they best apply. Like all descriptive terms their use is often loose and can only be defined to a limited extent. More detailed background to these terms may be found by consulting the General Index.

Terms have been grouped under the following four subheadings to aid understanding, but reference via the General Index may be necessary if an alphabetical enquiry is required: peatlands; special local terms denoting function and circumstance of formation; soil terms; and functional terms denoting processes.

Peatlands

Mire is now an internationally accepted term which includes the generally **ombrotrophic** types namely **bog** [*Hochmoor*, *Weissmoor* (in part) and *Reisermoor* (in part), German; *mosse*, Swedish] on the one hand, and the **minerotrophic** types such as **fen** (*Niedermoor*, *Flachmoor* and *Braunmoor*, German; *kärr*, Swedish) and **carr** (swamp, North American; *Sumpfwald* or *Bruchmoor*, German; *löv-kärr*, Swedish; *korpi*, Finnish) on the other. The literal translations of *Hochmoor* and *Niedermoor*, "high-moor" and "low-moor", are used in much of the older literature in English but they tend to be misleading and are falling into disuse.

General words for **mire** from countries having land of the relevant type include *veen* (Dutch), *Moor* (German), *myr* (Swedish and Norwegian), *mýri* (Icelandic), *suo* (Finnish), *boloto* (Russian), *soo* (Estonian), *tourbière* (French) and *muskeg* (Indian word used in Canada).

Swamp (*Sumpf*, German) and **marsh** imply eutrophic conditions but are less specific words in popular usage. The latter is often confined to wetlands with more or less mineral soils but, apart from the North American usage above, **swamp**, like **wetland**, is very widely used to include both mires and marshes.

Special local terms

Many terms coined in one country have subsequently been adopted elsewhere.

The following are in common usage in different parts of the mire literature in English. They often refer to circumstances of formation and by doing so elaborate on the general term **mire** defined above. In general they are given in alphabetical order according to the language of their first usage.

Aapa (Finnish): extensive watery sedge mires of northern Finland. A cold-climate variant of mires, having ridge-and-pool surfaces oriented along the contours. This kind of mire is widespread in the Boreal Zone.

Blanket bog, *terrainbedeckendes Moor* (German), **climatic bog**: an extensive mire type over undulating terrain, not confined to depressions and usually formed in response to a very humid climate.

Carr (synonyms: see above): fen with scrub or woodland.

Flark (Swedish), ***rimpi*** (Finnish), **mud-bottom (plant) communities**: area of exposed peat with an algal film, sometimes having a sparse cover of sedges. Such areas are often locally frequent; though usually fairly small, some may be up to 1 km in length.

Hochmoor (German), ***högmosse*** (Swedish), **raised bog**: classic "peat bog" of temperate northwestern Europe, western and north-central U.S.S.R., and elsewhere; often includes a mosaic of *Weissmoor* and *Reisermoor* (German) which are types of raised bog vegetation respectively dominated by sedges and by dwarf shrubs. These mires were initiated during the early Postglacial or later, developing by accumulation processes to a depth of several metres and often extending over hundreds of hectares. The term refers to the convex cupola of ombrotrophic peat raised a few metres above the level of the surrounding land. This is an important feature reflecting the water balance of the area (see pp. 14–17). However, many raised bogs are not domed, or only asymmetrically so, having developed eccentrically. Associated terms are:

Rand, ***Hochmoorrand*** (German): outwards-sloping margin at the periphery of a raised bog.

Lagg (Swedish): marginal zone outside the **Rand** containing fen vegetation and representing the transition between raised-bog peat and mineral soils.

Bog expanse: the inner part of a raised bog, within the *Rand*.

Lawn, ***fastmattsamhällen*** (Swedish): a uniform expanse of firm mire vegetation notably of wiry cyperaceous plants or grasses (e.g. *Molinia*), and to be distinguished from both hummocks and soft or mossy communities commonly found in mires. Usually lawns are local but may have considerable extent.

Mixed mire, ***blandmyr*** (Swedish): an aapa mire with ridges high enough to become ombrotrophic, with the rest, largely **flarks**, remaining minerotrophic.

Niedermoor, ***Braunmoor*** (German), ***kärr*** (Swedish), **fen**: minerotrophic mire usually having a wider range of species than bogs. In this respect, **poor fens** (acid) differ less from bogs than **rich fens** (from slightly acid to alkaline and often distinctly calcareous).

Palsa (Finnish): mound of peat with a permanently frozen core. Palsas often reach 3 to 6 or even 7 m in height and may sometimes be hundreds of metres in diameter. They are confined to winter-cold areas where, however, permafrost is not continuous.

Schwingmoor (German), ***gungfly*** (Swedish), **floating mat**: mire formation floating on the water of moraine hollows, "kettle holes", sink holes, etc., but most frequently developed along sheltered lake shores. Floating reed swamps in the Danube delta are termed *plaur*, and floating rafts of *Cyperus papyrus* in the Sudan are known as *sudd*.

Soak, ***dråg*** (Swedish), **water track**, **vattenbana** (Swedish), ***Wasserbahne*** (German): strip of fen with seepage of moving water, crossing bogs or separating bog areas from each other.

Strangmoor (German), ***strängmyr*** (Swedish), **patterned mire**, **string bog**: an aapa mire with hummock-and-hollow relief caused by abundant local flow of melt water. Ridges (*Stränge*, strings) and pools are oriented in a ladder pattern at right angles to the downslope direction. In large eccentric and concentric examples of *Hochmoore* a corresponding pattern is often developed (*kermi* bogs: Finnish). The most extensive patterned mires are found in the Boreal to Low Arctic Zones.

Valley bog, **headwater bog**: mire (usually fen rather than bog proper) formed in the valleys of slowly moving rivers, or in slightly concave water-gathering depression grounds. The term is often used for fairly extensive mire formations such as the flood-plain mires of Eastern Europe; but it is used in a British context for quite small mires (for example, see Taylor, Part B, Ch. 1).

Soil terms

Brief reference is made to the terms used for the organic matter of mire systems; these include **peat**, **turf**, **moss**, **muck** and **sod**. The term **turf** is more common in Irish than in English usage; in England it refers mostly to a lawn-grass substrate, i.e. sod. In other languages a linguistically related word is used in the sense of peat: *Torf* (German), *torf* (Russian and Polish), *turf* (Dutch), *turvas* (Estonian), *torv* (Swedish and Norwegian), *turve* (Finnish), *tourbe* (French). Peat itself is essentially any predominantly organic deposit formed in a wet environment, nearly always *in situ*; but definitions giving more functionally derived descriptions exist (e.g. Waksman, 1942). **Moss** is usually now restricted to meaning relatively undecomposed peat, especially including the obvious remains of *Sphagnum*. Occasionally the word has been used synonymously with **bog** but such usage seems potentially confusing. **Muck** is a term used in the U.S.A. and generally refers to peat or gyttja soils containing mineral matter, sometimes under agricultural or horticultural usage. **Mud**, ***gyttja*** (Swedish) and ***dy*** (Swedish) indicate partly organic bottom deposits in lakes or the sea containing remains of water life, etc., which predominate in the often pale-coloured *gyttja*, whereas the dark *dy* is partly deposited from colloidal suspensions. **Tephra** are layers of volcanic ash found, for instance, in the mires of Iceland and Japan (see pp. 19 and 23), and also in other volcanic areas (Alaska, Kamchatka, Patagonia).

The two main functional horizons of mires are distinguished as the **acrotelm**: the surface, active, and more water-permeable upper layer; and the **catotelm**: sub-surface, fossilized layers (Ingram, 1978).

These and other soil terms are discussed in more detail by Clymo (Ch. 4).

Functional terms

The following terms refer mainly to conditions and processes of mire formation. They are grouped according to a specific range of terms in each case and there are several cases of overlapping meaning between groups.

Allogenic: [change or development] produced by influences external to the mire itself.

Autogenic: [change] resulting from processes internal to the mire, such as the accumulation of peat in modifying the water regime. (The development of many raised bogs since Postglacial times, as revealed by stratigraphic studies of peat profiles, illustrates allogenic change in the earlier stages followed later by autogenic change and represented by the formation of the "raised" cupola.). Strictly speaking, even autogenic development includes allogenic influence, such as that of climate and the chemical constituents of precipitation and fall-out.

Allochthonous: [peat] of sedimentary origin, i.e. not formed *in situ*.

Autochthonous: [peat] formed *in situ*.

Limnetic, limnic: [peat formation] occurring on or in deep water by free-floating or deeply rooted plants.

Telmatic: [peat formation] at the water table due to plants growing under conditions of periodic flooding.

Terrestric: [peat formation] above the general water table.

Paludification, ***Versumpfung*** (German), ***försumpning*** (Swedish): formation of mire systems over previously forested land, grassland or even bare rock, due to climatic or autogenic processes. The literal meaning is "swamping".

Terrestrialization, ***Verlandung*** (German), ***igenväxning*** (Swedish): formation of a mire system by filling of a water body with organic remains, usually by gradual extension of peat-forming communities outwards from the shoreline of a lake.

Auxotrophic: becoming more eutrophic.

Eutrophic: mineral-rich.

Oligotrophic: mineral-poor, nearly synonymous with **dystrophic** but the latter more limited in application to conditions naturally toxic to some organisms.

Mesotrophic: intermediate.

Meiotrophic: becoming more oligotrophic.

Minerotrophic: [a supply of water to vegetation] originally derived from mineral soils or rocks but sometimes via lakes or rivers as intermediates; it may be eutrophic, mesotrophic or oligotrophic.

Ombrotrophic: [a supply of nutrients] exclusively from rain water (including snow and atmospheric fall-out), therefore making nutrition extremely oligotrophic often in an unbalanced way.

Ombrogenous: [peat or vegetation] formed under ombrotrophic conditions.

Geogenous (either **topogenous** or **soligenous**): **geogenous** refers to peat or vegetation formed under minerotrophic conditions; **topogenous** refers to a stagnant and essentially non-moving water variant while **soligenous** indicates a moving-water type.

Rheophilous: a term for peat vegetation growing under moving-water conditions which may be either mineral-rich or mineral-poor in origin. It is almost synonymous with soligenous (see above), though often used in the wider sense of geogenous (minerotrophic).

Ombrophilous: [vegetation] tolerating a predominantly rain-water influence.

Recurrence surface, *rekurrensytor* (RY) (Swedish): a term used in peat stratigraphy denoting a striking change in conditions of peat formation (see Frenzel, Ch. 2). Most commonly it is a sharp discontinuity from highly humified peat below, changing to relatively undecomposed *Sphagnum* above (Fig. 1.7) ["recurrence" from the fact that there is a repetition of a sequence of similar climatic events believed to be contemporaneous in different bogs (Granlund, 1932; Von Post, 1946).]

Grenzhorizont (German): a particular recurrence surface (RY III) associated with the major change of climate from the Subboreal to the Subatlantic (see Weber, 1926; Von Bülow, 1930; Schneekloth, 1968b), and dated by radiocarbon at about 500 B.C. or slightly earlier.

Regeneration complex [a term coined by Osvald (1923), based partly on earlier evidence from Von Post and Sernander (1910)]: an area on a bog expanse with a supposedly cyclic process of bog growth. **Hummocks** of *Sphagnum* are thought to develop from **hollows** due to semi-aquatic *Sphagnum* spp. growing more quickly, and new **hollows** to be produced by inundation of the previously existing and slower-growing vegetation (including lichens and dwarf shrubs) on the **hummocks**. Though such successions are easy to demonstrate from stratigraphic evidence, such a mechanism does not apply to all hollow-and-hummock complexes, because large, deep hollows or real bog pools have been stable in position over thousands of years (see Ruuhijärvi, Part B, Ch. 2).

Conclusion

The terms listed above are the more important ones used in mire studies. There are a large number of other terms which are composite expressions (see Ruuhijärvi, Part B, Ch. 2). These incorporate descriptive words appropriate to a particular mire area, for example; ***Waldhochmoor*** (German), **pine bog** (***tallmosse***: Swedish), or **spruce muskeg**, **swamp forest mire** and **tropical forest swamp**, all of which represent woodland mires in different circumstances; or **poor fen**, **weakly soligenous bog** or **transition bog**, all of which are ***Übergangsmoore*** (German) and which describe mesotrophic conditions with different degrees of emphasis. Use of such expressions is usually clear from the context, but in general care is needed to avoid unnecessary proliferation of terms, and Waksman's (1942) comment seems appropriate: "If all conditions are taken into consideration only very few types [of peat] need to be recognized and few terms employed".

Other local and special terms will be encountered in the text, and are explained where they are introduced. The General Index will enable the explanation to be found readily.

The helpful co-operation of Professor Hugo Sjörs in the preparation of this short glossary is gratefully acknowledged.

REFERENCES[1]

Anonymous, 1968.[1]

Belak, S., 1968. In: C. Lafleur and J. Butler (Editors), *Proc. 3rd Int. Peat Congr. Quebec, 1968*, p. 22.

Bernatowicz, S. and Pieczyńska, E., 1965. Organic matter production of macrophytes in the lake Tałtowisko [Mazurian lakeland]. *Ekol. Pol., Ser. A*, 13(9): 113–124.

Bernatowicz, S. and Radziej, J., 1965. Produkcja roczna makrofitów w kompleksie jeziora Mamry. [Annual production of macrophytes in the archipelago of Mamra.] *Pol. Arch. Hydrobiol.*, 12: 307–348.

Bernatowicz, S. and Zachwieja, J., 1966. Types of littoral found in the lakes of the Masurian and Suwałki lakelands. *Ekol. Pol., Ser. A*, 14(28): 519–545.

Bernatowicz, S., Pieczyńska, E. and Radziej, J., 1968. The biomass of macrophytes in the lake Śniardwy. *Bull. Acad. Pol. Sci.*, II, 16: 625–629.

Bick, W., Robertson, A., Schneider, R., Schneider, S. and Ilnicki, P., 1976. *Słownik torfoznawczy niemiecko — polsko — angielsko — rosyjki*. [*Glossary for Bog and Peat. German — Polish — English — Russian*.] Biblioteczka Wiadomości I.M.U.Z. No. 56. Instytut Melioracji i Użytków Zielonych, Warsaw, 178 pp.

Bjarnasson, O., 1954. The peat deposits of Iceland. In: *Proc. 1st Int. Peat Soc. Symp., Dublin, 1954*, pp. 1–8.

Bjarnasson, O.B., 1968. Chemical investigations of Icelandic peat. In: R.A. Robertson (Editor), *Proc. 2nd Int. Peat Congr., Leningrad, 1963*, 1: 69–73.

Bohn, H.L., 1971. Redox potentials. *Soil Sci.*, 112: 39–45.

Cajander, A.K., 1913. Studien über die Moore Finnlands. *Acta For. Fenn.*, 2(3): 1–208.

Clymo, R.S., 1978. A model of peat bog growth. In: O.W. Heal and D.F. Perkins (Editors), *Production Ecology of British Moors and Montane Grasslands*. Ecological Studies, 27. Springer, Berlin, pp. 187–223.

Conway, V.M., 1948. Von Post's work on climatic rhythms. *New Phytol.*, 47: 220–237.

Dahl, E., 1956. Rondane: mountain vegetation in South Norway and its relation to the environment. *Skr. Nor. Vidensk.-Akad. Oslo, Mat.-Naturv. Kl.* 1956 (3): 1–374.

Dahl, E., 1969. Teorier omkring myrkomplexenes dannelse. In: *Myrers Økologi og Hydrologi: Symposium om Myrer, Ås, 1969*. International Hydrological Decade, Oslo, pp. 20–24.

Dąmbska, I., 1961. Roślinne zbiorowiska jeziorne okolic Sierakowa i Międzychódu. [Plant communities of lakes in the region of Sierakow and Miedzychod.] *Poznan. Tow. Przyj. Nauk, Wydz. Mat.-Przyr., Pr. Kom. Biol.*, 23(4): 1–120.

Dąmbska, I., 1965. Roślinność litoralu jezior lobeliowych Pojezierza Kartuskiego. [The littoral vegetation of the "Lobelia" lakes in the Kartuzy Lake District.] *Pozn. Tow. Przyj. Nauk, Wydz. Mat.-Przyr., Pr. Kom. Biol.*, 30(3): 1–55.

Daniels, R.E., 1978. Floristic analyses of British mires and mire communities. *J. Ecol.*, 66: 773–802.

Dau, J.H.C., 1823. *Neues Handbuch über den Torf, dessen Natur, Entstehung und Wiedererzeugung. Nutzen im Allgemeinen und für den Staat*. J.C. Hinrichse Buchhandlung, Leipzig, 244 pp.

Dierssen, K. and Dierssen, B., 1978. The distribution of communities and community complexes of oligotrophic mire sites in western Scandinavia. *Colloq. phytosoc. Lille* 7 (*sols tourbeux*): 95–119.

Dobson, A.T., 1979. Mire types of New Zealand. In: *Proceedings of the International Symposium on Classification of Peat and Peatlands, Hyytiälä, 1979*. International Peat Society, University of Helsinki, Helsinki, pp. 82–94.

Du Rietz, G.E., 1949. Huvudenheter och huvudgränser i svensk myrvegetation. *Sven. Bot. Tidskr.*, 43: 274–309.

Einarsson, T., 1968. On the formation and history of Icelandic peat bogs. In: R.A. Robertson (Editor), *Proc. 2nd Int. Peat Congr., Leningrad, 1963*, 1: 213–216.

Eurola, S., 1962. Über die regionale Einteilung der südfinnischen Moore. *Ann. Bot. Soc. Vanamo*, 33(2): 163–243.

Ferda, J., 1968. Czechoslovakia: Peat bogs and their utilization in Czechoslovakia. In: C. Lafleur and J. Butler (Editors), *Proc. 3rd Int. Peat Congr., Quebec, 1968*, pp. 10–11.

Fosberg, F.R., 1936. Miscellaneous Hawaiian plant notes I. *Occas. Pap., B.P. Bishop Mus.*, No. 12: 15.

Froment, A., 1975. Les premiers stades de la succession végétale après incendie de tourbe, dans la reserve naturelle des Hautes Fagnes. *Vegetatio*, 29: 209–214.

Früh, J. and Schröter, C., 1904. Die Moore der Schweiz mit Berücksichtigung der gesamten Moorfrage. *Beitr. Geol. Schw. Geotechn. Ser.*, 3. Lief.

Godwin, H., 1946. The relationship of bog stratigraphy to climatic change and archaeology. *Proc. Prehistor. Soc. 1946*, 1: 1–11.

Godwin, H., 1952. Recurrence-surfaces. *Dan. Geol. Unders., II. Raekke*, 80: 22–30.

Godwin, H. and Conway, V.M., 1939. The ecology of a raised

[1]As indicated in the footnote on page 1, this reference list had not been compiled at the time of the author's death, and it was necessary to put it together from various sources. Some papers mentioned may have been incorrectly identified, and certain references could not be identified at all.

bog near Tregaron, Cardiganshire. *J. Ecol.*, 27: 313–363.

Goodwillie, R., 1980. European peatlands. *Counc. Europe, Nat. Environ. Ser.*, 19: 1–75.

Gopal, B., 1973. A survey of the Indian studies on ecology and productivity of wetlands and shallow water communities. *Pol. Arch. Hydrobiol.*, 20(1): 21–29.

Gore, A.J.P., 1961. Factors limiting plant growth on high-level blanket peat. II. Nitrogen and phosphate in the first year of growth. *J. Ecol.*, 49: 605–616.

Gore, A.J.P., 1975. An experimental modification of upland peat vegetation. *J. Appl. Ecol.*, 12: 349–366.

Gore, A.J.P. and Olson, J.S., 1967. Preliminary models for accumulation of organic matter in an *Eriophorum/Calluna* ecosystem. In: J.S. Olson and R. Williams (Editors), *Mathematical Models and Natural Systems: Analysis of Stability and Change of Ecosystems*. Oak Ridge Natl. Lab., Oak Ridge, Tenn., 20 pp.

Gorham, E., 1953. Some early ideas concerning the nature, origin and development of peat lands *J. Ecol.*, 41: 257–274.

Gorham, E., 1957. The development of peatlands. *Q. Rev. Biol.*, 32: 145.

Gorham, E., 1961. Water, ash, nitrogen and acidity of some bog peats and other organic soils. *J. Ecol.*, 49: 103–106.

Göttlich, K. (Editor), 1976. *Moor- und Torfkunde*. Stuttgart, 260 pp.

Granlund, E., 1932. De svenska högmossarnas geologi. *Sver. Geol. Unders. Årsb.*, 26: 1–193.

Heikurainen, L., 1977. *Peatland Terminology for Forestry. English, German, Russian, Swedish, Finnish*. I.U.F.R.O. Working Party S1.05.01 Report, Helsinki, 78 pp.

Heinselman, M.L., 1963. Forest sites, bog processes and peatland types in the glacial Lake Agassiz region, Minnesota. *Ecol. Monogr.*, 33: 327–374.

Hejný, S., 1973. Ecosystem study on wetland biome in Czechoslovakia. *Czech. Acad. Sci., Czech., I.B.P. Rep.*, No. 3, Třeboň.

Herbichowa, M., 1979. Roślinność atłantyckich torfowisk Pobrzeza Kaszubskiego. [Vegetation of Atlantic peatlands on the Kaszubian coast.] *Soc. Sci. Gedanensis Acta Biol.*, 5: 1–52.

Hill, M.O., Bunce, R.G. and Shaw, M.W., 1975. Indicator species analysis, a divisive polythetic method of classification, and its application to a survey of native pine woods in Scotland. *J. Ecol.*, 63: 597–613.

Hölzer, A., 1977. *Vegetationskundliche und ökologische Untersuchungen im Blindensee-Moor bei Schonach*. Diss. Bot., 36. Kramer, Vaduz, 195 pp.

Hornburg, P., 1972. National plan for preserving bogland in Norway. In: *Proc. 4th Int. Peat Congr., Otaniemi, 1973*, pp. 179–190.

Ilnicki, P., 1977. Moore in Polen. *Telma*, 7: 203–213.

Imhof, G. and Burian, K., 1972. *Energy-Flow Studies in the Wetland Ecosystem (the Reed-Belt of the Lake Neusiedlersee)*. Austrian Acad. Sci, Special Publication for I.B.P., Vienna, 15 pp.

Ingram, H.A.P., 1967. Problems of hydrology and plant distribution in mires. *J. Ecol.*, 55: 711–724.

Ingram, H.A.P., 1978. Soil layers in mires: function and terminology. *J. Soil Sci.*, 29: 224–227.

Jasnowski, M., 1972. Rozmiary i kierunki przekształceń szaty roślinnej torfowisk. [Extents and directions of changes of plant cover of the bogs.] *Phytocoenosis*, 1(3): 193–209.

Jasnowski, M., 1975. Torfowiska i tereny bagienne w Polsce. [Peatlands and swamps in Poland.] In N.Ya. Kats (Editor), *Bagna kuli ziemskiej*. P.W.N., Warsaw, pp. 356–390.

Jasnowski, M., 1977. Probleme und Methoden des Moorschutzes in Polen. *Telma*, 7: 215–240.

Jasnowski, M. and Pałcziński, A., 1976. Mire vegetation and types of peatlands in Poland. In: *Peatlands and their Utilization in Poland, 5th Int. Peat Congr., Poznań*. Warsaw, pp. 5–28.

Jenny, H., Gessel, S.P. and Bingham, F.T., 1949. Comparative study of decomposition rates of organic matter in temperate and tropical regions. *Soil Sci.*, 68: 419–432.

Jensen, U., 1961. Die Vegetation des Sonnenberger Moores im Oberharz und ihre ökologischen Bedingungen. *Natursch. Landschaftspfl. Niedersachsen*, 1: 73 pp.

Jones, H.E. and Gore, A.J.P., 1978. A simulation of production and decay in blanket bog. In: O.W. Heal and D.F. Perkins, with W.M. Brown (Editors), *Production Ecology of British Moors and Montane Grasslands*. Ecological Studies, 27. Springer, Berlin, pp. 160–186.

Kats, N.Ya. [Katz, N.Ya], 1971. *Bolota zemnogo shara*. [*Swamps of the Earth*.] Nauka, Moscow, 295 pp.

Kaule, G., 1974. *Die Übergangs- und Hochmoore Süddeutschlands und der Vogesen*. Diss. Bot., 27, Kramer, Vaduz, 247 pp.

Kępczyński, K., 1960. Zespoły roślinne Jezior Skąpskich i otaczających je łąk. [Plant groups of the lake district of Skepe and the surrounding peat-bogs.] *Stud. Soc. Sci. Torun.*, Suppl. 6: 1–244.

Kępczyński, K., 1966. Stosunki florystyczne i fitosocjologiczne litoralu jeziora Sukiel. [Floral and phytosociological relations in the littoral of Lake Sukiel.] *Zesz. Nauk. Wyzsz. Szk. Roln. Olsztynie*, 21: 757–775.

Kivinen, E. and Pakarinen, P., 1980. Peatland areas and the proportion of virgin peatlands in different countries. In: *6th Int. Peat Congr., Duluth, Minn.*

Kovács, M., 1962. *Die Moorwiesen Ungarns*. Die Vegetation ungarischer Landschaften 3. Akademia Kiado, Budapest, 214 pp.

Kroigaard, 1968.[1]

Kulczyński, S., 1939/40. *Torfowiska Polesia*. Kraków, 777 pp. (2 vols.).

Kulczyński, S., 1949. Peat bogs of Polesie. *Mem. Acad. Polon. Sci. Lett., Cl. Sci. Math. Nat., Ser. B. Sci. Nat.*, 15: 1–356. (English version of Kulczyński, 1939/40).

Kuntze, H., 1971. Moorböden Norddeutschlands. *Mitt. Dtsch. Bodenkundl. Ges.*, 13: 105–150.

Kvét, J., Svoboda, J. and Fiala, K., 1969. Canopy development in stands of *Typha latifolia* L. and *Phragmites communis* Trin. in South Moravia. *Hidrobiologia (Bucuresti)*, 10: 63–75.

Lemée, M.G., 1938. Recherches écologiques sur la végétation du Perche. *Rev. Gén. Bot.*, 50: 415–433; 671–690.

Løddesøl, Aa., 1948. *Myrene i næringslivets tjeneste*. Grøndal, Oslo, 330 pp.

Løddesøl, Aa., 1968. The present situation on bog reclamation

[1]See footnote, p. 30.

and peat production in Norway. In: C. Lafleur and J. Butler (Editors), *Proc. 3rd Int. Peat Congr., Quebec, 1968,* p. 18.

Lundqvist, G., 1951. Beskrivning till jordartskarta över Kopparbergs län. *Sver. Geol. Unders.*, Ca 21: 1–213.

McVean, D.N. and Ratcliffe, D.A., 1962. *Plant Communities of the Scottish Highlands. A Study of Scottish Mountain, Moorland and Forest Vegetation.* Monographs of the Nature Conservancy, 1. London, 429 pp.

Malmström, C., 1923. Degerö stormyr. *Medd. Statens Skogsförsöksanst.*, 20: 1–206.

Malmström, C., 1930. Om faran för skogsmarkens försumpning i Norrland. *Medd. Statens Skogsförsöksanst.*, 26: 1–162.

Marek, S., 1965. Biologia i stratygrafia torfowisk olszynowych w Polsce. [Biology and stratigraphy of alder peatlands in Poland.] *Zesz. Probl. Postepow Nauk Roln.*, 57: 5–505.

Masing, V., 1960. *Saksa–inglese–rootsi–soome–eesti–vene sooteaduslik oskussönaslik.* [*German–English–Swedish–Finnish–Estonian–Russian Mire Science Dictionary.*] Riikl. Ülik., Tartu, 110 pp.

Mejstřik, V., 1968. Present state of research in peat deposits in Czechoslovakia. In: R.A. Robertson (Editor), *Proc. 2nd Int. Peat Congr., Leningrad, 1963*, pp. 91–97.

Mentz, A., 1912. Studier over de danske mosers recente vegetation. *Bot. Tidsskr.*, 31: 177–463.

Michna, I. 1976. Roślinne zbiorowiska jeziorne Pojezierzy Drawskiego i Bytowskiego. [Lake plant communities of the Lake Districts Drawsko and Bytów.] *Poznan. Tow. Przyj. Nauk, Wydz. Mat.-Przyr., Prace Kom. Biol.*, 43: 1–7.

Misra, R., 1946. A study in the ecology of low lying lands. *Ind. Ecol.*, 1: 27–46.

Moen, A., 1973. Norwegian national plan for mire preservation. In: *Proc. Int. Peat Soc. Symp., Glasgow, Sept. 1973.* Int. Peat. Soc., Helsinki, 13 pp.

Moen, A., 1975. Myrundersøkelser i Rogaland. *K. Nor. Vidensk. Selsk. Mus. Rapp. Bot. Ser. 1975*, 3: 127 pp.

Moore, P.D. and Bellamy, D.J., 1974. *Peatlands.* Elek Science, London, 221 pp.

Moretti, A. and Balboni, A., 1968. Peat deposits of Italy. In: R.A. Robertson (Editor), *Proc. 2nd Int. Peat Congr., Leningrad, 1963*, 1: 49–58.

Næss, T., 1969. Östlandets myrområder — utbredelse og morfologi. *Nor. Kom. Int. Hydrol. Dec. Rapp.*, 1: 75–88.

Netherlands Soil Survey, 1961. *Bodemkaart van Nederland. Schaal 1:200.000.* [*Soil map of The Netherlands. Scale 1:200.000.*] Stichting voor Bodemkartering, Wageningen, 9 sheets.

Neuhäusl, R., 1972. Subkontinentale Hochmoore und ihre Vegetation. *Studie Č.S.A.V. Praha*, 13: 1–121.

Nordhagen, R., 1943. Sikilsdalen og Norges fjellbeiter. *Bergens Mus. Skr.*, 22: 607 pp.

Olaczek, R., 1967. Zespoly szuwarowe i turzycowe dolin Bziry i Zianu. [*Scirpus* and *Phragmites* stands in the Bziry and Zianu valleys.] *Zesz. Nauk. Uniw. Lódz., Ser. II.*, Z23: 75–199.

Olkowski, M., 1966. Najczęściej występujące zbiorowiska roślinne na gitiowiskach mazurskich. *Zesz. Probl. Postepow Nauk Roln.*, 66: 33–41.

Osvald, H., 1923. Die Vegetation des Hochmoores Komosse. *Sven. Växtsociol. Sällsk. Handl.*, 1: 1–436.

Osvald, H., 1925a. Zur Vegetation der ozeanischen Hochmoore in Norwegen. *Sven. Växtsociol. Sällsk. Handl.*, 7: 1–106.

Osvald, H., 1925b. Die Hochmoortypen Europas. *Veröff. Geobot. Inst. Eidg. Tech. Hochsch. Stift. Rübel, Zürich*, 3: 707–723.

Osvald, H., 1949. Notes on the vegetation of British and Irish mosses. *Acta Phytogeogr. Suec.*, 26: 1–62.

Oświt, J., 1973. Warunki rozwoju torfowisk w dolinie dolnej Biebrzy na tle stosunków wodnych [Conditions for development of peatlands in the lower Biebrza valley with regard to the water regime.] *Roczn. Nauk Roln., Ser. D*, 143: 1–80.

Overbeck, F., 1975. *Bøtanisch-geologische Moorkunde unter besonderer Berücksichtigung der Moore Nordwestdeutschlands als Quellen zur Vegetations-, Klima- and Siedlungsgeschichte.* Karl Wachholtz, Neumünster, 719 pp.

Pacowski, R., 1967. Biologia i stratygrafia torfowiska wysokiego Wieliszego na Pomorzu Zachodnim. [Biology and stratigraphy of the peatlands of high Wieliszego in western Pomerania.] *Zesz. Probl. Postepow Nauk Roln.*, 76: 101–196.

Pałczyński, A., 1972. Biologia, paleofitosocjologia i kierunki zagospodarowania Bagien Jaćwieskich (Pradolina Biebrzy). [Biology, paleophytosociology and management of the Jaćwieskie swamps (Biebrzy valley). *Zesz. Nauk Wyzsz. Szk. Roln. Wrocławin*, 98.

Palczyński, A., 1975. Bagna Jaćwieskie pradoliny Biebrzy. [The Jaćwiez bogs in the proglacial valley of the Biebrza River.] *Roczn. Nauk Roln., Ser. D*, 145: 1–232.

Pallis, M., 1916. The structure and history of plav: the floating fen of the delta of the Danube. *J. Limnol. Soc. (Bot.)*, 43: 233–290.

Polakowski, B., 1969. Zespól *Cladietum marisci* (All. 1922) Zobrist 1935 w północno-wschodniej Polsce. [*Cladietum marisci* in north-eastern Poland] *Fragm. Florist. Geobot. (Kraków*), 15(1): 86–90.

Polakowski, B. and Dziedzic, J., 1972. Zespól *Hydrocharitetum morsus-ranae* van Langendonck 1935 w północno-wschodniej Polsce. [*Hydrocharitetum morsus-ranae* in north-eastern Poland] *Fragm. Florist. Geobot.* (*Kraków*), 18(3/4): 353–358.

Pop, E., 1960. *Mlaştinile de turba din Republica Populara Romînă.* Biblioteca Biologica Vegetala III. Acad. Rep. Pop. Rom., Bucuresti.

Radforth, N.W., 1955. *Organic Terrain Organization From the Air. Handbook No. 1*, Defense Research Board, DR 95. Ottawa, Ont.

Radforth, N.W., 1958. *Organic Terrain Organization From the Air. Handbook No. 2*, Defense Research Board, DR 124. Ottawa, Ont.

Radforth, N.W., 1977. Muskeg hydrology. In: N.W. Radforth and C.O. Brawner (Editors), *Muskeg and the Northern Environment in Canada.* University of Toronto Press, Toronto, Ont.

Radforth, N.W. and Bellamy, D.J., 1973. A pattern of muskeg — a key to continental water. *Can. J. Earth Sci.*, 10: 1420–1430.

Rodewald-Rudescu, L., 1974. *Das Schilfrohr.* Phragmites communis *Trinius.* Die Binnengewässer, 27. Schweizerbart'sche Verlagsbuchhandlung, Stuttgart, 302 pp.

Rusińska, A., 1974. Zespoły szuwarowe i oczeretowe południowej części Puszczy Bukowej pod Szczecinem.

[Plant communities of the alliances of Magnocaricion and Phragmition in the southern part of Puszcza Bukowa and Szczecin.] *Bad. Fizjogr. nad Polską Zach.*, B26: 165–192.

Rycroft, D.W., Williams, D.J.A. and Ingram, H.A.P., 1975. The transmission of water through peat. I. Review. II. Field experiments. *J. Ecol.*, 63: 535–556; 557–568.

Rybníček, K., 1974. *Die Vegetation der Moore im südlichen Teil der Böhmisch–Mährischen Höhe.* Vegetace ČSSR A6. Academia, Prague, 243 pp.

Sakaguchi, Y., 1978. *Geology of Peatlands in Japan.* University Press, Tokyo.

Sakaguchi, Y., 1979. Distribution and genesis of Japanese peatlands. *Bull. Dep. Geogr. Univ. Tokyo*, 11: 17–42.

Schallinger, K.M., 1968. Peat soils in Israel. In: C. Lafleur and J. Butler (Editors), *Proc. 3rd Int. Peat Congr., Quebec, 1968*, p. 17.

Schneekloth, H., 1968a. Peat bog cartography in N.W. Germany. An example of modern research on natural deposits. In: R.A. Robertson (Editor), *Proc. 2nd Int. Peat Congr., Leningrad, 1963*, 1: 99–104.

Schneekloth, H., 1968b. The significance of the limiting horizon for the chronostratigraphy of raised bogs: results of a critical investigation. In: C. Lafleur and J. Butler (Editors), *Proc. 3rd Int. Peat Congr., Quebec, 1968*, p. 116.

Schothorst, C.J., 1977. Subsidence of low moor peat soils in the western Netherlands. *Geoderma*, 17: 265–291.

Schwaar, J., 1977. Humifizierungswechsel in terrainbedeckende Mooren von Gough Island/Südatlantik. *Telma*, 7: 77–90.

Schwaar, J., 1979. The conservation of mires in the Southern Hemisphere (Tierra del Fuego, Gough Island, Easter Island, Juan Fernandez, and others). In: *Proceedings of the International Symposium on Classification of Peat and Peatlands, Hyytiälä, 1979.* International Peat Society, University of Helsinki, Helsinki, pp. 329–331.

Selling, O.H., 1948. Studies in Hawaiian pollen statistics. Part III. On the Late Quaternary history of the Hawaiian vegetation. *Spec. Publ. Bernice P. Bishop Mus., Honolulu*, 39: 154 pp.

Shoham, D. and Levin, I., 1968. Subsidence in the reclaimed Hula swamp area. *Israel J. Agric. Res.* 18: 15–18.

Sjörs, H., 1948. Myrvegetation i Bergslagen. *Acta Phytogeogr. Suec.*, 21: 1–299.

Sjörs, H., 1959. Bogs and fens in the Hudson Bay lowlands. *Arctic*, 12(1): 1–19.

Sjörs, H., 1963. Bogs and fens on Attawapiskat River, northern Ontario. *Nat. Mus. Can. Bull.*, 186: 45–133.

Smith, R.I.L., 1979a. Peat forming vegetation in the Antarctic. In: *Proceedings of the International Symposium on Classification of Peat and Peatlands, Hyytiälä, 1979.* International Peat Society, University of Helsinki, Helsinki, pp. 58–67.

Smith, R.I.L., 1979b. Classification of peat and peatland vegetation on South Georgia in the Sub-Antarctic. In: *Proceedings of the International Symposium on Classification of Peat and Peatlands*, Hyytiälä, 1979. International Peat Society, University of Helsinki, Helsinki, pp. 96–108.

Sobotka, D., 1967. Roślinność strefy zarastania bezodplywowych jezior Suwałszczyzny. [Vegetation of the zone subject to overgrowth in endorheic lakes of the Suwałki Region.] *Monogr. Bot.*, 23(2): 175–259.

Soó, R., 1954. Die Torfmoore Ungarns in dem pflanzensoziologischen System. *Vegetatio*, 5–6: 411–421.

Stanek, W., 1977a. Ontario clay belt peatlands — are they suitable for forest drainage? *Can. J. For. Res.*, 7: 656–665.

Stanek, W., 1977b. Classification of muskeg. In: N.W. Radforth and C.O. Brawney (Editors), *Muskeg and the Northern Environment in Canada.* University of Toronto Press, Toronto, Ont., pp. 31–62.

Steindórsson, S., 1975. Studies on the mire-vegetation of Iceland. *Soc. Sci. Island.*, 41: 226 pp.

Suzuki, S. and coworkers, 1973. *Ecology of Mires.* Tokyo, 399 pp. (Quoted by Sakaguchi, 1979.)

Szafer, W. and Zarzycki, K. (Editors), 1972. *Szata roślinna Polski. [The vegetation Cover of Poland.]* A.W.N., Warsaw, Vol. 1, 615 pp.; Vol. 2, 347 pp.

Szczepański, A., 1978. The ecology of macrophytes in wetlands. *Pol. Ecol. St.*, 4(4): 45–94.

Tansley, A.G. (Editor), 1911. *Types of British Vegetation.* Cambridge University Press, Cambridge, 416 pp.

Tansley, A.G., 1939. *The British Isles and their Vegetation.* Cambridge University Press, Cambridge, 930 pp. (republished as second edition in 1953, in two volumes, with the same pagination).

Tolonen, K., 1979. Peat as a renewable resource: long-term accumulation rates in north European mires. In: *Proceedings of the International Symposium on Classification of Peat and Peatlands, Hyytiälä, 1979.* International Peat Society, University of Helsinki, Helsinki, pp. 282–296.

Tüxen, J., 1976. Über die Regeneration von Hochmooren. *Telma*, 6: 219–230.

Ullmann, H. and Stehlik, A., 1972. A moor of nordic type in the Alps. In: *Proc. 4th Int. Peat Congr., Otaniemi, 1972*, 1: 75–88.

UNESCO, 1973. *International Classification and Mapping of Vegetation.*

Urquhart, C. and Gore, A.J.P., 1973. The redox characteristics of four peat profiles. *Soil Biol. Biochem.*, 5: 659–672.

Vanden Berghen, C., 1951. Landes tourbeuses et tourbières bombées à Sphaignes de Belgique. *Bull. Soc. R. Bot. Belg.* 84: 157–226.

Von Bülow, K., 1925. *Moorkunde.* Walter de Gruyter, Berlin, Leipzig.

Von Bülow, K., 1929. *Allgemeine Moorgeologie. Einführung in das Gesamtgebiet der Moorkunde.* Bornträger, Berlin, 308 pp.

Von Bülow, K., 1930. Zur Frage des Grenzhorizontes. *Z. Dtsch. Geol. Ges.*, 82: 38–41.

Von Post, L., 1944. Pollenstatistiska perspektiv på jordens klimathistoria. *Medd. Stockh. Högsk.*, 2: 79–113.

Von Post, L., 1946. The prospect for pollen analysis in the study of the earth's climatic history. *New Phytol.*, 45: 193–217.

Von Post, L. and Granlund, E., 1926. Södra Sveriges torvtillgångar I. *Sver. Geol. Unders.*, C 335: 1–127.

Von Post, L. and Sernander, R., 1910. Pflanzen-physiognomische Studies auf einigen Torfmooren in Närke. *Livret-Guide Exc. Suède, 11ᵉ Congr. Géol.*, 14: 1–48.

Vorren, K.D., 1967. Evig tele i Norge. *Ottar*, 51: 1–26.

Waksman, S.A., 1942. *The Peats of New Jersey and their Utilization. Part A. Nature and Origin of Peat, Composition*

and Utilization. Department of Conservation and Development, Trenton, N.J., pp. 16–24.
Walker, D. and Walker, P.M., 1961. Stratigraphic evidence of vegetation in some Irish bogs. *J. Ecol.*, 49: 169–185.
Walter, H., 1968. *Die Vegetation der Erde in öko-physiologischer Betrachtung. Band II. Die gemässigten und arktischen Zonen.* Gustav Fischer-Verlag, Jena, 1001 pp.
Weber, C.A., 1902. *Über die Vegetation und Entstehung des Hochmoors von Augstumal in Memeldelta mit vergleichenden Ausblicken auf andere Hochmoore der Erde.* Paul Parey, Berlin, 252 pp.
Weber, C.A., 1908. Aufbau und Vegetation der Moore Norddeutschlands. *Englers Bot. Jahrb.*, 90 (Suppl.): 19–34.
Weber, C.A., 1926. Grenzhorizont und Klimaschwankungen. *Abh. Nat. Ver. Bremen*, 26: 98–106.
Wendelberger, G., 1973. Ein österreichischer Moorschutzkatalog. *Telma*, 3: 163–171.
Werenskjold, W., 1943. Högmossarnas välvning i södra Sverige. *Geol. Fören. Stockh. Förh.*, 65: 304–305.
Wickman, F.E., 1951. The maximum height of raised bogs and a note on the motion of water in soligenous mires. *Geol. Fören. Stockh. Förh.*, 73: 413–422.
Yamanaka, M., 1963. Ecological studies of the Takadayachi Moor III. Pollen analytical studies of Takadayachi Moor. *Ecol. Rev.*, 16: 27–32.
Yoshioka, K., 1963. Ecological studies of the Takadayachi Moor in the Hakkoda Mountains 1. General aspects of the environment and vegetation. *Ecol. Rev.*, 16: 13–26.
Yoshioka, K., 1974.[1]

[1]See footnote, p. 30.

Chapter 2

MIRES — REPOSITORIES OF CLIMATIC INFORMATION OR SELF-PERPETUATING ECOSYSTEMS?[1]

BURKHARD FRENZEL

INTRODUCTION

The various types of mire ecosystems (terminology: Cajander, 1913; Waldheim, 1944; Osvald, 1949; Tansley, 1953; Du Rietz, 1954; Schneekloth and Schneider, 1972; Overbeck, 1975), with their wealth of stored information on past ecological conditions, seem to possess promising clues for the understanding of the character, strength, and periodicity of climatic changes in the past. The reason for this is that the mire vegetation often accumulates organic material year after year without much disintegration or redeposition. So it may be expected that all steps in the evolution of the ecosystem investigated can be followed easily and dated accurately, especially if radio-carbon dating is used cautiously. Moreover all the pollen grains and spores which fall onto the surface of the mire vegetation are in general held to be perfectly stored and protected there against most of the deteriorating influences, so that the organic content of fossil and subfossil peat layers not only seems to portray the evolution of the mire itself but of its surroundings as well.

MAIN REQUIREMENTS FOR VALID INTERPRETATION

In evaluating the validity of the assumptions made, the principal prerequisites for the formation of peat and for its palaeoclimatological interpretation should be duly borne in mind:

(1) Production and accumulation rate of organic matter must be greater than its decay. This can be achieved either by a relatively large primary production or by a reduced rate of decay (e.g., by cold climate, by drought, by lack of oxygen, or by an acid soil solution). So, from a palaeoclimatological point of view, the accumulation rate of peat is equivocal.

(2) The accumulation of organic material must happen without interruptions, so that all palaeoclimatic information is perfectly preserved. This means that the degree of water saturation in the peat investigated should always have been roughly the same. If interruptions in peat formation should have happened, the palaeoecological information of the site studied must enable a reliably exact discrimination between climatic, geomorphologic, biogenic or anthropogenic reasons. But very often the responses of the mire vegetation to detrimental influences of the environment resemble each other strongly irrespective of the factor involved.

(3) Diagnostically essential features of plant or animal remains must be preserved perfectly, without selective destruction of some taxa only, causing a wrong interpretation of past conditions. Moreover, a general destruction of all the organic matter, leading to some type of mud, must strongly impede the palaeoecological analysis or may cause an ambiguity in its interpretation. Most detrimental for the preservation of plant and animal remains are repeated or long-lasting periods of oxidation. They may be caused either by climate or by tectonics, eustatic or isostatic changes in sea level, by geomorphological or biogenic processes in the mire under consideration, or by man. Last but not least, a rapid decay of plant material may originate from a surplus of precipitation causing a strong influx of soligenous water.

---

[1]Manuscript received July, 1979.

(4) The mire vegetation should be governed by macroclimate only, without being influenced by its own growth processes, and without itself influencing the microclimate, topoclimate and edaphic conditions of its surroundings. Whether autogenic growth processes or even feed-back mechanisms had ever governed the evolution of a mire under study is often very difficult to evaluate. While climatic or edaphic influences on the surroundings seem to be negligible in mountainous areas, their existence must be thoroughly checked in flat lowlands. Even in mountainous areas this prerequisite is perhaps not always met — for instance where blanket bogs are developing. Blanket bogs are further discussed by Taylor in Part B, Ch. 1.

(5) It should be possible to distinguish between changes caused by climate and those caused by seral development of the vegetation. This is in general felt to be possible, but only if long peat exposures or several borings in the same peat bog can be examined simultaneously. Past vegetation types must be identifiable, or those types of past vegetation which can be deduced from palaeobotanical investigations should have equivalents in modern vegetation. But in this respect some difficulties are met with (see, for instance Grosse-Brauckmann, 1963, 1975a; Kats and Kats, 1964; and Rybniček, 1973).

(6) The ecology of plant and animal taxa characteristic of the various types of mire vegetation should be unequivocal today and should have remained constant during Postglacial times at least. It is a well-known fact that, frequently, neither of these requirements appears to hold true (Früh and Schröter, 1904; Du Rietz, 1954; Kubitzki, 1960; Malmer, 1962; Kats and Kats, 1964; Müller, 1965; Aletsee, 1967). From this it follows that palaeoecological interpretations may become ambiguous. Good examples of this are *Scheuchzeria palustris* and *Sphagnum imbricatum*. The former occurs currently in northwestern Central Europe in actively growing bogs, and also in the quite different climate of northern Siberia. *Sphagnum imbricatum*, so characteristic today for the mire vegetation of northwestern Central Europe, has been found in peat deposits of central Russia, having grown evidently in eutrophic conditions of the Likhvin Interglacial (Kats and Kats, 1964; Levkovskaya et al., 1970). The same moss species occurs at the present day in a bog habitat of the Kolkhida region (Neĭshtadt, 1961b). Botch and Masing (Part B, Ch. 4) describe other examples of apparently different ecological requirements in a single species.

(7) Past changes in ecological conditions must be dated accurately, as has been done most elaborately by Nilsson (1964a, b), Hibbert et al. (1971), Aaby and Tauber (1974) and Aaby (1975). If short, yet striking inconsistencies really have existed in the curve of ^{14}C ages versus real ages, as Suess (1970) had pointed out and as has been confirmed by Suess and Becker (1977) for some periods of the European Postglacial, palaeoecological changes which are actually randomly distributed have become artificially grouped together, causing a non-random distribution of ^{14}C ages and so stimulating palaeoclimatic speculations.

The discussion so far has emphasized the principal difficulties and obstacles to palaeoclimatic interpretations of peat stratigraphy. The essential point is that, while it cannot be denied that climate strongly influences the growth of peat, several non-climatic agents can cause comparable effects.

A SHORT SURVEY OF POSTGLACIAL CLIMATE HISTORY

The present state of knowledge of Postglacial climate history is reported by Sawyer (1966), Grichuk (1969), Lamb (1972), Godwin (1975), Overbeck (1975), World Meteorological Organization (1975), Frenzel (1977) and others (see Fig. 2.1). Though there are some difficulties in reconstructions of temperature, it should be stressed that the evaluation of changes in rainfall, runoff, and humidity is still more prone to misinterpretation (see p. 40 et seq.). This can be overcome only by considering two sources of palaeobotanical evidence: the information given by the pollen flora of the peat; and the information given by macroremains and the peat itself. Both these items will be discussed at some length in the following pages. Before doing so several points should be made:

General trends of climate during the Postglacial, according to different sources of information

Temperature indicators. Previous work has led to a good general agreement that, during the last 10 300

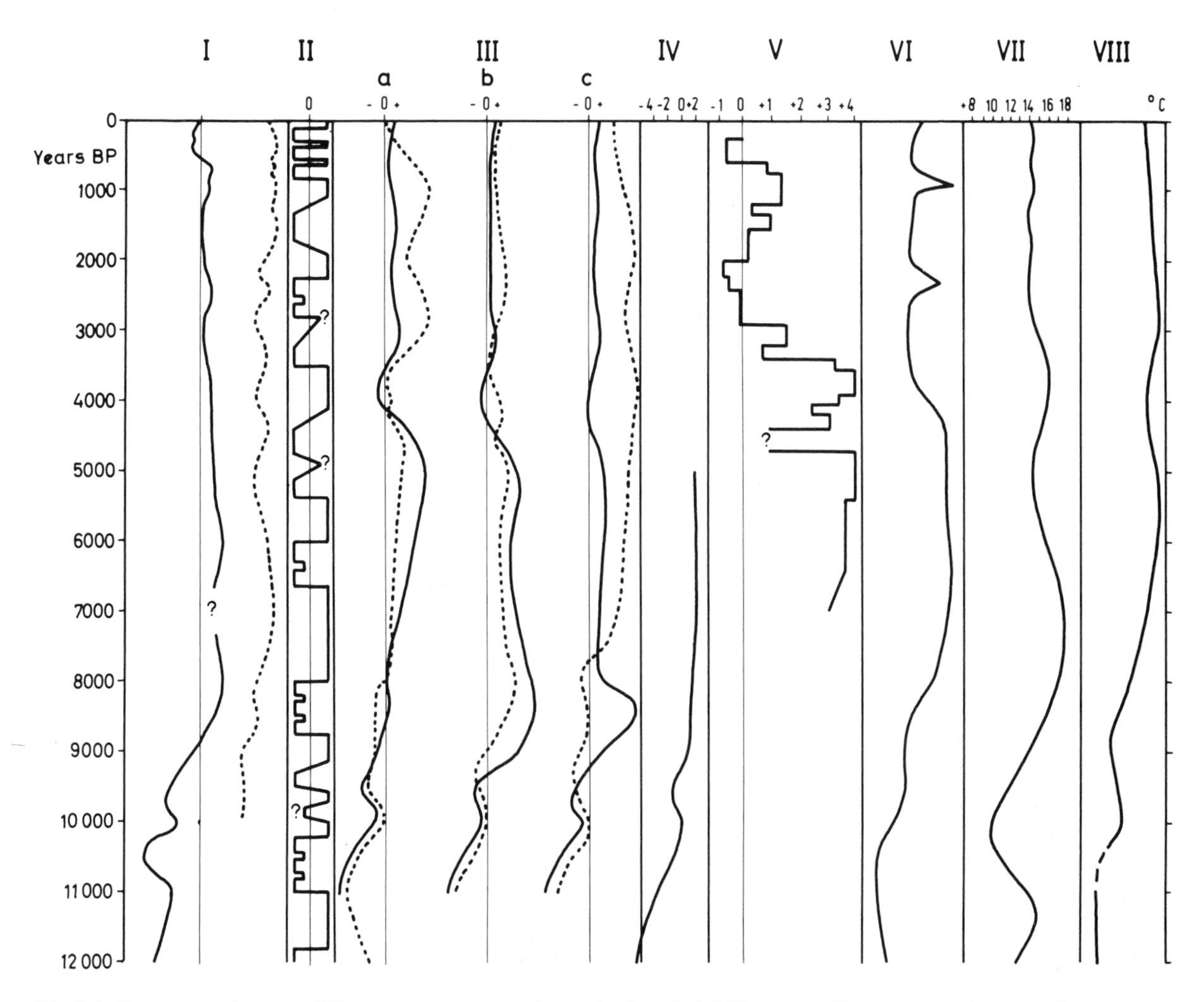

Fig. 2.1. Comparison between different attempts to evaluate the Postglacial history of climate. *I*=northwestern Central Europe (Overbeck, 1975); full line: temperature; dotted line: humidity; values increase to the right. *II*=Alps (Patzelt, 1977); warmer conditions to the right. **III** = northern Eurasia (Khotinskii, 1976a, b); full line: temperature; dotted line: humidity; *a*, Russian plain; *b*, Siberia; *c*, the Far East. *IV*=Japan (Tsukada, 1967): temperatures. *V*=northwestern Canada (Nichols, 1975): summer temperatures. *VI*=Arizona (Martin, 1963): summer precipitation; higher values to the right. *VII*=Sabana de Bogotà, Colombia (Van der Hammen and Gonzalez, 1960): mean annual temperature. *VIII* = Cherangani Hills, Kenya (Van Zinderen Bakker, 1964): temperatures; higher values to the right.

years or so, there has been a general and very rapidly acting improvement of climate. Perhaps the best example for this is the $^{18}O/^{16}O$ curve of Camp Century on the inland ice of Greenland (Dansgaard et al., 1970). Regrettably, this curve is equivocal for palaeoclimatological purposes because it is mainly a measure of the changing ice budget, not just a measure of temperature changes alone. Notwithstanding this, the general trend seems to be widely accepted: at first a rapid increase of temperature, followed by a long-lasting phase of roughly constant thermal conditions, with slight cooling possibly near the end of this period (see, for instance, Mörner and Wallin, 1977). The increase of temperature seems to have begun at between 13 500 and 14 000 B.P., causing phases of generally improved climates like the Susacá Interstadial of South America (Van der Hammen and Gonzalez, 1965) and of its equivalent in other parts of the world [northeastern North America: Lasalle (1966); East Africa: Van Zinderen Bakker (1972); western Scotland and adjacent areas: Pennington et al. (1972), Coope (1977a, b), etc.]. Even by this stage climatic history seems to have differed in some regions of the globe as compared with others, as Mercer (1969) had pointed out in respect of the

Allerød and Younger Dryas times in Northern and Central Europe, as compared with North America. This is a point of general importance which can be paralleled by other observations on the Postglacial history of climate (Frenzel, 1975; Williams and Van Loon, 1976). Another point of interest is that, occasionally, palaeoclimatic conclusions drawn from various "climate indicators" may differ strongly from each other. According to Coope (1977a, b), the carabid fauna of western England and western Scotland point to increasing warmth even prior to the famous Bølling Interstadial (13 250–12 400 B.P.), the greatest warmth having been felt during this Interstadial, whereas the temperature seems to have declined again during the ensuing Allerød Interstadial (11 850–10 800 B.P.). Yet the microclimate for the carabids investigated had changed at the same time: the open steppe-like plant communities with their warmth-indicating beetles were later replaced by juniper heath communities and even by birch forests, the local climate of which is much cooler during summer, and in general much more oceanic than that of steppe-like plant communities. This type of biogenic change in microclimate within the changing plant communities should be duly taken into consideration when broad trends in macroclimate are postulated. It will be seen that similar difficulties arise when changes in palaeoclimatic conditions for the developing vegetation of a single mire are considered.

"Pollen zones". During the Postglacial, climatic changes occurred repeatedly. Only a few years ago it was held in general that the transitions between the various "zones" in pollen diagrams were caused by synchronously acting changes of climate. Kats (1959a) and Heusser (1960) had shown however that these changes did not happen simultaneously along a S–N profile in the Northern Hemisphere. This view was corroborated by Ruddiman and McIntyre (1973) for the Atlantic Ocean. Comparable observations have been reported along an E–W profile by Watts and Wright (1966) in Nebraska, by Wright (1968) for the Great Lakes region in North America, by Hibbert et al. (1971) for the British Isles and their surroundings, and by Khotinskiĭ (1971a) for Central and Eastern Europe. Khotinskiĭ's view (1976a, 1977) that the Atlantic/Subboreal transition happened roughly synchronously all over northern Eurasia was not well founded since the pollen diagrams used for this cannot be compared exactly with one another, due to very divergent plant-geographical conditions.

From these difficulties it follows that the basal information for the study of changes in vegetation and therefore possibly in climate can only be deduced from pollen assemblages, such assemblages being characterized by a more or less homogenous pollen flora over a variable span of time, not by abstractions with much more definite implications, like "pollen zones" (West, 1970). Moreover, it should be stressed, as Watts (1973) has done, that the transition from one pollen assemblage to the following one very often lasted for a relatively long time, sometimes embracing a millennium or so. During such a period, there had been immigration of new competitors and several steps in the seral succession of vegetation types and soils could occur (Andersen, 1964; Iversen, 1964). This often means that a change in pollen assemblages cannot be as precisely dated as was thought previously when the impact of climate on the vegetation seemed to have been the governing factor.

Radioactive carbon (^{14}C). Nevertheless it must be admitted that climate changed several times, even if only to a minor extent (see below, pp. 39, 56), during the Postglacial. Nilsson (1964b), Aaby (1975), Aaby and Tauber (1974) and Van Geel (1976) pointed to changes with a period of about 150 to 260 years indicated by their research work in Sweden, Denmark and The Netherlands. Others have felt that local effects might have been overemphasized, so that Wendland and Bryson (1974) tried to solve this problem by a statistical approach (Fig. 2.2), and Frenzel (1975) has used a wealth of glaciological, oceanographical, geological, geomorphical, zoological, and botanical indicators of past climates, which were first thoroughly filtered in order to eliminate effects other than those due to changes of climate (Fig. 2.2). Even such methods, however, leave one problem unresolved: changes in palaeoecological conditions seem to have been accompanied by rapidly oscillating changes in the ^{14}C content of the atmosphere. It is thus possible that a non-random distribution of palaeoecological change may have been simulated, thus causing a fallacious impression, firstly of well-defined times of climatic change, and secondly of a connexion between changes in atmospheric ^{14}C and those of

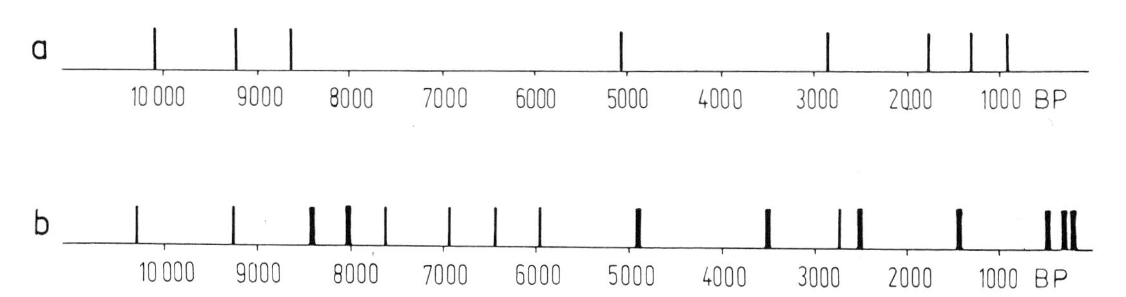

Fig. 2.2. Ages of Postglacial changes of climate. a. According to Wendland and Bryson (1974). b. According to Frenzel (1975).

climate (Frenzel, 1978). This problem needs more attention than has hitherto been devoted to it.

Atmospheric circulation effects

Up till now it has been generally held that climatic changes during the Postglacial happened simultaneously all over the world. This was the reason why names such as Preboreal, Subboreal, etc., have been so widely accepted (see Taylor in Part B, Ch. 1). Apart from changes in palaeoecological conditions which are known to have occurred at different times, there are changes in climate within the last few years which differ in their tendencies from one region to another, this being caused by the mode of atmospheric circulation (Krames, 1951, 1952; Røstad, 1955; Rodewald, 1970; Lamb, 1972). Clearly such sources of change could have had significance in the past also (Frenzel, 1976; Williams and Van Loon, 1976). This means that at a given date, climate may have ameliorated in one region of the temperature zones, whereas it worsened simultaneously elsewhere and remained constant in still other areas.

Amplitude of climatic change

It is now becoming evident that the amplitude of climatic change during the Postglacial was often on a relatively small scale only. This is deduced from the fact that the alpine timberline had already reached its present height in the central Alps by early Postglacial times, and since then has only oscillated by about 250 m or so at the most. From this it may be deduced that summer temperatures oscillated by only about $\pm 0.7°C$ during the whole of Postglacial times (Frenzel, 1966; Jelgersma, 1966; Patzelt, 1977; Zoller, 1977). If so a serious difficulty must arise, because ecosystems sometimes display relatively strong inertia and certain threshold values must be passed to initiate changes in their composition or area of distribution, as Smith (1965) convincingly pointed out.

The regulatory processes of ecosystems are sometimes strong enough to buffer even relatively important changes of climate. This is shown by the larch–spruce communities of eastern Siberia or by the fir–beech forest in southern Central Europe which thrived there for about 10 000 and 6000 years respectively, with almost no changes in composition (Khotinskiĭ, 1971a; Frenzel, 1978). This may create the fallacious impression that no climatic changes have taken place. On the other hand, small changes of climate may cause unexpectedly large consequences where the regulatory processes had created a labile equilibrium in the ecosystems investigated. A good example of this seems to be the initiation of blanket bog formation. Changes like this might be interpreted as having been caused by a larger change in climate than actually occurred.

There is still another point to be made. In contrast to the relatively low values of change in temperature just cited for the central Alps there are those which were calculated for northernmost Canada, for instance by Nichols (1967, 1969, 1975), Grichuk (1969) and Ritchie (1972), amounting roughly to between 3 and 5°C since the middle of the Holocene. While it is likely that they are only reflections of the various effects just quoted, the possibility is not completely ruled out that such changes are real ones. This would mean that the amplitude of climatic change might have differed greatly between various parts of the globe, as was demonstrated for recent climatic oscillations by Rodewald (1970) (see also Nichols, 1976). It is to be hoped that this very important palaeoclimatological problem will be investigated much more intensively soon.

Influence of man

As will be seen from the following sections, man appears to have caused serious changes in palaeoecological situations at different times and in various places, initiating tendencies in the development of ecosystems which might be interpreted erroneously as showing changes of climate.

Thus the picture of climatic history becomes increasingly more complicated, and this impression is further strengthened if the uncontrolled and often very delicately acting consequences of the impact of man are taken into consideration.

PALAEOECOLOGICAL SOURCES OF INFORMATION

Geographic differences less useful than changes in a single mire

Despite the hazards of interpretation, the use of peat as a repository of palaeoclimatic information is favoured by the vast distribution of Postglacial peat all over the world. The fact that mire provinces obviously governed by climate exist is further encouragement (Osvald, 1925a, b; Von Post, 1926; Dokturovskiĭ, 1928; Von Bülow, 1929; Kats, 1930, 1931, 1957, 1958, 1959b, 1961, 1969, 1972; Waldheim and Weimarck, 1943; Waldheim, 1944; Godwin, 1946; Sjörs, 1948; Schmitz, 1952; Lutz, 1956; Ruuhijärvi, 1960; Straka, 1960; Eurola and Ruuhijärvi, 1961; Neĭshtadt, 1961a, b, 1966, 1972, 1975; Eurola, 1962; McVean and Ratcliffe, 1962; Auer, 1965; Bellamy and Bellamy, 1966; Aletsee, 1967; Kulikova et al., 1971; Overbeck, 1975; Schwaar, 1976). Figs. 2.3 to 2.6 give some examples of the increasing resolution of such differences in space. Several authors (for instance, Von Bülow, 1929; Schmitz, 1952; Kats, 1957, 1960, 1961, 1972; Ruuhijärvi, 1960; Bower, 1961; Eurola and Ruuhijärvi, 1961; Eurola, 1962; Tallis, 1965; and Bellamy and Bellamy, 1966) have tried to evaluate the climatic factors governing this regional differentiation (see, for instance, Fig. 2.7), but it is still today nearly impossible to use them for palaeoclimatological reconstructions. In general not only one but several climatic factors, in addition to geological and edaphic ones, contribute to the establishment of plant-geographical zonations. This is further complicated by the fact that the relative importance of the various factors studied changes repeatedly in an uncontrolled way. It may be suggested that the best results are obtained by studying small areas and a restricted number of indicator plants only as was done by Boatman (1962) for *Schoenus nigricans* in Ireland and by Overbeck (1975) for *Sphagnum imbricatum* in northwestern Central Europe. Even in these examples the limiting factors are not well understood (see, for instance, Olausson, 1957).

If it is accepted that the different geographical types of mire vegetation in general cannot be used directly for palaeoclimatic reconstructions, the development in time and space of the most characteristic plant communities of the mire studied should be more useful. Of prime importance here is the discrimination between topogenous, soligenous, and ombrogenous mires and the changes in the position of the so-called *Mineralbodenwasserzeigergrenze* [mineral soil water limit — see Sjörs (1948, p. 280)]. As early as 1823, Dau had concluded that bogs are nourished by rain water only, whereas swamp, fen, and carr vegetation thrives on ground water, more or less rich in mineral nutrients. At the same time Dau had described successions from fen to carr, from fen to bog, and from carr to bog. This concept of seral development (see Tallis, Chapter 9) proved to be most stimulating for the understanding of the prevailing tendencies in the evolution of ecological conditions of the mire vegetation (see, for instance, Früh and Schröter, 1904; Cajander, 1913; Osvald, 1923, 1949; Du Rietz and Nannfeldt, 1925; Kats, 1926; Von Post, 1926; Dokturovskiĭ, 1927; Tansley, 1939; and Overbeck, 1975) though actually the mire vegetation seems often to remain nearly unaltered, without any remarkable seral changes, over several centuries (Grosse-Brauckmann, 1975a).

The *Grenzhorizont*

These ideas of Dau (1823) focused on the dynamics of plant communities at a given place under constant external conditions. On the other hand, it was Weber (1910) who stressed that these seral successions are part of minor oscillations only in the composition of the mire vegetation, climatic changes exerting a much stronger influence. The starting point for this hypothesis was Weber's observation in northwestern Central Europe of a strongly humified *Sphagnum* peat which was co-

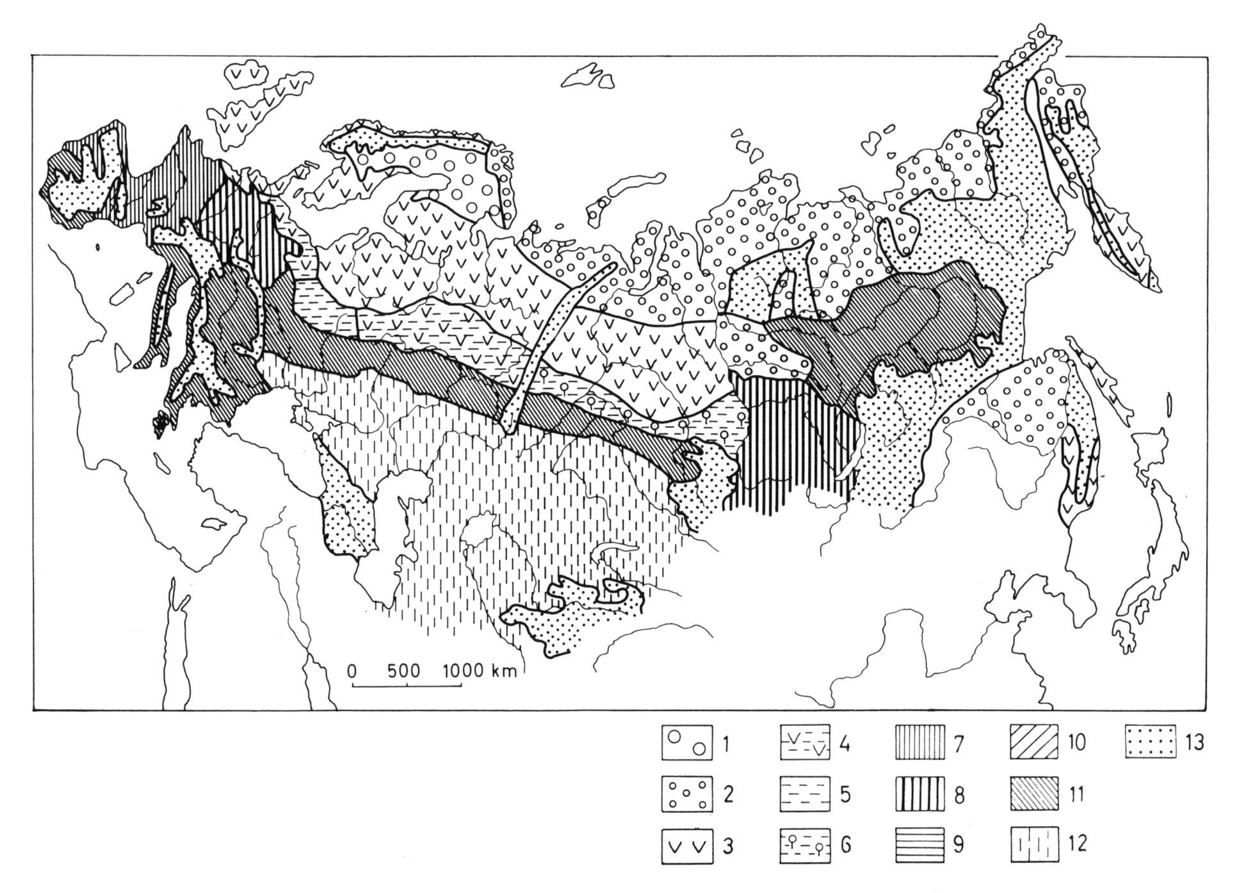

Fig. 2.3. Provinces of bog types in northern Eurasia (Kats, 1958). Legend: *1* = aapa bogs; *2* = fens and bogs of the cold climate zone with permafrost; *3* = blanket bogs and other types of ombrogenous bogs; *4* = fen and carr; typical bogs relatively rare; *5* = like *4*, but mostly characterized by a eutrophic herb and moss vegetation; *6* = widely distributed eutrophic moss and carr vegetation; *7* = various types of eutrophic fen and swamp vegetation; ombrogenous bogs very rare; *8* = mostly carr vegetation and other types of eutrophic swamps, overgrown by coniferous and deciduous trees; *9* = eutrophic fens covered by a rich broad-leaved tree vegetation; *10* = tree-covered eutrophic fen vegetation, rich in relics of Tertiary times, as for instance *Taxodium* and *Nyssa*; sometimes fen vegetation, rich in herbs; on the sea-shore often, in brackish water, including mangrove vegetation; *11* = dominating fen and swamp vegetation, rich in herbs; tree cover only very sparse; *12* = fresh-water fen vegetation with strongly humidified peat, this humification steadily increasing to the south; *13* = various types of topogenous fen and swamp vegetation in mountainous areas.

vered abruptly by a very slightly humified, ombrogenous *Sphagnum* peat. Weber called the sharp transition between these *Sphagnum* peat layers the *Grenzhorizont* and dated it in one place at between 1000 B.C. and 750 B.C. The strong humification of the older, black *Sphagnum* peat was held to originate from a long period of weathering without further peat growth, under a relatively warmer and drier climate. This view was based on observations made by Gradmann (1898) on the striking parallelism between the distribution of prehistoric settlements and the recent occurrence of steppe plants within the southern part of Central Europe, the so-called *Steppenheide-Theorie*. In this theory, Neolithic and Bronze Age peoples were thought to have sited their villages and fields on open steppe areas associated with a warm dry climate. The view was further favoured by Blytt's (1876) classical observation on the changing composition of subfossil plant remains in peat bogs of western Norway (see also Taylor, Part B, Ch. 1). These findings, and the ensuing observation of comparable, similar stages in the evolution of forest communities, led to the assumption that climatic change was of prime importance for the changes in vegetation. In Central Europe these stages are referred to as the "mitteleuropäische Grundsuccession" (Rudolph, 1931). This found its counterpart in several other

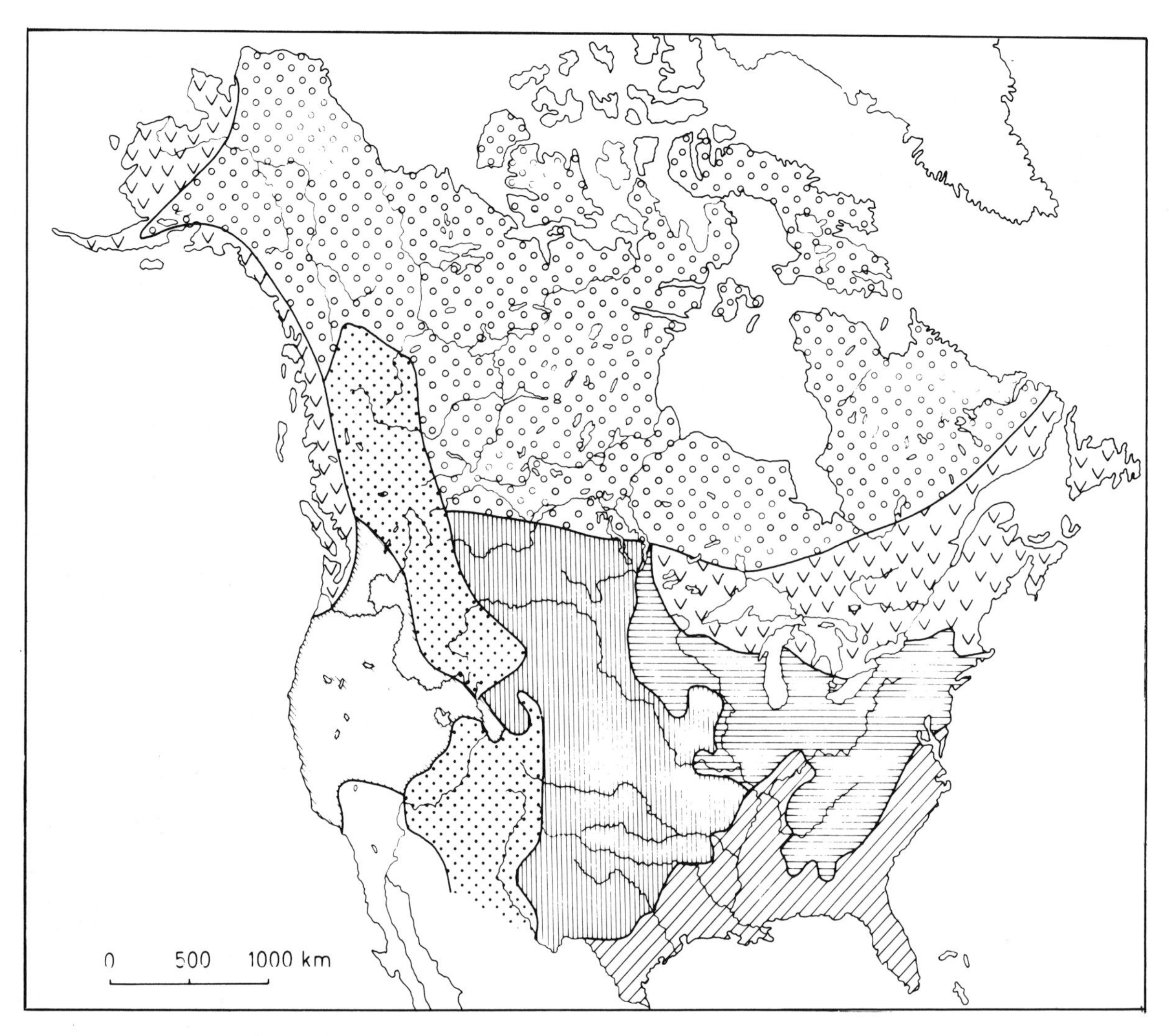

Fig. 2.4. Provinces of bog types in North America (Kats, 1958, 1959b). For explanation of symbols, see Fig. 2.3.

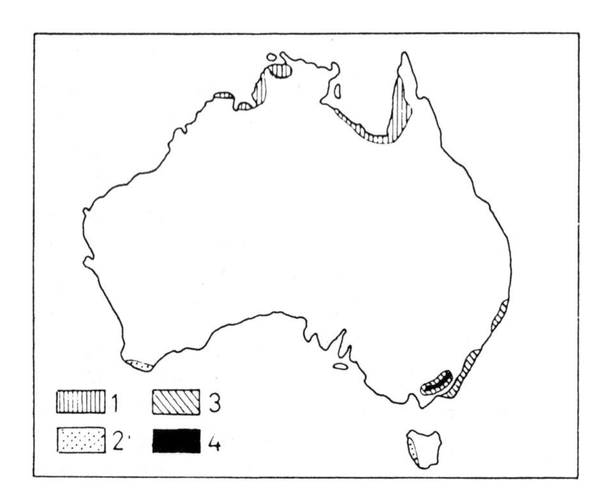

Fig. 2.5. The geographical distribution of fen and bog vegetation in Australia and Tasmania (Kats, 1969). Legend: *1* = fenland in lowlands on the sea shore, and mangrove vegetation; *2* = fen and carr vegetation, characterized by tall Cyperaceae like *Gymnoschoenus*, shrubs, and trees; *3* = moist sclerophyllous forests, sometimes on peat; *4* = peat bogs, sometimes consisting of various *Sphagnum* species.

regions of Europe or of the Northern Hemisphere in general (see, for instance, Von Post, 1924; Jessen, 1935; Deevey, 1939; Hansen, 1947; Overbeck, 1950; Godwin, 1956; Neĭshtadt, 1957). One of the reasons for general acceptance of the idea of climatic control was that the *Grenzhorizont* seemed to have developed over a very vast area [Central Europe:

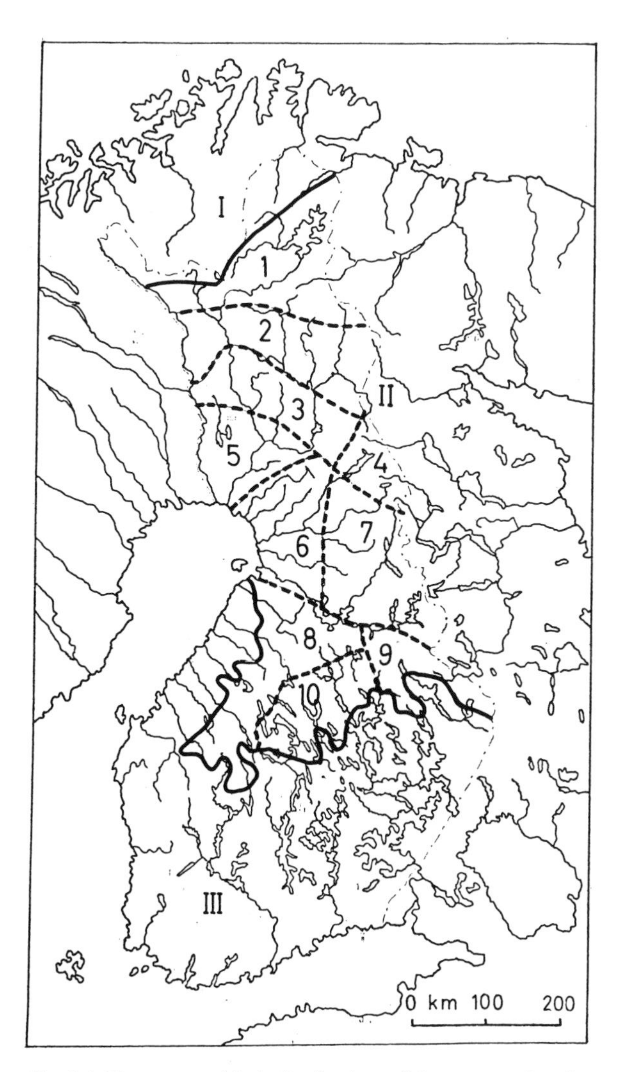

Fig. 2.6. The geographical distribution of bog vegetation in Finland (Ruuhijärvi, 1960). Legend: *I*=palsa bogs. *II*=aapa bogs: *1*=aapa bogs of the northernmost forest belt; *2*=aapa bogs of northern Peräpohjola; *3*=aapa bogs of southern Peräpohjola; *4*=bogs on mountain slopes of Kuusamo; *5*= eutrophic aapa bogs in northern Pohjanmaa; *6*=aapa bogs of northern Pohjanmaa; *7*=aapa bogs in Kainuu; *8*=aapa bogs in Suomenselka; *9*=aapa bogs in northern Karelia; *10*=aapa bogs in the lake region. *III*=ombrogenous bog vegetation of southern Finland.

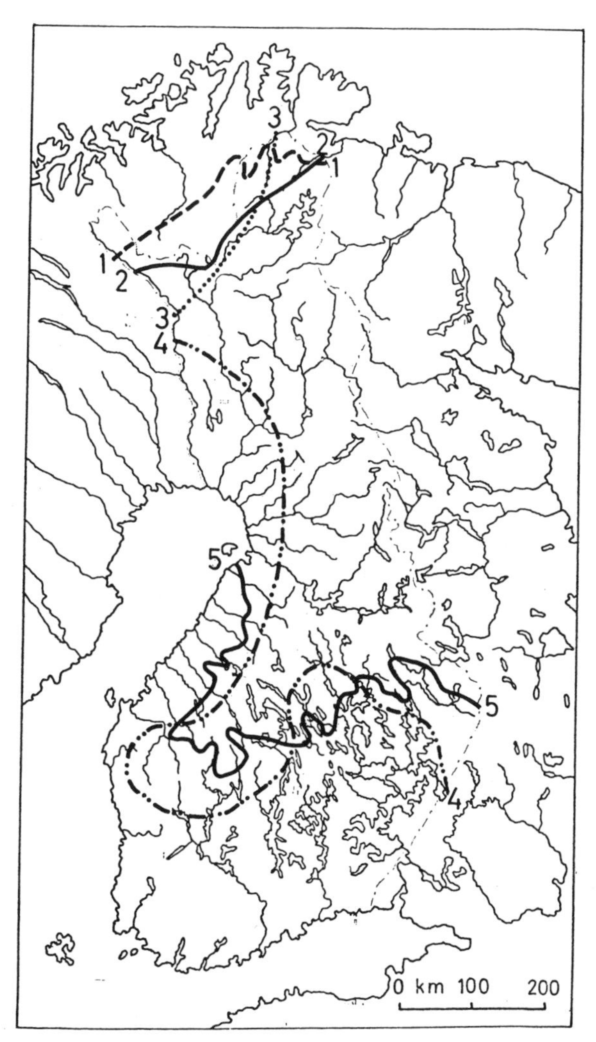

Fig. 2.7. Finnish bog types and climate (Ruuhijärvi, 1960). Legend: *1*=potential evapotranspiration, 360 mm per year; *2*=borderline between the palsa and aapa bog provinces; *3*=mean annual temperature of −1.0°C; *4*=hydrothermic quotient of 15 during the summer months; *5*=borderline between the aapa and the ombrogenous bog provinces.

Gross (1934), Overbeck (1975); Soviet Union: Dokturovskiĭ (1928), Thomson (1930), Neĭshtadt (1957), Khotinskiĭ (1971b, 1977), Kamyshev (1972); western Canada: Heusser (1960); Gough Island: Schwaar (1977); see also Gore, Chapter 1], allegedly at the same time. It should be noted that, even by the end of the last century and the beginning of the present one, several authors felt that the formation of the strongly humified peat and of the *Grenzhorizont* need not depend on climate only (Früh and Schröter, 1904; Von Post, 1910, 1926; Potonié, 1912) but might have been caused by autogenic processes in the bog itself. Without paying attention to these warnings, the climatic interpretation of the *Grenzhorizont* was widely accepted, though quite different things like a strongly humified ombrogenous *Sphagnum* peat in Central Europe and a strongly humified soligenous or even topogenous fen–carr peat in Russia and in western Siberia were lumped together. Moreover, the meaning of the word *Grenzhorizont* had undergone a serious change. Instead of its original meaning, Soviet scientists understood, by the same word, the whole layer of the strongly humified peat, irrespective of its ecological and plant-sociological situation (Khotinskiĭ, 1971b). The difficulties in understanding and interpretation arising from this change in meaning had already been shown convincingly by Gross in 1934. This paper seems to have been forgotten. Perhaps the enthusiasm for interpreting changes in humification of the peat layers and of their botanical constituents in terms of climate was enhanced by the ingenious hypothesis of Granlund (1922) (see Gore, Ch. 1). The widespread layers of retarded growth and marked humification were thought to be due to attainment of an equilibrium between effective rainfall, runoff, and the height of the bog cupola. Subsequent deposition of relatively unhumified peat depended on increased rainfall and humidity, obviously once more an indication of the reality and the periodicity of climatic change! These so-called recurrence surfaces (Swedish: *rekurrensytor*, RY) could be dated at RY I 1200 A.D., RY II 400 A.D., RY III 600 B.C., RY IV 1200 B.C., RY V 2300 B.C., RY VI 2800 B.C., RY VIII 3700 B.C. (but in some cases the estimated ages deviated greatly; see Brandt, 1948) and it was thought that the famous *Grenzhorizont* was the equivalent of RY III. It was only in 1952 that Godwin pointed out that to account for the great number of recurrence surfaces in Granlund's hypothesis (see also Olausson, 1957) it would be required that the climate should have become increasingly moist, either stepwise or at least continuously. Furthermore, such a process must have continued since 2700 B.C., the date of the earliest recorded recurrence surface. Evidence for this tendency in climatic history is lacking. Some years later several authors succeeded in demonstrating that the famous *Grenzhorizont* or its hypothetical equivalents varied in age even within a single peat bog. The point is discussed in more detail on pp. 53–56. It became clear that autogenic processes could account for the formation of recurrence surfaces, independently of climate.

From what has been discussed already, it follows that there are several problems in the investigation of climate history and of the formation of peat, yet the most elusive questions concern climatic control of peat growth.

THE DEVELOPING MIRE — SELF-PERPETUATING MECHANISMS VERSUS CLIMATE-CONTROLLED PROCESSES

A growing mire contains two sets of palaeoclimatic information — the pollen flora and the peat stratigraphy, representing successive stages of the participating plant communities. It will be remembered that it was the peat stratigraphy that stimulated palaeoclimatic speculations and hypotheses so much (cf. *Grenzhorizont*: Weber, 1910; recurrence surfaces: Granlund, 1922; retardation layers: Godwin, 1952). On the other hand, autogenic processes like *Stillstands-*, *Erosions-* and *Wachstumskomplex* (Osvald, 1923) can produce changes in the stratigraphy which look as if they have been induced climatically. This was stressed already by Früh and Schröter (1904), Von Post (1910), Weber (1911), Potonié (1912), and recently most convincingly by Casparie (1969). So it will be the aim of this section to discuss the problem of distinguishing between climate-induced changes in stratigraphy on the one hand, and consequences of autogenic or non-climatic allogenic processes on the other.

Processes of mire induction

Two types of mire formation occur (see also Tallis, Chapter 9) by overgrowth of lakes (*Verlandung* or terrestrialization) and by waterlogging of supra-aquatic terrestrial soils (*Versumpfung* or paludification).

Terrestrialization (*Verlandung*). The initiation and rate of overgrowth of lakes depend on lake water levels, on the trophic conditions of the lake itself and of its catchment area, on the plant-geographical situation, on the area and shape of the lake, and the topography of its bottom and its surroundings (Kiryushkin et al., 1967; Tamozhaĭtis, 1967). The seral development of the aquatic plant communities involved has been investigated repeatedly (see, for instance, Tansley, 1953; Vasari and Vasari, 1968; and Overbeck, 1975) and is discussed by Tallis (Ch. 9). Even so, allegedly clear examples of seral development are strongly governed by local historical factors (Wąs, 1965; Grosse-Brauckmann, 1976).

From the foregoing list of factors it is evident that climate plays but an indirect part in the initiation and early stages of the overgrowth of lakes. Because neither the catchment area nor the lake topography will have changed substantially during Postglacial times, the role of a number of important hydrological, geological and chemical influences can be assessed for any given mire-infilled lake. The character of lake sediments is also useful for the interpretation of past events on the catchment and in the lake itself. Trophic conditions can cause some problems because changes due to anthropogenic as opposed to spontaneous effects have to be distinguished. The catchment area is particularly important here. The excellent work of Mackereth (1966) and of Pennington and her colleagues (Pennington and Lishman, 1971; Pennington et al., 1972) has done most to shed some light on the difficulties met with, showing that changes caused by large-scale clearances strongly resemble those produced by spontaneous degradation of the prevailing vegetation or by soil development. Here the question of buffer capacity of the plankton and of the lake sediment itself is involved, as well as the turnover in nutrient content of the terrestrial biotopes; such matters as thresholds and inertia have also been referred to (Smith, 1965). So the velocity in the rate of overgrowth or a change in the contributing plant communities need not be directly correlated with the character of the disturbing agent nor with its strength.

Tallis (Ch. 9) describes several instances of active mire formation in English lakes. Other instances of seral change quoted by Tallis (Ch. 9) reveal the variety of sequences of plant communities which can occur during the overgrowth of lakes. Terrestrialization, at least in the early stages, is essentially a topographic process of mire formation, so that much of the stratigraphy of such mires can be ascribed to definable events other than climate. To this extent, from a palaeoclimatological point of view, the problems are relatively simple compared with those encountered in interpreting the stratigraphy of paludified terrestrial soils.

Paludification (*Versumpfung*). Flat basins or river valleys are of course prone to paludification if their bottom is clad by water-impermeable sediments and if they are periodically or nearly always oversaturated with slowly running or stagnant water. Other landforms that are very often covered by swamp, fen, carr or even by bog are riverine lowlands accompanying rivers but separated from them by dams of sand and gravel which were piled up by the river itself during floods. These lowlands are characterized by a high ground-water table which rises nearly continuously, in step with the piling up of the dams and with the rising of the bed of the river itself. Examples of this can be found repeatedly all over the world (see, for instance, Fig. 2.8). Here paludification begins when the river flow becomes stable and when the amount of water transported was sufficient to influence the accompanying lowlands by seepage. This seems to have often happened within the temperate zones by the onset of the Late Glacial (Fig. 2.9), but of course just the reverse was true for Arctic and Subarctic or North Temperate Zones when the ground hitherto perennially frozen had thawed, rendering the percolation of water to greater depths possible and so draining the Arctic riverine swamps previously existing. Thus, though both these processes were caused and governed by similar climatic events, they had exactly opposite consequences. Moreover, it should be stressed that the growth of these riverine mires does not depend on climate only, but on various geomorphological and anthropogenic factors as well, as will be shown later. This may serve as another example of the necessity, if palaeoclimatic conclusions based on peat stratigraphy are to be drawn, of not only studying the evolution of the mire proper but of analyzing in detail all the palaeoecologic conditions of the whole region, of which the mire may be only one small part.

The first example — flat basins or river valleys which are paludifying — is a well-known fact. Here the paludification is favoured by a thin layer of sand on top of impermeable clay or silt (Koshcheev, 1953) or by dunes encroaching onto the plain and impeding runoff (Borówko-Dłúżakowa, 1961). Of course the same may result from the activity of beavers, though the real importance of this "artificial" paludification seems to be unknown, and presumably it will be very difficult to demonstrate convincingly that beavers had even been of importance in the paludification of river systems in the past.

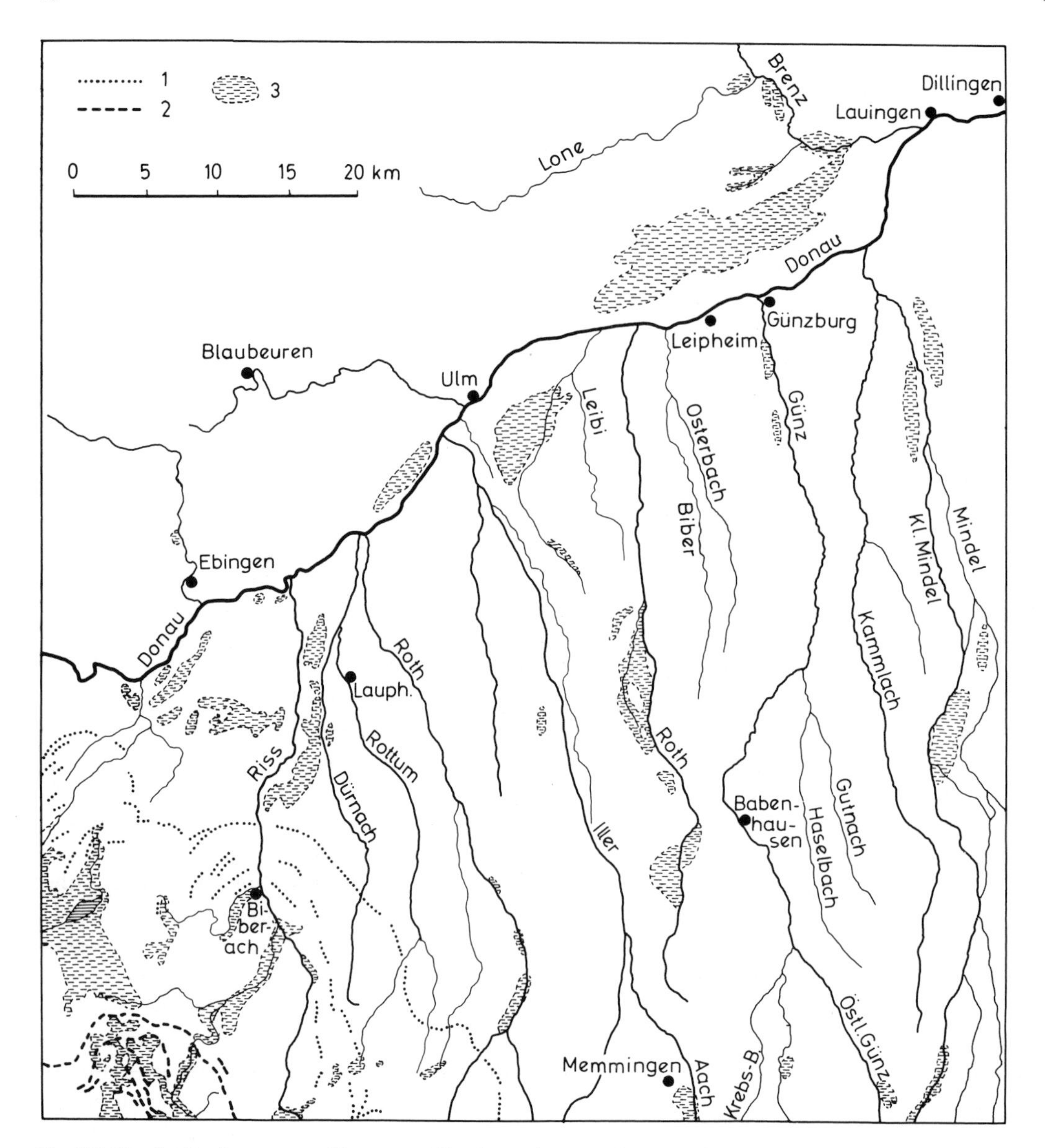

Fig. 2.8. The distribution pattern of fen, carr and bog in southwestern Germany. Legend: *1* = end moraines of the Riss glaciation; *2* = end moraines of the Würm glaciation; *3* = fen, carr and bog. Outside the end moraine systems fen and carr in general only occur on river terraces.

The paludification of flat-bottomed basins can be induced by tectonic movements without any influence of climate (Kiryushkin et al., 1967) or by eustatic and isostatic movements of sea level (see, for instance, Cajander, 1913; Du Rietz and Nannfeldt, 1925; Thomson, 1930; Aario, 1932; Jonas, 1935; Godwin, 1952; Von der Brelie, 1954, 1955; Ruuhijärvi, 1960; Fries, 1965; Oldfield, 1967; Behre, 1970; Schneekloth, 1970; Khotinskiĭ, 1971b, 1977; Val'k, 1971; Tooley, 1974, 1976; and Menke, 1976). Eustatic changes in sea level are often linked to climate (see, for instance, Fairbridge, 1961; Jelgersma, 1966; Mörner, 1971; Tooley, 1974; and Menke, 1976), but palaeoecological changes in terrestrial habitats caused by a rising sea level do not only depend on the height of mean sea level (i.e. indirectly on climate); other factors like difference in the height of tides, the formation of sandbars, etc., also contribute to an over-saturation of water in the low-lying areas near the coasts (see, for

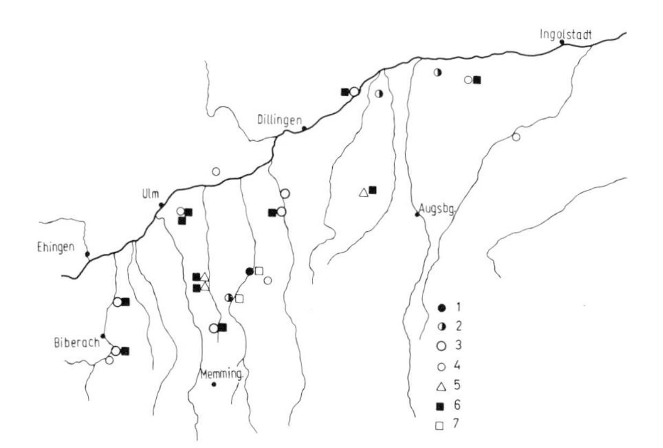

Fig. 2.9. The age of the beginning of fen and bog formation on river terraces in southwestern Germany. Legend: *1*=oldest Dryas time; *2*=younger Dryas time; *3*=Subboreal; *4*=Subatlantic; *5*=mediaeval times; *6*=charcoal at the bottom of the peat; *7*=charcoal lacking there.

instance, Menke, 1976). These factors render the palaeoclimatological interpretation of paludifying areas near the sea extremely difficult. During the formation of mires in such situations there exist peculiar interrelations between changes in sea level and rates of growth of the peat under consideration (Godwin, 1952; Khotinskiĭ, 1971b, 1977). This may be explained either by the direct damming up of ground water and the ensuing seepage of soligenous water onto the mire, or by an indirect increase in the humidity of the air due to enhancement of the transpiration and evaporation rates. Though an understanding of the mechanisms responsible is still far off, progress is being made (see, for example, Ingram, Chapter 3).

Another type of mire can be formed when water-permeable strata, like screes, terminate on water-impermeable sediments. Special types of fen communities with peat accumulation appear to be due directly to the amount of water available (Kukla, 1965) and this in turn seems to depend on climate. Possibilities for misinterpretation are illustrated by an example from the crest of the Hunsrück Mountains, western Germany. This consists of Lower Devonian quartzites, the screes of which encroached during the Pleistocene onto the surrounding Devonian clay schists. At the lower end of the screes, water sources nourish a beautiful fen and carr vegetation. The pollen diagram (Fig. 2.10) shows that, during long periods of the Postglacial, an alder (*Alnus*) forest, rich in birch (*Betula*) and willow (*Salix*) with a well-developed field layer of moisture-loving herbs, accompanied the brooklets and rivulets. In this vegetation the biomass must have been considerable, but in general peat did not accumulate. Obviously the production of organic matter and its decay were in balance. Suddenly, fen or carr peat formation began, and roughly at the same time alder was replaced by birch, forming a birch carr. This transition was characterized by the replacement of the hitherto predominant oak (*Quercus*) forests by beech (*Fagus*) and by the disappearance of ivy (*Hedera helix*). At the same time acidophilic plants like Ericaceae could spread (Fig. 2.10). In Central Europe, beech is a sub-oceanic plant, loving moist climates, and the flowering of ivy is prevented if the mean temperatures of the coldest month are less than $-1.5°C$. So the sequence of events seems to be unequivocal. Climate had become moist and cool. This had favoured the leaching of soils on the quartzites which are poor in mineral nutrients, rendering the expansion of an acidophilic shrub vegetation possible. On the other hand the moist and cool climate must have stimulated the accumulation of peat. Since beech immigrated there only sporadically at about 2100 B.C., its general expansion, accompanied by the events just discussed, seems to point to the Subatlantic deterioration of climate. Other evidence is conflicting, however. Though the area investigated is only a very small one, the sharp change in vegetation and in ecological conditions did not happen simultaneously — on the contrary it lasted for a long time, from 2000 B.C. to Roman or mediaeval times. Moreover, the transition cited was always characterized by rather extensive clearings. From this it follows that the triggering effect could have been the clearings and the ensuing agricultural activity of man. Oak was replaced at various times by beech, and overgrazing of the beech forests caused the spread of acidophilic ericaceous plants. The clearings could have changed the rate of transpiration and run-off, causing the accumulation of peat, and the light-demanding ivy was overcome by the shade-producing beech. Thus, the allegedly unequivocal evidence for the Subatlantic deterioration of climate could easily have been caused by man. The same sequence of events seems to have been fairly widespread in Central and Western Europe (see, for instance, Cajander, 1913; Koshcheev, 1953; Schneekloth, 1963b; Moravec and Rybničková, 1964; Rybničková, 1966; Jankovská, 1971; Szczepanek, 1971; Radke, 1972; Planchais, 1973; Turner and Kershaw, 1973; De Beaulieu, 1974; Frenzel, 1976; Grosse-Brauckmann and Streitz, 1976; and Peschke, 1977). Obviously the evidence from different locations varies so that interpretations differ, and others (Firbas and Broihan, 1937; Trautmann, 1952; Schlüter, 1964; Lange, 1967; Lange and Schlüter, 1972) have taken the view that climate is the primary cause of change and the activities of man only contribute to the effects. The matter is an intriguing one because in the examples cited a marked increase in the accumulation of various types of peat began simultaneously with the earliest indications of clearing by man or other agricultural activities.

It must be asked whether the formation of blanket bog in Western Europe, which is so characteristic of Ireland and Scotland and of large regions of England and Wales as well, was triggered exclusively by climatic change (Hafsten and Solem, 1976). As can be seen from Table 2.1 several blanket bogs are relatively young, and the formation of blanket peat began in the examples cited contemporaneously with the onset of man's activity (see Taylor, Part B, Ch. 1). This favours the view that the equilibrium between the formerly existing forests and their environment was unstable (Tallis, 1964; Pennington, 1965; Birks, 1971, 1972; Pennington and Lishman, 1971; Pennington et al., 1972; Tallis and McGuire, 1972; O'Sullivan, 1973, 1974; Turner et al., 1973). The role of forests in influencing soil moisture is now much better understood (see for example Ingram, Ch. 3), so it is possible that removal of an existing tree cover could have increased soil wetness so as to accelerate peat formation (see Pennington, 1965; Moore, 1977). In southern Germany vast fenland areas occur on the lowermost river terraces. These were most certainly caused by clearings at various times. Clearing of the forests from hill slopes adjacent to the rivers caused a downwash of fine silt onto the previously permeable gravels of the river terraces. The increasing superficial run-off and the ensuing floods appear to have caused the initiation of the peat which is so characteristic of the present-day conditions (Fig. 2.9). Apart from the increasing irregularity in run-off there was a greatly increased sediment load in the rivers, causing a considerable accumulation and redeposition of sand and gravel. By impeding drainage locally this led to a new phase of formation of fen and carr peat in the lowlands adjacent to the rivers. So here again the indirect effects of man could have been responsible for mire formation. This means that great care is needed if paludification of terrestrial soils is to be used for the

TABLE 2.1

Formation of blanket peat and its connexion with the impact of man in some European blanket bogs

Beginning of the formation of blanket peat or of cyperaceous peat	Locality	Devastations of the forests	Clearances	Agriculture
300–500 A.D.	northern Derbyshire	+	+	+
550 B.C.	Antrim, Ireland	+	+	+
750 B.C.	northern Pennines	+	+	+
750 B.C.	Antrim, Ireland	+		
950 B.C.	Antrim, Ireland	+	+	+
1125 B.C.	Antrim, Ireland	+	+	+
1350 B.C.	Antrim, Ireland	+		
1550 B.C.	Antrim, Ireland	+	+	+
1650 B.C.	Lancashire	+	+	+
1660 B.C.	Antrim, Ireland	+		
1750 B.C.	northwestern Scotland;	+	+	+
	Antrim, Ireland	+	+	+
1850 B.C.	Antrim, Ireland	+		
2215 B.C.	Northern Ireland	+	+	+
2000–2500 B.C.	Northern Ireland	+	+	+
2250 B.C.	western Scotland	+	+	+
3000 B.C.	southern Pennines	+	+	+

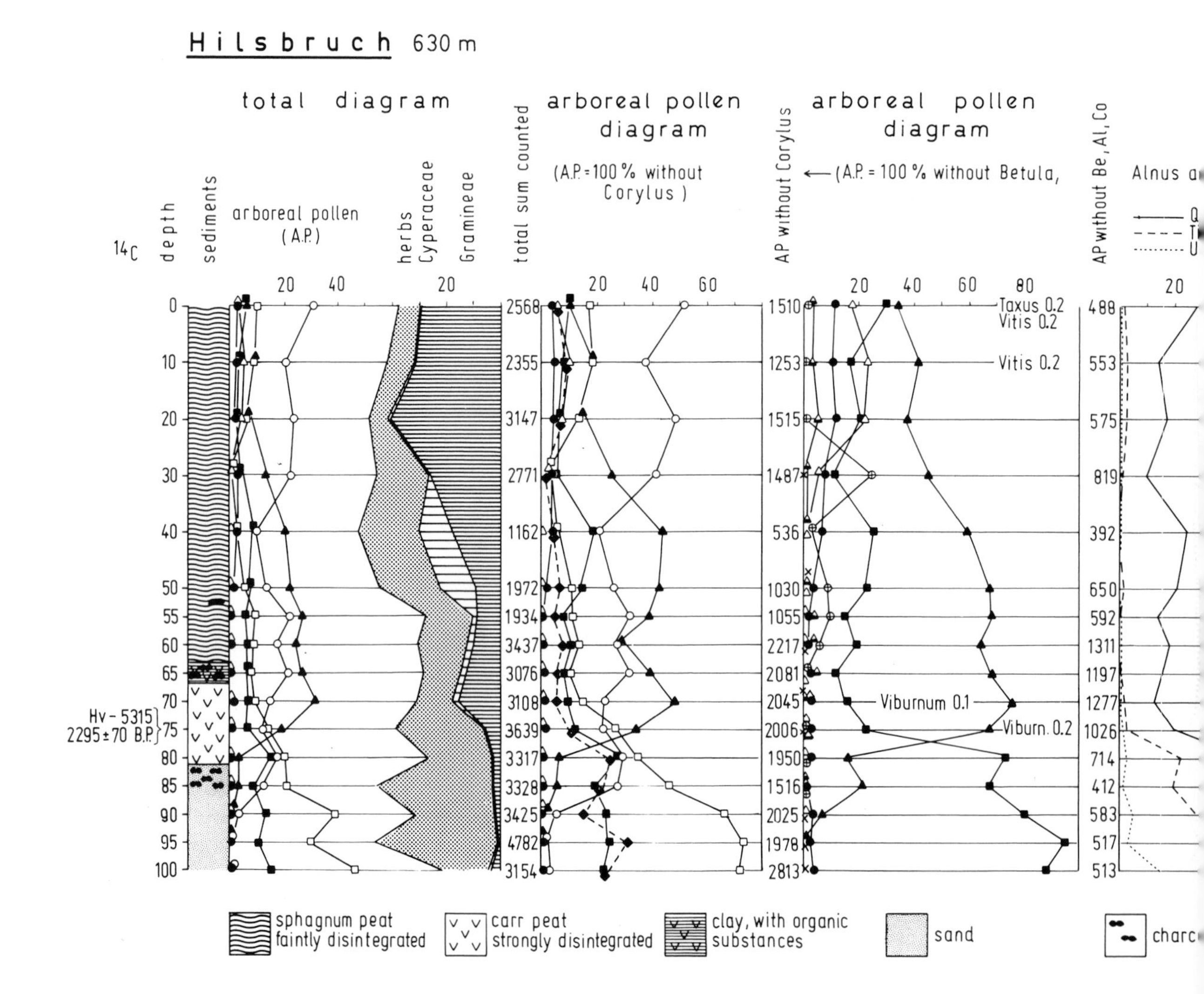

Fig. 2.10. Pollen diagram of the Hilsbruch, central Hunsrück Mountains, western Germany. "Arboreal pollen" includes both trees and shrubs. All data are expressed as percentages of the total (or partial total as indicated) for the horizon in question.

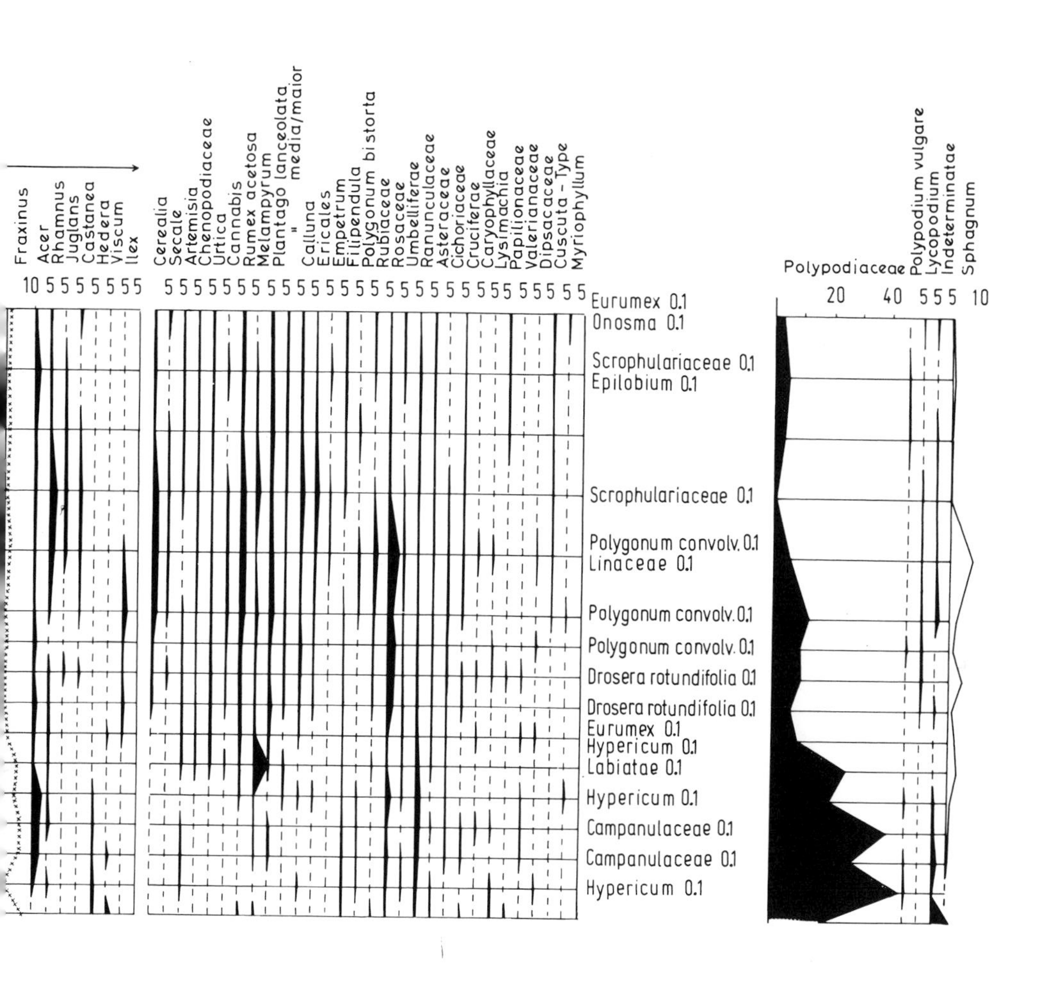

Fraxinus
Acer
Rhamnus
Juglans
Castanea
Hedera
Viscum
Ilex
Cerealia
Secale
Artemisia
Chenopodiaceae
Urtica
Cannabis
Rumex acetosa
Melampyrum
Plantago lanceolata
media/maior
"
Calluna
Ericales
Empetrum
Filipendula
Polygonum bistorta
Rubiaceae
Rosaceae
Umbelliferae
Ranunculaceae
Asteraceae
Cichoriaceae
Cruciferae
Caryophyllaceae
Lysimachia
Papilionaceae
Valerianaceae
Dipsacaceae
Cuscuta - Type
Myriophyllum
10 5 5 5 5 5 5 5
5 5 5 5 5 5 5 5 5 5 5 5 5 5 5 5 5 5 5 5 5 5 5 5 5 5 5 5 5
Eurumex 0.1
Onosma 0.1
Scrophulariaceae 0.1
Epilobium 0.1
Scrophulariaceae 0.1
Polygonum convolv. 0.1
Linaceae 0.1
Polygonum convolv. 0.1
Polygonum convolv. 0.1
Drosera rotundifolia 0.1
Drosera rotundifolia 0.1
Eurumex 0.1
Hypericum 0.1
Labiatae 0.1
Hypericum 0.1
Campanulaceae 0.1
Campanulaceae 0.1
Hypericum 0.1
Polypodiaceae
Polypodium vulgare
Lycopodium
Indeterminatae
Sphagnum
20
40
5 5 5
10

reconstruction of past climates. Another non-climatic factor which is able to initiate widespread paludification is the presence of certain layers of volcanic ash (Auer, 1965).

Often, though not always, the accumulation of peat is preceded by the formation of mor and of raw humus. In Fig. 2.11 the mean residence time of ^{14}C in the uppermost 10 to 15 cm of various soils is plotted against the geographical latitude. As can be seen, the residence time seems to increase with increasing latitude. This effect seems to be strengthened in dry climates (chernozem) as compared with moister ones (podzol). But of course the amount of organic matter accumulated and resting on podzolic soils is much greater than in chernozems. From Fig. 2.11 it follows that the relatively slow downward transport of organic matter in podzolic soils must cause an appreciable accumulation of organic matter, which can very easily lower the permeability, thus tending to favour paludification. This process is reinforced where a hard-pan is formed in the subsoil (Neĭshtadt, 1972) due to the precipitation of iron oxides. Thus, this climatically controlled type of pedogenesis may indirectly cause paludification — though, of course, only with a lapse of unknown duration after the triggering effect of climate had taken place. The accumulation of raw humus in the first place not only depends on climate (Iversen, 1958) but on the nutrient status of the weathering rocks, as well as on the history of vegetation and on man's activity (P'yavchenko, 1953; Sirén, 1955; Menke, 1963; Iversen, 1964; Dimbleby, 1965; Frenzel, 1978). However, once accumulation of raw humus has begun, the nutrient status, the water permeability (Cajander, 1913; Von Bülow, 1929; Steffen, 1931; Burges, 1951) and the heat conductivity (Tyrtikov, 1969) all change progressively. In cold-climate zones, this may cause an upward movement of the surface of perennially frozen ground, or even the initiation of permafrost, causing waterlogging with ensuing swamp and mire formation as well. So the palaeoclimatological significance of this type of paludification is again far from direct.

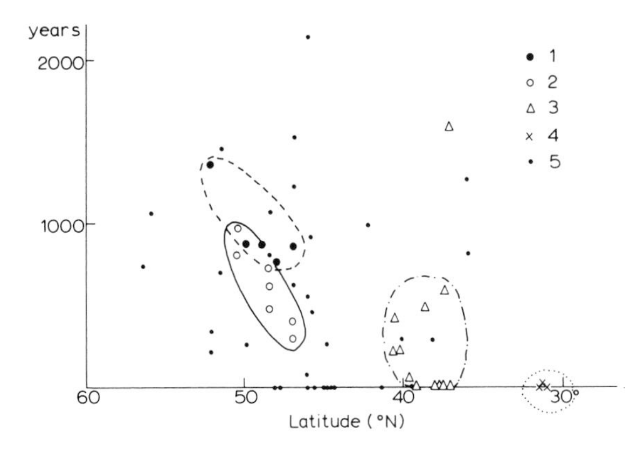

Fig. 2.11. Mean residence time of radiocarbon in the A horizon of various soils. Legend: *1* = chernozems; *2* = podzols; *3* = Vertisols; *4* = brunizems; *5* = brown earth, parabrown earth, pseudogley.

Another point of interest which should be mentioned is the self-intensifying effect of some types of paludification and of mire formation. As has been shown repeatedly, surplus water running from the growing mires, either superficially or as ground water (Cajander, 1913; Von Bülow, 1929; Steffen, 1931; Aario, 1932; Paasio, 1933; Waldheim, 1944; P'yavchenko, 1953; Dimbleby, 1965; Neĭshtadt, 1972; Frenzel, 1978), may have adverse effects on soil formation in the surroundings of the mire in question, accelerating paludification there. This may be intensified by the microclimate of the growing mire, without any change in macroclimate. A possible example of this is given in Table 2.2, where the area of peat-covered ground of the Bakchar mire in the Tomsk area of western Siberia is given for various periods of the Holocene. Tallis (Ch. 9) also discusses the work of Kulczyński (1949) and Heinselman (1963), who described the related phenomenon of upslope mire formation.

To sum up this section about climatically and non-climatically induced mire formation, it cannot be denied that climate has often influenced the formation of mire vegetation, but it may well have

TABLE 2.2

Area of mire-covered ground during the Postglacial in the Bakchar mire, western Siberia (from Neĭshtadt, 1971)

Approximate ages (years B.P.)	Area paludified during the period indicated. (ha)	Rate of increase in the paludified area (ha yr^{-1})
0–2000	53 000	26.5
2000–4000	73 100	36.5
4000–6000	64 000	32.0
6000–8000	33 000	16.7
8000–9000	3 200	3.2

been simulated by various non-climatic factors too (Von Post, 1926). These may act together or separately, rendering the evaluation of the agents responsible extremely difficult. As to the influence of geomorphological agents, "One could say that the climate determines the general limits within which landform determines the details" (Bower, 1961), and Romanova (1967) suggested that the geomorphological and geological conditions are more important for paludification and for the types of mires met with in the lowlands of western Siberia than is climate.

The transition to bog formation

In bog-forming areas (Figs. 2.3 and 2.4), once a lake is overgrown or paludification has started, generally any influence of eutrophic water will gradually decline and bog formation increase. Exceptions are found where a mire remains under the eutrophic influences, and Ruuhijärvi (Part B, Ch. 2) discusses the transition between raised bogs and aapa mires in southern Finland in this context. Overbeck (1975) has pointed out that the apparent coincidence of the transition from fen or carr to bog with the very beginning of the Atlantic period has been used as one of the best arguments for this event having been caused by climate. Overbeck (1975) found that the frequency distribution of this transition in Central Europe was as shown in Table 2.3.

According to Khotinskiĭ (1977) this transition happened in the Shuvalovo bog at about 8700 B.P., in the Kalina bog at about 5400 B.P., in the Leosaare bog at about 1900 B.P., and in the Berendeevo bog at about 3400 B.P. Comparably marked differences in the ages of ombrogenous peat are given by Weber (1967) for Estonia. Such dates hardly favour the view of only one phase of climatically induced formation of ombrogenous peat; it is more likely that this transition was largely caused by local conditions, the topography, mineral status and hydrology being of prime importance.

TABLE 2.3

Frequency distribution (out of 37 sites) of the beginning of bog formation (^{14}C ages)

Age (years B.P.)	Frequency of bog formation (beginning of ombrogenous peat)
2000–1000	1
3000–2000	7
4000–3000	4
5000–4000	5
6000–5000	5
7000–6000	4
8000–7000	7
9000–8000	4

Various authors have contributed to our knowledge of the processes leading from eutrophic topogenous swamp, fen or carr to oligotrophic ombrogenous bog. These results will not be repeated here (see, for instance, Weber, 1911; Osvald, 1923; Kats, 1926; Von Post, 1926; Rudolph, 1929; Tansley, 1953; Ruuhijärvi, 1963; Elina, 1969; Dickinson, 1973 and Overbeck, 1975). Clymo (Ch. 4) deals with the chemistry of the process in some detail. It is worth stressing that, within the climatic belt of bog formation, each part of a living fen or carr in which local growth proceeds more rapidly than the general rise of the ground-water table and in which the mean air humidity is high enough to render *Sphagnum* growth possible may serve as a starting point for bog formation. In this respect the divergent water requirements of the various *Sphagnum* species must be taken into consideration (Waldheim, 1944; Sjörs, 1948; Jensen, 1961; Malmer, 1962; Overbeck, 1975). Moreover, Grosse-Brauckmann (1963) found that, within former carr vegetation, the tree canopy seems to have been astonishingly open, so that heliophytes like most species of *Sphagnum* could thrive there easily, whereas they fail to grow under the relatively dense tree canopy found on some bogs at the present day. From this it was concluded that such bogs are no longer actively forming peat (Rybniček, 1977). Nevertheless, the transition from topogenous to ombrogenous conditions must have often lasted for long periods. Even if several decimetres of peat have accumulated, the chemical influence of the ground water from the underlying mineral soil is still observable (Kats, 1926; Du Rietz, 1954; Boatman, 1961; Eurola, 1962). Jensen (1961) showed that, in the Sonnenberger Moor, Harz, the influence of the ground water in the mineral soil terminated only when from 100 to 150 cm of peat had been accumulated. The thickness of isolating peat needed seems to increase where, during the time of snow-melt or during heavy rains, huge amounts of soligenous water are able to influence the mire's ecology. This is the basis of the previous reference to Ruuhijärvi (Part B, Ch. 2) concerning the transition zone between raised bogs and aapa fens in Finland. Such considerations may help to avoid the hypothesis of very fast transitions from fen or carr to bog in relation to sudden climatic changes. The importance of the substrate beneath the peat for the existing type of mire vegetation may be seen from Fig. 2.12 for the huge Vasyugan bog in the central part of western Siberia (Neĭshtadt, 1972, 1975; see also Part B, Figs. 4.35, 4.36).

Thus, within the climatic belt of bog formation, each fen or carr carries the potential for bog development; but whether this will take place or not depends on several more or less independent factors, the combination and relative importance of which may vary greatly. Among these factors change of macroclimate is only one.

The problems of the *Grenzhorizont* and of related phenomena

As has been stated earlier, the question as to the reasons for the formation of strongly humified older *Sphagnum* peat as compared with an only faintly humified younger *Sphagnum* peat, and of the transition between both these types of peat, the famous *Grenzhorizont*, has been of prime importance for the subdivision of the Holocene as well as for palaeoclimatological speculations. The uncritical acceptance of synchronism of the transition from each type of black and strongly humified peat to white and far less humified peat with the *Grenzhorizont* proper (Weber, 1911), without a careful analysis of the former plant communities involved, has proved to be misleading (Overbeck, 1950; Gehl, 1952; Lundqvist, 1962; Schneekloth,

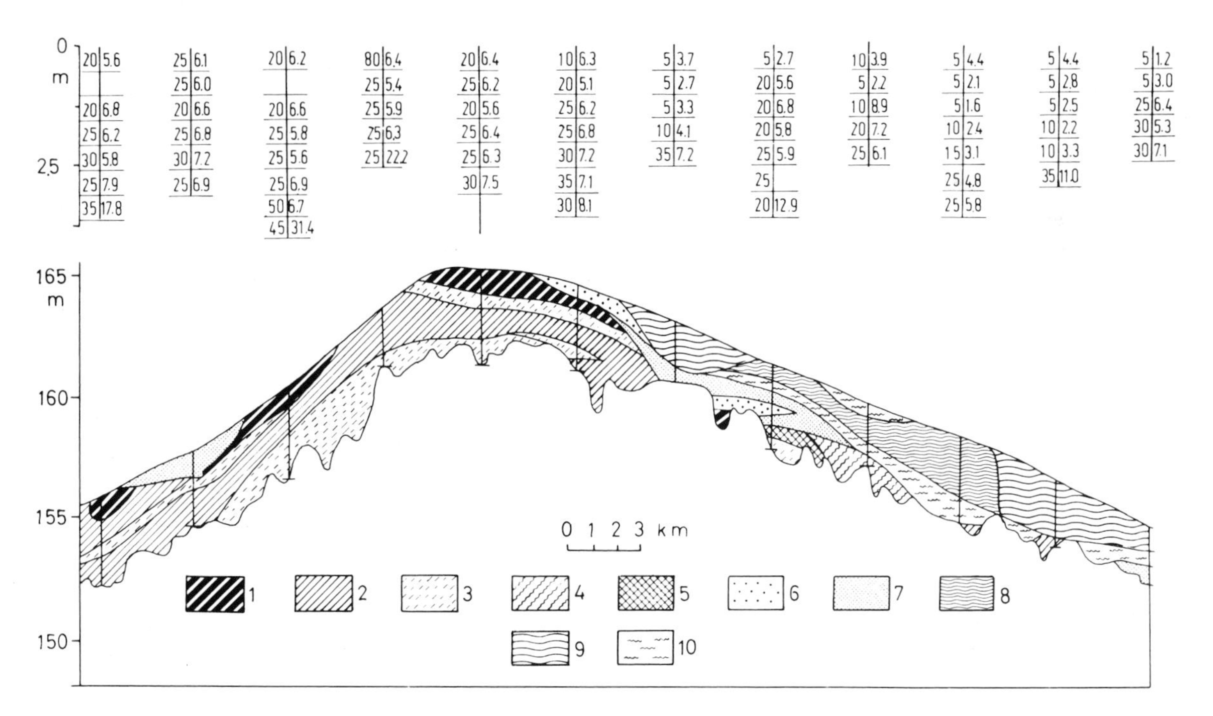

Fig. 2.12. The influence of the substrate on the fen and bog vegetation of the Vasyugan Bog, western Siberia, along a south (left) to north (right) profile (Neĭshtadt, 1972). Upper part: depth of borings indicated in the lower part of the diagram; to the left of the vertical lines: degree of disintegration of the plant material (%); to the right of the vertical lines: ash content (%). Legend: *1* = Cyperaceae peat; *2* = Hypnaceae fen peat; *3* = Cyperaceae–Hypnaceae fen peat; *4* = Poaceae fen peat; *5* = Poaceae–moss fen peat; *6* = Poaceae–moss transition peat; *7* = Cyperaceae–*Sphagnum* transition peat; *8* = various types of ombrogenous peat; *9* = *Sphagnum fuscum* peat; *10* = *Sphagnum* hollow and pool peat. The subsoil on the south-facing slope is rich in $CaCO_3$, Cl^- and SO_4^{2-}, whereas these substances are lacking in the substratum of the north-facing slope.

1963a, b, 1965, 1968, 1970; Schneider and Steckhan, 1963; Aletsee, 1967; Neĭshtadt, 1967a, b; Khotinskiĭ, 1971b, 1977; Neĭshtadt and Zelikzon, 1971; Neĭshtadt et al., 1974).

There are several reasons which may serve as explanations for this. Though the ecological and chemical status of soligenous and of topogenous peat differs strongly from that of ombrogenous peat, all these different types can be transformed by weathering into strongly humified, black peat layers. Even today (see Overbeck, 1975) there have only been very few thorough palaeobotanical analyses of the plant communities involved in the formation of these different types of peat. This situation is not improved by the fact that advanced humification and intensive decay of different types of plant and animal remains may have been selective, rendering the palaeoecological interpretation prone to mistakes. Very often it has only been the difference in colour and subjective assessment of humification that has been used as a criterion for the delimitation of the *Grenzhorizont*. The use of exact spectrophotometric measurements to distinguish black and white peat may change the picture considerably (Olausson, 1957; Overbeck, 1975). It must be regretted that, if areas with soligenous influences are excluded, no ecological situations seem to be known which under present-day conditions favour the formation of a typical black, strongly humified *Sphagnum* peat (see Overbeck, 1975). Weber has aimed at explaining the formation of the well-humified black *Sphagnum* peat by the assumption of a long-lasting dry period after the formation of an ombrogenous peat. This view was supported more recently by Neĭshtadt (1967b), who pointed to the differences in age between the uppermost layers of the black peat and the lowermost horizons of the overlying white peat in some of the European peat bogs. On the other hand, this does not hold true for the general case, which on the contrary points to an uninterrupted process of peat formation during this very transition (Overbeck, 1950; Van Zeist, 1955; Kubitzki, 1960). Last but not least, the typical *Grenzhorizont* seems to occur only in a very restricted area on the southern rim of the North Sea and on the south-eastern flank of the Baltic Sea (see, for instance, Aletsee, 1967; Overbeck, 1975; and also Gore, Ch. 1).

Taking this together with the appreciably diverging ages for the transition from black to white *Sphagnum* peat, it will be understood that, although there do exist some climatic background conditions directly responsible for the formation of the black, strongly humified peat, this influence can be much modified and camouflaged or changed by local conditions of various origin (see, for instance, Jonas, 1936; Olausson, 1957; Aartolahti, 1965; and Hayen, 1966).

How complicated the problems involved actually are may be seen from the following example. It is in general agreed that the accumulation rate of organic matter during the formation of the black *Sphagnum* peat seems to have been much smaller than during the formation of the white peat. However, Weber (1967) and Val'k (1971) have calculated that the accumulation rate must either have remained approximately constant in Estonia since the beginning of the Atlantic period or that it was roughly the same during Atlantic and Subatlantic times. Conversely in the Schwarzwald (Black Forest) of southwestern Germany, despite the absence of a typical *Grenzhorizont*, a transition between slowly and rapidly accumulating *Sphagnum* peat can be traced.

In continuation of my previous work on the chemistry and the absorption of the present-day aerosols by various plant communities I have investigated the problem of prehistoric clearings (by burning forests) and their possible contribution to the aerosols of the time. It could be shown that, synchronously with the extensive Neolithic, Bronze Age or Iron Age clearings in the Rhein-Graben region, much ash was transported by the prevailing westerly winds to the mountains of the adjacent Schwarzwald. The growth rate of peat increased, presumably in response to the fertilizing effect on the moss communities, and declined again when the ash content diminished following a diminution of man's activity in the nearby western lowlands (Fig. 2.13). Comparable correlations have been found by other authors (for instance, Olausson, 1957; and Schneekloth, 1963a). This raises the possibility that man may have been one of the agents responsible for the formation of fast growing white *Sphagnum* peat. Schwaar (1977) has pointed to comparable conditions for the formation of the white peat on Gough Island (see Gore, Ch. 1). However, in discussing this problem it should be borne in mind that the present-day

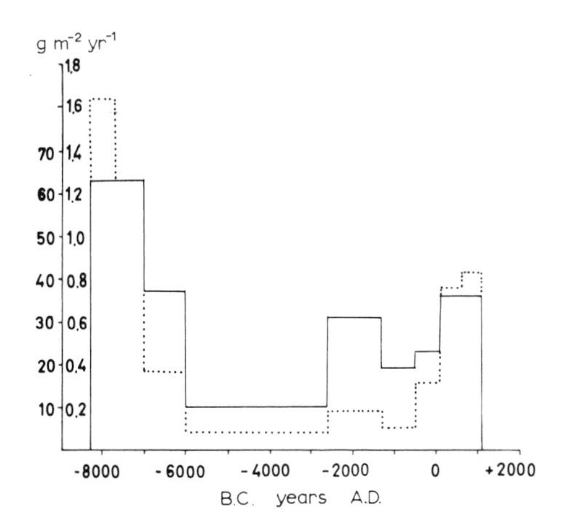

Fig. 2.13. Ash content and peat growth during Postglacial times in the Wildseemoor, near Kaltenbronn, Black Forest, southwestern Germany (Keitel, 1978). Full line: accumulation of dry matter; dotted line: accumulation of ash substances.

moss communities may contain appreciably variable amounts of ash and of mineral nutrients (see, for instance, Witting, 1948; Malmer, 1962; Bellamy and Bellamy, 1966; Aletsee, 1967; Grosse-Brauckmann, 1975b; Rump et al. 1977; and the literature cited in these papers). Furthermore, in order to keep a sense of proportion, it is worth referring to Clymo (Ch. 4) who summarizes the current views on rates of peat formation.

As an alternative to Granlund's hypothesis (see p. 44), Godwin (1952) and Walker and Walker (1961) suggested that it was not the continuous, or at least stepwise, increase in precipitation that has controlled the growth rate of the contributing species of *Sphagnum*, thus leading indirectly, through hydrological changes, to phases of desiccation and weathering, but that the climate had oscillated, sometimes favouring, sometimes retarding growth on vast areas of the bog. Walker and Walker (1961; see Fig. 2.14) convincingly demonstrated such a sequence of events over vast areas of a growing bog and thus also replaced the formerly held view of a cyclic evolution of hummocks with hollows in between them (see Osvald, 1923; Kulczyński, 1949; Jensen, 1961; and Fig. 2.15). Recent investigations, over a period of more than twenty years, on the active development of hummocks and hollows in a Bavarian bog (Schmeidl, 1977 — see Fig. 2.16), have supported the view held by Walker and Walker (1961) and by Godwin (1952) that, during the active growth of the bog, the topographical position of hummocks and ponds remains constant.

If renewed growth above the retardation layer depends solely on an increase in water saturation, the question remains whether such an increase was caused by climate only. Walker and Walker (1961) were cautious in attributing changes to climate alone, and did not exclude the role of local conditions. According to Casparie (1969) such local conditions of drainage, with simultaneous oversaturation with water in other parts of the mire, were of prime importance for the changing growth

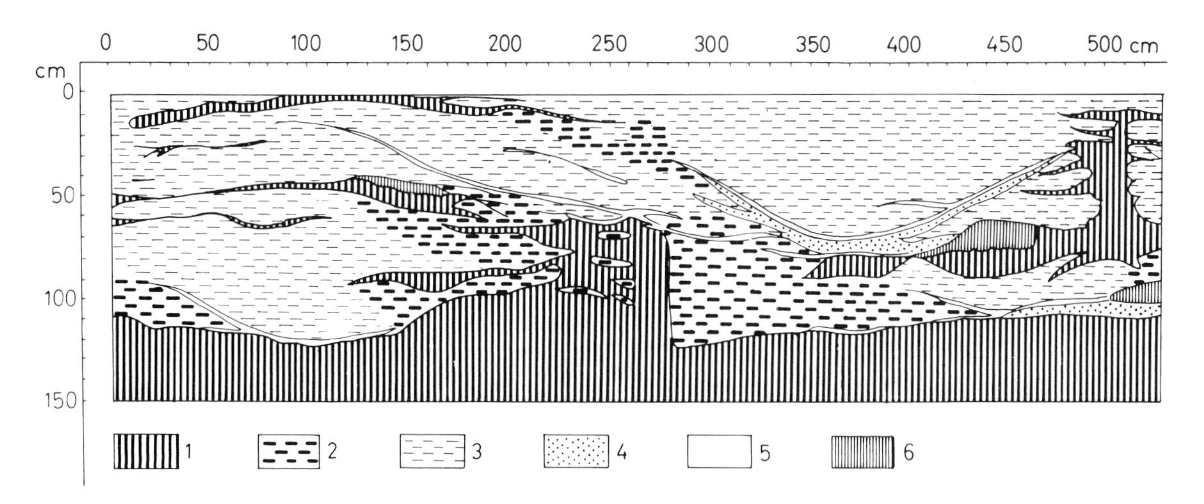

Fig. 2.14. Stratigraphy of a peat exposure at Fallahogy, northwestern Ireland (Walker and Walker, 1961). Legend: *1* = *Calluna–Sphagnum* peat; *2* = *Sphagnum–Calluna* peat; *3* = *Sphagnum* peat; *4* = *Sphagnum cuspidatum* peat; *5* = *Sphagnum cuspidatum* mud; *6* = *Eriophorum vaginatum*.

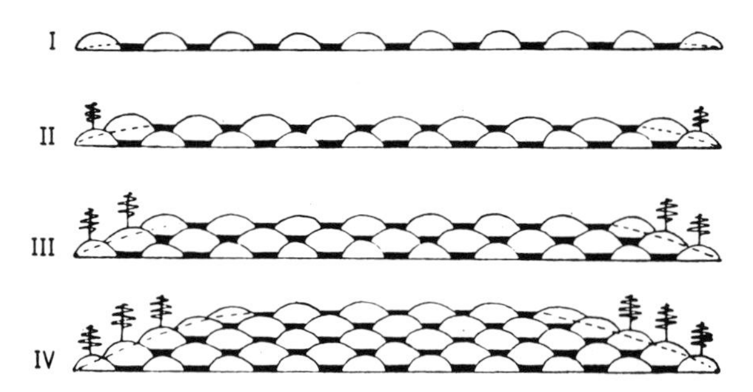

Fig. 2.15. Schematic vertical section through a growing ombrogenous bog (Kulczyński, 1949), showing for four stages the theoretically deduced history of hummocks and hollows (black). Height of the ground-water table indicated by the dotted lines.

and humification rates. That retardation layers and actively growing parts may occur simultaneously in a single mire (Olausson, 1957) is a well-known fact (see, for instance, Fig. 2.17). Moreover, the age of a particular retardation layer may differ in various parts of any one peat bog by some hundreds of years (Van Zeist, 1955; Lundqvist, 1962; Aartolahti, 1965; Schneekloth, 1965; Overbeck, 1975). This leads to the general conclusion that, according to our present knowledge, retardation layers and recurrence surfaces may originate either from climatic triggering effects plus local conditions or from local conditions only (see Hayen, 1966). In any case there are likely to be long delays between the allegedly acting climatic factor and the reaction of the bog investigated. This is complicated even further by the observation of Kuntze and Burghardt (1977) that very dry summers, the mean precipitation of which may amount to only 45 to 50% of normal mean values, obviously influence only the uppermost very thin layers of the bogs. One must agree with Gross (1934), Walker and Walker (1961), Schneekloth (1965, 1968, 1970), Casparie (1969), Overbeck (1975) and others that conclusions as to whether climatic factors had influenced the growth rate of peat bogs should be based on a comprehensive analysis of long horizontal exposures only, in order to avoid misinterpretation of local conditions. Moreover, only the oldest date available for any particular retardation layer should be looked upon as indicating the suggested date of climatic change (Overbeck, 1975). Furthermore, only several of these investigations, done on one peat bog and compared with several others, may help to establish the history of climate reliably (Godwin, 1961; Overbeck, 1975).

An example of a transition from one type of ombrogenous bog to another which may be more reliably governed by climate is thought to occur with the commencement of hummock ridge, or string, formation. This has been used by Frenzel (1966) for palaeoclimatic reconstruction since it was held that this special type of mire is caused by freezing processes.

There is some doubt however as to whether typical string (*kermi*) mires are the effects of differential freezing processes or whether they are caused mainly by large amounts of water running from the bog cupola (Sjörs, 1946). Further reference to this aspect of *kermi* mire formation is made by Ruuhijärvi (Part B, Ch. 2). In general, *kermi* mires are confined today to the Boreal coniferous forest belt or to the northern part of mixed broad-leaved and coniferous forests in continental regions of Europe and North America. In contrast to what has been thought previously, the onset of hummock ridge formation may have happened in southern zones earlier than in northern or more continental ones (Table 2.4). Presumably, within the North Temperate Zone, the effect of a layer of *Sphagnum* peat thick enough to influence the local heat flux was at least as important as macroclimate. This implies that the autogenic initiation of freezing can cause the formation of hummock ridges in *kermi* mires — without any accompanying change of macroclimate. The onset of string development in this way could occur at different times in different parts of the same mire province.

CONCLUSION

The central aim here has been to show that the development of mires is governed not only by climate but by a wealth of internal and external factors, complicating enormously the evaluation of past influences of climate. The main reason for this situation seems to be that climate did not change markedly during Postglacial times. If it is permissible to extrapolate from the changes in temperature calculated from the oscillating alpine timberline ($\pm 0.7°C$, according to Patzelt, 1977) and from the polar timberline in Northern Europe (Hustich, 1958), with approximately corresponding values, it becomes evident why all the other internal and external factors could exert such strong influences

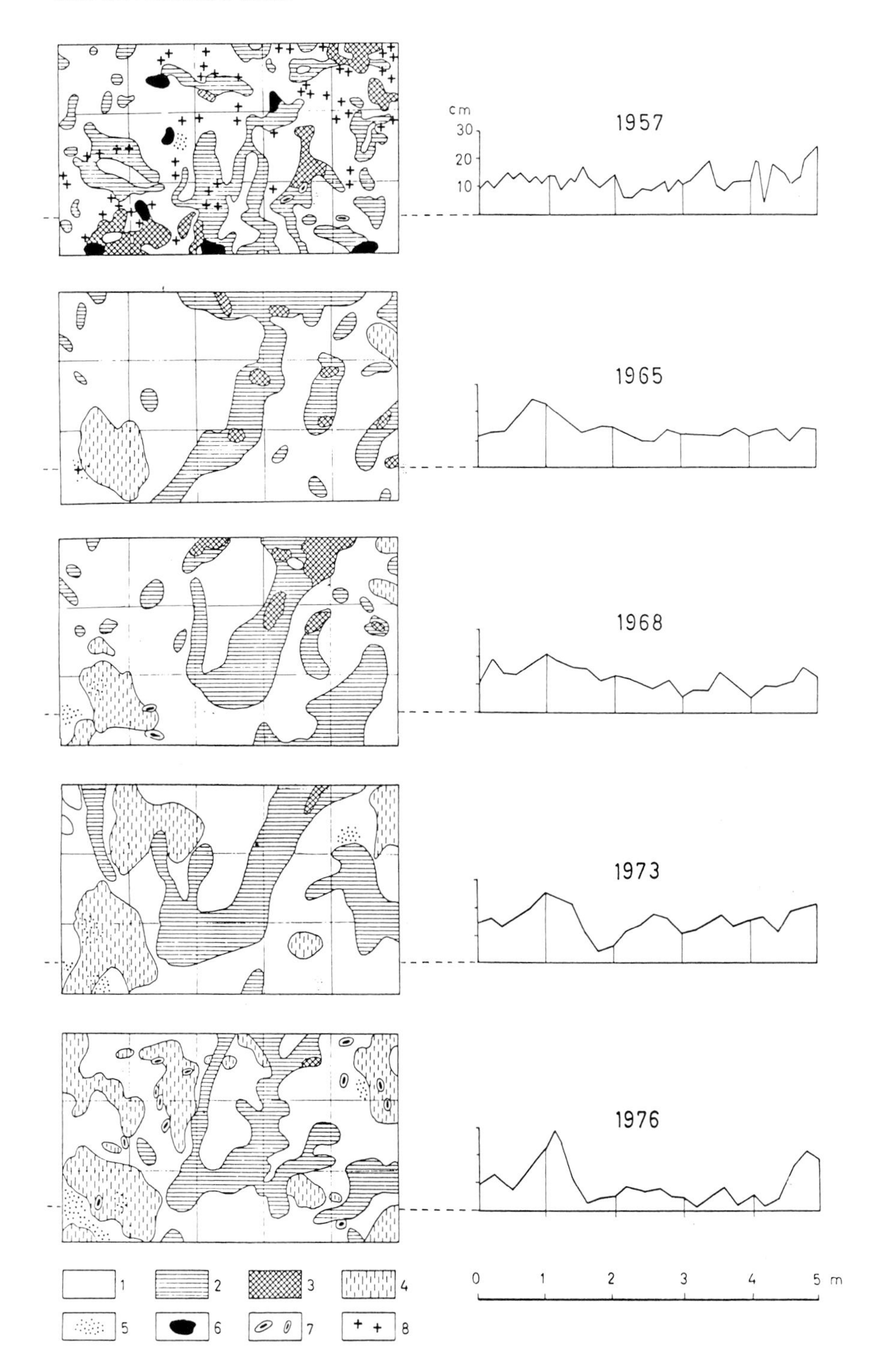

Fig. 2.16. The microrelief of a Bavarian ombrogenous bog (*Sphagnetum medii*, subassociation *Rhynchospora alba*) and its development during twenty years (Schmeidl, 1977). Left: maps showing the topography and plant associations involved; right: vertical sections at the places indicated by the dotted lines, through the uppermost horizons of the peat bog, showing the changes in topography of the bog surface. Legend: *1* = red-coloured hummock association; *2* = red-coloured hollow association; *3* = hollows with *Sphagnum cuspidatum*; *4* = *Polytrichum strictum*; *5* = *Calluna vulgaris*; *6* = pools; *7* = bare hollows; *8* = *Rhynchospora alba*.

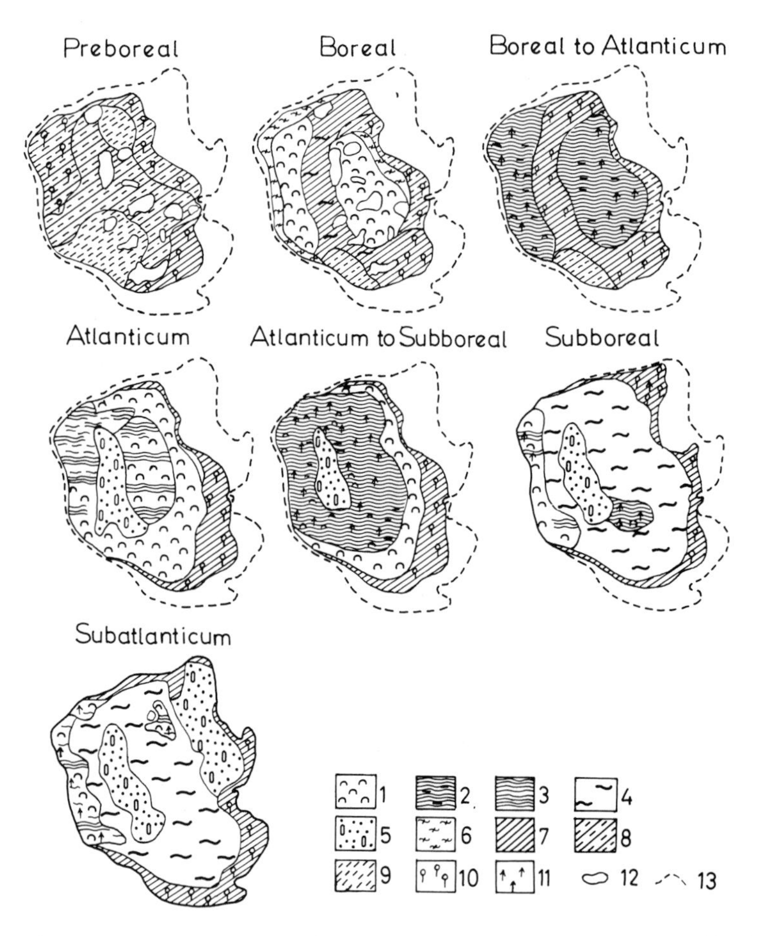

Fig. 2.17. The changing distribution pattern of fen and bog plant communities during the Holocene in the Shuvalovo peat bog, northwestern U.S.S.R. (Khotinskiĭ, 1977). Legend: *1* = *Sphagnum fuscum* association; *2* = *Sphagnum–Eriophorum* association; *3* = *Eriophorum* association; *4* = *Sphagnum teres* association; *5* = *kermi* complex; *6* = *Pinus–Eriophorum* complex; *7* = Cyperaceae association; *8* = Cyperaceae–Hypnaceae association; *9* = Hypnaceae association; *10* = *Betula*; *11* = *Pinus*; *12* = fossil lakes; *13* = present-day outline of the bog complex.

TABLE 2.4

Approximate date for the beginning of hummock ridge formation in various parts of northern and eastern Europe

Area	Date of beginning of hummock ridge formation	Author
Southwestern Häme, Finland	after 2400 B.P.	Aartolahti (1965)
Northern Satakunta, Finland	after 4000 B.P.	Aartolahti (1965)
Russian Karelia	second part of the Subatlantic	Aartolahti (1965)
Northern Finland	presumably Subatlantic period	Ruuhijärvi (1963)
Shuvalovo bog near Leningrad	end of the Atlantic or beginning of Subboreal times	Khotinskiĭ (1971b)

on the developing mires. Of course there are indications of much larger changes in climate during Postglacial times but even these are sometimes conflicting. This can be seen from the values given for the Canadian and Alaskan Arctic and Subarctic by Nichols (1967, 1969, 1975), Grichuk (1969), Rampton (1971) and Ritchie (1972) as compared with those of Marr (1948), who could not find there any clearcut influence of changing climates (see also Hustich, 1965).

Perhaps areas of the globe exist over which the relative importance of climate, and of the other contributing factors in the formation of mires, are different from one another. Possibly, where climate is of utmost importance for the growth of plants, as in subarctic, subalpine or steppe regions, the spontaneous evolution of mires is governed most of all by climate. It is in these places, however, that secondary disturbances of the spontaneous equilibrium by overgrazing, agriculture and so on are likely to influence the growth rate of the mires much more than might be expected from the climatic changes only.

On the other hand, where climate is optimal for plant growth, as in the Temperate Zone and the moist tropics, factors other than climate may be of prime importance for the growth of mires. Unfortunately, it is in these regions that man has most intensively changed the ecosystems and the edaphic and geomorphological conditions surrounding the growing mires.

In practice much can be done to overcome the difficulties described. It is clearly essential to study the development of the various types of mire vegetation and their antecedents most accurately by the use of extended peat exposures. The use of mires as convenient traps for pollen grains is but one of the means of investigation and the use of modern chemical and physical techniques should also help to overcome some of the more obvious pitfalls of misinterpretation. Such investigations must necessarily be accompanied by well-based palaeoecological studies of the whole area of which the mire proper is only one small part. The mire reflects the palaeoecological history of the whole catchment area or of a large geomorphological unit, and its development can only be understood against the background of the changing palaeoecological situation of this larger area.

REFERENCES

Aaby, B., 1975. Cykliske klimavariationer de sidste 7500 år påvist ved undersøgelser af højmoser og marine transgressionsfaser. *Dan. Geol. Unders., Årb.*, 1974, 91–104.

Aaby, B. and Tauber, H., 1974. Rates of peat formation in relation to degree of humification and local environment as shown by studies of a raised bog in Denmark. *Boreas*, 4: 1–17.

Aario, L., 1932. Pflanzentopographische und paläogeographische Mooruntersuchungen in Nord-Satakunta. *Fennia*, 55(1): 179 pp.

Aartolahti, T., 1965. Oberflächenformen von Hochmooren und ihre Entwicklung in Südwest-Häme und Nord-Satakunta. *Fennia*, 93(1): 286 pp.

Aletsee, L., 1967. Begriffliche und floristische Grundlagen zu einer pflanzengeographischen Analyse der europäischen Regenwassermoorstandorte (mit einer Diskussion der floristischen Gradienten innerhalb der europäischen Hochmoorvegetation). *Beitr. Biol. Pflanzen*, 43: 117–160.

Andersen, S.Th., 1964. Interglacial plant successions in the light of environmental changes. *Rep. VI Int. Congr. Quaternary, Warsaw, 1961*, 2: 359–368.

Auer, V., 1965. The Pleistocene of Fuego-Patagonia; part IV: Bog profiles. *Ann. Acad. Sci. Fenn., Ser. A III, Geol. Geogr.*, 80: 160 pp.

Behre, K.E., 1970. Die Entwicklungsgeschichte der natürlichen Vegetation im Gebiet der unteren Ems und ihre Abhängigkeit von den Bewegungen des Meeresspiegels. *Probl. Küstenforsch. Südl. Nordseegeb.*, 9: 13–47.

Bellamy, K. and Bellamy, R., 1966. An ecological approach to the classification of the lowland mires of Ireland. *Proc. R. Ir. Acad. Sci. Sect. B*, 65: 237–251.

Birks, H.H., 1971. Studies in the vegetational history of Scotland. I. A pollen diagram from Abernethy Forest, Inverness-Shire. *J. Ecol.*, 58: 827–846.

Birks, H.H., 1972. Studies in the vegetational history of Scotland. II. Two pollen diagrams from the Galloway Hills, Kirkcudbrightshire. *J. Ecol.*, 60: 183–217.

Blytt, A., 1876. Forsög till en theori om invandringen af Norges flora under vexlende regfulde og törre tider. *Nyt Mag. Naturvet.*, Kristiania.

Boatman, D.J., 1961. Vegetation and peat characteristics of blanket bogs in County Kerry. *J. Ecol.*, 49: 507–517.

Boatman, D.J., 1962. The growth of *Schoenus nigricans* on blanket bog peats. 1. The response to pH and the level of potassium and magnesium. *J. Ecol.*, 50: 823–832.

Borówko-Dłużakowa, Z., 1961. Historia flory Puszczy Kampinoskiej w późnym glacjale i holocenie. *Przegl. Geogr.*, 33: 365–382 (with Russian and English summaries).

Bower, M.M., 1961. The distribution of erosion in blanket peat bogs in the Pennines. *Trans. Inst. Br. Geogr.*, 29: 17–30.

Brandt, A., 1948. Über die Entwicklung der Moore im Küstengebiet von Süd-Pohjanmaa am Bottnischen Meerbusen. *Ann. Bot. Soc. Zool.-Bot. Fenn. "Vanamo"*, 23(4): 134 pp.

Burges, A., 1951. The ecology of the Cairngorms. III The *Empetrum-Vaccinium* zone. *J. Ecol.*, 39: 271–284.

Cajander, A.K., 1913. Studien über die Moore Finnlands. *Acta Forest. Fenn.*, 2: 298 pp.

Casparie, W.A., 1969. Bult- und Schlenkenbildung im Hochmoortorf (zur Frage des Moorwachstums-Mechanismus). *Vegetatio*, 18: 146–180.

Coope, G.R., 1977a. Fossil coleopteran assemblages as sensitive indicators of climatic changes during the Devensian (last) cold stage. *Philos. Trans. R. Soc. Lond.*, B 280: 313–340.

Coope, G.R., 1977b. Coleoptera as clues to the understanding of climatic changes in North Wales towards the end of the last (Devensian) glaciation. *Cambria*, 4(1): 65–72.

Dansgaard, W., Johnsen, S.J., Clausen, H.B. and Langway, C.C., 1970. Ice cores and paleoclimatology. In: I.U. Olsson (Editor), *Radiocarbon Variations and Absolute Chronology*. Almqvist and Wiksell, Stockholm, pp. 337–348.

Dau, J.H.Chr., 1823. *Neues Handbuch über den Torf, dessen Natur, Entstehung und Wiedererzeugung. Nutzen im Allgemeinen und für den Staat.* J.C. Hinrichschen Buchhandlung, Leipzig, 244 pp.

De Beaulieu, J.L., 1974. Evolution de la végétation sur la bordure montagneuse cévenole au postglaciaire, d'après les pollens. *Bull. Soc. Languedoc. Géogr.*, 8: 347–358.

Deevey, E.S., 1939. Studies on Connecticut lake sediments I. A postglacial climatic chronology for Southern New England. *Am. J. Sci.*, 237: 691–724.

Dickinson, W., 1973. The development of the raised bog complex near Rusland in the Furness district of North Lancashire. *J. Geol.*, 61: 871–886.

Dimbleby, G.W., 1965. Post-glacial changes in soil profiles. *Proc. R. Soc. Lond.*, B 161: 355–362.

Dokturovskiĭ, W.S., 1927. Die Sukzession der Pflanzenassoziationen in den russischen Torfmooren (Materialien zur Vergleichung der skandinavischen und russischen Torfmoore). *Veröff. Geobot. Inst. Eidg. Techn. Hochsch. Stift. Rübel Zürich*, 4: 123–143.

Dokturovskiĭ, W.S., 1928. Über die Grenzen der *Sphagnum*-moore und über Moorgebiete in USSR (Rußland). *Bot. Not.*, 1928: 54–62.

Du Rietz, G.E., 1954. Die Mineralbodenwasserzeigergrenze als Grundlage einer natürlichen Zweigliederung der Nord- und Mitteleuropäischen Moore. *Vegetatio*, 5–6: 571–585.

Du Rietz, G.E. and Nannfeldt, J.A., 1925. Ryggmossen und Stigsbo Rödmosse, die letzten lebenden Hochmoore der Gegend von Upsala. Führer für die vierte I.P.E. *Sven. Växtsociol. Sällsk. Handl.*, 3: 21 pp.

Elina, G.A., 1969. O razvitii bolot v glubokikh vpadinakh na severe Karelii. In: M.I. Neĭshtadt (Editor), *Golotsen*. Nauka, Moscow, pp. 165–171 (English summary).

Eurola, S., 1962. Über die regional Einteilung der südfinnischen Moore. *Ann. Bot. Soc. Zool.-Bot. Fenn. "Vanamo"*, 33(2): 243 pp. (Finnish summary).

Eurola, S. and Ruuhijärvi, R., 1961. Über die regionale Einteilung der finnischen Moore. *Arch. Soc. Zool.-Bot. Fenn. "Vanamo"*, 16: 49–64.

Fairbridge, R.W., 1961. Eustatic changes in sea level. In: L.H. Ahrens, F. Press, K. Rankama and S.K. Runcorn (Editors), *Physics and Chemistry of the Earth, 4*. Pergamon Press, London, pp. 99–185.

Firbas, F. and Broihan, F., 1937. Das Alter der Trockentorfschichten im Hils. *Planta*, 26: 291–302.

Frenzel, B., 1966. Climatic change in the Atlantic/Sub-boreal transition on the Northern Hemisphere: botanical evidence. In: J.S. Sawyer (Editor), *World Climate From 8000 to 0 B.C. Proceedings of the International Symposium held at Imperial College, London, 1966*, pp. 99–123.

Frenzel, B., 1975. The distribution pattern of Holocene climatic change in the Northern Hemisphere. In: *Proceedings of the WMO/IAMAP Symposium on Long-Term Climatic Fluctuations, Norwich, 1975.* World Meteorological Organization, Geneva, Nr. 421.

Frenzel, B., 1976. Die Missenmoore des Nordschwarzwaldes; Spiegelbild der Umweltbelastungen? *Daten Dok. Umweltsch.*, 19: 99–107.

Frenzel, B., 1977. Postglaziale Klimaschwankungen im südwestlichen Mitteleuropa. In: B. Frenzel (Editor), *Dendrochronologie und postglaziale Klimaschwankungen in Europa. Erdwiss. Forsch.*, 13: 297–322.

Frenzel, B., 1978. Landschaftsgeschichte und Landschaftsökologie des Kreises Freudenstadt. In: *Der Kreis Freudenstadt.* Konrad Theiss Verlag, Stuttgart, pp. 52–76.

Frenzel, B., 1979. Dendrochronologie und Landschaftsökologie. *Allg. Forstzeitschr.*, 49: 1355–1359.

Fries, M., 1965. The late-Quaternary vegetation of Sweden. *Acta Phytogeogr. Suec.*, 50: 269–284.

Früh, J. and Schröter, J., 1904. *Die Moore der Schweiz, mit Berücksichtigung der gesamten Moorfrage.* Beiträge zur Geologie der Schweiz, Geotechnische Serie, III. Lieferung, Bern, 750 pp.

Gehl, O., 1952. Die Hochmoore Mecklenburgs, nebst einem Beitrag zur Waldgeschichte des Küstenraumes zwischen Elbe und Oder. *Beih. Z. Geol.*, No. 2: 99 pp.

Godwin, H., 1946. The relationship of bog stratigraphy to climatic change and archaeology. *Proc. Prehist. Soc.*, 1946, No. 1: 1–11.

Godwin, H., 1952. Recurrence-surfaces. *Dan. Geol. Unders., II. Raekke*, Nr. 80: 22–30.

Godwin, H., 1956. *The History of the British Flora.* Cambridge University Press, Cambridge, 384 pp.

Godwin, H., 1961. Radiocarbon dating and Quaternary history in Britain. *Proc. R. Soc. Lond., Ser. B*, 153: 287–320.

Godwin, H., 1975. *A Factual Basis for Phytogeography.* Cambridge University Press, Cambridge, 541 pp.

Gradmann, R., 1898. *Das Pflanzenleben der Schwäbischen Alb, 1 und 2.* Tübingen, 470 pp.; 351 pp.

Granlund, E., 1922. De svenska högmossarnas geologi. *Sver. Geol. Unders.*, C 373: 193 pp.

Grichuk, V.P., 1969. Opyt rekonstruktsii nekotorykh elementov klimata severnogo polusariya v atlanticeskiĭ period golotsena. In: M.I. Neĭshtadt (Editor), *Golotsen.* Nauka, Moscow, pp. 41–57.

Gross, H., 1934. Zur Frage des Weberschen Grenzhorizontes in den östlichen Gebieten der ombrogenen Moorregion. *Beih. Bot. Centralblatt*, 51(II): 305–353.

Grosse-Brauckmann, G., 1963. Über die Artenzusammensetzung von Torfen aus dem nordwestdeutschen Marschen-Randgebiet (eine pflanzensoziologische Auswertung von Großrestuntersuchungen). *Vegetatio*, 11: 325–341.

Grosse-Brauckmann, G., 1975a. Einige allgemeine Ergebnisse von Torf-Großrestuntersuchungen. *Telma*, 5: 39–42.

Grosse-Brauckmann, G., 1975b. Über Beziehungen zwischen chemischen Merkmalen von Torfen und ihrem botanischen Charakter. In: *Moor und Torf in Wissenschaft und Wirtschaft*, pp. 145–155.

Grosse-Brauckmann, G., 1976. Zum Verlauf der Verlandung bei

einem eutrophen Flachsee (nach quartärbotanischen Untersuchungen am Steinhuder Meer). II. Die Sukzessionen, ihr Ablauf und ihre Bedingungen. *Flora*, 165: 415–455.

Grosse-Brauckmann, G. and Streitz, B., 1976. Das Wiesbüttmoor: Über die Pflanzendeck eines kleinen Naturschutzgebietes im Spessart, Teil 1. Mitteilungen Lochmühl Nr. 17. *Aufsätze Reden Senckenberg. Naturforsch. Ges.*, 28: 103–108.

Hafsten, U. and Solem, Th., 1976. Age, origin, and palaeoecological evidence of blanket bogs in Nord-Trøndelag, Norway. *Boreas*, 5: 119–141.

Hansen, H.P., 1947. Postglacial forest succession, climate, and chronology in the Pacific Northwest. *Am. Philos. Soc. Trans.*, 37: 1–130.

Hayen, H., 1966. Moorbotanische Untersuchungen zum Verlauf des Niederschlagsklimas und seiner Verknüpfung mit der menschlichen Siedlungstätigkeit. *Neue Ausgrabungen und Forschungen in Niedersachsen*, 3: 280–307.

Heinselman, M.L., 1963. Forest sites, bog processes and peatland types in the glacial Lake Agassiz region, Minnesota. *Ecol. Monogr.*, 33: 327–374.

Heusser, C.J., 1960. Late-pleistocene environments of North Pacific North America. *Am. Geogr. Soc., Spec. Publ.*, No. 35: 308 pp.

Hibbert, F.A., Switsur, V.R. and West, R.G., 1971. Radiocarbon dating of Flandrian pollen zones at Red Moss, Lancashire. *Proc. R. Soc. Lond., Ser. B*, 177: 161–176.

Hustich, I., 1958. On the recent expansion of the Scotch pine in Northern Europe. *Fennia*, 82(3): 25 pp.

Hustich, I., 1965. Correlation of tree-ring chronologies of Alaska, Labrador, and Northern Europe. *Acta Geogr.*, 15(3): 3–26.

Iversen, J., 1958. The bearing of glacial and interglacial epochs on the formation and extinction of plant taxa. *Uppsala Univ. Årsskr.*, 1958, No. 6: 210–215.

Iversen, J., 1964. Retrogressive vegetational succession in the Postglacial. *J. Ecol.*, 52 (*Suppl.*): 59–70.

Jankovská, V., 1971. The development of vegetation on the western slopes of the Bohemian–Moravian uplands during the late Holocene period: A study based on pollen and macroscopic analysis. *Folia Geobot. Phytotaxon.*, 6: 281–302.

Jelgersma, S., 1966. Sea-level changes during the last 10 000 years. In: J.S. Sawyer (Editor), *World Climate From 8000 to 0 B.C. Proc. Int. Symposium held at Imperial College, London, 1966*, pp. 54–71.

Jensen, U., 1961. Die Vegetation des Sonnenberger Moores im Oberharz und ihre ökologischen Bedingungen. *Natursch. Landschaftspfl. Niedersachsen*, 1: 73 pp.

Jessen, K., 1935. Archaeological dating in the history of North Jutland's vegetation. *Acta Archaeol.*, 5(3).

Jonas, F., 1935. Die Vegetation der Hochmoore am Nordhümmling. *Repert. Spec. Nov. Regni Veget.*, Beih. 78, 1: 143 pp.

Jonas, F., 1936. Das Grenzhorizontproblem. *Beih. Bot. Centralblatt*, 54B: 370–376.

Kamyshev, N.S., 1972. Sravnitel'naya kharakteristika sfagnovykh bolot Oksko-Donsko nizmennosti. *Byull. Mosk. O.-Va. Ispyt. Prir., Otd. Biol.*, 77(3): 88–99.

Kats, N.Ya. [Katz, N.Ya.; Kac, N.J.], 1926. *Sphagnum* bogs of Central Russia: Phytosociology, ecology and succession. *J. Ecol.*, 14: 177–202.

Kats, N.Ya., 1930. Zur Kenntnis der Moore Nordosteuropas. *Beih. Bot. Centralblatt*, 46B: 297–394.

Kats, N.Ya., 1931. Zur Kenntnis der oligotrophen Moortypen des europäischen Rußlands. *Beih. Bot. Centralblatt*, 47B: 177–210.

Kats, N.Ya., 1957. O granitsakh torfyano-bolotnych pochv, o rastitel'nykh arealakh i ikh formirovanii v svyazi s klimatom i geologicheskoi istoriei. *Pochvovedenie*, 1957, No. 5: 12–21.

Kats, N.Ya., 1958. O tipakh bolot i ikh razmeshchenii v kholodnoi i umerennoi zonakh severnogo polushariya. *Pochvovedenie*, 1958, No. 6: 13–20.

Kats, N.Ya., 1959a. Termicheskaya volna pozdnego pleistotsena i razvitie rastitel'nosti *Izv. Akad. Nauk S.S.S.R., Ser. Geogr.*, 1959, No. 6: 77–88.

Kats, N.Ya., 1959b. O bolotakh i torfyanikakh severnoi Ameriki. *Pochvovedenie*, 1959, No. 10: 44–52.

Kats, N.Ya., 1960. Über die Moortypen und ihre Verbreitung in der nördlichen und gemäßigten Zone der nördlichen Halbkugel. In: *VII Mezinárodni sjezd pro vseobecný výzkum raselin, Františkovy Lázně*, 3 pp.

Kats, N.Ya., 1961. O vypuklikh bolotakh poberezhiĭ moreĭ na zapadnykh granitsakh S.S.S.R. *Byull. Mosk. O.-Va. Ispyt. Prir., Otd. Biol.*, 66(2): 44–64.

Kats, N.Ya., 1969. Bolota Avstralii i Tazmanii. *Byull. Mosk. O.-Va. Ispyt. Prir., Otd. Biol.*, 1969(2): 106–116.

Kats, N.Ya., 1972. O rasprostranenii torfyanikov na zemnom share, o tipakh ikh i pri. *Bot. Zh.*, 57: 198–210.

Kats, N.Ya. and Kats, S.V., 1964. Die Eigentümlichkeiten der pleistozänen Pflanzengesellschaften. *Rep. VI Int. Congr. Quaternary, Warsaw, 1961*, 2: 439–445.

Keitel, A., 1978. *Untersuchungen zum Chemismus der Moor- und Seeablagerungen im nördlichen Schwarzwald.* Diplomarbeit, Botanisches Institut der Universität Hohenheim, Hohenheim, 150 pp.

Khotinskiĭ, N.A., 1971a. Palinologickeskie materialy k problemam paleogeografii golotsena tichookeanskoi okrainy SSSR. In: M.I. Neĭshtadt (Editor), *Palinologya golotsena.* Akad. Nauk S.S.S.R., Inst. Geogr., Moscow, pp. 171–187 (English summary).

Khotinskiĭ, N.A., 1971b. *Appendix to Guide for Field Route Nr. 1-B. The Problem of Boundary Horizon, With Special Reference to the Shuvaloff Peat Bog. III.* International Palynological Conference, Novosibirsk, 41 pp.

Khotinskiĭ, N.A., 1976a. Theoretische Aspekte bei der Erforschung der Paläogeographie des Holozäns. *Petermanns Geogr Mitt.*, 120: 192–196.

Khotinskiĭ, N.A., 1976b. Archäologisch-paläogeographische Forschungen im Zentrum der Russischen Ebene. *Z. Archäol.*, 10: 161–172.

Khotinskii, N.A., 1977. *Golotsen severnoi Evrazii. Opyt transkontinental'noi korrelyatsii etapov razvitiya rastitel'nosti i klimata.* Nauka, Moscow, 198 pp.

Kiryushkin, V.N., Starichenkov, I.P., Tikhomirov, L.I., 1967. Vliyanie geologo-geometeorologicheskikh uslovii mestnosti na formirovanie bolot (na primere bolot severo chasti Archangel'skoi oblasti). In. A.A. Nitsenko (Editor), *Priroda bolot i metody ikh issledivanij.* Nauka, Leningrad, 11–15.

Koshcheev, A.L., 1953. Die Vermoorung ganzer Rodungen auf

Sandböden und ihre Bekannte maßnahmen. *Tr. Inst. Lesa Akad. Nauk S.S.S.R.*, 13: 10–50.

Krames, K., 1951. Die Korrelation zwischen Temperatur- und Niederschlagsanomalien im Winter der Nordhemisphäre. *Geogr. Ann.*, 33: 210–229.

Krames, K., 1952. Die Korrelation zwischen Temperatur- und Niederschlagsanomalien im Sommer der Nordhemisphäre. *Geogr. Ann.*, 34: 238–260.

Kubitzki, K., 1960. Moorkundliche und pollenanalytische Untersuchungen am Hochmoor "Esterweger Dose". *Schr. Naturwiss. Ver. Schleswig-Holstein*, 30: 12–28.

Kukla, St., 1965. Rozwój torfowisk źródliskowych na terenach pólnocno-wschodniej Polski. *Zesz. Probl. Postepów Nauk Roln.*, 57: 395–483 (with Russian and English summaries).

Kulczyński, St., 1949. Torfowiska Polesia. *Mém. Acad. Polon. Sci. Lettr., Cl. Sci. Math. Nat., Sér. B, Sci. Nat.*, No. 15: 356 pp.

Kulikova, G.G., Liss, O.L., Predtechenskiĭ, A.V., Skobeeva, E.I. and Tjuremnov, S.N., 1971. Rastitel'nyi pokrov torfyanykh bolot srednego Priob'ya i zakonomernosti ego razmeshcheniya. *Vestn. Mosk. Univ., Ser. Biol., Pochvoved., Geol. Geogr.*, 1971, No. 2: 53–57.

Kuntze, H. and Burghardt, W., 1977. Die Entwicklung der Bodenfeuchte deutscher Hochmoorkulturen in einem Trockenjahr. *Telma*, 7: 55–54.

Lamb, H.H., 1972. *Climate. Present, Past and Future, 1. Fundamentals and Climate now.* Methuen, London, 613 pp.

Lange, E., 1967. Zur Vegetationsgeschichte des Beerberggebietes im Thüringer Wald. *Repert. Spec. Nov. Regni Veget.*, 76: 205–219.

Lange, E. and Schlüter, H., 1972. Zur Entwicklung eines montanen Quellmoores im Thüringer Wald und des Vegetationsmosaiks seiner Umgebung. *Flora*, 161: 562–585.

Lasalle, P., 1966. Late Quaternary vegetation and glacial history in the St. Lawrence lowlands, Canada. *Leidse Geol. Meded.*, 38: 91–128.

Levkovskaya, G.M., Kind, N.V., Zavel'skiĭ, F.S. and Forova, V.S., 1970. Absolyutnyĭ vozrast torfyanikov rayona g. Igarka i raschlenenie golotsena zapadnoi Sibiri. *Byull. Kom. Izuch. Chetvertichn. Perioda, Akad. Nauk S.S.S.R.*, 37: 94–101.

Lundqvist, G., 1962. Geological radiocarbon datings from the Stockholm Station. *Sver. Geol. Unders., Ser. C, No. 589, Årsb.*, 56(5): 23.

Lutz, J.L., 1956. Spirkenmoore in Bayern. *Ber. Bayer. Bot. Ges. Erforsch. Heim. Flora*, 31: 58–69.

Mackereth, F.J.H., 1966. Some chemical observations on postglacial lake-sediments. *Philos. Trans. R. Soc. Lond., Ser. B 250*, No. 765: 165–215.

McVean, D.N. and Ratcliffe, D.A., 1962. *Plant Communities of the Scottish Highlands. A Study of Scottish Mountain, Moorland and Forest Vegetation.* Monographs of the Nature Conservancy, 1. London, 429 pp.

Malmer, N., 1962. Studies on mire vegetation in the Archaean area of southwestern Götaland (South Sweden). I. Vegetation and habitat conditions on the Åkhult mire. *Opera Bot.*, 7(1): 322 pp.

Marr, J.W., 1948. Ecology of the forest–tundra ecotone on the east coast of Hudson Bay. *Ecol. Monogr* 18: 117–144.

Martin, P.S., 1963. *The Last 10 000 Years. A Fossil Pollen Record of the American Southwest.* The University of Arizona Press, Tucson, Ariz., 87 pp.

Menke, B., 1963. Beiträge zur Geschichte der Erica-Heiden Nordwestdeutschlands. *Flora*, 153: 521–548.

Menke, B., 1976. Befunde und Überlegungen zum nacheiszeitlichen Meeresspiegelanstieg (Dithmarschen und Eiderstedt, Schleswig-Holstein). *Probl. Küstenforsch. Südl. Nordseegeb.*, 11: 145–161.

Mercer, J.H., 1969. The Allerød Oscillation: A European climatic anomaly? *Arct. Alp. Res.*, 1: 227–234.

Moore, P.D., 1977. Vegetational history. *Cambria*, 4(1): 73–83.

Moravec, J. and Rybničková, E., 1964. Die *Carex davalliana*-Bestände im Böhmerwald-Vorgebirge, ihre Zusammensetzung, Ökologie und Historie. *Preslia*, 36: 376–391.

Mörner, N.A., 1971. Eustatic and climatic oscillation. *Arctic Alpine Res.*, 3: 167–171.

Mörner, N.A. and Wallin, B., 1977. A 10 000-year temperature record from Gotland, Sweden. *Palaeogeogr., Palaeoclimatol., Palaeoecol.*, 21, 113–138.

Müller, K., 1965. Zur Flora und Vegetation der Hochmoore des nordwestdeutschen Flachlandes. *Schr. Naturwiss. Ver. Schleswig-Holstein*, 36: 30–77.

Neishtadt, M.I., 1957. *Istoriya lesov i paleogeografiya SSSR v golotsene.* Akad. Nauk S.S.S.R., Moscow, 403 pp.

Neĭshtadt, M.I., 1961a. Uchet bolot i torfyanykh resursov v SSSR. *Izv. Akad. Nauk S.S.S.R., Ser. Geogr.*, 1961, No. 2: 46–52.

Neĭshtadt, M.I., 1961b. Zakonomernosti geograficheskogo raspredeleniya torfyanykh bolot i ikh tipov na territorii SSSR. *XIX Mezhdunar. Geogr. Kongr. Stokgolme, Akad. Nauk S.S.S.R.*, pp. 199–207.

Neĭshtadt, M.I., 1966. Prinzipien zur Rayonierung der Moore in der UdSSR. *Ann. Acad. Sci. Fenn., Ser. A, Geol.-Geogr.*, III, No. 89: 29 pp.

Neĭshtadt, M.I., 1967a. Ob absolyutnom vozraste torfyanykh bolot Zapadnoi Sibiri. *Rev. Roum. Biol., Sér. Bot.*, 12(2–3): 181–186.

Neĭshtadt, M.I., 1967b. Stratigrafiya torfyanykh mestorozhdenii v svete dannykh absolyutnogo vozrasta. In A.A. Nitsenko (Editor). *Priroda bolot i metody ikh issledovaniya.* Nauka, Leningrad, pp. 90–95.

Neĭshtadt, M.I., 1971. Mirovoi prirodnyi fenomen — zabolochennost' zapadno-sibirskoi ravniny. *Izv. Akad. Nauk S.S.S.R., Ser. Geogr.*, 1971, No. 1: 21–34.

Neishtadt, M.I., 1972. Bolota Ob'-Irtyshskogo mezhdurech'ya. In: *Prirodnye usloviya osvoeniya Ob'-Irtyshskogo mezhdurech'ya, Moscow*, pp. 322–346.

Neĭshtadt, M.I., 1975. Die Holozäntorfablagerungen von Westsibierien und die sie bildenden Pflanzenkomplexe. *Biul. Geol.*, 19: 173–181.

Neĭshtadt, M.I. and Zelikzon, E.M., 1971. Neue Angaben zur Stratigraphie der Torfmoore Westsibiriens. *Acta Agral. Fenn.*, 123: 29–32.

Neĭshtadt, M.I., Firsov, L.V., Orlova, L.A. and Panychev, V.A. 1974. Some peculiarities of Holocene processes in Western Siberia. *Geoforum*, 17: 77–83.

Nichols, H., 1967. The post-glacial history of vegetation and climate at Ennadai Lake, Keewatin, and Lynn Lake, Manitoba (Canada). *Eiszeitalter Gegenw.*, 18: 176–197.

Nichols, H., 1969. The late Quaternary history of vegetation and climate at Porcupine Mountain and Clearwater Bog, Manitoba. *Arct. Alp. Res.*, 1: 155–167.

Nichols, H., 1975. The time perspective in northern ecology: Palynology and the history of the Canadian boreal forest. *Proc. Circumpolar Conf. Northern Ecology, 1975, Ottawa, Ont.*, 1: 157–165.

Nichols, H., 1976. Historical aspects of the northern Canadian treeline. *Arctic*, 29: 38–47.

Nilsson, T., 1964a. Standardpollendiagramme und C^{14}-Datierungen aus dem Ageröds Mosse im mittleren Schonen. *Lunds Univ. Årssk., N.F., 2*, 59(7): 52 pp.

Nilsson, T., 1964b. Entwicklungsgeschichtliche Studien im Ageröds Mosse, Schonen. *Lunds Univ. Årsskr., N.F., 2*, 59(8): 34 pp.

Olausson, E., 1957. Das Moor Roshultsmyren. Eine geologische, botanische und hydrologische Studie in einem südwestschwedischen Moor mit exzentrisch gewölbten Mooselementen. *Lunds Univ. Årsskr., N.F., 2*, 53(12): 72 pp.

Oldfield, F., 1967. The paleoecology of an early Neolithic waterlogged site in Northwestern England. *Rev. Palaeobot. Palynol.*, 4: 67–70.

O'Sullivan, P.E., 1973. Pollen analysis of mor humus layers from a native Scots pine ecosystem, interpreted with surface samples. *Oikos*, 24: 259–272.

O'Sullivan, P.E., 1974. Two Flandrian pollen diagrams from the east-central highlands of Scotland. *Pollen Spores*, 16: 33–57.

Osvald, H., 1923. Die Vegetation des Hochmoores Komosse. *Sven. Växtsociol. Sällsk. Handl.*, I: 436 pp.

Osvald, H., 1925a. Die Hochmoortypen Europas. *Veröff Geobot. Inst. Eidg. Tech. Hochsch. Stift. Rübel Zürich*, 3: 707–723.

Osvald, H., 1925b. Zur Vegetation der ozeanischen Hochmoore in Norwegen. *Sven. Växtsociol. Sällsk. Handl.*, VII: 106 pp.

Osvald, H., 1949. Notes on the vegetation of British and Irish mosses. *Acta Phytogeogr. Suec.*, 26: 62 pp.

Overbeck, F., 1950. Die Moore. In: K. Gripp, F. Dewes, F. Overbeck, *Geologie und Lagerstätten Niedersachsens, 3: Das Känozoikum in Niedersachsen (Tertiär, Diluvium, Alluvium und Moore), 4. Abteilung.* Bremen–Horn, 112 pp.

Overbeck, F., 1975. *Botanisch-geologische Moorkunde unter besonderer Berücksichtigung der Moore Nordwestdeutschlands als Quellen zur Vegetations-, Klima- und Siedlungsgeschichte.* Karl Wachholtz, Neumünster, 719 pp.

Paasio, I., 1933. Über die Vegetation der Hochmoore Finnlands. *Acta Forest. Fenn.*, 39(3): 210 pp. (Finnish summary).

Patzelt, G., 1977. Der zeitliche Ablauf und das Ausmaß postglazialer Klimaschwankungen in den Alpen. In: B. Frenzel (Editor), *Dendrochronologie und postglaziale Klimaschwankungen in Europa. Erdwiss. Forsch.*, 13: 248–259.

Pennington, W., 1965. The interpretation of some post-glacial vegetation diversities at different Lake District sites. *Proc. R. Soc. Lond., Ser. B*, 161: 310–323.

Pennington, W. and Lishman, J.P., 1971. Iodine in lake sediments in Northern England and Scotland. *Biol. Rev.*, 46: 279–313.

Pennington, W., Haworth, E.Y. Bonny, A.P. and Lishman, J.P., 1972. Lake sediments in Northern Scotland. *Philos. Trans. R. Soc. Lond.*, B 264: 191–294.

Peschke, P., 1977. Zur Vegetations- und Besiedlungsgeschichte des Waldviertels (Niederösterreich). *Mitt. Komm. Quartärforsch. Österr. Akad. Wiss.*, 2: 84 pp.

Planchais, N., 1973. Contribution à l'analyse pollinique des sols de l'Aigoual, étage du hêtre et pelouse sommitale. *Pollen Spores*, 15: 293–309.

Potonié, H., 1912. Die rezenten Kaustobiolithe und ihre Lagerstätten, III. Die Humusbildungen (2. Teil) und die Liptobiolithe. *Abh. K. Preuß. Geol. Landesanst., N.F.*, 55(III): 322 pp.

P'yavchenko, N.I., 1953. Zabolachivanie lesov v baseyne reki Sheksna. *Tr. Inst. Lesa Akad. Nauk S.S.S.R.*, 13: 51–76.

Radke, G.J., 1972. Landschaftsgeschichte und -ökologie des Nordschwarzwaldes. *Hohenheimer Arb.*, 68: 121 pp.

Rampton, V., 1971. Late Quaternary vegetational and climatic history of the Snag-Klutlan Area, Southwestern Yukon Territory, Canada. *Geol. Soc. Am. Bull.*, 82: 959–978.

Ritchie, J.C., 1972. Pollen analysis of late Quaternary sediments from the arctic treeline of the Mackenzie River Delta region, Northwest Territories, Canada. *Nordia*, No. 2: 253–271.

Rodewald, M., 1970. Die Ozeane als Vorzugsgebiete klimatischer Veränderlichkeit der atmosphärischen Zirkulation. *Beilage zur Berliner Wetterkarte*, 80/70; So 22/70, 26 pp.

Romanova, E.A., 1967. Nekotorye morfologicheskie charakteristiki oligotrofnykh bolotnykh landshaftov zapadnosibirskoi nizmennosti kak osnova ikh tipologii i rayonirovanii In: A.A. Nitsenko (Editor), *Priroda bolot i metody ikh issledovaniya.* Nauka, Leningrad, pp. 63–67.

Røstad, A., 1955. *On Long Range Temperature Waves in Europe.* Akademisk Forlag, Oslo, 75 pp.

Ruddiman, W.F. and McIntyre, A., 1973. Time-transgressive deglacial retreat of polar waters from the North Atlantic. *Quaternary Res.*, 3: 117–130.

Rudolph, K., 1929. Die bisherigen Ergebnisse der botanischen Mooruntersuchungen in Böhmen. III. Die Stratigraphie und Entwicklungsgeschichte der böhmischen Moore. *Beih. Bot. Centralblatt*, 45B: 139–180.

Rudolph, K., 1931. Grundzüge der nacheiszeitlichen Waldgeschichte Mitteleuropas (bisherige Ergebnisse der Pollenanalyse). *Beih. Bot. Centralblatt*, 47B: 111–176.

Rump, H.H., Van Werden, K. and Herrmann, R., 1977. Über die vertikale Änderung von Metallkonzentrationen in einem Hochmoor. *Catena*, 4: 149–164.

Ruuhijärvi, R., 1960. Über die regionale Einteilung der nordfinnischen Moore. *Ann. Bot. Soc. Zool.-Bot. Fenn. "Vanamo"* 31(1): 360 pp.

Ruuhijärvi, R., 1963. Zur Entwicklungsgeschichte der nordfinnischen Hochmoore. *Ann. Bot. Soc. Zool.-Bot. Fenn. "Vanamo"*, 34(2): 1–40.

Rybniček, K., 1973. A comparison of the present and past mire communities of Central Europe. In: H.J.B. Birks and R.G. West (Editors), *Quaternary Plant Ecology*, Blackwell, Oxford, pp. 237–261.

Rybniček, K., 1977. Die Ursachen der Änderung und des Unterganges der Moorvegetation in Mitteleuropa. *Telma*, 7: 241–249.

Rybničková, E., 1966. Pollen-analytical reconstruction of vegetation in the upper regions of the Orlické Hory mountains, Czechoslovakia. *Folia Geobot. Phytotaxon.*, 1: 289–310.

Sawyer, J.S. (Editor), 1966. *World Climate from 8000 to 0 B.C. Proceedings of the International Symposium held at Imperial College London, 18 and 19 April 1966*. Royal Meteorological Society, London, 229 pp.

Schlüter, H., 1964. Zur Waldentwicklung im Thüringer Gebirge, hergeleitet aus Pollendiagrammen, Archivquellen und Vegetationsuntersuchungen. *Arch. Forstwes.*, 13: 283–305 (Russian and English summaries).

Schmeidl, H., 1977. Veränderung der Vegetation auf Dauerflächen eines präalpinen Hochmoores (vorläufige Mitteilung). *Telma*, 7: 65–76.

Schmitz, H., 1952. Moortypen in Schleswig-Holstein und ihre Verbreitung. *Schr. Naturwiss. Ver. Schleswig-Holstein*, 26: 64–68.

Schneekloth, H., 1963a. Das Hohe Moor bei Scheeßel (Kreis Rotenburg/Hannover). *Beih. Geol. Jahrb.*, 55: 1–104.

Schneekloth, H., 1963b. Das Weiße Moor bei Kirchwalsede (Kreis Rotenburg/Hannover). *Beih. Geol. Jahrb.*, 55: 105–138.

Schneekloth, H., 1965. Die Rekurrenzflächen im Großen Moor bei Gifhorn — eine zeitgleiche Bildung? *Geol. Jahrb.*, 83: 477–496.

Schneekloth, H., 1968. Altersunterschiede des Schwarz-/Weißtorfkontaktes im Kehdinger Moor. *Geol. Jahrb.*, 85: 135–146.

Schneekloth, H., 1970. Das Ahlen-Falkenberger Moor; eine moorgeologische Studie mit Beiträgen zur Altersfrage des Schwarz-/Weißtorfkontaktes und zur Stratigraphie des Küstenholozäns. *Geol. Jahrb.*, 89: 63–96.

Schneekloth, H. and Schneider, S., 1972. Vorschlag zur Klassifizierung der Torfe und Moore in der Bundesrepublik Deutschland. *Telma*, 2: 57–63.

Schneider, S. and Steckhan, H.U., 1963. Das Große Moor bei Barnstorf (Kreis Grafschaft Diepholz). *Beih. Geol. Jahrb.*, 55: 139–192.

Schwaar, J., 1976. Die Hochmoore Feuerlands und ihre Pflanzengesellschaften. *Telma*, 6: 51–59.

Schwaar, J., 1977. Humifizierungswechsel in terrainbedeckenden Mooren von Gough Island/Südatlantik. *Telma*, 7: 77–90.

Sirén, G., 1955. The development of spruce forest on raw humus sites in Northern Finland and its ecology. *Acta Forest. Fenn.*, 62(4): 408 pp. (Finnish summary).

Sjörs, H., 1946. Myrvegetationen i övre Långanområdet i Jämtland. *Ark. Bot.*, 33 A (6): 1–96.

Sjörs, H., 1948. Myrvegetationen i Bergslagen. *Acta Phytogeogr. Suec.*, 21: 299 pp. (English summary).

Smith, A.G., 1965. Problems of inertia and threshold related to postglacial habitat changes. *Proc. R. Soc. Lond.*, B 161: 331–342.

Steffen, H., 1931. Vegetationskunde von Ostpreußen. *Pflanzensoziologie*, 1: 406 pp.

Straka, H., 1960. Literaturübersicht über Moore und Torfablagerungen aus tropischen Gebieten. *Erdkunde*, 14.

Suess, H., 1970. Bristlecone pine calibration of the radiocarbon time scale from 5400 B.C. to the present. In: I.U. Olsson (Editor), *Radiocarbon Variations and Absolute Chronology*. Almqvist and Wiksell, Stockholm, pp. 303–312.

Suess, H. and Becker, B., 1977. Der Radiocarbongehalt von Jahrringproben aus postglazialen Eichenstämmen Mitteleuropas. In: B. Frenzel (Editor), *Dendrochronologie und postglaziale Klimaschwankungen in Europa. Erdwiss. Forsch.*, 13: 156–170.

Szczepanek, K., 1971. Kras Staszówski w świetle badań paleobotanicznych. *Acta Palaeobot.*

Tallis, J.H., 1964. The pre-peat vegetation of the Southern Pennines. *New Phytol.*, 63: 363–373.

Tallis, J.H., 1965. Studies on Southern Pennine peats. IV. Evidence of recent erosion. *J. Ecol.*, 53: 509–520.

Tallis, J.H. and McGuire, J., 1972. Central Rossendale: The evolution of an upland vegetation. I. The clearance of woodland. *J. Ecol.*, 60: 721–737.

Tamoshaĭtis, Yu.S., 1967. Lozha bolot kak odin iz osnovnykh boloto-obrazovatel'nykh faktorov. In: A.A. Nitsenko (Editor), *Priroda bolot i metody ikh issledovaniya*. Nauka, Leningrad, 27–30.

Tansley, A.G., 1939. *The British Isles and Their Vegetation*, Cambridge University Press, Cambridge, 930 pp.

Thomson, P.W., 1930. Die regionale Entwicklungsgeschichte der Wälder Estlands. *Acta Comm Univ. Tartu. (Dorpatensis), A, Math. Phys., Med.*, 17(2): 87 pp.

Tooley, M.J., 1974. Sea-level changes during the last 9000 years in north-west England. *Geogr. J.*, 140(pt.1): 18–42.

Tooley, M.J., 1976. Flandrian sea-level changes in West Lancashire and their implications for the "Hillhouse Coastline." *Geol. J.*, 11(pt. 2): 137–152.

Trautmann, W., 1952. Pollenanalytische Untersuchungen über die Fichtenwälder des Bayerischen Waldes. *Planta*, 41: 83–124.

Tsukada, M., 1967. Vegetation and climate around 10 000 B.P. in Central Japan. *Am. J. Sci.*, 265: 562–585.

Turner, J. and Kershaw, A.P., 1973. A late- and post-glacial pollen diagram from Cranberry Bog, near Beamish, County Durham. *New Phytol.*, 72: 915–928.

Turner, J., Hewetson, V.P., Hibbert, F.A., Lowry, K.H. and Chambers, C., 1973. The history of the vegetation and flora of Widdybank Fell and the Cow Green Reservoir basin, Upper Teesdale. *Philos. Trans. R. Soc. Lond., Ser. B*, 265: 327–408.

Tyrtikov, A.P., 1969. Die Versumpfung von Gewässern in der nördlichen Taiga Westsibiriens und die Dynamik der ewigen Gefrornis. *Vestn. Mosk. Univ.*, 24(1): 51–54.

Val'k, U.A., 1971. Ob izmeneiyakh klimata Estonii v golotsene (po materialam izucheniya torfyanykh bolot). In: M.I. Neĭshtadt (Editor), *Palinologiya golotsena*. Akad. Nauk S.S.S.R., Geogr. Inst., Moscow, pp. 43–52 (English summary).

Van Geel, B., 1976. *A Paleoecological Study of Holocene Peat Bog Sections, Based on the Analysis of Pollen, Spores and Macro- and Microscopic Remains of Fungi, Algae, Cormophytes and Animals*. Hugo de Vries-Laboratorium, University of Amsterdam, Amsterdam, 75 pp.

Van der Hammen, T. and Gonzalez, E., 1960. Upper Pleistocene and Holocene Vegetation of the "Sabana de Bogota" (Colombia, South America). *Leidse Geol. Meded.*, 25: 261–315.

Van der Hammen, T. and Gonzalez, E., 1965. A late-glacial and Holocene pollen diagram from Cienaga del Visitador (Dept. Boyaca, Colombia). *Leidse Geol. Meded.*, 32: 193–201.

Van Zeist, W., 1955. Pollen analytical investigations in the Northern Netherlands with special reference to archaeology. *Acta Bot. Neerl.*, 1–81.

Van Zinderen Bakker, E.M., 1964. A pollen diagram from Equatorial Africa. Cherangani, Kenya. *Geol. Mijnb.*, 43: 123–128.

Van Zinderen Bakker, E.M., 1972. Late Quaternary lacustrine phases in the southern Sahara and East Africa. In: E.M. van Zinderen Bakker (Editor), *Paleoecology of Africa and of the Surrounding Islands and Antarctica*, 6. A.A. Balkema, Cape Town, pp. 15–27.

Vasari, Y. and Vasari, A., 1968. Late- and post-glacial macrophytic vegetation in the lochs of Northern Scotland. *Acta Bot. Fenn.*, 80: 1–120.

Von Bülow, K., 1929. *Allgemeine Moorgeologie. Einführung in das Gesamtgebeit der Moorkunde.* Bornträger, Berlin, 308 pp.

Von der Brelie, G., 1954. Transgression und Moorbildung im letzten Interglazial. *Mitt. Geol. Staatsinst. Hamburg*, 23: 111–118.

Von der Brelie, G., 1955. Die Küstentorfe Ostfrieslands und ihre marine Beeinflussung. *N. Jahrb. Geol., Paläontol. Mineral., Monatsh.*, 1955: 201–217.

Von Post, L., 1910. Das Skagershultmoor. In: L. von Post and R. Sernander, *Pflanzenphysiognomische Studien auf Torfmooren in Närke. Livret-guide des excursions en Suède du XIe Congres Géologique Internationale, Excursion A7*, No. 14: 1–24.

Von Post, L., 1924. Some features of the vegetational history of the forests of southern Sweden in post-arctic time. *Geol. För. Stockh. Förh.*, 46: 83–128.

Von Post, L., 1926. Einige Aufgaben der regionalen Moorforschung. *Sver. Geol. Unders., Ser. C, No. 337, Årsb.*, 19(4): 41 pp.

Waldheim, St., 1944. Die Torfmoorvegetation der Provinz Närke. *Lunds Univ. Årsskr., N.F., 2*, 40(6): 91 pp.

Waldheim, St. and Weimarck, H., 1943. Bidrag till Skånes Flora. 18. Skånes myrtyper. *Bot. Not.*, 1943: 1–40.

Walker, D. and Walker, P.M., 1961. Stratigraphic evidence of vegetation in some Irish bogs. *J. Ecol.*, 49: 169–185.

Wąs, St., 1965. Geneza, sukcesje i mechanizm rozwoju warstw mszystych torfu. *Zesz. Probl. Postepów Nauk Roln.*, 57: 305–393 (Russian and English summaries).

Watts, W.A., 1973. Rates of change and stability in vegetation in the perspective of long periods of time. In: H.J.B. Birks and R.G. West (Editors), *Quaternary Plant Ecology*. Blackwell, Oxford, pp. 195–206.

Watts, W.A. and Wright, H.E., 1966. Late-Wisconsin pollen and seed analysis from the Nebraska sandhills. *Ecology*, 47: 202–210.

Weber, C.A., 1910. Was lehrt der Aufbau der Moore Norddeutschlands über den Wechsel des Klimas in postglazialer Zeit? *Z. Dtsch. Geol. Ges.*, 62: 2.

Weber, C.A., 1911. Das Moor. *Hannov. Geschichtsbl.*, 14: 255–270.

Weber, K.Yu., 1967. Vozrast bolot i prirost torfa po dannym sporovo-pyl'cevykh diagramm severo-vostochnoi Estonii. In A.A. Nitsenko (Editor), *Priroda bolot i metody ikh issledovaniya*. Nauka, Leningrad, pp. 103–107.

Wendland, W.H. and Bryson, R.A., 1974. Dating climatic episodes of the Holocene. *Quaternary Res.*, 4: 9–24.

West, R.G., 1970. Pollen zones in the Pleistocene of Great Britain and their correlation. *New Phytol.*, 69: 1179–1183.

Williams, J. and Van Loon, H., 1976. The connection between trends of mean temperature and circulation at the surface: part III. Spring and autumn. *Mon. Weather Rev.*, 104: 1591–1596.

Witting, M., 1948. Preliminärt meddelande om fortsätta katjonbestämningar i myrvatten sommaren 1947. *Sven. Bot. Tidskr.*, 42, 1: 116–134.

World Meteorological Organization, 1975. *Proceedings of the WMO/IAMAP Symposium on Long-term Climatic Fluctuations, Norwich, 18–23 August 1975.* W.M.O., Nr. 421, Geneva, 503 pp.

Wright, H.E., 1968. The roles of pine and spruce in the forest history of Minnesota and adjacent areas. *Ecology*, 49: 937–955.

Zoller, H., 1977. Alter und Ausmaß postglazialer Klimaschwankungen in den Schweizer Alpen. In: B. Frenzel (Editor), *Dendrochronologie und postglaziale Klimaschwankungen in Europa. Erdwiss. Forsch.*, 13: 271–281.

Chapter 3

HYDROLOGY[1]

H.A.P. INGRAM

INTRODUCTION

The ecosystems which form the subject of this monograph have in common one feature: their soil spends much of its time in a waterlogged state. This, indeed, is the basic characteristic of mires wherever they occur, and is the justification for devoting a separate chapter of this volume to their hydrology.

Hydrology has achieved the status of a distinct scientific discipline. It was, therefore, with some trepidation that, having myself no training in this discipline, I accepted the late Tony Gore's invitation to contribute this chapter. My justification must be that this is a monograph aimed principally at ecologists and that, while hydrologists like other students of the natural environment must in a certain sense be reckoned ecologists, few western practitioners have yet chosen to study the peculiar water relations of mires in the depth that confers insight into the ecologist's point of view and lends understanding to the questions about mires which an ecologist might ask. The adjective "western" is used advisedly, because in the Soviet Union there has developed a considerable mutual understanding between those who study mire hydrology and those with a wider interest in the ecology of these ecosystems, so that my other great problem in compiling this review has been the extent and importance of the literature emanating from non-English speaking countries, especially from the U.S.S.R.; in this connexion, further reference may be made to Ch. 4 in Part B by Botch and Masing.

While I do not pretend to have surveyed this literature in its entirety, I have tried to provide here an introduction to it and so, for the first time, to write a treatment in English which is reasonably comprehensive. This is not, however, the first western review of any kind. Important aspects of my subject have been dealt with by Baden and Eggelsmann (1963), Ingram (1967), Dooge (1975), Rycroft et al. (1975a), Eggelsmann and Schuch (1976), Linacre (1976), Boelter and Verry (1977) and Goode et al. (1977), and I have been very glad to have their reviews at my disposal. Moreover, much of this chapter has been written while editing an English edition of K.E. Ivanov's (1975a) treatise. This is the most advanced work on the hydrological aspects of mire ecology yet to have appeared and should be studied by anyone wishing to take the subject further.

Most of the world's mires are raised bogs of one sort or another and, fortunately, my own experience has mainly been gained on mires of this type. Regrettably, knowledge of the hydrology of other kinds of mire is less extensive and for some, such as the riverine swamps of the tropics, scarcely any hydrological information seems to be available. Naturally, I should be very grateful if anyone who notices significant omissions or misconceptions here would write to the University of Dundee to let me know, since this will help to correct bias in future comprehension of the subject.

Whilst acknowledging the need to be critical, I am very conscious that one must temper criticism with admiration. No one with experience of field work in mire hydrology can fail to appreciate the very great effort required to obtain any worthwhile results in an environment accessible only with difficulty, remote from power supplies and laboratory facilities, and rendered hostile by its exposure, its humidity, its insect fauna and its rough, wet and unstable ground surface. I make no apo-

[1] Manuscript completed March, 1981.

logy for mentioning these problems, because failure to understand them merely adds to the difficulties. A research-funding agency recently suggested that we borrow a device which had been developed for installing lysimeters in mineral soils. We abandoned our application after learning, from other sources, that the device in question weighed several tons so that, even if we could without irreparable damage to the ecosystem have transported it to the right place, it must surely have sunk into the mire beyond hope of recovery. With such practical obstacles as these to overcome, mire hydrology has some remarkable achievements to its credit.

In this chapter a main section is devoted to each major hydrological process, namely recharge, evaporation, storage and discharge, with a summary of conclusions after all except the first. There is an initial brief summary of the hydrological cycle which links these processes together; also a final account of some ideas for building "ecosystem models" of mires on the foundation of their hydrology. This is one of several places where aspects of theoretical telmatology are touched upon, but this chapter does not present a rigorous theoretical treatment, for which Ivanov's (1975a) book should be consulted.

Notation

The mathematical notation (in which vectors are not distinguished typographically) and location of the definitions of mathematical terms, and their dimensions, are specified as follows:

A	angle	Fig. 3.24	–
B	parameter of Ivanov's equation for hydraulic conductivity in layers	eq. 3.37	dimensionless
C	flux density of open channel flow	eq. 3.2	$L\,T^{-1}$
D	flux density of total discharge	eq. 3.4	$L\,T^{-1}$
E	flux density of evapotranspiration	eq. 3.2	$L\,T^{-1}$
F	coefficient of groundwater	p. 116	dimensionless
G	ground heat flux density	eq. 3.6	$H\,L^{-2}T^{-1}$
H	sensible heat flux density	eq. 3.6	$H\,L^{-2}\,T^{-1}$
I	flux density of interception loss	p. 76	$L\,T^{-1}$
J	elevation of raised bog surface	eq. 3.40	L
K	hydraulic conductivity	eq. 3.30	$L\,T^{-1}$
L	width or radius of raised bog	eqs. 3.41, 3.42, 3.44	L
N	flux density of total recharge	eq. 3.3	$L\,T^{-1}$
P	flux density of precipitation	eq. 3.2	$L\,T^{-1}$
Q	flux density of diffuse surface flow	eq. 3.2	$L\,T^{-1}$
R	flux density of radiation	eq. 3.6	$H\,L^{-2}\,T^{-1}$
T	temperature	p. 77	θ
U	flux density of seepage	eq. 3.2	$L\,T^{-1}$
V	flux density of pipe flow	eq. 3.2	$L\,T^{-1}$
W	stored water	eq. 3.1	L
Y	specific yield (aquifer yield)	eq. 3.24	dimensionless
Z	elevation of local water table	eq. 3.22	L
a	area of microtope	eq. 3.39	L^2
b	Wickman's evapotranspiration constant	eq. 3.40	$L\,T^{-1}$
c	Romanov's empirical coefficient for evapotranspiration	eq. 3.18	$L\,T^{-1}$
d	depth of water stage below local mire surface	eq. 3.21	L
e	vapour pressure of water	eq. 3.8	$M\,L^{-1}\,T^{-2}$
f	general function; parameter of Penman's equation	eq. 3.9	dimensionless
h	hydrostatic pressure (piezometric head)	eq. 3.20	$M\,L^{-1}\,T^{-2}$
k	Wickman's coefficient	eq. 3.40	$L\,T$
l	latent heat of vaporization of water	eq. 3.6	$H\,L^{-3}$ or $H\,L^{-1}$
m	parameter of Ivanov's equation for hydraulic conductivity in layers	eq. 3.37	dimensionless
q	volume of water	eq. 3.14	L^3
r	vapour diffusion resistance	eq. 3.16	$L^{-1}\,T$
s	radius of pore space	eq. 3.14	L
t	time	p. 120	T
u	wind speed	eq. 3.8	$L\,T^{-1}$
w	water stage above local datum	eq. 3.21	L
x	horizontal distance	eq. 3.42	L
z	elevation of local datum point	eqs. 3.20, 3.22	L
α	Romanov's evapotranspiration coefficient	eq. 3.18	$H^{-1}\,L^3$
β	Bowen ratio	eq. 3.11	dimensionless
γ	psychrometric constant	eq. 3.13	$M\,L^{-1}\,\theta^{-1}\,T^{-2}$
η	error in water balance computation	eq. 3.2	$L\,T^{-1}$
θ	volumetric water content	p. 100	dimensionless
ρ	spectral reflection coefficient (albedo)	eq. 3.7	dimensionless

σ	surface tension of water	eq. 3.15	$M\,T^{-2}$
τ	matric suction	eq. 3.15	$M\,L^{-1}\,T^{-2}$
ϕ	water potential	eq. 3.20	$M\,L^{-1}\,T^{-2}$ or L
∇	Laplacian operator	eq. 3.31	–

THE HYDROLOGICAL CYCLE AND THE WATER BALANCE

As an approach to understanding ecosystems, one possible way of looking at hydrology is to regard it as a form of accountancy in which the currency is water (Ward, 1975; Miller, 1977). In hydrology, notions of continuity and the conservation of matter have the force of axioms. The idea that one should be able to account for all the water in a system is basic. This means that all the water which enters a system must sooner or later leave it. At the boundaries of the system, the entry and exit of water can be measured as a series of fluxes. Within the system, inequalities of influx and efflux cause variation with time in the quantity of water stored. If the stored water is symbolised W and is regarded as increasingly positive the greater the amount, one may write:

$$\text{influx} - \text{efflux} - \Delta W = 0 \qquad (3.1)$$

The various flow and storage processes for part of a water catchment containing a mire are shown schematically in Fig. 3.1. It is helpful to consider the system as comprising four compartments, namely the atmosphere, the mire itself, the mineral soils and their parent rock adjacent to the mire, and the local system of surface streams. Since the ecologist may be interested in hydrological phenomena at various different scales it is important to identify not only the fluxes across the mire boundary but also the various flow processes occurring within the mire. In fact it seems probable that most of the boundary fluxes are associated with internal processes, but the scheme becomes complicated in places and any pictorial representation is bound to involve simplifications and assumptions of varying dubiety. Most of these will be examined later in this

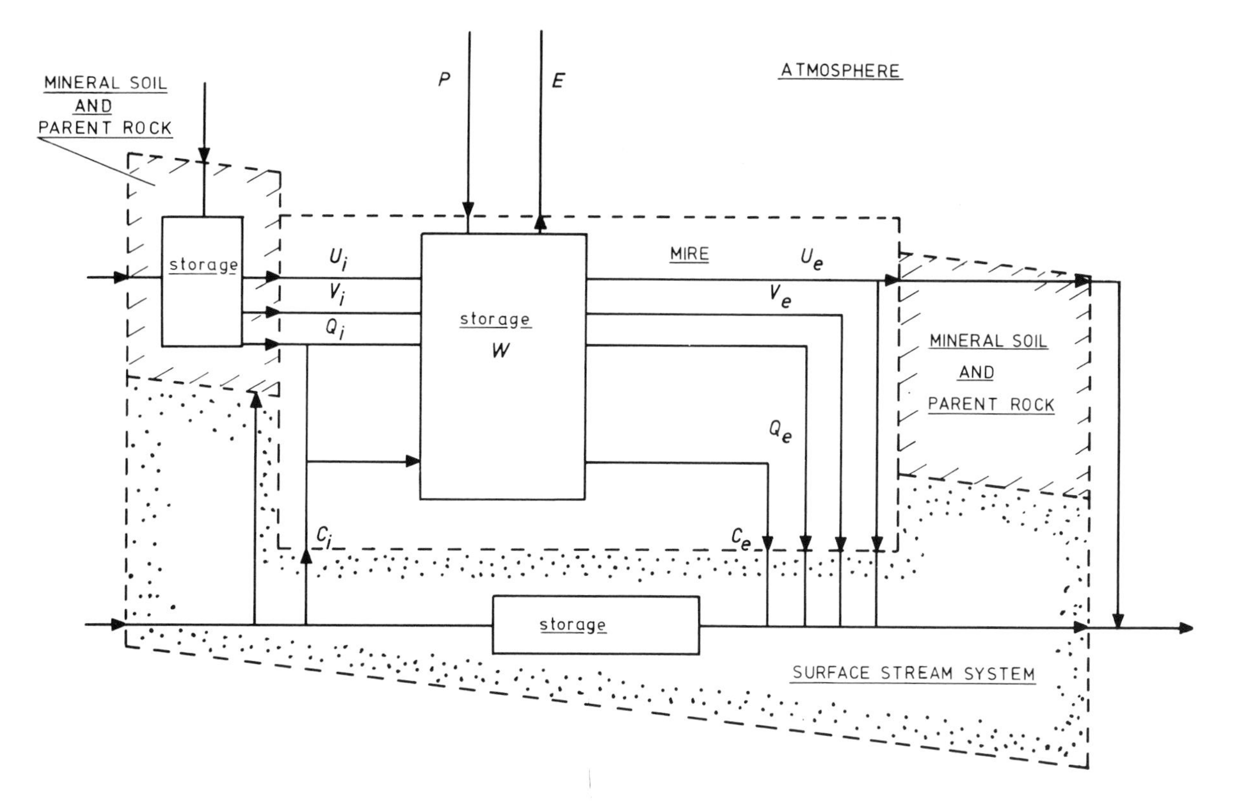

Fig. 3.1. Schematic diagram of the water relations of a mire, showing the compartments in which storage occurs and the fluxes of water involved in their recharge and discharge.

account. At this point the reader's attention is drawn to four of them:

(1) The local system of surface streams is thought of as distinct from the mire itself. Some mires develop their own systems of open channels in which discharge water collects on its way towards the mire boundary. The involvement of open channel flow as an influx process occurs when one is considering the input of water from another part of a mire complex, or when these channels become involved in carrying flood water onto a mire from the local surface streams. Floods are an important source of water in some mires, but when the mire channels can no longer accommodate the flood water it becomes a diffuse surface flow.

(2) With the possible exception of some of the seepage, all the effluxes eventually join the local surface streams. The scheme represents them as doing so separately, but this is a simplifying assumption which may be invalid in some mires and at certain seasons.

(3) Water transfer in ecosystems is complex. The scheme suggested here assumes that it is helpful to consider it as the outcome of six basic processes and that these are sufficiently different to be worth separate study. These processes are the precipitation of atmospheric moisture, the seepage of liquid water through peat considered as a porous medium, flow in pipes and fissures which are not directly open to the atmosphere, diffuse surface flow or sheet flow, unconfined flow in directed channels and evapotranspiration. It is assumed that other processes, such as porous-medium vapour flow, are relatively unimportant or that, like snowmelt, dew formation or interception loss, they can safely be regarded as special cases of processes already identified.

(4) The scheme assumes that in general all the processes involve distinct pathways disposed, on the whole, in parallel. Again this is a simplifying assumption. Its limitations should not be overlooked.

For the purposes of this account it has been necessary to devise a new expression for the water balance outlined in eq. 3.1. This expression reflects the view that the processes of water transfer are more basic to an understanding of the subject than the pathways involved. Although the formulation is based upon that of Ivanov (1957; summarized by L.G. Bavina in Sokolov and Chapman, 1974), it incorporates certain processes which that author ignored. The expression is:

$$P+U_i+V_i+Q_i+ \\ +C_i-U_e-V_e-Q_e-C_e-E-\Delta W-\eta=0 \quad (3.2)$$

Regrettably it does not seem possible to reconcile the symbolisms conventionally used in all the biological and geographical disciplines relevant to this account. Here: P=precipitation; U=seepage; V=pipe flow; Q=diffuse surface flow; C=open channel flow; E=evapotranspiration; and η=error. The first five terms are influxes (input, recharge), while the next five are effluxes (output, discharge), distinguished where appropriate by the suffixes i and e, respectively.

The water balance concept is applicable not only to whole mire ecosystems but also to portions of these systems such as lysimeter vessels or experimental plots delimited by ditches. Eq. 3.2 lends itself to simplification, and the various alternative forms of it which are appropriate to different conditions are given later. A convenient general simplification is to combine the various surface and sub-surface influxes in a total surface and sub-surface recharge term, N, where:

$$N=U_i+V_i+Q_i+C_i \quad (3.3)$$

while using a corresponding term D for the total surface and sub-surface discharge, where:

$$D=U_e+V_e+Q_e+C_e \quad (3.4)$$

By classifying the various fluxes in this way we obtain, from eq. 3.2, the useful expression:

$$P+N-D-E-\Delta W-\eta=0 \quad (3.5)$$

THE WATER INPUT TO MIRES: RECHARGE

In the general water balance equation (3.5) the total input of water or recharge is equal to $P+N$. The existence of two types of recharge, the meteoric supply, P, derived from the atmosphere and the telluric supply, N, derived from surrounding rocks and soils is of fundamental importance in the hydrology and ecology of mires. These two types differ not only in their origin but also in their chemical quality, seasonality and amenability to measurement. This latter aspect is troublesome to the investigator and implies that the body of evidence is of very varied quality.

Telluric and meteoric recharge

Hydrologically, mires can be divided into two categories according to the nature of their water supply. Fens are mires developing in valleys or topographic basins. Part of their water recharge comes from atmospheric precipitation, but the remainder is telluric and it is this part which has the greater ecological effect. Due to their low-lying physiographic situation, water which has been in contact with mineral particles and with the soils of the surrounding catchment enters these mires by seepage (U_i), pipe flow (V_i), by diffuse surface flow or run-on (Q_i) or from points upstream through the open channel network (C_i) (see Fig. 3.1 and eq. 3.3, also Figs. 3.27 and 3.28).

Most bogs, by contrast, are recharged mainly by meteoric water and their surface vegetation is largely isolated from telluric influence. In raised mire systems (generally referred to as raised bogs in this chapter), the central parts of the mire massif all lie above the level to which the mineral surroundings drain. Despite occasional assertions that matric forces ("capillary action") are capable of raising this drainage water from the eutrophic mire margin onto the mire expanse[1] (e.g. Gosselink and Turner, 1978, fig. 6), this seems most improbable on physical grounds, since E. Granlund showed that the capillary rise was too small (Wickman, 1951) and any peat sufficiently compact to generate enough matric suction to attract water to the highest parts of raised bogs would be too impermeable to sustain the required rates of flow. Indeed the only evidence for such an influx into any raised bog is that from the Steinhuder Meer near Hannover, where a study of permeability and potential distribution enabled the lake outflow beneath adjacent bogs to be calculated using Darcy's law (see p. 121). The result and Eggelsmann, 1961; Eggelsmann, 1962) was $N \simeq 23$ mm yr^{-1}: an insignificant proportion of the local mean annual precipitation $\bar{P} = 601$ mm yr^{-1}. In general it seems likely that the oligotrophic expanses of bogs are fed almost entirely by precipitation and that, for these, input $= P$. Unsaturated flow and infiltration into peat soils have been studied by Bartels and Kuntze (1973) and by Renger et al (1976). In this account it has been assumed that infiltration does not limit the recharge of mires by precipitation, except when they are frozen (p. 140) or have been intensively burned and a gelatinous aquiclude[2] has formed at the surface (p. 136).

Chemical and thermal indicators of recharge water

Various suggestions have been made for distinguishing mire waters derived from telluric and meteoric recharge. Bellamy (1959) defined ombrotrophic waters as having $[Ca^{2+}]/[Mg^{2+}] < 1$, few bases and low pH. In this he agreed with Gorham (1956, 1957), who suggested that so long as some telluric water reaches the mire surface the reaction of the water remains above pH 4.5. An extensive body of geochemical data on Swedish mires was collected by Witting (1947), Malmer and Sjörs (1955), Olausson (1957) and Malmer (1962b). During the course of this work the influence of rain storms and the seasonal state of plant growth became apparent, and the possibilities of measuring electrical conductance were explored. Various authors, including Chapman (1964), Gore (1968) and Boatman et al. (1975), have discussed aspects of the geochemistry of some British mires. Similar studies have been carried out in recent years in Minnesota (Hawkinson and Verry, 1975; Boelter and Verry, 1977) to distinguish lakes and mires which participate in the regional ground-water flux from those which are "perched" above it and do not. The perched-mire water had $[Ca^{2+}] < 4$–5 mg l^{-1}, pH 3–4 and specific conductance < 80 μS cm^{-1} and supported bog vegetation, while the ground-water mires were fens containing water with $[Ca^{2+}] > 15$ mg l^{-1}, pH 4–8 and specific conductance > 100 μS cm^{-1}. In these latter mires further evidence of telluric recharge was provided in winter where the warmer water upwelling through the base of the peat prevented surface freezing. Similar upwellings have been noted at the Stordalen mire near Abisko,

---

[1] **Mire expanse**: the central area of a raised bog, as distinct from the peripheral **mire margin** (Sjörs, 1948).

[2] An **aquiclude** is a layer which may contain water without measurably transmitting it.

northern Sweden, but here the ground frost was so intense and persistent that recharge with water from the neighbouring higher ground was effectively prevented by an artesian aquiclude over most of the mire (Skartveit et al., 1975). According to Alekseevsky and Tereshchenko (1975), artesian upwellings not involving ground frost are responsible for the recharge of headwater fens in the Pripyat' catchment in Poles'ye. Artesian recharge is a particularly troublesome instance of the general problem of identifying telluric recharge in mires — a problem noted by Egglesmann and Schuch (1976).

The general problems of assessing the chemical quality of the recharge water once it has entered the mire were emphasized by Summerfield (1974). Whilst studying the mineral nutrition of *Narthecium ossifragum* he found considerable variations in water quality over short distances and with progress of the growing season. Similar variations have been noted by other authors, and are reviewed by Tervet (1976). The spatial variation was related to the position and rooting depth of vascular plants, and became apparent over distances as small as 10 cm. Summerfield warned that water samples taken from wells or mire pools may differ profoundly from those adjacent to the roots of nearby plants.

Recharge and mire distribution

Meteoric water may be precipitated in a variety of forms in addition to rain. Snow, hail, fog or dew may also be involved, the latter two being distinguished as "occult precipitation". Problems arise in measuring these recharge fluxes. Many conventional rain gauges generate aerodynamic anomalies at the collecting orifice, and the resulting underestimates vary with the gauge design (and hence with the political administration of the gauge network) and the windiness of the climate (see Ward, 1975; and p. 76). Few rain gauges are effective as snow gauges, and there is no general agreement on how occult precipitation should be measured, although Schlüter (1970) thought it was important in determining the distribution of plant communities in the Thüringer Wald.

Various attempts have been made to explain the distribution of mires in terms of patterns of precipitation. According to Ivanov (1975a), bogs are absent from the zone of deficient moisture in the Soviet Union. In this climatic zone mean potential evapotranspiration exceeds mean precipitation, and the only mires are eutrophic fens, formed where surface saturation is maintained by telluric water. For oligotrophic bogs, Kats (1971) discerned a replacement series mediated by climatic humidity which increases westwards from the Urals to the Atlantic seaboard of Europe. In the east the raised bogs support pine forest, which towards the west becomes increasingly restricted to the *rand* or marginal slope where drainage is more sharp. The convex bogs of eastern Sweden with their concentrically elongated hummocks (ridges or kermis) and pools (flarks) give way westwards to flatter surfaces and eventually to the blanket bogs of Norway, northern Britain and Ireland, in which the humidity is high enough to support the spread of mire on to sloping terrain. A similar morphological trend, but with very different plant species, is seen in Tierra del Fuego and the neighbouring South American mainland where, according to Pisano Valdes (1973), the precipitation falls abruptly from $P > 2000$ mm in the west to < 350 mm yr^{-1} on the Atlantic side of the continent, sufficient only to support minerotrophic mires.

More local climatic effects on mire distribution have been recorded from various countries, and the effects of precipitation have sometimes been deduced with some precision. Southern Africa is a region not noted for its bogs, yet in the highlands of Lesotho, from which many rivers of the arid and semi-arid surrounding areas take their source, mires of various types are found (Van Zinderen Bakker and Werger, 1974). Minerotrophic fens are fed by water draining from basalt screes. Later these become ombrotrophic and acid. Their water supply is ensured by three factors: high annual rainfall ($P > 1600$ mm locally); rainfall maxima in the summer season; and the frequent presence of cloud during summer days, which gives rise to occult precipitation while evaporation is curtailed. Indeed it is clear that variations in evaporation (considered below) strongly modify the geographical effects of precipitation on mires. The concept of "effective rainfall" is discussed by Gore (Ch. 1). In temperate regions evaporation often decreases as clouds increase and radiation diminishes with altitude (Harding, 1979), while the annual receipt of solar radiation also falls off with increasing latitude. Discussing the effects on British mires, Taylor (1976) notes that in the south (Wales) ombrotrophic mires are found in an area delimited either by the 1270-mm (50-inch) average annual isohyet, or by the 305-m (1000-ft) contour or both. But he also quotes R.A. Robertson as suggesting that, in the north (Scotland), 1000 mm (40 inch) average annual precipitation is sufficient to sustain ombrotrophic mire, with an even lower limit in the extreme north (Caithness). Still more detailed studies have been made in Finland. According to Solantie (1974) the mean July value of the monthly potential water surplus ($P-E_t$: see section on evaporation below) is negative in the south of that country but positive in the north, and the line of equality is an important biogeographical limit. For mires it marks the limit north of which peatlands exceeded 30% of total land area in 1953, and coincides approximately with the southern limit of aapa fens (see Ruuhijärvi, Part B, Ch. 2).

Finally one must recall that the form of precipitation can profoundly affect mire development. Sonesson (1970) showed that among the bogs and nutrient-poor fens of the Lake Torneträsk area of northern Sweden a series of chionophobous to chionophilous vegetation types could be discerned, their distribution depending on the distribution of snow. The deepest and longest-lying snowfalls occur in the west of the area, while in the east bog hummocks are more sheltered and less frozen (Sonesson, 1969). As to the presence of individual mire species, *Vaccinium myrtillus* was positively correlated with deep snow, while *Andromeda polifolia* and *Empetrum nigrum* subsp. *hermaphroditum* were negatively correlated. These findings agree with Schlüter's (1970) study of mires in the Thüringer Wald of the German Democratic Republic. There, raised bogs covered with *Piceo-Vaccinietum* were the last to clear of snow in spring, while the moss-rich spring fens enjoyed the longest growing season due to continued recharge with warmer, telluric water during the winter.

Clearly there are important general correlations between mire distribution and precipitation, with fens predominating in areas of low rainfall, raised bogs in wetter conditions, and blanket mires in the wettest parts. But the complexities mentioned above make it impossible to discern precise relationships except over small areas. Moreover, the experience at Komosse should be salutary. In this large and much-studied southern Swedish mire, Johansson (1974) found that total precipitation for the growth season (late April to mid-September) of 1973 varied from less than 280 mm to more than 340 mm in a consistent pattern from one part of the mire complex (macrotope) to another. Future publications from this project will be awaited with interest to see how variation in precipitation from place to place compares with its variation year to year.

Unfortunately the literature contains scarcely any reliable data on N, the actual amount or rate of supply of telluric water to mires. It is possible that, in time, advances in lysimeter techniques (pp. 76–96) or geohydrology may remedy this, but meanwhile it is a most serious gap in our knowledge.

Faced with these difficulties, the course adopted here is to quote authors' original data on local precipitation where appropriate, as the best available means of indicating the climate of mire development.

EVAPORATION

A large proportion of the water recharge which mires receive is subsequently discharged by evaporation. In their evaporative behaviour mires show several interesting features, some of which may be ascribed to special properties of their vegetation, others to their physical peculiarities. The literature is extensive, the methods diverse and the data sometimes conflicting.

Theoretical considerations: potential evapotranspiration

Evaporation from terrestrial ecosystems is the sum of three processes. By **interception loss** is meant the evaporation of water from plant surfaces which have been wetted by rain, dew or fog. By **transpiration** is meant the evaporation of water from within the plants themselves, which generally involves the movement of further supplies from storage in the soil to the sites of evaporation in the aerial shoot systems. The third process is **direct evaporation** from the non-living soil or plant litter which forms the floor of the community and in which the vascular plants are rooted. Because it is difficult — and not always very relevant — to measure transpiration and direct evaporation separately they are frequently considered as a single process termed evapotranspiration.

In recent years the physical and biophysical mechanisms involved in evaporation have become fairly clearly understood, though processes occurring at the plant/atmosphere interface seem clearer than those in the soil. For background and theoretical detail the reader is referred to the accounts by Konstantinov (1966), Rose (1966), Slatyer (1967), Budyko (1958, 1974), Monteith (1973, 1975), Ward (1975) and Miller (1977).

The magnitude of the evaporative flux at any moment is controlled by three factors, namely the energy available, the "sink strength" and the water supply. If the energy conversion in photosynthesis is ignored, the input R_n of energy to the site in question is dispersed partly as sensible heat, H, when the atmosphere is warmed by contact with the heated surfaces of the soil or vegetation; partly as the latent heat content, lE, of water vapour evaporated from these surfaces (l being the latent heat of vaporisation of water and E the evaporative flux); and partly as heat, G, conducted into the ground. These processes may be summarized by means of an energy balance equation:

$$R_n - H - lE - G = 0 \qquad (3.6)$$

in which all the terms are flux densities. The term R_n is the net radiation or radiation surplus. It is the energy which remains after a fraction ρ of the total short-wave radiation, R_s, impinging on the site has been reflected, while a quantity R_l has been re-radiated at longer wave-lengths due to the low black-body temperature of the earth's surface. Thus one has:

$$R_n = R_s(1-\rho) - R_l \qquad (3.7)$$

where the reflection coefficient, ρ, is known as the albedo.

The sink strength is the ability of the atmosphere to take up the evaporated moisture, and is a reflexion of atmospheric humidity and the aerodynamics of the system. Uptake is favoured by a steep gradient of water vapour pressure between the evaporating surface and the bulk air, and by rapid air movements which cause turbulence and a quick replacement of moisture-laden air by drier air. If the vapour pressures at the evaporating surface and in the bulk air assume the values e_0 and e_a respectively, the sink strength effect is most simply expressed by the Dalton equation:

$$E = f(\bar{u})(e_0 - e_a) \qquad (3.8)$$

in which it is assumed that the transfer coefficient for water vapour is dependent only on $\bar{u}$, the mean wind speed.

Penman (1948, 1963) proposed a method which combines the surface energy balance with the aerodynamic approach to permit the estimation of potential evaporation from more or less standard and readily available meteorological data. The potential evaporation, E_0, is the evaporation from an extensive open water surface at the site in question. Alternatively it is possible by using a seasonally variable function, f, to compute E_t, the potential evapotranspiration, as:

$$E_t = f\, E_0 \qquad (3.9)$$

Penman has defined potential evapotranspiration as the evapotranspiration from a short, physiologically active, green crop, completely covering an extensive area of ground kept sufficiently well moistened to prevent moisture stress. His method therefore permits the estimation of the combined evaporative effect of the energy climate and atmospheric conditions, which are factors external to the ecosystem, upon one of two terrestrial ecosystem types with water-supply characteristics that are standardized — in the first more uniform and therefore preferable type because a free water surface is involved, and in the second type because the effects of soil moisture, vegetation type and physiological state are carefully defined. Estimates of E_0 or E_t therefore serve as bases for comparison with the actual evaporation, E, from any particular ecosystem. Evapotranspiration estimates for particular sites gain greatly in general significance when comparative potential evaporation estimates for the same sites are available.

Penman's method for potential evaporation has become accepted in many countries with a wide range of climates because of its value in predicting the irrigation requirements of crops and the run-off from water catchments. It is, however, only one of several approaches with similar aims. For present purposes, four other approaches are significant. The first is that of Budyko and his associates, widely used in the U.S.S.R. (Budyko, 1958, 1974; Budyko et al., 1962; Thornthwaite and Hare, 1965; Konstantinov, 1966; Barry, 1969; Kuzmin, 1972), which is similar to that of Penman. Thornthwaite's (1948) method has also found wide favour, especially in the U.S.A. It involves a simpler and more empirical approach, based on mean air temperature and hours of daylight, and is reputed to work best in the interior of a large, continental land mass (Ward, 1975). Haude's (1952, 1955) method appears to be virtually unknown outside Germany. It is based on single daily observations of the saturation deficit of the atmosphere and on an empirically determined function which varies with season and depth of water table. Using Haude's method, Uhlig (1954) reported results which correlated well with E_t estimates made by the Penman method, although the slope of Uhlig's regression suggested that Haude's approach tends to underestimate low values while overestimating high ones. Finally, potential evaporation may be estimated by means of an open water evaporating device such as the U.S. Weather Bureau Class A pan. Theoretically this device is highly unsound, with its small surface area, its low thermal capacity and its liability to heat advection, to aerodynamic disturbance caused by its raised position and the rim of the open vessel, and to anomalous exchanges of heat across its walls and floor. Nevertheless it is cheap and easy to use, except in freezing conditions. It was used by Kohler et al. (1955) to estimate lake evaporation, and adopted by the World Meteorological Organization (W.M.O.) for use during the International Geophysical Year (Chang, 1965). Tests carried out in the Israeli Negev (Stanhill, 1961), in the U.S.A. (Nordenson and Baker, 1962) and in Finland (Mustonen, 1964) suggest that it yields results which compare closely with Penman's E_0 under a wide range of climatic conditions. Konstantinov (1966) has discussed at length the properties of evaporation pans and has compared the performance of the Class A type with Soviet designs.

The Class A pan is generally considered to evaporate at about 1.4 times the rate of an open lake surface under the same conditions (Kohler et al., 1955; Linacre, 1976). If Penman's $E_0 \simeq$ pan E_0 and if Penman's E_t/E_0 is taken at the Western European value (for the equinoctial months) of 0.7 on a year-round basis or 0.8 for the main part of the growing season (Penman, 1963), the following approximations emerge, which will be found useful in interpreting what follows: Penman's $E_t \simeq$ lake E_0 for complete year; $E_t \simeq 1.1$ (lake E_0) for growing season; $E_t \simeq 0.7$ (pan E_0) for complete year; $E_t \simeq 0.8$ (pan E_0) for growing season.

Maps of estimates of potential evaporation or potential evapotranspiration are available for several large geographical areas. N.N. Ivanov's (1957) map covers the whole world. For the U.S.A., maps of average annual lake evaporation and Class A pan evaporation were published by Kohler et al. (1959), while more recent maps appeared in the *Water Atlas of the United States* (Geraghty et al, 1973). Maps for the U.S.S.R. have been drawn by Borisov (1965). For parts of Europe, maps have been given by Penman (1954), Mohrmann and Kessler (1959) and Eggelsmann (1978: Haude's method) while, for the British Isles, Ward (1975) has published a map of pan evaporation, and county-by-county tabulations by L.P. Smith, using Penman's method, are also available (Anonymous, 1967).

Mention must finally be made of atmometers, which are instruments for measuring potential evapotranspiration over short periods of time and often over, or within, small stands of vegetation. These instruments usually measure the rate of water loss from the standardized surface of a saturated porous material. Data from them have frequently been used as a basis for comparison with the results of ecophysiological studies of water loss from stands of particular species or plant communities (Walter, 1960). In this context the Piche evaporimeter is the most relevant. Its evaporating surface is a disc of white or green filter paper (Steubing, 1965; Slavík, 1974). De Vries and Venema (1954) noted that it underestimates the effect of radiation whilst overestimating the influence of wind. On the other hand, Dilley and Helmond (1973), in the course of a detailed study of atmometry, suggested that evaporation from Piche evaporimeters might in some circumstances be described by energy balance–aerodynamic combination formulae of the Penman type.

The estimation and measurement of evapotranspiration in mires

Although technical developments now permit direct estimation of the evaporative flux density, at least during short-term experiments at experimental stations, most of the estimates for mires have been made by indirect methods. The following remarks should assist interpretation of the data quoted below which result from applying these methods.

Catchment hydrology

Approaches through soil physics or catchment hydrology are based on the principles of continuity and conservation of matter expressed in the form of eq. 3.5.

The evapotranspiration term, E, includes I, the evaporation of water intercepted by plant surfaces. Increases in storage (W) over the period in question are reckoned positive when inserted in this equation and, like E, W may be considered to have various components:

$$W = W_p + W_r + W_s + W_u, \quad (3.10)$$

where W_p is water detained on plant surfaces; W_r is water ponded or stored as snow, etc., above the soil surface; W_s is soil moisture in the unsaturated zone and capillary fringe; and W_u is ground water stored beneath the water table.

Any device for estimating E by drawing up a water balance in an artificially isolated soil monolith is called here a lysimeter. This device actually measures liquid discharge (D in eq. 3.5), enabling E to be estimated as the residual term in the water balance. The monolith should ideally mimic exactly the hydrology of its surroundings, but frequently fails to do so. Common faults include disturbance to the airflow and radiation due to gaps or upstanding walls round the exposed surface of the monolith; discontinuities in vegetation cover between monolith and surroundings caused by restrictions in rooting depth or by root disturbance at installation; toxicity of lysimeter components; interference with aeration or mineral nutrition because the walls interrupt lateral water movement; discrepancies in soil surface level; distortion of the soil thermal profile by the walls of the lysimeter; differences in soil moisture profile and elevation of the water table between monolith and surroundings. Short-term lysimeter estimates demand measurement of W. This is best done by weighing, which in theory at least gives a measure of all the components on the right of eq. 3.10. Some lysimeter designs rely on measurements of water-table height or combinations of these with the specific yield profile of the monolith, but these strictly give one component, W_u, only. The two relevant forms of liquid water discharge, Q (diffuse surface flow) and U (ground-water discharge: see pp. 69, 70), are generally combined by controlling both processes at the wall of the lysimeter.

When applied at the scale of a whole catchment or naturally isolated catchment compartment a different set of difficulties is encountered. Catchment boundaries may be hard to identify (p. 130). Deep leakage of ground water past measuring weirs may occur. Storage changes are difficult to measure and some new techniques, such as the neutron probe, are hard to apply in peat soils. In large catchments the accounting process may be hampered by the time taken for precipitation to concentrate and appear at the outflow point.

As noted above, precipitation, P, is hard to measure accurately (Robinson and Rodda, 1969; Rodda, 1972; Bochkov and Struzer, 1972). The requirements for gauging P as a water balance component are more stringent than for conventional meteorology.

Furthermore, because E has to be estimated as the residual term in all water balance studies, its deviations from the true value will be the algebraic sum of the deviations of measurements from true values for all the other terms. Except for the unusually detailed and interesting comparative study of Baden and Eggelsmann (1964), evapotranspiration estimates derived from water balance studies of catchments are not considered here, partly because the technical problems outlined above are so formidable, partly because few catchments are entirely covered by mire and fewer still by mire of a single type.

Atmospheric physics

Alternatively several approaches to estimating E through applications of atmospheric physics are available. Although the eddy correlation technique proposed by Swinbank (1951) has been tried on Russian mires (Romanov, 1968a, b) and in tall helophyte swamps (Linacre et al., 1970), most of the more successful work has either used the Dalton equation (3.8) or the energy-balance approach. The former method has found favour with workers interested in tall helophyte communities (Linacre, 1976). The latter has been used in other types of mire, notably in the Soviet Union. Starting from the energy balance equation (3.6) it may be shown that:

$$E = \frac{R_n - G}{l(1+\beta)} \quad (3.11)$$

where β, the Bowen ratio, is the ratio of the flux density of sensible heat to that of latent heat:

$$\beta = \frac{H}{lE} \quad (3.12)$$

This ratio is most conveniently determined from measurements of temperature, T, and water vapour pressure, e, made with ventilated psychrometers at two levels above the evaporating surface. Assuming equality of the transfer coefficients for sensible heat and water vapour, it may also be shown that:

$$\beta = \gamma \frac{\Delta T}{\Delta e} \quad (3.13)$$

where γ is the psychrometric constant, in units of kPa $^{\circ}C^{-1}$. The general precautions and limitations of the method are set out by Slatyer and McIlroy (1961), while details of practical application to mires will be found in the official Soviet instruction manual (Anonymous, 1961). It is recommended that levels for the atmospheric sensors be kept as low as possible, while Romanov (1968a) has emphasized the need to avoid heat advection and eddy disturbance by selecting homogeneous areas (known as microlandscapes or microtopes) that are as large as possible, siting the equipment at least 100 m inside the boundary and avoiding areas with forest or a more-or-less continuous cover of dwarf shrubs.

A major advantage of the energy-balance approach is that by focusing attention on the details of energy partition at the evaporating surface its use affords greater insight into the mechanisms controlling the varying flux density of evapotranspiration.

More complete guides to the evaluation of evapotranspiration are provided by Slatyer and McIlroy (1961), by Konstantinov (1966) and by Sokolov and Chapman (1974).

Ecophysiological aspects

The supply of water for evaporation from a mire which is covered by vegetation depends largely upon features of the plants themselves. As noted elsewhere in this volume, mire vegetation is unusual in several respects, and several taxonomic groups of plants are especially characteristic of mires. These phytosociological and floristic facts contribute to the distinctive character of mire evaporation. While little seems to be known about the ecophysiology of families like the Restionaceae and Epacridaceae, which are important in mires in the Southern Hemisphere (see Campbell, Part B, Ch. 5), more information is available from Holarctic mires.

In this context it is the raised bogs of the northern deciduous and coniferous forest zones that are of greatest interest. In floristics and life-forms, their vegetation is characterized by a variable tree cover, but above all by the abundance of evergreen chamaephytes (dwarf shrubs) of the family Ericaceae; of graminoid herbs, especially Cyperaceae; and of bryophytes, *Sphagnum* being predominant while species of the suborder Jungermannineae (leafy liverworts) are inconspicuous but very abundant associates. Some of these raised bogs are also very rich in fruticose members of the lichen genus *Cladonia*.

Effects of community architecture

The community structure is generally very patchy. In treeless raised bogs the mire expanse generally comprises a mosaic in which level areas ("lawns") alternate with hollows and hummocks (see Sjörs, Part B, Ch. 3). The hollows may contain permanent or seasonal pools. They support mainly bryophytes and algae, and represent the wet end of a sequence of sites in which the species of *Sphagnum* replace one another and vascular plants become increasingly conspicuous (Ratcliffe and Walker, 1958) until, on the tallest, driest hummocks, moribund *Sphagnum* spp. may be found in the shade of a dense dwarf shrub layer. Many of the dwarf shrubs, such as *Calluna vulgaris* which is abundant in the raised bogs of Western Europe, therefore show a

TABLE 3.1

Daily temperatures at 20 cm above various raised bog communities (1 and 2 July, 1957) and evaporation from Piche atmometers (means of six successive 24-h periods, September, 1956) at the Dářko fishponds (data of Rybníček and Rybníčková, after Neuhäusl, 1975)

Community	Mean air temperature (°C)		Daily amplitude (°C)	Atmometer evaporation (cm^3 day^{-1})
	maximum	minimum		
Sphagnetum magellanici	33.4	9.0	24.4	9.5
Pino rotundatae–Sphagnetum, *Vaccinium-Calluna* phase	29.8	11.7	18.1	4.4
Vaccinio uliginosi–Pinetum	30.3	12.2	18.1	3.6

markedly clumped pattern in which mutual shelter and shading create a distinctive local microclimate. Similar effects occur in the stands of graminoids, since several of the mire-expanse Cyperaceae are tussock-formers (*Eriophorum vaginatum*, *Scirpus cespitosus*). This point was discussed by Firbas (1931). On 3 August, 1929 the air temperature at 1 m over the Schwarzes Moor in the Rhön hills east of Frankfurt (Main) in Germany varied between 16.1° and 17.4°C during the noon period (11:00–14:00 h). During this same period the temperature 1 cm below the soil surface varied in the *Calluna–Eriophorum* association between −2.4° and +3.9° (mean +1.4°C) while at 1 cm depth in hummocks of *Sphagnum fuscum* it varied between +8.7° and +11.6° (mean 10.1°C). When evaporative losses from Piche atmometers were measured on three similar days (11–13 August, 1929) the mean reduction in the loss at 2 cm above the ground compared with the loss at 30 cm was 30% in the *Eriophorum–Sphagnum capillifolium* association, compared with 43% in the more sheltered and shaded *Calluna–Eriophorum* association. Neuhäusl (1975) reported microclimatological observations in various raised bog communities in the Bohemian–Moravian upland of Czechoslovakia. Here also the soil surface of *Sphagnum* stands displayed temperature fluctuations of great amplitude, and again the amplitude was much diminished by the presence of dwarf shrubs. Very high rates of evaporation from Piche atmometers occurred over the *Sphagnetum magellanici*. The rates were much reduced by the presence of trees or dwarf shrubs, as in the *Pino rotundatae–Sphagnetum* and the *Vaccinio uliginosi–Pinetum*. These results are summarized in Table 3.1.

The *Sphagnum* carpet

The high Piche evaporation values which occur over *Sphagnum* communities on dry, sunny days in summer are thought to be due to other ecophysiological peculiarities of mire vegetation, as are the high air temperatures. Species of the genus *Sphagnum* possess a very uniform and highly characteristic structure which is illustrated in Fig. 3.2. In aquatic forms, such as *S. cuspidatum*, *S. recurvum* and *S. auriculatum* var. *auriculatum*, a population comprises a mat of more-or-less prostrate stems, often largely submerged except for their capitula (apical tufts of branches). In the terrestrial species (*S. magellanicum*, *S. capillifolium*) the distal portions of the stems are erect and form more or less dense cushions. The erect portions are roughly coterminous with the living parts of the plants. The older, proximal parts which have died are less rigid and tend to lie prostrate (Overbeck and Happach, 1957). The densest part of the cushion in a terres-

Fig. 3.2. *Sphagnum*. A. Ribbed hyaline cells of the stem cortex in *S. molle*, with bases of detached leaves. B. Lawn of *S. recurvum*, showing prostrate dead and erect living portions of the plants. C. Capitulum and side branches of the aquatic *S. cuspidatum*, in which no branches invest the stem. D. Capitula and side branches of the terrestrial *S. magellanicum*, showing squarrose or patently downcurved side branches, together with pendulous appressed branches which invest the stem. E. Side branch of *S. cuspidatum* with most leaves removed, showing the porose "retort cells" of the cortex. F. Portion of abaxial surface of branch leaf in *S. capillifolium*, showing hyaline cells strengthened by arching wall fibrils, and pores, many of which are opposed. G. Overlapping branch leaves of *S. magellanicum*, showing architecture of the extra-cellular storage spaces in the capitulum and their relation to the leaf-cell pores. A, E, F and G by Olivia M. Bragg; B by H.A.P. Ingram; C and D by H.A.P. Ingram and I. Tennant.

A
B
10 cm
100 μm
C
D
1 cm
1 cm
E
100 μm
F
10 μm
G
100 μm

trial *Sphagnum* species is formed by the surface layer of capitula in which the stem internodes have not yet elongated, leaving the branches densely crowded together. The older but still living portions of the stems bear fascicles of branches, spaced apart by internodes which have elongated. Whereas in the aquatic species the side branches are all similar and spreading, in the terrestrial species the fascicles comprise branches of two kinds: elongate, flagelliform branches directed downwards either pendulously and free of the stem or else appressed to it and joining with others to form a loose envelope; or shorter, somewhat ascending, divergent branches which radiate outwards from the stem (Parihar, 1962). In effect, therefore, the structure of a *Sphagnum* cushion comprises a dense and finely porous "roof" of side branches supported on a much less dense layer of vertical columns interspersed by much larger spaces and obtaining some lateral bracing from the occasional divergent side branches. The curves of plant volume versus depth presented by Romanov (1961) show minima not at the surface but at 3 to 9 cm below it (Fig. 3.3). Similar curves are given by Ivanov (1953). Porosity, on the other hand, attains its maximum value in this same layer (Fig. 3.4). The integrity of the "roof" is maintained by the even elongation of stems within a stand. It is more marked in pure stands (single-species populations) and among the hummock-forming species (see Clymo, Ch. 4).

Sphagnum species appear to be devoid of water-conducting tissue and there are no rhizoids. Water is therefore absorbed directly through the weakly cutinized cells of the leaves and the stem, retained in the plants by matric forces and conducted through networks of various kinds of capillary spaces. The leaves of *Sphagnum* are most abundantly and typically developed on the branches. They are single-layered sheets of cells (Fig. 3.2) in which the small, photosynthetic chlorophyllous cells alternate regularly with much larger hyaline cells which are dead at maturity. In time of drought these latter cells are empty of water. They then impart a whitish appearance to the *Sphagnum* tussocks, which have a high albedo and reflect more incident radiation

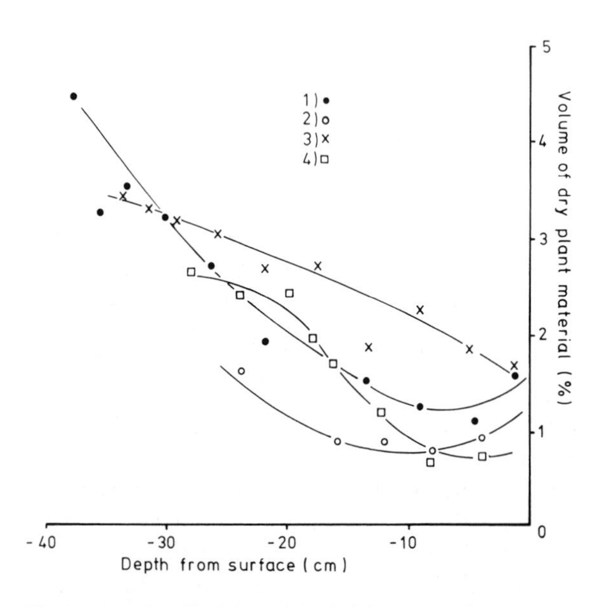

Fig. 3.3. Relationship between depth below the surface and the volumetric fraction occupied by plant material in the surface layers of *Sphagnum* bogs in the northwestern U.S.S.R. Legend: *1*=hummock from the mire expanse at Lammin-Suo Bog; *2*=the same from Shirinskoye Bog; *3*=shallow hollows from the mire expanse at Lammin-Suo Bog; *4*=the same from Shirinskoye Bog. (After Romanov, 1961.)

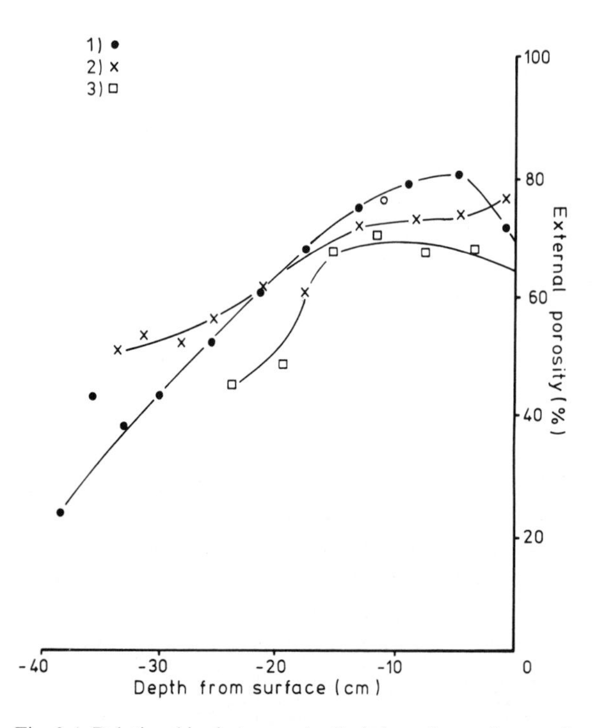

Fig. 3.4. Relationship between depth below the surface and external porosity in the surface layers of *Sphagnum* bogs in the northwestern U.S.S.R. Legend: *1*=hummocks from the mire expanse at Lammin-Suo; *2*=shallow hollows from the same; *3*=shallow hollows from the mire expanse at Shirinskoye. (After Romanov, 1961.)

than when damp (see below). In wet weather they absorb water rapidly due to matric forces, and in some species their walls are strengthened against inward collapse by fibrous ribs of cellulose and perforated by pores. The hyaline cells thus confer a high water storage capacity on the capitula, reported by Romanov (1968a) to attain 27.0% of volume in the surface 2.9 cm of *S. fuscum* cushions, but it is likely that extra-cellular storage capacity in interstices between the overlapping branch leaves and between the leafy comal branches themselves is still greater. For this Romanov (1968a) has reported 71.6% of volume for the surface 2.9 cm of *S. fuscum*. In some raised bogs, this species forms the tallest hummocks. Along with *S. capillifolium* and others of the *Acutifolia* section it belongs at the fine-leaved end of a range of species with successively larger leaf cells and branch leaves, with longer branches and with a coarser overall structure. The coarsest members of this series are classed in the *Palustria* group and include *S. magellanicum* and *S. papillosum*, which form low hummocks or cover the lowest slopes of the taller ones. It is tempting to correlate the habitats of members of this series with the increasing matric forces developed in the finer comal capillaries, enabling those species which grow furthest from the water table to retain water most strongly.

However, conflicting evidence on this point was presented by Overbeck and Happach (1957), who considered that living *Sphagnum magellanicum* had a greater capacity to retain water against gravity than had *S. fuscum*. This finding appears to be confirmed by the results of Päivänen (1973; see p. 112) who studied the decline in water content of various very slightly humified *Sphagnum* materials between saturation ($\phi = 0$) and pF 1 ($\phi = -0.001$ MPa) and found differences in the range 20 to 23% of volume where *S. fuscum* was dominant, compared with only about 3% of volume where *S. magellanicum* was dominant. In fact, since *S. magellanicum* retained more water than *S. fuscum* over the whole range of water potentials investigated by Päivänen down to about -1.5 MPa, there is clearly need for more work on water retention by *Sphagnum* before the ecophysiological differences between the hummock and hollow species can be clearly understood.

Ribbed hyaline cells are found also in the surface layer of the stem cortex of some species (e.g. *S. palustre*). They form a water transfer pathway according to Oltmanns (1885, 1887), who used suspended carmine particles as tracers. By this means he also found that a pathway was formed by the spaces between stems and appressed side branches (Fig. 3.2). Since these extracellular spaces are wider than the intracellular system they are more efficient, as would be predicted from a crude physical model based on Poiseuille's law:

$$\frac{dq}{dt} = f(s^4) \tag{3.14}$$

where dq/dt is the flux rate and s is the radius of the space through which the water passes. On the other hand, since water is retained in these spaces by a matric force, τ, which results from the surface tension, σ, of the water but is inversely related to radius:

$$\tau = \frac{2\sigma}{s} \tag{3.15}$$

it is the largest spaces which empty first under conditions of moisture stress. Hence, unlike a stand of vascular plants, there is no sustained supply of water to the surface of a *Sphagnum* cushion under strongly evaporative conditions. In relation to the architecture of an entire cushion it is worth noting that in hummock-forming *Sphagnum* species the largest spaces between the elongated stem columns are too big to develop any significant matric attraction. Since also it is beneath hummocks that the water table is furthest from the surface, these spaces are almost invariably empty of water and can contribute to the maintenance of surface dampness only by the highly inefficient processes of vapour transfer which occur in the gas phase. A more detailed theoretical treatment of the water relations of *Sphagnum* stands is given by Romanov (1968a).

Early experiments on evaporation were reported by Leick (1929), while experimental data on capillary rise in *Sphagnum* were given by Overbeck and Happach (1957). Their material was collected with minimal disturbance, packed at its natural density into vertical cylinders, air-dried and supplied with water from a water table maintained 12 cm below the surface. After 48 h the capitula of the aquatic *S. recurvum* had risen to only about 100% of dry weight, while their surfaces seemed to be drying. By contrast the capitula of *S. magellanicum* and *S. capillifolium* had attained contents in excess of 1000% d.w. Neuhäusl (1975) obtained similar results. In the course of experiments on the water relations of various surface materials from mires he found that on hot summer days evaporation from the capitula of *S. recurvum* and *S. cuspidatum* was so intense that they lost water even when their surfaces were only 2 cm above the water table. These findings explain the rapid development of dry atmospheres over *Sphagnum* lawns and account for the high atmometer readings found there.

Leafy liverworts

Morphologically, the leafy liverworts of raised bogs are for the most part small and inconspicuous inhabitants of the surfaces of *Sphagnum* stands. Their cell surfaces are thought to be even more weakly cutinized than are those of *Sphagnum*, and the commoner genera (*Calypogeia, Cephalozia, Cephaloziella, Mylia, Odontoschisma*) are entirely lacking in the special water-storage organs found in some liverworts of drier habitats. No physiological information on their water relations seems to be available.

Lichens

The lichens of raised bogs either occur as epiphytes on the older branches of dwarf shrubs (*Hypogymnia physodes*) or, in the case of the widespread genus *Cladonia*, as appressed squamules and more or less erect podetia which may be trumpet-shaped scyphi or fruticose, resembling miniature bushes. Again, little ecophysiological information is available on water relations of mire lichens, although in other habitats different *Cladonia* species have been reported to dry out at different rates (Lechowicz and Adams, 1974). It seems, however, likely that they add to the interception capacity of mire communities and they may influence the aerodynamics in some conditions by increasing surface roughness.

Dwarf shrubs (chamaephytes)

Several of the dwarf shrubs which are most characteristic of the mire expanse in Western European raised bogs are important in other ecosystems as well. *Calluna vulgaris, Erica tetralix* and *Empetrum nigrum* are examples, while *Vaccinium myrtillus* is more characteristic of the marginal slope. *Erica tetralix* is also found on gleyed mineral soils with a tendency to waterlogging. *Calluna, Vaccinium myrtillus* and *Empetrum* flourish in well-drained soils; and *Calluna* is a species of economic importance as a grazing plant in northern England and eastern Scotland, so that its biology has been intensively studied (Gimingham, 1960, 1972). Other dwarf shrubs such as *Vaccinium uliginosum, V. oxycoccos, Andromeda polifolia* and *Ledum palustre* are more strictly confined to oligotrophic mires (Oberdorfer, 1962) and are less clearly understood biologically. *Rubus chamaemorus*, a species characteristic of Norwegian and British blanket bogs, has edible fruits and has been studied in great detail. The water relations of this and several other mire species have been considered by Rakhmanina (1970).

Rubus chamaemorus is a deciduous species. *Vaccinium myrtillus* and *V. uliginosum* are also deciduous, though *V. vitis-idaea*, a related species of mineral soils, is not. *V. oxycoccos*, sometimes placed in a distinct genus as *Oxycoccus palustris*, is however evergreen, as are all the other species mentioned. The fact that all the evergreen mire shrubs are sclerophyllous has occasioned much comment in the literature, even though plants with persistent leaves tend to be sclerophyllous under widely varying conditions. The so-called "xeromorphism of raised bog plants" will not be discussed here in detail since Bradbury and Grace deal with aspects of the subject in Chapter 8.

Firbas (1931) emphasized that the strongest xeromorphism was shown by the evergreen species, which showed an unusual development of the vascular system comparable to that of desert plants, but coupled with high stomatal densities. *Calluna* and *Empetrum* were exceptions, and hypertrophied vascular systems appeared more marked in plants from hollows than from hummocks. The effect was especially marked in the evergreen *Vaccinium* species, which Firbas therefore also considered to be xeromorphic.

The root penetration of these dwarf shrubs varies in response to waterlogging, but in general they tend to be much more shallow-rooted than their herbaceous, graminoid associates, and have root systems virtually confined to the acrotelm[1] (Firbas,

[1] **Acrotelm**: the surface layer of a mire soil, overlying the **catotelm** (Table 3.6).

1931; Metsävainio, 1931; Heath and Luckwill, 1938). Using a radioactive tracer technique, Boggie et al (1958) found that on deep peat *Calluna* was markedly surface-rooting, and that absorption of phosphorus and rubidium was mainly in the 0 to 15 cm layer in this species as well as in *Erica tetralix*, so that it seems reasonable to conclude that water also is likely to be supplied mainly from the surface layer. Leiser (1968) found that the young roots of many Ericaceae, including *Calluna*, develop a hydrated mucilaginous sheath. This may counter the adverse effects of low water tables on these shallow root systems.

Early ecophysiological work on ericoid dwarf shrubs in heath habitats, reported by Gimingham (1972), had revealed the relatively low water content of live plants. Firbas (1931) found this to be equally true in the mire species, where it was accompanied by low osmotic potentials (high osmotic pressures) in the sap expressed from turgid leaves or young leafy shoots. In *Calluna* the osmotic potentials ranged from about −1.5 MPa down to −3.0 MPa. The results from *Vaccinium uliginosum* and *V. oxycoccos* mostly lay within the same range, as did those from *Empetrum nigrum*. The experiments of Bannister (1964b) showed that *Calluna* experiences substantial water deficits in winter. This suggests that high osmotic pressure of the cell sap in its young shoots may be correlated with high tolerance to dehydration and the low pressure (turgor) potentials of water in the leaf cells which ensue. *Erica tetralix*, which in bogs is generally found in wetter sites than *Calluna*, showed a similar annual march of water potential, but its leaf cells remained more turgid in winter than those of *Calluna* and it was inferred to be less tolerant of desiccation.

While bryophytes and lichens have no physiological mechanisms regulating the loss of water vapour, vascular plants possess vapour barriers in the form of cuticles perforated by stomata of variable aperture. Firbas (1931) studied transpiration in the xeromorphic raised bog species by the detached shoot method of L.A. Ivanov (see Slavík, 1974), recognizing that this method is apt to overestimate the rate. Even so, he concluded that for all the species studied the rate, calculated either per unit fresh weight, per unit dry weight, or per unit of leaf surface area, was low in comparison to mesomorphic sun-plants. Average values for the noon hours of sunny summer days ranged from 15.5 mg water (g fresh weight)$^{-1}$ min^{-1} [16 mg (dm leaf)$^{-2}$ min^{-1}] in *Vaccinium uliginosum* down to 7.4 mg g^{-1} min^{-1} (10 mg dm^{-2} min^{-1}) in *Empetrum nigrum*, both growing amongst *Sphagnum*. The Ivanov method has in recent years been criticized for shortcomings not recognized by Firbas, and it is clear that its results demand the most cautious interpretation (Slatyer, 1967).

Biophysical studies of the detailed mechanism of vapour loss and of the resistances to vapour flow in and around the leaf have suggested that control of transpiration by stomata is likely to be most efficient in plants with small, or at least narrow, leaves. The prevalence of such plants on the mire expanse of raised bogs gains interest from such findings. Bannister (1964a) showed that, in wet sites, water loss from *Calluna* was restricted by stomatal closure when the water potential in transpiring shoots fell to about −1.8 MPa. At this point the shoot water content was 76% of its value at full turgidity, while *Erica tetralix* tolerated a water loss of only 10% before its stomata closed. *Vaccinium uliginosum*, on the other hand, transpired freely down to a relative water content of 64%, while *V. myrtillus* closed its stomata at 79%, equivalent to a leaf water potential of −2.3 MPa (Bannister, 1976). The evidence for stomatal control of transpiration in these plants is therefore clear, although it mainly relates to plants from heaths, and the *Vaccinium* data make it hard to interpret these findings in relation to the known facts of mire plant autecology. Indeed Firbas (1931) maintained that, although the raised bog xeromorphs show maximum stomatal apertures in dull weather, their stomata remain fairly wide open even on sunny days at noon. Under these conditions he also noted that the stomata of *Vaccinium uliginosum* closed further and earlier in plants growing on *Calluna*-covered hummocks than in *Sphagnum*-rich hollows. In view of his data on osmotic potentials (see above) it seems possible that the water deficits imposed in the experiments reported by Bannister are seldom encountered in mires, except on hummocks.

More recent studies include the work of Small (1972), who reported spot determinations of leaf water potential and stomatal resistance for various chamaephytes from the Mer Bleue raised bog near Ottawa, Canada. Resistance values lay between 1 and 3 s cm^{-1}. Then Hinshiri (1973) obtained results suggesting that stomatal restriction in *Calluna* does not become effective until the water potentials in leafy shoots fall to values between −2.5 and −3.5 MPa — in general some 1 to 1.7 MPa below the point suggested by Bannister, who also used detached shoots. Using the same Fukuda–Hygen approach (Slavík, 1974) as Bannister to interpret the loss of water as a function of time, Ashmore (1975) identified the stomatal closure point at water contents equivalent to −2.2 MPa, much closer to Bannister's closure water potentials than to most of Hinshiri's. Ashmore, however, argued on grounds of physiological theory that the act of cutting off the shoot would be expected to lower both water content and potential at the point when the stomatal resistance increases. Thus all these results, including those obtained using the Ivanov method, could be misleading. A laboratory comparison of cut shoots with intact plants confirmed this prediction: the differences in resistance were highly significant in the expected direction.

Ashmore's (1975) study of the ecophysiology of *Calluna* is probably the most significant yet undertaken in the mire context. He worked on blanket bog at Moor House near Alston in northern England and used a sophisticated approach. He solved the technically formidable problem of constructing and operating a steady-state diffusion porometer at a remote field site by methods which have been outlined by Incoll (1977) and derive from the work of Parkinson and Legg (1972). Attached leafy shoots were briefly enclosed in a gas-tight polypropylene chamber[1], so that the release of water vapour into an initially dry air stream could be followed. The total resistance, r, encountered by vapour diffusing from the mesophyll cells in the leaves is the sum:

$$r = r_s + r_a \qquad (3.16)$$

of r_s, the stomatal resistance of the leaves and r_a, the resistance of the atmospheric boundary layer beyond the leaf surface. The value of r_a for the stirred atmosphere of the chamber was determined in calibration tests with moist filter paper, and r_s could then be obtained by subtracting r_a from r values calculated on the basis of Fick's law. This assumes that the rate of outward diffusion of water vapour is a linear function of the fall in vapour concentration along the pathway controlled by the stomata, and that the source concentration is given by the saturation vapour pressure at the shoot temperature, while the sink concentration can be determined by a humidity sensor in the chamber atmosphere. It also assumes equality in shoot and air temperature, an assumption justified by observations of Grace (1970) and Hinshiri (1973) that for *Calluna* the difference never exceeds 1°C, which enabled Ashmore to calculate that stomatal resistances might be underestimated by up to 25%, though usually by much less.

Ashmore also measured shoot water potentials by the Dixon–Scholander pressure bomb technique (Slavík, 1974). The *Calluna* plants began the day with water potentials close to zero, declining towards noon to values between −0.7 and −1.3 MPa on cloudy or sunny summer days respectively. Concurrently stomatal resistance was found to increase from minimum values between 2 and 4 s cm^{-1} to maxima between 10 and 20 s cm^{-1} on sunny days, the changes on a cloudy day being erratic. On the basis of these field studies, Ashmore placed the stomatal closure point between −0.8 and −1.0 MPa, which is appreciably higher than the cut-shoot experiments of others had suggested, indicating that stomatal control of transpiration might be effective on most days during the growing season. Some results for one of the sunny days appear in Fig. 3.5. Among other advantages, this work fills a conspicuous gap in the data of Firbas (1931) and one is left in no doubt that physiological mechanisms in the vascular plants can indeed control transpiration from mires.

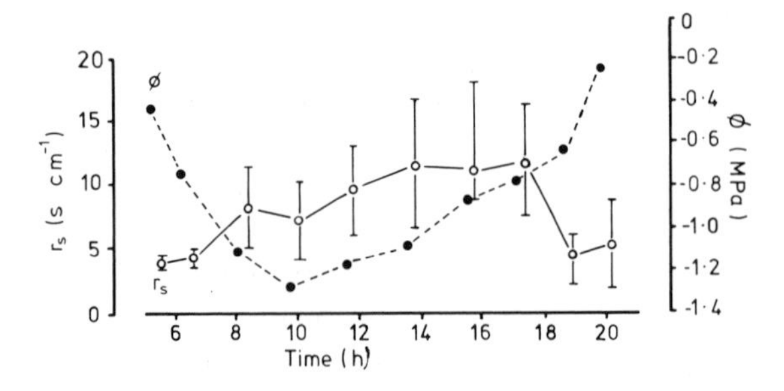

Fig. 3.5. Diurnal march of the water relations of *Calluna vulgaris* on blanket bog at Moor House (England) on a sunny day, 21 May 1974: stomatal resistance r_s and shoot water potential ϕ Vertical bars: 95% confidence limits. (After Ashmore, 1975.)

[1]Claimed to have negligible effect on the shoot energy balance during the test periods.

Graminoids

Some of the cyperaceous graminoids which accompany dwarf shrubs on the raised-bog mire expanse also display xeromorphism. *Eriophorum vaginatum*, for instance, bears setaceous, triangular leaves less than 1 mm broad. Their substomatal cavities contain remarkable hypodermal plates which are cutinized in continuity with the epidermal cells (Firbas, 1931). Similar hypodermal plates occur beneath the stomata in *Scirpus cespitosus*, but in this species there are no foliage leaves and the stomata are borne on the slender, ridged stems. In *Eriophorum angustifolium* the leaves are also very narrow but have a distinct, channelled lamina below, resembling those of *Rhynchospora alba* in which, however, hypodermal plates are absent. All these plants are summer-green. Their aerial organs gradually die in autumn and presumably cease to transpire, though since they may not collapse until the following season they continue to furnish surfaces from which intercepted water can evaporate in winter.

Unlike the dwarf shrubs in which the physiologically active root system is a perennial structure, these graminoids have adventitious root systems which are repeatedly renewed by the outgrowth of new roots from perennating rhizomes at the soil surface, often from points close to the buds from which the aerial parts are formed. In *Eriophorum vaginatum* the outgrowth of new roots is not a continuous process but shows two maxima, the greater production occurring in June–July when the roots reach greater depths, the lesser production of shallower roots occuring in winter (Wein, 1973).

Some idea of rooting depth in the graminoids can be obtained from the radioisotope experiments of Boggie et al. (1958) already mentioned. Phosphorus absorption by *Scirpus cespitosus* and *Eriophorum vaginatum* was fairly evenly distributed, but declined towards the maximum isotope placement depth of 61 cm (24 in). Absorption by *Eriophorum angustifolium* increased down to this depth. This conforms with the evidence of Heath and Luckwill (1938), whose morphological studies showed that these plants are relatively deep-rooting. *Narthecium ossifragum*, however, absorbed mostly from the surface layer, again in conformity with its shallow-rooting habit.

Firbas (1931) was unable to detect much stomatal closure in *Eriophorum vaginatum* or *E. angustifolium*. He found the stomata at their widest on cloudy days, narrowing somewhat in sunny conditions and at noon. His results were obtained by liquid penetration and alcohol fixation methods which do not permit valid comparison between different species (Slavík, 1974).

More recent studies at Barrow, Alaska (Miller et al., 1978) have involved measuring the water loss of bundles of cut leaves by weighing in a chamber of known ambient temperature and humidity. This enabled stomatal resistances to be obtained for comparison with water potentials determined by the Dixon–Scholander pressure-bomb method. The resulting curve for *Eriophorum vaginatum* is reproduced at Fig. 3.6, from which it may be seen that, while the resistance increases steadily with falling water potential, it only rises rapidly below about −0.7 MPa. In the field at mid-summer, leaf water potentials in *Eriophorum* seldom declined below −0.5 MPa at midday. It therefore seems unlikely that transpiration in this species encoun-

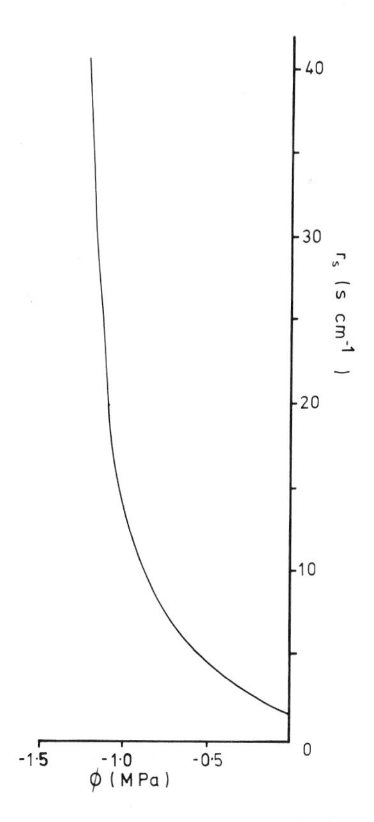

Fig. 3.6. Dependence of stomatal resistance r_s upon water potential ϕ in leaves of *Eriophorum vaginatum* from Barrow, Alaska, U.S.A. (After Miller, et al., 1978.)

ters much physiological control, at least in the wet, Arctic tundra. It should be noted, however, that since two subspecies of *E. vaginatum* occur in Alaska (Wein, 1973) it may be unwise to apply these conclusions elsewhere.

According to Firbas, the only raised-bog species to show strong stomatal movements was the insectivorous *Drosera rotundifolia* — also the only species which he regarded on general grounds as mesomorphic. Fibras also compared the transpiration from *Eriophorum vaginatum* plants growing in different situations, again using the detached-leaf method of L.A. Ivanov. He found that on a fresh-weight basis the plants from low-lying *Sphagnum* communities transpired about 20% less than those in *Calluna* hummocks under similar weather conditions. This result, for a deep-rooted species well supplied with water, is what one might expect in view of the more exposed situation of hummock plants. Nevertheless it is worth noting that Firbas found exactly the converse situation when he made the same comparison for *Calluna* and *Vaccinium uliginosum*. It is tempting to conclude that the development of water deficits in these more shallow-rooted species imposes upon them a somewhat different role in evapotranspiration from bogs, and that during the growing season they may well contribute less to the vapour flux than might be supposed from their contribution to the total standing crop. Further examination of this would require the stomatal or "canopy" resistances of several species to be evaluated on a ground-area basis, as was done by Rutter (1967) for a plantation of *Pinus sylvestris*. Pending this one can only speculate as to the quantitative importance of plant mechanisms in regulating mire evaporation on the basis of broader data — data in which variety of species and site has been achieved at the expense of physiological precision. These data are examined in the following two sections.

Evapotranspiration from bryophyte–chamaephyte–Cyperales mires (raised bogs) in the Holarctic

Eggelsmann (1964b) has given a set of tabular summaries indicating the approximate limits within which the daily average evapotranspiration from various European mire communities may be expected to lie. The results, all obtained during the summer months, lay between about 1.5 mm day^{-1} for open sedge–*Sphagnum* communities in south-central Finland (Huikari, 1963) and about 6 mm day^{-1} for *Sphagnum* hummocks in the Belgian Hautes Fagnes (Tinbergen, 1940, see below). A selection of studies will now be considered in detail.

A useful body of broadly based data on evaporation from various mire communities is presented in Neuhäusl's (1975) report (see above). It relates to the series of raised bogs and other mires at Dářko near Žd'ár on the river Sázava. They are situated at an altitude of 600 to 650 m, in the Žd'árská Vrchy Hills on the borders of Moravia and Bohemia (Czechoslovakia). Most of the work to be described was carried out by Neuhäusl and his colleagues during a dry period in the summer of 1957.

Transpiration was determined by the much-criticised detached-organ method of L.A. Ivanov (see above). The daily march of transpiration and of water content per unit dry weight was determined for the eighteen commonest vascular plants. Samples were removed and quickly weighed on a torsion balance located at a central measuring point. Simultaneous Piche atmometer measurements at sampling and measuring points formed the basis of a series of correction factors. These varied between 0.56 for *Vaccinio uliginosi-Pinetum* and 1.18 for *Trifolio-Festucetum rubrae*, but Neuhäusl considered the whole procedure very approximate.

From the curves of daily march of transpiration and water content and from measurements of fresh-weight biomass a set of estimates of daily "stand transpiration" was prepared. The stands fell into four groups, as follows:

(1) Stands with low transpiration (<2 mm day^{-1}; $E < E_0$)[1]: *Eriophorum vaginatum–Sphagnum recurvum* stage of secondary succession in old peat cuttings; *Sphagnetum magellanici* on the primary, open mire expanse.

(2) Stands with negligible transpiration (→0 mm day^{-1}) but high evaporation ($E = E_0$) from exposed water surfaces: *Sphagnum cuspidatum* and *S. recurvum* stages of secondary succession in old peat cuttings.

(3) Stands with moderate transpiration (2–4 mm day^{-1}; $E > E_0$): *Caricetum fuscae*, *Caricetum dioi-*

---

[1] E is the evapotranspiration of the stand, E_t the potential evapotranspiration, and E_0 the evaporation from an extensive open water surface (see p. 74).

cae, *Carici-Festucetum rubrae caricetosum paniceae* of meadows and fens; *Carex rostrata* stage of secondary succession; *Pino rotundatae-Sphagnetum* and mountain peaty forest of *Betula pubescens*, *Pinus sylvestris* and *P. mugo* ssp. *rotundata* on the mire expanse. Evaporation from the ground layer was here low.

(4) Stands with high transpiration (>4 mm day^{-1}; $E \gg E_0$): *Carici-Festucetum rubrae* meadows with *Carex rostrata*; meadows and pastures with *Trifolio-Festucetum rubrae* and *Nardo-Festucetum capillatae*; the retrogressive, degraded successional stage following drainage and deforestation of the mire expanse; *Vaccinio uliginosi-Pinetum* on the mire expanse; plantations of *Picea abies* with *Vaccinium myrtillus*. The effects of good water supply and/or ventilation (the "clothes-line effect") can perhaps be discerned here.

The device of multiplying transpiration estimates for fragments of vegetation by standing-crop estimates to obtain estimates for stands poses problems because it takes no account of the effect of stand structure on the evaporation process. However, in qualitative terms the results from the Dářko mires are generally what would be expected, in that rapid evapotranspiration was associated either with high water tables or with tall vegetation.

Experiments with similar objectives but employing a soil-physical, rather than an atmospheric, approach were described by Tinbergen (1940). The site was a small saddle raised bog in a col at 620 m altitude at the source of the river Roer (Rur), which rises east of Liège near the northern edge of the Belgian Ardennes. Though surrounded within 200 to 300 m by patches of coniferous woodland and *Polytricheto-Salicetum*, the site itself was open, and communities dominated by *Sphagnum* and dwarf shrubs were a much more important feature than at Neuhäusl's site.

Evapotranspiration was determined from the water balance of soil monoliths of surface area 20 × 20 cm and depth 25 cm, with intact vegetation. The monoliths were weighed in a tent with an accuracy sufficient to permit estimates of evapotranspiration to be made to ±0.1 mm. Tinbergen compared his evapotranspiration data with data on open water evaporation obtained with an evaporation pan (compare Prytz, 1932). By evaluating the quotient E_{sample}/E_{pan} for days when both were measured, E_{sample} values were estimated for those days (just over one quarter of the total) for which measurements of E_{pan} only were available. It was found that the value of the quotient varied between 1.6 and 0.8. The observations of evapotranspiration were made from 28 June to 27 August, 1938.

Interest centred on comparing evaporation from two principal communities on the mire expanse. In the first type *Sphagnum papillosum* was dominant. The second community lacked *Sphagnum* and was dominated by *Calluna vulgaris* and *Erica tetralix*. During the two months of observations, evaporation from pure *Sphagnum* stands, totalling 183.6 mm, differed little from evapotranspiration in the chamaephytes which totalled 173.5 mm. Total free water evaporation was 162.3 mm.

Schmeidl et al. (1970) have described a hydrological investigation of the Chiemseemoore. These mires form a series on the south shore of the Chiemsee near Bernau, Bavaria, at an altitude of about 530 m in the northern foothills of the Alps. They are located some 300 km southwest of Neuhäusl's (1975) site at Žd'ár. Precipitation in the summer (140 mm month^{-1}) was appreciably greater than in winter (80 mm month^{-1}). The vegetation was characteristic of many raised bogs, with *Calluna* and other woody species at the rand and *Sphagnum* on the mire expanse.

Evapotranspiration was studied with two types of lysimeters: initially the Popov apparatus (see Steubing, 1965); later a larger design similar to that of Ivitskiĭ (1938). Using the first method, it was found that evaporation from lawns of *Sphagnum magellanicum* and *S. capillifolium* was generally rather less than estimates of potential evapotranspiration made by the Thornthwaite method. The difference was greatest in July or August, when E/E_t fell to about 0.6. In April and May however, values of E/E_t as high as 1.4 were noted. For the season as a whole E/E_t was about 0.9. To illustrate the difficulty of this approach, the regression of E on E_t is reproduced in Fig. 3.7. Although the regression was highly significant, giving $E = 1.45\ E_t$, the failure of the relationship to pass through the origin is unsatisfactory. It implies that, for evaporation to occur, potential evapotranspiration must first reach a threshold below which evaporation from the lysimeter was negligible and above which the rate was unexpectedly sensitive to environmental conditions.

The larger lysimeters were therefore installed as

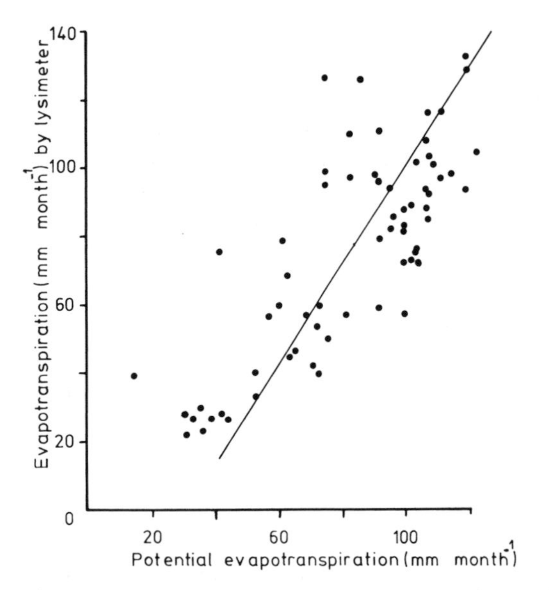

Fig. 3.7. Regression of evapotranspiration from the intact mire expanse of the Chiemseemoore, Bayern (Bavaria, Federal Germany), measured with the Popov lysimeter, on estimates of potential evapotranspiration by Thornthwaite's method. Monthly totals, May–October 1959–1968. (After Schmeidl et al., 1970.)

an alternative approach. The results are given in Table 3.2, together with data from the smaller lysimeters for the same days. It should be noted that the measurements were not made with the vegetation (hummock and hollow *Sphagnum* spp.) in its natural position. The total evaporation for the fourteen days on which the larger lysimeters were used lay very close to the total potential evapotranspiration computed by Haude's method, while the Popov lysimeter results were about 24% lower. Schmeidl et al. considered that the larger lysimeters gave better estimates of evaporation from *Sphagnum* lawns because in them the water table declined through evaporation only, and in a similar manner to that in the surrounding mire expanse, while in the Popov design the water in the monolith was discharged by drainage as well.

A somewhat similar, though deeper, lysimeter arrangement was used by Ivitskiĭ (1938) to compare evaporation from bare peat with that from peat sown with a sward of *Phleum pratense*.

Eggelsmann (1954) also used lysimeters to measure evaporation at the Königsmoor, a raised bog situated beside the river Wümme between Bremen and Hamburg in the lowlands of northwestern Germany. The Königsmoor lies at an altitude of 38 to 40 m. The mean annual precipitation was

TABLE 3.2

Comparison of evapotranspiration results obtained by different methods on a series of dry days at the Chiemseemoore, 1968 (after Schmeidl et al., 1970)

Month	Date	Potential evapotranspiration (Haude's method) (mm)	Evapotranspiration		Duration of bright sunshine (h)
			Popoff small lysimeter (mm)	0.2 m^2 lysimeter (mm)	
July	2	7.1	5.2	7.6	14.0
	3	6.5	2.4	6.1	12.8
	4	2.3	2.0	2.9	1.7
	5	4.9	3.0	6.6	10.7
	8	5.1	4.2	6.1	11.0
	30	5.0	5.2	5.8	12.2
August	1	4.4	4.2	4.5	7.4
	2	4.9	4.0	3.9	2.3
September	2	4.7	4.0	3.9	10.4
	3	3.5	2.5	3.0	4.8
	7	1.4	1.0	1.2	0.0
	8	3.0	2.4	3.0	5.9
	9	4.0	2.6	3.0	9.3
	10	3.5	2.9	3.3	10.6
Total	14 days	60.3	45.6	60.9	113.1

645 mm, 351 mm falling in the summer half-year, May to October (Baden and Eggelsmann, 1964, 1970). The mire expanse was treeless. Its vegetation had been sporadically disturbed by burning, a little drainage and peat extraction since the middle of the last century, and in these respects it resembled a great many raised bogs in Western Europe. The upper peat layers were mainly *Sphagnum* of the *Cymbifolia* and *Acutifolia* sections, but at the time of these investigations the surface vegetation was a *Callunetum* in which *Calluna vulgaris*, *Erica tetralix* and *Eriophorum vaginatum* were associated with *Andromeda polifolia*, *Vaccinium oxycoccus* and a number of *Sphagnum* species.

In Egglesmann's lysimeters the monolith was accommodated in a steel cylinder of depth 100 cm and had a surface area of 500 cm^2. Each installation comprised a pair of lysimeters, a Hellman rain gauge exposed at ground level (but without an in-splash prevention device) and a shelter containing a balance for weighing the monolith cylinders. The balance was sufficiently accurate to permit storage to be measured to ± 0.2 mm equivalent depth. Precautions were taken to reduce the effect of an annular gap between the lysimeter monolith and the surrounding concrete access cylinder.

At installation, the monolith was removed in 20-cm sections, using a cylindrical coring punch, and reassembled in its cylinder so as to preserve the natural stratigraphy.

Observations were made daily during May to October and at longer intervals in other seasons. Water-table measurements in the monolith were compared with those in the surroundings and adjusted to match them.

Baden and Eggelsmann (1964) compared evapotranspiration estimates from the semi-natural mire expanse at the Königsmoor with three other sets of data:

(1) evaporation from a neighbouring area of *deutsche Hochmoorkultur*, tile-drained since 1912, fertilized and supporting a grass sward used either as pasturage or as hay meadows;

(2) estimates of evapotranspiration computed as the residual term in the catchment water balance;

(3) potential evapotranspiration estimated by Haude's method. [As Linacre (1976) has pointed out, no direct measure of E/E_0 was made in this study. However Prytz (1932), working in similar circumstances in Jutland, obtained the quotient $E/E_0 = 1.25$, based on comparisons between data from large lysimeters and sunken evaporation pans.]

(1) The regression of monthly totals of evapotranspiration from *deutsche Hochmoorkultur* on that from the uncultivated raised bog is reproduced in Fig. 3.8. With a correlation coefficient of 0.944 it was highly significant. However, the slope was less than unity, so that if E is the evapotranspiration from the intact raised bog expanse and E_r is that from the reclaimed area:

$$E_r = 0.83\ E + 6\ \text{mm}$$

This regression suggests that under conditions of low evaporative demand the intact raised bog lost water vapour more slowly than the reclaimed area, but that the converse obtained under conditions of high demand. However the eight-year mean value for E of 506 mm differed little from E_r, with a mean of 491 mm for the same period, while the mean May to October rate for E was 2.01 mm day^{-1} compared with 2.15 mm day^{-1} for E_r (Egglesmann, 1964b). These are very small differences.

(2) Having determined the correlation between water-table height and soil moisture content (the specific yield) for the Königsmoor, Eggelsmann

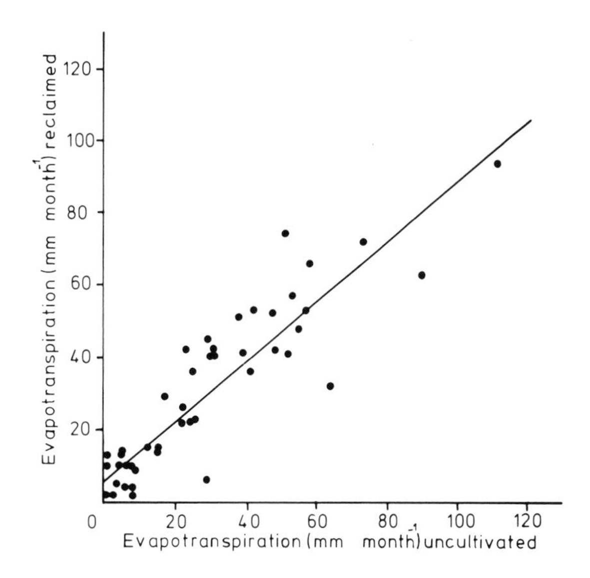

Fig. 3.8. Eggelsmann's data on the relation between evapotranspiration from unreclaimed and reclaimed parts of the Königsmoor raised bog in Niedersachsen (Lower Saxony, Federal Germany). Monthly totals, 1951–1958. (From Baden and Eggelsmann, 1964.)

(1963) was able to use water-level fluctuations to compute changes in soil moisture storage. These results were then inserted into a water-balance equation of the form:

$$P-D-E_{\mathrm{m}}-\eta=0 \tag{3.17}$$

(compare eq. 3.5) in order to compute monthly evaporation from the mire catchment, E_{m}, as the residual term in a budget drawn up for the catchment as a whole. When these results were correlated with lysimeter evaporation, E, for the uncultivated mire expanse the results were as shown in Fig. 3.9. Here:

$$E=0.996\ E_{\mathrm{m}}+1.3\ \mathrm{mm}$$

with a correlation coefficient of 0.674, again highly significant. The scatter of points about the regression line is an interesting indication of the variability to be expected in a comparison of this kind. Even if it were possible to measure run-off, D, with equal accuracy at all rates, which could not have been achieved with the V-notch weirs employed in this study, there remains the problem that all catchment water-balance studies are subject to "accounting errors" which arise due to the time lag between the arrival of a rainstorm and the resulting run-off. Such errors may indeed have been negligible in such a small catchment, but their effect on hydrological investigations in mires has seldom been considered. Nevertheless, the fact that the regression line passes so close to the origin and that its gradient is so near unity is perhaps the best available evidence that lysimetry is an appropriate technique for measuring evapotranspiration in mires.

(3) The regression of lysimetric evapotranspiration from the uncultivated Königsmoor raised bog on the potential evapotranspiration, E_{t}, by Haude's method is shown in Fig. 3.10, from which:

$$E=0.617\ E_{\mathrm{t}}+14.2$$

with a correlation coefficient of 0.871 (highly significant). here the intercept and the slope of the regression line (appreciably less than unity) reflect the fact that, under conditions of low evaporative demand, actual evapotranspiration tended to exceed potential evapotranspiration, whereas when evaporative demand was high the converse effect occurred. Furthermore, Eggelsmann's (1963) diagrams showing the march of evaporation for the eight-year period confirm that the winters were

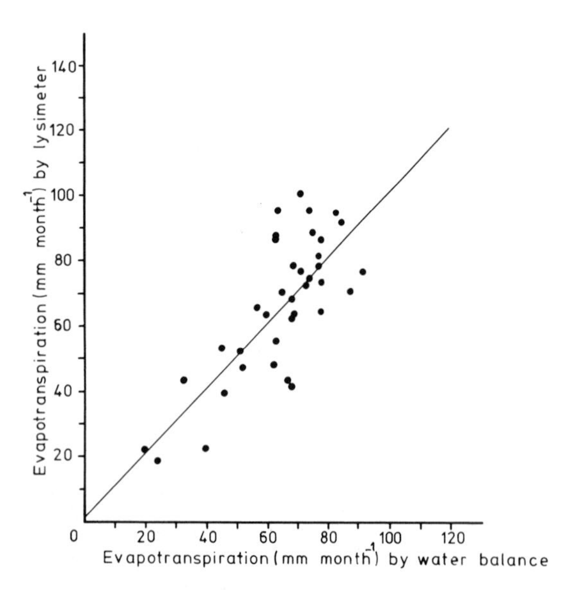

Fig. 3.9. Monthly totals of evaporation from a raised bog, obtained as residual term in the catchment water balance, compared with lysimeter estimates. Königsmoor, Lower Saxony, 1951–1958. (From Eggelsmann, 1963).

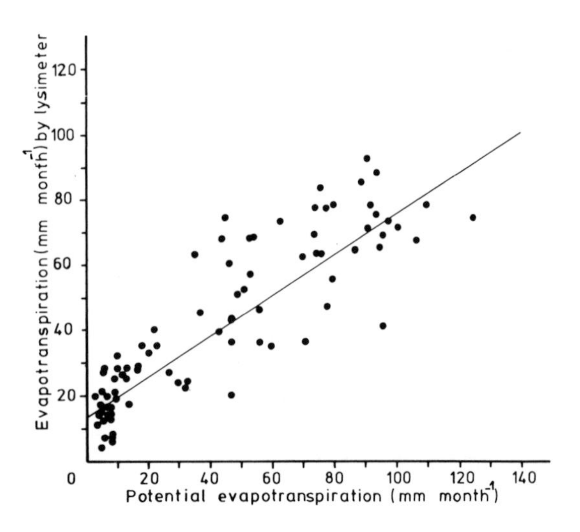

Fig. 3.10. Regression of lysimetric estimates of evapotranspiration on potential evapotranspiration determined by Haude's method. Monthly totals for the uncultivated expanse of the Königsmoor raised bog, Lower Saxony, 1951–1958. (From Eggelsmann, 1963.)

marked by periods of 3 to 7 months for which $E > E_t$, while for up to six months in summer $E < E_t$, with the quotient E/E_t falling to 0.43 in unusually dry spring weather. From the eight-year summary curve given by Eggelsmann (1964b) it appears that the average situation was as follows:

September to March: $E/E_t \simeq 1.5$
April to August: $E/E_t \simeq 0.9$
complete year: $E/E_t \simeq 1.0$

Eggelsmann (1964b, 1975) considered that evapotranspiration from *Sphagnetum* in northwestern Germany would exceed that from mire *Callunetum* by a factor which would raise the annual value of the quotient E/E_t to about 1.1 or greater. However, his data for *Sphagnetum* were estimates rather than lysimetric observations. Baden and Eggelsmann (1966) concluded generally that raised bogs display a higher rate of evapotranspiration than comparable catchment areas without mire development. They considered that low evapotranspiration from *Sphagnum* during dry summer weather was due to the drying out of the capitula in the absence of efficient upward transport of soil water, and that this explained the seasonal contrast with communities dominated by vascular plants.

Extensive research into mire hydrology has been carried out by the State Hydrological Institute in Leningrad. Regarding evaporation, detailed reports in English have been given by Romanov (1968a, b) and Bavina (1967). The approach was based upon a combination of lysimetry and micrometeorology.

Two models of lysimeter were used in these studies (Anonymous, 1961; Romanov, 1968a, b). The GGI B-1000 lysimeter resembled Eggelsmann's device. It embodied a cylindrical monolith of surface area 1000 cm^2. Storage was estimated to 0.2 mm by weighing, and the device was considered suitable for determining evapotranspiration totals for relatively short periods of the order of a day. In the GGI B-500 model, the surface area was 500 cm^2. The cylindrical monolith could not readily be removed from the mire for weighing. Instead a constant water table was maintained in the monolith. Because natural storage changes could therefore not be measured the B-500 model was only considered suitable for reckoning evapotranspiration over relatively long periods of the order of one month. The monolith cylinders were of galvanized iron in both models, and Romanov (1968a) calculated that, in the B-1000 model, errors due to temperature effects could increase evaporation by 0.1 mm h^{-1}. On the other hand, he expected that poor conditions for mineral nutrition and root aeration in the stagnant soil water of the isolated monolith would depress transpiration along with other physiological activity.

Romanov and his collaborators also made extensive use of the energy-balance method (eq. 3.11) as a micrometeorological technique for estimating evapotranspiration from mires. The net radiation, R_n, was estimated using a reversible pyranometer (Yanishevskiy, 1957: see Šesták et al., 1971) for the short-wave component and calculations based on air temperature, humidity, cloudiness and surface temperature for the long-wave component. Estimates of G, the ground heat flux, were based on profiles of soil temperature and on studies of soil thermal capacity using the loss of heat from electrically heated, buried spheres. The psychrometer heights recommended by Romanov (1968a) for determining the Bowen ratio, β, were rather large compared with those suggested by Slatyer and McIlroy (1961).

The Soviet studies were based on sets of observations carried out at four stations: Zelenogorsk near Leningrad, Tooma near Tartu in Estonia, Kemeri near Riga in Latvia and Krestunovo near Pinsk in Poles'ye. The first three were intact raised bogs. Krestunovo was a fen. All are located in the most westerly part of the Soviet Union.

Measurements using the B-1000 lysimeter in the central parts of raised bogs were reported for the season of plant growth (Bavina, 1967). The mean seasonal totals lay between 417 and 459 mm and varied only slightly with site and vegetation type. Of the evaporative flux, 70% occurred in May, June and July, 16% in August and 13 to 15% in September.

For the Krestunovo fen the mean evaporation for the five growing seasons 1961–65 was 594 mm, with 493 mm and 672 mm as extreme values. Compared with the central parts of raised bogs, the water table here fluctuated much more widely in level between +20 cm (flooded) at the start of the season and −70 cm or lower (below the base of the lysimeter monolith) towards its end.

For the remainder of the year the mire surfaces

was snow-covered. Bavina (1967) tentatively suggested a value of 25 mm for the annual evaporation from snow, plus an additional 21 to 26 mm to account for the snow-free periods outside the May to September season of plant growth.

One therefore arrives at the following average values for mire evaporation: bogs 416 to 510 mm; fens averaging 642 mm but liable to greater variation from year to year. Using Borisov's (1965) estimate of 340 to 475 mm potential evapotranspiration for the region, one obtains for bogs a ratio E/E_t of about 1.1 and an equivalent for fens of about 1.4. Bavina (1972), however, doubted whether such a difference exceeds the limits of computational error.

Romanov (1953) suggested a method of estimating evapotranspiration from mires, using the relationship:

$$E = \alpha R_n + c \tag{3.18}$$

in which α and c are empirically determined coefficients. The method is discussed in detail by Romanov (1956, 1968a, b) and Bavina (1967, 1972), and has been widely quoted (Kuzmin, 1972; Linacre, 1976; Goode et al., 1977). Since, however, the values of E used in computing α and c were themselves determined from energy-balance observations, rather than by independent methods such as lysimetry, they are largely governed by R_n. It is thus impossible to check the reliability of these estimates of E or to judge the worth of the maps of mire evapotranspiration "norms" given by Romanov and Bavina.

Komosse is a complex of mires of various types interspersed with moraines and lakes. It lies at an altitude of 320 to 350 m on the border between the provinces of Halland and Jönköping in southern Sweden (Osvald, 1923). It was chosen as a representative site for the International Hydrological Decade, and early results have been reported by Johansson (1974). Evapotranspiration was studied using lysimeters of the GGI B-1000 pattern, and compared with estimates of potential evaporation made with the Class A pan, pans of the Soviet GGI-3000 pattern (Konstantinov, 1966) and by the Penman method. Komosse is a fairly large mire complex, roughly 10 km in length and some 5 km maximum width, with a conserved area of about 24 km^2 of which 77% is covered with peat. Twelve open pan sites, two of which were also lysimeter stations, were laid out in order to estimate variability over the mire area. The ranges of total E_0 for the season May to September were 333 to 429 mm in 1972 and 409 to 490 mm in 1973. High values of E_0 were recorded at stations in the open, central parts of the mire, but values for the more sheltered stations in the western parts of the complex were mostly 10 to 20% lower. From Fig. 3.11 (for one of the central stations) it is estimated that E/E_0 for 1972 was 0.64 (Class A pan) or 0.85 (GGI-3000 pan). Compared with results from other mires these quotients are small and suggest E/E_t values of 0.8 to 0.9.

By contrast, some results obtained in the U.S.A. are remarkably high. Relatively little work on evapotranspiration from mires has been carried out in North America. An example is reported by Bay (1966a), for a forested, raised bog near Grand Rapids, Minnesota. Bay used a compact layer of highly humified herbaceous peat some 75 cm below

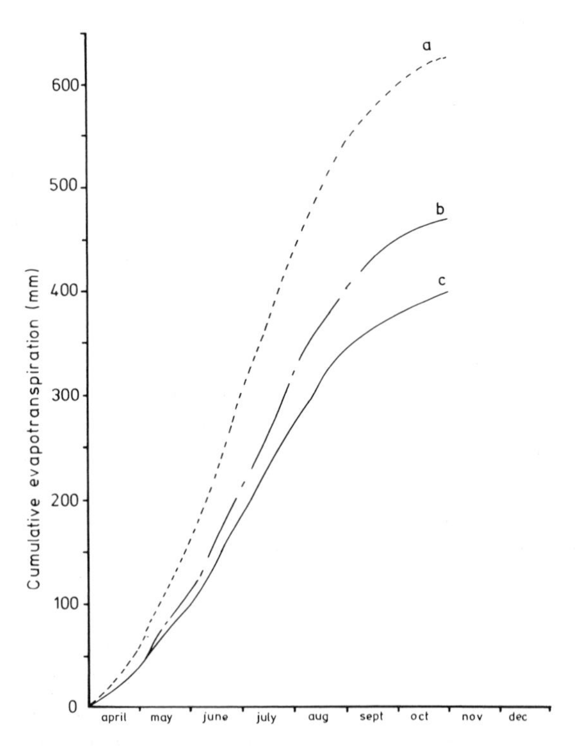

Fig. 3.11. Cumulative data on evapotranspiration from the centre of the Komosse mire complex, south Sweden, obtained during 1972: a = evaporation from USWB Class A evaporation pan; b = from sunken GGI 3000 pan; c = evapotranspiration estimated with a GGI 1000 lysimeter. (From Johansson, 1974.)

the surface as the impermeable floor of a bottomless lysimeter tank 3 m in diameter. This approach had the virtue of causing scarcely any disturbance to the soil profile during installation. Storage changes were estimated at weekly intervals. Changes in E followed changes in E_0, measured with the Class A pan, fairly closely. Over the period 13 July to 2 November 1964, E/E_0 was 1.13. Assuming E_0 estimates made with the Class A pan are similar to those made by Penman's method and applying the appropriate seasonal adjustments suggested by Penman (1956, 1963), one obtains $E/E_t = 1.45$. Bay maintained his internal water tables between 2.5 and 10 cm higher than those outside, and this may account for such a high value compared with data from European mires.

Some information on the effect of tree species on evapotranspiration comes from a pilot study by Heikurainen (1963) on mires with *Betula*, *Picea* and *Pinus* at Hyytiälä near Helsinki. The method adopted was akin to that of Bay (1966a, see above). It involved the use of 20 × 20 m sample lysimeter monoliths isolated from their surroundings by galvanized steel walls and lower peat layers of small permeability. Since the inner and outer water tables were closely similar, leakages of water were deemed to be slight.

Water tables were found to be very sensitive to transpiration, showing in some cases two periods of decline each day (Fig. 3.12), corresponding with the forenoon and afternoon periods of stomatal opening. Beneath *Betula*, these fluctuations ceased at leaf fall. Water tables rose in response to precipitation. The relationship between rainfall and water-table rise was used to determine a coefficient, which could then be used to convert water-table recession into estimates of evapotranspiration.

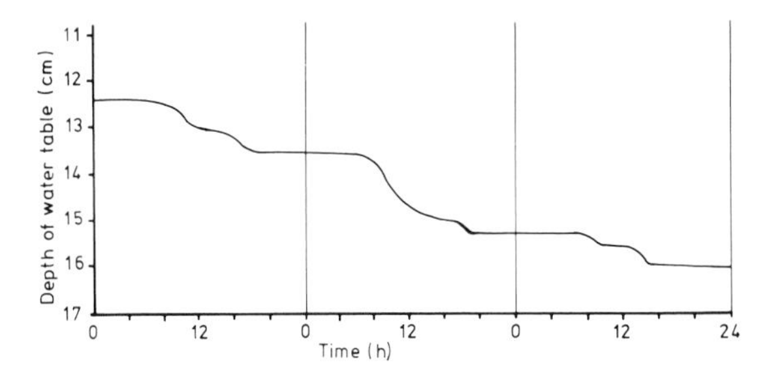

Fig. 3.12. March of declining depth of water table over a dry three-day period in a wooded mire near Helsinki, Finland, in July 1962. (After Heikurainen, 1963.)

Heikurainen concluded that *Betula* transpired as much as the conifers even though inferior both in standing crop and in growth increment. This conflicts with the evidence of P'yavchenko (1960) and Molchanov (1963a, b), who found appreciable differences between one tree species and another in the Soviet Union. Interception losses were not measured in Heikurainen's study.

Belotserkovskaya (1975) also found different rates of evapotranspiration from different types of forested mire near Tomichevo in Poles'ye, White Russia (Belorussiya). Using an energy-balance approach with improved net radiometers, an attempt was made to account separately for the contribution of different layers of vegetation. In *Sphagnum* communities with stunted *Pinus* the moss layer contributed about 30% of the total evapotranspiration, which in this case included interception loss. The march of daily moisture consumption for the growing season is shown for the two main types of forested mire in Table 3.3. Although the contribution of the moss layer in the *Pinus–Sphagnum* forest was so great, total evapotranspiration was higher in the *Betula–Vaccinium* forest because, while in leaf, *Betula* transpired almost twice as much water as the stunted *Pinus* (for the whole growing season 461 mm, compared with 264 mm).

Bay (1967a, b, 1968) also worked on forested bogs. His sites were situated near Grand Rapids in northern Minnesota, where the mires occurred in inorganic depressions. Here a ground layer of *Sphagnum* and other bryophytes, with a field layer of ericaceous dwarf shrubs, grew beneath almost pure stands of the markedly helophytic tree *Picea mariana*. Evaporation was estimated as the residual item in the catchment water balance, and compared with estimates of potential evapotranspiration made by the Thornthwaite method. Because of the complex geology of these catchments, estimates of mire evapotranspiration were available only for dry periods when ground-water recharge from mineral aquifers could be assumed negligible. During such periods, E always fell short of E_t, with E/E_t sometimes falling as low as 0.5. Bay ascribed the low rates of evapotranspiration to water-table recession beyond −30 cm (see pp. 94 and 106). Perhaps the significant facts here are, firstly, that interception losses do not occur in dry weather so that the effect must be entirely due to soil physical processes, and

TABLE 3.3

Monthly march of mean total daily evapotranspiration (mm) in mires near Tomichevo supporting 38-year tree stands, computed from their energy balance (after Belotserkovskaya, 1975)

Vegetation type	Depth of water table (cm)	May	June	July	August	September	October
Sphagnum with stunted *Pinus*	20–86	3.3	4.0	3.8	2.5	1.6	1.0
Betula with *Vaccinium*	41–149	3.4	4.3	4.1	3.0	1.9	0.9

secondly, that *Picea mariana* is known to develop a very shallow root system when it grows in peat liable to waterlogging (Heikurainen, 1964).

The effects of water table on evapotranspiration

This section illustrates the interactive nature of hydrological processes in mires and the difficulties which this creates in the measurement, for comparative purposes, of individual components of the water balance. Where water tables are adjusted experimentally, estimates of evapotranspiration suffer from an "oasis effect" if the areas involved are small, and the adjusted water tables are atypical of the surroundings, as may occur in lysimeters. Artificial effects apart, differences of E may be due to changes in the albedo (the reflection coefficient of the mire surface), to changes in length of the upward path of soil water, or to interception losses from the canopy, and all these are liable to vary depending on the time of year. Even if the experimenter is successful in attaining measures of these components, they still have to be related to large-scale variation in conditions in real mires before convincing comparisons can be made, for instance between bog and fen.

Bavina (1967) quoted data on the albedo (ρ in eq. 3.7) of various mire surfaces. For the *Sphagnum*–chamaephyte microtopes of the raised-bog expanse the average value during the growing season was 16%, there being a regular seasonal march from 12% in April to 20% in October at Tooma (see above). At Krestunovo the sedge–*Hypnum* microtopes generally gave values of 19 to 20% in mid-summer, with lower values early and late in the season — a similar pattern of seasonal variation to that found by Berglund and Mace (1972) for a *Sphagnum*–sedge–chamaephyte bog in Minnesota. As Bavina suggests, it seems probable that fen and especially bog vegetation reacts to lowering of the water table in summer by an alteration in the colour of the mosses as they dry out. In turn, this leads to a greater reflection of short-wave radiation so that the proportion of incident radiation available to cause evapotranspiration will diminish the more evaporative output exceeds water recharge. Negative feedback is therefore a feature of evapotranspiration in these communities. Indeed Williams (1970), quoting evapotranspiration measurements by unspecified techniques at the Mer Bleue bog near Ottawa, indicated that E with the water table at -30 cm was only 75% of its rate with the water table at the surface. The point was also discussed by Belotserkovskaya et al. (1969). On the basis of work on the energy balance of a string bog complex in the Lammin-Suo raised bog near Leningrad, they concluded that evaporation and evapotranspiration from pools exceeded evapotranspiration from ridges by 21 to 33% under average mid-summer conditions. In very dry years, however, hollows showed lower evapotranspiration than ridges because, under conditions of very low water table (-25 cm) the pool *Sphagnum* species (*S. balticum*, *S. cuspidatum*, *S. majus*) turned very white and developed a high albedo of about 27% compared with 17 to 19% for the albedo of *S. fuscum* and dwarf shrubs on the ridges.

A general study of the effects of water table on evapotranspiration was carried out by Virta (1966) for the months of June, July and August from 1959 to 1962. Three sites were involved: a string fen at Korvanen near Naarsaapa, in Finnish Lappland (see also Virta, 1960); a *Sphagnum papillosum* fen surrounding a rimpi fen at Mösky, near Pohjoisneva in Vaasa; and a raised bog at Loppi near Luutasuo in Häme. The last site is some 300 km north-northwest of the nearest Soviet site (Tooma). At Korvanen, the dominant plants were *Carex limosa*, *Menyanthes trifoliata*, *Calliergon stramineum* and *Drepanocladus* spp.; at Mösky, *Carex limosa*, *Rhynchospora alba*, *Sphagnum compactum*, *S. papillosum* and *S. recurvum* var. *tenue*; at Loppi, *Calluna vulgaris*, *Eriophorum vaginatum*, *Scirpus cespitosus*, *Sphagnum balticum*, *S cuspidatum*, *S. fuscum* and *S. capillifolium*. All sites were treeless.

Evapotranspiration was measured with cylindrical brass lysimeter tanks 50 cm deep and 120 cm in diameter (exposed surface area 1.13 m^2). No provision was made for measuring storage. Instead each tank was provided with an overflow at which all free water above a predetermined level was discharged into a collecting vessel, where the amount (D_m) could be measured each morning. The sample monolith was artificially recharged each evening with a known quantity (N_e) of water. Thus, by morning the quantity of water stored in the profile would be equal to that 24 h previously. It was assumed that the moisture profile in the zone above the water table was not disturbed by precipitation or evapotranspiration during the night. The basic water balance equation (3.5) could, for these experiments, therefore be simplified to:

$$P - E - (D_m - N_e) = 0 \qquad (3.19)$$

For the estimation of E_0, evaporation pans of diameter 120 cm and depth 25 cm were installed on the mire surface. Estimates of E_0 and E_t by various methods including Penman's were also made.

Virta's monthly results for the three sites are summarized in Table 3.4. [In this Table, Bavina's (1967) lysimeter data from Tooma are given for comparison. Exact agreement is not to be expected because of the differing ground-water regimes.] Virta thought that the somewhat greater values from the northern site (Korvanen) might have been due to abnormal weather. When results for lysimeters with the water table at −2 cm were compared with potential evapotranspiration, the value of E/E_t for Penman's E_t differed from unity by 5% at most, except at Mösky in 1961. However the natural water-table depths at Virta's sites were greater than this, and the season of measurement was very brief. It is thus probable that Virta's lysimeter results with water tables maintained at

TABLE 3.4

Evapotranspiration (mm) during the summer months measured with lysimeters at four mire stations in the eastern Baltic seaboard, with internal water tables maintained at various depths (*indicates that more than 10% of the value was estimated by comparison) (data for Tooma after Bavina, 1967; remaining data after Virta, 1966)

Station	Nature of site	Year	Depth of water table (cm)	Month			Total
				June	July	August	
Korvanen	spring	1959	2	138*	112	74	324
	fen	1960	2	105	136	63	304
Mösky	*Sphagnum*	1960	2	126	104	86	316
	papillosum	1961	2	92*	69	59*	220*
	fen		15	66	64	43	173
Loppi	raised	1962	2	104	84	56	244
	bog		4	77	69	46	192
			11	51	51	39	141
			16	57	60	40	157
Tooma	raised	1960	ambient	102	94	67	263
	bog with	1961	ambient	99	95	61	255
	Pinus	1962	ambient	84	85	82	251

-2 cm would be subject to an "oasis effect", and would therefore exceed the rate of evaporation from a mire where the water table happened naturally to be at that depth.

Evaporation values from lysimeters with the water table at -4, -11 and -16 cm were compared with the values at -2 cm. The results are shown in Fig. 3.13. They suggest an initially steep decline in evapotranspiration as the water table recedes towards -11 cm. Beyond this the rate levels off at 60% of that at -2 cm. No doubt part of the decline in rate can be ascribed to effects of desiccation on the albedo of mosses as mentioned above. However, as the water table recedes, the path length for capillary rise increases. At the same time the hydraulic conductivity declines as water is withdrawn from the larger pores. The position is greatly modified by the presence of higher plants, due to the enhanced exploitation of soil-moisture resources by their vascularized root systems.

Theoretical treatments of the effect of water-table depth on evaporation from bare soil have been attempted by Gardner (1958), Wind (1960) and Kastanek (1973), while Cowan (1965) has analyzed the effect of rooted vascular plants.

The curve in Fig. 3.13 appears to conform quite well to the shape predicted by Gardner's model. Similar results were obtained in other Finnish studies by Huikari (1959, 1963), who made use of a series of compartments in some mires at Liesneva in southern Finland which were being drained for forestry. The compartments were bounded by drains and were of different widths, so that during the summer the mean water-table depths varied from -7 to -31 cm. After allowing for groundwater recharge or discharge, evapotranspiration was estimated from the water balance as the difference between rainfall and run-off. In unplanted compartments where the vegetation was mainly *Sphagnum* or *Polytrichum*, the evapotranspiration declined from about 60% to under 5% of rainfall with increasing water-table depth (Table 3.5). The wooded compartments supported a standing crop of 45 m^3 ha^{-1}, mainly as *Pinus sylvestris* up to 8.6 m tall. Here a decline from 75 to 55% of rainfall was recorded as water-table depth increased. Although it is tempting to ascribe this reduced water-table effect to greater root development, it should be noted that interception loss is likely to account for a considerable part of the total evaporative flux from forests. This was clearly recognized by Huikari, but its measurement did not form part of his study.

Romanov (1968b) considered the relationship between evaporation and the water table in relation to the rooting habit of dwarf shrubs. His findings for raised bogs in the European U.S.S.R. are summarized in Fig. 3.14. He found that the roots of the shrubs penetrated to between -10 and -20 cm, and that the "capillary fringe" (saturated zone with negative water potential) was between 15 and 20 cm thick. He interpreted the abrupt decline in evaporation when the water table receded below -40 cm as the result of diminished supply to roots which had lost contact with the capillary fringe,

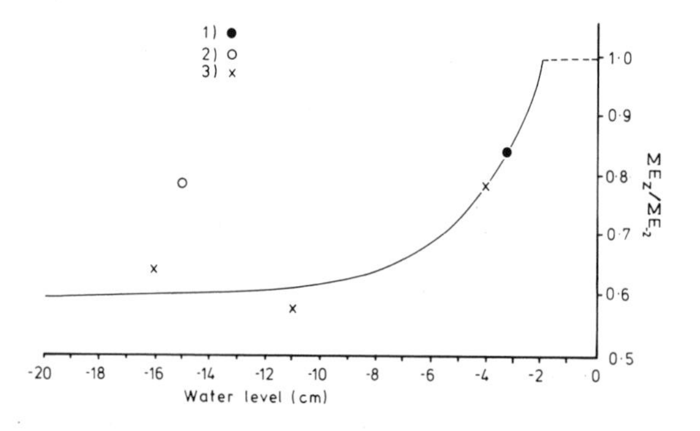

Fig. 3.13. Evaporation from Finnish mires, estimated with lysimeters during June, July and August, while water tables were maintained at various depths, z. Evaporation is expressed as a quotient of (total with water table at depth z)/(total at $z = -2$ cm) and compared with z. Legend: *1* = string fen in Lappland, 1959; *2* = *Sphagnum papillosum* fen in Vaasa; *3* = raised bog in southern Häme, 1962. (After Virta, 1966.)

TABLE 3.5

The effect of drain spacing on mean water-table depth and evapotranspiration between 30 June and 12 September, 1954 in a mire at Liesneva, Finland (after Huikari, 1959)

Distance between drains (m)	Mean depth of water-table (cm)	Evapotranspiration as fraction of precipitation (%)
100	17	65
80	7	57
60	9	58
40	11	53
20	19	37
10	29	12
5	29	2

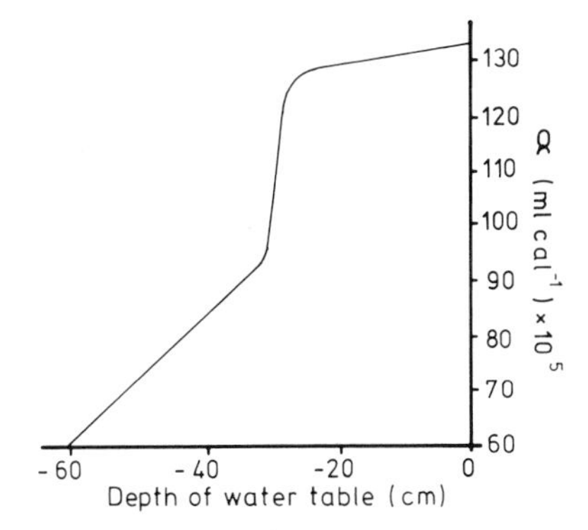

Fig. 3.14. Romanov's coefficient of evaporation, α, as affected by water-table depth during June, July and August. α is the slope of the curve of E vs R_n (eq. 3.18). (After Romanov, 1961.)

which seems to accord with the predictions of Cowan's model. The linear decline in evapotranspiration with water-table depth which Romanov found to be a feature of sedge fens was presumably due to deeper penetration by the annually renewed, aerenchymatous, adventitious root systems of tall graminoid helophytes.

Evapotranspiration from communities of tall helophytes (reed swamps, etc.)

Tall helophytes are conspicuous and ecologically significant components of many fresh-water ecosystems and often form important stages in the hydrosere successions which lead to the formation of mires (see Tallis, Ch. 9). Nevertheless they probably do not represent a very significant fraction of the mires of the world. Moreover, their aquatic habit and the water resources available to them imply that their hydrology differs in various respects from that of other mire communities, and is better considered as an aspect of limnology, which is treated in a companion volume in this series. For these reasons, and also because evaporation from these communities has recently been the main subject of a careful review by Linacre (1976), they will not be considered here in great detail.

The term "tall helophyte" implies a plant whose aerial parts exceed 1 m in stature and whose root systems are adapted to waterlogged conditions. Some of the plants defined in this way are woody, for example *Taxodium distichum* of subtropical North American swamps and *Pandanus helicopus* of the riverine swamps of lowland Malesia, but there is hardly any published information on evapotranspiration in these conditions. Most tall helophytes are monocotyledonous herbs. Important genera include *Arundo*, *Calamagrostis*, *Carex*, *Cladium*, *Cyperus*, *Juncus*, *Phalaris*, *Phragmites*, *Scirpus*, *Typha* and *Zizania*. They form a vegetation type, generally known as swamp or marsh, found near the shores of lakes or the banks of large rivers in soils subject to permanent or periodic inundation with shallow water.

One species, *Phragmites australis* (= *P. communis*) is almost cosmopolitan in distribution and of great economic importance. This species especially has been studied ecophysiologically (Rodewald-Rudescu, 1974) and information on water potential, stomatal behaviour and other aspects of its transpiration physiology has been collected (e.g. by Tuschl, 1970).

At the community level, Linacre (1976) in his review considered at some length the question of the effect of tall helophytes on lake evaporation, which is of great interest to water engineers. From considerations of the greater turbulence induced by the roughness of the vegetated surface and of the more rapid radiative heating of emergent plant parts compared with open water, it might be thought that tall helophytes would enhance evaporation from lakes. This conclusion appeared to be confirmed by the early work of Gel'bukh (1963, 1964) on *Phragmites* communities in the Ila Delta and in lakes in northern Kazakhstan, though Bernatowicz et al. (1976) have mentioned still earlier work by Novikova (1963) on sparse stands of *Phragmites* and *Typha angustifolia* in the Kengirdam Reservoir, also in Kazakhstan, but showing the converse effect.

On the other hand, tall helophytes have a much higher albedo than open water, especially in winter when their aerial parts are dead but still standing. For instance, Bray (1962) found albedos of 22% for the growing season and 34% for the whole year in a *Typha* marsh near Bethel, Minnesota. These results may be compared with figures of 5 to 10% quoted for open water by Rodda et al. (1976). It follows that more of the incident radiation should be retained by open water, while water surfaces beneath emergent vegetation will be screened from radiation and sheltered from atmospheric turbulence. That these effects are probably of decisive importance is suggested by the more recent study conducted by Sjeflo (1968; see also Eisenlohr, 1975) on the shallow ponds ("potholes"[1]) of the Coteau du Missouri plateau of North Dakota. While some ponds were clear of vegetation, others supported tall helophyte vegetation in which *Carex atherodes*, *Scirpus acutus*, *Scolochloa festucacea* and three species of *Typha* were especially important (Stewart and Kantrud, 1972). Using an improved technique based on the Dalton equation (3.8) and the water balance, Sjeflo concluded that evaporation from a clear pothole was about 11% greater than evapotranspiration from vegetated ones ($E/E_0 = 0.9$) for the year as a whole. Seasonally, at one site the ratio E/E_0 rose from 0.7 or 0.8 in May to maxima about 1.0 in mid-summer, before declining towards 0.5 in October in accordance with a steep fall in evapotranspiration.

One possible cause of this late-season decline may have been interruption of the transpiration stream with the onset of senescence and death in the autumn. This was also suggested by Šmíd (1975) who worked on vigorous stands of *Phragmites australis* in a fish pond in southern Moravia. Evapotranspiration was estimated by the energy-balance method and compared with lake evaporation measured with a floating pan. For a series of calm days with bright, anticyclonic weather the ratios E/E_0 were 1.0 (1 June), 1.9 (27 June), 1.4 (11 August), 0.9 (5 October). At the height of the growing season Bowen ratios (β, eqs. 3.11 and 3.12) approached zero and were taken to imply negligible moisture stress, no stomatal closure and no physiological control of evaporation — an interesting contrast to the situation in bogs. These results do not permit seasonal estimates to be made, but tend to make one cautious of accepting Linacre's (1976) conclusion that tall helophytes reduce lake evaporation.

It should be remembered that these swamps are azonal, and that stands of tall helophytes often occur in environments which are much drier than those surrounding bogs and many types of fen. Linacre (1976) has stressed the modifying effect upon evapotranspiration which dry weather may cause in these conditions, due to the advection of sensible heat. It should also be noted that the quotient E/E_0 will be greater when evaluated on a lake basis than on the basis of readings from a Class A pan.

In their evaporative behaviour, reed swamps and allied systems therefore differ profoundly from the majority of true mires. Although, taking the year as a whole, they may show E/E_t quotients (perhaps 1.3 for the prairie potholes) which fall between those of bogs and fens, this annual value conceals important contrasts. The negative feedback characteristic of bryophyte mires is absent here. Instead, at least in herbaceous swamps, the phenology of the helophytes curtails the input of heat by radiation in winter, while heat advection in summer may cause a temporary rise of E/E_t above 2.5 in arid surroundings.

[1]**Potholes**: semi-permanent shallow lakes with wetland development along the shoreline.

Concluding remarks on evapotranspiration

The lack of detailed agreement in the data presented in this section is not surprising in view of the variety of methods employed to measure both actual and potential evapotranspiration, and of the criticisms of this work that can be made. However, certain features of the process seem sufficiently well established to encourage tentative generalization.

Firstly, there is good evidence that rates of evapotranspiration depend on the vegetation, even within one type of mire, and this is especially clear where trees are present.

Secondly, it seems probable that the actual evapotranspiration from bogs is approximately equal to potential evapotranspiration, while on limited evidence that from fens is greater. Thus E/E_t for treeless bogs lies between 1.0 and 1.1, while for fens the quotient is about 1.4 or a little less.

Thirdly, on a seasonal basis and in relatively snow-free areas where the comparison is relevant, the ratio E/E_t undergoes an annual cycle on treeless bogs, with $E > E_t$ in winter and $E < E_t$ in summer. This can be ascribed to an element of negative feedback in the partition of energy at the surface, so that a greater proportion of the high incident radiation at mid-summer is dispersed as sensible rather than latent heat. This, in turn, is the outcome of three processes: increase in the albedo of the moss surface as it dries out due to lack of vascular tissue and inefficient water supply; stomatal control of transpiration by vascular plants, especially dwarf shrubs, when retreating water tables deprive their shallow root systems of moisture adequate to maintain high leaf water potentials; and soil-physical effects produced by declining water tables. It would help one to understand evapotranspiration from mires if more were known about the relative magnitudes of these three effects. The relationship between the climate, the vegetation and the soil is of great interest here, and the prospects for further analysis seem very promising.

Further it is clear that one is justified in regarding the tall helophyte swamps as special cases, which differ from mires in this as in so many aspects of their ecology.

For the future, promising lines of enquiry might well include: improvements in the design of lysimeters and precipitation gauges; combined studies by lysimetry and atmospheric physical approaches; ecophysiological studies of diffusion resistances in leaves, canopies and whole communities from a wider range of mire ecosystems, including the forested mires of the tropics and the aapa fens and palsa mires of the subarctic; evaluation of Soviet studies by increased effort in comparable climates elsewhere; more attention to mires in cool regions free from prolonged snow-lie and in parts of South America, Malesia, New Zealand and northern Canada where mires from which there are no data occur under very varied climates.

WATER STORAGE

General theoretical and practical considerations

It frequently happens that, over short and finite periods of time, the inputs of water to an ecosystem are not equal to the outputs. When this occurs, the principle of continuity is satisfied by changes in storage (ΔW in the water balance equation, eq. 3.2). Excessive input results in an increase in storage, while excessive output reflects a decrease in stored water.

Terrestrial ecosystems store water in a variety of states and locations. Water is detained as surface films on the wetted aerial organs of vegetation during the interception of rain, dew or fog. This is generally a very temporary form of storage. More long-lasting storage occurs when atmospheric moisture droplets form ice or rime on plant surfaces which have cooled by radiation on cloudless winter nights. In windy climates, the presence of vegetation is also associated with the lodgement of wind-driven or drifted snow as air currents are arrested by the plants. The amount of water stored in this way depends on the density of the vegetation and on the stature and habit of its component plants.

Surface storage of water occurs in hollows in uneven terrain. It may result from the presence of scarcely permeable soil layers near the surface, or it may reflect the emergence of a high water table. "Ponding" of water at the surface may also occur immediately after drought during which the drying surface soil layers develop very low hydraulic conductivities. Water is stored as snow on mire surfaces just as it is elsewhere. Some effects are considered below (pp. 140–145).

But the most important location for water storage in terrestrial ecosystems is generally the soil. In mires and other ecosystems with partially waterlogged soils there is an important hydraulic index known as the water table. This is often erroneously believed to mark the top of the zone of saturated soil pores. In fact, its position relates entirely to the free energy status or potential of the soil water, and only indirectly to the degree of saturation. The water table is the surface at which the hydrostatic pressure, h, of the soil water is zero (i.e., equal to atmospheric pressure). Below the water table h becomes positive, due to gravitational effects, and is measured as the rise in level in a tube or piezometer, open at both ends. Above the water table h values are negative and are associated with the interfacial forces of matric suction. Here, pressures are measured as the fall in water level in the open limb of a U-shaped tensiometer tube, closed at the other end by a porous plate across which hydrostatic equilibrium is established. The hydrostatic pressure is related to ϕ, the water potential, by:

$$\phi = h + z \tag{3.20}$$

where z is the distance, above an arbitrarily defined reference datum plane, of the point in the soil where ϕ is to be measured. Eq. 3.20 implies that ϕ is measured in the same units (generally cm) as h and z, which follows when potential is defined per unit weight of water. When it is defined per unit volume of water the units of ϕ are those of pressure (generally megapascals, MPa). At 20°C, 100 cm head of water is equivalent to 0.0098 ($\simeq$0.01) MPa.

The development of matric suction and of negative values of h is usually associated with the presence of unsaturated soil pores, but not invariably so. The exception is a region immediately above the water table where the pores are saturated even though h is below zero. This region is the "capillary fringe". It is the upper surface of the capillary fringe, rather than the water table, which therefore marks the limit of saturation in a soil. Capillary fringes between 20 and 40 cm thick have been reported in mires (Boelter, 1966, 1974b; Päivänen, 1973). For reclaimed mires the phenomenon has been exhaustively studied by Renger et al. (1976).

The volumetric water content, θ, of a given soil sample is expressed as water volume per unit volume of soil and attains its maximum value, θ_{sat}, when the sample is saturated and water has displaced air from all the pores. In the unsaturated part of a soil profile, variations in θ indicate the storage of varying amounts of water. The amount stored is related to the hydrostatic pressure or water potential. The relationship, which is curvilinear, is called the "moisture characteristic" and is determined for each soil sample in laboratory tests using tension tables, pressure plates or pressure membranes. That θ is not a single-valued function of h or ϕ will be apparent from Fig. 3.15. Because of the operation of matric forces at their narrow entrances, emptying of soil pores occurs at lower pressures than refilling. The result is a hysteresis loop, between θ values determined on drying the sample (decreasing h: upper curve), and those determined on wetting it (increasing h: lower curve). Being simpler to conduct, drying tests are more usually reported in the literature.

Recent years have seen the development of a variety of devices for measuring water potentials in unsaturated soils. For use above shallow water tables the most appropriate device under field conditions is the soil moisture tensiometer. This comprises a porous ceramic thimble or "filter candle" which is buried in the soil, filled with air-free water and attached to a manometer. Matric forces in the fine ceramic pores prevent the intrusion of air while allowing water to enter or leave. In this way the water in the thimble attains the same

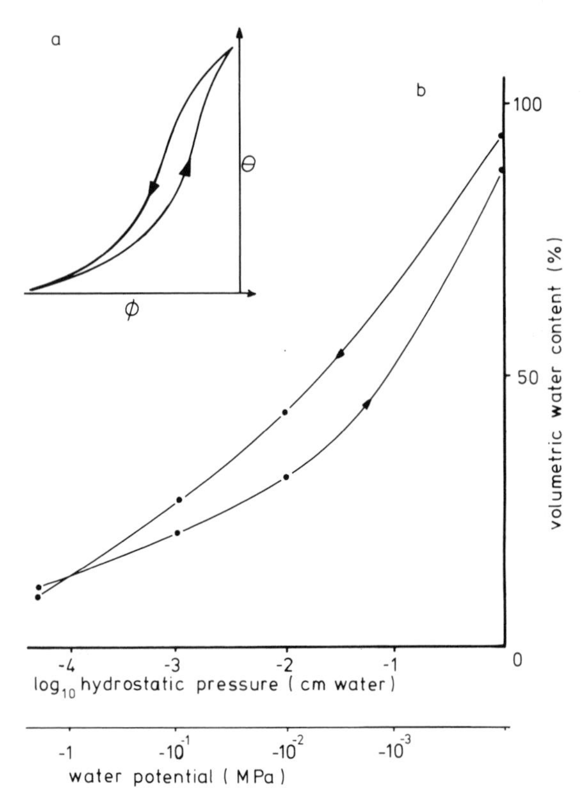

Fig. 3.15. Moisture characteristic curves, showing hysteresis: a, schematic; b, for a sample of raised bog peat (after Kuntze, 1965). In b the volumetric water content θ is shown plotted against matric suction (or negative hydrostatic pressure) h and water potential ϕ in the wetting mode (lower curve) and in the drying mode (upper curve). Note that the effect of hysteresis upon the results was most marked when this sample had a water potential in the vicinity of -0.01 MPa, corresponding to the situation in a surface layer in equilibrium with a water table at 100 cm depth. Matric suction experimentally imposed by centrifugation.

potential as that in the surrounding soil, which can therefore be read off directly from the manometer (see, for instance, Webster, 1966). Knowing the moisture characteristic of the soil, it is then theoretically possible to infer the volumetric water content. Indeed, this has often been done in mineral soils, and also proves possible in drained peat, as shown by Päivänen (1973), Ahti (1974) and by Mr D.G. Pyatt of the U.K. Forestry Commission (pers. comm., 1976). So far, however, it has not been successfully applied in intact mires (Boelter, 1974b), due to the difficulty of using conventional manometry at low pressures. However, Miss O.M. Bragg, based at the writer's laboratory, has recently made promising attempts to overcome this by replacing mercury by manometric fluids of lower density.

Field estimates of the soil moisture content may also be achieved by the neutron-scattering technique. This depends on the interaction between neutrons emitted by a buried radioactive source and the moveable protons (hydrogen nuclei) present in the soil moisture, and has been widely used as a technique for measuring water storage in mineral soils (for example Bell, 1973). Unfortunately, in peat soils the carbon skeletons of the solid matrix also contain high concentrations of fixed protons. Irwin (1966) described an early attempt to apply the technique in mires. He experienced difficulty in calibrating his instrument, but thought that the fixed proton problem might in time be overcome. Boelter (1974a) referred to similar early misfortunes when the allied technique of gamma ray attenuation was applied to peat, but recently this has been used as a non-destructive laboratory technique with notable success by Mr P.M. Hayward and Dr R.S. Clymo of Westfield College, London (Hayward, 1978).

The relationship between ϕ, h and z enables the moisture profile in an unsaturated monolith at equilibrium to be defined by the moisture characteristic, as shown in Fig. 3.16 in which the capillary fringe is also illustrated.

Hydrologists conventionally distinguish between ground water, which is water stored in the saturated zone, and soil moisture, which is water in the unsaturated zone. Some of the soil moisture occurs in the vapour state, and ice may be present at low temperatures.

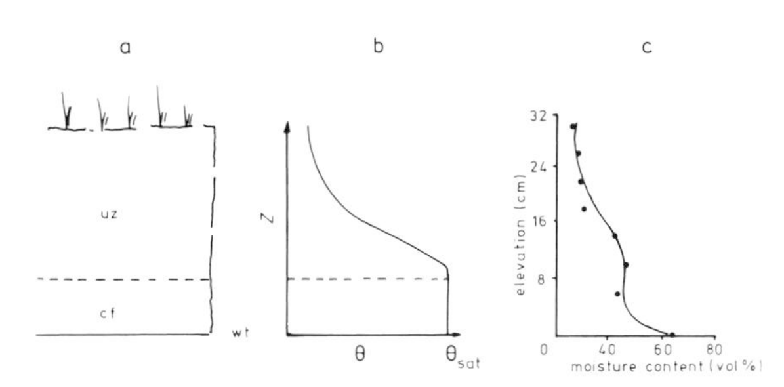

Fig. 3.16. Equilibrium soil moisture profiles. a. Schematic diagram of monolith showing relationship between unsaturated zone (*uz*), capillary fringe (*cf*) and water table (*wt*). b. Moisture profile associated with a as a plot of volumetric moisture content θ against elevation Z above an arbitrary datum plane. Since, at equilibrium, the water potential $\phi = -Z$, the moisture profile in the unsaturated zone has the same shape as the relevant portion of the moisture characteristic curve. c. Example of an equilibrium moisture profile, determined in the laboratory on a monolith from a ridge in the mire expanse of the Shiriniskoye raised bog near Leningrad, with its water table maintained at the datum level 32 cm below the mire surface. (After Romanov, 1961.)

More extensive treatment of the points discussed above will be found in textbooks dealing with soil physics (e.g., Rose, 1966; Luthin, 1966; Slatyer, 1967; Childs, 1969; Hillel, 1971; Marshall and Holmes, 1979).

Several authors have discussed the particular situation of soil water in peat and the forces affecting its energy status. Malmström (1939) proposed a classification of soil water in mires which took account of colloidal behaviour and other physico-chemical processes as well as matric and gravitational effects. The binding of water onto peat particles was further considered by Volarovich and Churaev (1966) and by Romanov (1968a), who recorded that, in the middle to late stages of humification, between 11 and 30% of the total water content might be bound by surface forces: more being bound in moderately than in highly humified peat. This could be due to an increase in the proportion of lignin (and other residues with hydrophobic properties) with advancing decomposition, as suggested by Kay and Goit (1977). After thermodynamic analysis these authors likened the water adsorption of peat to that of dispersible clay and humate extracts (sodium montmorillonite and sodium humate) suggesting that, in its reactivity to water, peat behaves like a colloid with a large specific surface. Specific surfaces of 1560 $cm^2\ g^{-1}$ were calculated by Romanov (1968a) for unhumified *Sphagnum–Eriophorum* peat, rising with humification to over 36 000 $cm^2\ g^{-1}$.

The physical behaviour of water in soils differs fundamentally according to its state. The foregoing remarks apply to liquid water stored in the soil. However, storage can also occur in the form of ice, and, in mires, this can lead to the development of highly anomalous structures known as palsas. These will be considered later.

Storage mediated by living plants

Since water accounts for most of the fresh biomass of the organisms in any community, it follows that some of the water stored by terrestrial ecosystems is to be found in the tissues of living animals and, much more importantly, plants. As mentioned above, most terrestrial mesophytes possess structural devices and physiological mechanisms which conserve and replace water, with the result that in these plants the water content generally varies mainly with the biomass of standing crop, but is relatively little affected by changes in water availability due to weather. Following the terminology of O. Stocker, Bannister (1976) describes these plants as being isohydric, in contrast to anisohydric forms which are subject to large fluctuations in their water content. As mentioned on pp. 82 to 86, the xeromorphism shown by many of the vascular plants of Holarctic mires may be accompanied by tolerance of rather widely fluctuating water content. In the ericoid dwarf shrubs the water content is low in any case, compared with that of mesophytes, as was shown by Stocker in 1923, but this is in diametric contrast to the situation in certain angiosperms from mires in the Southern Hemisphere.

In most New Zealand mires, *Sphagnum* is the dominant peat-forming genus, as in many other regions. In some bogs however other plants are more important. Of special interest is the monocotyledonous angiosperm family Restionaceae, which is mainly found in the drier regions of southern Africa and Australia. According to Cockayne (1967) and Hupkens van der Elst (1972), several restiad genera are widespread mire plants in humid areas in New Zealand. *Empodisma lateriflora* is perhaps the most typical species. Its role in the vegetation of certain raised bogs near Hamilton (North Island) has been described by Campbell (1964; and Part B, Ch. 5). In this area, which is mild but windy, *Empodisma* is associated with *Sporadanthus traversii* of the same family, and the two species are co-dominants of a mire expanse community in which bryophytes are unimportant. The root systems of *Empodisma* are very remarkable. Its primary roots are stout and grow horizontally outwards from the vertical rhizomes. They bear very numerous, fine, secondary roots, many of which grow upwards above the mire surface, especially round the bases of the sheltering *Sporadanthus* plants, the stem bases of which thus become invested with conical masses of material with a felted consistency, due to the dense covering of persistent root hairs found on all the *Empodisma* roots. Campbell describes these root masses as being sponge-like, with a water content up to fifteen times their dry weight. She considers them to play a part in the water economy of the ecosystem similar to that of the *Sphagnum* cushions of Holarctic raised bogs, although, on a dry-weight basis, the water content of this latter material is frequently much greater. In view of the anomalous botanical nature of the surface water reservoir in these bogs it is interesting to recall that, throughout the major portion of their geographical range, the Restionaceae are xerophytes (Engler and Gilg, 1912). Even the helophytes mentioned here mainly store water by matric forces developed in external capillaries beyond the influence of the epidermal apparatus of the plant.

Even in New Zealand, such anomalous water reservoirs in angiosperm helophytes do not seem to be very common. Generally, it is the living and dead tissues of bryophytes which comprise the bulk of the matrix in which variable water storage occurs. In contrast with the situation in vascular plants, the ecologically dominant gametophyte generation lacks both vascular tissues and root systems which replace losses efficiently, and an apparatus of cuticle and stomata which regulates and restricts evaporation. The external pathways of water movement which some mosses possess have been shown to be of little consequence in conditions of rapid potential evaporation (see above, pp. 78–82). Under such conditions, water evaporated from these plants is replaced only slowly. The bryophytes, therefore, undergo marked fluctuations in water content and this is especially significant in Holarctic mires, where they are often the dominant element in the standing crop. The reservoir which the bryophytes of mires south of the tundra zone provide probably varies more in its water content over short intervals than botanical reservoirs in any other ecosystem, because most of these plants belong to the genus *Sphagnum*. The special water storage properties conferred by the hyaline cells in the stem cortex and branch leaves of this genus are immediately apparent when a saturated handful is gathered and squeezed. Water then runs out as though from a sponge. The detailed anatomy of these plants is discussed above (pp. 78–80). According to Romanov (1968a), almost 100% of the volume of a hummock of *Sphagnum fuscum* comprises pores which might, in some conditions, become waterlogged. Normally, in damp conditions, only the intracellular pores would be filled in the upper parts of the hummock. These comprise roughly 20% of the total volume in the living stems just below the capitula, rising to over 70% as the external pores become compressed in dead material at depths of 40 cm. At these latter depths the external porosity approaches that of mineral soils, whereas close to the hummock surface the external porosity, at 70 to 80%, is exceedingly high, compared even with the best-structured mineral soils. This is due to the peculiar biologically produced pore architecture of *Sphagnum*. Moreover, the exceedingly sparing use of plant material in fabricating the solid matrix of the pores is reflected in very low bulk densities. Romanov (1968a) recorded values as low as 0.017 g cm^{-3} near the surfaces of hummocks of *Sphagnum fuscum*. From the practical view-point (Nys, 1954; Boelter and Blake, 1964; Boelter, 1970, 1976) it has been repeatedly emphasized that, when water contents of peat and peat-forming materials are discussed, the minute amount of solid matter is useless as a basis for expressing the results. These must accordingly be given per unit volume, and not per unit dry weight as in conventional soil-testing practice.

Hydrology and soil layers: acrotelm and catotelm

Expressing the results of water storage tests is only one of several interrelated problems which result from the unique structure and mode of formation of mire soils. All such soils have a very high organic content, often approaching 100% of dry matter. Sometimes new organic material is added in the form of dead plant litter, as in tall helophyte swamps (reed swamps in the broadest sense) and many fens but, in other fens, and especially in the vast *Sphagnum* bogs of the Holarctic, the new soil-forming material is alive and grows *in situ*. In these mires it is hydrologically meaningless to maintain any distinction between the living surface layer of *Sphagnum* and the dead material below. The part of the soil profile in which most storage changes occur contains both a living, growing layer of plants and a dead layer undergoing diagenesis by humification. With increasing depth, anaerobic conditions prevail and the process of diagenesis becomes imperceptibly slow. The anaerobic conditions are associated with permanent waterlogging and, consequently, with a constant water content.

The distinction between the upper, periodically aerated and partly living soil layer and the lower, anaerobic layer which is dead except for the aerenchymatous roots of helophytic angiosperms, is an important concept, certainly fundamental to any clear understanding of the hydrology and ecology of mires. Elsewhere (Ingram, 1978) it has been argued that the concept is also fundamental to mire pedology — at least to the pedology of intact mires. Its importance to mire hydrology concerns water transmission through peat (pp. 123–137 below) as well as water storage, and was first given explicit recognition by Soviet mire hydrologists (Ivanov, 1953). They use the terms "active layer" for the upper layer which is periodically aerated and "inert layer" for the lower, anaerobic layer. In English, however, the use of these terms leads to confusion: the first is pre-empted by students of permafrost to refer to a totally different situation, while the second creates a misleading impression of the part played by the lower layer in the overall water relations of mire ecosystems (see p. 149 below). As preferable alternatives, meaningless without definition, the terms **acrotelm** and **catotelm** have been suggested (Ingram, 1978). They are defined in Table 3.6, which is based mainly on the original Soviet definitions of Ivanov (1953) and on their subsequent amplification by Romanov (1968a).

Storage in the acrotelm

The collection of water-table data

The water tables in mires are generally located by "dip wells". These are unlined holes dug, augered or cored into the acrotelm and provided with a fixed index at the surface. The water level in the well is measured in relation to the index with a graduated tape, point gauge or hook gauge. If the index is located at the soil surface (the top of the acrotelm), the quantity being estimated is d, the local depth of the water table. For certain purposes it is more convenient to convert this to w, the local water stage (Fig. 3.17), which is the elevation of the water table above some arbitrarily chosen datum point such as the foot of a dip well (Virta, 1966). The various local datum points are generally chosen to lie on a surface which is parallel to, but slightly lower than, the mire surface. Its elevation accordingly varies from place to place on the mire, and at any given point it will have the value z with respect

TABLE 3.6

Soil layers in mires, based on the concepts of Ivanov (1953), Romanov (1968a) and Ingram (1978)

Position	Upper	Lower
English name	acrotelm	catotelm
Russian name	*deyatel'nyĭ sloĭ, aktivnyĭ sloĭ*	*inertnyĭ sloĭ*
Exchange of energy	rapid	slow
Exchange of matter	rapid	slow
Water table	present	absent
Water content (vs time)	variable	constant
Permeability (vs depth)	widely variable, highest at the surface	relatively constant, low
Water transmission	Darcyan	non-Darcyan
Aeration	periodically aerated	anaerobic
Activity of peat-forming aerobic microbes	high	nil (general level of microbial activity low)
Macroflora	matrix of living plant material	dead, except for a few roots
Upper surface	upward limit of matric forces	lower limit of rapidly variable characteristics

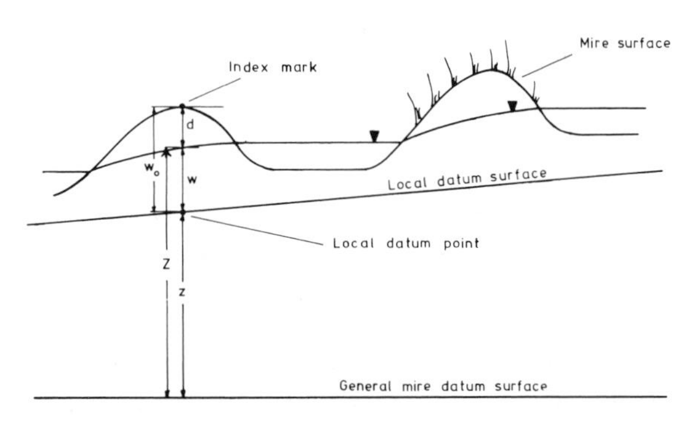

Fig. 3.17. A scheme of measured and derived quantities to specify the position of a mire water table in a sectional elevation. ▼ indicates the water surface in a flooded hollow or flark and the water table in a neighbouring hummock or ridge. (Modified after Virta, 1966.)

to the general datum plane for the mire as a whole. For hydraulic analyses involving comparisons of water table elevation at two or more points the elevations, Z, are computed with respect to this general datum plane. These relationships are shown in Fig. 3.17, whence it is apparent that, if w_0 is the stage corresponding to the mire surface,

$$w = w_0 - d \quad (3.21)$$

and

$$Z = w + z \quad (3.22)$$

Problems arise due to the reported tendency of the elevation, $w_0 + z$, of the mire surface to vary somewhat. This is a seasonal effect known in Germany as *Mooratmung* ("mire breathing"), first reported from Store Vildmose in Jutland by Prytz (1932) and later from the Ersterweger Dose (Uhden, 1967) and the Königsmoor (Baden and Eggelsmann, 1964), which are both raised bogs in Lower Saxony (Niedersachsen). The uncultivated surface of the Königsmoor showed an annual oscillation of 15 mm amplitude in 1956 and 30 mm in 1957. A detailed examination of such oscillations, based mainly on Soviet experience, was made by Aref'eva (1963), who considered how they might be allowed for during observations of the water stage in mires. While its effects have yet to be explored fully, the phenomenon may indicate changes in total soil storage capacity (see below) and it certainly necessitates caution in computing estimates of Z from those of d.

Other problems of a practical nature arise. If a dip well is lined to prevent its walls collapsing, the lining must be perforated or the well becomes a piezometer. As such it responds, not to fluctuations of the water table, but to variations in hydrostatic pressure at the lower end. Only when there is a complete absence of vertical water movement through the peat will the hydrostatic pressure, h, at this point be equal to w, the water stage with reference to the same point. Vertical water movements imply inequalities of potential with a vertical component, causing the piezometric level to vary with the depth of insertion of the tube, as reported by Rutter (1955) from wet heaths in southern England.

Due to the lower storage capacity of the drained soil compared with a neighbouring dip well, a given amount of water recharge or discharge causes a greater alteration in the elevation of the water table than in the well used to measure it. In practice this means that dip wells show time lags in their response to water-level movements. Since equilibrium takes longer to establish in a broad well, such wells yield the least reliable records and narrow wells are to be preferred even if they require perforated liners. (In many Holarctic bogs, liners are not essential in wells above about 5 cm radius unless recording is to continue for many years. Below this diameter, liners become increasingly necessary to locate the well and to prevent the water-surface detector becoming fouled with peat.) In broader wells the point of contact between a steel tape or point gauge and the water surface can readily be detected by eye at the water table depths commonly encountered in intact Holarctic mires. Narrower wells require a detector of some kind. Various electrical contact devices have been suggested for use in mires and one of these is in routine use in the Soviet Union (Anonymous, 1961). A reliable, simple and cheap method is to blow down a narrow tube as it is lowered down the well. As the end of the tube approaches the water level a rising note followed by bubbling is heard and in this way the position of the water level can readily be established to within 1 mm, in lined wells of the order of 1 cm diameter. In narrower wells than this, matric forces are liable to affect the result. The blow-pipe method works less easily in windy weather.

For the continuous recording of water-table fluctuation two approaches are available. The first uses a counter-weighted float and tape to draw a trace on a paper chart or to actuate a transducer and electronic data-logging system. Such recorders resemble those commonly employed in stream gauging, so several accurate and robust designs are available. Installations of this type are used in the Soviet Union (Anonymous, 1961). They have the advantage of being direct-reading, and the zeroing and calibration can be checked with auxiliary manual instruments in the field. Their chief disadvantage is that they require a well of large diameter (30 cm in the Soviet design) to accommodate the float, so their response is subject to time delay. The second type of recorder uses a flexible bulb, responsive to changes in the surrounding hydrostatic pressure (Toebes and Ouryvaev, 1970). Pressure changes in the bulb are transmitted either pneumatically or hydraulically to a metal bellows which actuates the recorder pen or transducer (Fig. 3.18). This arrangement can be robust and is reliable, provided the tube connecting the bulb to

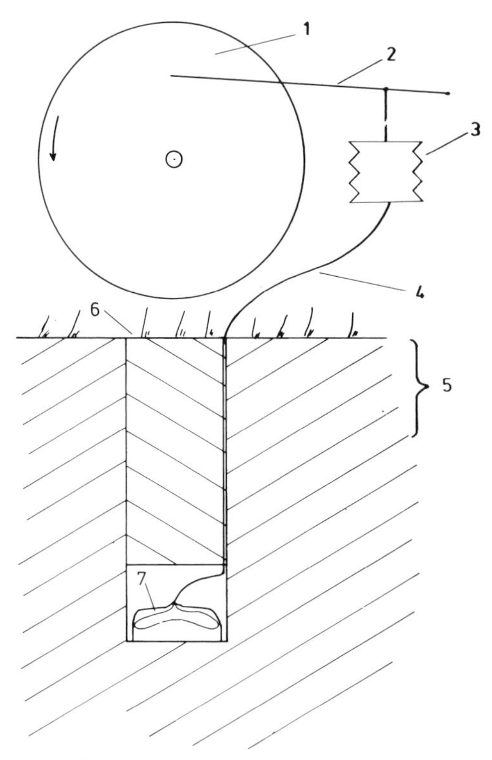

Fig. 3.18. Diagram of "Barnesbury" pressure bulb recorder (Negretti and Zambra Ltd.) installed to monitor water-table movements at Dun Moss. Legend: *1* = disc chart rotated weekly by spring-driven clock; *2* = pen arm carrying tube conveying ink to porous-tipped writing point from reservoir (not shown); *3* = metal bellows communicating pneumatically through *4* = a thin lead tube with *7* = rubber pressure bulb in inverted cup; *5* = range of variation in water stage; *6* = vegetated plug of peat of depth between 45 and 60 cm occupying upper part of well. (After Rycroft, 1971; and Mott, 1973.)

the recorder is kept as short as possible, otherwise it responds to ambient temperature. Its chief advantage is that, alone of all the methods discussed, the well can be eliminated if the bulb is placed in a cavity below the minimum anticipated water-table position. Response lags are then also eliminated, at least in theory. The disadvantage is that, without a well, zeroing and calibration cannot be checked in the field. Equipment supplied for monitoring liquid levels in storage tanks is commercially available.

Users of both types of recorder often forget that their response is liable to lag behind the changes being measured due to mechanical friction within the instrument. A plot of indicated water level against true level therefore shows hysteresis between the rising and falling limb of a complete cycle of oscillation. Such a plot (Fig. 3.19) is termed a Van der Casteel diagram. Each instrument should

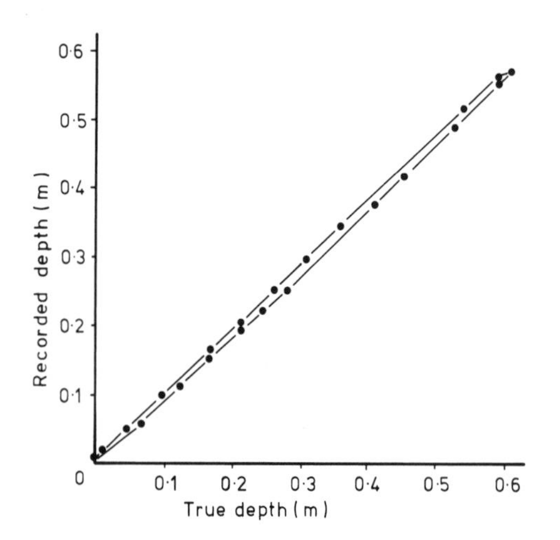

Fig. 3.19. Van der Casteel diagram constructed from laboratory tests on a pressure bulb stage recorder of the type shown in Fig. 3.18. Variable water depths imposed in a tank and measured with a micrometer point gauge. (After Mott, 1973.)

be tested in this way before installation, and periodically thereafter (Ven Te Chow, 1964).

Soviet workers (Anonymous, 1961) and many others (e.g. Virta, 1966) have noted the disturbance to water tables, and to water levels in wells, which results when observers walk on the mire surface near the observation point, especially in mires with soft surfaces. These undesirable effects are best eliminated by working from duckboarding on post piles. The official Soviet manual (Anonymous, 1961) suggests duckboarding at least 5 m long, on posts driven at least 1.5 m deep or, in shallow mires, driven through to the underlying mineral deposits. The duckboarding and its supports must not touch the well or the supports of the measuring or recording devices. In the absence of such facilities, Virta (1966) suggests strict adherence to a timed routine at each visit. Alternatively, Dr David Goode of the United Kingdom Nature Conservancy Council has devised a portable measuring instrument with electrical contact water-level detection that can be placed on the well, operated and read by an observer standing some distance away on the mire surface (Wadsworth, 1968, p. 271).

Modern developments in pneumatic and electronic pressure transduction seem not to have been applied to the study of mire water tables. Expense and lack of robustness in a harsh, wet environment may have been discouraging factors. Also such devices add little to the investigator's capability unless they can be combined, either with automatic data processing, with centralized and therefore of necessity telemetric logging, or with both. The remoteness of many mires adds to the problems of calibrating and maintaining these facilities. An early attempt at telemetry using electromechanical synchro-devices was described by Yakovlev (1969).

Water-table behaviour

Continuous records of water-table fluctuations have been published since the early days of modern systematic studies of mire ecology.

For fens, the early water-table studies of Godwin (1931, 1932) are still of considerable interest. They were conducted at Wicken Fen near Cambridge, on a surviving fragment of the fenlands of eastern England between the lower reaches of the rivers Witham and Ouse. The records were made by a float recorder in a well. They showed clearly the diurnal pattern of decline in water level associated with transpiration, the decline averaging between 3 and 4 cm per day in high summer (see p. 97).

They also showed the irregular pattern of rise associated with rainstorm recharge. Over most of the fen, precipitation appeared to be the main source of recharge water and since in June, July and August rainfall amounted to little more than 40% of estimated evapotranspiration, these months were marked by a steep decline in the water table. Only in the vicinity of the artificial drainage ditches and canals was the decline mitigated by ground-water recharge.

The high water tables generally met with in mires were noticed by their earliest students. Steele remarked in 1826 that "The surface of a peat-bog is ... even in the midst of summer, always wet and spongy". The detail revealed by studies like those of Godwin strongly supported this conclusion. At Wicken Fen the minimum recorded water table was at −48 cm, reached in the dry summer of 1929. But for the most part levels were high, generally between −20 and −4 cm from November to early June. Even the lowest levels, however, far exceed those occurring during the growing season in mineral soils of agricultural quality, where depths greater than 1.5 m are commonly met with (Eggelsmann, 1978). This comparison confirms the high water table as the single most distinctive hydrological and ecological feature of the mire landscape (see Fig. 3.20).

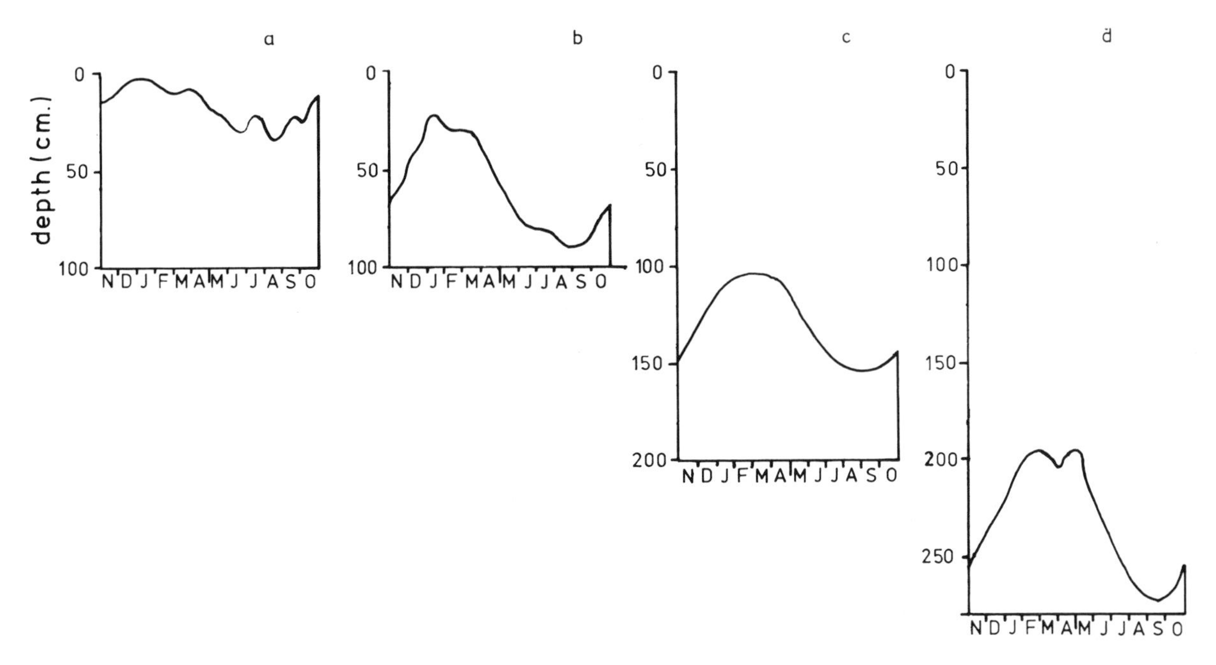

Fig. 3.20. Annual march of the water table during various periods of observation at four contrasting sites in north Germany. a. Pre-drained but unreclaimed part of the Königsmoor raised bog (1951–56). b. Drained and cultivated raised bog pasture grassland, Königsmoor (1951–56). c. Mixture of arable, woodland and pasture grassland on loams, sandy soils and peats at Fuhrberg near Hannover (Aller Basin) (1916–60). d. Mixed arable, woodland and pasture grassland on sandy loams and sands at Ummeln (upper Ems) (1921–40). (After Eggelsmann, 1971.)

Comparisons of maximum and minimum water-table depths may be significant for the ecology of helophytes and vital for the correct design of lysimeter tanks and other research equipment, but they yield at best a crude summary of this aspect of mire hydrology. By dividing the acrotelm into a series of layers of 5 or 10 cm depth, the total time for which the water table resides in each layer can be computed. This approach was adopted at the Königsmoor raised bog in Lower Saxony, where Baden and Eggelsmann (1964) used daily readings from dip wells to compile frequency diagrams which show at a glance the general character of the water-table regime, and cumulative duration curves from which the time spent by any part of the profile either above or below the water table can be read off (Fig. 3.21). The frequency diagrams are the basis of estimates of mean (time-weighted) water-table depth, while the duration curves enable modal depths (which are exceeded or not reached on equal numbers of days) to be read off. The annual frequency diagram for the uncultivated *Callunetum* is strongly skewed with its mode 8 cm above its mean, reflecting high winter water tables. On the other hand, the annual frequency diagram for the raised-bog grassland shows two peaks, one reflecting a winter mode, the other a summer mode 53 cm deeper. This summer mode is 8 cm below the corresponding mean, due to skewness and a predominance of water tables which were, relatively, very low.

From Table 3.7 the very marked effect of the agricultural treatment of the grassland on the water table is apparent. The grassland had been drained 39 years prior to the start of observations by means of tile drains at 20-m spacings, which lay at depths of 80 to 90 cm during the observation period. By this time the mean downward displacement of the water table was 31 cm in winter, 63 cm in summer and 50 cm for the year as a whole, a finding of great significance in discussions of mire conservation (see below) because of what it suggests about the effect of land reclamation on total storage capacity.

A similar though less detailed approach was adopted by Vidal and his colleagues to the analysis of ten years of water-table data from the Chiemsee mires (Schmeidl et al., 1970). In the damper climate of the Alpine foothills of Bavaria (Bayern) where

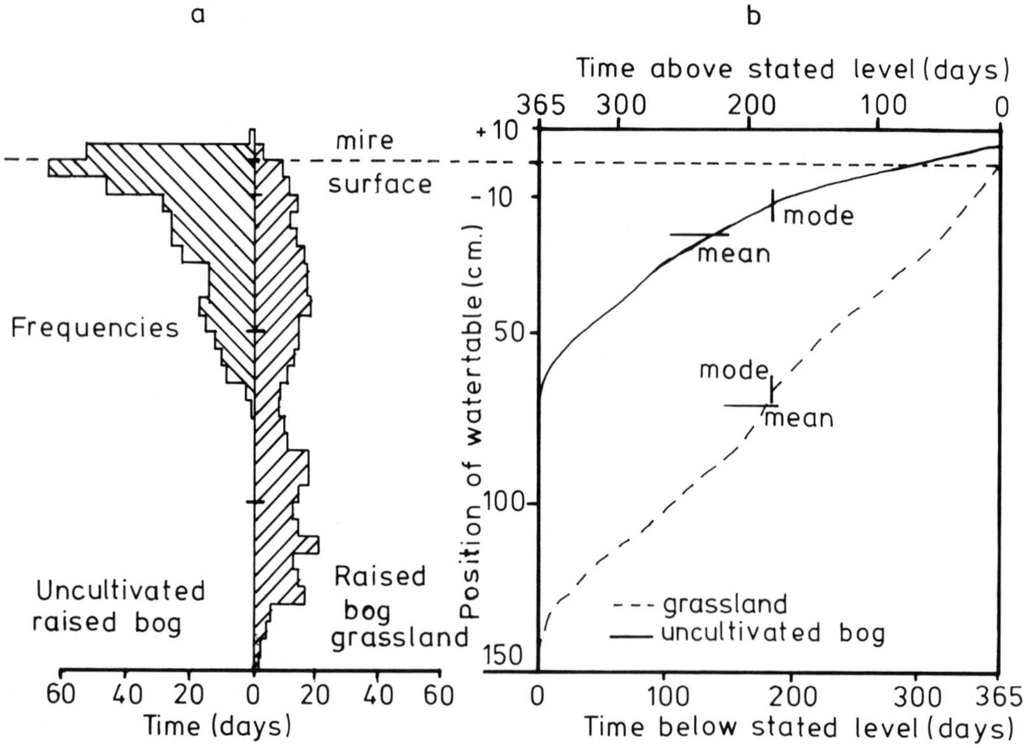

Fig. 3.21. Water table regime at the Königsmoor, Lower Saxony, during 1951–58, plotted (a) as frequency histograms of residence time in 5 cm depth zones and (b) as cumulative duration curves. (From Baden and Eggelsmann, 1964.)

TABLE 3.7

Statistical summary of water-table data from the Königsmoor for 1951–1958, taken from Baden and Eggelsmann (1964) and Fig. 3.21

	Elevation of water-table relative to mire surface (cm)	
	raised-bog *Callunetum*	raised-bog grassland
Annual mean	−21	−71
Annual mode	−13	−70
Winter (Nov.–Apr.) mean	−12	−43
Winter (Nov.–Apr.) mode	−5	−37
Summer (May–Oct.) mean	−29	−92
Summer (May–Oct.) mode	−25	−100

precipitation is roughly double that in Lower Saxony, water tables in raised bogs tend in general to be higher than those in the Königsmoor (Eggelsmann and Schuch, 1976). Nevertheless the same skewness towards high water tables is apparent in the frequency distribution from the unreclaimed parts of the Chiemsee mires, while data from reclaimed parts are more symmetrically distributed, again reflecting the action of drains in removing the excess when the water balance was in surplus.

Considering the general water balance (eq. 3.5) it should be possible to account for the behaviour of the mire water table in terms of water inputs and outputs if its position is an adequate indication of the amount of water stored. With this in mind, Schmeidl et al. (1970) attempted to examine the relationships between changes in water storage, precipitation and run-off by applying linear correlation techniques to the simultaneous monthly mean values. The clearest correlations between water stage and precipitation and between water storage and precipitation less run-off were obtained for the summer half-year (May–October). Those for the winter half-year were much weaker and indeed, when winter run-off was taken into account, the correlation coefficients were negative. However, this analysis showed several anomalous features which were apparent to the authors themselves, and it is worth noting that attempts to correlate data such as stage residence times are very likely to mislead unless the data are first transformed towards a more nearly normal distribution.

An alternative approach to the description and analysis of mire water-table behaviour is that of Mott (1973). It was based on a six-year series of records from seven pressure-bulb recorders installed in 1967 on various parts of the expanse of Dun Moss, a small raised bog in the foothills of the southeastern Scottish Grampians. Data from continuous chart logs were transcribed at synchronized six-hourly time intervals and filed in a digital computer for easy retrieval and analysis. The charts were read to the nearest 0.01 ft (approximately 3 mm). Smoothed curves of residence time versus depth are shown in Fig. 3.22 for two sites, one (Site 5) near the apex of the domed mire surface, the other (Site 2) near the crest of the marginal slope or rand. The curves are of similar form. They show a peak at or near the mire surface, indicating that the water tables resided near the bottoms of the hollows for about half the period studied. The local datum surface was chosen to follow the hollow bottoms, and at the site near the apex the peak of the curve coincided exactly with this datum ($w=0$). At the crest of the rand, the peak lay 0.08 ft below the datum ($w=-24.4$ mm) and was broader, suggesting that sloping sites experience slightly lower and more variable water-table regimes than more level sites. The remaining half of the data points formed a "tail", showing that in periods of water deficit the stage declined to $w=-0.85$ ft(-25.9 cm) at each site. Little difference between the "tails" is discernible, suggesting that evapotranspiration affected the water table similarly everywhere. There is much scope for further application of the techniques of time series analysis to the study of water tables and other aspects of mire hydrology.

Factors controlling water-table movement

From these studies it follows that water-table fluctuation in mires can be interpreted through the effect of their water balance on storage. This suggests that prediction of the water-table regime from water-balance data should be possible. The engineering problems presented by mires have stimulated several attempts to explore this possibility. Novikov (1964) described a procedure based upon a climatically determined "index of aridity" which could be computed from meteorological records. The index summarized the main variables affecting water input and evaporative discharge as a ratio of incident radiation to precipitation. The liquid discharge was computed from data on the

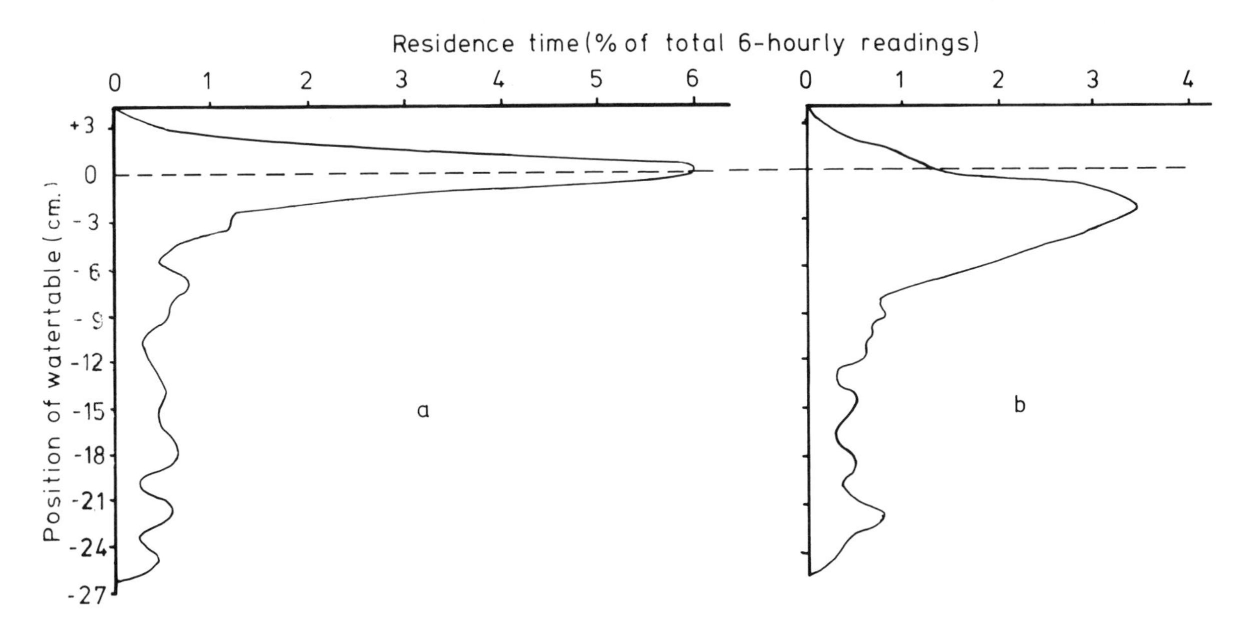

Fig. 3.22. Smoothed curves of residence time vs depth of water table at Dun Moss near Blairgowrie, Scotland, during 1970–72, based on six-hourly readings of charts from pressure bulb recorders. Local zero datum level at bottom of hollows in the mire expanse. a. From a site near the centre of the domed mire surface. b. From a site near the crest of the marginal slope. (After Mott, 1973.)

shape and size of the mire, its surface run-off characteristics and the depth and permeability of the acrotelm, all these parameters being estimated photogrammetrically or from maps, and from the known hydrological characteristics of the various microlandscapes previously studied empirically by K.E. Ivanov (1957). Starting from a known water stage, Novikov was then able to predict subsequent stages from the movements of the index of aridity, and his predictions and observations differed by more than 5 cm on only 4% of the test days. The problem of predicting water stages at the start of the growing season was also considered by K.E. Ivanov (1957). In mires in the forest zone of the U.S.S.R. the water input at this season is determined by the rate of snowmelt. Nekrasova (1973) tested Ivanov's method and again found good agreement between prediction and observation. In view of the great refinement in evaluating meteorological factors achieved by Penman, Budyko and others (pp. 74–75) and of improvements in techniques for predicting the liquid water discharges from mires (Anonymous, 1971; see below, pp. 133–134), further improvements in water-table prediction are certainly within our grasp. Meanwhile the success of these early attempts shows that most of the processes which control water-table movements have already been identified, at least in the typical raised-bog situation.

Nevertheless, some raised bogs show anomalous behaviour. An example was described by Bay (1966b, 1968, 1970, 1973) from Minnesota, where certain bogs have formed in depressions in permeable fluvio-glacial sands. In typical raised bogs the water table in the centre of the mire expanse is perched, sometimes several metres, above the regional water table, but in these the mire water table was continuous with that in the surrounding sands, and the surroundings clearly influenced the water regime. In the case described, the effect of groundwater recharge from the surroundings was to reduce the annual amplitude of water-table oscillation compared with normal raised bogs in the area. The richer vegetation and more mineralized ground water of these bogs suggests analogy with some of the transition mires of Europe (Kulczyński, 1949), so it might be better not to regard them as raised bogs in the strict sense.

In fens, the movements of the water table are complicated by the water supply. Ombrogenous bogs are fed by precipitation alone, but the ecological character of a fen is dominated by an influx of ground water, often mineral-enriched by contact with surrounding rocks and soils. The effect of this was clearly apparent in the early work at Wicken (see above). Its consequence is that water-table behaviour in fens is not dependent on meteorological variables and mire parameters alone, but also on the hydrological nature of the surrounding catchment. These additional influences make predictive generalization difficult, largely because these catchments are so variable in their topographical and geological details. Based on extensive studies of the Poles'ye mires, Kulczyński concluded that the main contrast was between large catchments with long detention times and small catchments with short detention times. The former contained fens with slowly oscillating water tables remaining at high levels for long periods. The latter produced the opposite effect: their fens had rapidly oscillating water tables with long periods of low water stage. The problem was later studied by Bavina (1966), working in the same area.

A discussion of water recharge in fens will be found above, pp. 70–72.

The water table in relation to mire vegetation

Although the work of Mott (1973) described above showed differences in water-table behaviour between the mire expanse and the sloping rand of a raised bog, the variations of stage (w, Fig. 3.17) were small in comparison to the variation in local datum level (z) which, for the whole raised bog which Mott studied, exceeded 8 m. Findings of this kind are widespread. However, they do not invalidate the general conclusion that, in consequence of the shallowness of the acrotelm and of water tables in most intact mires, the water table resembles the mire surface in its overall shape.

Nevertheless, small differences in stage occur consistently, season by season, in different parts of mires — sometimes, as in the blanket bog studied by Boatman and Tomlinson (1973), with little apparent reference to the shape of the surface. Such differences are of great ecological significance and appear to be concerned in many of the contrasts in vegetation which occur from place to place in the same mire. These contrasts are considered more fully elsewhere (e.g. Crawford, Ch. 7; Tallis, Ch. 9; Speight and Blackith, Ch. 10; Sjörs, Part B, Ch. 3;

Botch and Masing, Part B, Ch. 4; Junk, Part B, Ch. 9). Here it is only necessary to indicate some underlying hydrological patterns.

In raised bogs, the deeper water tables beneath the rand are a consequence of the greater slope, the steeper hydraulic gradient and the more rapid drainage in that part of the mire (pp. 135–136 below). Conversely, Kulczyński (1949) has pointed out that the phenomenon of *Mooratmung* referred to earlier has the effect of reducing both the depth and the oscillation of water tables in the mire expanse. Thus, it happens that dwarf shrubs which perform better with good drainage predominate on the rand, while the mire expanse is the typical site of the regeneration complex, in which the hummock-and-hollow-forming *Sphagnum* species can co-exist in the presence of high water tables throughout most of the year, but in the absence of the complete inundation to which the marginal fens are periodically subject. Water tables in floating mire mats (*Schwingmoor*) may show marked self-regulation, becoming almost static, with respect to the living plants, towards the free edge of the mat (Buell and Buell, 1941; Vitt and Slack, 1975).

The detailed conformation of the water table in the mire expanse is in itself of considerable interest. Some authors (e.g. Moore and Bellamy, 1973) have suggested that matric forces within the hummocks draw up the water table so that its microrelief is a subdued image of the mosaic of hummocks and hollows. This view is based on a misinterpretation of the processes involved. Matric forces have no direct effect upon the water table. In the static situation its position is determined by gravity. When water moves, the water table also moves to satisfy the storage requirements of the local water balance. Where the acrotelm has high permeability the water-table movements are homogeneous and it remains flat; but where the permeability of the acrotelm is low, and fluxes associated with interception or evaporation differ between hummocks and hollows, an undulating water table results. Recent work at Dun Moss (Scotland) (Bragg, 1978) has revealed some of the details of this complex and fascinating situation. Flat water tables were associated with the highly permeable *Sphagnum magellanicum* (see p. 133 below), but the permeability of *S. capillifolium* was lower, and during the warm part of the year water tables beneath hummocks of this species were nearly always concave, depressed below those in neighbouring hollows. Our earlier studies had shown summer depressions of water tables of the order of 5 cm beneath tall hummocks of *S. fuscum*.

Numerous studies indicate that the distribution of vegetation in mires depends not only on the mean water stage but also on the fluctuations of the water table (Sjörs, 1948; Kulczyński, 1949; Ratcliffe and Walker, 1958; Romanova, 1961; Malmer, 1962a; Dierschke, 1969; Balátová-Tuláčková, 1972; Rybníček, 1974; Bell and Tallis, 1974, etc.) It has been suggested (Ingram, 1967) that water-table movements affect root aeration and the mineral nutrition of plants, while Nye and Tinker (1977) have argued that it is the oscillation of water tables, rather than high water tables in themselves, which adversely affect the performance of mesophytes.

The water table and water storage: specific yield

Chapman (1964) has pointed out that, in mires of the *Schwingmoor* type, with floating rafts of peat, water storage is unrelated to the local water stage measured from the peat surface. Otherwise the amount of water stored in the mire acrotelm depends on the depth of the water table. Eggelsmann (1957) first showed how close this relationship can be in intact mires with water tables shallower than 150 cm. His work was confirmed by Heikurainen et al. (1964), but conflicts with Ahti's (1974) results for drained peat in summer. The available storage space is related to the porosity of the acrotelm, but the relationship is too complex to be of great practical or quantitative use. Regions of the acrotelm with a high proportion of large, readily drainable pores give up water more readily than regions with a low total porosity, or a high proportion of small pores in which water is strongly retained by matric forces. It follows that the capacity of any part of the acrotelm to serve as a temporary storage reservoir is related, not only to its porosity (proportion of total volume occupied by "voids"), but also to the pore size distribution and pore architecture, which determine the proportion of pores which drain at a given (negative) water potential. Only in the coarsest of mineral materials, in which the capillary fringe is of negligible thickness, would the notion of all pores emptying as the water table passes down through them be a valid approximation. In other media, including peats, which tend to possess very high total porosities exceeding 80% of the bulk volume (Boelter, 1974a), the pore architecture may be so complex that there are too many variables for an analytical approach to have much practical significance. Accordingly attempts to quantify the relation between storage and the water table generally involve recourse to the concept of specific yield which is expressed as an empirically determined parameter. This may conveniently be thought of as the equivalent, for a profile with a water table, of the well-known concept of field capacity for a well-drained soil with a water table too deep to be of consequence. Much discussion of specific yield arises from ground-water engineering practice in aquifers below the soil. Mire water tables are so shallow that here the concept has immediate ecological interest.

The specific yield, Y, also known as the aquifer yield (Marshall and Holmes, 1979), is commonly defined as the discharge of ground water which takes place when the water table is lowered through unit distance. If for simplicity it is supposed that there is a complete absence of influx and of pipe, surface or channel efflux, the terms, P, U_i, V, Q, C and E in the water balance (eq. 3.2) equate to zero and one may write:

$$U_e = \Delta W_s + \Delta W_u \qquad (3.23)$$

where W_s, W_u represent water stored above and below the water table respectively, signifying that ground-water discharge U_e is a distraint upon storage in the soil profile only; thus:

$$Y = \frac{\Delta W_s + \Delta W_u}{\Delta w} \qquad (3.24)$$

and, if quantities of water are measured in equivalent depths and in the same units as the water stage, w, then Y will be a dimensionless quantity with a value less than unity.

The specification of storage in terms of two components emphasizes that water may be stored in the soil profile both above the water table and below it. If the discharge of stored water involved the latter component only, there would be no difficulty, since all pores would drain completely as ϕ becomes negative and Y would be numerically equal to θ_{sat}. The problem in interpreting specific yield lies in the fact that as the water table falls and ϕ becomes negative the drainage of pores is commonly incomplete. The discharge of water in the capillary fringe and unsaturated zone has been discussed by Childs (1969). When the water-table movement is relatively fast, as in some of the well-pumping tests which are conventionally used to estimate specific yield, the results for a given soil sample may vary widely, and the same variability may arise when the water table is close to the surface, as so often in mires. Implicit in the above definition of specific yield is the notion that the water given up by the profile is only that which drains away under gravity, so that the yield concerned represents the difference between the water content at saturation, θ_{sat}, and at field capacity, θ_{fc}. Such a definition may provide adequate insight into some processes, such as those connecting mire water tables with the water balance in winter. However, they may be quite irrelevant when the point in question is the effect of water balance on water tables in summer, when evapotranspiration is a major item in the water balance, depleting the moisture content in the upper part of the profile well below field capacity.

Water storage in various peats from Minnesota was tested in the laboratory by Boelter (1964, 1970, 1974a), using samples removed from the field with minimal disturbance. Boelter assumed that lowering of the water table generally decreased the water potential of the horizon in question from zero to -0.01 MPa ($=-0.1$ bar or pF2, the potential or tension usually accepted as corresponding to field capacity). Using tensiometers in Finnish peats, Ahti (1974) showed that, even after drainage, the water potential near the surface did not fall below -0.02 MPa, so Boelter's assumption appears justified as a useful working hypothesis. Boelter found that the "water yield coefficient" (Y) of living *Sphagnum*, defined in this way, was 0.85, diminishing with depth, compression and humification to values appreciably below 0.5, and even to 0.08 in highly humified material [see further comments by Boelter (1976) and similar work by Galvin (1976)]. Boelter (1969) quoted the following ranges of values of Y for peats with different gravimetric content of tissue fragments (fibres) larger than 0.1 mm across: fibric (fibres $>67\%$, slightly humidified): $Y>0.42$; hemic (fibres 33–67%, moderately humified): $Y=0.15$–0.42; sapric (fibres $<33\%$, highly humified): $Y<0.15$.

Päivänen (1973) conducted a similar though more detailed series of tests of the water-retaining capacity of the peat from various types of mire in central Finland. He used the standard methods of soil-physical laboratories (tension cells; pressure plate and pressure membrane apparatus) to follow the relationship between declining water potential and water content in samples from depths of 10 to 15 cm in intact and drained sites. He also found that water yield coefficients diminished rapidly with increasing humification, which was in turn closely correlated with bulk density, thus:

Bulk density ($g\ cm^{-3}$)	Approximate humification (Von Post)	*Y*
0.05	H1	0.60
0.10	H4	0.36
0.15	H6	0.22
0.20	H9	0.18

Where intact mires are concerned, interest centres on the yield of water from weakly humified peat near the surface. Table 3.8 summarizes results calculated from various pertinent investigations, with the emphasis on unreclaimed sites, from which water yield coefficients for peat, as distinct from live *Sphagnum*, vary between about 0.2 and 0.7, raised-bog peats generally showing higher values than fen peats. Galvin's low value of 0.14 for Irish blanket peat is interesting. Presumably it reflects the highly humified nature of much peat of this kind.

From the standpoint of ecosystem hydrology the results of this approach to specific yield must be interpreted with caution. The water content of peat at $\phi = -0.01$ MPa relates to a matric suction of 100 cm water. However, when the water table subsides through a given layer of the acrotelm the potential does not immediately decline to this value. In theory, only when the water table has fallen 100 cm below the layer in question would its water potential fall to -0.01 MPa. In practice, an evaporative demand at the surface complicates the situation, so that a linear relationship between

TABLE 3.8

Specific yields of surface layers in different types of mire, based on the laboratory data of various authors

Locality	Material	Humification	Specific yield (*Y*)	Authority for original data
Raised bog				
Swan River, Minn., U.S.A.	*Sphagnum*	slight	0.60	Boelter (1972)
Hyytiälä, Finland	*Eriophorum–Sphagnum*	H2	>0.5	Heikurainen (1963)
Hyytiälä, Finland	*Sphagnum*	H2	0.83–0.56	Heikurainen (1963)
Ireland	*Sphagnum*	H2	0.42	Galvin (1976)
Transition mire (fen→bog)				
Grand Rapids, Minn., U.S.A.	*Sphagnum*–chamaephyte	slight	0.57	Boelter (1972)
Fen				
Poland	small sedge–moss	slight	0.24	Okruszko and Szuniewicz (1975)
Poland	tall sedge	moderate	0.16	Okruszko and Szuniewicz (1975)
Blanket bog				
Ireland	?	H8	0.14	Galvin (1976)

Except for those based on Heikurainen (1963), whose results were obtained by testing the effect of artificial recharge and discharge upon water stage, the estimates were obtained by subtracting the volumetric water content at $\phi = 0.01$ MPa from that at $\phi = 0$, and therefore represent the yield of the surface layer while the water table declines from the surface to a depth of 100 cm (Boelter, 1970) water potential and distance to the water table is often not maintained (Kuntze, 1974; Pyatetsky, 1974). Using tensiometers, Päivänen (1973) showed that, in a chamaephyte-dominated mire with *Pinus* that was undergoing drainage, the water potential at 5 to 10 cm declined linearly with the water table only to 62 cm, after which water potential fell much more steeply (Fig. 3.23). It is therefore clear that data on water-yield coefficient are chiefly of value in the notions of variability and relative magnitude which they afford in comparing different samples. In general they seriously over-estimate the specific yield itself. Indeed, it should be emphasized that the data from which Table 3.8 was compiled are derived from the extensive literature on peat moisture characteristics; that most of this literature refers to reclaimed mires; and that the original authors generally collected their data with other ends in view.

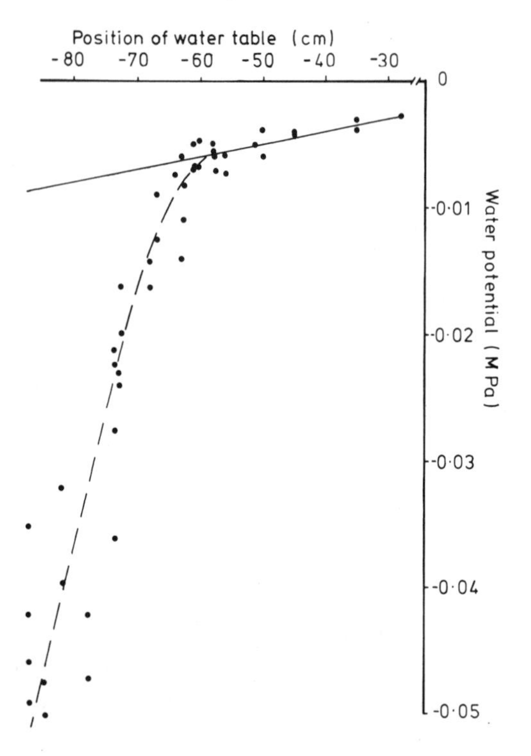

Fig. 3.23. Relationship between water potential and position of the water table in the *Eriophorum–Calluna* peat of a chamaephyte-dominated drained mire with *Pinus sylvestris* in the south Finnish province of Häme. Water potentials were determined close to the surface with tensiometers operating from 5 to 10 cm depth. The unbroken curve shows the theoretical linear relationship between water potential and depth to water table. The broken curve was fitted to the data by hand. (After Päivänen, 1973.)

The most successful attempt to link laboratory studies of soil moisture retention with the specific yield is probably still that of Vorob'ev (1963). It concerned the very extensive tracts of mire in the lowlands near Barabinsk in western Siberia which are mostly fens, though interrupted by small islets of raised bog known as *ryams*. Vorob'ev first used "capillarimeters" (Anonymous, 1961; Romanov, 1953, 1968a) to determine the moisture characteristics of samples from successive layers, 5 cm thick, throughout the deposit. He then computed the response of each layer to the lowering of the water table in a sequence of 5-cm steps, by summing the water lost at each step. Summation over all the layers above the water table gave the total loss per unit fall in the water table — in other words, the specific yield. The results were predictably dependent on humification and botanical composition. For *Hypnum*-sedge microlandscapes the specific yield of the acrotelm had a maximum value in the region of 0.34, decreasing fairly rapidly to 0.15 at 30 cm and below, where sedge remains were dominant. For graminoid microlandscapes, dominated mainly by sedges, rushes and tall grasses, the smallest values of about 0.12 were obtained near the surface, while most of the upper 50 cm of profile gave values about 0.15. The method was not applied to the *ryams*. Vorob'ev's results confirm the over-estimation inherent in cruder interpretations of laboratory data. They are a little higher than those which he obtained by field methods (see below).

Several workers have used field methods to estimate specific yields in mires. Their strategy has generally involved following the reaction of the water table to precipitation input from storms of known intensity, while the value of other items in the water balance was known or could be inferred. Prytz (1932) made an early attempt to determine specific yields for a raised bog in Jutland by relating water-table movements to precipitation less evapotranspiration, but this approach is hampered by the difficulty of determining evapotranspiration (see above, section on evaporation). An alternative approach is to suppose that evapotranspiration is negligible while it is raining, so that specific yields in the acrotelm can be computed from the reaction of the water table to precipitation during a series of individual storms. This method was first advocated by K.E. Ivanov (1957) and later applied by several investigators in Finland, the Soviet Union and the United States. It avoids some of the difficulties in interpreting laboratory studies, but cannot conveniently be applied unless enough suitable rainstorms are experienced during the period of field observation and, less objectionably for the present purpose, it often does not permit investigation below the acrotelm, which may be desirable in planning drainage works. If the water table is deep the specific yield is liable to be overestimated due to detention of water in the surface layers to form a moisture profile with only tenuous contact with the capillary fringe, so that less water reaches the water table than might otherwise be the case, especially in a light storm.

Nevertheless, in cases where it has been possible to derive estimates of specific yield from both field and laboratory measurements (Vorob'ev, 1963; Bay, 1968) similar sites have tended to give similar results, and the work of Päivänen (1973) suggests that hysteresis may account for some at least of the recorded discrepancies.

However when the water table is shallow the water which infiltrates after rain may only be detained very briefly. This was suggested by the laboratory experiments of Heikuranen et al. (1964) who followed the changes in weight of the top 10 cm of a woody sedge–*Sphagnum* profile (humification H3 on the Von Post scale) to which the equivalent of 10 mm of rainfall had been applied from above. With the water table at a depth of 20 cm, half of the added water had left the top layer within the first 54 min, and it took only 16 h for 90% of the added water to disperse.

Vorob'ev (1963) maintained that, in the presence of interception by vegetation, large amounts of precipitation are necessary if specific yield estimates with small errors are required, but pointed out that rainfall of low intensity and long duration is partly dispersed as run-off. Only 15 to 20% of the storms recorded in his study proved suitable. Vorob'ev's field method is explained in Fig. 3.24. He divided the acrotelm into a series of layers, supposing that in each layer the change in water stage, Δw, would be a linear function of the infiltrating water if the layer was thin enough, a supposition subsequently confirmed by Heikurainen et al. (1964; see below). He further supposed that the infiltrating water was the precipitation, P, less W_d, the water detained in

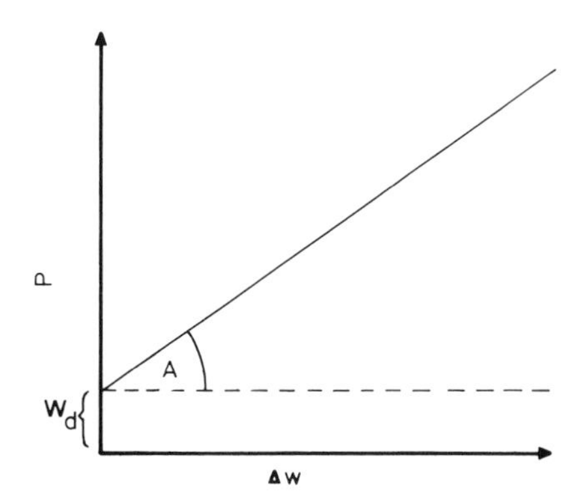

Fig. 3.24. Diagram to illustrate Vorob'ev's (1963) field method for determining the specific yield of the individual soil layers in a mire. See text for explanation.

the overlying layers, and that for any layer the value of W_d remained constant between storms. (It may be noted that, for the topmost layers with shallow water tables, this supposition appears to conflict with the findings of Heikurainen et al. noted above). Repeated rises in the water table in reaction to successive storms permitted a linear plot of P against Δw to be obtained for each layer, and the value of W_d (in reality an average value) was read off as the intercept on the ordinate. Thus detention could be eliminated from the calculation and specific yield computed as:

$$Y = \tan A \tag{3.25}$$

Vorob'ev's field results for the specific yield of fens in the lowlands near Barabinsk are shown plotted against depth in Fig. 3.25, and for bogs near Barabinsk and in Karelia in Fig. 3.26. At the surface of the fens highly humified hypnoid moss peats showed the lowest yields (about 0.3), while those of the graminoid and rush peats were somewhat higher (about 0.4). In the *Sphagnum*–shrub microlandscapes of the raised bogs the specific yields of the surface layers were still higher (about 0.45) according to the data obtained from the Lammin-Suo raised bog in the Karelian isthmus near the borders of Russia and southern Finland. In all mires, the yield values at greater depths converged towards 0.08 to 0.10, although the pattern of variability was similar to that revealed in the corresponding laboratory tests described above. This may have been due to the combined effect of interception losses in light rainstorms and surface run-off in heavy storms.

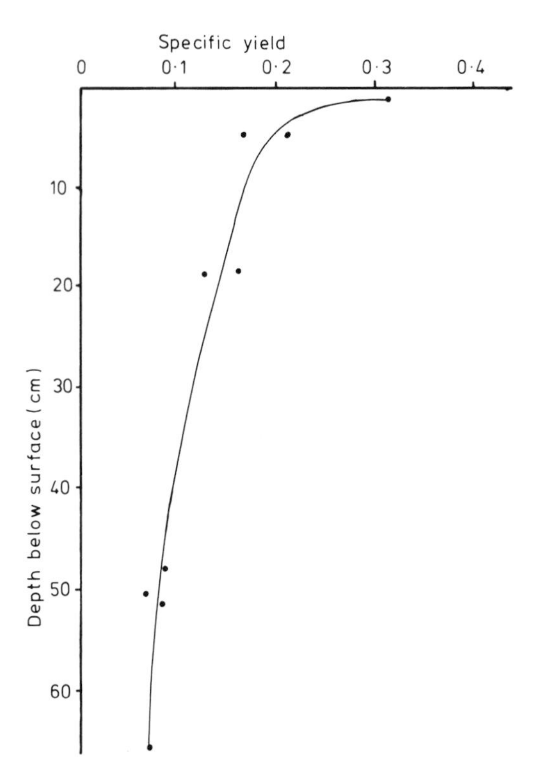

Fig. 3.25. Specific yield in relation to depth in hypnoid-moss peats from the Tarmanskoye Fen in the lowlands near Barabinsk (southern part of the West Siberian depression). (Field data of Vorob'ev, 1963.)

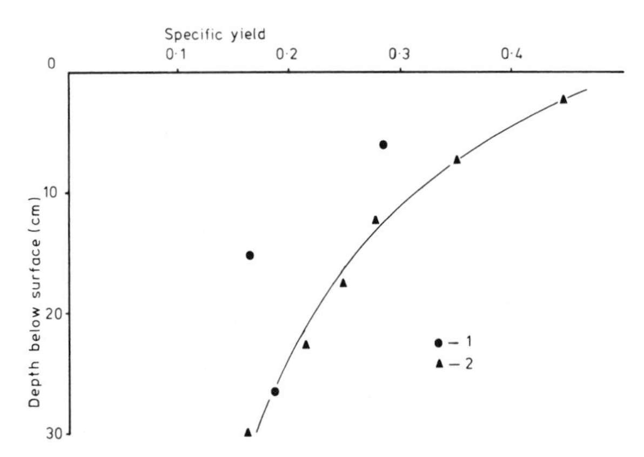

Fig. 3.26. Specific yield in relation to depth in Soviet raised bogs. Legend: *1*=for a "ryam" in western Siberia; *2*=for the Lammin-Suo bog in Karelia. (Field data of Vorob'ev, 1963.)

Heikurainen (1963) used rainstorm records and water-table observations to compute "coefficients of groundwater", *F*, where $F=1/Y$. Using 20×20 m plots, separated from their surroundings by sheet steel piling to prevent run-off (see p. 93), he determined values for *Y* ranging from 0.2 to 0.4 for sites in central Finland, but gave no details of the nature of the peat. In an elaboration of this work, Heikurainen et al. (1964), working on woody sedge–*Sphagnum* peat with humification H3 to H6, used cylindrical samples replaced *in situ* after being divided into three layers (0–15, 15–25 and 25–40 cm) by horizontal mesh screens. The samples could be weighed periodically and their weights compared with the water stage measured in a dip well in the vicinity. Within each layer, water content and stage showed a close linear correlation (coefficient seldom <0.85, generally >0.9), but here again the values of the specific yield changed with depth, being greatest (about 0.4) for the topmost layer and least (about 0.2) for the deepest. Moreover, the relationship between specific yield and depth appeared to be approximately linear. Bearing in mind the presence of moderately humified peat in the top layer these results are in good accord with Päivänen's (1973) subsequent laboratory determinations for peat from the same area.

The most promising field approach to the determination of storage and specific yield is the weighable lysimeter, which enables water stages and the weights of stored water to be simultaneously determined in the same monolith at any instant without any requirement for measurements of precipitation or evaporation. Although advocated by Heikurainen and Laine (1974), the potential benefits of this method have so far not been fully exploited.

Storage in the catotelm

If the catotelm remains saturated, as it should in an intact mire, it might be expected to store a quantity of water which, while it would increase exceedingly slowly with peat accumulation, would remain sensibly constant over any normal period of study. This, however, may well be an oversimplified view. From the work of Heikurainen et al. (1964) with three layers of peat (see above) there was some evidence that water content below the water table was not constant, but that the water content of the bottom layer continued to increase as the water table rose, even when that layer was completely submerged. On the hypothesis of a linear variation in specific yield with depth, the water table would continue to influence the water content of submerged peats down to a depth of 53 cm. This finding is of interest in view of more recent evidence that water storage in the catotelm may indeed depend upon the local hydrostatic pressure (p. 126 below). This evidence came to light during work on the anomalies which arise when seepage tubes are used to test the permeability of the catotelm. As described later, there are features of the time course of these tests in humified peat which are most readily explained by supposing that water enters or leaves the experimental cavity, not only as a result of flow down potential gradients in the surrounding peat, but also, and perhaps more rapidly, in response to pressure changes in the cavity. Thus if, for instance, the test is set up by an initial downward displacement of water in the tube, this discharges part of the water stored in the vicinity of the cavity. Recharge takes place if water is subsequently poured back into the tube, when the gradient-induced flux and the recharge flux are in opposite directions. The nature of this "isopinal" storage is not yet understood. Since peat is not a rigid material, elasticity or plasticity of the aquifer cannot be ruled out and may also be implicated in the phenomenon of *Mooratmung* (see above).

It therefore seems probable that short-term changes in catotelm water storage may take place. However the phenomenon has scarcely been recognized as yet, so it is at present impossible to assess its importance and the mechanism is not at all understood.

Any such short-term changes would be superimposed upon the slow upward trend in storage as the overall size of the catotelm increases through time. The rate of bog growth is reported to be of the order of 1 mm yr^{-1} (for general discussion see Clymo, Ch. 4; Tallis, Ch. 9). Even if most of this is water, its effect on the annual water budget is, in practical terms, negligible.

On the other hand, the water content of the catotelm is known to decrease with increasing humification of peats of widely differing botanical origin (Grebenshchikova, 1956, 1957; Yamazaki et al., 1957; Romanov, 1968a). No doubt this results from the increasingly close packing of solid residues as decomposition proceeds. Enhanced by the effects of increasing overburden pressure, the decrease in water content of the lower layers of peat in a developing mire might eventually completely offset the increase in water content near the surface, causing the long-term trend in overall water content to reach a maximum.

Conclusions: the general characteristics of water storage in mires

Most of the water in mire ecosystems is stored beneath the ground in the catotelm. Despite recent evidence of storage changes related to hydrostatic pressure, it seems probable that the water content of the catotelm remains sensibly constant over long periods of time. Most of the changes in water storage in mires accompany water-table movements in the acrotelm. But the acrotelm of an intact mire is generally only a thin layer of soil. While it may be freely-draining at the top, especially in the interstices between the organs of mosses and other living plants, many of the most characteristic mire plants have morphological peculiarities which promote moisture retention close to the upper surface of the acrotelm. Furthermore, decay within the acrotelm rapidly increases moisture retention towards its lower surface. According to Eggelsmann (1971), the storage changes accompanying the movement of water tables generally amount to no more than 3 to 10% by volume. Even if one accepts the wider range indicated by the laboratory tests summarized in Table 3.8, the very thinness of the acrotelm to which the water-table movement is confined implies that most mires have a limited capacity either to store or to release water in the short term. A comparison of the overall hydrology of mires with catchments without mire development (Fig. 3.20) showed that the unreclaimed part of the Königsmoor raised bog had the smallest short-term storage capacity (36 mm), even though it had once undergone preliminary drainage. By contrast, some catchments with mineral soils in Germany possessed a short-term storage capacity amounting to 96 mm. Drained parts of the Königsmoor resembled mineral-soil catchments in their storage behaviour.

The low capacity of intact mires for the temporary storage of water is confirmed by studies on many different sites, including forested mires, in many countries (for example, Nys, 1957; Ferda and Pasák, 1969; Schmeidl et al., 1970; Boelter and Verry, 1977). The idea that intact mires act as useful water storage reservoirs seems first to have been put forward by Alexander von Humbolt (1769–1859) and is sometimes still advanced as an argument for their conservation (e.g. Crawford, 1969). In its simplest form this argument is clearly wrong: drained mires are better reservoirs and mineral-soil ecosystems are better still. The case for mire conservation should involve more subtle hydrological arguments concerning, among other matters, water quality; the effects of afforestation, grassland improvement or peat extraction on items of the water balance; and the long-term consequences of oxidation and wastage of drained areas of mire. Parts of this problem are closely involved with run-off from mires and will be considered below; but before turning to this aspect of mire hydrology an anomalous form of storage remains to be discussed.

Anomalous water storage: palsas

While most of the ecologically significant water storage in mires occurs in the liquid state, storage in the form of ice is claimed to be important in the formation and growth of one particular mire feature, namely the palsa. In the Northern Hemisphere, palsa mires are found in the southern part of the permafrost zone, where frozen subsoil is discontinuous. The palsas themselves are peat-covered mounds which generally support oligotrophic vegetation. According to Moore and Bellamy (1973) the mounds may be of great size, averaging 3 m in height and 20 to 100 m in breadth and length. Plateau-like eminences can arise through the coalescence of adjacent palsas. A review of the fairly extensive European and North American literature prior to 1974 was published by Brown (1977).

Juvenile palsas appear in shallow mire pools beneath which no permafrost is initially present. It is thought that these pools freeze to the bottom in winter, and that the resultant mechanical stresses are relieved by local upheavals of the saturated peat forming the floor of the pool, accompanied, it is suggested, by the development of an ice lens between the peat and the mineral substrate. On emerging from the pool, the peaty bottom materials dry out by evaporation in summer. Their insulating effect then causes the ice lens to persist. The mound therefore also persists, although its peat is no thicker than that of the surrounding pond floor and it is supported entirely by ice. In autumn re-wetting of the peat followed by freezing causes its thermal conductivity to rise. The mound therefore cools more rapidly than its surroundings and this effect is later enhanced by its elevated surface, exposure to wind and thinner snow cover. The ice lens then grows by the centripetal migration of moisture through the now unsaturated peat (see p. 140) down gradients of thermodynamic potential, and the palsa enlarges as increasing amounts of ice are stored within it. It appears that, although *Sphagnum*, chamaephytes and even trees become established on older mounds, the formation of peat contributes little or nothing to their growth. Indeed, oxidation and wastage of the insulating peat layer eventually exposes the ice. Thawing then sets in and the palsa collapses. The physics of palsas is difficult to study due to the fragility of the terrain, and undoubtedly more details of their development remain to be discovered. For the moment they appear to represent a singular type of medium-term water storage with extraordinary ecological consequences.

LIQUID WATER DISCHARGE

The efflux or output of water by evapotranspiration was considered in detail above (section on evaporation), where the discharge of vapour to the atmosphere above a mire was discussed. All the remaining effluxes are discharges of water in the liquid state. In this section one is concerned not only with discharge across the mire boundary into the surrounding landscape but also with processes occurring within the mire which regulate the flow towards the boundary.

Classification of liquid discharge processes

As indicated above (pp. 69–70), the movement of water through mire ecosystems is best analyzed as a combination of several simultaneous but distinct processes. In classifying these processes one must aim to group them in a way which gives the clearest understanding of their total result. Generally this means separating processes which can be recognized as having distinct physical characteristics. In this way one can draw on the whole experience of geophysics and bring this efficiently to bear upon the solution of problems in mire ecology. The need for a careful classification is nowhere greater than in this instance, for the various processes involved in liquid discharge from mires are not well documented in the general hydrological and ecological literature, and yet many of them are individually quite well understood by geophysicists.

The scheme of classification adopted here is based on that of K.E. Ivanov (1957). Ivanov's later treatment (1975a) is referred to only briefly here, and the emphasis differs from his. For instance, my experience leads me to think that seepage in the catotelm may be more significant than Soviet workers would generally concede. It would be possible to devise a flow classification based on

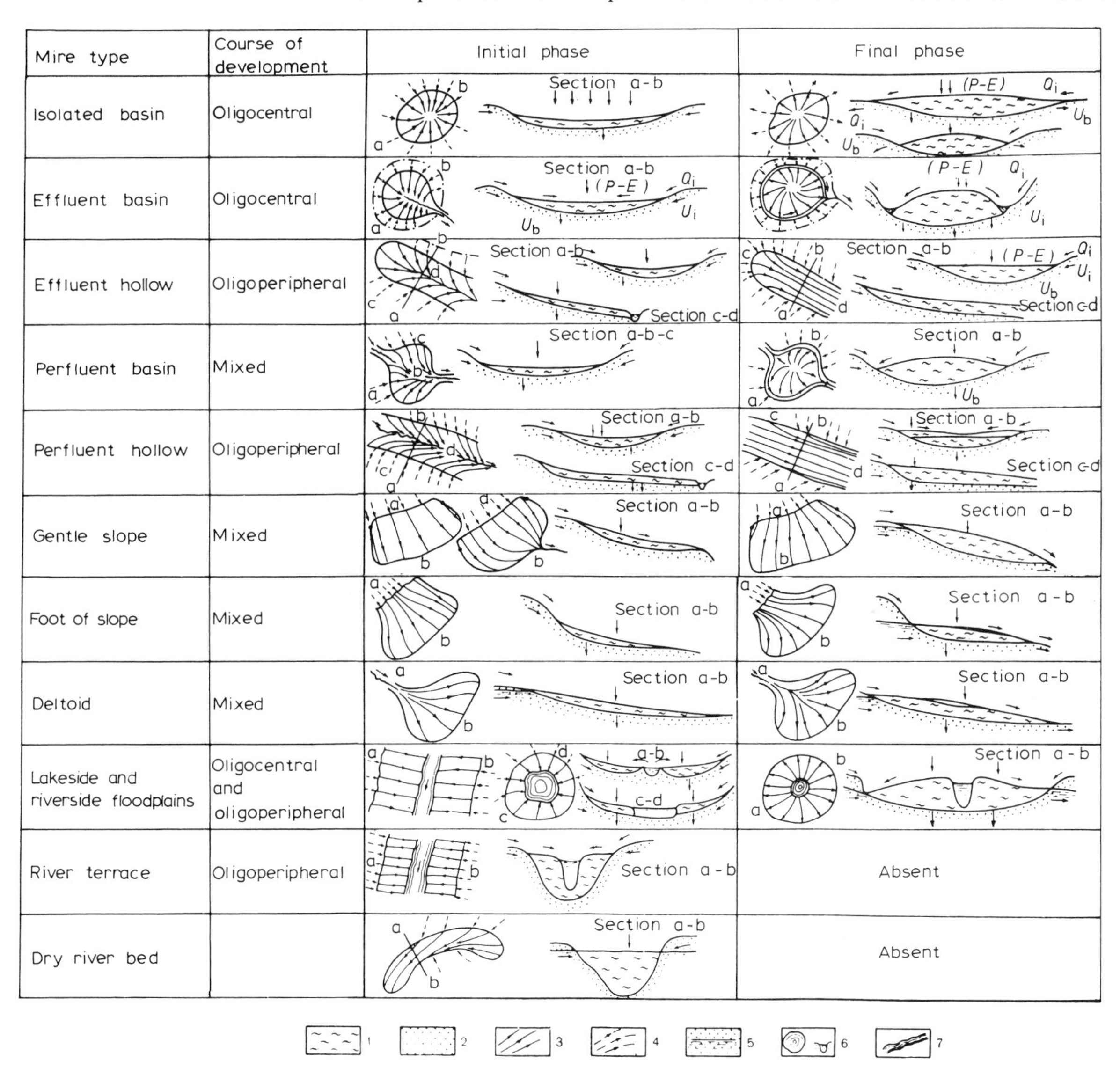

Fig. 3.27. K.E. Ivanov's diagrammatic classification of recharge, throughflow and discharge processes during the initial and final phases of development of various types of mire (mesotope classes). Concave surface relief in the initial phase is linked with a telluric recharge component and entirely eutrophic vegetation; the convex surface of the final phase isolates most of the vegetation from telluric influence and it therefore becomes oligotrophic. Where oligotrophic vegetation first becomes established in the central parts of the mire, development is said to take an oligocentral course typical of raised bogs; but where, as in valley bogs, oligotrophic communities first appear at the edge, the course of development is said to be oligoperipheral. Legend: P=precipitation; E=evapotranspiration; Q_i=diffuse surface recharge; U_i=subsurface recharge; U_b=discharge by deep seepage; *1*=peat; *2*=mineral environment of mire; *3*=throughflow; *4*=telluric recharge; *5*=water table in mineral environment; *6*=mire lake; *7*=stream. After Ivanov (1975a).

numerous different types of mire at various stages of development, as Ivanov (1975a) has done to a certain extent (Fig. 3.27). For brevity, the present scheme is illustrated only by a typical raised bog of the saddle type. This is chosen because it exemplifies most of the processes which seem to be important in mires generally and because I am especially familiar with it. However, certain processes in blanket bogs or mires with central lakes may not be adequately comprehended.

The discharge processes in a raised bog are shown schematically in Fig. 3.28. In the plan view, diffuse surface flows (Q_e) are drawn with three arrows. Single arrows denote concentrated, directed flows. These occur either as directed overland flow (Q_e) in shallow marginal water tracks, as pipe flow (V_e), or as open channel flow (C_e) in lagg streams or rills on the mire expanse. Subsurface discharge processes are shown in the section. Three forms of seepage discharge are distinguished, namely seepage through the acrotelm towards the margin (U_a), seepage through the catotelm towards the margin (U_c), and deep seepage (U_b) — the leakage of water through the base of the mire. These together comprise the total seepage discharge (U_e); thus:

$$U_e = U_a + U_c + U_b \tag{3.26}$$

The part played by these various processes in the overall water balance of the mire is expressed mathematically by substituting N from eq. 3.3 for the surface and subsurface recharge terms of eq. 3.2, giving:

$$P + N - U_e - V_e - Q_e - C_e - E - \Delta W - \eta = 0 \tag{3.27}$$

and the details of discharge are then fully described by substituting the components of seepage (U_e) from eq. 3.26:

$$P + N - U_a - U_c - U_b - V_e - \\ - Q_e - C_e - E - \Delta W - \eta = 0 \tag{3.28}$$

It should be noted that the example of Fig. 3.28 refers to a raised bog which is mature, though still actively growing. In terms of Ivanov's (1975a, fig. 4) analysis it is in the third and final phase of its development. The contribution to the total water balance of the various terms in eq. 3.28 would have differed, at least quantitatively, in the two preceding phases.

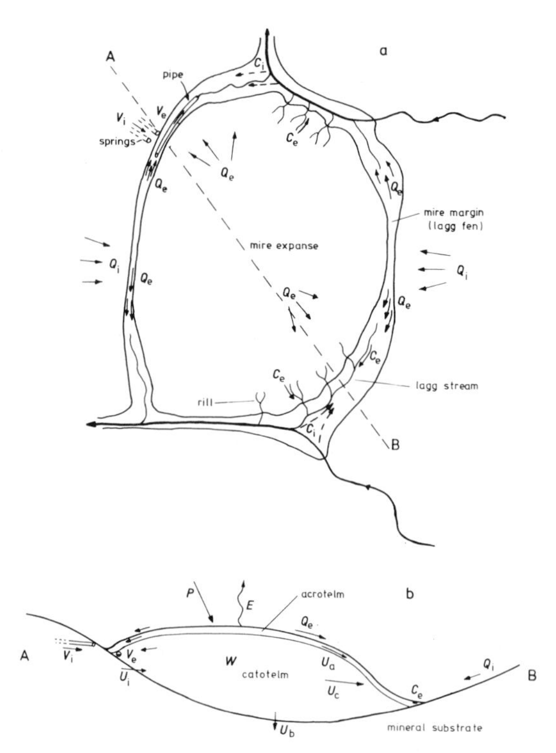

Fig. 3.28. Scheme of water fluxes associated with a raised bog in a col (an eccentrically domed saddle raised bog). a. Plan. b. Section along AB. The eccentricity has arisen because the hillside to the right of the mire presents the larger catchment. See text for explanation. Diagrammatic.

Physical characteristics of liquid discharge processes

Here the liquid discharge terms of eq. 3.28 will be discussed in turn to explain the physical nature of the processes involved and how each can be described parametrically. Various accounts of the relevant geophysics are available, including, for porous medium flow, those of Muskat (1937), Scheidegger (1960), Polubarinova-Kochina (1962), Luthin (1966), Rose (1966), Bear et al. (1968), Childs (1969) and Hillell (1971); for pipe flow and surface flow those of Kirkby (1969) and Ward (1975); and for open channel flow those of Chebotarev (1962) and Ven Te Chow (1964).

Seepage

The mass transport of liquid through porous media is conveniently known as seepage or filtration flow. It is a process of great interest to geophysicists because of its practical importance in oil and mining technology, in the water supply and sewage disposal industries, and in the irrigation and drainage of land used for farming and forestry. There is an extensive literature which dates back at least to 1800, and theoretical analysis is closely linked with approaches to the analogous phenomena of heat conduction and electrical currents in solid media. Hitherto the analysis of water movement in soils has provided the greatest impetus to practical studies of seepage in peat.

Theoretical considerations. Water moves through soils in response to various different forces which only fortuitously act in the same direction. In order to avoid the cumbersome procedures of vector algebra it has become conventional to regard all these forces as contributing in their various ways to the potential energy status of water, otherwise known as the water potential, ϕ. Water then moves through the system from regions where its potential is high to those where it is low. The speed of movement, $\mathrm{d}q/\mathrm{d}t$, is a function of the steepness of the gradient, grad ϕ, (which is dimensionless if potentials and distances are expressed in the same units — see p. 99), and the direction is that of steepest slope and is normal to surfaces of equal potential (isopotentials). Thus:

$$\frac{\mathrm{d}q}{\mathrm{d}t} = -f\,(\mathrm{grad}\ \phi) \qquad (3.29)$$

The loss of water potential which takes place as the water moves down the potential gradient is associated with the expenditure of energy in overcoming forces which tend to oppose the flow. These forces may be of various kinds. Some are due to the frictional drag between layers of water molecules sliding past each other at different speeds. Some are associated with the deflection of molecules from the straight paths of movement imposed by inertia. Some have to do with the ordered arrays of molecules maintained by electro-osmotic forces in the neighbourhood of fixed ionic charges in the porous matrix.

The frictional drag in a liquid is its viscosity. Different liquids move through the same porous medium at rates determined by their viscosities. Where only one liquid, namely water, is of interest the practical significance of viscosity is its variation with temperature, which can be found in standard laboratory reference manuals. From the theoretical standpoint, it can be shown that where laminar flow occurs, so that the forces resisting seepage are exclusively of viscous origin, the relationship between flow rate and potential gradient is represented by a simple model of the form:

$$\frac{\mathrm{d}q}{\mathrm{d}t} = -K\ \mathrm{grad}\ \phi \qquad (3.30)$$

where K is a parameter, characteristic of the porous medium, whose value depends only on the temperature. The same analysis shows that, if large pores are combined with potential gradients steep enough to induce turbulence in the interstices of the porous medium, some energy is dispersed in overcoming inertia. Thus an increase in potential gradient induces a less than proportional increase in flow rate, and the simple, linear model of eq. 3.30 fails to describe the relationship between gradient and flow. Moreover, if electro-osmotic forces are involved, the model again fails, but in this instance increases in hydraulic gradient induce a more than proportional increase in rate of flow as increasing numbers of water molecules become detached from ordered arrays. On this view, the model of eq. 3.30 is merely the simplest of a range of possible models describing the flow of liquid water through porous media down a potential gradient. It is known as Darcy's Law, and the parameter K is termed the hydraulic conductivity. The flow rate $\mathrm{d}q/\mathrm{d}t$ is not the actual rate of flow through the pores but the rate at which water enters or leaves the porous medium. If it is expressed as volume per unit cross-section of aquifer per unit time and, if the potentials are expressed gravimetrically in units of length (see above) so that grad ϕ is a dimensionless number, the units of K are those of length per unit time, or velocity. Experiments using permeameters have shown that Darcy's Law adequately describes gravitational water flow through a variety of porous materials, to which values of hydraulic conductivity can accordingly be ascribed.

In most soils gravity is the major component of the gradients of water potential which cause mass discharge of liquid water, and the gradients themselves are generally well below unity. Furthermore, the pore-size distribution of soils is within too fine a range for turbulence to be probable at these low gradients. Consequently there are many situations in soils where Darcy's Law is useful as a parametric description of seepage. Indeed, such situations are so numerous that much of the literature embodies the uncritical assumption that all water seepage in soils may be described in this way. However, it is also known that substantial electro-osmotic effects are implicated in other soil processes, such as cation exchange and the shrinkage of clays on de-watering without air entry. It is therefore scarcely surprising to find accumulating evidence of soils in which Darcy's simple linear model fails because the flux increase is more than proportional to gradient (Swartzendruber, 1966).

Liquid-water discharge down potential gradients induced by gravity takes place mainly in the saturated zone of a soil. This is for two reasons. Firstly, the discharge flux is mainly horizontal in direction. The gravitational component of potential gradient causes vertical movement in the unsaturated zone; only in the saturated zone can it cause water to flow sideways. Secondly, although Darcy's Law can serve as a model for flow in the unsaturated zone, it does so only with such modified values of hydraulic conductivity as are appropriate to the moisture content of each unsaturated layer. As the moisture content diminishes and continuous water films retreat into channels that are increasingly narrow, tortuous and disconnected, the conductivities fall. For soils of average porosity K varies inversely and sharply with θ; thus hydraulic conductivity in the unsaturated zone is comparatively low, and so this zone is of more interest in relation to infiltration (pp. 70–73) and evaporation (pp. 73–99) than to liquid discharge.

The nature of the potential gradients inducing seepage in saturated mire aquifers requires comment. As has been shown, these gradients are an expression of the energy which drives the water through the peat. But in addition to conforming with principles of energy conservation the flow process must also satisfy the law of conservation of matter, which underlies the principle of continuity that is so basic to the water-balance concept in hydrology. Where water flows through a homogeneous aquifer of varying cross-sectional area, continuity demands that the flux density shall be high in the narrow sections and low in the broad ones. In mires this situation arises, for instance, when marginal peat deposits thin out against the rising surface of a less permeable substratum (Fig. 3.28b). Since by definition the aquifer is homogeneous, there is no increase in permeability in its narrow sections, so continuity is here satisfied by steepening potential gradients.

Where an incompressible fluid moves through an isotropic medium in response to a gradient of potential energy, and where the relationship of flow rate to potential gradient is of some such linear form as Darcy's Law expresses, it may be shown that the conditions of matter and energy conservation will be satisfied by the expression:

$$\nabla^2\phi = 0 \tag{3.31}$$

where ∇ ("del") represents the sum of the partial derivatives of grad ϕ, taken in the directions of the axes of a coordinate system. For a system of Cartesian coordinates in three dimensions,

$$\nabla = \frac{\delta\phi}{\delta x} + \frac{\delta\phi}{\delta y} + \frac{\delta\phi}{\delta z} \tag{3.32}$$

and analogous statements can be written in cylindrical or spherical coordinates. Eqs. 3.31 and 3.32 together constitute a statement of the Laplace equation.

It is often difficult to derive a solution of Laplace's equation that would be appropriate to a particular problem of ground-water flow. Instead an approximation is used. In soil physics the best known of these is the Dupuit–Forchheimer approximation, the predictions of which give satisfactory agreement with solutions of Laplace's equation for cases where these happen to be available. The approximation is based upon the definition of water potential given in eq. 3.20:

$$\phi = h + z \tag{3.33}$$

The water potential at a point below the water table in an aquifer is due partly to the elevation, z, of that point above an arbitrary zero datum plane and partly to the pressure, h, or head of water above the point, which is equal to the depth of its submergence below the water table. In the absence of water movement the potential at all such points is by definition the same (i.e. the water body is an isopotential) and must be equal to the elevation, Z, of the water table, since also:

$$h+z=Z \tag{3.34}$$

hence:

$$\phi=Z \tag{3.35}$$

and so the water table is horizontal. If, conversely, the water table is tilted as shown in Fig. 3.29, only those points underlying lines of equal Z (i.e. water-table contours) will be isopotential, thus the isopotentials take the form of surfaces. In essence the Dupuit–Forchheimer approximation assumes that these surfaces are vertical and that their water potentials are equal to the elevations of the water table along their upper edges, whence:

$$\text{grad } \phi=\text{grad } Z \tag{3.36}$$

This, of course, implies that flow is everywhere horizontal beneath a tilted water table, which is clearly impossible since the water table is itself one of the confining boundaries and water in its vicinity flows parallel to it. The situation is more realistically sketched in Fig. 3.30, which shows lines of flow converging on a lagg fen. Since flow takes place in the direction of steepest decline in potential, flow lines must be normal to the isopotentials, which are therefore not vertical but curved. The approximate character of this approach is most clearly seen at the right-hand side of the sketch.

There is not a large amount of experimental evidence on the nature of water seepage through

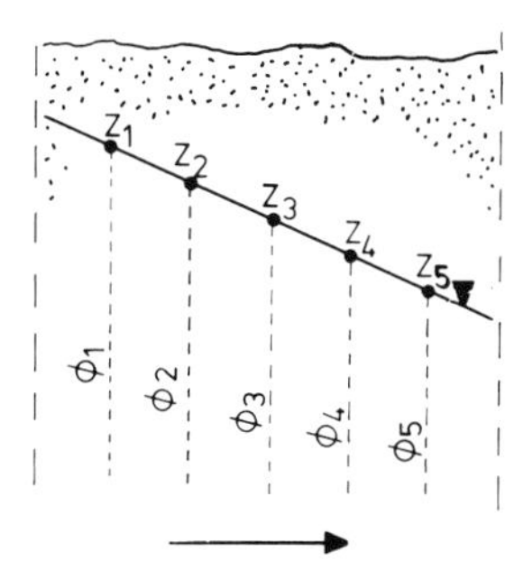

Fig. 3.29. Diagram of section through a porous aquifer containing a uniformly sloping water table (▼). The plane of the section is the direction of (maximum) slope of the water table. The isopotential surfaces $\phi_1>\phi_2>\ldots>\phi_5$ are drawn according to the conventions of the Dupuit-Forchheimer approximation and are "labelled" by the corresponding water table heights $Z_1>Z_2>\ldots>Z_5$. Arrow shows direction of flow.

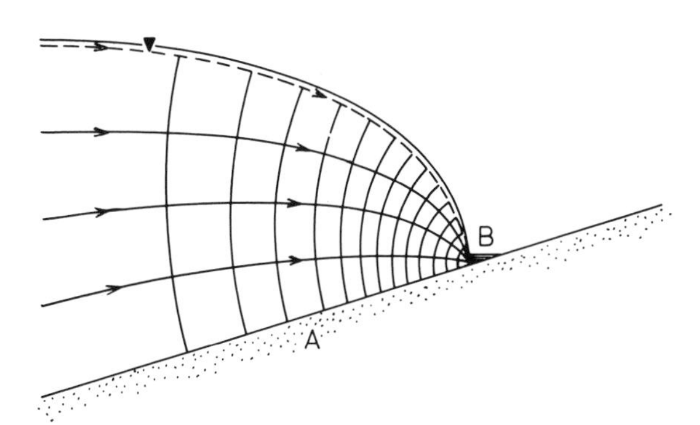

Fig. 3.30. Sketch of flow net in a vertical radial section through the marginal slope and lagg of a raised bog abutting against a rising slope of relatively impermeable mineral strata *A*. *B* is the lagg fen at the mire margin; ▼ is the water table; the flow lines are arrowed; the isopotentials crossing them are spaced at equal increments of potential. Diagrammatic and with exaggerated vertical scale.

mires. Most western authors have been content to assume Darcyan linearity, while Soviet investigators have not been very interested in the circumstances which give rise to most difficulty.

Seepage in the acrotelm. As noted above (p. 103), the site of most rapid water seepage is the acrotelm. Here, Romanov (1968a) has quoted velocities as high as 2.5 cm s^{-1}. Because of these very high velocities, most Soviet evidence on acrotelm seepage has been obtained from laboratory experiments. These have used soil monoliths cut from the surface layers of mires during winter and transported to the laboratory while still frozen in order to avoid disturbance. Horizontal flow has then been imposed by enclosing each monolith in a "filtration flume" which can be tilted to vary the potential gradient. The natural direction of flow is maintained by ensuring that each monolith is cut out with its long axis normal to the surface contours of the mire. Downhill flow along this axis is then maintained by the side walls of the flume. Technical details are given by K.E. Ivanov (1953, 1957; Anonymous, 1961). No deviations from Darcyan proportionality were reported during the course of laboratory tests of this kind, using acrotelm monoliths of widely varied origin (K.E. Ivanov, 1953, 1957, 1975a; Romanov, 1968a).

Rycroft (1971) has described field tests on acrotelm material from a Scottish raised bog. The material was a mixture of *Sphagnum magellanicum* with *S. cuspidatum*, slightly humified (H1–H2 on Von Post's 10-point scale) and forming the recent infill of an abandoned drainage ditch. The seepage tube method of Luthin and Kirkham (1949) was used to impose potential gradients in the part of the acrotelm surrounding an unlined cavity at the foot of the seepage tube. In this method, which is described by Rycroft et al. (1975a), the water level in the seepage tube recovers towards its initial equilibrium position at a rate which dies away logarithmically with time, provided the soil behaves in a Darcyan manner. This was indeed the pattern of recovery found in Rycroft's acrotelm tests (Ingram et al., 1974; Rycroft et al., 1975b), not only in the *Sphagnum* material described above but also in the acrotelm of a marginal fen in the same mire, with *Carex rostrata* and hypnoid mosses (Fig. 3.31). Because the gradients of water potential imposed in these seepage-tube tests are initially steeper than those used in Ivanov's "filtration flume", and because they vary in space along the convergent flow lines as well as diminishing with time, it seems that Darcy's Law may be accepted as a robust and useful model of water transmission in the saturated part of the acrotelm.

Nature of seepage in the catotelm: laboratory tests. Numerous studies of seepage in the catotelm have been carried out as an aid to mire drainage. Various seepage-tube techniques and auger-hole methods have been successfully developed for determining

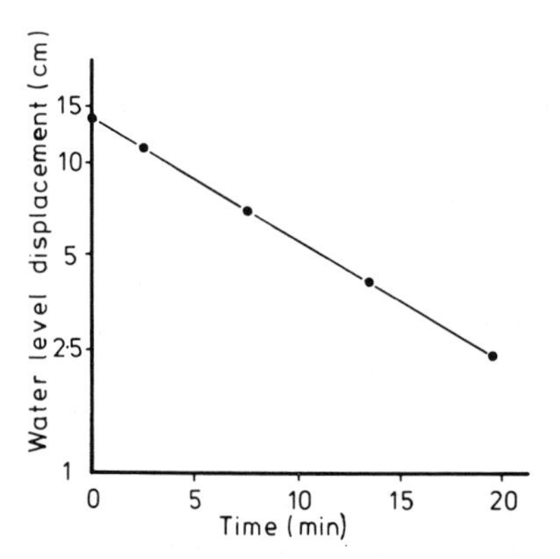

Fig. 3.31. Time course of water table recovery (log scale) for seepage tube installed in weakly humified acrotelm peat (*Sphagnum magellanicum* and *S. cuspidatum*) in a choked drainage ditch at Dun Moss, Perthshire (Scotland). Test in depletion mode. Hydraulic conductivity $K = 3.1 \times 10^{-3}$ cm s^{-1}. (From Rycroft, 1971.)

hydraulic conductivity values to be used in the solution of practical drainage problems in mineral soils (Rycroft et al., 1975a). This is possible because Darcy's Law holds throughout the range of potential gradients and pore-size distributions likely to be encountered in agricultural soils of good quality, as may readily be confirmed in experiments with laboratory permeameters. However, when the same methods were first tried out in mires the validity of Darcy's Law in the very different conditions of the catotelm was merely assumed. The earlier workers carried out no laboratory tests to check the validity of the law, even though the assumption that it applied underlay all their field methods. Only Eggelsmann (1964a) seems to have been worried by this. He used piezometers to obtain the distribution of isopotentials and streamlines between mole and tile drains in reclaimed raised bogs in Niedersachsen and concluded that Darcy's Law applied in the catotelm. Eggelsmann pointed out that laboratory experiments on seepage in mires are of doubtful validity because the delicate structure of peat is bound to be disturbed during sampling. So far as it goes, this is a fair and important comment and undoubtedly accounts for the reluctance of many mire hydrologists to carry out laboratory tests. On the other hand, one should be careful to distinguish between experiments to determine the physical mechanism of a process and tests to evaluate its parameters. It is for the latter purpose that laboratory methods are unsuitable. However, they may be more appropriate for studying the mechanism of water transmission through peat because, in the laboratory, the investigator has more control over the flow geometry. Most field techniques involve flow lines that either converge or diverge and may, as in Eggelsmann's (1964a) studies, do both. For reasons of continuity, these flow lines are associated with potential gradients that are non-uniform in space, and may also vary in time as in the recovering seepage tube described above. Where Darcy's Law applies, such lack of uniformity is immaterial. Where it does not, implying a non-linear flux–gradient relationship, the actual values of potential gradients imposed in tests must be unambiguously known if the extent of the departure from linearity is to be assessed.

Boelter (1965) carried out permeability tests in the laboratory. He used replicated samples in unconfining permeameters (in which the samples were not enclosed at the upper surface). For a variety of peats, horizontal conductivities exceeded those measured in a vertical direction, but the differences were not statistically significant. No departures from linearity were reported, but it is likely that the potential gradients used varied little from test to test.

Rycroft (1971) conducted a short series of laboratory tests on humified *Sphagnum–Eriophorum–Calluna* peat from the catotelm of Dun Moss, a Scottish raised bog. Like Boelter's, the tops of his samples were unconfined, but his permeameter system permitted a variety of potential gradients and mean water potentials to be imposed on his samples. Water temperatures and pH were not controlled, and were appreciably higher than those experienced in the field. The rate of flow appeared to be independent of the mean water potential in the sample, but showed a more-than-proportional increase with the potential gradient across the sample, suggesting departure from Darcyan linearity.

Further laboratory studies of vertical water transmission through peat samples from the Dun Moss catotelm were made by J.M.B. Brown and J. Waine (Waine, 1976). They developed a fully confining permeameter system, with close temperature control at typical field temperatures, and chemically inert sample cells to prevent drift in pH. Water from the mire was used, but it was found that over long periods of testing the rates of flow declined, apparently due to pore blockage following hydrogen sulphide evolution as a result of anaerobic bacterial reduction. This aging effect, which is not found in field tests, was countered by the addition of 10 p.p.m. Cu^{2+} ions to the water. Potential gradients (grad ϕ; see p. 120) ranging from 1.5 to 17.6 were imposed in these experiments. Increasing gradient induced a similar more-than-proportional rise in flux to that observed by Rycroft (Fig. 3.32). Again there was no evidence of an effect of mean water potential. It is important to note that the gradient range between 0 and 1 was inaccessible due to the technical problems of measuring the very low potential differences and flow rates which invariably arise in such tests. This means that it is still uncertain how the system would behave in response to those gradients most frequently encountered in the field.

Similar non-linear effects have been reported for other fine-grained soils (Mitchell and Younger, 1967), especially in swelling soils (Russell and Swartzendruber, 1971). Several possible mechanisms could be responsible, but the addition of peat to the list of anomalous substances suggests that the explanation must be sought in the interaction between water and the solid matrix. The high proportion of physically bound water reported in humified peats by Volarovich and Churaev (1966, 1968), Volarovich et al. (1972) and Kay and Goit (1977) supports this suggestion.

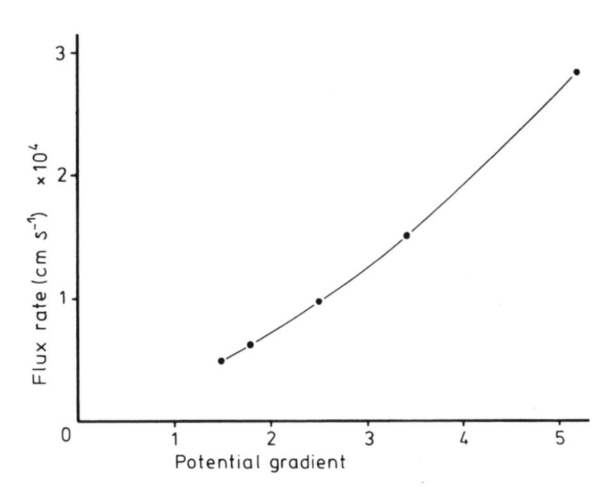

Fig. 3.32. Relationship between potential gradient and flux rate in a series of laboratory permeameter tests on a sample of humified *Sphagnum–Eriophorum–Calluna* peat from the mire expanse at Dun Moss, Perthshire. (From Waine, 1976.)

The simplest assumption about the situation portrayed in Fig. 3.32 is that the curve extrapolates to zero. If this is valid, it implies that the non-linearity over the gradient range 0 to 1 is weak. Lacking further evidence, it may therefore be best at present to treat seepage in the catotelm as being approximately linear, remembering that this is an assumption which still lacks experimental proof.

As to the use of laboratory permeameters to examine the effects of average water potential, Galvin and Hanrahan (1967) conducted a series of such tests on highly humified blanket peat in Ireland. They showed that permeabilities rose with increasing water potential from values in the neighbourhood of $K = 6.4 \times 10^{-6}$ cm s^{-1} at $\phi = -0.0275$ MPa to $K = 1.2 \times 10^{-5}$ cm s^{-1} at $\phi = 0.0824$ MPa. Galvin and Hanrahan considered the blockage of pores when entrapped air expanded at lower water potentials to be responsible for this effect, and it is interesting to note that pore blockage by gaseous decomposition products has recently been suggested as the cause of a falling off in seepage with time in some peats (Bodarenko et al., 1975). However, these average potential effects were not found by Rycroft (1971) or by Waine (1976).

Seepage in the catotelm: field tests. Boelter (1965) and Päivänen (1973) found that laboratory tests yielded considerably higher permeability values than field tests on the same peat, an effect which they ascribed to disturbance of the structure of laboratory samples or leakage at the permeameter wall. Most parametric descriptions of seepage in the catotelm are based on field permeability tests in which such disturbance is thought to be minimized. The available methods and results of their application to peat have been reviewed by Rycroft et al. (1975a). The methods are of two kinds. In each of these the permeability characteristics are inferred by displacing the water level from a known equilibrium position (corresponding to the water table) in a well, and recording the ensuing return towards equilibrium, the rate of which diminishes as equilibrium approaches, but is also related to the permeability. In the first group of methods the well is unlined. These are called "auger-hole methods". The second group uses wells which are lined with close-fitting tubes, leaving only the floor of the well or a shallow cavity at the bottom for water to enter or leave. These "seepage-tube methods" yield results which in theory relate to a smaller volume of peat. They have also been termed "piezometer methods" or "infiltration pipe methods".

The auger-hole methods have been much used on the continent of Europe (Eggelsmann, 1978) — especially, perhaps, on mires already subject to some drainage. Rycroft et al. (1975b) reported difficulty in applying this approach to an intact raised bog because the acrotelm was so permeable as to be the dominant influence on the results. However, if a seepage tube is used it prevents interference from the acrotelm.

The course of a seepage-tube test in the acrotelm has already been discussed and is shown in Fig. 3.31. Unfortunately, the method has given rise to much difficulty when applied to catotelm peat (Ingram et al., 1974; Rycroft et al., 1975b), but for quite different causes than those which affect auger holes. The problem is exemplified by the curves shown in Fig. 3.33. These relate to tests in the catotelm of Dun Moss, where the peat concerned

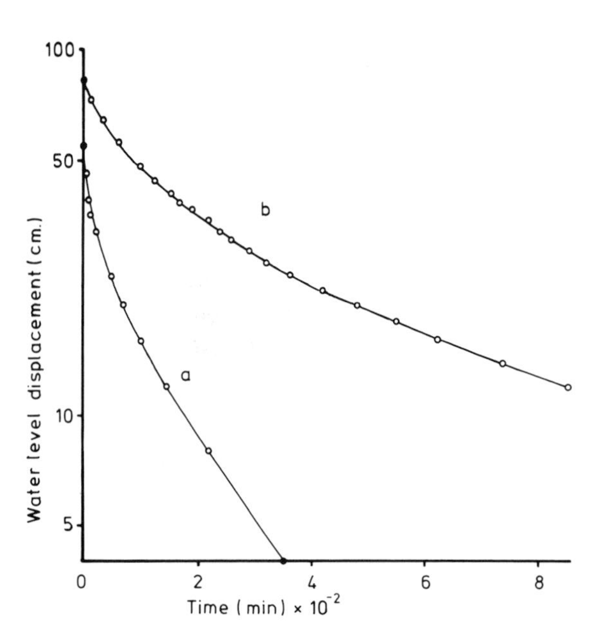

Fig. 3.33. Time course of water-level recovery (log scale) for seepage tube installed in humified catotelm peat (*Sphagnum–Eriophorum–Calluna*) on the mire expanse at Dun Moss, Perthshire. *a*, test in recharge mode; *b*, test in depletion mode. (After Ingram et al., 1974.)

was mainly *Sphagnum*, together with *Eriophorum vaginatum* and *Calluna vulgaris*, of humification H3 to H6. Here, there was no linear relationship between level displacement and time. Similar anomalies were reported by Yamamoto (1970) and Dai and Sparling (1973), using seepage tubes in Japanese and Canadian mires, respectively.

The two curves of Fig. 3.33 relate to the two possible initial modes for a seepage-tube test. "Depletion" implies that water was removed to start a test; "recharge" implies that water was first poured into the tube. These modes should generate flow of identical geometry but opposite direction, and hence recovery curves of identical shape. The difference in shape between these two curves arose from the fortuitous circumstance that it took longer to establish the initial conditions in the depletion mode than in the recharge mode. The importance of preparation time was further established by a series of seepage-tube tests described by Rycroft et al., (1975b), which also indicated that in general the immediately previous history of the system influenced the course of recovery.

The explanation of these findings seems to lie in the phenomenon of isopinal storage described above (p. 117). It will be recalled that water storage in the catotelm appears to vary with hydrostatic pressure or water potential. Starting a seepage-tube test involves altering the water potential at the face of the cavity. If, as in a depletion test, the potential is lowered, water is then released from storage and enters the cavity. The rate at which it does so depends on the fall in potential. It also depends upon the condition of the store. If the store has not been completely refilled following a previous test, or if the initial draw-down takes so long that partial adjustment of the contents occurs before the test begins, the flux of stored water into the cavity will be less rapid than if prolonged freedom from disturbance had permitted complete replenishment or the draw-down had been instantaneous. This potential-induced water flux is superimposed upon the gradient-induced flux of seepage water originating at the water table. Indeed, as described earlier, it is possible to arrange for the two fluxes to be in opposite directions during a seepage-tube test, so that, temporarily, the water level in the tube either moves away from its equilibrium position or else becomes stationary when the magnitudes of the opposing fluxes become equal. In a normal test, however, the fluxes are additive throughout. Thus, the initial recovery rate is enhanced well beyond its value in the absence of isopinal storage. If, therefore, observations at successive levels are used to compute hydraulic conductivity values for each interval between observations, the result is a series of values diminishing with time.

As the test proceeds, the potential gradient inducing seepage declines. Since the peat is known to be weakly non-linear from laboratory tests, it follows that the computed hydraulic conductivity values must decline during the test for this reason also. In the strict sense it is misleading to speak of hydraulic conductivity, which, as the Darcy function, is constant by definition. On the assumption of near-linearity at low potential gradients, the term should be used only as an approximation, and then computed only from data obtained as time $t \rightarrow \infty$ in the latter part of a test, for only at these latter stages will the storage flux become negligible and the potential gradients low. This is the technique recommended by Rycroft et al. (1975b), who suggested that seepage-tube tests be continued for periods of the order of 10^3 min in order to obtain sensibly constant values. It seems probable that a similar technique should be followed in auger-hole tests. Inspection of the course of individual recharge tests reported by Rycroft et al. (1975b) shows that failure to follow this recommendation could cause "hydraulic conductivities" to be overestimated by more than two orders of magnitude.

Pipe flow

In 1912, Elgee suggested that blanket peat in the north Yorkshire moors in northern England was being eroded by mechanical and chemical processes occurring beneath the surface. "Subcutaneous erosion" and "suffosion" are alternative names for these processes. Their effects on other blanket bogs have been described by Moss (1913), Osvald (1949) and Overbeck (1975). Here they often produce systems of tunnels, shafts and sink-holes, reminiscent of karst erosion features (Ingram, 1967). The sound of running water can often be heard, and thus these features are frequently detectable along the junction between bog margins and steep rock slopes or along the central axes towards which fen areas drain. Pearsall (1950) identified the collapse of the roofs of these tunnels as the initial stage in gully erosion of blanket bog on Rannoch Moor in Scotland, a conclusion confirmed during work in Wales by Taylor and Tucker (1970).

Similar features are also found in raised bogs and have been described from various parts of Germany by Rudolf and Firbas (1927), Dittrich (1952) and Eggelsmann (1960), and from northern England by Nicol and Robertson (1974).

Water discharge through suffosion tunnels, or soil pipes as they are now termed, has been widely studied and the literature is reviewed by Jones (1981). So far as mires are concerned, the only soil pipes in peat whose hydrological function has been given detailed attention are those in the Institute of Hydrology experimental catchment on the upper River Wye, on the eastern flanks of Plynlimon in Wales (Newson, 1976; Newson and Harrison, 1978; Gilman and Newson, 1980). Here the pipes occurred on slopes between plateau blanket bogs and valley floor mires, either in peaty podzols or in solifluction "head" or amorphous peat beneath *Juncus effusus* flushes. Their depth was controlled by soil structure, and they were generally found where less cohesive material rested upon a cohesive horizon of low permeability. Their mean cross-section was 67.5 cm^2, but it varied rapidly over short distances, being sometimes great enough for an adult to crawl through. Floors were smoothly rounded, but roofs were rough and often ruptured by vertical shafts, which occurred at 2.6-m intervals on average. The mean density of pipes was 42.3 km km^{-2}, rising to 180 km km^{-2} over the area which the pipes actually drained. They formed dendritic networks with 5000 nodes km^{-2} on average (21 000 nodes km^{-2} over the area drained). They were thought to develop from shrinkage cracks caused by drought, which then enlarged due to the eluviation or mechanical suffosion of finer material by water, seeping under pressure towards the phreatic surface of the wall of the crack. Eventually, this removed all but a few fragments of the most cohesive aggregates — nodules a few centimetres in diameter which clung to the roof on rootlets but were finally flushed out during storms. On this interpretation, a combination of steep potential gradients and widely variable water content is necessary for pipe formation in peat. This may help to explain the prevalence of the process at the margins of bogs. Pipes appeared to develop rapidly (10–20 years), but their active life was commensurately brief due to roof collapse, blockage and short-circuiting.

Overland flow

The mass movement of water over the land surface without defined channels is termed overland flow or sheet flow. In recent years controversy has arisen among hydrologists about the nature and importance of this process (Ward, 1975). The older view is associated with the name of R.E. Horton (1933), who suggested that overland flow was a consequence of soil physical constraints. According to this concept any process, such as saturation of the profile or slow infiltration, which causes ponded water to appear at many points on the surface, will lead to overland flow under intense snow melt or precipitation, when the water in neighbouring hollows overflows to become a continuous network or sheet, moving downslope towards the nearest stream. The more recent view is mainly due to J.D. Hewlett (see Hewlett and Hibbert, 1967), who discounted infiltration as a limiting process. In his view, the storm hydrograph peaks which Horton had ascribed to rapid overland flow are in reality due to rapid flow in the channel network. The size or "contributing area" of this network varies with time. During times of rain storm or snow melt it expands. Thus parts of the catchment nearest to the permanent streams become flooded and involved in overland flow as well as in the much slower flow of ground water.

Parametric description of this form of discharge involves identifying the contributing areas and establishing how they vary with the weather and the hydrological state of the system. So far little quantitative work of this kind seems to have been attempted on mires.

Open channel flow

General nature and significance. Mire recharge which is not dispersed by evapotranspiration or deep seepage and does not contribute to the long-term increase in the water stored in a growing mire must ultimately appear as a discharge flux in surface channels: rills, streams or rivers. This flux has various interests for the ecologist. In the first place it modifies the environment for plant growth by creating peculiar local conditions of mineral nutrient concentration, substrate aeration and enhanced water supply. Here the vegetation differs from that of the hinterland, the helophytes giving way to hydrophytes of variously amphibious habit and systematic position, ranging from Rhodophyceae (*Batrachospermum* spp.) through bryophytes (*Fontinalis*) to angiosperms (*Potamogeton*, *Callitriche*, etc.), most of which are also found in moving water outside mire ecosystems. Secondly, the surface streams in mires may change with time due to erosion or other processes and are therefore involved in long-term ontogenetic changes, often with profound ecological and hydrological consequences. Thirdly, the open-channel discharge from mires has certain unusual characteristics which affect the engineering problems of the water supply industry and downstream river management. Because these characteristics may be altered by human interference involving drainage or afforestation, they are important in the conservation and applied ecology of mires.

Classification. Hydrologists are accustomed to distinguish two forms of open-channel discharge. Quickflow is contributed by overland flow, pipe flow and rapid seepage throughflow as well as by direct precipitation onto the channel itself. It is the major component of discharge during rainstorms or snow-melt. Baseflow by contrast is a form of sustained discharge that generally also continues under other conditions. In mire-free areas it results from ground-water discharge below the basal soil horizon or, supposedly, from unsaturated seepage in steeply sloping soils. Ward (1975) has quoted the opinion that about 80% of river flow in the eastern U.S.A. is baseflow. However, the quickflow–baseflow classification seems to have arisen more in response to the practical demands of the engineer than as a consequence of detailed scientific appreciation of the mechanisms involved. Some scepticism over imprecise definitions may not be out of place.

This classification gives rise to a further classification of surface streams into ephemeral, intermittent and perennial types. The first category experience no baseflow because they are perched above the water table and are therefore inaccessible to ground water. The second experience some baseflow in the wet season where there is a marked annual alternation of periods with contrasting weather. Perennial stream beds are always below the water table so baseflow makes a continuous contribution.

So far as mires are concerned, perched water is unlikely and surface streams fall into the two latter categories, with rills, erosion gullies (groughs) and the headwaters of lagg streams and fen watercourses being intermittent, while larger streams are perennial.

Measurement. Open channel discharge is very commonly measured by water authorities and supply undertakings, so that experience of the technology is extensive [see for example Ven Te Chow (1964) and Wilson (1974)]. Three approaches are commonly adopted. In dilution gauging an aqueous solution of a known concentration of a readily detectable tracer is introduced into the stream and the discharge rate is inferred from the dilution which then occurs, measured by quantitative analysis of water samples obtained downstream. This method has not commended itself to mire hydrologists because of the ever-present risk of adsorption of the tracer at ion-exchange sites in the peaty channel walls (see below). This, by exaggerating the dilution effect, would cause discharge rates to be overestimated. In the velocity–area method, discharge is estimated as the product of mean channel flow velocity and channel area, the latter being a function of channel geometry and the level, or stage, of the water surface. Floats or current meters are used to measure velocity, and the thermoelectric current meter of Daniels et al. (1977) offers promise for use in rills (see below). The results can be used to determine the relationship of stage to discharge and thus form a basis for continuous recording, always provided that the channel geometry does not change with time. (Such changes are common in some types of soft sediments and may create severe practical difficulties.) On the other hand, the velocity–area method causes least disturbance to the hydrological regime under study. In practice, the method is a convenient means of reconnaissance, enabling permanent gauging structures of appropriate capacity to be designed. Gauging structures permit the most accurate discharge measurements to be made. These are rigid installations inserted in the channel. They are made of resistant materials so that their geometry remains constant through time. In addition, the gauging structure imposes a constriction on the channel either by raising its floor, as in the sharp- or broad-crested weir, or by narrowing the channel width to produce a standing wave. Broad-crested weirs and width constrictions are conveniently accommodated in flumes. The purpose of the structure is to cause ponding of water on the upstream side to a level or stage which is unaffected by conditions on the downstream side. The relationship between stage and discharge is either determined by calibration tests in the field or hydraulics laboratory, or else taken from published accounts of standard designs. Knowing this relationship, discharge rates can be inferred from continuous records of water stage and the amount of discharge calculated by integration with respect to time.

Gauging structures have the disadvantage of raising water tables on the upstream side. This may interfere with the water balance in various ways and may enhance the leakage of discharge water seeping along the channel banks past the gauging point, so that a screen of sheet piling must be installed across the banks (Bay, 1967a). Properly installed gauging structures are also very expensive and require individual engineering design to meet the special requirements of each gauging point. Some of the problems arising in mires are briefly mentioned in the official Soviet manual (Anonymous, 1961).

Hydrograph analysis. A record of open-channel discharge as it varies with time is known as a hydrograph, and in hydrology it is the practice to regard the discharge/time curve of each rain storm or snow melt event as to some extent distinct from those which precede and follow it. Much hydrograph analysis has traditionally depended upon an empirical generalization: the unit hydrograph or hydrograph generated by a unit equivalent depth (say 25 mm) of direct surface discharge, produced by a short and uniform rain storm of unit duration (say, 2 h). The form of the unit hydrograph affords a basis for comparing the channel discharge characteristics of catchments with differing topography, geology, soils, vegetation and land use. Typically a hydrograph curve shows a rapid initial rise, a peak and then a fall or recession at diminishing rate. In comparing unit hydrographs from different catchments interest centres on three features of the curve: (1) the mean travel time, concentration time or time elapsing between the start of the storm (or the point when the flux density of precipitation exceeds a given value) and the attainment of maximum discharge; (2) the value of the maximum discharge rate; and (3) the rate of recession following this maximum. These features characterize aspects of the hydrography of water catchments that are important, not only in the design of gauging installations, but also in evaluating the catchment as a water resource, a source of flooding hazard downstream, a system evolving by erosion, a source of sediment and dissolved ions, or a system capable of responding to engineering operations and other changes in land use.

An additional significant feature is the total open-channel discharge flux which the catchment generates over a given period of time. Since flux totals may be obtained by integrating hydrograph curves with respect to time they are clearly related to the three hydrograph characteristics previously mentioned. They are, for instance, closely dependent on discharge maxima and steepness of recession. For small catchments, like many mires, the flux total is an important item, of use in drawing up or checking a short-term water balance. But for large catchments, characterized by long mean travel times, the delay between precipitation or snow melt in headwater areas and the arrival of this recharge water at the gauged exit point may be so long that flux totals can only be used in very long-term water budget computations.

Rainfall duration may affect the hydrograph where storage capacity is limited. On the Hewlett model (see above), the whole catchment will only contribute to storm discharge if the storm lasts longer than the mean travel time. The intensity and movement of storms also affect the hydrograph. These meteorological factors interact with catchment physiography. Since many raised bogs are approximately isodiametric and lie on or near water partings, mean travel times tend to be brief and high peak discharges occur, due to the roughly synchronous efflux of water from all parts of the mire. By contrast, fens and valley bogs tend to be long and narrow, lying downstream from water partings. Compared with a similar area of raised bog, the arrival of headwater precipitation at the outlet of a fen is delayed, and discharge from its usually larger catchment builds up more slowly to a peak representing less efflux per unit area of mire (Ivanov, 1953).

These effects are predictable from general hydrological experience, which also suggests that the form of the channel network depends on other catchment characteristics. Thus, catchments whose soil-physical peculiarities generate much overland flow compared with throughflow possess a denser pattern of surface streams than those in which throughflow predominates. In this connexion the absence of an open-channel network from the mire expanse of many raised bogs, despite their high prevailing water tables, is clearly of great significance, emphasizing the role of the acrotelm as an aquifer capable of sustaining high flux densities of discharge without erosion.

Catchment delineation and heterogeneity. Hydrograph analysis in mires raises the basic problem of catchment delineation. In fens and raised bogs, and even in certain types of blanket bog, the surface relief is very gentle. Although the water table in bogs is often so close to the surface that lack of coincidence between hydraulic and topographic divides is seldom a problem, the topographic divide generally lies in the most nearly horizontal area of mire expanse where contours are few and hard to plot and where identification of the catchment boundary is attended by great uncertainty. In fens formed in basins of porous mineral material the ground water is concealed, and it may again be difficult to estimate the area contributing to recharge (Boelter and Verry, 1977). Sander (1976) considered this difficulty in relation to a fen catchment and suggested an ingenious solution, using an electrical analogue model. Since evaluation of stream discharge in units of equivalent depth demands accurate knowledge of catchment size, the importance of the problem is very evident. A related and equally serious problem is that of catchment heterogeneity. A fen or raised bog is generally surrounded by areas of mineral soil supporting mesophytic vegetation. It is therefore hard to find a site for a gauging structure on a stream whose catchment lies exclusively on the mire. In Fig. 3.28 the only streams fulfilling this condition are the rills, and these are the streams whose catchment areas are least easy to delineate and measure. All the other streams draining the mire also drain the surrounding mineral terrain. The importance of this problem is especially obvious when one is interested in comparing the hydrology of mires with that of mineral soils. For this purpose comparisons between catchments in closest proximity would be most informative; but there are few places where comparable catchments occur side-by-side, one being entirely mire-covered, the other entirely devoid of mire. There are certainly many catchments completely covered with blanket bog, but here the problem is to find comparable mineral terrain. In addition the geographical distribution of blanket bog is limited to regions with particular climatic regimes, so the results of such comparisons would not necessarily apply elsewhere.

General remarks on catchment behaviour. Experience of worldwide catchment observations and of such experimental manipulation of the vegetation as was seen in the ecosystem study of Hubbard Brook, New Hampshire (Bormann and Likens, 1979) suggest a few relevant generalizations. Small catchments, in which a sparse standing biomass of vegetation is rooted in shallow, skeletal or eroded soils overlying impermeable strata, respond strongly and rapidly to recharge. They have brief concentration times, steep recession curves and discharge maxima which are high relative to dry-weather flow. By contrast, large catchments, or catchments forested with a dense standing biomass generating a thick but unwaterlogged surface layer of organic material or rooted in deep, well-drained soils overlying porous and permeable strata, respond weakly and slowly. Their concentration times are long, their recession curves gentle and their discharge maxima relatively low. From such catchments the yield of sediment, colloidal matter and dissolved minerals is also comparatively low, and it is often suggested that the proportion of recharge appearing as channel discharge at the exit may be low also, although this is more debatable.

Ontogeny of open channels. Like the other components of a mire ecosystem, open channels are structures whose nature and disposition changes as the ecosystem develops. A discussion of their function in the ecosystem therefore demands brief reference to the developmental history of open channels in mires.

Channel development in mires differs essentially from corresponding processes elsewhere because the peaty material of the channel bank is mainly an autochthonous sediment, formed by biological processes in the hinterland rather than by deposition of water-borne material. Whereas, in areas of low relief, allochthonous sedimentation may be intimately bound up with the stream channel from which the sediment spreads, the presence of a stream may also either have no direct influence on mire growth in its vicinity or else may actually inhibit peat formation by the local effect of aeration and mineral availability in enhancing decomposition. Ivanov (1975a) regarded mire development as being mainly counter-erosive. Thus, the formation of a fen in a shallow lake generally involves the progressive constriction of a channel which at first was very broad (Kulczyński, 1949). This biologically driven constrictive type of channel formation has no exact counterpart in other ecosystems. The influence of mire growth on channel formation is also evident in raised bogs. Here, the channel of the lagg stream must shift centrifugally to accommodate the spread of the bog outwards from the depression in which it first developed. In this case, mire development also causes an increase in the altitude of the channel bed (Fig. 3.34). This aggradation might steepen the stream in the direction of flow, so that the same discharge flux could be accommodated in a narrower channel. Since it seems possible that more vigorous erosion in steeper, faster lagg streams might eventually set a limit to the centrifugal spread of a raised bog, such pro-

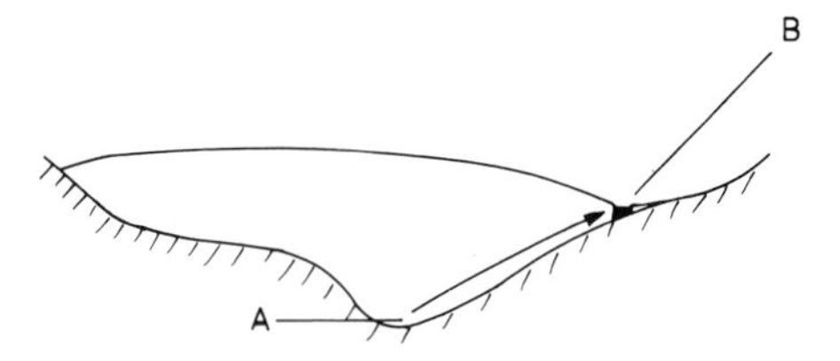

Fig. 3.34. Diagram illustrating migration of the main drainage channel during development of a saddle raised bog (a raised bog in a saddle or col between two neighbouring hills). A = Position of antecedent fluvioglacial channel; B = position of stream in main lagg fen at the present day.

cesses clearly offer scope for study in relation to the problems of growth, shape, size and stability in these ecosystems.

In many mires, by contrast, the channel system appears to enlarge at the expense of the intact peat. In raised bogs, discharge at the rand can become canalized into a series of small erosion channels. In German these are called *Rüllen*, a term for which the English "rills" is an appropriate equivalent. *Rüllen* are described in standard works (Overbeck, 1975; Kaule and Göttlich, 1976), and Eggelsmann (1967) considered them from the hydrological standpoint. Many of the rills have been destroyed by agricultural reclamation of the German mires during the past sixty years, but they were formerly characteristic of the large raised bogs in the northern lowlands, where the upslope catchment on the mire expanse generates large volumes of discharge at the rand. A map of 1897 recorded rills well over 1 km in length in the Grosses Moor near Uchte in Lower Saxony. The main channels were third-order streams, their prevailing direction centrifugal and their branching markedly dendritic. Erosion and the inhibition of *Sphagnum* growth were important characteristics. It is interesting to note that rills are seldom observed in the smaller raised bogs near the Alpine foothills of Bavaria. Eggelsmann and Schuch (1976) ascribed this partly to the smaller extent of catchment presented by the mire expanse in these bogs, and partly to suppression of erosion by stands of *Pinus mugo* which grow on them.

Extensive areas of erosion are frequently encountered in blanket bog (Moss, 1913; Conway and Millar, 1960; Bower, 1961; McVean and Ratcliffe, 1962). Sometimes the uneroded portions ("hags") are divided by rills with peat-floored channels. In other cases the original bog is divided into entirely separate remnants by deep "groughs", floored in some places with redistributed peat, in others with the mineral bedrock. Although such features are now very characteristic of the plateau blanket bogs of British uplands, they may be comparatively new. With the aid of a stratigraphic and palynological study, Tallis (1965) found that a typical erosion gully in the southern English Pennines began to form about 200 years ago. Later (Tallis, 1973) he measured rates of sediment transport which suggested 200 to 250 years or somewhat longer for the formation of a grough 330 m in length. While pipe formation and collapse are also thought to initiate blanket-bog erosion (see above), Tallis concluded that at his study site muirburn, which is a common form of upland management in Britain, was the original cause (see below). Frost-heaving and drought were important erosive agents, but substantial erosion was observed during snow-melt and heavy rains. These two latter conditions were also specified by Eggelsmann and Schuch (1976) as the main agents of erosion on upland raised bogs in Central Europe (Harz, Solling, Krkonoše, etc.) where rills are often met with.

Thus, the various discharge processes discussed above are not only related in the short term by the exigencies of continuity, but also historically through their effect on the evolution of the mire catchment.

Observations of discharge fluxes in open channels draining mires are described below (pp. 140–145).

Rates of discharge processes

Ground-water flow

Hydrological opinion currently emphasizes the importance to discharge of a process termed "interflow". This is the lateral downslope seepage of water through soils, which occurs either as an unsaturated flow or else as a saturated flow below the water table in a layer of saturation that is "perched" above the true ground-water table. Apart from occasional references by Romanov (1968a) to the occurrence of lateral downslope flow above the water table in laboratory tests, scarcely anything seems to be known about interflow in mires. Since the process is considered to be most important on steep hillsides while becoming less important, relative to ground-water flow, on gentle slopes, it seems best to proceed meanwhile as if it were of negligible account in mires, whilst keeping an open mind for the future.

Nys (1955) used aniline blue and other dyes to demonstrate seepage between pools in raised bogs, and there have been a few attempts to measure the rate of the process "directly" by means of tracers. Some of these have entailed the use of dissolved materials, although the physical complications created by the organic matrix and the technical difficulties of such an approach have long been known (Kaufmann and Orlob, 1956). Daniels et al. (1977) sought to avoid these problems by using, in effect, heat as a tracer. They were however unable to calibrate their modified form of Poppendiek thermoelectric anemometer except in a freely flowing water body without any solid matrix, and at flow rates higher than those which they associated with the least compact acrotelm materials. Despite superficial attraction, these "direct" measurement techniques have hitherto proved unattractive to mire hydrologists. This is because such methods introduce further elements of complexity into the physics and physical chemistry of a system which is already far from simple. In order to take full account of the complexities it would be necessary to calibrate the chosen technique for every variety of peat matrix in which it was to be used. Thus, the technique would not even provide measurements which were "direct" in the restricted sense here implied. In the writer's view, a parametric approach, which measures the forces at work and the resistances to be overcome, not only provides more insight into the nature of the flow process but also confers practical possibilities of generalization and prediction offering the greatest advantage when dealing with hydraulic systems which are so heterogeneous in space and so variable in time. The fact that in these systems both Darcy's Law and the Dupuit–Forchheimer theory are to some extent approximations does not affect the basic validity of this approach, as is shown by its success in predicting discharge from intact mires (Ivanov, 1975a) and by its usefulness to mire drainage engineers (Eggelsmann, 1978).

Permeability in the Acrotelm. In Rycroft's (1971) seepage-tube tests in the acrotelm of Dun Moss (Ingram et al., 1974; Rycroft et al., 1975b; see pp. 123–124) the result of $K = 3.1 \times 10^{-3}$ cm s^{-1} obtained for the overgrown ditch was fairly typical of hydraulic conductivities in marginal situations; but Romanov (1968a) reported very much higher values in the acrotelms of bogs in the Soviet Union. He quoted hydraulic conductivities in the range 10^1 to 10^2 cm s^{-1}, decreasing with depth d (cm) according to an empirical formula proposed by K.E. Ivanov (1957):

$$K_d = \frac{B}{(d+1)^m} \tag{3.37}$$

in which B was found to vary between 2290 and 41 700 and m between 2.4 and 3.8. The values of parameters B and m were found to vary in regular manner with the type of microtope (or micro-landscape: Romanov, 1968a; Anonymous, 1971), as in the examples quoted in Table 3.9. They have been used in the U.S.S.R. to compute rates of seepage through the acrotelms of mires. Discharge predictions of this kind require only a knowledge of the microtopes involved and the behaviour of the water table, since these two characteristics are sufficient to establish the potential gradients, the permeability characteristics of the acrotelm and the thickness of material involved in water transmission. According to the Dupuit–Forchheimer approximation, the potential gradient may be taken as equal to the gradient of the water table, which is itself a characteristic of the microtope (for a discussion of mire microtopes see Ivanov (1953), Romanova (1961), pp. 146–148 and Botch and Masing, Part B, Ch. 4). Ivanov (1975a) mentions that some of the least compact acrotelm materials become supersaturated in wet conditions so that the whole structure expands to admit intrusions of free water. This applies especially to carpets of plants like *Sphagnum cuspidatum*. Romanov (1968a) associated this process especially with waterlogged hollows in *Sphagnum–Eriophorum* or *Sphagnum–Scheuchzeria* microtopes, in which the bulk density and hydraulic conductivity of the acrotelm underwent regular seasonal variation.

Romanov (1968a) also deplored the lack of information on the variation of hydraulic conductivity of fen acrotelms with depth, and the situation appeared to have improved little during the thirteen years preceding the Eberswalde symposium (see, for instance, Wertz, 1975). Romanov quoted results for *Carex–Hypnum* peats from

TABLE 3.9

Values of the parameters B and m in eq. 3.37 for seepage throughflow in the acrotelms of bogs in the forest zone of the U.S.S.R. (based on K.E. Ivanov, 1957; after Romanov, 1961) (The limiting depth d specifies the upper surface of the soil horizon to which the quoted values apply.)

Microtope	B	m	d (cm)	Level at which $d=0$
Sphagnum–chamaephyte with dwarf *Pinus*	3670	2.38	5	mean elevation of hummocks and hollows
Sphagnum–Eriophorum–chamaephyte with *Pinus*	2290	3.0	5	mean elevation of hollows
Pinus with *Sphagnum* and chamaephytes	41 700	3.8	5	mean elevation of hollows
Ridges: *Sphagnum*–chamaephyte with *Pinus*	3670	2.67	8	mean elevation of ridges
Pools: *Sphagnum*–*Eriophorum*	2290	2.67	3	mean elevation of pool surfaces

Belorussiya (White Russia) showing progressive decline in hydraulic conductivity from 3.1×10^{-3} cm s^{-1} at 0 to 50 cm depth to 6×10^{-5} cm s^{-1} in the 100 to 150 cm layer. From the Soviet work it seems that the hydraulic conductivity of bog acrotelm materials may exceed anything found in fen acrotelms by more than four orders of magnitude. However, experience from Germany suggests that slightly humified (H3) fen peats are more permeable than bog peats of the same humification (Baden and Eggelsmann, 1963; Rycroft et al., 1975b; Eggelsmann, 1978). Nevertheless the discrepancy may be more implied than real, since the more eutrophic conditions in fens probably favour more rapid decomposition (Given and Dickinson, 1975; Codarcea, 1977). This ensures that the coarse-textured and scarcely humified moss materials in the acrotelms of raised bogs have little counterpart in fens.

Permeability in the catotelm. As in the acrotelm, so in the catotelm the permeability of peat is correlated with various other attributes. The factors which influence it have been discussed by many authors, notably Baden and Eggelsmann (1963), Boelter (1965), Päivänen (1973) and Eggelsmann (1978). The subject was reviewed by Rycroft et al. (1975a). Peats from Holarctic mires in the middle stages of humification generally show hydraulic conductivity values between 10^{-4} and 10^{-2} cm s^{-1}. The actual value obtained depends upon:

(1) the botanical composition — *Sphagnum* (least permeable) $<$ hypnoid mosses $<$ *Carex* $<$ *Phragmites* (most permeable);

(2) the degree of humification — the least humified peats of given botanical composition are the most permeable;

(3) the bulk density — this is strongly and positively correlated with humification (Boelter, 1969; Päivänen, 1969), so that permeability varies inversely with bulk density, though Päivänen (1973) found that humification on the scale of Von Post accounted better for the variation in permeability than did bulk density;

(4) the fibre content — this is inversely related to humification (Farnham and Finney, 1965), so that Boelter (1970) found that hydraulic conductivity was positively correlated with fibre content;

(5) the substance volume — this is the European term for the porosity of peat, obtained by dividing the bulk density by the specific gravity of the solid matrix, and is thus directly correlated with humification and inversely correlated with hydraulic conductivity (Baden and Eggelsmann, 1963);

(6) the drainable porosity or water yield coefficient (pp. 111–116) — the most readily drainable pores also present the least resistance to water transfer through a saturated aquifer, so that Boelter (1976) and Galvin (1976) found a close correlation between the logarithm of K and drainable porosity; and

(7) the surface loading — this diminishes per-

meability through its effect in increasing the bulk density of this highly compressible medium (Wechmann, 1943; Hanrahan, 1954).

A summary table of field data on peat permeability from various sources has been given elsewhere (Rycroft et al., 1975a). It lists "hydraulic conductivites" ranging from 6×10^{-8} cm s^{-1} for highly humified blanket peats up to 5×10^{-3} cm s^{-1} for slightly humified fen peat. Päivänen's (1973) results, which were omitted from this table, lay approximately in the range 6×10^{-6} cm s^{-1} (*Sphagnum*, H8–H10) to 10^{-3} cm s^{-1} (*Sphagnum*, H3; *Carex*, H3–H5, brushwood, H3–H6). These latter data have the advantage that the test method was described in some detail. The seepage-tube method of Luthin and Kirkham (1949) was used, with readings taken only during the first two thirds of water-level recovery which, as shown above, probably led to the conductivity being overestimated. It is less easy to judge the worth of many of the data used in compiling the table of Rycroft et al. (1975a) because some workers gave few details of method. The results of Rycroft et al. (1975b) were also omitted from that table. For a series of seepage-tube tests in the *Sphagnum–Eriophorum–Calluna* peat (H3–H6) of the Dun Moss mire expanse their values generally lay in the range 10^{-5} to 10^{-4} cm s^{-1}. In contrast to those of Päivänen, these data refer to the latter parts of tests ($t \rightarrow 10^3$ min).

In interpreting these data, it should be remembered that they generally relate only to the top of the catotelm. The writer knows of no permeability tests in mires carried out at depths below those examined by Galvin and Hanrahan (1967), who imposed a maximum draw-down to 2.3 m.

It is, however, clear from the data quoted above that peat in the catotelm is generally less permeable than in the acrotelm, the difference in permeability between these two layers varying from, at one extreme, as little as a single order of magnitude in some fens with moderately humified catotelm peat to, at the other extreme, as much as eight orders of magnitude in actively growing raised bogs with highly humified catotelms. For Holarctic raised bogs in which *Sphagnum* species have long dominated the mire expanse it seems likely that an acrotelm with K typically averaging about 1 cm s^{-1} would overlie a catotelm with average K about 10^{-4} cm s^{-1} — a difference of about four orders of magnitude.

Potential gradients. From the tendency discussed above (pp. 106–109) for mire water tables to remain near the surface, it appears that water-table and surface gradients are very similar in mires. If, on the basis of the Dupuit–Forchheimer approximation (eq. 3.36), one equates water-table gradients with potential gradients, it follows that from surface gradients one can infer the potential gradients associated with seepage parallel to the surface (the "throughflow" of hydrologists). This is the approach used in Ivanov's (1975a) analysis of mire hydrodynamics.

Various aspects of surface gradient in relation to water discharge from raised bogs were considered by Eggelsmann (1967) and Eggelsmann and Schuch (1976), who used old maps to reconstruct generalized profiles for the period before human interference. On the basis of 64 raised bogs in Lower Saxony and 28 in Bavaria, they presented the idealized profiles shown in Fig. 3.35. It will be noted that the so-called "cupolas" or "domes" of peat in these mires do not slope evenly, neither are their profiles arcuate. Instead the mire expanse is almost horizontal over a relatively large area, but the rand or marginal slope becomes rapidly steeper from its crest towards the lagg. Thus, Eggelsmann and Schuch aptly describe these profiles as having the shape of a watch-glass. Similar, though flatter, profiles are described for southwestern Finland by Aartolahti (1965); for the dry continental interior of Europe by Kulczyński (1949); and for the extensive forested raised bogs of the Southeast

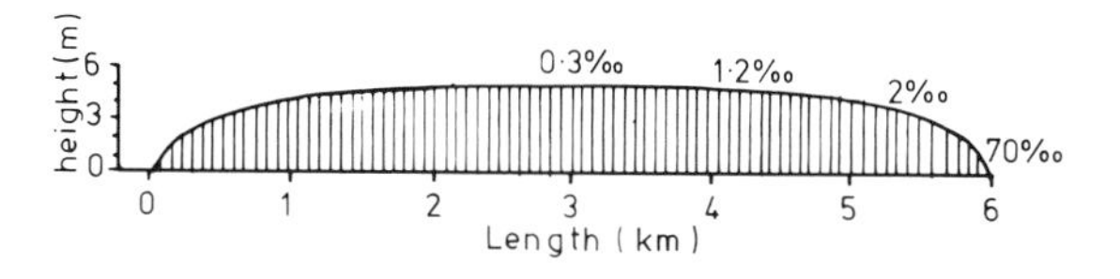

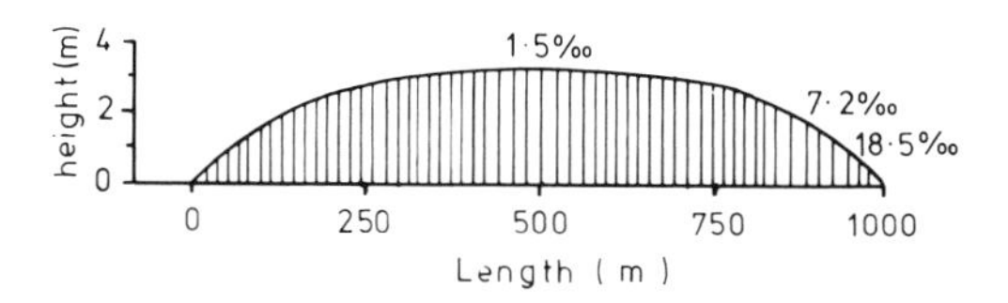

Fig. 3.35. Generalized profiles of raised bogs in Germany. Above: mean of 64 bogs in Lower Saxony, vertical scale exaggerated ×125. Below: mean of 28 selected bogs in Bavaria, vertical scale exaggerated ×50. Surface heights above lowest altitude of lagg stream. Figures above the bog surface are slopes in parts per thousand. (After Eggelsmann and Schuch, 1976.)

Asian tropics by Anderson (1964) and Andriesse (1974). In Lower Saxony, average surface gradients attained 70‰ near the lagg. The gradient relationships of raised bogs are of great significance for their stability, as explained below.

Large areas of blanket bog may also be almost level, but the capacity of this form of mire to develop on hillsides causes many examples to show marked surface gradients throughout their history. Evidence of surface gradients in Scottish blanket bog is given by McVean and Ratcliffe (1962), who recorded the following mean gradients beneath various vegetation noda (maxima in brackets):

Trichophoreto–Eriophoretum typicum[1]	50‰	(180‰)
Empetreto–Eriophoretum	70‰	(270‰)
Trichophoreto–Callunetum[1]	100‰	(270‰)
Calluneto–Eriophoretum	100‰	(470‰)
Trichophoreto–Eriophoretum caricetosum[1]	100‰	(530‰)
Molinia–Myrica nodum	180‰	(470‰)
Molinieto–Callunetum	250‰	(470‰)

The unevenness of slope in blanket bog causes ground-water flow to converge towards the steepest gradients that are locally available, where it becomes canalized into "water tracks" but seldom appears on the surface (Ingram, 1967).

Fens which develop during the course of hydrosere succession in mineral-rich ponds or lakes (Godwin, 1978) or in broad lowland river valleys (Balátová-Tuláčková, 1972) seldom show appreciable surface gradients. Fens and peat flushes in the uplands may slope steeply (McVean and Ratcliffe, 1962; Birks, 1973; Rybníček, 1974), and so may the lagg fens which comprise the marginal water tracks of raised bogs (Aletsee, 1967).

Flux densities of ground-water flow. Too little is known about the variations of peat permeability to generalize about seepage fluxes in intact mires. Moreover the history of the subject makes comment capricious in its emphasis.

For raised bogs, interest centres on the relative magnitude of seepage fluxes through the acrotelm and the catotelm. Suppose 8 m of catotelm with $K=10^{-4}$ cm s^{-1} underlie 25 cm of water-filled acrotelm with $K=0.7$ cm s^{-1} [an estimate for *Sphagnum–Eriophorum* microtopes based on eq. 3.37 and given by Romanov (1968a)] one finds, by assuming Darcy's Law and applying eq. 3.30, that the ratio $U_a/U_c=220$. Thus, the greater depth of aquifer in the catotelm is more than counteracted by its much lower permeability, and the significance of the acrotelm for ground-water flow is confirmed. General considerations of acrotelm architecture expressed in eq. 3.37 imply that heavier precipitation or more rapid snow melt involve progressively shallower and more permeable regions of the acrotelm in flow as the water table rises. Thus, the system responds by more rapid total seepage (see below), and the acrotelm is seen to be an aquifer with a high capacity to transmit storm- or melt-water, protecting the ecosystem from the erosion that would result from run-off over a loose and unstable surface (see above). The contribution of the catotelm to ecosystem stability will be considered below. Meantime, one may note that the same values of grad ϕ apply in the acrotelm as in the underlying catotelm.

In Europe, blanket bogs attain their greatest development in the northern and western parts of the British Isles. In the Outer Hebrides, for example, the Isle of Lewis is almost entirely covered by mire of this kind, either intact or reclaimed. Most British blanket bogs are rather frequently burnt over in a crude attempt to enhance their grazing value (in Scotland the practice is termed "muirburn"). In consequence the acrotelm often disappears and may in extreme cases be replaced by a gelatinous surface layer of algal origin which has very low permeability (Conway and Millar, 1960; McVean and Lockie, 1969). The sheet flow of storm water over these haplotelmic[2] mires may be an important contributory cause of peat erosion, such as the sheet erosion of blanket bog described by Bower (1961). Since pipe formation is often also involved (see above), it is scarcely surprising to find many areas of blanket bog in a highly eroded condition. In less extreme cases most liquid-water discharge may still be accommodated as seepage flow in the catotelm. This is a feature of blanket bogs along the northwestern seaboard of Scotland, where the extremely rugged relief of the mineral substrate produces very numerous bogs, each of small area, with very variable surface gradients and an abundance of water tracks. Ingram (1967) found that the more steeply inclined water table in a

---

[1] *Scirpus cespitosus* has been called *Trichophorum caespitosum*.
[2] **Haplotelmic:** lacking an acrotelm (Ingram, 1978).

Molinia–Myrica water track was accompanied by higher peat permeability, with the result that seepage velocities in the track exceeded those beneath *Trichophoreto-Eriophoretum typicum* in the surrounding mire expanse by a factor of 25. Probably the distinctive vegetation of these water tracks owes its existence to the difference in mineral nutrient availability that accompanies the difference in seepage rate. The literature is full of such suggestions, and the very use of the term "rheophilous" in mire descriptions is extremely suggestive.

The effect of seepage flow on vegetation and mire development is likely to be of particular interest in fens. Godwin and Bharucha (1932) showed that the rate of recession of water table at Wicken Fen was inversely related to distance from the open drainage canals, thus clearly showing the importance of ground-water flow. Kulczyński (1949) considered in detail the role of fen peats as aquifers in the flat region of Poles'ye, between Brest-Litovsk and Pinsk, which is drained by the rivers Bug and Pripyat'. Here reeds and tall sedges were sometimes found associated with short sedges and mosses, growing in a diplotelmic[1] soil with well-demarcated acrotelm and catotelm layers. The tall plants were rooted in the catotelm, the short in the acrotelm. In some cases rising water tables caused the mossy acrotelm to float clear of the catotelm. This allowed a free flow of water to occur in the intervening gap, without submerging the acrotelm surface. Kulczyński called this condition "dysaptic". In other cases, termed "cryptodysaptic", the acrotelm and catotelm remained connected and were less well differentiated. Here the acrotelm merely expanded to accommodate a rising water table. In the dysaptic fens the complete mechanical independence of the two layers was sometimes accompanied by contrasting mineral nutrient status, leading to a mixture of tall, eutrophic species such as *Phragmites australis* or *Rumex hydrolapathum* with short, more oligotrophic plants such as *Scheuchzeria palustris* and *Sphagnum* sect. *Cuspidata* in sites with a slow flow of water. Rapid water movements and widely oscillating water tables eliminated the upper layer, forming the haplotelmic tall helophyte fens characteristic of river flood terraces in which surface flow was the salient hydrological feature. However a very sluggish flow enabled the small sedges to assume dominance and ultimately to overwhelm the taller species and form a more oligotrophic dwarf-sedge fen.

It is clear from this that seepage-flow characteristics have much to do with the differentiation of fen vegetation. However, despite the smaller range of permeabilities characteristic of fen peat, there have been few attempts to explore this in greater detail except in the context of mire reclamation.

[1]**Diplotelmic**: having both acrotelm and catotelm (Ingram, 1978).

Pipe flow

Newson and Harrison (1978) considered that hydraulic analysis of pipe flow (see above) would be difficult due to the irregularity of the pipes. The relationship between flow velocity and discharge is shown in Fig. 3.36. On Plynlimon, water flow in pipes often ceased altogether in dry weather, but at moderate discharges the pipes were one quarter to half full, and dye tests showed velocities of about 0.1 m s^{-1}. Heavy storms induced water jets at vertical shafts, and subterranean leakage was also apparent. Hydrologically, this study suggests that pipe flow is best regarded as a discharge process which is in series with others. Seepage controls the rate of entry of water into the pipes, but once it has

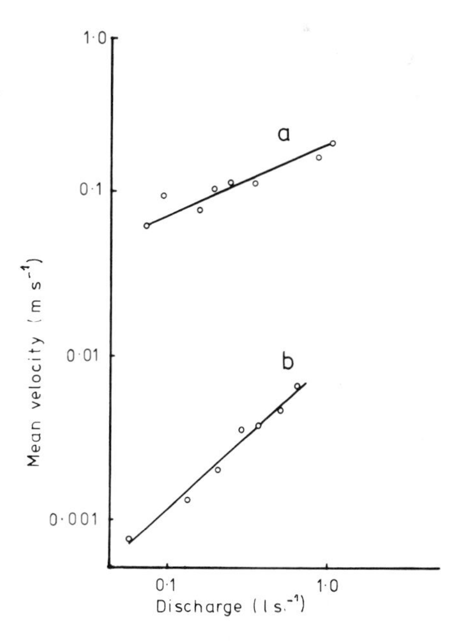

Fig. 3.36. Velocity/discharge relationships (log scales) for (*a*) pipe flow and (*b*) overland flow (surface runoff) in a bog at Plynlimon, mid-Wales. (Brine irrigation experiments by Miss P. Newman; from Newson and Harrison, 1978.)

reached them it is removed much more rapidly than it would be in their absence. This, in turn, affects the rising limb of the storm hydrograph and shortens the concentration time of the catchment. Jones (1981) thought pipes might prevent "bog bursts", but it is not yet known what causes these.

Overland flow

The more local character of Hewlett's overland flow (see above) accords much better with experience of mires than do Horton's ideas. In raised-bog ecosystems, for example, one does not observe sheet flow on the mire expanse when the acrotelm is intact. Storms may cause pools to enlarge and, indeed, to overflow their normal margins. However, as air photographs have strikingly and repeatedly shown, bog pools form a regular pattern, the long axes of elongated pools lying parallel to the surface contours (Fig. 3.37). This pattern is less remarkable than is sometimes claimed; water would run out of elongated pools in any other orientation! This preferred orientation is especially obvious in the aapa mires of northern latitudes where such pools were first called "flarks", and it does much to preclude their participation in overland flow. Moreover the large, steep-sided, dystrophic, permanent pools or "dubh lochans" that occur in Scottish raised and blanket bogs have bottoms of catotelm peat of low permeability (Boatman and Tomlinson, 1973). Thus, their enlargement and overflow merely brings them into hydraulic continuity with a greater depth of the acrotelm. Hence they are best regarded as local components of the ground-water flow system. In this respect, their hydraulic conductivity is infinitely great. However, since their water surfaces are sensibly horizontal, they are also virtually isopotentials and presumably have little influence on the overall ground-water flux, whose magnitude is still controlled by the permeabilities and potential gradients in the surrounding matrix of plants and peat. Ivanov (1975a, b) has treated the hydrology of mire pools in great detail. Theoretical and practical studies of their life history and hydraulics have led him to conclude that their enlargement during wet periods and contraction in dry periods favour the long-term stability of raised bogs by buffering them against climatic fluctuations.

In the west there has been little relevant experimental work on surface run-off, but recent studies in mid-Wales are a useful beginning. Newson and Harrison (1978) described experiments at Plynlimon by Miss P. Newman, who induced overland flow in a convex valley bottom *Sphagnum–Eriophorum* bog by artificial upslope irrigation with brine. These trials confirmed the high infiltration capacity of mire soils and communities already referred to (pp. 70–71). They also suggest that overland flow occurred, not as a result of ponding above unsaturated layers of soil on the Horton model, but only in consequence of a rise in the ground-water table itself and its appearance above the surface. Indeed, it proved difficult to separate overland flow from acrotelm seepage through *Sphagnum* hummocks. At low discharge rates, overland-flow velocities were less than 10^{-3} m s^{-1} — about two orders of magnitude lower than pipe-flow velocities at the same discharge rate. However, overland-flow velocities rose more steeply with discharge than did pipe flow (Fig. 3.36). The irrigation rates required to generate overland flow were equivalent to very high rates of precipitation.

For blanket bog, some estimates of the circumstances causing overland flow were given by Tallis (1973). They relate to the eroding summit plateau of the southern English Pennines where the mire vegetation is dominated by *Eriophorum vaginatum*, associated with *E. angustifolium* and various chamaephytes. Tentatively, in view of his incomplete hydrological data, Tallis concluded that up to 20% of a year's precipitation of 1060 mm was dispersed into the groughs by overland flow. As in the study reported by Newson and Harrison, the process seemed to start above a saturated profile. It was not initiated by a shortfall of infiltration capacity in the Horton manner. Thus, the meteorological conditions for overland flow depended on antecedent events affecting the water table, and ranged from an estimated mean surplus ($P-E$) of about 0.5 mm day^{-1} for a water table at -6.5 cm up to about 50 mm day^{-1}. A mean surplus of as little as 0.7 mm day^{-1} was thought capable of producing overland flow from an initial water table at -25.5 cm during a nine-day period.

Circumstances are quite different in the marginal lagg fen of a raised bog. Here the lagg stream may be invisible in dry weather. At other times it swells and may in some cases become a torrent involving most of the area of the lagg. After the storm the

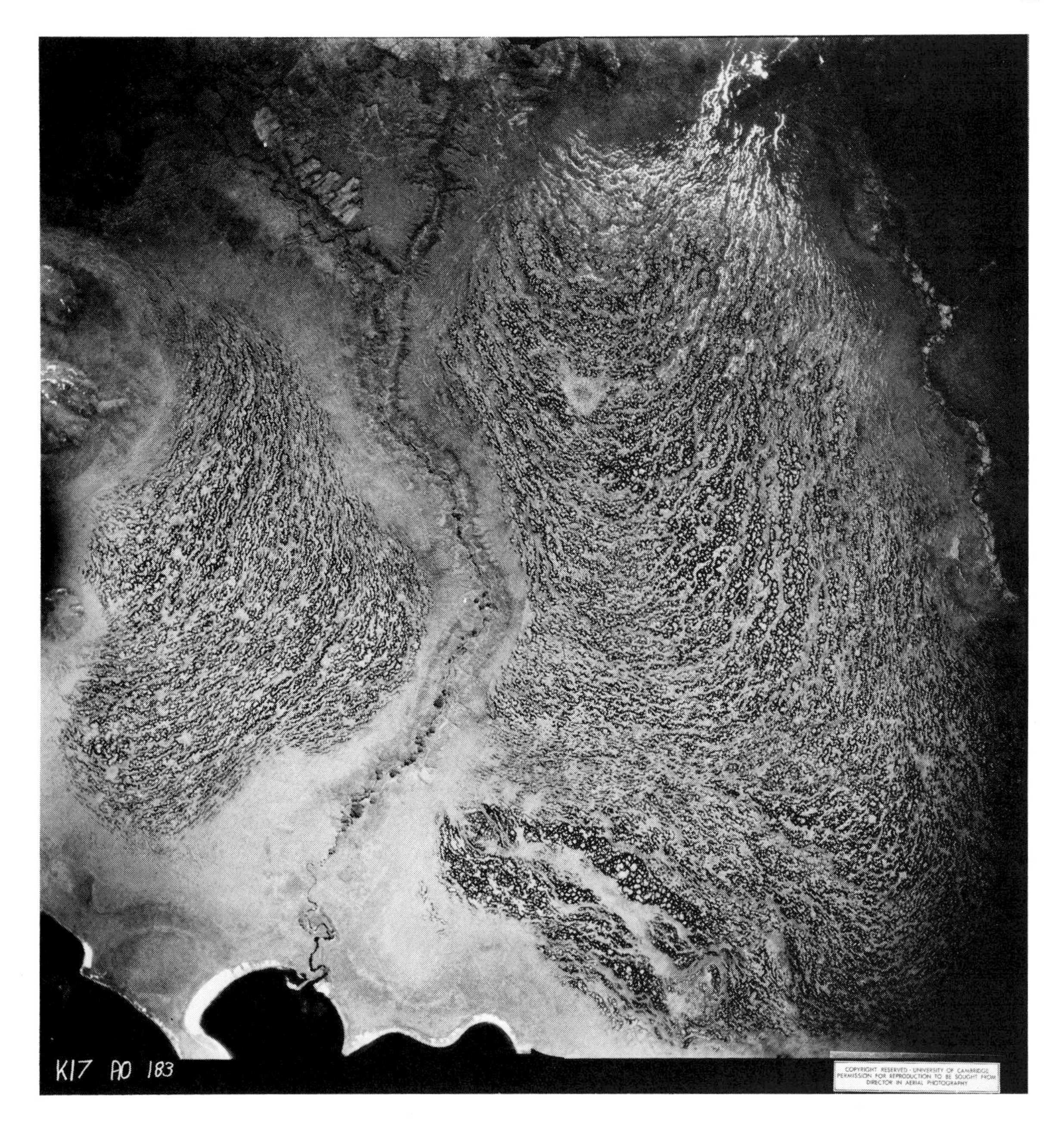

Fig. 3.37. Air photograph of Claish Moss, Sunart in west Scotland. This mire lies beside the shore of the fresh-water Loch Shiel, 10 km inland from the Sound of Arisaig. The lake shore forms its northern margin, from which there is an overall upward gradient towards the southern margin (the brightly illuminated area towards the top of the photograph), where the mire abuts against a low range of rugged hills. On the basis of the stratigraphy, surface vegetation and overall morphology of the mire, Moore (1977) compared it with the eccentric raised mires of Scandinavian authors. The pattern of pools or flooded flarks alternating with hummocks or low ridges is typical of many raised bogs in continental Eurasia. The two sections flanking the stream would be described by Ivanov (1975a) as mesotopes and the whole mire as a macrotope, following the proposals of E.A. Galkina. The length of the stream, from its confluence to the lake shore, is approximately 1 km. Photograph by J.K. St. Joseph, Cambridge University collection, copyright reserved.

leaves of tall sedges and other less rigid structures remain bent over in the direction of the current, while leaf litter is left stranded against tussocks of *Juncus* and other stiff species. These, however, are strictly local effects. On the Hewlett view, raised bogs are catchments in which the marginal fen marks the limit of extension of the contributing area. Ecologically, this view gives fresh insight into the differentiation of mire margin and mire expanse. For, while the prostration of tall herbs in the lagg fen protects the smaller and more delicate plants from the swollen flow of surface water, it is unlikely that the sparser, stiffer and shorter angiosperms of the mire expanse would protect the loose carpet of rootless *Sphagnum* from the erosive effects of sheet flow. Overland flow and erosion in blanket bogs have already been discussed.

The expansion of open channels over neighbouring fen communities was described by Kulczyński (1949) in the vast fens of the Central European watershed around Pinsk. The most frequently flooded fens with the most rapidly oscillating water levels were those with more steeply sloping surfaces or which occupied the shallowest basins. In these, aerobic decomposition during summer water-level minima produced dense peat, liable to sink beneath the water at other seasons. This tendency was increased by the deposition of dense and slimy allochthonous sediment during flooding. Erosion of the dense, slightly permeable peat led to canalization of the drainage into channels which gradually became hydraulically isolated from the hinterland, provided routes for rapid run-off, and shortened the periods of inundation of the surrounding fen with an overland flow of nutrient-rich water. As a result, these "fluvial fens" became almost pure *Phragmites* swamps with a few floating-leaved hydrophyte associates, but without a ground layer of mosses. The more retentive fens of deeper basins with more slowly oscillating water levels developed differently, with flow mainly beneath the surface as described above (p. 137). Here, as always in mires, differences in hydrological behaviour are intimately connected with the ecology of the vegetation.

Open channel flow

This section is concerned with results of measuring channelled surface discharge within, and in particular from, mires and with the evaluation of the fluxes denoted by C_e in Fig. 3.28.

Winter. Most mires have developed in the cool interiors of the northern continental land masses. Here winter precipitation falls as snow which reflects almost all incident radiation, and polar air-masses at high pressure prevent the incursion of the warmer oceanic air streams which cause intermittent melting along the seaboard. Thus, snow accumulates during the winter, little melt-water is released into the surface-channel network and the months pass unmarked by sudden hydrological events.

In a study of the hydrology of a small bog of the raised type in northern Minnesota, Bay (1969; see also Boelter and Verry, 1977) found virtually no discharge in the open drainage channel from the catchment between late November or early December and late March or early April, despite the occurrence of 20% of annual precipitation during these months. The bog concerned was covered with *Sphagnum* spp. and scattered dwarf shrubs (Bay, 1970). It bore a poorly grown stand of *Picea mariana*, and the data relate to a five-year period from 1962. Similar results were obtained in many of the raised bogs of the Soviet Union for which hydrographs were presented by K.E. Ivanov (1957).

The process of infiltration into frozen mineral soils has been studied by Stoekeler and Weitzman (1960) and by Willis et al. (1964). Similar studies on peat were initiated by Pavlova (1970). When surface layers are saturated in autumn, freezing produces a relatively impermeable layer, and little infiltration can occur. Initially unsaturated layers remain permeable, an aspect of importance in the growth of palsa mounds (see p. 118). During the winter period this has the consequence that water tables fall steadily in response to an upward thermodynamic flow of moisture towards the descending frost front. However, in late winter the trend is reversed, as liquid melt water becomes available for infiltration. Whatever the detailed mechanism, the result at the end of a continental winter is a diminishing blanket of melting snow and ice overlying a saturated soil, a situation which also obtains in mires. In consequence the onset of spring is marked by exceedingly high discharges recorded in the channel network, and hydrographs from mires in the Soviet Union and North America invariably show maxima at that season. In Minnesota, April and May are normally the peak discharge months. They are also fairly rainy, and Bay (1969) found that, for the five years from 1962, a mean of 57% of total annual discharge through streams draining his raised bog site (see above) occurred before the beginning of June. Very high spring discharges are also seen in Soviet mire hydrographs (K.E. Ivanov, 1957), following negligible flows in winter.

The winter discharge pattern from mires in oceanic climates is quite different. Here the incursion of warm air-masses is liable to cause snowmelt at any time, and this is quickly reflected in channel hydrographs. Reviewing data from various German mires, Eggelsmann and Schuch (1976) showed that numerous high discharge peaks occurred throughout the winter, and that, in north Germany, about 79% of annual discharge occurred during the winter period November to March, which experienced about 42% of annual precipitation. These data relate to the eight years from 1951, and were collected from the unreclaimed Königsmoor raised bog (see p. 88 above for the site details). Even from the Chiemsee mires in the foothills of the Bavarian Alps, Eggelsmann and Schuch noted discharges for the November–March period which amounted to 38% of the annual total, despite the relatively high precipitation which occurs during summer in that area. In both these German studies, which are described in detail by Baden and Eggelsmann (1964, 1970) and by Schmeidl et al. (1970), catchment delimitation was achieved artificially by means of open drains. Similar experience in northern Germany was reported by Uhden (1967) and in Central Europe by Ferda and Pasák (1969).

At the Komosse raised bog complex in Sweden, the three-year hydrograph given by Johansson (1974) showed a winter discharge pattern intermediate between the continental and oceanic types. During the winters of 1970 and 1972 there were long periods with little flow, followed by a well-marked spring peak, but in 1971 there occurred substantial midwinter discharge and the spring peak was correspondingly lower.

The most oceanic raised bog so far subjected to intensive hydrological study is probably Blacklaw Moss, Lanarkshire, which lies at an altitude of about 230 m on the northwestern flank of the Southern Uplands of Scotland (Robertson et al., 1968). The plant cover was mainly *Calluna vulgaris* and *Eriophorum vaginatum* with discontinuous *Sphagnum*, suggesting some light drainage and possibly burning in former times. The experimental plot was delineated by a system of open perimeter drains, and discharge was measured with a V-notch weir over the three years 1959 to 1961. The mean annual rainfall was 813 mm (see also Holland, 1967), with substantial falls occurring in all months, maxima in late summer to early winter and minima in late spring and early summer. Snow was not important. Here the march of monthly discharge totals followed closely the march of precipitation during the winter months of November to February. There was no spring discharge peak. The data therefore resemble those from northern Germany and form a link with results from blanket bogs in wetter climates.

Since blanket bogs develop only in the mild wet oceanic climates of the continental margins, one expects their winter discharge patterns to differ greatly from those of the typical raised bogs of the continental interior. Conway and Millar (1960) analyzed the discharge over V-notch weirs on streams draining variously treated areas of deep blanket peat at altitudes exceeding 500 m in the upper catchment of the River Tees in the northern English Pennines. Precipitation, falling mostly as short-lying snow in winter, amounted to a mean annual total of about 2000 mm. It was rather evenly distributed seasonally, though with a slight tendency for the wettest months to occur in autumn and early winter, with spring and early summer somewhat drier. Winter discharge over all the weirs closely followed precipitation, there being little prolonged storage as either snow or ice even at this altitude.

Similar work on artificially delimited blanket-bog sub-catchments at Glenamoy in County Mayo, western Ireland, was described by Burke (1975). Mean annual precipitation was 1265 mm, this being a low-altitude site. The seasonal distribution resembled that in the northern Pennines, although snow was unimportant. Again the winter surface discharge, from open drains surrounding an otherwise undrained plot with its natural vegetation intact, followed closely the pattern of precipitation, wet days generating high flows and *vice versa* (Fig. 3.38).

In the windy climates of the western seaboard of Europe the flux density of open-channel discharge from bogs falls noticeably short of catchment input in the form of precipitation, even in midwinter. For instance, Conway and Millar (1960) found a short-

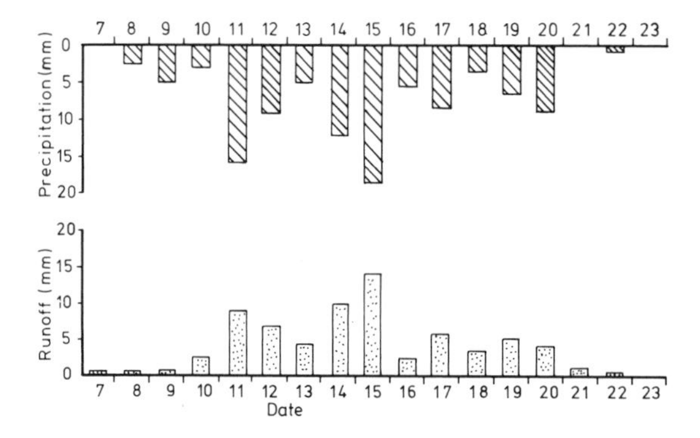

Fig. 3.38. Daily totals of precipitation for February, 1971 at Glenamoy, in northwestern County Mayo, Ireland, compared with surface run-off from blanket bog. Total precipitation 102.2 mm; total runoff 72.3 mm. (After Burke, 1975.)

fall of 10 to 25%. Similar effects are apparent in the data of Robertson et al. (1968) and Burke (1975). In view of the rather different methods of catchment delimitation used in these studies, it is unlikely that the discrepancy is entirely due to leakage past the stream gauges, especially since rain gauges in such climates are liable to underestimate precipitation (pp. 72 and 76). At a season when inputs of radiant energy are low the explanation more probably lies in the high flux density at which sensible heat is made available by atmospheric turbulence, causing intercepted water to evaporate quickly from wet plants.

Very little information is available on the surface-water discharge characteristics of fens, whether drained or otherwise. The area of mire reclaimed for agriculture at Crossens near Southport in northwestern England unfortunately includes peats of various origins from limnic to oligotrophic bog, so little of specific reference to fens can be deduced from experiences of pumping operations recorded by Hall and Prus-Chacinski (1975). However, Bavina (1975) has given data for a nine-year period from an intact flood-plain fen in the southern part of the forest zone of the western U.S.S.R. Surface and subsurface discharges were added to obtain these data. If precipitation and evapotranspiration are ignored, discharge was exceeded by recharge in five of the years during the early winter period of November to February, but the net decrease in storage due to these non-meteoric fluxes was small, amounting to an overall mean of only 5 mm during this snowbound season. Snow melt during the spring period of March and April produced the expected peak of liquid-water discharge in most years, with the mean excess over recharge amounting to 68 mm — more than half the annual total.

Summer. Peat is formed by plants which are typically helophytes or hydrophytes. Because these plants require abundant moisture for growth, the circumboreal mires seldom develop in climates with scanty precipitation during the warm season of growth, and even in lower latitudes growth is interrupted during the dry season (Moore and Bellamy, 1973). Most of the mires for which hydrological data are available therefore experience moderate to abundant summer rainfall. In consequence their discharge behaviour varies less with geographical location during the warm season than it does in winter, and generalization is somewhat easier. Useful reviews are given by Boelter and Verry (1977) and by Goode et al. (1977), especially the section of the latter entitled "The Water Balance of Peatlands".

Hydrographs from raised and blanket bogs usually show rapid responses to precipitation. The concentration time during which open-channel discharge builds up to its peak value is often very short. This is partly due to the tendency for these mires to be situated in the headwaters of the streams, so that the catchment is of limited area and the mean travel time is short. This feature is apparent in most of the studies cited above. Another feature is the close dependence of the height attained by the discharge peak upon the antecedent moisture conditions. This is to be expected: high water tables leave less room for the temporary storage of storm water in the soil profile. It is well illustrated by the work of Bay (1969) in Minnesota (Fig. 3.39).

There is a suggestion that bogs which are entirely free from human interference, especially those with an intact *Sphagnum* carpet, produce lower and more delayed peak flows, followed by long and shallow recession curves. This is apparent from the comparison of variously managed sub-catchments in Pennine blanket bog by Conway and Millar (1960). It also emerged from an analysis of Bay's (1969) raised-bog results by Boelter and Verry (1977). The relatively high storage capacity of the intact moss acrotelm and the level surface relief were suggested

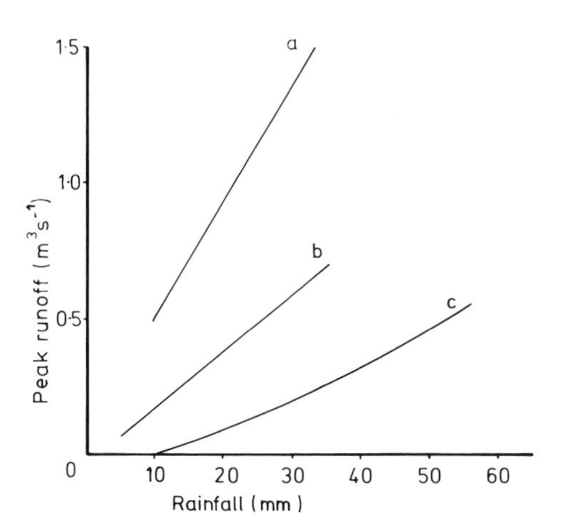

Fig. 3.39. Response of liquid water discharge to storm rainfall for a mire of the raised bog type (area 9.7 ha) in Minnesota: a = high water tables, fluctuating above the average moss surface and within the hummock–hollow microtophography; b = medium water tables; c = low water tables, more than 15 cm below the average hollow floor. (Work of R. Bay, after Boelter and Verry, 1977.)

causes, though the latter is not a characteristic of many blanket-bog areas.

However, prolonged recession curves are certainly not characteristic of summer hydrographs from raised bogs in general. Indeed, the most typical feature of warm-season hydrographs from the majority of the studies referred to above is the tendency for discharge to fall to low values or to cease completely after relatively brief dry periods. Two weeks without rain may often suffice to produce this effect, which was found in the Königsmoor in northwestern Germany by Baden and Eggelsmann (1964), on the Chemutovka catchment in the Krušné Mountains (Erzgebirge) of Czechoslovakia by Ferda and Pasák (1969), in various parts of the forest zone of the U.S.S.R. (K.E. Ivanov, 1953, 1957; Bavina, 1975), by Robertson et' al. (1968) at Blacklaw Moss in Scotland, by Bay (1969) in Minnesota, and at Komosse in Sweden by Johansson (1974). Of the intact raised bogs so far studied, only the Chiemseemoore in Bavaria were an exception (Schmeidl et al., 1970). Here, in the Alpine foothills, mean annual precipitation over the ten-year study period exceeded 1300 mm; over 840 mm of this fell in the summer half-year, and dry episodes were very rare. Such unusual conditions are bound to generate anomalous run-off patterns.

The low summer flows from most raised bogs always occur when water tables are low, and there are two reasons for them. The first is concerned with storage. As described above (pp. 106–111), most temporary variations in water storage in mires are accommodated by movements of the water table. These movements are, however, restricted by the limited depth of the acrotelm, so that the storage capacity of mire soils is much smaller than that of many mineral soils. The second reason is concerned with hydraulic conductivity in the acrotelm which is the effective aquifer for most of the contributing seepage. As shown above, this decreases sharply with depth (eq. 3.37). Therefore, as the water table falls during a dry spell, not only is the depth of effective aquifer increasingly diminished but throughflow is also increasingly confined to the least conductive parts of the acrotelm.

The diminution in channel flow resulting from these two effects is clearly shown by the curves in Fig. 3.40, which resemble the relationships obtained by Chapman (1965) and Bay (1973). Chapman also suggested ways in which the relationship might be used in water-budget estimating. A similar body of techniques has been used for predicting raised-bog discharge in the Soviet Union by empirical methods, based only on data about climate and water-table behaviour (Anonymous, 1971).

Discharges from the blanket bogs studied by Conway and Millar (1960) in northern England and by Burke (1975) in western Ireland, at the sites described above, were much reduced in summer. However, there were no months with negligible discharge from intact blanket bog. As in the Bavarian raised bogs of Schmeidl et al. (1970), this is not because conditions in the acrotelm differ from other mires, but because of the wetness of the summer climate. Burke, for instance, reports that on average rain fell on three days in four at Glenamoy, and was only slightly more frequent in winter than in summer. On the other hand one of the sub-catchments of Conway and Millar was extensively eroded by groughs, and continued to yield large quantities of silt which had to be removed from the weir stilling pool during the study. This area had also been severely burnt and its *Sphagnum* cover was negligible. Several periods occurred during which run-off in the stream draining this haplotelmic mire ceased entirely, and one of them lasted six weeks.

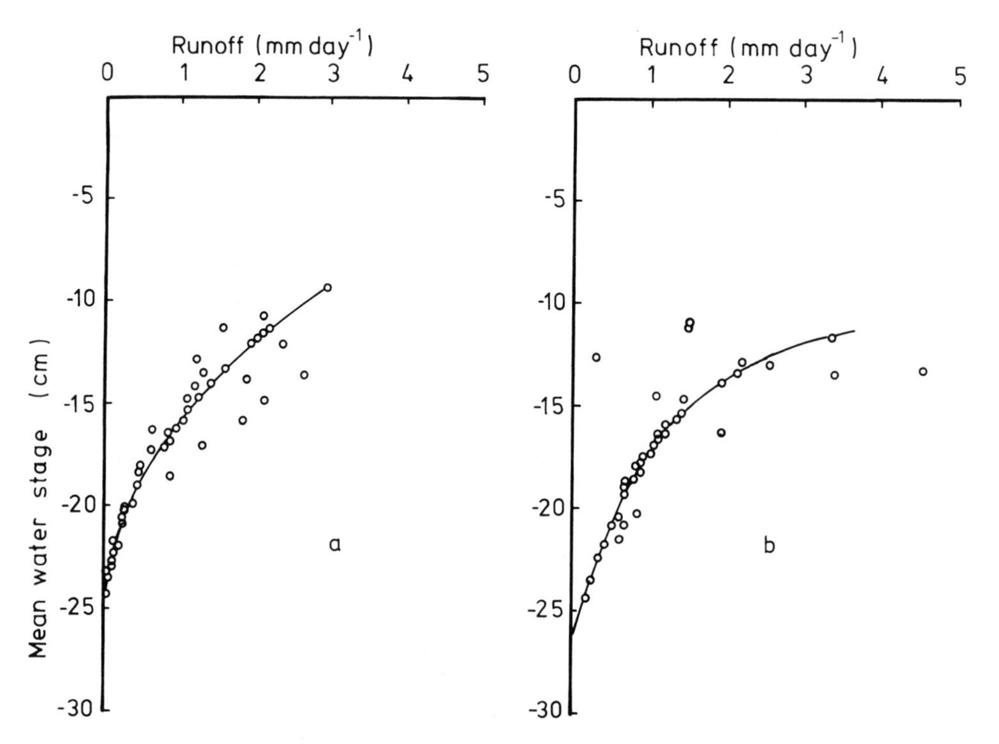

Fig. 3.40. Relationship between computed run-off and water stage at the Lammin-Suo raised bog near Leningrad. a. Based on two dip wells in central part of mire. b. Based on four dip wells in convex part. (After Romanov, 1961.)

Low summer discharge fluxes in open channels which drain bogs are generally explained by the rapid decline of water tables towards the least permeable parts of the acrotelm, this decline being a response not only to throughflow, pipe flow, etc., but especially to the high evaporative flux which prevails at this season whenever water tables approach the surface and potential evapotranspiration is high (see pp. 94 and 98).

Judging from the scanty evidence available, fens and bogs present an interesting contrast in their discharge behaviour during the summer season. Although fens show higher rates of evapotranspiration than bogs in similar conditions, discharge in their channel networks is much less variable during the summer. This is apparent in fens in the southwestern forest zone of the Soviet Union (Bavina, 1975), Germany (Eggelsmann and Schuch, 1976) and Minnesota (Boelter and Verry, 1977). These fens are not buffered against drought by frequent rainfall, by large capacity for water storage or by any other internal mechanism, but by their external sources of water from the hinterland. While bogs develop as independent hydraulic systems, the ground-water and soil-water aquifer of a fen is continuous with that of the mineral surroundings which drain into it (pp. 71–72) and from which most of its water is derived. Of the nine years' data reported by Bavina (1975) only those for one year, the driest, showed a marked difference in discharge between the fen and its surroundings. Goode et al. (1977) thought this might be due to the high evaporation rates which are sustained in fens — a conclusion according well with that above (p. 98).

The capacity of the fen hydraulic system to sustain an abundant supply of moving water, laden with minerals derived from the hinterland and often well aerated, is of fundamental ecological significance. In all except the most oceanic climates, the oligotrophic vegetation of bogs must withstand a low nutrient input exacerbated by periods of ground-water stagnation and drought which coincide with warm and sunny weather that would otherwise be best for plant growth. Fens, with their more permeable catotelms and greater proportion of deeply rooting species, can support eutrophic vegetation barely hampered by physiological stress during the season of growth (Ingram, 1967; Conner and Day, 1976; Gosselink and Turner, 1978; Richardson et al., 1978).

Conclusions on liquid-water discharge from mires

Several discharge processes operate in mires. They differ markedly in their physical nature and in their relationship to the hydrological system as a whole. Most are highly significant for the ecology of mire vegetation and all are closely involved with the growth and degeneration of whole mire ecosystems. They show interesting geographical variations which, although clearly related to climate and to the major features of mire distribution, require more widespread and long-term study before they can be clearly understood. Concentration on mires of raised-bog type means that there is relatively little evidence on how those processes operate in blanket bogs and fens. Nevertheless many of the features of mire-discharge hydrographs can be understood at least qualitatively by reference to evapotranspiration, to storage and to processes occurring in preceding stages of the hydrological cycle. There is still a notable lack of good hydrological comparison between mire catchments of different sorts or between mires and neighbouring mineral soils. But the evidence available lends little support to the notion, so important in arguments about wildlife conservation, that mires in the natural state are more useful water reservoirs and regulators of river flow than other ecosystems. They are not. Indeed, from the engineering point of view they present serious problems in these respects.

MIRE WATER RELATIONS AND THEORETICAL TELMATOLOGY

In the introduction to this chapter, reasons were given for supposing that hydrological factors and processes might play a uniquely significant part in the functioning of mire ecosystems. That no originality can be claimed for this idea is made clear by the character and extent of the literature reviewed above and by the frequency with which writers on mire ecology in general have suggested hydrological bases for the phenomena which they describe. The ecologist, faced with all this, should consider the activities of Elijah on Mount Carmel (I Kings 18) and recall that the significance of a concept ought to be measured, not by the frequency with which it is invoked, but by the insights to which it leads and its effectiveness in solving problems.

In a general way it is clear that many of the phenomena of mire distribution, mire development and mire ecology can be understood in hydrological terms. These include the association of different types of mire with climatic wetness noted above all by Kats (1971), and the interaction between development, morphology and the mineral nutrition of the plant cover which has been understood, in outline at least, since the time of Weber (1908). Such notions of causality are too basic to modern telmatology to be relegated to a specialized chapter and are therefore mainly dealt with elsewhere in this volume (Gore, Ch. 1, Tallis, Ch. 9).

The plan of the earlier sections of this chapter is based on the water-balance equation (eq. 3.2). It has been seen that, so far as mires are concerned, the terms of this equation embody inputs by precipitation and ground-water recharge that may not be well quantified and are certainly highly variable geographically. These interact with local physiography and geology to produce patterns of mire development that may also be complex on a local scale. Evapotranspiration seems to be more consistent, apparently varying in a rather regular manner so that, water bodies apart, the wettest types of mire generate the densest flux. Most characteristically, water storage is uniformly high, varying within rather narrow limits. This, coupled with the peculiar diplotelmic soil structure of intact mires, ensures that an input of storm water is quickly followed by a flood peak in the drainage system, oscillation in the efflux being somewhat reduced by compensating variations in evaporation rate. Except in fens, where the influx of telluric water predominates, the hydrograph of drainage from intact, diplotelmic mires is highly characteristic. It shows a series of peaks, each associated with a single storm and comprising a rapid rise as saturation is established in the upper, least humified and most permeable, regions of the acrotelm, followed by a rather steep recession as water is carried off through these layers. Dry periods, even fairly short ones, are marked by very low discharge rates because the bottom of the acrotelm and the entire catotelm are humified and of much lower permeability.

Despite substantial gaps in knowledge of hydrological processes in mires, it is therefore clear that a view of the evidence is now beginning to emerge which has three merits: it integrates several salient features of function and structure in these ecosystems; it provides a practical basis for assessing their significance in land management; and it suggests fruitful directions for future enquiry. The converse question now arises of how this view of mire hydrology bears upon understanding of the detailed nature of mire ecosystems themselves. Recent years have seen several attempts to answer this question quantitatively — to provide, that is, some theoretical justification for the intuitive notion of so many observers that water is important for mires and that therefore hydrology ought to provide helpful insights in mire ecology.

An essay at providing such a theoretical framework was recently made by Wildi (1978). It was based on a study of some thin blanket mires developed among clay-covered ground moraines on a valley floor in Switzerland. Wildi identified a network of causality in mire development. He regarded peat accumulation as the combined result of the production and decay of vegetation. These depended upon soil aeration and mineral nutrient supply which, in their turn, were governed by the throughflow of water. Finally, throughflow was controlled by the water balance and the slope of the substrate. Wildi's contribution has the merit of identifying hydrological processes as paramount among several functional attributes which have long been thought significant in this context (e.g. Ingram, 1967). In accordance with the current fashion in ecology he then proceeded to construct a computer simulation model, which combined a series of differential equations describing these processes. In my view this part of the exercise is much less helpful. For instance, the equations concerned with peat accumulation and mineral nutrition represent a naïve conception of these processes, neglecting for instance the important contributions of Clymo (1978; see also Clymo and Reddaway, 1973) to existing understanding of them. On throughflow, Wildi not only ignored all the large body of recent work on hillslope hydrology (e.g. Warrick, 1970) but tried to interpret the process with the aid of such concepts as the centre of gravity, important in the mechanics of rigid bodies but quite alien to the domain of fluid dynamics. The major structural feature which Wildi's scheme fails to comprehend is the diplotelmic nature of the substrate.

The most carefully considered theoretical approach currently available has been worked out since 1946 in the Soviet Union. Although it pays much more detailed attention to hydrological processes than to others in Wildi's scheme, its impact deserves to be greater because these aspects are so fundamental and because it incorporates such extensive experience of them. This approach is chiefly the work of a team based in Leningrad and has been described in a number of publications, most recently by Ivanov (1975a, b), whose writings should be consulted for a thorough introduction. Here there is space only for a brief outline.

Ivanov's work relates mainly to mires of raised bog type which cover large areas in the northwestern European Soviet Union and in western Siberia. At the present time these mires are at a stage of development in which the central parts form an oligotrophic mire expanse covered by various types of vegetation, each of which, together with its characteristic surface microtopography and acrotelm conditions, forms a particular mire microtope (microlandscape), a geotope (landscape element) at the lowest taxonomic level (Russian: *mikrolandshaft*). The bog microtopes of the mire expanse, together with endotelmic[1] pools, flarks or lakes and the fen microtopes of the mire margin, form a mesotope (Russian: *mesolandshaft*) equivalent to a single, isolated raised bog. Where raised bogs originating at separate centres of peat accumulation have coalesced during development, the result is a geotope at the highest taxonomic level: a macrotope (Russian: *makrolandshaft*).

A study of the vegetation and physiography of individual mire mesotopes or macrotopes by ground-controlled air photogrammetry permits the investigator to draw plans of each mire massif. Besides showing the distribution of microtopes it is possible to deduce the horizontal directions of the streamlines of throughflow and represent these also on the plans. These streamlines describe throughflow in the acrotelm only, and their trend in a vertical plane therefore follows the surface gradient of the mire. As mentioned earlier in this chapter,

[1]**Endotelmic** (Russian *vnutribolotnyi*): within the mire.

Soviet experience has confirmed that the acrotelm and the vegetation that has formed it in each mire microtope has a distinct hydrological regime. It loses water by evapotranspiration at a certain mean rate which can be predicted from local meteorological norms. Assuming negligible seepage U_c in the catotelm or loss U_b through the base of the peat deposit, and introducing a new term U_m for the local subsurface input from the adjacent microtope up the streamline, it becomes possible to draw up a local water balance for each microtope. In terms of eq. (3.28) this is written:

$$\bar{P}+\bar{U}_m-\bar{U}_a-\bar{E}=0 \qquad (3.38)$$

since, over the periods for which these mean values apply, one can neglect changes in storage. This would be the simplest form, applicable to microtopes with no pipe flow, no diffuse surface run-off and no surface streams. Along with analogous expressions for hydraulically more complex microtopes it permits the computation of a value for $\bar{U}_a$, the mean net flux density of water supply to the microtope which is dispersed by seepage in the acrotelm. Thus, the mean net water supply $\bar{q}$ to each microtope of area a can be calculated as:

$$\bar{q}=a\,\bar{U}_a=a(\bar{P}+\bar{U}_m-\bar{E}) \qquad (3.39)$$

from which the mean efflux of water per unit length of the downstream boundary of the microtope can be determined. This forms the basis for evaluating $\bar{U}_m$ for the next microtope downstream.

The actual seepage fluxes or rates of throughflow along streamlines can be related to the nature and situation of each microtope as a function of three parameters, namely water level, acrotelm hydraulic conductivity, and surface slope. Using the Dupuit–Forchheimer approximation (eq. 3.36) the slope gives the potential gradient in Darcy's Law (eq. 3.30) which expresses the dependence of the flux on hydraulic conductivity. The depth of the water table controls the cross-sectional area involved in transmission. As outlined by Romanov (1968a), Soviet workers have found that all three parameters are highly characteristic of the different types of mire microtope. Since the overall pattern of throughflow in any mire massif must satisfy the principle of continuity, it follows that the disposition of microtopes must not only provide the correct set of local water balances through appropriate resistances to evapotranspiration, but it must also be such as to transmit the correct seepage flux at each point on the surface.

This analysis provides an immediate theoretical explanation for the coexistence of various types of vegetation on the surfaces of large mire systems, and for the finding that these types are distributed in a regular manner, repeated on different massifs.

Ivanov (1975a) has also offered an explanation of the tendency for many mire microtopes to take the form of mosaics of intimately related yet contrasting component elements: hollows with hummocks, ridges or strings with flarks, and the like. It is clear from evidence preserved in the peat itself that mires are persistent ecosystems, the lifespans of which are to be reckoned in thousands of years. It is also known that the period of the Flandrian over which most of the circumboreal mires have developed was marked by a complex series of climatic oscillations which have caused changes, recorded in the stratigraphy of peat deposits, that are synchronous over wide areas. These changes (Aaby, 1976) are generally taken to indicate shifts in the water balance. (For a contrary view, see Raikes, 1967.) At first sight, these shifts seem incompatible with the delicately integrated hydraulics of the mire ecosystem suggested by Ivanov's theory. In fact, however, the system is self-regulating. The microtopes react to wet episodes by relative expansion of the more transmissive elements, while the less transmissive elements expand in drier periods. For this to be possible the elements must alternate along the streamlines. More than half the total area of the intact circumboreal mires is occupied by ridge-flark microtopes with the elements elongated parallel to the contour, and consequently alternating in the direction of flow. Ivanov (1975a) argues that their abundance is due to their stability in the face of climatic change, the flarks being the elements of more rapid throughflow, and also of more rapid evapotranspiration.

The theory also encompasses the formation and development of endotelmic pools and lakes. These can be regarded as the microtopes or microtope elements of highest transmissivity and lowest resistance to evaporation. Thus, they begin their existence as flooded hollows or flarks. They are, however, unique among the major surface features of mires in being prone to erosion. This causes them to enlarge and spread at the expense of neighbouring helophytic microtopes. The process may continue for some time without major disturbance to the surrounding water levels, but pools may eventually link up with rills or endotelmic streams which are themselves eroding by cutting back their beds. This causes a lowering of the pool water level, sometimes by 2 to 2.5 m, and an abrupt destabilizing of the surrounding microtopes because their water levels fall below the permissible limits for sustained growth and the hydraulic continuity at the pool bank is broken. Eventually the water table is lowered over large areas of the eroding mire; the pools, now much shallower, are overgrown by vegetation; peat formation, slowed down as the pools expanded, accelerates.

These concepts, described here in outline only, provide for the first time a coherent body of theory sufficient to account for the integrated juxtaposition of a great many well-known features of mire ecosystems. To the pure ecologist they raise questions about the evolution of whole communities and complex ecosystems. To the applied ecologist, they offer a guide to the consequences of various alternative prescriptions for land management.

The Soviet theories outlined above provide useful insight into the mechanisms which govern details of the vegetation and other features of the mire surface, and of their mutual relationships. They do not, however, set out to account for the shape of that surface and the morphology of the mire as a whole. Peat accumulation and the growth of a mire is in total effect a counter-erosive process, through which the original vegetated surface becomes progressively changed while its level is raised, its relief becomes subdued, and its water supply and drainage are transformed. The new shape may thus be ascribed to the obtrusion of a catotelm of increasing thickness between the inorganic bedrock and the peat-producing layer of the acrotelm.

These processes occur during the formation of fens and blanket mires, but the most striking changes in physiography are those which accompany the development of raised bogs. The classical process of terrestrialization, as originally described from the lowlands round the southern Baltic, involves the most profound transformation, from a concave basin with centripetal drainage and canalization of the nutrient supply in a central stream or water track, to a convex "cupola" with centrifugal drainage and concentration of nutrients in the peripheral lagg.

The question of how these peat domes maintain their saturation is one which has long puzzled telmatologists, some of whom still fail to perceive that, if water could be "drawn up" from the lagg, as indicated by Gosselink and Turner (1978, fig. 6), its minerotrophic influence would appear in the vegetation of the marginal slope and the central mire expanse (see also pp. 70–71). In fact the dome remains saturated because its drainage is impeded. Catastrophic onset of prolonged drought would lead to its gradual emptying of water. This was clearly understood by C.A. Weber as early as 1908. Weber wrote, "Der auf diese Weise zu namhafter Höhe über dem ehemaligen Seespiegel aufgehäufte Torf stellt einen undurchlässigen Boden dar, der in unserm niederschlagsreichen Klima zu einer erneuten Versumpfung Anlass gibt. [The peat, piled up in this manner to a considerable height above the original lake level, produces an impermeable soil which, in our rainy climate, brings about renewed paludification.]"

The notion that raised-bog saturation is maintained by a dynamic equilibrium between supply and drainage led Granlund (1932) to postulate a connexion between raised-bog convexity and rainfall, since the mire expanse is here fed entirely by meteoric water. At the time, Granlund's paper aroused less interest than it deserved, but his work has since been favourably reconsidered by Wickman (1951) and by Dahl (1969). Granlund showed that, in southern Sweden, it was possible to discern a regular relationship between the breadth of the raised bogs, the height of their peat domes and the annual rainfall, the most convex bogs occurring in the wettest areas. Similar regular relationships between the maximum height and breadth of more than 100 raised bogs were later discerned by Aartolahti (1965) in two areas of southwestern Finland. On the basis of Granlund's data, Wickman showed that:

$$J_m^2 = k(\bar{P} - b) \tag{3.40}$$

where J_m is the maximum height of the cupola above the lagg stream, $\bar{P}$ is the mean annual rainfall, b is a constant = 480 mm which represents the mean annual evapotranspiration, and k is a function whose value depends on the size class of each bog. Wickman also suggested that, under the same mean annual rainfall:

$$J_m^2 = f(L) \tag{3.41}$$

where L is a linear measure of the size of the bog in plan. In this instance, however, there were serious and consistent disagreements between theory and reality in the larger size classes.

In their theoretical analysis of Granlund's data, Wickman and Dahl both assumed that the peat through which precipitated water drains away is homogeneous. This is not so. In an intact diplotelmic mire most of the flow occurs through the acrotelm, as has been shown above (pp. 132–137). Moreover the system is so adjusted that the acrotelm, despite its thinness, accommodates the variations in flux density, associated with storms and drought, by seepage beneath the oscillating water table. In such mires the water table never descends below the base of the acrotelm. If it were to enter the catotelm or, more strictly, if the surface of its associated capillary fringe were to do so, the colloids in the humified peat would undergo irreversible desiccation and the system would begin to disintegrate (Hooghoudt, 1950).

From this it is clear that the condition for continuing growth and stability in raised bogs is the maintenance of the catotelm in a saturated condition, even under the most adverse (i.e. the driest) conditions which a mire is likely to experience. The following treatment is based on that of Ingram (1977).

The consequence of the stability condition set out in the previous paragraph is that the catotelm should be continuously occupied by a body of ground water that is coextensive with it. Such water bodies have been termed "ground-water mounds" by Marino (1974). They arise in many situations where water is supplied from above to a porous aquifer which impedes its dispersal by horizontal seepage. Using the Dupuit–Forchheimer approximation (p. 122), a series of relationships may be predicted (Childs, 1969) between the water supply U, the hydraulic conductivity K of the aquifer and the height Z of the water table, or surface of the mound, at various specified points. For an aquifer with a rectilinear flow towards two parallel open channels at distance $2L$ apart, containing water of negligible depth, we have:

$$\frac{U}{K} = \frac{Z^2}{2Lx - x^2} \tag{3.42}$$

where x is the distance of the point in question from the nearest channel. The highest point of the mound is located as a line midway between the channels, and its elevation Z_m above them is given by:

$$\frac{U}{K} = \frac{Z_m^2}{L^2} \tag{3.43}$$

The corresponding solutions for a system with a circular peripheral channel are:

$$\frac{U}{K} = \frac{2Z^2}{L_c^2 - x_c^2} \tag{3.44}$$

where L_c is the radius of the channel and x_c the radial distance from the centre to the point in question; and:

$$\frac{U}{K} = \frac{2Z_m^2}{L_c^2} \tag{3.45}$$

If the thickness of the acrotelm is taken to be negligible compared with the dome as a whole, eqs. 3.43 and 3.45 predict that the shape of a raised bog in profile will be elliptical, and not parabolic as claimed by Werenskjold (1943). At Dun Moss, a small Scottish raised bog, the profile was found to be a good fit to this prediction. In this bog the flow approximates to the rectilinear case and Z_m corresponds to a value for the quotient U/K close to 10^{-3}. The year May 1972 to April 1973 was also the driest in a decade at this site. Water-balance computations for that year gave a value of $U_{min}/K = 1.2 \times 10^{-3}$, assuming the approximate applicability of Darcy's Law (p. 125) and taking a value of K determined at low hydraulic gradients near to the top of the catotelm. The agreement is encouraging, and the theory suggests a basis for the functional relationship of eq. 3.40, provided $\bar{P}$ is linearly related to U_{min}. Clearly, more work is required on the long-term water balances of raised bogs whose catotelm permeabilities are known in some detail.

It seems, therefore, that at least for some mires it is possible to account for the major features of physiography and plant distribution using hydraulic models. These treat the mire as a system of two layers in which the recharge can be studied in terms of a water balance, while the discharge is treated by the ordinary methods of soil physics. That this should be so provides further justification for the view that the role of water is central to the functioning of mire ecosystems.

ACKNOWLEDGEMENTS

Many people have helped me in the preparation of this chapter by their kindness in furnishing me with literature and notes, and by their willingness to discuss the problems of mire hydrology; I am especially grateful to the following: Bent Aaby, Danish Geological Survey, Copenhagen; Michael Ashmore, Imperial College, London; Peter Ashton, Arnold Arboretum, Jamaica Plain, Mass.; Rudiger Bartels and Rudolf Eggelsmann, Ausseninstitut für Moorforschung, Bremen; John Birks, Cambridge Botany School; Don Boelter and Sandy Verry, United States Department of Agriculture Forest Service; Jim Brown and Walter McNicoll, University of Dundee; A. Garkin, International Peat Society; Kevin Gilman and Malcolm Newson, Institute of Hydrology, Staylittle; Konstantin Ivanov, University of Leningrad; Anthony Jones, University College, Aberystwyth; Viktor Masing, University of Tartu; Donald MacKerron, Scottish Horticultural Research Institute, Invergowrie; John Monteith, University of Nottingham; Robert Neuhäusl and Kamil Rybníček, Czechoslovak Academy of Sciences; M. Novák, Water Management Research Institute, Prague; Hans Olson, University of Lund; Edmundo Pisano, Instituto de la Patagonia, Punta Arenas; Manfred Renger, Niedersächsisches Landesamt für Bodenforschung, Hannover; Allan Robertson, Macaulay Institute, Aberdeen; Fred Schlegel, Valdivia; Heinz Schlüter, Jena; Josef Votruba, Land Reclamation Research Institute, Zbraslav; Roy Ward, University of Hull; Otto Wildi, University of Western Ontario, London, Ont.; David Williams, Swansea Department of Chemical Engineering; Edward Youngs, Rothamsted. To my former colleagues Peter Mott, David Rycroft, David Tervet and John Waine, and to their successors Olivia Bragg and Andrew Coupar, I am grateful for much enlightening comment and for permission to quote unpublished results. I thank Ellis Armstrong, Iain Tennant, Angela Petrie, Melvin Daft and Jonathan Weyers, who have eased my task in various ways, also my father, A.W.K. Ingram of Rugby, who has helped to prepare the index entries. I am both grateful and fortunate to have enjoyed the patient, encouraging and meticulous editorial assistance of David Goodall and of the late Tony Gore, who had been my friend for nearly twenty years.

REFERENCES

Aaby, B., 1976. Cyclic climatic variation in climate over the past 5,500 yr reflected in raised bogs. *Nature*, 263: 281–284.

Aartolahti, T., 1965. Oberflächenformen von Hochmooren und ihre Entwicklung in Südwest-Häme und Nord-Satakunta. *Fennia*, 93: 1–268.

Ahti, E., 1974. Measuring seasonal moisture variation of drained peatlands by using tensiometers. In: *International Symposium on Forest Drainage, Jyväskylä–Oulu, 1974*. Soc. For. Finland, Helsinki, pp. 81–86.

Alekseevsky, V.E. and Tereshchenko, K.P., 1975. Sources of incoming water in bogs of the Ukrainian Pripyat Polessie as a result of reclamation. In: *Hydrology of Marsh-Ridden Areas. Proc. IASH Symp., Minsk, 1972*. IASH/UNESCO, Paris, pp. 109–116.

Aletsee, L., 1967. Begriffliche und floristische Grundlagen zu einer pflanzengeographischen Analyse der europäischen Regenwassermoorstandorte. *Beitr. Biol. Pflanz.*, 43: 117–283.

Anderson, J.A.R., 1964. The structure and development of the peat swamps of Sarawak and Brunei. *J. Trop. Geogr.*, 18: 7–16.

Andriesse, J.P., 1974. Tropical lowland peats in South-East Asia. *Comm. Dep. Agric. Res., R. Trop. Inst. Amsterdam*, No. 63: 63 pp.

Anonymous, 1961. *Manual of Hydrometeorological Observations in Swamplands*. Instructions for Hydrometeorological Stations and Posts, 8. Meteorol. Transl., 6, Atmosph. Environ. Serv., Downsview, Ont., 120 pp.

Anonymous, 1967. Potential transpiration. *Tech. Bull. Minist. Agric. Fish. Fd.*, No. 16: 77 pp.

Anonymous, 1971. *Ukazaniya po raschetam stoka s neosushennykh i osushennykh verkhovykh bolot*. [*Instructions for Computation of Flow from Reclaimed and Unreclaimed Raised Bogs*.] Gidrometeoizdat, Leningrad, 84 pp.

Aref'eva, A.I., 1963. Sezonnye kolebaniya poverkhnosti sfagnovykh bolot pod vliyaniem gidrometeorologicheskikh faktorov. *Tr. Gos. Gidrol. Inst.*, 105: 80–108. [Seasonal fluctuations of the surface of *Sphagnum* swamps under the influence of meteorological factors. *Sov. Hydrol.*, 1963: 309 (abstract)].

Ashmore, M.R., 1975. *The Eco-Physiology of* Calluna vulgaris

(L.) Hull in a Moorland Habitat. Dissertation, University of Leeds, Leeds, 145 pp.

Baden, W. and Eggelsmann, R., 1961. Moorhydrologische Untersuchungen am Westrand des Steinhuder Meeres zur Feststellung eines unterirdischen Seeabflusses. *Wasser Boden*, 13: 403–410.

Baden, W. and Eggelsmann, R., 1963. Zur Durchlässigkheit der Moorböden. *Z. Kulturtechn. Flurbereinig.*, 4: 226–254.

Baden, W. and Eggelsmann, R., 1964. Der Wasserkreislauf eines nordwestdeutschen Hochmoores. *Schriftenr. Kurat. Kult., Hamburg*, 12: 1–156.

Baden, W. and Eggelsmann, R., 1966. Diskussionsbeitrag. *Dtsch. Gewässerkdl. Mitt.*, 10: 22–24.

Baden, W. and Eggelsmann, R., 1970. The hydrologic budget of the highbogs in the Atlantic region. *Proc. 3rd Int. Peat Congr., Quebec, 1968*, pp. 206–211.

Balátová-Tuláčková, E., 1972. *Flachmoorwiesen im mittleren und unteren Opava-Tal (Schlesien)*. Vegetace ČSSR, A4. Academia, Prague, 201 pp.

Bannister, P., 1964a. Stomatal responses of heath plants to water deficits. *J. Ecol.*, 52: 151–158.

Bannister, P., 1964b. The water relations of certain heath plants with reference to their ecological amplitude III. Experimental studies: general conclusions. *J. Ecol.*, 52: 499–509.

Bannister, P., 1976. *Introduction to Physiological Plant Ecology*. Blackwell, Oxford, 273 pp.

Barry, R.G., 1969. Evaporation and transpiration. In: R.J. Chorley (Editor), *Introduction to Physical Hydrology*. Methuen, London, pp. 82–97.

Bartels, R. and Kuntze, H., 1973. Torfeigenschaften und ungesättigte hydraulische Leitfähigkeit von Moorböden. *Z. Pflanzenernähr., Düng, Bodenkd.*, 134: 125–135.

Bavina, L.G., 1966. Vodnyi balans nizinnykh bolot Polesskoi vlagimennosti. [Water balance of fens in the Poles'ye lowlands.] *Tr. Gos. Gidrol. Inst.*, No. 135: 181–196. (Quoted by Dooge, 1975).

Bavina, L.G., 1967. Refinement of parameters for calculating evaporation from bogs on the basis of observations at bog stations. *Sov. Hydrol.*, 1967: 348–370.

Bavina, L.G., 1972. Water balance of swamps and its computation. In: *World Water Balance. Proc. IASH Symp., Reading, 1970, 2.* IASH/UNESCO, Paris, pp. 461–466.

Bavina, L.G., 1975. Water balance of swamps in the forest zone of the European part of the USSR. In: *Hydrology of Marsh-Ridden Areas. Proc. IASH Symp., Minsk, 1972.* IASH/UNESCO, Paris, pp. 297–303.

Bay, R.R., 1966a. Evaluation of an evapotranspirometer for peat bogs. *Water Resour. Res.*, 2: 437–442.

Bay, R.R., 1966b. Factors influencing soil-moisture relationships in undrained forested bogs. In: W.E. Sopper and H.W. Lull (Editors), *Forest Hydrology*. Pergamon, Oxford, pp. 335–343.

Bay, R.R., 1967a. Techniques of hydrologic research in forested peatlands, U.S.A. *Proc. 14th IUFRO Congr., Munich, 1967*, 1: 400–415.

Bay, R.R., 1967b. Ground water and vegetation in two peat bogs in northern Minnesota. *Ecology*, 48: 308–310.

Bay, R.R., 1968. Evaporation from two peatland watersheds. *Rep. Disc. IASH, Gen. Assembl. Bern, 1967*, pp. 300–307.

Bay, R.R., 1969. Runoff from small peatland watersheds. *J. Hydrol.*, 9: 90–102.

Bay, R.R., 1970. The hydrology of several peat deposits in northern Minnesota. *Proc. 3rd Int. Peat Congr., Quebec, 1968*, pp. 212–218.

Bay, R.R., 1973. Water table relationships on experimental basins containing peat bogs. In: *Results of Research on Representative and Experimental Basins. Proc. IASH Symp., Wellington, 1970.* IASH/UNESCO, Paris, pp. 360–368.

Bear, J., Zaslavsky, D. and Irmay, S., 1968. *Physical Principles of Water Percolation and Seepage*. Arid Zone Res. Ser. 29. UNESCO, Paris, 465 pp.

Bell, J.N. and Tallis, J.H., 1974. The response of *Empetrum nigrum* L. to different mire water regimes, with special reference to Wybunbury Moss, Cheshire and Featherbed Moss, Derbyshire. *J. Ecol.*, 62: 75–95.

Bell, J.P., 1973. *Neutron Probe Practice*. Inst. Hydrol., Wallingford, (Rep. 19), 63 pp.

Bellamy, D.J., 1959. Occurrence of *Schoenus nigricans* L. on ombrogenous peats. *Nature*, 184: 1590–1591.

Belotserkovskaya, O.A., 1975. The water and heat balance of the forests of Byelorussian Polessie. In: *Hydrology of Marsh-Ridden Areas. Proc. IASH Symp., Minsk, 1972.* IASH/UNESCO, Paris, pp. 321–331.

Belotserkovskaya, O.A., Largin, I.F. and Romanov, V.V., 1969. Investigation of surface and internal evaporation on high-moor bogs. *Sov. Hydrol.*, 1969: 540–554.

Berglund, E.R. and Mace, A.C., 1972. Seasonal albedo variation of black spruce and *Sphagnum* — sedge bog cover types. *J. Appl. Meteorol.*, 11: 806–812.

Bernatowicz, S., Leszczynski, S. and Tycznska, S., 1976. The influence of transpiration by emergent plants on the water balance in lakes. *Aquat. Bot.*, 2: 275–288.

Birks, H.J.B., 1973. *Past and Present Vegetation of the Isle of Skye: a Palaeoecological Study*. Cambridge University Press, London, 415 pp.

Boatman, D.J. and Tomlinson, W.R., 1973. The Silver Flowe I. Some structural and hydrological features of Brishie Bog and their bearing on pool formation. *J. Ecol.*, 61: 653–666.

Boatman, D.J., Hulme, P.D. and Tomlinson, W.R., 1975. Monthly determinations of the concentrations of sodium, potassium, magnesium and calcium in the rain and in pools on the Silver Flowe National Nature Reserve. *J. Ecol.*, 63: 903–912.

Bochkov, A.P. and Struzer, L.R., 1972. Estimation of precipitation as a water balance element. In: *World Water Balance. Proc. IASH Symp., Reading, 1970, 1.* IASH/UNESCO, Paris, pp. 186–193.

Bodarenko, N.F., Danchenko, O.I. and Kovalenko, N.P., 1975. Issledovanie prirody nestabil'nosti fil'tratsionnogo potoka v torfakh. [Studying the nature of filtration flow instability in peats.] *Pochvovedenie*, 1975(10): 137–140.

Boelter, D.H., 1964. Water storage characteristics of several peats *in situ*. *Soil Sci. Soc. Am. Proc.*, 28: 433–435.

Boelter, D.H., 1965. Hydraulic conductivity of peats. *Soil Sci.*, 100: 227–231.

Boelter, D.H., 1966. Hydrologic characteristics of organic soils in Lake States watersheds. *J. Soil Water Conserv.*, 21: 50–53.

Boelter, D.H., 1969. Physical properties of peats as related to degree of decomposition. *Soil Sci. Soc. Am. Proc.*, 33: 606–609.

Boelter, D.H., 1970. Important physical properties of peat materials. *Proc. 3rd Int. Peat Congr., Quebec, 1968*, pp. 150–154.

Boelter, D.H., 1972. Water table drawdown round an open ditch in organic soils. *J. Hydrol.*, 15: 329–340.

Boelter, D.H., 1974a. The hydrologic characteristics of undrained organic soils in the Lake States. In: A.R. Aandahl, S.W. Buol, D.E. Hill and H.H. Bailey (Editors), *Histosols: their Characteristics, Classification and Use*. Soil Sci. Soc. Am., Madison, Wisc., pp. 33–46.

Boelter, D.H., 1974b. The ecological fundamentals of forest drainage. Peatland hydrology: peatland water economy. In: *International Symposium on Forest Drainage — Coordinators' Papers and Discussions, Jyväskylä–Oulu, 1974*. Soc. For. Finland, Helsinki, pp. 35–46.

Boelter, D.H., 1976. Methods for analysing the hydrological characteristics of organic soils in marsh-ridden areas. In: *Hydrology of Marsh-Ridden Areas. Proc. IASH Symp., Minsk, 1972*. IASH/UNESCO, Paris, pp. 161–169.

Boelter, D.H. and Blake, G.R., 1964. Importance of volumetric expression of water contents of organic soils. *Soil Sci. Soc. Am. Proc.*, 28: 176–178.

Boelter, D.H. and Verry, E.S., 1977. Peatland and water in the northern Lake States. *Gen. Tech. Rep. U.S. Dep. Agric. For. Serv.*, NC-31: 22 pp.

Boggie, R., Knight, A.H. and Hunter, R.F., 1958. Studies of the root development of plants in the field using radioactive tracers. *J. Ecol.*, 46: 621–639.

Borisov, A.A., 1965. *Climates of the USSR*. C.A. Halstead (Editor), R.A. Ledward (Translator). Oliver and Boyd, Edinburgh, 255 pp.

Bormann, F.H. and Likens, G.E., 1979. *Pattern and Process in a Forested Ecosystem*. Springer, New York, N.Y., 253 pp.

Bower, M.M., 1961. The distribution of erosion in blanket peat bogs in the Pennines. *Publ. Inst. Br. Geogr.*, 29: 17–30.

Bragg, O.M., 1978. Water relations of *Sphagnum* communities of a Scottish raised bog. *Bull. Br. Ecol. Soc.*, 9(4): 11 (abstract).

Bray, J.R., 1962. Estimates of energy budgets for a *Typha* (cattail) marsh. *Science*, 136: 1119–1120.

Brown, R.J.E., 1977. Muskeg and permafrost. In: N.W. Radforth and C.O. Brawner (Editors), *Muskeg and the Northern Environment in Canada*. Toronto University Press, Toronto, Ont., pp. 148–163.

Budyko, M.I., 1958. *The Heat Balance of the Earth's Surface*. (Translated by N.A. Stepanova.) OTS 131692, U.S. Dep. Commerce, Washington, D.C., 259 pp.

Budyko, M.I., 1974. *Climate and Life*. D.H. Miller (Editor). Academic Press, New York, N.Y., 508 pp.

Budyko, M.I., Yefimova, N.A., Aubendok, L.I. and Strokina, L.A., 1962. The heat balance of the surface of the earth. *Sov. Geogr.*, 1962 (5): 3–16.

Buell, M.F. and Buell, H.F., 1941. Surface level fluctuation in Cedar Creek Bog, Minnesota. *Ecology*, 22: 314–321.

Burke, W., 1975. Effect of drainage on the hydrology of blanket bog. *Ir. J. Agric. Res.*, 14: 145–162.

Campbell, E.O., 1964. The restiad peat bogs at Motumaoho and Moanatuatua. *Trans. R. Soc. N.Z.*, 2: 219–227.

Chang, J.-H., 1965. On the study of evapotranspiration and water balance. *Erdkunde*, 19: 141–150.

Chapman, S.B., 1964. The ecology of Coom Rigg Moss, Northumberland II. The chemistry of peat profiles and the development of the bog system. *J. Ecol.*, 52: 315–321.

Chapman, S.B., 1965. The ecology of Coom Rigg Moss, Northumberland III. Some water relations of the bog system. *J. Ecol.*, 53: 371–384.

Chebotarev, N.P., 1962. *Theory of Stream Runoff*. A. Wald (Translator), M. Diskin (Editor). Israel Progr. Sci. Transl., Jerusalem, 464 pp.

Childs, E.C., 1969. *An Introduction to the Physical Principles of Soil Water Phenomena*. Wiley, London, 493 pp.

Clymo, R.S., 1978. A model of peat bog growth. In: O.W. Heal and D.F. Perkins (Editors), *Production Ecology of British Moors and Montane Grasslands*. Springer-Verlag, Berlin, pp. 187–223.

Clymo, R.S. and Reddaway, E.J.F., 1973. A tentative dry matter balance sheet for the wet blanket bog on Burnt Hill, Moor House N.N.R. In: *Aspects of the Ecology of the Northern Pennines, 3*, Nature Conserv. Counc., London, pp. 1–15.

Cockayne, L., 1967. *New Zealand Plants and Their Story*. E.J. Godley (Editor). Government Printer, Wellington, 4th ed., 269 pp.

Codarcea, F., 1977. Some considerations on the peat formation process. *Proc. 5th Int. Peat Congr., Poznan 1976*, 2: 90–98.

Conner, W.H. and Day, J.W., Jr., 1976. Productivity and composition of a baldcypress — water tupelo site and a bottomland hardwood site in a Louisiana swamp. *Am. J. Bot.*, 63: 1354–1364.

Conway, V.M. and Millar, A., 1960. The hydrology of some small peat-covered catchments in the northern Pennines. *J. Inst. Water Eng.*, 14: 415–424.

Cowan, I.R., 1965. Transport of water in the soil–plant–atmosphere system. *J. Appl. Ecol.*, 2: 221–239.

Crawford, R.M.M., 1969. Landwirtschaftlische Nutzpflanzen: grössere Überflutungstoleranz. *Umschau*, 17: 535–539.

Dahl, E., 1969. Teorier omkring myrkomplexenes dannelse. In: *Myrers Økologi og Hydrologi: Symposium om Myrer, Ås, 1969*. Norwegian Comm. Int. Hydrol. Decade, Oslo, pp. 20–24.

Dai, T.S. and Sparling, J.H. 1973. Measurement of hydraulic conductivity of peats. *Can. J. Soil Sci.*, 53: 21–26.

Daniels, R.E., Pearson, M.C. and Rydén, B.E., 1977. A thermal-electric method for measuring lateral movement of water in peat. *J. Ecol.*, 65: 839–846.

De Vries, D.A. and Venema, H.J., 1954. Some considerations on the behaviour of the Piche evaporimeter. *Vegetatio*, 8: 225–234.

Dierschke, H., 1969. Grundwasser-Ganglinien einiger Pflanzengesellschaften des Holtumer Moores östlich von Bremen. *Vegetatio*, 17: 372–383.

Dilley, A. C. and Helmond, I., 1973. The estimation of net radiation and potential evapotranspiration using atmometer measurements. *Agric. Meteorol.*, 12: 1–11.

Dittrich, J., 1952. Zur natürlichen Entwässerung der Moore. *Wasser Boden*, 4: 286, 288.

Dooge, J., 1975. The water balance of bogs and fens. In: *Hydrology of Marsh-Ridden Areas. Proc. IASH Symp., Minsk, 1972*. IASH/UNESCO, Paris, pp. 233–271.

Eggelsmann, R., 1954. Über die Bestimmung der Verdunstung vom bewachsenen Moorboden. In: *Festschr. 100-Jahre Bauschule Suderburg*. Uelzen, pp. 67–74.

Eggelsmann, R., 1957. Zur Kenntnis der Zusammenhänge zwischen Bodenfeuchte und oberflächennahem Grundwasser. *Wasserwirtschaft*, 47: 283–287.

Eggelsmann, R., 1960. Über den unterirdischen Abfluss aus Mooren. *Wasserwirtschaft*, 50: 149–154.

Eggelsmann, R., 1962. Durchlässigkeit und Grundwasserströmung am Beispiel der Moore am Steinhuder Meer und eines Moormarschpolders im Bremer Blockland. *Ber. Landesanst. Bodennutz. Schutz*, 3: 1–8.

Eggelsmann, R., 1963. Die potentielle und aktuelle Evaporation eines Seeklimahochmoores. *IASH Publ.*, No. 63: 88–97.

Eggelsmann, R., 1964a. Verlauf der Grundwasserströmung in entwässerten Mooren. *Mitt. Dtsch. Bodenkdl. Ges.*, 2: 129–139.

Eggelsmann, R., 1964b. Die Verdunstung der Hochmoore und deren hydrographischer Einfluss. *Dtsch. Gewässerkdl. Mitt.*, 8: 138–147.

Eggelsmann, R., 1967. Oberflächengefälle und Abflussregime der Hochmoore. *Wasser Boden*, 19: 247–252.

Eggelsmann, R., 1971. Über den hydrologischen Einfluss der Moore. *Telma*, 1: 37–48.

Eggelsmann, R., 1975. The water balance of lowland areas in north-western regions of the F.R.G. In: *Hydrology of Marsh-Ridden Areas. Proc. IASH Symp., Minsk, 1972*. IASH/UNESCO, Paris, pp. 355–367.

Eggelsmann, R., 1978. *Subsurface Drainage Instructions*. (Translation edited by Int. Comm. Irrig. Drain.) Parey, Berlin, 283 pp.

Eggelsmann, R. and Schuch, M., 1976. Moorhydrologie. In: K. Göttlich (Editor), *Moor- und Torfkunde*. Schweizerbart'sche, Stuttgart, pp. 153–162.

Eisenlohr, W.S., 1975. Hydrology of marshy ponds on the Coteau du Missouri. In: *Hydrology of Marsh-Ridden Areas. Proc. IASH Symp., Minsk, 1972*. IASH/UNESCO, Paris, pp. 305–311.

Elgee, F., 1912. *The Moorlands of North-Eastern Yorkshire*. A. Brown, London, 361 pp.

Engler, A. and Gilg, E., 1912. *Syllabus der Pflanzenfamilien*. Borntraeger, Berlin, 7th ed., 387 pp.

Farnham, R.S. and Finney, H.R., 1965. Classification and properties of organic soils. *Adv. Argon.*, 7: 115–162.

Ferda, J. and Pasák, V., 1969. *Hydrologic and Climatic Function of Czechoslovak Peat Bogs*. Výzkumný ústav meliorací, Zbraslav, 358 pp. (Czech with English summary).

Firbas, F., 1931. Untersuchungen über den Wasserhaushalt der Hochmoorpflanzen. *Pringsheims Jahrb. Wiss. Bot.*, 74: 459–696.

Galvin, L.F., 1976. Physical properties of Irish peats. *Ir. J. Agric. Res.*, 15: 207–221.

Galvin, L.F. and Hanrahan, E.T., 1967. Steady state drainage flow in peat. *Natl. Res. Counc. Highways Res. Bd. Res. Rec.*, 203: 77–90.

Gardner, W.R., 1958. Some steady state solutions of the unsaturated moisture flow equation with application to evaporation from a water table. *Soil Sci.*, 85: 228–232.

Gel'bukh (Gelboukh), T.M., 1963. Evapotranspiration from overgrowing reservoirs. *Rep. Disc. IASH, Gen Assembl. Berkeley, 1963*, p. 87.

Gel'bukh, T.M., 1964. Evaporation from reed fields in water bodies. *Sov. Hydrol.*, 1964: 363–382.

Geraghty, J.J., Miller, D.W., Van der Leeden, F. and Troise, F.L., 1973. *Water Atlas of the United States*. Water Inf. Cent., Port Washington, N.Y., 122 maps with text on reverse.

Gilman, K. and Newson, M.D., 1980. Soil pipes and pipeflow — a hydrological study in upland Wales. *Br. Geomorphol. Res. Group Res. Monogr. Ser.*, No. 1: 110 pp. (Geo Books, Norwich).

Gimingham, C.H., 1960. Biological flora of the British Isles. *Calluna vulgaris*, (L.) Hull. *J. Ecol.*, 48: 455–483.

Gimingham, C.H., 1972. *Ecology of Heathlands*. Chapman and Hall, London, 266 pp.

Given, P.H. and Dickinson, C.H., 1975. Biochemistry and microbiology of peats. In: E.A. Paul and A.C. McLaren (Editors), *Soil Biochemistry*. Dekker, New York, N.Y., pp. 124–211.

Godwin, H., 1931. Studies in the ecology of Wicken Fen I. The ground water level of the fen. *J. Ecol.*, 19: 449–472.

Godwin, H., 1932. Water levels in Wicken Sedge Fen. In: J.S. Gardiner (Editor), *The Natural History of Wicken Fen, 6*. Bowes and Bowes, Cambridge, pp. 615–625.

Godwin, H., 1978. *Fenland: its Ancient Past and Uncertain Future*. Cambridge University Press, Cambridge, 196 pp.

Godwin, H. and Bharucha, F.R., 1932. Studies in the ecology of Wicken Fen II. The fen water table and its control of plant communities. *J. Ecol.*, 20: 158–191.

Goode, D.A., Marsan, A.A. and Michaud, J.-R., 1977. Water resources. In: N.W. Radforth and C.O. Brawner (Editors), *Muskeg and the Northern Environment in Canada*. Toronto University Press, Toronto, Ont., pp. 299–331.

Gore, A.J.P., 1968. The supply of six elements by rain to an upland peat area. *J. Ecol.*, 56: 483–495.

Gorham, E., 1956. The chemical composition of some bog and fen waters in the English Lake District. *J. Ecol.*, 44: 142–152.

Gorham, E., 1957. The development of peat lands. *Q. Rev. Biol.*, 32: 145–166.

Gosselink, J.G. and Turner, R.E., 1978. The role of hydrology in freshwater wetland ecosystems. In: R.E. Good, D.F. Whigham and R.L. Simpson (Editors), *Freshwater Wetlands*. Academic Press, New York, N.Y., pp. 63–78.

Grace, J., 1970. *The Growth Physiology of Moorland Plants in Relation to their Aerial Environment*. Dissertation, University of Sheffield, Sheffield, 250 pp.

Granlund, E., 1932. De svenska högmossarnas geologi. *Sver. Geol. Unders., Ser. C*, No. 373: 1–193.

Grebenshchikova, A.A., 1956. O vlagoemosti torfov. [On the water capacity of peats.] *Pochvovedenie* 9: 102.

Grebenshchikova, A.A., 1957. [The moisture content of small virgin peat areas.] *Byull. Nauchno-Tekh. Inf. Tsentr. Torfobolotnoi Opytn. Stn.*, 1: 19–22. [Abstract: *Soil Fert.*, 22: 441 (1959).]

Hall, M.J. and Prus-Chacinski, T.M., 1975. Forecasting run-off volumes for the drainage of peat lands. *J. Agric. Eng. Res.*, 20: 267–278.

Hanrahan, E.T., 1954. An investigation of some physical properties of peat. *Geotechnique*, 4: 108–123.

Harding, R.J., 1979. Radiation in the British uplands. *J. Appl. Ecol.*, 16: 161–170.

Haude, W., 1952. Zur Möglichkeit nachträglicher Bestimmung der Wasserbeanspruchung durch die Luft und ihrer Nachprüfung an Hand von Topfversuchen und Abflussmessungen. *Ber. Dtsch. Wetterdienstes U.S. Zone*, 32: 27–33.

Haude, W., 1955. Zur Bestimmung der Verdunstung auf möglichst einfache Weise. *Mitt. Dtsch. Wetterdienstes*, 2(11): 1–24.

Hawkinson, C.F. and Verry, E.S., 1975. Specific conductance identifies perched and ground water lakes. *U.S. Dep. Agric. For. Serv. Res. Pap.*, NC-120: 5 pp.

Hayward, P.M., 1978. Water and the growth of *Sphagnum*. *Bull. Br. Ecol. Soc.*, 9(4): 12 (abstract).

Heath, G.H. and Luckwill, L.C., 1938. The rooting systems of heath plants. *J. Ecol.*, 26: 331–352.

Heikurainen, L., 1963. On using ground water table fluctuations for measuring evapotranspiration. *Acta For. Fenn.*, 76(5): 1–15.

Heikurainen, L., 1964. Improvement of forest growth on poorly drained peat soils. *Int. Rev. For. Res.*, 1: 39–113.

Heikurainen, L. and Laine, J., 1974. Estimating evapotranspiration in peatlands on the basis of diurnal water table fluctuations. In: *International Symposium on Forest Drainage, Jyväskylä–Oulu, 1974.* Soc. For. Finland, Helsinki, pp. 87–96.

Heikurainen, L., Päivänen, J. and Sarasto, J., 1964. Ground water table and water content in peat soil. *Acta For. Fenn.*, 77(1): 1–18.

Hewlett, J.D. and Hibbert, A.R., 1967. Factors affecting the response of small watersheds to precipitation in humid areas. In: W.E. Sopper and H.W. Lull (Editors), *Forest Hydrology*. Pergamon, Oxford, pp. 275–290.

Hillell, D., 1971. *Soil and Water: Physical Principles and Processes*. Academic Press, New York, N.Y., 288 pp.

Hinshiri, H.M., 1973. *Field and Experimental Studies of the Water Relations of* Calluna vulgaris *(L.) Hull with Special Reference to the Effect of Wind.* Dissertation, University of Aberdeen, 174 pp. (Quoted by Ashmore, 1975.)

Holland, D.J., 1967. Evaporation. *Br. Rainfall, 1961*, No. 3: 5–34 (H.M.S.O., London.)

Hooghoudt, S.B., 1950. Irreversibly desiccated peat, clayey peat and peaty clay soils; the determination of the degree of reversibility. *Trans. 4th Int. Congr. Soil Sci., Amsterdam*, 2: 31–34.

Horton, R.E., 1933. The rôle of infiltration in the hydrologic cycle. *Trans. Am. Geophys. Union*, 14: 446–460.

Huikari, O., 1959. Metsäojitettujen turvemaiden vesitaloudesta. [On the water management of forest drained peat soils.] *Comm. Inst. For. Fenn.*, 51(2): 1–45.

Huikari, O., 1963. Effect of distance between drains on the water economy and surface runoff of *Sphagnum* bogs. *Trans. 2nd Int. Peat Congr., Leningrad, 1963, 2.* H.M.S.O., Edinburgh, pp. 739–742.

Hupkens van der Elst, F.C.C., 1972. Nutrient requirements for establishment and maintainance of good pastures on high moor peat soils in New Zealand. *Proc. 4th Int. Peat Congr., Otaniemi, 1972*, 3: 21–36.

Incoll, L.D., 1977. Field studies of photosynthesis: monitoring with $^{14}CO_2$. In: J.J. Landsberg and C.V. Cutting (Editors), *Environmental Effects on Crop Physiology*. Academic Press, London, pp. 137–155.

Ingram, H.A.P., 1967. Problems of hydrology and plant distribution in mires. *J. Ecol.*, 55: 711–724.

Ingram, H.A.P., 1977. Some hydrological problems in intact peat deposits. *Rep. Scott. Hydrol. Group Inst. Civ. Eng.*, (abstract). 1976, 3 pp.

Ingram, H.A.P., 1978. Soil layers in mires: function and terminology. *J. Soil Sci.*, 29: 224–227.

Ingram, H.A.P., Rycroft, D.W. and Williams, D.J.A., 1974. Anomalous transmission of water through certain peats. *J. Hydrol.*, 22: 213–218.

Irwin, R.W., 1966. Progress report on the application of a neutron soil moisture meter to organic soil. *Proc. 11th Muskeg Res. Conf., Natl. Res. Counc. Can., Tech. Mem.*, No. 87: 45–54.

Ivanov, K.E., 1953. *Gidrologiya bolot.* [*Hydrology of Mires.*] Gidrometeoizdat, Leningrad, 296 pp.

Ivanov, K.E., 1957. *Osnovy gidrologii bolot lesnoĭ zony i raschety vodnogo rezhima bolotnykh massivov.* [*Rudiments of the Hydrology of Mires in the Forest Zone and Computations of the Water Regime of Mire Massifs.*] Gidrometeoizdat, Leningrad, 500 pp.

Ivanov, K.E., 1975a. *Vodoobmen v bolotnykh landshaftakh.* Gidrometeoizdat, Leningrad, 280 pp. [*Water Movement in Mirelands.* A. Thomson and H.A.P. Ingram (Translators). Academic Press, London, 277 pp. (1981).]

Ivanov, K.E., 1975b. Hydrological stability criteria and reconstruction of bogs and bog-lake systems. In: *Hydrology of Marsh-Ridden Areas. Proc. IASH Symp., Minsk, 1972.* IASH/UNESCO, Paris, pp. 343–353.

Ivanov, N.N., 1957. *Mirovaya karta isparyaemosti.* [*World Map of Potential Evaporation.*] Gidrometeoizdat, Leningrad, 38 pp.

Ivitskiĭ, A.I., 1938. Isparenie s bolot v zavisimosti ot klimata, osusheniya i kul'tury. *Gidrotekh. Melior.*, 1938(1/2): 62–66. (Evaporation from bogs, depending on climate, drainage and cultivation. TT 67-51386, U.S. Dep. Commerce, Washington, D.C.)

Johansson, I., 1974. Hydrologiska undersökningar inom myrkomplexet Komosse. *Rap. K. Tek. Högsk., Sekt. Lantmät., Inst. Kulturtek.*, 3(17): 1–161.

Jones, J.A.A., 1981. The nature of soil piping: a review of research. *Br. Geomorphol. Res. Group Res. Monogr. Ser., 3.* Geo Books, Norwich, 301 pp.

Kastanek, F., 1973. Calculation of vertical moisture flow in a soil body during evaporation, infiltration and redistribution. *Tech. Bull. Inst. Land Water Manage. Res. Vienna*, No. 86: 49–58.

Kats, N. Ya., 1971. *Boloto zemnogo shara.* [*Mires of the Terrestrial Globe.*] Nauka, Moscow, 295 pp.

Kaufman, W.J. and Orlob, G.T., 1956. Measuring ground water movement with radioactive and chemical tracers. *J. Am. Water Works Assoc.*, 48: 559–572.

Kaule, G. and Göttlich, K., 1976. Begriffsbestimmungen anhand der Moortypen Mitteleuropas. In: K. Göttlich (Editor), *Moor- und Torfkunde.* Schweizerbart'sche, Stuttgart, pp. 1–21.

Kay, B.D. and Goit, J.B., 1977. Thermodynamic characterisation of water adsorbed on peat. *Can. J. Soil Sci.*, 57: 497–501.

Kirkby, M.J., 1969. Infiltration, throughflow and overland flow. In: R.J. Chorley (Editor), *Introduction to Physical Hydrology*. Methuen, London, pp. 108–121.

Kohler, M.A., Nordenson, T.J. and Fox, W.E., 1955. Evaporation from pans and lakes. *Res. Pap. U.S. Dep. Comm.*, No. 38: pp. 21.

Kohler, M.A., Nordenson, T.J. and Baker, D.R., 1959. Evaporation maps for the United States. *Tech. Pap. U.S. Weather Bur.*, No. 37: 13 (5 plates). (U.S. Dep. Commerce, Washington, D.C.)

Konstantinov, A.R., 1966. *Evaporation in Nature*. L. Shichtman (Translator). Israel Progr. Sci. Transl., Jerusalem, 523 pp.

Kulczyński, S., 1949. Peat bogs of Polesie. *Mém. Acad. Pol. Sci.*, B15: 1–356.

Kuntze, H., 1965. Physikalische Untersuchungsmethoden für Moor- und Anmoorboden. *Landwirtsch. Forsch.*, 18: 178–191.

Kuntze, H., 1974. Effects of drainage. In: *International Symposium on Forest Drainage — Coordinators' Papers and Discussions, Jyväskylä–Oulu, 1974.* Sor. For. Finland, Helsinki, pp. 111–124.

Kuzmin, P.P., 1972. Methods of estimating evaporation from land applied in the USSR. In: *World Water Balance, Proc. IASH Symp., Reading, 1970, 1.* IASH/UNESCO, Paris, pp. 225–231.

Lechowicz, M.J. and Adams, M.S., 1974. Ecology of *Cladonia* lichens II. Comparative physiological ecology of *C. mitis*, *C. rangiferina* and *C. uncialis*. *Can. J. Bot.*, 52: 411–422.

Leick, E., 1929. Zur Frage der Wasserbilanz von Hochmooren. *Mitt. Naturwiss. Ver. Neu-Vorpommern*, 52/56: 146–174.

Leiser, A.T., 1968. A mucilaginous root sheath in Ericaceae. *Am. J. Bot.*, 55: 391–398.

Linacre, E., 1976. Swamps. In: J.L. Monteith (Editor), *Vegetation and the Atmosphere, 2.* Academic Press, London, pp. 329–347.

Linacre, E.T., Hicks, B.B., Sainty, G.R. and Grauze, G., 1970. The evaporation from a swamp. *Agric. Meteorol.*, 375–386.

Luthin, J.N., 1966. *Drainage Engineering*. Wiley, New York, N.Y., 250 pp.

Luthin, J.N. and Kirkham, D., 1949. A piezometer method for measuring permeability of soil *in situ* below a water table. *Soil Sci.*, 68: 349–358.

McVean, D.N. and Lockie, J.D., 1969. *Ecology and Land Use in Upland Scotland.* Edinburgh University Press, Edinburgh, 134 pp.

McVean, D.N. and Ratcliffe, D.A., 1962. *Plant Communities of the Scottish Highlands.* H.M.S.O., London, 445 pp.

Malmer, N., 1962a. Studies on mire vegetation in the Archaean area of southwestern Götaland (south Sweden) I. Vegetation and habitat conditions on the Åkhult mire. *Opera Bot. Soc. Bot. Lund.*, 7(1): 1–322.

Malmer, N., 1962b. Studies on mire vegetation in the Archaean area of southwestern Götland (south Sweden) II. Distribution and seasonal variation in elementary constituents on some mire sites. *Opera Bot. Soc. Bot. Lund*, 7(2): 1–67.

Malmer, N. and Sjörs, H., 1955. Some determinations of elementary constituents in mire plants and peat. *Bot. Not.*, 108: 46–80.

Malmström, C., 1939. Methoden zur Untersuchung der Wasserverhältnisse von Torfböden. In: E. Abderhalden (Editor), *Handbuch der biologischen Arbeitsmethoden, 11(4).* Urban and Schwartzenburg, Berlin, pp. 373–390.

Marino, M.A., 1974. Growth and decay of groundwater mounds induced by percolation. *J. Hydrol.*, 22: 295–301.

Marshall, T.J. and Holmes, J.W., 1979. *Soil Physics*. Cambridge University Press, Cambridge, 345 pp.

Metsävainio, K., 1931. Untersuchungen über das Wurzelsystem der Moorpflanzen. *Ann. Bot. Soc. Zool.-Bot. Fenn.*, 1: 1–417.

Miller, D.H., 1977. *Water at the Surface of the Earth.* Academic Press, New York, N.Y., 557 pp.

Miller, P.C., Stoner, W.A. and Ehleringer, J.R., 1978. Some aspects of water relations of arctic and alpine regions. In: L.L. Tieszen (Editor), *Vegetation and Production Ecology of an Alaskan Arctic Tundra.* Springer, New York, N.Y., pp. 343–357.

Mitchell, J.K. and Younger, J.S., 1967. Abnormalities in hydraulic flow through fine-grained soils. *Am. Soc. Test. Mater., Spec. Tech. Publ.*, No 417: 106–139.

Mohrmann, J.C.J. and Kessler, J., 1959. Water deficiencies in European agriculture. *Publ. Int. Inst. Land Reclam. Impr., Wageningen*, 5: 1–60 (10 maps).

Molchanov, A.A., 1963a. *The Hydrological Role of Forests.* A. Gourevitch (Translator). Israel Progr. Sci. Transl., Jerusalem, 407 pp.

Molchanov, A.A., 1963b. Regulation of hydrological regime by biological methods in areas undergoing swamping. In: *The Increase of Productivity of Swamped Forests. Tr. Inst. Lesa Akad. Nauk S.S.S.R.*, 49: 47–56. (OTS 61-31225, U.S. Dep. Commerce, Washington, D.C.)

Monteith, J.L., 1973. *Principles of Environmental Physics.* Arnold, London, 241 pp.

Monteith, J.L. (Editor), 1975. *Vegetation and the Atmosphere, 1: Principles.* Academic Press, London, 278 pp.

Moore, P.D., 1977. Stratigraphy and pollen analysis of Claish Moss, North-West Scotland: significance for the origin of surface pools and forest history. *J. Ecol.*, 65: 375–397.

Moore, P.D. and Bellamy, D.J., 1973. *Peatlands.* Elek, London, 221 pp.

Moss, C.E., 1913. *Vegetation of the Peak District.* Cambridge University Press, London, 235 pp.

Mott, P.J., 1973. *On the Development of a Raised Peat Bog.* Dissertation, University of Dundee, Dundee, 134 pp.

Muskat, M., 1937 (repr. 1946). *The Flow of Homogeneous Fluids through Porous Media.* Edwards, Ann Arbor, Mich., 763 pp.

Mustonen, S.E., 1964. Potentiaalisen evapotranspiraation määrittämisestä. [Estimating potential evapotranspiration.] *Acta Agral. Fenn.* 102(2): 1–24.

Nekrasova, I.V., 1973. Comparison of observed bog water levels with data obtained by computer. *Sov. Hydrol.*, 1973: 305–310.

Neuhäusl, R., 1975. *Hochmoore am Teich Velké Dářko.* Vegetace ČSSR, A9. Academia, Prague, 267 pp.

Newson, M.D., 1976. Soil piping in upland Wales: a call for more information. *Cambria*, 3(1): 33–39.

Newson, M.D. and Harrison, J.G., 1978. Channel studies in the Plynlimon experimental catchments. *Inst. Hydrol. Wallingford Rep.*, No. 47: 61 pp.

Nicol, A.T. and Robertson, R.A., 1974. *Bolton Fell, Cumbria.*

Report on Peat Resources and Their Development. Unpublished memorandum, Macaulay Institute, Aberdeen. 8 pp.

Nordenson, T.J. and Baker, D.R., 1962. Comparative evaluation of evaporation instruments. *J. Geophys. Res.*, 67: 671–679.

Novikov, S.M., 1964. Computation of the water-level regime of undrained upland swamps from meteorological data. *Sov. Hydrol.*, (1964), 1–22.

Novikova, Y.V., 1963. O transpiratsii gidrofitnykh rastenii i ikh razhode vody na ispareniye iz Kengirskogo vodokhranilshcha. [Transpiration of hydrophytes and their part in the total evaporative discharge of water from the Kengirdam reservoir.] *Tr. Inst. Bot. Akad. Nauk Kaz. S.S.R.*, 16: 118–135. (Quoted by Bernatowicz et al., 1976.)

Nye, P.H. and Tinker, P.B., 1977. *Solute Movement in the Soil-Root System.* Blackwell, Oxford, 342 pp.

Nys, L., 1954. La capacité pour l'eau et le pH des tourbières bombées. *Ann. Soc. Géol. Belg.*, 77B: 289–296.

Nys, L., 1955. La circulation de l'eau dans les tourbières bombées. *Ann. Soc. Géol. Belg.*, 78: 463–467.

Nys, L., 1957. Tourbières hautes et débits de rivières. *Bull. Soc. R. For. Belg.*, 64: 217–229.

Oberdorfer, E., 1962. *Pflanzensoziologische Exercursionflora für Süddeutschland.* Ulmer, Stuttgart, 2nd ed., 987 pp.

Okruszko, H. and Szuniewicz, J., 1975. Porositätsdifferenzierung und damit verbundene Luft- und Wasserverhältnisse in den Niedermoorböden. In: *Beiträge der internationales Symposium zu Problemen der Wasserregulierung auf Niedermoor, Eberswalde, 1974.* Akad. Landwiss. D.D.R., Berlin, pp. 227–250.

Olausson, E., 1957. Das Moor Roshultsmyren. Eine geologische, botanische und hydrologische Studie in einem südwestschwedischen Moor mit exzentrisch gewölbten Mooselementen. *Acta Univ. Lund, N.S. 2*, 53(12): 1–72.

Oltmanns, F., 1885. Zur Frage nach der Wasserleitung im Laubmoosstämmchen. *Ber. Dtsch. Bot. Ges.*, 3: 58–62.

Oltmanns, F., 1887. Ueber die Wasserbewegung in der Moospflanze und ihren Einfluss auf die Wasservertheilung im Boden. *Cohns Beitr. Biol. Pflanz., Breslau*, 4: 1–49.

Osvald, H., 1923. Die Vegetation des Hochmoores Komosse. *Sven. Växtsociol. Sällsk. Handl.*, 1: 1–436.

Osvald, H., 1949. Notes on the vegetation of British and Irish mosses. *Acta Phytogeogr. Suec.* 26: 1–62.

Overbeck, F., 1975. *Botanisch-geologische Moorkunde.* Wachholtz, Neumunster, 719 pp.

Overbeck, F. and Happach, H., 1957. Über das Wachstum und den Wasserhaushalt einiger Hochmoorsphagnen. *Flora, Jena*, 144: 335–402.

Päivänen, J., 1969. The bulk density of peat and its determination. *Silva Fenn.*, 3(1): 1–19.

Päivänen, J., 1973. Hydraulic conductivity and water retention in peat soils. *Acta For. Fenn.*, 129: 1–70.

Parihar, N.S., 1962. *An Introduction to Embryophyta, 1. Bryophyta.* Central Book Depot, Allahabad, 4th ed., 377 pp.

Parkinson, K.J. and Legg, B.J., 1972. A continuous flow porometer. *J. Appl. Ecol.*, 9: 669–675.

Pavlova, K.K., 1970. Phase composition of water and thermophysical characteristics of frozen peat in the study of infiltration. *Sov. Hydrol.*, 1970: 138–159.

Pearsall, W.H., 1950. *Mountains and Moorlands.* Collins, London, 312 pp.

Penman, H.L., 1948. Natural evaporation from open water, bare soil and grass. *Proc. R. Soc., Lond.*, A193: 120–145.

Penman, H.L., 1954. Evaporation over parts of Europe. *IASH (Gen. Assy. Rome), Publ.*, 62(3): 168–176.

Penman, H.L., 1956. Evaporation: an introductory survey. *Neth. J. Agric. Sci.*, 4: 9–29.

Penman, H.L., 1963. *Vegetation and Hydrology.* Commonw. Agric. Bur., Farnham Royal, 124 pp.

Pisano Valdes, E., 1973. Fitogeografia de la peninsula de Brunswick, Magellanes I. Comunidades meso-higromorficas e higromorficas. *An. Inst. Patag.*, 4: 141–206.

Polubarinova-Kochina, P.Ya., 1962. *Theory of Groundwater Movement.* J.M.R. de Wiest (Translator). Princeton University Press, Princeton, N.J., 613 pp.

Prytz, K., 1932. Der Kreislauf des Wassers auf unberührtem Hochmoor. *Ingeniörvidensk. Skr., Ser. A*, 33: 1–126.

Pyatetsky, G.Y., 1974. The drainage rate and time for providing it on the bogs of Karelia. In: *International Symposium on Forest Drainage, Jyväskyla–Oulu, 1974.* Soc. For. Finland, Helsinki, pp. 117–126.

P'yavchenko, N.I., 1960. Forest swamp science and forest land draining in the U.S.S.R. In: *Questions of Forestry and Forest Management. Papers for 5th World Forestry Congress.* Akad. Nauk S.S.S.R., Moscow, pp. 294–299. (Quoted by Bay, 1966b.)

Raikes, R., 1967. *Water, Weather and Prehistory.* John Baker, London, 208 pp.

Rakhmanina, A.T., 1970. Vodnyi rezhim rastenii. [The water regime of plants.] In: B.N. Norin (Editor), *Ekologiya i biologiya rastenii vostochnoevropeiskoi lesotundry, 1.* Nauka, Leningrad, pp. 253–302.

Ratcliffe, D.A. and Walker, D., 1958. The Silver Flowe, Galloway, Scotland. *J. Ecol.*, 46: 407–445.

Renger, M., Bartels, R., Strebel, O. and Giesel, W., 1976. Kapillarer Aufstieg aus dem Grundwasser und infiltration bei Moorböden. *Geol. Jahrb.*, F3: 9–51.

Richardson, C.J., Tilton, D.L., Kladec, J.A., Chamie, J.P.M. and Wentz, W.A., 1978. Nutrient dynamics of northern wetland ecosystems. In: R.E. Good, D.F. Whigham and R.L. Simpson (Editors), *Freshwater Wetlands.* Academic Press, New York, N.Y., pp. 217–241.

Robertson, R.A., Nicholson, I.A. and Hughes, R., 1968. Runoff studies on a peat catchment. In: *Trans. 2nd Int. Peat Congr., Leningrad, 1963, 1.* H.M.S.O., Edinburgh, pp. 161–166.

Robinson, A.C. and Rodda, J.C., 1969. Rain, wind and the aerodynamic characteristics of raingauges. *Meteorol. Mag.*, 98: 113–120.

Rodda, J.C., 1972. On the question of rainfall measurement and representativeness. In: *World Water Balance. Proc. IASH Symp., Reading, 1970, 1.* IASH/UNESCO, Paris, pp. 173–186.

Rodda, J.C., Downing, R.A. and Law, F.M., 1976. *Systematic Hydrology.* Newnes–Butterworths, London, 399 pp.

Rodewald-Rudescu, L., 1974. *Das Schilfrohr* Phragmites communis *Trinius.* Die Binnengewässer, 27. Schweizerbart'sche, Stuttgart, 302 pp.

Romanov, V.V., 1953. Issledovanie ispareniya so sfagnovykh

bolot. [Research into evaporation from *Sphagnum* mires.] *Tr. Gos. Gidrol. Inst.*, 39: 116–135.

Romanov, V.V., 1956. Gidrofizicheskie metody rascheta vodnogo balansa bolotov. *Pochvovedenie*, 1956(8): 49–56. [Hydrophysical methods of calculating the water balance of bogs. OTS 60-21121, U.S. Dep. Commerce, Washington, D.C.]

Romanov, V.V., 1961. *Gidrophysika bolot.* Gidrometeoizdat, Leningrad, 359 pp. (Translated as Romanov, 1968a).

Romanov, V.V., 1968a. *Hydrophysics of Bogs.* N. Kaner (Translator), Heimann (Editor). Israel Progr. Sci. Transl., Jerusalem, 299 pp.

Romanov, V.V., 1968b. *Evaporation from Bogs in the European Territory of the U.S.S.R.* N. Kaner (Translator), Heimann (Editor). Israel Progr. Sci. Transl., Jerusalem, 183 pp.

Romanova, E.A., 1961. *Geobotanicheskie osnovy gidrologicheskogo izlucheniya verhovykh bolot (c ispol'zovaniem aerofotos'emki).* [*Radiation (Utilising Air Photogrammetry) in Relation to the Geobotanical and Hydrological Elements of Raised Mires*]. Gidrometeoizdat, Leningrad, 244 pp.

Rose, C.W., 1966. *Agricultural Physics.* Pergamon, Oxford, 230 pp.

Rudolf, K. and Firbas, F., 1927. Die Moore des Riesengebirges. *Beih. Bot. Zentbl.*, 43(2): 69–144.

Russell, D.A. and Swartzendruber, D., 1971. Flux–gradient relationships for saturated flow of water through mixtures of sand, silt and clay. *Soil Sci. Soc. Am. Proc.*, 35: 21–26.

Rutter, A.J., 1955. The composition of wet-heath vegetation in relation to the water table. *J. Ecol.*, 43: 507–543.

Rutter, A.J., 1967. An analysis of evaporation from a stand of Scots pine. In: W.E. Sopper and H.W. Lull, (Editors), *Forest Hydrology.* Pergamon, Oxford, pp. 403–417.

Rybníček, K., 1974. *Die Vegetation der Moore im südlichen Teil der Böhmisch-Mährischen Höhe.* Vegetace ČSSR, A6. Academia, Prague, 243 pp.

Rycroft, D.W., 1971. *On the Hydrology of Peat.* Dissertation, University of Dundee, Dundee, 268 pp.

Rycroft, D.W., Williams, D.J.A. and Ingram, H.A.P., 1975a. The transmission of water through peat I. Review. *J. Ecol.*, 63: 535–556.

Rycroft, D.W., Williams, D.J.A. and Ingram, H.A.P., 1975b. The transmission of water through peat II. Field experiments. *J. Ecol.*, 63: 557–568.

Sander, J.E., 1976. An electric analogue approach to bog hydrology. *Ground Water*, 14: 30–35.

Scheidegger, A.E., 1960. *The Physics of Flow Through Porous Media.* Toronto University Press, Toronto, Ont., 313 pp.

Schlüter, H., 1970. Vegetationskundlich-synökologische Untersuchungen zum Wasserhaushalt eines hochmontanen Quellgebietes im Thüringer Wald. *Wiss. Veröff. Geogr. Inst. Dtsch. Akad. Wiss., N.F.*, 27: 23–146.

Schmeidl, H., Schuch, M. and Wanke, R., 1970. Wasserhaushalt und Klima einer kultivierten und unberuhrten Hochmoorfläche am Alpenrand. *Schriftenr. Kurat. Kult., Hamburg*, 20: 1–171.

Šesták, Z., Čatský, J. and Jarvis, P.G. (Editors), 1971. *Plant Photosynthetic Production. Manual of Methods.* W. Junk, The Hague, 818 pp.

Sjeflo, J.B., 1968. Evapotranspiration and the water budget of prairie potholes in North Dakota. *U.S. Geol. Surv. Prof. Paper*, No. 585-B: 1–49.

Sjörs, H., 1948. Mire vegetation in Bergslagen, Sweden. *Acta Phytogeogr. Suec.*, 21: 277–290.

Skartveit, A., Rydén, B.E. and Kärenlampi, L., 1975. Climate and hydrology of some Fennoscandian tundra ecosystems. In: F.E. Wielgolaski (Editor), *Fennoscandian Tundra Ecosystems, I.* Springer-Verlag, Berlin, pp. 41–53.

Slatyer, R.O., 1967. *Plant–Water Relationships.* Academic Press, London, 366 pp.

Slatyer, R.O. and McIlroy, I.C., 1961. *Practical Microclimatology.* UNESCO/CSIRO, Canberra, A.C.T., 308 pp.

Slavík, B. (Editor), 1974. *Methods of Studying Plant Water Relations.* Springer-Verlag, Berlin, 449 pp.

Small, E., 1972. Water relations of plants in raised *Sphagnum* peat bogs. *Ecology*, 53: 726–728.

Šmid, P., 1975. Evaporation from a reedswamp. *J. Ecol.*, 63: 299–309.

Sokolov, A.A. and Chapman, T.G. (Editors), 1974. *Methods for Water Balance Computation.* UNESCO Press, Paris, 127 pp.

Solantie, R., 1974. The influence of water balance in summer on forest and peatland vegetation and bird fauna and through the temperature on agricultural conditions in Finland. *Silva Fenn.*, 8: 160–184 (Finnish with English summary).

Sonesson, M., 1969. Studies on mire vegetation in the Torneträsk area, northern Sweden II. Winter conditions of the poor mires. *Bot. Not.*, 122: 481–511.

Sonesson, M., 1970. Studies on mire vegetation in the Torneträsk area, northern Sweden III. Communities of the poor mires. *Opera Bot. Soc. Bot. Lund.*, 26: 1–120.

Stanhill, G., 1961. A comparison of methods of calculating potential evapotranspiration from climatic data. *Isr. J. Agric. Res.*, 11: 159–171.

Steele, A., 1826. *The Natural and Agricultural History of Peat-Moss or Turf-Bog.* Laing and Black, Edinburgh, 401 pp.

Steubing, L., 1965. *Pflanzenökologisches Praktikum.* Parey, Berlin, 262 pp.

Stewart, R.A. and Kantrud, H.A., 1972. Vegetation of prairie potholes, North Dakota, in relation to quality of water and other environmental factors. *U.S. Geol. Surv. Prof. Paper*, No. 585-D: 1–36.

Stocker, O., 1923. Die Transpiration und Wasserökologie nordwestdeutscher Heide- und Moorpflanzen am Standort. *Z. Bot.*, 15: 1–41.

Stoekeler, J.H. and Weitzman, S., 1960. Infiltration rates in frozen soils in northern Minnesota. *Soil Sci. Soc. Am. Proc.*, 24: 137–139.

Summerfield, R.J., 1974. The reliability of mire water chemical analysis data as an index of plant nutrient availability. *Plant Soil*, 40: 97–106.

Swartzendruber, D., 1966. Soil–water behaviour as described by transport coefficients and functions. *Adv. Agron.*, 18: 327–370.

Swinbank, W.C., 1951. The measurement of vertical transfer of heat and water vapour by eddies in the lower atmosphere. *J. Meteorol.*, 8: 135–145.

Tallis, J.H., 1965. Studies on southern Pennine peats IV. Evidence of recent erosion. *J. Ecol.*, 53: 509–520.

Tallis, J.H., 1973. Studies on southern Pennine peats V. Direct observations of peat erosion and peat hydrology at

Featherbed Moss, Derbyshire. *J. Ecol.*, 61: 1–22.

Taylor, J.A., 1976. The peat deposits of the British Isles: their location and evaluation. *Proc. 5th Int. Peat Congr., Poznan, 1976*, 4: 228–243.

Taylor, J.A. and Tucker, R.B., 1970. The peat deposits of Wales: an inventory and interpretation. *Proc. 3rd Int. Peat Congr., Quebec, 1968*, pp. 163–173.

Tervet, D.J., 1976. *Geochemical Investigations at Dun Moss.* Dissertation, University of Dundee, Dundee, 309 pp.

Thornthwaite, C.W., 1948. An approach toward a rational classification of climate. *Geogr. Rev.*, 33: 55–94.

Thornthwaite, C.W. and Hare, F.K., 1965. The loss of water to the air. *Meteorol. Monogr.*, 6: 163–180

Tinbergen, L., 1940. Observations sur l'évaporation de la végétation d'une tourbière dans les Hautes-Fagnes de Belgique. *Mém. Soc. R. Sci. Liège*, 4: 21–76.

Toebes, C. and Ouryvaev, V., 1970. *Representative and Experimental Basins. An International Guide for Research and Practice.* UNESCO, Paris, 348 pp.

Tuschl, P., 1970. Die Transpiration von *Phragmites communis* Trin. im geschlossenen Bestand des Neusiedler Sees. *Wiss. Arb. Burgenld.*, 44: 126–186.

Uhden, O., 1967. Niederschlags- und Abflussbeobachtungen auf unberuhrten, vorentwässerten und kultivierten Teilen eines nordwestdeutschen Hochmoores, der Esterweger Dose am Küstenkanal bei Papenburg. *Schriftenr. Kurat. Kult., Hamburg*, 15(1): 1–99.

Uhlig, S., 1954. Berechnung der Verdunstung aus klimatologischen Daten. *Mitt. Dtsch. Wetterdienstes*, 1954(6): 1–24.

Van't Woudt, B.D. and Nelson, R.E., 1963. Hydrology of the Alakai Swamp, Kauai, Hawaii. *Bull. Hawaii Agric. Exper. Stn.*, No. 132: 1–30.

Van Zinderen Bakker, E.M. and Werger, M.J.A., 1974. Environment, vegetation and phytogeography of the high-altitude bogs of Lesotho. *Vegetatio*, 29: 37–49.

Ven Te Chow, 1964. Runoff. In: Ven Te Chow (Editor), *Handbook of Applied Hydrology, Section 14.* McGraw-Hill, New York, N.Y., 14-1–14-54.

Virta, J., 1960. Evapotranspiration measurements in a string fen in northern Finland. *IASH (Gen. Assy. Helsinki), Publ.*, 53: 438–441.

Virta, J., 1966. Measurement of evapotranspiration and computation of water budget in treeless peatlands in the natural state. *Comment. Phys.-Math. Soc. Sci. Fenn.*, 32(11): 1–70.

Vitt, D.H. and Slack, N.G., 1975. An analysis of the vegetation of *Sphagnum*-dominated kettle hole bogs in relation to environmental gradients. *Can. J. Bot.*, 53: 332–359.

Volarovich, M.P. and Churaev, N.V., 1966. Effect of surface forces on transfer of moisture in porous bodies. In: B.V. Deryagin (Editor), *Research in Surface Forces, 2.* Consultants Bureau, New York, N.Y., pp. 212–219.

Volarovich, M.P. and Churaev, N.V., 1968. Application of the methods of physics and physical chemistry to the study of peat. In: *Trans. 2nd Int. Peat Congr., Leningrad, 1963, 2.* H.M.S.O., Edinburgh, pp. 819–831.

Volarovich, M.P., Gamayunov, N.I. and Lishtvan, I.I., 1972. Investigations of physical and physico-chemical properties of peat. *Proc. 4th Int. Peat Congr., Otaniemi, 1972*, 4: 219–226.

Vorob'ev, P.K., 1963. Investigations of water yield of low-lying swamps of western Siberia. *Sov. Hydrol.*, 1963: 226–252.

Wadsworth, R.M. (Editor), 1968. *The Measurement of Environmental Factors in Terrestrial Ecology.* Blackwell, Oxford, 314 pp.

Waine, J., 1976. *The Hydraulic Conductivity of Humified Peat.* Dissertation, University of Dundee, Dundee, 210 pp.

Walter, H., 1960. *Standortslehre. Einführung in die Phytologie, 3(1).* Ulmer, Stuttgart, 2nd ed., 566 pp.

Ward, R.C., 1975. *Principles of Hydrology.* McGraw-Hill, Maidenhead, 2nd ed., 367 pp.

Warrick, A.W., 1970. A mathematical solution to the hillside seepage problem. *Soil Sci. Soc. Am. Proc.*, 34: 849–853.

Weber, C.A., 1908. Aufbau und Vegetation der Moore Norddeutschlands. *Bot. Jahrb., Suppl.*, 90: 19–34.

Webster, R., 1966. The measurement of soil water tension in the field. *New Phytol.*, 65: 249–258.

Wechmann, A., 1943. Die Durchlässigkeit von Moorböden — das Ergebnis einer Umfrage. *Dtsch. Wasserwirtsch.*, 38: 136–138.

Wein, R.W., 1973. Biological flora of the British Isles. *Eriophorum vaginatum* L. *J. Ecol.*, 61: 601–615.

Werenskjold, W., 1943. Högmossarnas välvning i södra Sverige. *Geol. Fören. Stockholm Förhandl.*, 65: 304–305. (Quoted by Dahl, 1969.)

Wertz, G., 1975. Bemessungsgrundlagen zur Wasserregulierung auf Niedermoorstandorten. In: *Beiträge der internationales Symposium zu Problemen der Wasserregulierung auf Niedermoor, Eberswalde, 1974*, pp. 329–346.

Wickman, F.E., 1951. The maximum height of raised bogs and a note on the motion of water in soligenous mires. *Geol. Fören. Stockholm Förhandl.*, 73: 413–422.

Wildi, O., 1978. Simulating the development of peat bogs. *Vegetatio*, 37: 1–17.

Williams, G.P., 1970. The thermal regime of a *Sphagnum* peat bog. *Proc. 3rd Int. Peat Congr., Quebec, 1968*, pp. 195–200.

Willis, W.O., Parkinson, H.L., Carlson, C.W. and Haas, H.J., 1964. Water table changes and soil moisture loss under frozen conditions. *Soil Sci.*, 98: 244–248.

Wilson, E.M., 1974. *Engineering Hydrology.* Macmillan Press, London, 2nd ed., 232 pp.

Wind, G.P., 1960. Capillary rise and some applications of the theory of moisture movement in unsaturated soils. *Versl. Meded. Comm. Hydrol. Onderz. T.N.O.*, 5: 1–15.

Witting, M., 1947. Katjonbestämningar i myrvatten. *Bot. Not.*, 1947: 287–304.

Yakovlev, V.I., 1969. Remote ground-water level meter for bog stations. *Sov. Hydrol.*, 1969: 442–449.

Yamamoto, S., 1970. Study on permeability of soils, especially of peat soil. *Mem. Fac. Agric. Hokkaido Univ.*, 7: 307–411 (in Japanese with English summary).

Yamazaki, F., Soma, K. and Furuya, C., 1957. (On the pF curve of peat soil). *J. Agric. Eng. Soc. Jap.*, 25: 214–217. [Abstract: *Soil Fert.*, 22: 355 (1959).]

Yanishevskiĭ, Y.D., 1957. *Aktinometricheskie pribory i metody nablyudeniĭ.* [*Actinometric Instruments and Observation Methods.*] Gidrometeoizdat, Leningrad, 414 pp. (Quoted by Šesták et al., 1971.)

Chapter 4

PEAT[1]

R.S. CLYMO

PEAT STRUCTURE

Peat is the accumulated remains of dead plants. There is no clear break in the continuum between a mineral soil with organic matter in it — as, for example, in the surface of a podzol — and an almost pure *Sphagnum* peat of which more than 99% is organic matter. Most peats contain less than 20% of unburnable inorganic matter, but some soil scientists allow up to 35%, and commercial standards may allow up to 55%. In Iceland, volcanic ash is widespread and peat commonly contains more than 20% ash (Bjarnason, 1968). The depth of peat deposits is similarly unbounded, though it is common to take an arbitrary minimum depth of about 30 cm. Few deposits in the Temperate and Boreal Zones exceed 15 m in depth[2]. Peat accumulations in subtropical and tropical regions are mostly coastal or at high altitude, but may once have been more extensive than Temperate and Boreal Zone peats (Given and Dickinson, 1975). For example Anderson (1964) records 13 m depth of swamp forest peat accumulated during 4500 years on the coast of Sarawak, and Whitmore (1975), who summarized Anderson's work, noted that as much as 20 m of peat had been recorded. Tropical and subtropical peat may even now be more abundant than is commonly realised (Anderson, Part B, Ch. 6). For example, Coulter (1950, 1957) records about 800 000 ha of peatland in Malaya (about the same as the area in Scotland, where peat covers 10% of the land: Robertson, 1968), and Anderson calculates about 1.5×10^6 ha of peatland in Sarawak.

Peat is not a single homogeneous substance either in space or time. It starts as recently dead plant matter and begins a series of changes which are usually fast at first but slower later. Collectively these changes are often called decay, decomposition, breakdown or humification. The meaning of these terms is rather vague, but the processes encompassed by them, with varying emphasis, are: (1) loss of organic matter, as gas or in solution, as a result of leaching and of attack by animals and micro-organisms; (2) loss of physical structure; and (3) change of chemical state — for example the production of new types of molecules by micro-organisms.

Peat is of interest to workers in many fields — agriculture, horticulture, chemistry, power generation, civil engineering, medicine and mining — as well as in ecology; so there is a great deal of published information about it. Much of this information may initially be of interest to workers in one particular field. For example, consolidation under load was of interest, originally, to engineers. More recently it has become of interest to ecologists too (Clymo, 1978; Hutchinson, 1980). These different interests are displayed in the four-yearly International Peat Congress. The Proceedings of these Congresses form a useful entrée to the literature, but are, with one exception, a bibliographer's nightmare, having neither editors nor publication date. They are quoted here with the year in which the Congress was held (see p. 224). A very useful account of those parts of peat science which have been of interest in Germany is that of Göttlich (1976).

Peat has numerous characters, most of which

[1]Manuscript completed July, 1980.

[2]Florschütz et al. (1971) described a deposit near Padul, southern Spain, which had twelve alternating bands of peat and lake marl to a depth of 72 m. About half was peat, though of what botanical origin was not recorded. The peat at 10 m depth was about 4600 years old.

may be used in classification. Amongst the more generally important are:

1. Botanical composition
2. State of decomposition

Other characteristics which may be important for particular purposes are:

3. Concentration of inorganic matter (soluble and insoluble)
4. Concentration of inorganic solutes (especially calcium, potassium, ammonium and hydrogen ions, nitrate and phosphate)
5. Cation exchange capacity
6. Activity of micro-organisms
7. Bulk density (dry-matter mass per unit volume of peat)
8. Water content
9. Gas content
10. Drainage and water-retaining properties
11. Proportion of fibre (defined as a size class, and not fibre in the generally understood sense)
12. Structure (the arrangement of materials of different size class and type)
13. Heat of combustion
14. Colour
15. Age

Finally, there are three characters of the peat-forming system as a whole:

16. Topography and hydrology of the area which formed the peat
17. Morphology of the peat-forming system
18. History of the peat-forming system

These characteristics are independent in the sense that there is no compelling physical reason why most of them should not vary independently. In practice, however, there is high correlation between characteristics, so that knowledge of one or a few of them enables one to predict many others with considerable success. For example, the knowledge that a peat is composed mainly of undecomposed *Sphagnum* allows one to predict that it is probably light to medium brown, comes from raised bog (*Hochmoor*), has bulk density about 0.05 g cm^{-3}, holds (at field capacity) about fifteen times the dry mass of water and has about 50% gas space, has about 90% "fibre" (material retained by a 1.5-mm mesh sieve), a cation exchange capacity of about 1 mmol g^{-1}, a calorific content of about 17 to 20 kJ g^{-1}, an ash content of 1 to 2%, nitrogen concentration about 10 to 20 mg g^{-1}, phosphorus concentration about 1 mg g^{-1}, and so on.

For many practical purposes the chemical state and the origin of the peat can be predicted from botanical composition, and the physical properties from the state of decomposition. These two "key characters" may, to a considerable extent, be determined in the field. Colour often changes rapidly on exposure to air, probably as a result of oxidation, so it must be recorded in the field; delicate greens, blues, yellows, browns, oranges and reds visible in peat freshly exhumed from anaerobic conditions often change before one's eyes, and within an hour the peat is a uniform dark brown or black.

The other characteristics are normally measured in a laboratory, and more detailed determination of botanical composition and state of decomposition may need to be made there too. These characteristics are now considered in more detail.

Attributes of peat

Botanical composition

A broad distinction of three categories — **moss**, **herbaceous** (predominantly grasses and sedges) and **wood** — has been suggested (Kivinen, 1977).

There is a good case for subdividing the "moss" category into "*Sphagnum*" and "other" (mainly hypnoid) mosses, because hypnoid moss peat is commonly formed from plants supplied with water which has flowed through mineral soil, and frequently has a higher concentration of solutes than does *Sphagnum* moss peat, which commonly forms in ombrotrophic and oligotrophic conditions (see, for example, Mörnsjö, 1969). A few types of non-*Sphagnum* moss peat are oligotrophic (and mainly ombrotrophic). Examples are the subantarctic peats formed from the mosses *Chorisodontium aciphyllum* and *Polytrichum alpestre* (Collins, 1976; Fenton, 1978, 1980) and the *Racomitrium* peats of Andøya, northern Norway.

Identification of mosses to family or genus may be attempted in the field, but a microscope is usually necessary to confirm the identification and to extend it to lower levels. This may be of value. For example, the identification of *Sphagnum* to the level of Section *Acutifolia* (usually hummock species), *Sphagnum* (commonly lawn species) and *Cuspidata* (commonly pool species) made possible the reconstruction of the succession of communities on north temperate bogs, and gave rise to the constructively misleading hypothesis, commonly attributed to Osvald (1923), about cyclic growth of hummocks and hollows (Tansley, 1939; Watt, 1947; Walker and Walker, 1961; Tallis, Chapter 9, this volume). Another example is the use of the distinctive comb-fibrils of *S. imbricatum* leaf cells to show that for nearly 7000 years this species dominated the peat in the bog at Malham Tarn, England (Pigott and Pigott, 1959), but that it rather suddenly vanished from the peat in recent times, and is now extremely rare in the surface vegetation of the whole region.

The "herbaceous" category of peat may usefully be divided too. In north temperate regions grasses and sedges predominate, and this type of peat (typically a product of Magnocaricion vegetation) has usually formed under the influence of mineral soil water. As Kulczyński (1949) pointed out the plants are rooted (unlike mosses) and the leaves fall onto the waterlogged peat surface. Such peats are usually denser than moss peats and have a higher concentration of ash and of inorganic solutes. The ecological conditions which give rise to *Phragmites* peat, *Carex* peat, and *Cladium* peat differ considerably, though from the gardener's point of view the differences may be unimportant. But the distinction between these and peat formed from the remains of *Eriophorum vaginatum* (cotton grass) is likely to be important to almost everyone using peat because *Eriophorum* grows, usually, in oligotrophic conditions.

In north temperate and arctic peat deposits, "wood" peat is usually found in one of three positions: at the base of a deposit, above a layer of herbaceous peat, or as a layer in moss peat. Wood peat at the base of a deposit is commonly of *Betula* or *Pinus*, the trees probably being the last generation rooted in mineral soil. Wood peat above herbaceous peat may be formed from fen carr, and commonly contains *Alnus*, *Betula* and some *Salix*. Wood layers in temperate moss peat often contain *Betula* or *Pinus* and formed at times when a bog surface dried temporarily. In the arctic, wood layers are commonly of birch — often *Betula michauxii*. In the tropics and subtropics complete deposits may be of wood peat — for example the coastal Sarawak peats (Anderson, 1964; and Part B, Ch. 6).

Many peats are an intimate mixture of moss, herbaceous and wood types. Yet others are so decomposed that their botanical composition is not, or not easily, identifiable.

State of decomposition

This character, usually called the state of humification, is partly assessed by the extent to which plant structure is visible, and partly by colour (though this may change rapidly). Field assessments are usually made using the H scale devised by Von Post and Granlund (1926). This is similar to the Beaufort scale of wind velocity in that it replaces an entirely subjective description by a numerical scale based on specific, more objective, criteria. These include the colour of the fluid expressed when peat is squeezed in the hand, and the proportion and character of the material which remains in the hand after squeezing the peat (Table 4.1).

Other methods of assessing the extent of humification rely on chemical extractants and measurement of the density of brown colour. One of the simplest is to shake equal volumes of peat and 5% w/v potassium hydroxide, or a saturated solution of sodium pyrophosphate, allow a few drops of the liquid to spread on a filter paper, and compare the colour with arbitrary standard colours, preferably Munsell chart colours.

More complex procedures are used too. For example, Bahnson (1968, translated in Aaby and Tauber, 1974), dries and grinds the samples, then boils 0.2 g with 100 ml 0.5% sodium hydroxide for 1 h, filters, and measures absorbance with a yellow filter (EEL 626). Other methods use neutral sodium pyrophosphate at room temperature for 24 h as extractant (Schnitzer and Kahn, 1972), or dissolve all but "humus" with acetyl bromide (Overbeck and Schneider, 1940). Von Naucke (1976) discussed these and other methods of assessing humification.

When comparisons of methods are made (for example, Aaby and Tauber, 1974) there is broad general agreement, though in detail the agreement is poor. Fig. 4.1 shows the correlation between the H scale and a colorimetric estimate. A more extensive series of 614 samples was found by Karesniemi (1972) to give an approximately linear relation between H and a (different) colorimetric estimate. The lack of detailed agreement is not surprising: peat is a mixture of still largely unknown chemical substances (Schnitzer and Kahn, 1972), and the techniques used for assessing humification (e.g. Schnitzer, 1973) are still comparatively crude. Nevertheless they can, used in a relative way, give useful insights into some of the processes of peat formation.

TABLE 4.1

Humification scale, translated and tabulated from Von Post and Granlund (1926, pp. 29–30)

Scale number	Description	Proportion of "dy"	Plant structure	Expressed fluid	Peat lost	Peat retained in the hand	
						consistency	colour
H1	completely unhumified	none		colourless, clear			
H2	virtually unhumified	none		yellow-brown, clear			
H3	little humified	small		noticeably turbid	none	not porridgey	
H4	poorly humified	modest		very turbid		somewhat porridgey	
H5	fairly humified, structure distinct	fair	plain, but somewhat obscured	strongly turbid	some	very porridgey	
H6	fairly humified, structure less distinct	fair	indistinct, but still clear		up to 1/3	very porridgey	
H7	quite well-humified	considerable	much still visible		about 1/2	gruel-like	very dark
H8	well-humified	large	vague		2/3	only fibrous matter and roots remain	
H9	almost completely humified	most	almost none visible		almost all	homogeneous	
H10	completely humified	all	none visible		all	porridge	

The often quoted Von Post (1924) reports the existence of this scale but gives no details. Peat is squeezed in the hand. The fluid which escapes is examined and so are the nature and amount of peat substance remaining in the hand.
The term "dy" introduced by Von Post has no direct equivalent in English. It is nowadays used for amorphous (colloidal) dark brown "humic" matter from which all trace of macroscopic plant-structure has disappeared.

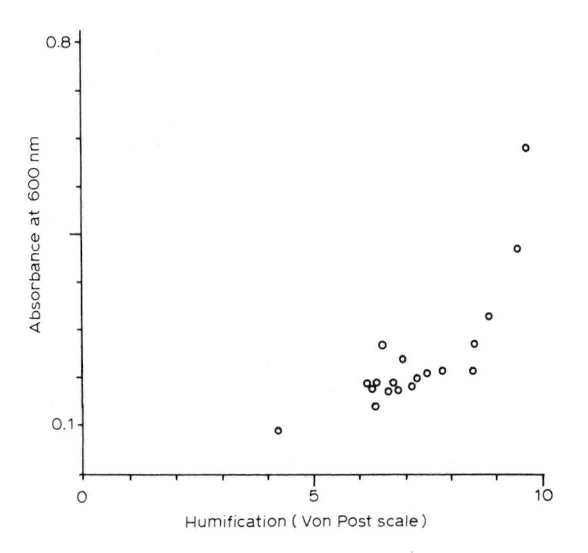

Fig. 4.1. Correlation of humification on the H scale with absorbance at 600 nm of an Na_2HPO_4/NaF extract. Data of Mattson and Koutler-Andersson (1954).

In *Sphagnum* peats in Finland there is a positive and approximately linear regression between bulk density and the Von Post H scale value (summarized by Päivänen, 1969). The relationship is approximately:

$$\rho = 0.01\ H + 0.04$$

where ρ = bulk density (g cm^{-3}); H = Von Post scale number (0–10). A remarkably similar relationship:

$$\rho = 0.01\ H - 0.05$$

may be calculated from the results shown by Karesniemi (1972).

Humification may also be related to void proportion (see below).

Concentration of inorganic matter

This is easily measured after dry ashing (at 450–550°C) or after wet ashing with concentrated strong acids, particularly nitric acid mixed with perchloric acid. Wet ashing is usually used only when chemical analyses are to be made, and the presence of perchloric acid is potentially dangerous. The mineralogy of the inorganic fraction may be a useful characteristic (Finney and Farnham, 1968), particularly the distinction between soil-derived quartz and plant-produced opaline silica (Chapman, 1964b).

Concentration of inorganic solutes

It is convenient to distinguish solutes in the water from those, probably nearly in equilibrium, in "exchangeable" forms. "Exchange" includes a variety of processes, including simple exchange, chelation and (for H^+) dissociation. Peat water may be expressed by pressure (Malmer, 1962b, p. 8), and "exchangeable" solutes extracted by 1*M* acetic acid, 1*M* ammonium acetate, or 0.5*M* barium acetate, at pH 7 (Puustjärvi, 1957) or by some other concentrated acid or salt. Different extractants give different results (Table 4.2A). Most of the H^+, Na^+, K^+, Mg^{2+} and Ca^{2+} in ombrogenous peat is exchangeable (Gore and Allen, 1956; see also Table 4.2B).

The concentration (or, strictly, the activity) of H^+ in peat water is an especially important character. It is commonly reported as "pH", and is useful because, as Pearsall (1954) pointed out, it is often highly correlated with other characters. Taken together with the concentration of Ca^{2+}, or even with electrical conductivity (if the influence of coastal sea spray is not important), it can be used to assess the relative importance of rain water and mineral soil water.

Measurements of pH made by pressing a glass electrode into damp peat are usually lower than those of peat water, and are difficult to interpret. Measurements of pH in concentrated potassium chloride solution mixed with peat may be useful: essentially, the method measures total exchangeable H^+, so the relative amounts of solution and dry peat mass should be recorded too (but rarely are).

Many chemical species change state as a result of the activities of micro-organisms or from purely chemical changes, especially between anaerobic and aerobic conditions. Examples are: Fe^{2+} and Fe^{3+}, Mn^+ and Mn^{2+}, S^{2-} and SO_4^{2-}. Investigation of these very important changes *in the field* has hardly begun.

TABLE 4.2

A. Concentration (mmol dm^{-3}) in peat from Åkhult mire, southern Sweden, of exchangeable cations extracted with 1*M* CH_3COONH_4 (Am) and corresponding value as % difference (D) extracted by 1*M* CH_3COOH. Selected from Malmer (1962a, pp. 304–307) (nd = not determined)

Am (mmol dm^{-3})				D (%)			
Na^+	K^+	Mg^{2+}	Ca^{2+}	Na^+	K^+	Mg^{2+}	Ca^{2+}
Replicates from "mud-bottom", H3–H4							
0.57	0.22	3.1	4.2	−18	9	nd	−2
0.40	0.33	3.0	3.2	28	9	nd	16
0.40	0.11	2.7	3.7	−10	0	nd	−11
0.52	0.22	3.5	5.2	8	−14	nd	−4
0.50	0.16	3.2	4.5	48	131	nd	0
Wet area, *Eriophorum vaginatum* zone							
0.48	0.79	2.3	1.8	−2	−5	−22	−22
0.40	0.28	2.5	2.4	15	57	0	−37
0.32	0.17	1.9	1.9	−9	0	nd	−16
0.52	0.31	2.8	2.4	−17	16	4	−25
Wet area, *Menyanthes trifoliata* zone, H5–H6							
0.49	0.79	3.1	9.2	33	−10	nd	−4
0.59	0.64	2.5	12.7	−3	6	nd	2
0.82	nd	2.1	7.8	−2	nd	nd	−19
Wooded bog, H2							
0.83	0.88	2.4	2.3	−10	−41	17	0

B. Cation concentration in a *Sphagnum–Eriophorum* peat (mmol dm^{-3}) using four extractants, compared with total ($HClO_4$–HNO_3 wet ashing) (from Boatman and Roberts, 1963; bulk density assumed to be 0.1 g cm^{-3})

	(Extractant pH)	Na^+	K^+	Mg^{2+}	Ca^{2+}
Wet ashed		1.2	0.7	3.1	3.1
Extracted with:					
HCl (50 mmol dm^{-3})	*c.* 1.2	1.2	0.3	4.3	3.0
CH_3COOH (1 mol dm^{-3})	*c.* 2.4	1.0	0.4	2.2	1.7
CH_3COOH (1 mol dm^{-3}, intermittent leaching)	*c.* 2.4	1.3	0.5	4.9	3.2
$BaCl_2$ (triethanolamine)	8.1	1.1	0.5	–	2.5

Cation exchange capacity

The term "cation exchange" is often used loosely, as it is here, as a description of experimental results. Besides cation exchange in the strict sense, there are other processes which can in general be described by equilibrium equations based on the law of mass action and with a specific coefficient. Dissociation and chelation are such processes. One may set up mathematical models with sets of, say, three such equations for each cation, and use experimental results to get best estimates of the coefficients for each process. These are useful for predictive purposes (Clymo, 1967). Most workers assume, however, that they are dealing with a single process. If they work with Na^+ or K^+ they will probably call it "exchange", but if with Pb^{2+} they may prefer to speak of chelation. There is usually very little evidence adduced on which the distinction can be supported (but see Sikora and Keeney, Ch. 6), and indeed there may in reality be a mixture of processes, or ones of intermediate character. "Exchange" is used here in the broad sense.

Sphagnum has a relatively high cation exchange capacity (Skene, 1915) as a result of abundant long chain polymers of uronic acids (Theander, 1954), but *Eriophorum vaginatum*, for example, has an unusually low cation exchange capacity (Knight et al., 1961; Clymo, 1963). The cation exchange capacity of bog peats may vary noticeably, therefore, over quite short distances both horizontally and vertically (Fig. 4.2). The chemical *potential* of a cation is related to concentration (or, strictly, to activity) in peat water, but the *capacity* of the peat to supply that ion is related to the cation exchange capacity (and to the concentration of solids such as calcium carbonate if present) and also to the concentration of other cations. Table 4.3 (from Malmer, 1962b) shows that, in accordance with expectation (Clymo, 1967), up to half the total (monovalent) Na^+ is in the peat water, but only 1 to 2% of the (divalent) Ca^{2+}.

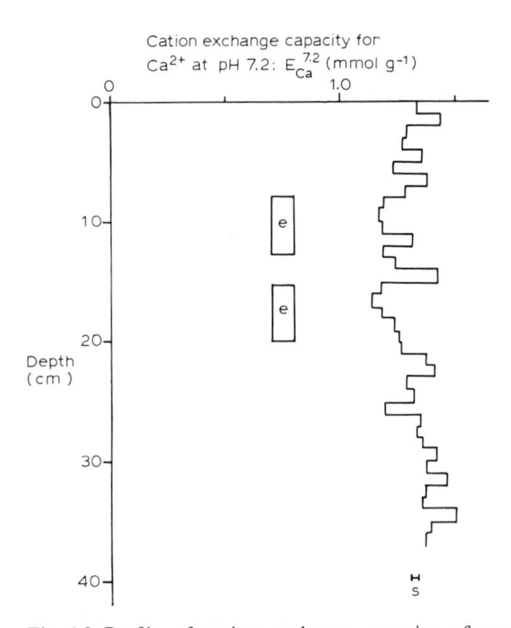

Fig. 4.2. Profile of cation exchange capacity of peat from Stordalen, Abisko, northern Sweden. Peat type *Sphagnum fuscum*, with some *Eriophorum* at points marked *e*. The standard deviation of the method is shown at *s*. (Clymo, unpubl.)

Cation exchange capacity is commonly expressed per unit dry mass of peat, and measured by saturating the exchange sites with a polyvalent cation at a pH sufficiently high to ensure that the uronic acids are dissociated, followed by washing in distilled water to remove cations in solution, then displacing the saturating cation with strong acid or alkali and measuring the amount thus displaced. The washing step is unsatisfactory because some of the saturating cation comes out into the water as the system moves toward a new equilibrium. It is preferable therefore to use a different procedure. The peat is pressed as dry as possible, weighed and transferred to the extracting solution. From the pressed dry weight and the final dry weight the mass of solution transferred with the peat is calculated. The cation concentration in the saturating solution is measured, as well as in the extracting solution, and then the amount of cation transferred with the peat is subtracted. With well-pressed peat this correction is usually less than 10% of the total exchangeable cations.

TABLE 4.3

Proportions of peat cations (as % of total) in peat water in southern Swedish mires; the water was extracted with a pressure-plate apparatus at 14.5×10^5 Pa for 24 h (from Malmer, 1962a).

Site	Na^+	K^+	Mg^{2+}	Ca^{2+}
Mud bottom				
Bog, Åkhult	44	17.0	3.4	2.6
Poor fen, Åkhult	45	11.5	1.7	0.8
Rich fen, Nåthult	40	11.3	3.5	1.6
Hummock				
Bog, Åkhult	27	4.7	2.6	2.3
Narthecium ossifragum community				
Åkhult	33	2.5	4.5	1.0
Kopporås	38	23.0	1.9	0.9
Tjörnarp	51	11.4	6.3	3.5

Because the COOH and OH groups dissociate as the pH is raised, the cation exchange capacity is very dependent on pH. Fig. 4.3 (redrawn from Belkevich and Chistova, 1968) illustrates this point. If E is the observed cation exchange capacity (mmol g^{-1}), and p is the pH, then: $E \simeq 0.3\ (p-3)$. For this reason it might be preferable to use cation exchange ability ($E_{Ca}^{7.2}$) to represent the measured capacity at a particular pH (7.2 in this example) and with a particular cation (calcium in this example), and to reserve the term "cation exchange capacity" for the maximum reached when all the exchange groups are ionized. The results depend on the concentration of the exchanger cation too (Clymo, 1963). Where detailed comparisons are to be made there is much to be said for allowing all samples, enclosed in individual net bags, to equilibrate together in the same bath of solution. The results in Fig. 4.2 were obtained in this way, and with the pressed-dry correction described above. The saturating solution was $CaCl_2 + Ca(OH)_2$ at pH 7.2, and the displacing solution was $0.1M$ HNO_3. The cation Ba^{2+} has commonly been used, in spite of the chances of precipitation of barium carbonate or barium sulphate if the pH is not controlled. Puustjärvi (1956) used H^+ as saturating cation, and

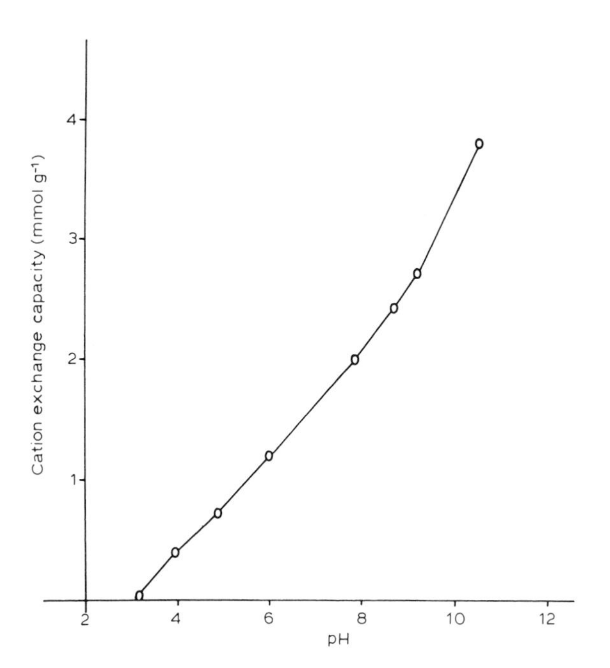

Fig. 4.3. Cation exchange capacity for Ba^{2+} of an *Eriophorum–Sphagnum* peat as a function of pH. Redrawn from Belkevich and Chistova (1968).

$0.25M$ barium acetate as displacer. He found values of $E_{Ba}^{6.5}$ from 0.07 mmol g^{-1} in peat from a *rimpi* fen composed of *Carex* plus non-*Sphagnum* mosses up to 1.5 (or, in isolated cases, 1.7 and 1.9) mmol g^{-1} in *Sphagnum fuscum* peat.

Activity of micro-organisms

The nature and activities of micro-organisms (mainly fungi and bacteria) have been receiving much more attention since 1950, following a long period of neglect after the early work of Waksman and Tenney (1928) and Waksman and Stevens (1929c).

In blanket bog, Collins et al. (1978) found that estimates of biomass of bacteria by different methods differed by one to two orders of magnitude, and of fungi by about one order of magnitude. Within these broad limits there was no clear indication that either group was dominant. The abundance of both groups declined in the anaerobic zone, though this total masked conspicuously different behaviour by groups defined by physiological function. Even obligately anaerobic bacteria were less abundant in the anaerobic zone than they were in the surface layers, which contain a lot of water and must have a large total volume of microanaerobic sites scattered through the generally aerobic zone.

The biomass of bacteria and fungi is of course not a good indicator of their physiological activity. That such activities are localized can be suggested by enrichment cultures (Clymo, 1965) or, better, by direct measurement of the products of micro-organism activity, for example sulphide (Benda, 1957; Clymo, 1965; Urquhart and Gore, 1973), or carbon dioxide and methane (Clymo and Reddaway, 1971). In general, however, there is still relatively little known about the activities of micro-organisms and particularly about microbial interactions in peat. Reviews of the microbiology of peat have been made by Given and Dickinson (1975) and by Dickinson in this volume (Ch. 5). Here it is sufficient to point out that the activity (or inactivity) of micro-organisms is one of the main causes of the very existence of peat. At the other end of the scale, the excessive activity of micro-organisms perhaps especially of thermophilic fungi (see, for instance, Küster and Locci, 1964) can raise the temperature of milled peat to 70°C, at which point, if air is admitted, chemical reactions become so fast that the peat burns spontaneously[1] (Strygin, 1968).

Bulk density, water content, gas content

There are many methods of making these measurements. Puustjärvi (1968, 1969) gives one example. In principle, four quantities must be known (Skaven-Haug, 1972).

A defined volume of peat is weighed. The peat is oven-dried and weighed again. A fourth quantity — the density of the peat dry matter — is needed. This may be determined by weighing the peat, suspended under water, after evacuation under water to remove gas (Clymo, 1970). Let:

V = volume of whole bulk of peat	V_d = volume of the dry peat substance
W_w = wet weight in air	M_w = wet mass in air
W_s = weight submerged in water	
W_d = dry weight in air	M_d = dry mass in air
ρ_w = density of water	

Then ρ_d, the density of peat dry matter, is given by:

$$\rho_d = \frac{M_d}{V_d} = \frac{\rho_w W_d}{(W_d - W_s)} \tag{1}$$

The dry matter density is defined in terms of *mass* per unit volume, but the simplest method for determining it depends on measuring the quotient of *weights*. The method should be accurate at any position on the earth's surface, but would be impossible if attempted in a condition of weightlessness. This is not likely to be a severe limitation in practice.

The *bulk* density, ρ, is given by:

$$\rho = M_d / V \tag{2}$$

The water content on a mass basis, ϕ_d, is given by:

$$\phi_d = \frac{(M_w - M_d)}{M_d} \tag{3}$$

[1]Spontaneous combustion of peat should be distinguished from the phenomenon known as "will-o'-the-wisp" (= *ignis fatuus* = Jack-o'-lantern) which anecdotal accounts record from the surface of peat bogs. It is tempting to speculate that this may be bubbles of methane ignited by rapid oxidation of inorganic or organic sulphides.

and the water content on a volume basis, ϕ_v, by:

$$\phi_v = \frac{(M_w - M_d)}{\rho_w V} \tag{4}$$

The void proportion, E, is given by:

$$E = 1 - \frac{M_d}{V \rho_d} \tag{5}$$

Finally, the gas content per unit volume, G, is given by:

$$G = 1 - \phi_v - \frac{M_d}{V \rho_d} \tag{6}$$

If such measurements are made for ecological purposes, it may be important to avoid sample compaction. Hiller borer samples are virtually useless. Those collected with a "Russian pattern" borer (West, 1968), with which a half-cylinder of metal is rotated to trap the sample, gave results for bulk density which were consistently about 5% below those for samples taken by horizontal excavation of monoliths from a cut-back peat face (Clymo, unpubl.). The monolith samples were each about fifty times the volume of the "Russian pattern" samples.

For detailed investigations of the top 50 to 100 cm, then, a rectangular sampler (Fenton, 1978) may be useful. One side of the sampler runs in grooves and can be slid out lengthwise. The three-sided box, with a sharp-sloping end, is forced down into the peat, cutting down three sides but leaving the fourth attached. This attachment causes the whole block to resist compaction. The sharp-ended fourth side of the cutter is then thrust down, running in its grooves, to complete the vertical isolation of a peat block. The whole must then be excavated with a spade. Any movement of the surface can be seen, so faulty blocks may be rejected.

Other workers have used cylindrical samplers. One such (Clymo, unpubl.) allows one to draw a cutter plate across the sampler base, thus retaining the core in the tube while both are extracted. The natural profile of water and of gas space (Fig. 4.4) may be fairly easily and accurately captured in this way — something which is difficult to manage by

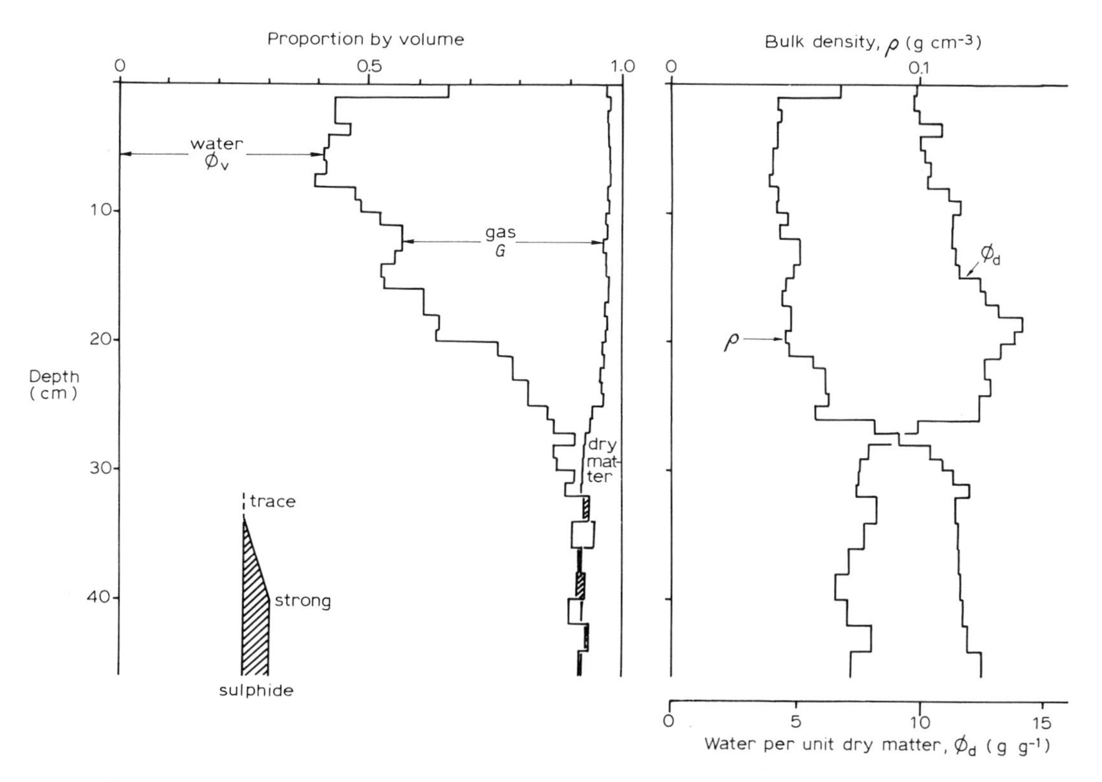

Fig. 4.4. Physical properties of a *Sphagnum fuscum* peat profile from Nordmjele, Andøya, northern Norway (Clymo, unpublished). The 20 cm diameter core was collected in a tube which was then sealed to prevent water loss. The core was later ejected, 1 cm at a time, and sliced with a sharp knife.

other methods. Again, faulty samples may easily be recognized and rejected.

Of the quantities defined above, bulk density and water content are determined routinely for commercial purposes (for example, Puustjärvi, 1968, 1969) but not as commonly as might be hoped for scientific purposes, and rather rarely on profiles (as opposed to isolated samples). An illustrative set of profiles is shown in Fig. 4.4. The void proportion (or quotient), E, is especially important in studies of the mechanical properties of peat (see, for example, Berry and Poskitt, 1972) and in studies of hydraulic conductance. The bulk density profile is important in studies of the rate of peat accumulation (Tallis and Switsur, 1973; Jones and Gore, 1978; Clymo, 1978) and ought to be measured whenever ^{14}C dates are determined.

The general features of *Sphagnum*-dominated profiles are that the bulk density in the top 1 cm is relatively high, but that immediately below the tightly packed capitula of the mosses it falls to about 0.02 g cm^{-3} (mosses of the Section *Sphagnum*) or about 0.04 g cm^{-3} (Section *Acutifolia*). Below this ρ increases gradually, and then, at the point where the weakened moss can no longer support the load above (commonly 10–30 cm deep), ρ increases rapidly to about 0.1 g cm^{-3}. At greater depth there may be perhaps a 20% further increase (for example, see Jones and Gore, 1978). Highly humified peat (Päivänen, 1969; Tallis and Switsur, 1973) and *Eriophorum vaginatum* peat (Tallis and Switsur, 1973) have greater bulk density — about 0.12 to 0.15 g cm^{-3}. The values calculable from the data of Bellamy and Rieley (1967) — between 0.18 and 0.22 g cm^{-3} in *Sphagnum fuscum* peat — seem remarkably high for natural peat, though artificial means can raise ρ to 1.3 g cm^{-3} (Freistedt, 1968). The specific gravity of 0.2 to 0.8 reported by Mattson and Koutler-Andersson (1954), and density from 0.35 to 0.96 reported by Walsh and Barry (1958), probably refer to wet-mass. A similar pattern: $\rho = 0.15$ g cm^{-3} at the surface, declining to 0.09 g cm^{-3} at 15 cm depth, then increasing again deeper in the profile, is reported by Fenton (1980) for Antarctic peats (though the explanation here is probably rather different — see p. 215). In all these peats the contribution of inorganic matter is negligible.

It is remarkable that in these peats no more than 3 to 10% by volume is organic matrix: the rest is water or gas. The proportions of water and gas are inversely related. The water-holding capacity of small peat samples which have been saturated and allowed to drain is closely related to bulk density (see, for example, Puustjärvi, 1968, 1972). In the *field* however the relationship is complicated by capillary phenomena. Apart from the top 1 cm, the surface peat contains 50% (*Acutifolia* peat) to 90% (*Sphagnum* peat) of gas, decreasing to zero at the water table. This has obvious consequences for the roots of vascular plants. The interrelationship of water-table, water-content profile, bulk density and decay is best discussed later.

For a wide range of commercial peats, Olsen (1968) has shown that humification H (on the Von Post scale) is related to ρ (g cm^{-3}) and E by:

$$\left.\begin{array}{l} H=(\rho-0.023)/0.015 \\ E=1.01-0.67\rho \end{array}\right\} \text{(valid between } \rho=0.03 \text{ and } 0.20)$$

Karesniemi (1972) and others report similar relationships.

The profile, *in situ*, of water content on a dry weight basis, ϕ_d, is affected by changes both in ϕ_v and in ρ, and quite complex shapes (for example, see Fig. 4.4) may result.

Drainage and water-retaining capacity

Relatively undecomposed peats have a high hydraulic conductance, and water flow through them follows Darcy's Law quite closely (Ingram et al., 1974; Rycroft et al., 1975a, b). The conductivity is about 10^{-1} to 10^{-3} cm s^{-1}. In decomposed peat the conductivity is much lower — perhaps 10^{-6} cm s^{-1}. Korpijaakko and Radforth (1972) reported an approximately linear relationship between conductivity, K_p (in cm min^{-1}) and the logarithm of humification on the Von Post scale:

$$K_p \simeq -0.5 \log_{10} H + 0.6$$

Rycroft et al. (1975b) showed that flow in humified peat deviates from Darcy's Law, so it is not clear how far the results of Korpijaakko and Radforth can be relied on. During periods of heavy precipitation (or snow melt) much of the water can flow vertically and then horizontally through layers of high conductance, but at other times, of low water table, movement may be very slow (e.g. Chapman, 1965).

The relationship of water content, ϕ_v, and equilibrium water potential, ψ, in peat is important, and depends primarily on the extent of humification and compaction. An example of the water potential capacity curve for a moderately humified peat is shown in Fig. 4.5 (capacity as ϕ_d rather than ϕ_v). Most of the water is removed between a potential corresponding to that in circular fully wettable water-filled pores between 1 and 20 μm diameter. This corresponds to the range of pore sizes commonly measured in peats, and one may conclude that most of the water in little-humified peat is held by capillary forces. "Wilting point" (at -20×10^5 Pa)[1] is well below this point (though the actual wilting point of most peat-inhabiting plants is not known), and is unlikely to be reached in natural conditions. Capillary forces in this range begin to overlap with other (stronger) chemical forces. In highly humified peats — H8 to H10 — the proportion of water with ψ below -20×10^5 Pa is larger (Boelter, 1968). In less humified peat (Fig. 4.5) the whole curve is lower and extends to higher water

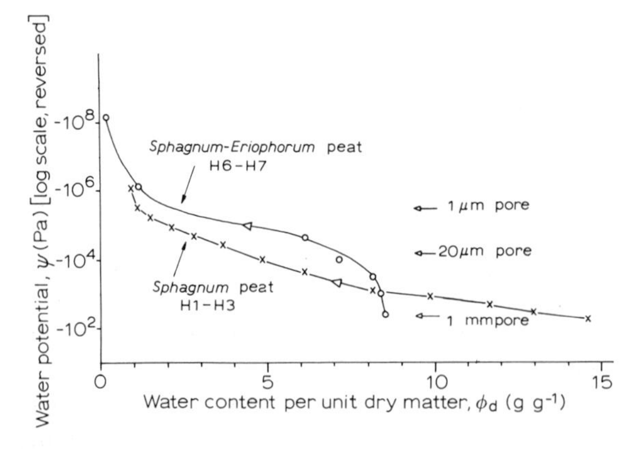

Fig. 4.5. Water potential/water content curves (drying, for *Sphagnum–Eriophorum* peat, H6–H7, and for *Sphagnum* peat, H1–H3). Redrawn from Van Dijk and Boekel (1968) and Puustjärvi (1968). The ψ value for cylindrical water-filled pores of 1 and 20 μm diameter is shown.

[1]The SI unit of pressure, the Pascal, is related to the Newton and other units by: 10^5 Pa $= 10^5$ N m$^{-2} \simeq 1$ bar $\simeq 1$ atm.

content because there are more and larger capillary spaces.

The exact shape of water potential/capacity curves varies with pretreatment. Changing the physical structure by freezing and thawing for example (Van Dijk and Boekel, 1968) has a marked effect, as does drying. If results are plotted on a volumetric basis the curves are similar, but shifted as a result of shrinkage on drying. In peats with some structure but low strength this shrinkage may be caused by the reduced hydrostatic pressure inside capillary films. The pressure is closely related to water potential in this case. This may account for the linear relationship between water loss and shrinkage (Irwin, 1968). In humified peats the process is more akin to the shrinking of a drying jelly. In unhumified peat in which the plant structures on the scale of 1 to 5 mm are preserved there is much less shrinkage on drying. Further information is given in Chapters 3 and 8.

Proportion of fibre

"Fibre" is the category of organic matter which fails to pass through a sieve of mesh size 1.5 mm. "Unrubbed" and "rubbed" fibre are distinguishable. Fragments of wood larger than 2 cm are excluded in this definition, but leaves are included (Day, 1968). The proportion of "fibre" must, ultimately, be measured, but with practice visual estimates on vertical (not horizontal) cleavage planes in the field may become accurate to within 10%. Technical details are given by Farnham (1968). In practice the bulk of material classified as "fibre" is not fibre in the sense of the botanical cell type, but consists of leaves and stems of *Sphagnum*, other bryophytes, segments of sedge leaves, fragments of wood from stems of dwarf shrubs, and so on. The main exception is undecomposed shoot bases of *Eriophorum* which contain a large proportion of fibres, and are fibrous in the technical and common senses of the word.

The proportion of fibre is usually closely linked with the extent of humification, and hence with the mechanical and hydraulic properties already discussed.

Structure

Structural features of primary importance are the size and distribution of pores, and the orientation of those components (if any) which have retained structural integrity. These features have influence on the mechanical properties of peat (compressibility and creep behaviour) and on the hydraulic properties. MacFarlane and Radforth (1968) suggest from microscopic examination that the structure of moss peat may be approximated by a series of parallel cylinders, each with an incompressible core (the stem) surrounded by a compressible sheath (leaves). Ohira (1962) suggested a more abstract model of loosely packed spheres, with the space between them filled by smaller particles, gas and water. Some features of the pore size and pattern may also be deduced from the pattern of absorption and elution of non-adsorbed radioactive tracers (Volarovich and Churaev, 1968).

Heat of combustion

This character (also known as "calorific content" or "energy content") may be measured by the standard techniques of bomb calorimetry. Such measurements are made routinely for commercial purposes but are rarely published and are not usually related to the processes of peat genesis. The results in Fig. 4.6 show that in both profiles there is

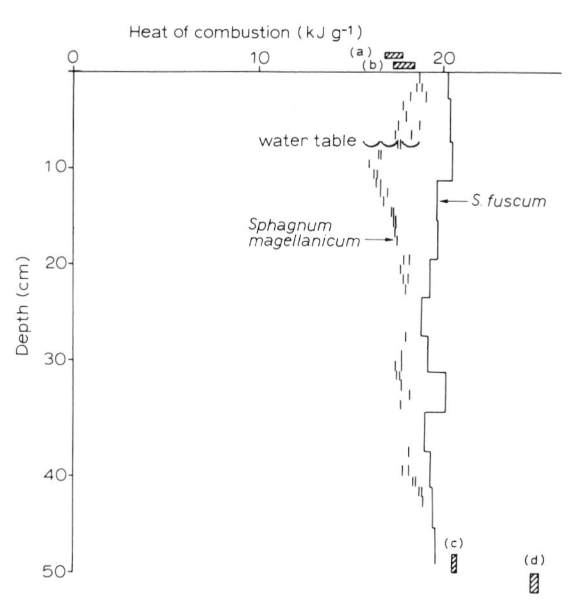

Fig. 4.6. Profiles of heat of combustion (="energy content" ="calorific value"). *Sphagnum fuscum* profile (Bellamy and Rieley, 1967) from northern England. *S. magellanicum* profile (Clymo, unpublished) from Cranesmoor, southern England. The four blocks are: (*a*) "*Sphagnum*" (Gorham and Sanger, 1967); (*b*) *S. capillifolium*, *S. papillosum*, *S. cuspidatum*, *S. recurvum* (E.D. Ford, quoted in Clymo, 1970); (*c*) "peat" (Duane et al., 1968; Kamula, 1968); (*d*) coal (average).

at first a decrease downwards in the heat of combustion, followed by an increase. This is clear in the *Sphagnum magellanicum* profile, where there are sufficient observations to make the trend plain. The *S. fuscum* profile shows the same trend (excepting the high value at 31 to 35 cm), though it is less well marked and the trough is at greater depth. The reversal may be connected with the change from aerobic to anaerobic conditions. The even higher values for commercial peat (used for fuel) and coal, and the results of Salmi (1954), suggest that the process continues.

Colour

Soil scientists have long recognised the need for objective descriptions of colour *in the field*, and commonly use standards such as Munsell charts. There are a surprising variety of delicate colours to be seen in freshly exposed peat, and these may be of ecological significance. For example, the distinction of upper "black-brown" peat above a layer described as "green" which in turn overlies "red-brown" peat has been associated with microbiological differences, which are, in turn, probably connected with the position and fluctuation of the water table (Collins et al., 1978). Reference to standard colours is rarely made however, probably because a complete set of such standards is expensive and only occasionally of use to ecologists.

Age

The age of peat is needed for the reconstruction of vegetation history and in studies of peat growth. For ages in the range from 200 to 20 000 yr the ^{14}C technique is satisfactory for ombrogenous peats unless there has been an abnormal amount of deep root growth, but may be less so for peat formed from plants growing with mineral soil water, particularly if this is highly calcareous. There is uncertainty about the calibration of ^{14}C ages (Pearson et al., 1977). There is probably little vertical movement of carbon-containing substances, though it may be that some substances present in small amounts, for example, amino acids (Swain et al., 1959) and pollen (Mackay, unpublished, 1976), are mobile in the top 20 cm of the peat.

Ages to 200 years cannot at present be measured accurately, though one particular event, the 1963 peak in ^{137}Cs produced by nuclear bomb tests, may be useful. Aaby et al. (1979) and Oldfield et al. (1979) doubt this and prefer the amount of ^{210}Pb, believing that the mobility of lead is low (but see Figs. 4.17 and 4.18). Attempts have been made to use the cumulative total of magnesium or aluminium to provide a continuous scale (Clymo, 1978), but it is evident that many inorganic elements are surprisingly mobile in the surface layers of peat (Damman, 1978; Mackay, unpublished, 1976), so such methods are approximate at best.

The only other method in use at present is painstaking correlation of pollen abundance with historical records of tree planting, industrial activity (causing deposition of soot), changes in cereal abundance, and so on (Lee and Tallis, 1973; Livett et al., 1979).

Attributes of peat-forming systems

The three characters which follow differ from the earlier ones in being characters of the peat-*forming* system as a whole, rather than of the peat *substance*. The three characters are often closely interlinked.

Topography and hydrology of the area in which peat formed

Because of their dependence on water, peat-forming systems may be very sensitive to topography. In regions of continuous high humidity such as Ireland and Newfoundland, peat may form direct on slopes up to 20° or more, although such "blanket peats" on steep slopes are likely to "flow" or "burst". In drier climates, basins or valleys may be essential to give the necessary amplification or stabilization of the water supply. The peat itself affects the hydrology. These matters are considered in much detail elsewhere in this volume.

Morphology of the peat-forming system

Structures may be seen in peat-forming systems at all scales, but three scales are most conspicuous. First is the scale of the individual plant, centred on about 1 to 10 cm. Second is a scale of 1 to 20 m. This corresponds to pool and hummock topography. There is a great variety of structures from the extraordinary "strings" of the northern "string-bogs" which may be 2 m across, 1 m tall, and hundreds of metres long, to the approximately circular hummocks of more southerly bogs on the west coast of Europe. Finally, there is the large-scale structure of a whole peat-forming system, such as raised-bog, eccentric-domed bog, *Schwingmoor*, etc. The scale here is hundreds or thousands of metres. Again, these structures are described in detail elsewhere in this volume.

History of the peat-forming system

The history of a peat-forming system is preserved, with distortion, in the peat substance. Except where cryoturbation has disturbed the sequence, the deepest peat is the oldest. Macrofossils — leaves, stems and roots — show what plants lived at that particular point. Microfossils — pollen, spores, etc. — may be local or regional. Different plants decay at different rates (see, for instance, Coulson and Butterfield, 1978; and Dickinson, Chapter 5, this volume) so it is not in general possible to reconstruct the vegetation history in detail, but major changes from, for example, *Phragmites* to *Sphagnum* (see, for instance, Chapman, 1964a) are of value as indicators of a change in the nature of the peat-forming vegetation. Sometimes the changes are surprising and would not have been expected from consideration of present topography and hydrology. For example, the sequence *Sphagnum* peat, *Cladium* peat, *Sphagnum* peat at Shapwick Heath, Somerset (Godwin, 1956, Fig. 14) probably resulted from flooding of a rainwater-dependent surface by calcareous mineral-soil water.

Classification of peat and peat-forming systems

There are a large number of classifications in use and disuse. A primary distinction is between classifications of peat *substance* and those of the peat-forming *system*. The general problems are well presented in the account of an International Peat Society Symposium (Kivinen et al., 1979).

Peat is of economic importance, so there are national commercial classifications of peat substance (summarized by Farnham, 1968) in, *inter alia*, the Soviet Union, Finland, Canada, the United States, Germany, Great Britain, Sweden, Poland, Norway and The Netherlands. [The order is that of the area, reported by Olenin (1968), of exploitable peat in each country.] Attempts are being made to produce an internationally agreed commercial classification (Kivinen, 1977) and this is shown in Appendix I.

Some classifications of peat-forming systems make use, in the main, of the peat substance characters and their vertical variation. An example of such a classification is that contained within the United States Department of Agriculture 7th Approximation soil classification. The relevant parts are shown in Appendix II. A similar classification is given by Farnham and Finney (1965), who give examples of other classifications too. These are "special classifications" in the same sense that a telephone directory is. They may be very good for one purpose, but useless for all others. The arbitrary definition of a "control section", 130 or 160 cm deep in the U.S.D.A. 7th Approximation, makes the classification of little value to those whose interest lies in the ecology or history of the peat-forming system, as Ingram (1978) points out.

Ecologists are likely to find the more "general" or so-called "natural" classifications of more use. There are very many of these, mostly making use of five main elements: topography and hydrology, present surface vegetation, morphology, water chemistry, and history. Amongst the earliest were those of Weber (1908) and Potonié (1908), and an influential later one is that of Sjörs (1948). An example of such a classification is shown in Appendix III, and others are given by Moore and Bellamy (1974) and elsewhere in this volume. Of particular interest are the classifications used in the Soviet Union, which contains perhaps 60% of the world's peat. Rather vague examples are given by Bradis and Andrienko (1972) and Elina (1972), and a more detailed description, but using Scandinavian and German terms, by Walter (1977); further discussion will be found in Chapter 4 of Part B.

Most of these schemes are intended for use within a limited geographic range: few of them would cope easily with the sub-Antarctic moss-peats (Fenton, 1978). Some interesting and very early "proto-classifications" are reviewed by Gorham (1953, 1957).

Most classifications recognize the fundamental importance of two factors: plant nutrition and source of water. The significance of these was first emphasized by Scandinavian ecologists, and is discussed elsewhere in this volume. Two extremes of each factor may be recognized. The water coming to the system may have a high concentration of plant "nutrients" (or a lower concentration but a constantly renewed supply, or both) in which case conditions may, inaccurately but commonly, be called eutrophic. Water which has a low concentration of "nutrients" or which is not often renewed, causes oligotrophic conditions. The present surface vegetation may be entirely dependent on precipitation for its water supply ("ombrotrophic"), or most of its water may have flowed through rock or mineral soil or both ("minerotrophic"). The terms "bog" and "fen", with qualifiers, are freely used in classifications. Their relationship to nutrition and source of water is shown in Table 4.4.

Precipitation alone does not cause eutrophic conditions, and peat-forming systems which depend entirely on precipitation are almost universally recognized (nowadays) as "bog". The vegetation often contains much *Sphagnum*, which can make the water acid (Clymo, 1967), and a limited number of vascular plant species, particularly *Eriophorum* and dwarf shrubs, which can tolerate acid, oligotrophic, wet conditions. Water which has flowed through sedimentary rocks and acquired a high concentration of solutes usually gives conditions suited to a wider range of vascular plants, mostly of non-bog species, and is recognized as "fen". Classifications differ in their treatment of peat-forming systems which receive most of their water from weathering-resistant rock or mineral soil, and which are therefore oligotrophic and commonly have a vegetation similar to that of "bog". Sometimes the vegetation character is considered paramount, sometimes the origin of the water. On the whole such mires — a general term — have been called "bog" in Britain, but "fen" in Scandinavia where the "mineral soil water limit" is considered of great importance.

TABLE 4.4

Broad outline of ecological classifications of northern peat-forming systems

Source of water	Nutrition	
	eutrophic	oligotrophic
Rock or soil	fen (a)	bog or fen (b)
Precipitation	–	bog (c)

Approximate equivalence of special terms:

	Weber (1908)	Potonié (1908)	Sjörs (1948)	Tansley (1939)
(a)	*Niedermoore*	*Flachmoore*	rich fen	fen
(b)	*Übergangsmoore*	*Zwischenmoore*	poor fen	valley bog
(c)	*Hochmoore*	*Hochmoore*	moss	raised bog; blanket bog

These terms are only approximately equivalent because different authors placed emphasis on different characters.

During its development a mire may have passed from fen to bog. Some classifications recognize this. Others ignore the history of the peat-forming system, but attach importance to morphology.

In general it is as well to remember that classifications are artificial, they are intended for *convenience*, and they should change to accommodate new knowledge. The quest for "the" classification of peat-forming systems has proved as difficult and as self-revealing (though not yet as deadly) as that for the Holy Grail, but there are now so many classifications in use (or disuse) that it should be possible to find one that is adequate for any specific limited purpose.

PEAT CHEMISTRY

Many factors affect the chemistry of peat. Amongst the most important are the nature of the original plant matter, the supply of inorganic solutes, the activities of plants and animals and of micro-organisms, environmental conditions (particularly temperature and the extent of waterlogging) and, finally, the age and history of the peat. These factors are not independent: the type and extent of microbiological activity depends on temperature and the extent of waterlogging; the nature of the original plants may change as eutrophic fen is succeeded by oliogtrophic (ombrotrophic) bog, and so on.

There are marked vertical patterns in chemistry associated with gradients in these factors, and significant horizontal patterns within any one peat-forming community: one need only consider the "string bogs" of arctic regions (Ruuhijärvi, Part B, Ch. 2) to appreciate this.

It is convenient to consider the organic compounds in peat separately from the inorganic ones. Nitrogen and phosphorus do not fit easily into this scheme, and are considered at convenient points in both sections.

Organic constituents of peat

Peat begins to form when any one of a large number of plant species dies. At the outset therefore it contains the full range of chemical compounds found in the parent plants, dominated in quantity by the structural cell-wall carbohydrates. As the peat ages, its chemistry changes. Some substances are selectively removed. For example, plant pigments do not survive in acid peat, though they do in lake sediments (Gorham, 1961a). Other substances are formed *de novo*; the development of brown colour cannot be explained simply by removal of non-brown matter. Micro-organisms and, in the surface layers, small animals are known to be active (Coulson and Butterfield, 1978; see also Chapters 5, 10 and 11 of this volume), and it is tempting to suppose that these are the main agents of change. The extent to which slow chemical reactions, taking hundreds or thousands of years to become obvious, are involved is not known. Some animal activities are unexpected: the production of wax by the aphid *Colopha compressa* living on the roots of *Eriophorum* is an example (Wheatley et al., 1975). The wax is in interlinked fibres and contains paraffins, with some carbohydrate and secondary amides. It forms white aggregates up to 3 mm diameter which can be found to 6 m depth at least. Presumably the aggregates at lower depth are fossil. Those at 1 m depth show fewer fine fibres and less carbohydrate and amides. This example is sufficient to show that the chemistry of peat is likely to be very complex.

The usual approach to the organic chemistry of peat has been to extract the peat with a variety of solvents, and attack it with other chemicals to make it soluble, then to apply more precise methods to these fractions.

In general our knowledge of the organic composition of peat, and of the changes which occur, is fragmentary. Specialists have detected the presence, often in unmeasured concentration, of a wide range of organic substances. It is rarely possible to relate such studies to one another, however, because too little is recorded about the general properties, such as botanical composition, humification, depth or age of the sample. Notable exceptions are the works of Waksman and Stevens (1928a, b, 1929a, b, c), of Mattson and Koutler-Andersson (1954), and especially of Theander (1954): all fractionated peat from known positions in a profile. Some of their results are shown in Table 4.5 and Fig. 4.7.

The material soluble in non-polar solvents is often called peat "wax". The proportion of "wax" increases with age and depth, or perhaps more specifically with humification, in the examples in both Table 4.5 and Fig. 4.7. This might be simply a result of selective survival; it does not seem to be associated with the presence of macroscopic remains of cuticularised plants, and there is no evidence for a concentration of aphid activity. The general problem illustrated here pervades the whole study of peat: to what extent is a difference observed now a record of historical differences in vegetation, and to what extent has it resulted from changes applied to the same material but over different time spans? Frenzel (Ch. 2) discusses this issue too.

The peat "wax" is a complex mixture of true waxes, "asphalt", and "resins" (Howard and Hamer, 1960) and includes at least some ligneous matter. It is also called "bitumen". Of these terms only wax (esters of fatty acids with alcohols other than glycerol; Fieser and Fieser, 1944) is at all clearly defined. For example, Gilliland and Howard (1968) made a further separation of peat "wax", using column chromatography. They recognised nine fractions, each of which was still a mixture, and identified, *inter alia*, alcohols with chain lengths of even numbers of carbon atoms from C_{20} to C_{30}, and acids with chain lengths from C_{17} to C_{33} (with marked dominance of odd number lengths).

The compounds extracted from peat in polar solvents (Table 4.5: NH_3 or KOH and H_2O) were constant in proportion at different depths at about 14% of dry mass, although the proportion in *Sphagnum* (probably dead for a few years) was about 10%, of which at least 3% of the total dry mass was fructose, and a further 3% other sugars (galactose, glucose, mannose, arabinose, and xylose). In the lower peats about 1% was in these sugars, but no fructose was detectable. The concentration of free sugar is surprisingly high; media used for growing micro-organisms commonly contain 2 to 5% of sugars, and there is no shortage of active micro-organisms in the surface layers of peat, so it may be that these "solvents" attack the peat chemically too.

Further comparison of the fractions in Table 4.5 and Fig. 4.7 is far from easy. Waksman and Stevens (1928a, b, 1929a, b, c), and Mattson and Koutler-Andersson (1954) used two acid-hydrolysis stages. The first was mild, and may be assumed to have broken some polymeric carbohydrates into shorter sections. The second hydrolysis was more severe,

TABLE 4.5

Fractionation of organic compounds in surface *Sphagnum* and peat from southern Sweden (Theander, 1954)

Material	Depth (cm)	Age (yr)	Humification (Von Post scale)	Ash (%)	Successive extractions[1] (%)						Residue	Lost[2]	Last four columns = holocellulose
					C_6H_6	CH_3OH	H_2O	bleach	NH_3 then H_2O	KOH then H_2O			
S. fuscum	5	1–10	–	1.5	1.7	6.3	4.6	3.7	5.6	3.7	72.5	1.9	83.7
S. imbricatum	5	1–10	–	1.8	1.7	4.5	3.6	3.7	6.9	3.6	73.4	2.6	86.5
Peat A													
S. imbricatum	17	200–400	3	1.0	4.0	3.6	1.5	26.3	8.0	5.6	48.5	2.5	64.6
Peat B													
S. tenellum, *S. recurvum*	47	700–1000	4	1.1	3.1	4.5	2.0	24.7	8.5	5.5	47.8	3.9	65.7
Peat C													
S. imbricatum	57	900–1200	6	1.7	5.9	4.6	1.6	43.9	8.8	6.5	27.8	6.6	49.7
Peat D													
S. imbricatum	113	1900–2200	3	0.7	3.3	3.9	1.0	22.1	8.3	5.1	53.8	2.5	69.7
Peat E													
Sphagnum spp.	295	4300–4800	6–7	1.7	6.7	4.9	1.1	57.1	8.8	5.2	20.4	5.4	39.8

[1]The fractionation procedure was: (1) C_6H_6 for two days; (2) CH_3OH five days; (3) dry and extract with H_2O 9 h; (4) Na hypochlorite + CH_3COOH at 60–65°C, 2–4 h; (5) swell in liquid ammonia, extract with H_2O; (6) 5% KOH under N_2, 8 h, extract with H_2O. Stages 1–3 are simple extractions causing little or no chemical modification. Stage 4 is a mild oxidation.
[2]Proportion lost from the holocellulose fraction during the preceding bleaching. This was estimated separately, and is excluded from the "bleach" fraction.

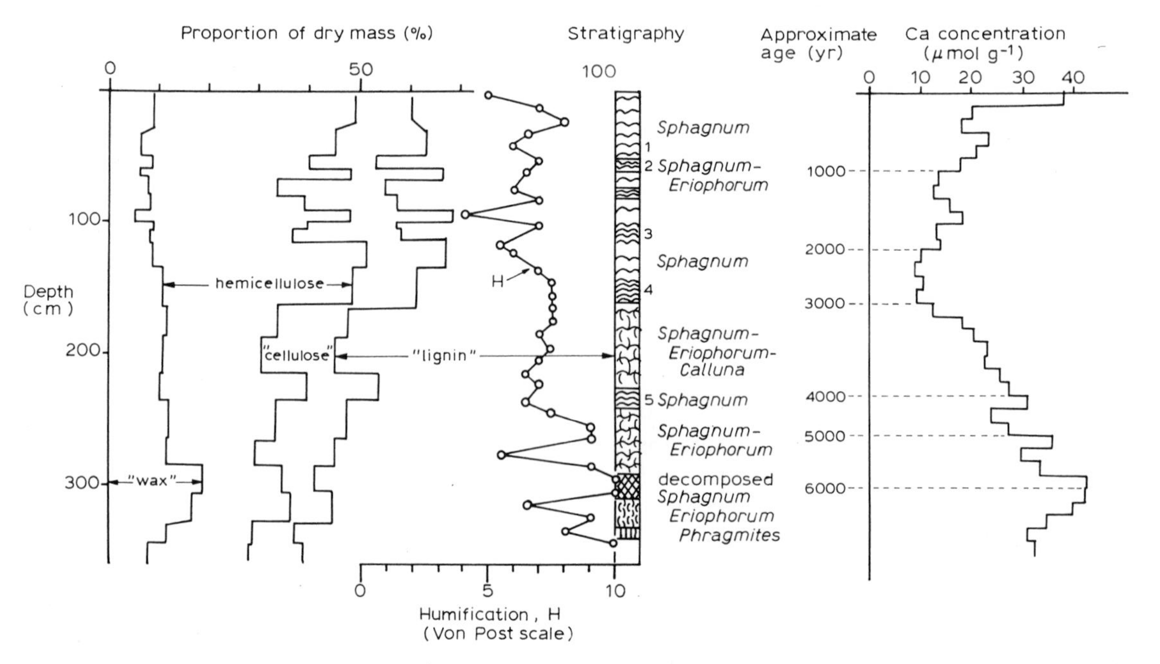

Fig. 4.7. Stratigraphy, botanical composition, humification (Von Post scale), age, exchangeable Ca concentration and organic fractions in Ramna bog, southern Sweden. The fractionation procedure was: (1) reflux in C_2H_5OH and C_6H_5 ("wax"); ; (2) 2% HCl, boil 1 h ("hemicellulose"); (3) 80% H_2SO_4 *in vacuo* $2\frac{1}{2}$ h, dilute, boil 5 h ("cellulose"); (4) residue ("lignin"). Redrawn from Mattson and Koutler-Andersson (1954). The numbers *1–5* in the stratigraphy column are recurrence horizons.

but not exceptionally so. The fractions thus isolated they called "hemicellulose", "cellulose", and (the residue) "lignin". "Hemicellulose" is nowadays usually isolated by hydrolysis with alkali and after removal of lignins. The terms "cellulose" and "lignin" are valid only in the most general sense.

Theander (1954) used a mild hypochlorite oxidation to remove lignins before alkaline hydrolysis, which should then give some indication of hemicellulose proportions (Bonner, 1950).

If these equivalences are accepted, then the two sets of data do agree quite closely. Again it is the degree of humification which seems important; "lignin" forms about 30% of the total at H3, rising to 50 to 60% at H6–7. The amounts of "hemicellulose" (Fig. 4.7, Table 4.5) differ, but whether the difference lies in the peat or in the methods cannot be determined. The proportion of "hemicellulose" +"cellulose" (Fig. 4.7) and of KOH-solubilized+ residue (Table 4.5) behave in much the same way, decreasing with humification from initial values of about 75% in newly dead *Sphagnum* to about 50% in the early stages of peat formation and later to about 25% in highly humified peat. Again it is impossible to assess the relative importance of historical difference and of development, but the drift of values over the middle section of Fig. 4.7, for the peat between 2000 and 4000 years old, which has little change in botanical composition, argues for continued slow changes.

The composition of the holocellulose (alkali-solubilized+residue) is shown in Table 4.6. In newly dead *Sphagnum* nearly 60% of the total mass could be positively identified. Most of the sugars and uronic acids declined in the same way, but mannose increased for the first thousand years. The identity of about 20% of the total dry mass, all in the holocellulose fractions, remained unknown. This proportion did not appear to depend on humification or age. Theander (1954) suggested that it might be non-carbohydrate material, but the recorded amounts of uronic acids seem to be lower than those measured by other workers (Clymo, 1963; Spearing, 1972) and lower by 5 to 10%, on a dry-weight basis, than those necessary to account for the cation-exchange properties of *Sphagnum* (Clymo, 1963) and peat (Puustjärvi, 1956). These cation exchangers — polymers containing uronic acids — are of great importance in peat chemistry. They provide not only the possibility of a large reserve *capacity* for cations, but also reduce the capacity for anions because the concentration of anions in the charged phase is reduced (Donnan, 1911).

The standard procedure for further fractionating the fraction of soil organic matter soluble in alkali or neutral sodium pyrophosphate ("humic substances") is to add acid. The material which remains in solution is "fulvic acid". That which precipitates is treated with ethanol. Insoluble material is "humic acid", soluble is "hymatomelanic acid". Schnitzer (1973) summarizes further procedures, and Schnitzer and Khan (1972) record hundreds of specific compounds isolated from soils by various procedures. These methods seem to have been applied to peat only infrequently. It is worth noting therefore (Table 4.6) that, in deeper and

TABLE 4.6

Sugars and uronic acids in acid hydrolysates of holocellulose fractions in Table 4.5; figures are proportion of *total* dry mass of peat (A = alkali-soluble (NH_3 and KOH solubilized); R = insoluble residue; ++ = present but less than 2%; + = present but less than 0.5%)

Material	Humification	Uronic acids		Galactose		Glucose		Mannose		Arabinose		Xylose		Rhamnose		Peat proportion accounted for here	Holocellulose unaccounted for here
		A	R	A	R	A	R	A	R	A	R	A	R	A	R		
S. fuscum	1	3.3	15.6	+	5.2	+	26.8	+	++	+	–	+	3.9	+	+	58.2	23.6
S. imbricatum	1	3.4	14.5	+	5.3	+	28.4	+	++	+	–	+	2.6	+	+	57.6	26.3
Peat A	3	3.3	7.8	+	2.6	++	21.8	+	2.2	+	–	++	2.2	+	+	42.9	19.2
Peat B	4	3.8	7.3	+	3.0	++	18.9	+	3.0	+	–	++	3.0	+	+	42.0	19.8
Peat C	6	2.5	3.7	+	++	++	13.6	+	++	+	–	++	++	+	+	25.9	17.2
Peat D	3	3.1	8.7	+	3.4	++	22.3	+	2.4	+	–	++	2.9	+	+	45.3	21.9
Peat E	6–7	++	++	+	++	++	11.4	+	++	–	–	+	+	+	+	18.1	16.3

older horizons, 60% falling to 20% seems to consist of polymers of a few sugars — predominantly glucose — and uronic acids, and 5% rising to 15% of a variety of complex non-polar compounds ("wax"). Only 20%, and possibly less, falls in the class of "humic substances".

Peat also contains, or can be treated to yield, a great variety of phenolic compounds. Many of these may be extracted in the "lignin" fraction forming 4% rising to 60% of the total mass. *Sphagnum* itself contains unusual lignins (Bland et al., 1968), and so does peat (Morita, 1968). In hot-water extracts of *Sphagnum–Eriophorum* peat, Wildehain and Henske (1965) found vanillin; *p*-hydroxybenzaldehyde; 4-hydroxy-2-methoxyacetophenone; 2,4,6 trihydroxyacetophenone; 3,4 dihydroxypropiophenone; glucoacetovanillone; *p*-hydroxybenzoic acid; vanillic acid; fernilic acid; protocatechuic acid; caffeic acid; *p*-hydroxyphenylpyruvic acid; 4-hydroxy-3-methoxyphenylpyruvic acid; and dehydrodivanillin. Such work is still at the stage of "finger printing" but eventually the amounts and origin of these substances must be known if the crudely derived bulk chemical properties of peat "lignin" are to be explained.

There are many reports that peat and *Sphagnum* extracts affect the growth of bacteria, fungi, algae and vascular plants. Sometimes growth is increased; in other cases it is decreased (for examples see Kvét, 1955; and Given and Dickinson, 1975). Very often there is an unsupported suspicion that phenolic compounds are having toxic effects. Control experiments with extracts of plants and humus from mineral soils, where decay is more complete, are the exception. It would be surprising if peat extracts did not have *some* effects. The important question is "How big are these effects in natural conditions?". It is difficult to answer this question.

The "lignin" fractions (insoluble in non-polar solvents, water, or sulphuric acid) as defined by Waksman and Stevens (1928a, b, 1929a, b, c) and by Mattson and Koutler-Andersson (1955) have some interesting gross chemical properties. For example, the proportion of lignin is closely correlated (Fig. 4.8) with the ability of the peat to

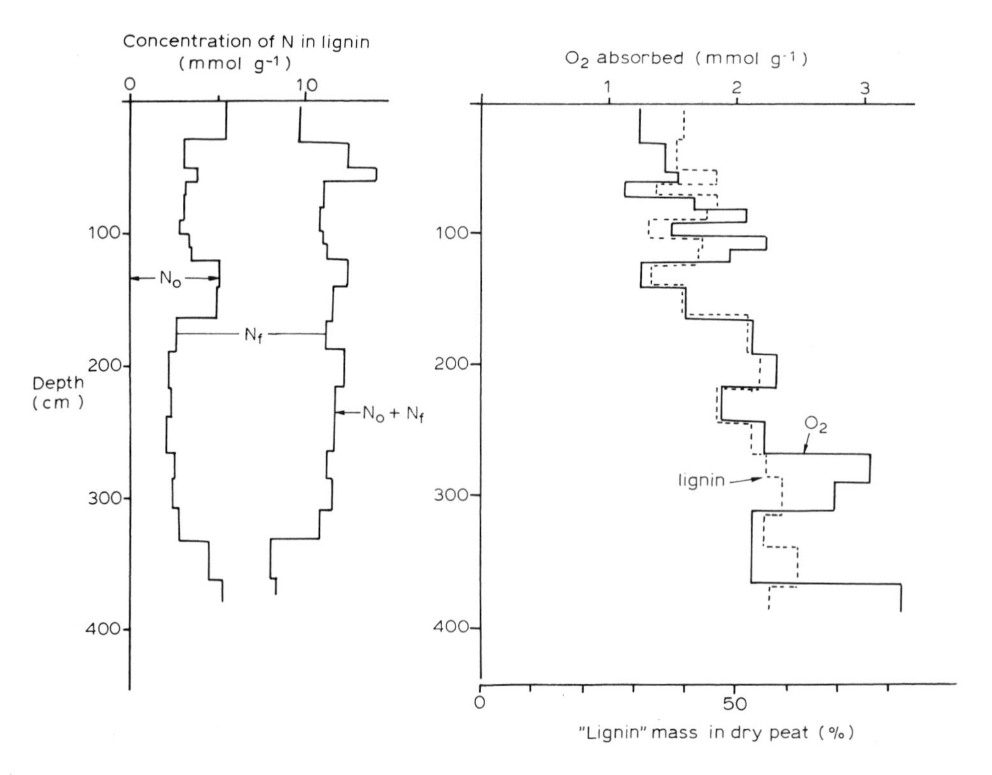

Fig. 4.8. Proportion of lignin, oxygen-absorbing capacity of peat, and properties of the "lignin" fraction of peat in Ramna bog, southern Sweden. N_0 = nitrogen concentration in "lignin"; N_f = concentration of nitrogen absorbed into non-exchangeable form from dilute ammonia. Redrawn from Mattson and Koutler-Andersson (1955).

TABLE 4.7

Fractionation of minerotrophic and ombrotrophic peat, and of newly dead plants (values are % of dry mass)

Depth (cm)	Peat type	Fraction[1]							Lost
		1 "wax"	2 or 3	4 "hemi-cellulose"	5 "cellulose"	6 "lignin"	7 "protein"	8 "ash"	
A. Raised bog, Cherryfield, Me (U.S.A.)[2]									
1–8	*Sphagnum*	2.4	1.5	26.5	16.9	27.2	4.1	2.0	19.4
8–20	*Sphagnum*	2.6	1.9	25.2	14.7	29.2	4.3	1.1	21.0
20–30	*Sphagnum*	2.8	1.8	24.6	16.0	28.9	5.1	1.0	19.8
30–46	*Sphagnum*	2.6	2.1	22.3	13.7	32.2	5.2	0.9	21.0
46–61	*Sphagnum*	3.0	3.2	18.5	14.7	33.2	4.8	1.1	21.5
183–214	*Sphagnum*	4.0	3.2	15.9	15.6	37.4	4.4	1.0	18.5
460–480	*Sphagnum*	4.9	4.3	12.7	11.9	44.8	4.7	1.1	15.6
550–580	sedge	6.0	5.1	6.0	5.1	54.1	11.5	2.8	9.4
B. Ramna bog, southern Sweden[3]									
60–65	*Sphagnum*	6.0		42.1	18.0	38.4	6.4[4]	0.8[4]	
67–79	*Sphagnum*	7.7		25.5	21.2	36.6			
190–216	*Sphagnum*	10.3		19.8	15.1	48.4	5.0	1.5[4]	
217–238	*Sphagnum*	10.0		29.6	14.0	39.9			
C. Lowmoor, Newton, N.J. (U.S.A.)[2]									
12	?	0.7		10.3	0	38.4	22.5	13.2	
18	?	1.1		9.0	0	50.3	18.7	10.1	
160–180	?	0.5		7.0	0	57.8	14.8	10.2	
D. Everglades, Cladium *peat*[2]									
0–65	surface	4.3	1.2	6.9	0.3	43.7	22.8	12.1	8.8
65–26	upper fibrous	4.7	1.1	6.4	0.3	46.1	23.1	10.0	8.4
26–40	pure *Cladium*	4.8	1.3	8.0	0.4	44.9	22.3	6.9	11.4
40–50	lower fibrous	4.7	1.4	7.6	0.4	47.6	21.3	8.1	9.0
50–62	upper colloidal	4.1	0.5	2.2	0	19.3	9.0	59.6	5.4
62–70	lower colloidal	2.3	0.7	2.6	0	28.5	13.0	42.3	10.6
110–120	lower fibrous	2.4	1.2	4.3	0	48.4	20.4	15.1	8.1
E. Newly dead plant material									
	Carex leaves	2.5	12.6	18.4	28.2	21.1	7.1	3.3	6.9
	Carex rhizomes	1.7	3.2	20.9	11.8	41.7	14.6	4.6	1.6
	Cladium leaves	1.1	6.9	21.5	28.3	29.1	7.2	3.9	2.1
	Cladium rhizomes	0.9	5.2	20.8	30.7	30.9	3.8	3.6	4.1
	Hypnum (moss)	4.6	8.4	18.9	24.8	21.1	4.2	4.3	13.7
	Sphagnum	1.6	1.6	24.5	15.9	19.2	1.9	19.9	13.5
	Pinus strobus leaves	11.4	7.3	19.0	16.4	22.7	2.2	2.5	6.0
	Oak leaves	4.0	15.3	15.6	17.2	29.7	3.5	4.7	10.1

[1]Extractants and fractions: (1) ether; (2) cold and hot water; (3) C_2H_5OH (boiling 95%), 1–2 h; (4) 2% HCl, boil, 5 h = "hemicellulose"; (5) 80% H_2SO_4 cold, 2 h then dilute and reflux, 5 h = "cellulose"; (6) residue, excluding (7) and (8) = "lignin"; (7) Kjeldahl N × 6.25 = "protein"; (8) inorganic ash.
[2]From Waksman and Stevens (1928a, b, 1929a, b, c).
[3]From Mattson and Koutler-Andersson, (1954).
[4]Approximate mean for the range.

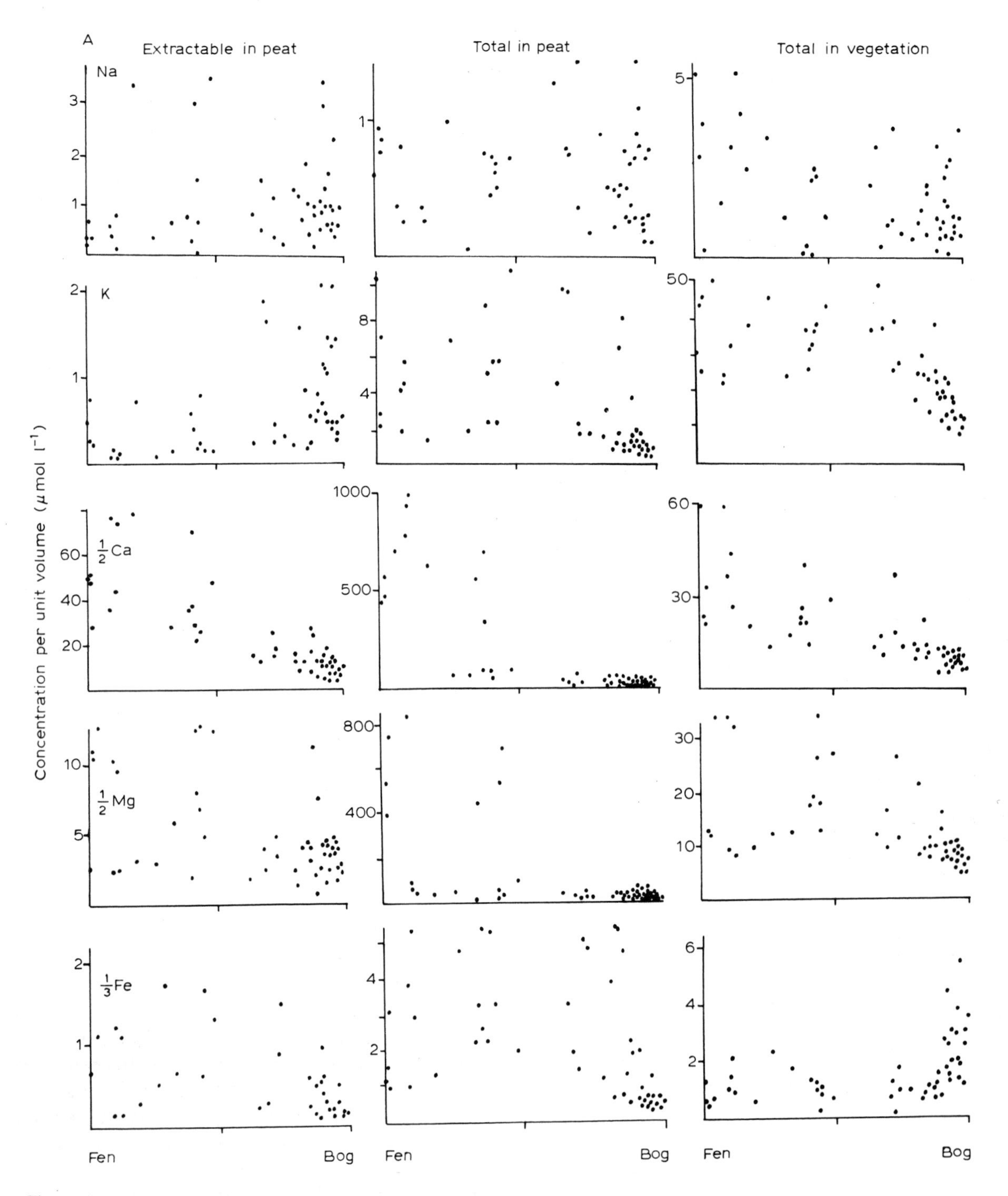
A
Extractable in peat
Total in peat
Total in vegetation
Na
K
$\frac{1}{2}$Ca
$\frac{1}{2}$Mg
$\frac{1}{3}$Fe
Concentration per unit volume (μmol l^{-1})
Fen
Bog
Fen
Bog
Fen
Bog

Fig. 4.9A.

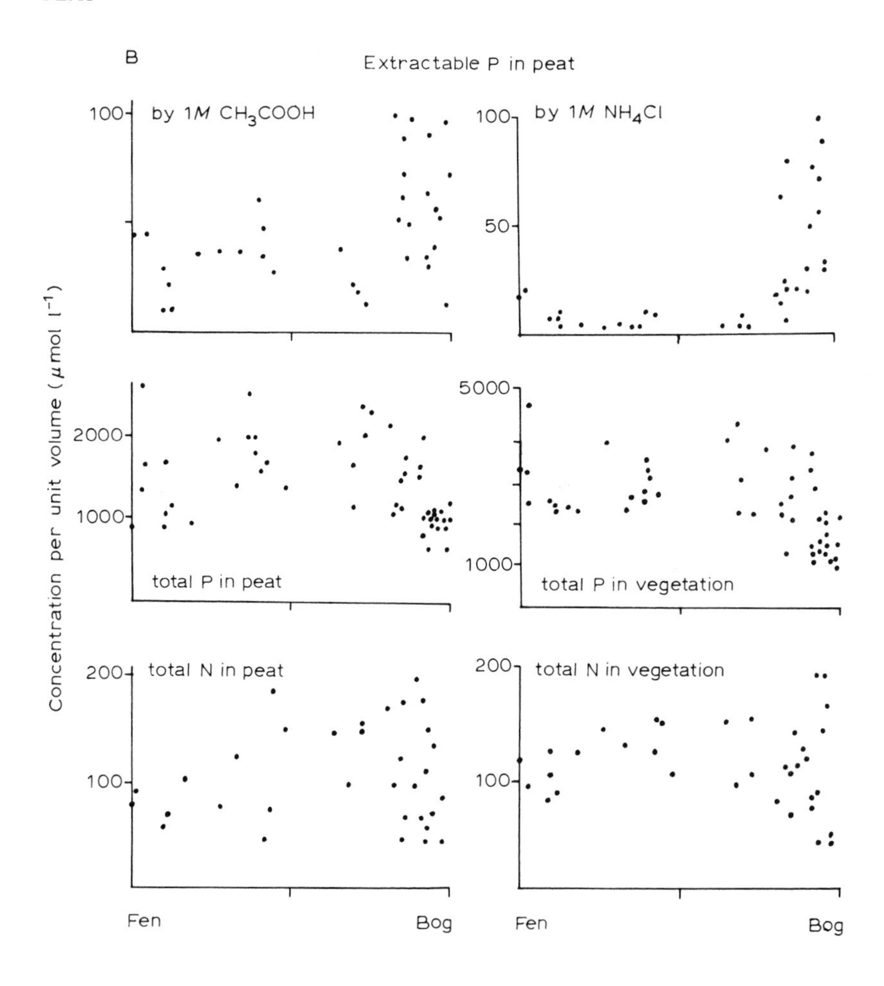

Fig. 4.9. Exchangeable and total concentration of elements in peat, and total in vegetation, in mires in southern Germany. The sites for samples were arranged using reciprocal averaging (Hill, 1973) of the vegetation. Fen is to the left, bog to the right. Na, K, Ca, Mg, Fe in Fig. A; N and P in Fig. B. Redrawn from Waughman (1980) assuming that peat and plants have bulk density of 0.1 g cm^{-3}.

combine chemically with O_2. The total Kjeldahl-nitrogen content of the "lignin", N_0, varies by a factor of three, but the Kjeldahl-nitrogen, after shaking with 12% ammonia solution in an atmosphere of oxygen, then washing with 0.1 *M* HCl and with water (i.e. the non-exchangeable N) is remarkably constant except for the top 50 cm and the base. It seems as if the "lignin" has a nearly constant "nitrogen capacity" which is filled to a greater or lesser extent at different depths. The combination of ammonia with the peat is accompanied by an oxidation. Mattson and Koutler-Andersson (1955) also found N_0 to be correlated with the pH of electrodialyzed peat, and inversely with "excess base", but the correlation with the proportion of "lignin" in the peat is almost as high.

Much of this account has been concerned with *Sphagnum*-dominated peat. It is of interest to compare these peats with other types. Table 4.7 shows some of the results obtained by Waksman and Stevens (1928a, b, 1929a, b, c) and Mattson and Koutler-Andersson (1954). The freshly dead plant materials differ in composition, but far less than the minerogenic and ombrogenic peats. In particular the minerogenic peats contain smaller proportions of "hemicellulose" and "cellulose", even after a relatively short time. On the other hand they have more "lignin" and "protein" (and ash). Again, it is impossible to separate the effects of original composition from those of development, but it seems clear that the changes during decay are at a different rate, or perhaps even of a different kind, in minerogenic peat and ombrogenous peat (see Sikora and Keeney, Ch. 6). The fractionation procedures are unable to distinguish readily metabolizable from resistant chemicals in the different fractions.

Inorganic constituents of peat

Some idea of the range of concentration of elements in peat can be obtained from Fig. 4.9, assembled from Waughman (1980). The peat samples on which these analyses were made came from the top 20 cm of sites in mires in southern Germany. The sites were ordinated by reciprocal averaging using vegetation composition, and are arranged from fen at the left to bog at the right. The vegetation ordination provides a quasi-objective means of positioning the sites on the fen–bog axis. The interrelation of vegetation and mire chemistry is thereby revealed. At the bog end occur species such as *Sphagnum* spp. and *Eriophorum vaginatum*; in the centre are *Menyanthes trifoliata*, *Carex rostrata* and *Equisetum fluviatile*; and at the fen end *Schoenus ferrugineus*, *Juncus subnodulosus*, and *Scorpidium scorpioides*. In fens, with pH about 7.5, the dominant cations in peat are Ca^{2+} ($Ca^{2+} \simeq 500$ $\mu mol\ l^{-1}$, of which about 1/20 is easily extractable) and Mg^{2+} ($Mg^{2+} \simeq 300$ $\mu mol\ l^{-1}$, of which about 1/50 is easily extractable). In bogs, with pH about 4.0, the total concentration is only one-tenth that in fens, or less, though the concentration of extractable Ca^{2+} and Mg^{2+} is not reduced by quite so much. The concentration of iron and phosphorus is also low in bogs, and higher towards the fen end of the continuum, but with some indications of lower concentrations again in rich fens. Sodium and nitrogen show no clear pattern, but total potassium shows a pattern similar to that of iron, though extractable potassium and phosphorus are present in smaller concentration in fens than in bogs. [The concentration of extractable phosphorus depends considerably on the extractant (Fig. 4.9B), and the concentration of phosphorus in peat water is much lower than either "extractable" or total phosphorus — see p. 187 below.] On the whole the concentration in the vegetation reflects the total concentration in peat, with the exception of iron which is more concentrated in bog vegetation than it is in bog peat, and less in fen vegetation than in fen peat.

The majority of analyses of inorganic constituents of peat have been made on north-temperate and boreal peatlands (but see Anderson, Vol. B, Ch. 6). Any account of the inorganic chemistry of peat must therefore be biased. It is convenient to consider first the processes affecting the profile of concentration of inorganic constituents, next examples of the vertical distribution in deep profiles and the pattern of horizontal variation, and thirdly the details in the top 50 cm — the so-called "active layer" (Romanov, 1968).

Processes affecting the inorganic constitution of peat

There are several processes which may combine to determine the concentration of an inorganic substance in peat.

Initial concentration in the plant or in the water in the peat. The biggest difference is between plants growing under the influence of mineral soil water, which usually has a relatively high concentration of, *inter alia*, calcium, iron, aluminium and manganese, and plants dependent on precipitation and dry deposition. The concentration in precipitation of elements such as sodium, magnesium and chlorine, which originate principally in sea spray, increases markedly towards coasts (for examples see Gorham, 1958; Boatman, 1961; and Sonesson, 1970). Other elements in precipitation and dry deposition, such as aluminium, manganese and iron, probably derive from soil dust (Peirson et al., 1973). Other distinguishable sources are industrial and domestic gases (sulphur dioxide, ammonia), smoke, industrial dusts, road dust (Tamm and Troedsson, 1955), volcanic ash, and extra-terrestrial particles (meteorites). Volcanic ash is of local importance, in Iceland for example. Meteorites are generally of negligible importance, though locally they can be catastrophic.

The supply of industrial and domestic gases, particles and solutes has undoubtedly increased and the increases, in a general way, are abundantly documented. Attempts to demonstrate increases in, for example, lead in Arctic and Antarctic snow and ice have posed extremely difficult analytical problems because of the low concentration (Murozumi et al., 1969), but the "concentration" of such elements by mosses including *Sphagnum* has made such work much simpler with bryophytes (Rühling and Tyler, 1971, 1973). The recent increase in acidity of precipitation (attributable to oxidation of sulphur dioxide) in Scandinavia has been documented (Granat, 1972; Brackke, 1976), and may have had some marked effects on fish populations. Earlier increases are suspected but are less well established. For example, Conway (1949) noted that a change of peat type at Ringinglow in the Pennines preceded the appearance of layers of soot, and drew attention to the high concentration of sulphur dioxide in the air and the low pH (about 3.1) of the present bog surface.

The concentration of inorganic constituents in dry deposition and in precipitation varies with place and fluctuates with season (Cawse, 1974) and, for example, with agricultural practice (Boatman et al., 1975). Long-term changes are superimposed on these fluctuations. Generalization is therefore difficult. There are numerous measurements of the concentration of solutes in precipitation, notably those in the network of stations (principally Scandinavian) described by Eriksson (1955) and in subsequent quarterly reports in *Tellus*. There are considerable technical difficulties in making such measurements (Paterson and Scorer, 1973) and in interpolating between stations (Granat, 1975), so there are few cases where precipitation chemistry can be compared directly with peat chemistry. One example, from the Pennine hills of England, is shown in Table 4.8. The peat came from the top 10 cm (A.J.P. Gore, pers. comm., 1979). The last line shows that the apparent effectiveness of trapping is $N > P > Ca \simeq Mg > K \gg Na$. The order for the last four is approximately that to be expected if these elements are trapped on cation exchange sites, and indeed, for these four elements, most *is* exchangeable (Gore and Allen, 1956; Smith, undated; Sonesson, 1970). Very little of the nitrogen and phosphorus is exchangeable, consistent with their presence in organic combination.

Nitrogen may be supplied in significant amounts as ammonia gas, which may be absorbed directly by acid peat (Ingham, 1950), or in special cases by "fixation" of atmospheric nitrogen (assessed by acetylene reduction) by root nodules of *Myrica gale* (Sprent et al., 1978) or — at a rate independent of pH between 4.5 and 7.5 — by blue-green algae associated with *Sphagnum* or other semi-aquatic bryophytes (Basilier, 1973; Basilier et al., 1978; Dickinson, Ch. 5).

Decay and loss as gas of peat organic matter. This would cause a non-selective increase in concentration of all inorganic solutes.

Relocation of inorganic constituents by physico-chemical processes. The hydraulic conductivity of humified peat is low (Rycroft et al., 1975b) and in permanently waterlogged peat of this kind the mass movement of water is very slow (Knight et al., 1972), so there is probably little mass redistribution of inorganic solutes below the surface 50 cm of peat. Diffusion may occur. There are few measurements of the diffusion rate of solutes in saturated peat. Giles (1977) found the diffusion coefficient of labelled (^{32}P) phosphate in saturated peat was about 10^{-6} cm^2 s^{-1}, compared with about 10^{-5} cm^2 s^{-1} in water and much lower values — around 10^{-8} to 10^{-10} cm^2 s^{-1} — in unsaturated soils (Nye and Tinker, 1977). There seem to be no records of diffusion coefficients of cations in saturated peat, but one might expect the value to be lower than for phosphate because the high cation exchange capacity effectively enlarges the volume of the peat.

Given the diffusion coefficient and concentration gradient, then Fick's Law may be used to calculate the amount of solute moved. For example, if the diffusion coefficient were 10^{-6} cm^2 s^{-1} and the concentration gradient that of calcium in Fig. 4.12 below (assuming bulk density of 0.1 g cm^{-3} gives a gradient of about 0.3 μmol cm^{-1} in the region just above the *Carex*–moss peat), then the flux would be about 10 μmol cm^{-2} yr^{-1}. The total calcium in the 250 cm above the *Carex*–moss peat is about 400 μmol cm^{-2}, representing about forty years of diffusion at this rate. The real elapsed time since the *Carex*–moss boundary is probably nearer to 4000 years, suggesting that, if diffusion is indeed the main process affecting the calcium profile in these conditions, then the diffusion coefficient must be about 10^{-8} cm^2 s^{-1} — only about one hundredth that of phosphate. This would be consistent with experimental measurements on cation exchange resins (Helfferich, 1962). If diffusion were the principal controlling factor one would also expect that the profile would be a hollow curve (which it is) and that trivalent cations would have even lower diffusion coefficients. The behaviour of iron in Fig. 4.12 is consistent with this. One would also expect there to be little change in concentration associated with changes in humification. This too

TABLE 4.8

Influx of solutes and particles (wet and dry deposition) on blanket bog at Moor House, England, and concentration of the corresponding elements in the top 10 cm of peat; the peat contains *Sphagnum*, *Calluna*, and *Eriophorum*

A. Influx in precipitation

Year	Precipitation (mm)	Influx[b] (μmol cm^{-2} yr^{-1})					
		NH_4+NO_3	PO_4	Na	K	Mg	Ca
1959–60	1790	–	0.11	14.8	1.4	1.0	2.6
1960–61	1910	–	0.30	12.2	1.2	–	3.3
1961–62	2060	7.6	0.09	12.9	0.9	–	3.1
1962–63	1870	8.9	0.60	24.1	1.5	–	5.7
1963–64	1800	10.3	0.19	10.2	0.8	–	2.2
1964–65	1770	13.3	0.35	17.2	1.6	1.6	2.8
Mean		10.0	0.27	15.2	1.2	1.3	3.3

B. Concentration in peat

		Concentration (μmol cm^{-3})					
		total N	total P	Na	K	Mg	Ca
Site 2[a]				1.0	1.3	2.6	3.0
Site 3[a]				0.7	1.0	3.2	3.0
Ten sites[c]	lowest	58	0.9	0.3	0.2	0.7	2.4
	highest	136	1.7	0.8	0.7	5.2	10.5

C. Approximate proportion of annual influx in 1 cm^3 of peat (median for top 10 cm)

	N	P	Na	K	Mg	Ca
	10	5	0.05	0.7	4.0	4.0

[a]From Gore and Allen (1956). [b]From Gore (1968), assuming peat bulk density is 0.1 g cm^{-3}. [c]From Smith (undated).

seems to be generally true (for example calcium in Fig. 4.7). But there is a wide gap between the conclusion that diffusion *could* account for part of the profile, and the conclusion that it *does*. Some elements in Fig. 4.12 — sodium, potassium, magnesium and aluminium — do not fit this hypothesis at all. If diffusion is important in these cases it must be superimposed on other processes.

The rate of relocation may be much influenced by changes of chemical state. Phosphate may be released into solution if peat is frozen then thawed (Sæbø, 1969); iron and manganese change valence between oxidized and reduced forms (aerobic and anaerobic peat); some elements (for example iron, manganese, lead and copper) can form relatively insoluble sulphides, and probably become less mobile in consequence when they pass into the anaerobic zone where sulphide is produced. They may also form relatively stable complexes with organic constituents in the peat (Sikora and Keeney, Ch. 6). The elements nitrogen and phosphorus are present in relatively low concentration but in a wide range of compounds which may be actively metabolized; the rate of "mineralization" of nitrogen is an important character if peat is used commercially as a plant substrate, and is related to the level of the water table (probably through the aerobic/anaerobic transition and microbial activity) in a complex way (Williams, 1974).

Relocation by washing through the surface layers is considered later.

Relocation of inorganic constituents by biological processes. The main processes are absorption of solutes by plant roots, then movement up into the rhizomes and leaves, and downward movement when roots grow down into the peat. Bog species differ in the vertical distribution of their roots, and more particularly in the profile of radioactive solute uptake (Boggie et al., 1958; Giles, 1977). Most species cycle solutes through the top 20 cm of peat, but the bog species *Eriophorum* may be effective to a depth of 60 cm or more. The roots of the fen plant *Cladium mariscus* extend to this depth (Conway, 1937), as probably do some roots of *Calamagrostis* spp. (Luck, 1964) and of the larger species of *Carex*. In *Sphagnum fuscum*-dominated bogs in the Abisko area of northern Sweden, *Rubus chamaemorus* is common. The rhizomes produce two types of roots: fine, surface-ramifying ones, and large (1–2 mm diameter) vertical roots at a density of about one per centimetre of rhizome. These vertical roots drill straight down to near the water table — perhaps from 10 to 60 cm down — with few or no laterals, then explode in a dense network of very fine branches. Stavset (1973) reports *R. chamaemorus* roots down to 180 cm in peat on Andøy in northern Norway.

The quantitative importance of roots of various bog plants in relocating phosphate has been examined by Giles (1977). Phosphate was used as a convenient tracer — ^{32}P has a half-life of about ten days and produces moderately energetic β radiation — but the concentration of *soluble* phosphorus in peat is but a tiny fraction of the total (see later). The exploitation of peat by a plant root is probably diffusion-limited, because the deep roots (at least) of bog plants appear to lack the mycorrhizal associations (Dickinson, Ch. 5) which might effectively extend out from the roots the zone from which rapid transport is possible (Sparling and Tinker, 1978a, b, c). It is, therefore, the *length* of root produced which is the most important measure of exploitation potential. Giles, taking into account the length of life and period of activity of roots, calculated an effective depletion cylinder of about 4 mm for most species. Table 4.9 shows the present proportions of species on a site in a valley bog in southern England, and the time needed for 95% of the peat volume at various depths to have been within the exploitation range of a root. Peak above-ground biomass is a fair estimator of productivity in this community, and is about the same as litter fall. Above-ground production is about six times the production of roots, and *Eriophorum* has a greater proportion of its roots at greater depths than other species. It is the main contributor in this community to relocation of phosphate from greater depths, as shown by the time for $P=0.95$ depletion. Given the usual rate of peat "growth" of about 1 cm yr^{-1} at the surface decreasing to about 1 mm yr^{-1} at 50 cm depth, it is apparent that there is a high probability that soluble phosphate will be kept in circulation at the surface. For cations, with smaller diffusion coefficients, exploitation even as close as 10 cm to the surface is probably much lower. The last line of Table 4.9, corresponding to a depletion cylinder of negligible thickness, gives some indication of the chance of exploitation of a totally immobile substance.

In general, one might expect there to be significant biological relocation of anions to a depth of 50 cm, but to a much smaller depth for cations. There is a great deal of internal cycling between root, rhizome and shoot — in some bog species at least (Sæbø, 1968, 1970, 1973, 1977; Chapin et al., 1975, 1978, 1979; Shaver et al., 1979). This internal cycling is considered in detail elsewhere in this volume.

The relocation of inorganic solutes by vascular plants is not surprising, but relocation in the surface layers by bryophytes is more so. *Sphagnum papillosum* plants supplied with $H^{14}CO_3^-$ incorporate ^{14}C, about 80% of which is in the alcohol-insoluble fraction after 24 h (Clymo, 1965); yet three months later, after the plants had grown on average 3.2 cm, as much as 35% of the ^{14}C in the plants after 24 h was found in the capitulum formed since the original treatment, and only 5% at and below the original treatment level. (Of the rest, 25% was in stem and branches which were probably part of the capitulum at the time of treatment.) This movement may have been partly outside the plant (as carbon dioxide respired and then reabsorbed),

TABLE 4.9

Shoot and root productivity, and the calculated time for 95% of peat to be exploited by plant roots, at a site in the valley bog at Cranesmoor, southern England

	Eriophorum angustifolium	*Molinia caerulea*	*Narthecium ossifragum*	*Rhynchospora alba*	Ericaceae	All five taxa
Biomass and dry mass productivity						
Maximum above-surface live biomass[1] ($g\ m^{-2}$)	7	12	16	19	8	61
Root productivity ($mg\ dm^{-3}\ yr^{-1}$)						
At depth 10 cm	3.6	14.2	18.4	8.9	1.6	47
20 cm	4.2	3.3	6.9	1.3	0.1	16
35 cm	0.5	0.5	0.02	<0.1	<0.1	1.2
Root productivity summed to 60 cm depth ($g\ m^{-2}\ yr^{-1}$)	0.9	2.6	3.7	1.5	0.3	9.0
Root growth in length						
Rate of root growth ($m\ dm^{-3}\ yr^{-1}$)						
At depth 10 cm	0.29	1.18	1.58	0.55	0.35	3.9
20 cm	0.47	0.15	0.42	0.15	0.01	1.2
35 cm	0.04	0.02	0.01	<0.01	<0.01	0.1
Rate of root growth summed to 60 cm depth ($m\ dm^{-3}\ yr^{-1}$)	110	200	290	100	54	750
Time (yr) for the depletion zone of roots to sum to 95% of the peat volume						
At depth 10 cm	180	50	30	100	210	13
20 cm	110	360	130	370	7500	44
35 cm	1250	2700	5500	>8000	>8000	750

The five taxa shown grow intermixed and account for about 98% of the biomass of rooted plants. *Sphagnum* is the other major component of the vegetation. Biomass and productivity of rhizomes are not included. Data of Giles (1977).

[1] Minimum was near zero, so these values approximate productivity. Litter-fall was 56 $g\ m^{-2}$ (cf. 61 $g\ m^{-2}$ maximum live biomass).

but it seems probable that most was within the moss.

The relocation of ^{137}Cs is another example. The radioactive ^{137}Cs produced during nuclear bomb tests reached a sharp peak in concentration in deposition in 1963. This is a useful date marker now well established in use on lake sediments (Ravera and Premazzi, 1972; Pennington et al., 1973, 1976). It has been used on peat profiles too (Clymo, 1978, and unpublished). Examples are shown in Figs. 4.10 and 4.18 below. The most notable feature is that most of the ^{137}Cs which has fallen is retained in the top 1 or 2 cm of live surface moss. In these cases vascular plants are hardly involved because the samples were from a nearly pure *Sphagnum* lawn.

In almost all cores there is a lower peak at a depth consistent with it having been laid down in 1963 and subsequently moved down and become spread over a greater depth. This lower peak in concentration of soluble ^{137}Cs may be not at all conspicuous in a single profile, as Fig. 4.10 shows, but it appears consistently. The concentration of insoluble ^{137}Cs, perhaps in glassy particles, is usually more conspicuous and rather higher up, perhaps because the peak has moved less.

These relocation processes are summarized in Fig. 4.11.

Deep peat-profiles

As an example, Fig. 4.12 (from data of Mörnsjö, 1968) shows the variation of concentration in a variety of substances at Fjällmossen, southern Sweden, in a concentrically domed peat bog with *Sphagnum* peat overlying *Carex*–moss peat over *Alnus* wood peat. There are several other profiles of the concentration of several elements on samples taken at intervals of from 10 to 5 cm to a depth of more than 2 m, *inter alia*, those of Sillanpää

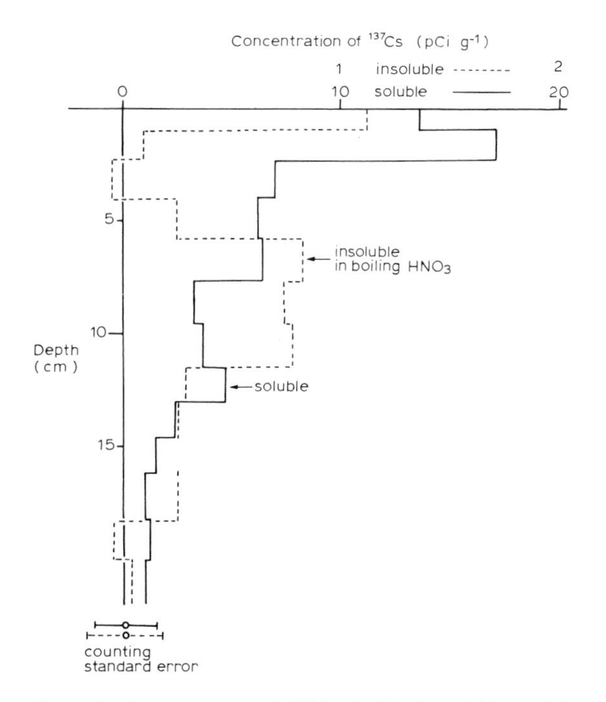

Fig. 4.10. Concentration of ^{137}Cs in HNO_3-soluble and insoluble fractions of peat collected in 1973 from Bohult mire, southern Sweden. Note the ten-fold difference in scale. Both fractions have a surface peak and a smaller peak further down which may be connected with the 1963 peak in ^{137}Cs produced by nuclear bomb tests. Results of Clymo (unpublished).

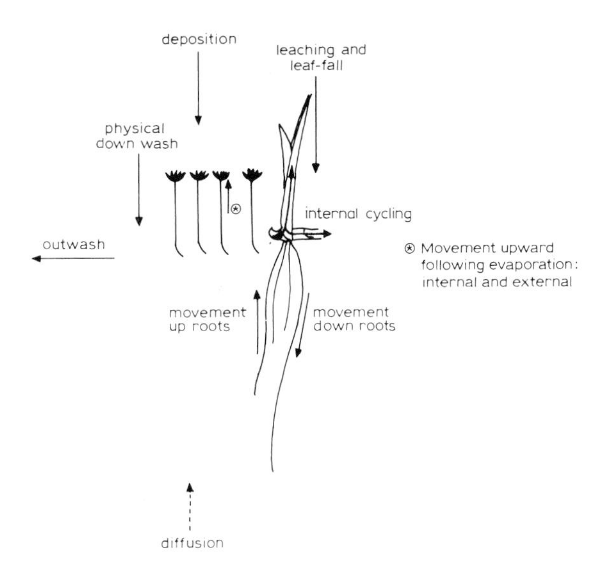

Fig. 4.11. Summary of processes causing movement of inorganic solutes (and particles) in peat. The surface contains stylized *Sphagnum* and a vascular plant.

(1972) from Finland; Mattson and Koutler-Andersson (1955), Sonesson (1970) and Damman (1978) from Sweden; Schneider (1968) from Germany; Chapman (1964b) and Green and Pearson (1977) from England; and Walsh and Barry (1958) from Ireland. Excluding the *Schwingmoor* and blanket bog the profiles all have features in common. Their stratigraphy shows a transition from fen communities (with or without trees) to bog. The basal and surface peats have high concentrations of silica ("SiO_2", or "insoluble ash"), calcium and iron, but the middle peat has lower concentrations. The same pattern, though less conspicuous, is shown by aluminium and manganese, while potassium has a particularly pronounced surface peak. Other elements such as magnesium are more variable, but sodium and hydrogen ions tend to increase in concentration from the bottom to the top of the profile. Nitrogen and phosphorus have been more rarely measured. They have high concentrations at the surface, but remain fairly constant below that (Mattson and Koutler-Andersson, 1955; Damman, 1978).

The high concentration in the top 30 to 50 cm may reasonably be connected with the relocating activities of live plants and animals, and with the change from aerobic to anaerobic conditions. The details of the processes operating in this horizon are considered later. The transition from high concentrations at the base to lower ones in the mid-section is commonly associated with the change from a eutrophic, mineral-soil-water-dependent, fen peat to an oligotrophic, precipitation-dependent, bog peat. For some substances the decline in concentration is abrupt: iron, aluminium and "SiO_2" are examples. For calcium, however, the decline is more gradual and occupies a metre or more. The possible role of diffusion in creating this shape has already been mentioned. In course of time either diffusion or biological relocation would smear an initially sharp transition. Mattson et al. (1944) drew attention to the usefulness of the Ca : Mg ratio as an indicator of the limit of influence of mineral soil water. In atomic units, the concentration of calcium in soil water usually exceeds that of magnesium, but in precipitation the sea-derived magnesium exceeds calcium in concentration, though the ratio depends on dust supply and on distance from the sea (for example, see Sonesson, 1970). Mattson and Koutler-Andersson

(1954) adopted the ratio of 1:1, though it is evident from their figure 7 that the erratic behaviour of the magnesium profile makes this nothing more than a rule of thumb. Chapman (1964b) used the same ratio, though perhaps with mass instead of atomic units of concentration, and found it a very good guide. The same atomic ratio is reached in Mörnsjö's Fjällmossen profile (Fig. 4.12) about 100 cm above the minerotrophic *Carex*–moss peat limit. It is exactly on the wood peat limit in Walsh and Barry's (1958) Cloncreen profile, but is 200 cm above the fen carr peat in their west coast raised bog at Kilmacshane. At Vassijaure (Sonesson, 1970) the ratio is more than 1:1 at all depths. It seems clear that no universal rule exists, and in view of the different sources of these two elements one would not expect one.

Where aluminium and iron have both been measured (Mattson and Koutler-Andersson, 1955; Chapman, 1964b; Mörnsjö, 1968; Sonesson, 1970; Damman, 1978; Clymo, unpublished) they usually show the pattern of Fig. 4.12. At the base, the atomic concentration of iron exceeds that of aluminium, and both are high. In the bulk of the ombrotrophic peat the concentration is low and both elements have about the same concentration. In the surface layers aluminium exceeds iron. As will be shown later, the thick 10 or 20 cm slices average out a much greater range of concentration in thin layers near the surface. The concentration of iron at the base of a peat deposit may be so high that iron-containing minerals such as siderite and vivianite (Casparie, 1972) are formed, as well as more amorphous deposits of "bog iron" (Newbould, 1960). The basal iron and aluminium are probably derived from the mineral soil water. In ombrotrophic conditions the iron and aluminium are probably derived from soil dust (Peirson et al.,

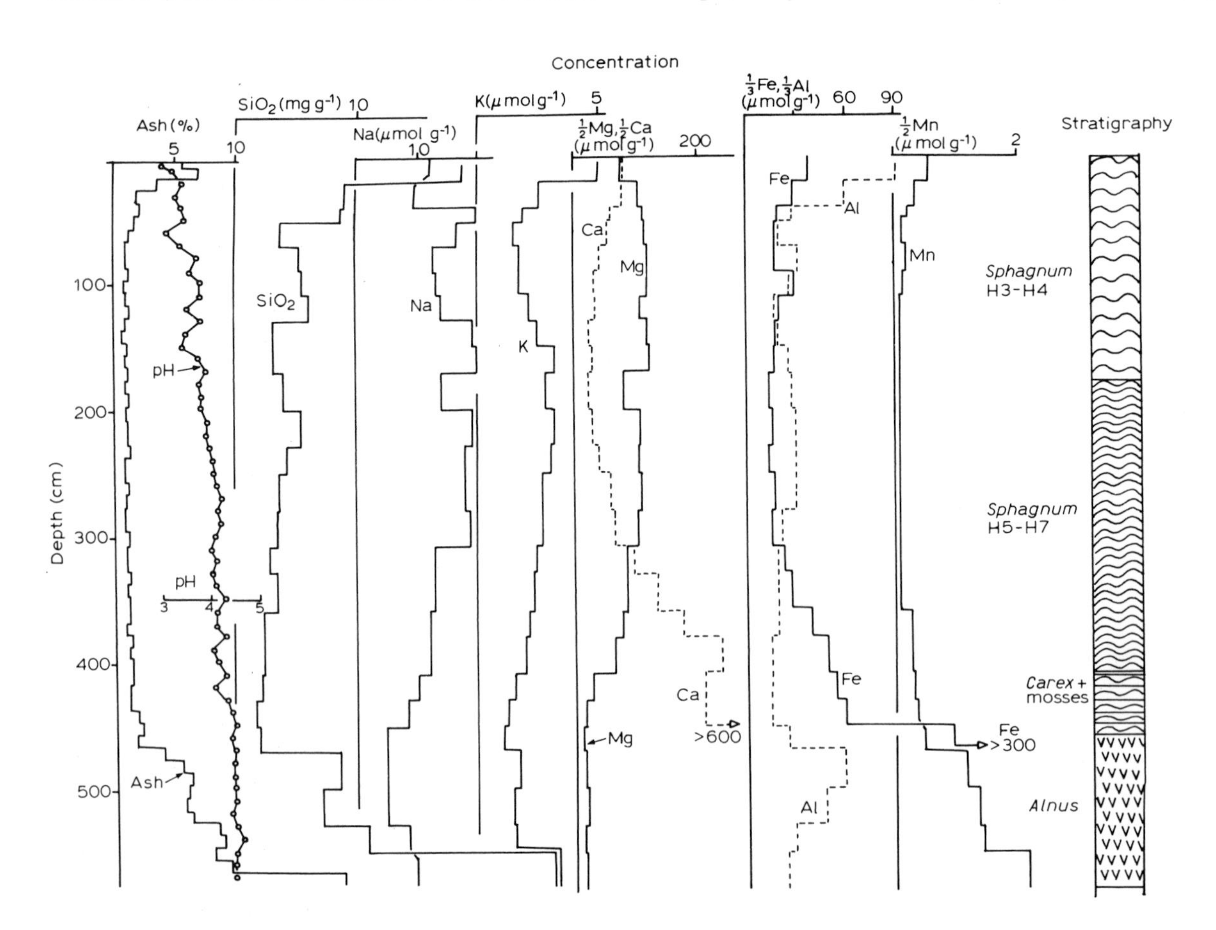

Fig. 4.12. Concentration of inorganic constituents in a peat profile from Fjällmossen, southern Sweden. The bog is (today) concentrically domed. Redrawn from Mörnsjö (1968).

TABLE 4.10

Concentration (per unit dry mass) of inorganic constituents in the mid-section of deep profiles (the sites are in north–south order; values are reliable only within a factor of 2)

Site	Concentration (μmol g^{-1})									Source
	Na	K	Mg	Ca	Mn	Fe	Al	"SiO_2"	pH	
Vassijaure (arctic Sweden)	4	1	30	50	0.1	20	30	0.2	4.0	Sonesson (1970)
Ramna[1] (central Sweden)	–	–	60[2]	20[2]	0.2	2	15	0.05	–	Mattson and Koutler-Andersson (1954)
Fjällmossen (southern Sweden)	20	3	40	20	0.05	5	10	0.01	4.0	Mörnsjö (1968)
Traneröds Mosse (southern Sweden)	15	2	40	25	0.2	4	10	0.3	–	Damman (1978)
Cruden Moss (northeastern Scotland)	50[3]	2	80	20	–	20	–	0.2	–	Stewart and Robertson (1968)
Coom Rigg Moss (northern England)	20	5	–	40	–	2	20	0.2	4.2	Chapman (1964b)
Cloncreen (Ireland)	20	4	60	30	–	4	–	0.1	4.7	Walsh and Barry (1958)

[1]Original data on air-dry mass. Assumed oven-dry mass is 80% of air-dry mass. [2]Exchangeable. [3]The Na profile is inverted, with the highest concentration in the mid-section.

1973). If this is so, then some of the differences in their profiles probably result from differences in the amount of relocation. The concentration of most of these elements in the central "fossil ombrotrophic" section of many profiles is surprisingly similar. Some results are collected in Table 4.10. The most variable of the elements commonly measured is iron, with a range from about 2 to 20 μmol g^{-1}. Total cation concentration is about 100 to 150 μmol g^{-1}. It is possible that this seeming uniformity is imposed by the cation exchange properties of the peat.

Isolated measurements of total nitrogen and phosphorus in peat have been frequently reported, but *profiles* have been less commonly measured. The relative expense of the methods rather than the importance of the subject must account for this. The chemical state of nitrogen and phosphorus is even more vaguely known than is that of the metals. Thus, it is possible to measure "total nitrogen" and "total phosphorus" or the exact species NH_4^+, NO_3^-, NO_2^- and the various forms of orthophosphate. The complex (probably organic) forms predominate. The range of concentration of inorganic phosphate in suction water from peat from the Åkhult mire (Malmer and Sjörs, 1955; Malmer, 1962b) was from 0.003 to 0.035 mmol l^{-1}. Sæbø (1968) recorded similar values. The concentration of extractable phosphate depends on the extractant. Malmer and Sjörs (1955) and Malmer (1962b) recorded, per unit volume of wet peat, from 0.03 to 0.37 mmol dm^{-3} extractable by 1*M* acetic acid, and Waughman (1980; see Fig. 4.9) recorded a range from 0.01 to 0.1 mmol dm^{-3}, the lower values being in fen peats. Waughman also showed that 1*M* ammonium chloride extracted less than acetic acid from fen peat, but about the same from bog peats. The total concentration of phosphorus in peat is much greater; Malmer and Sjörs (1955) and Malmer (1962b) found 0.8 to 3.2 mmol dm^{-3}, Smith (undated) recorded 2 to 5 mmol dm^{-3} in Pennine blanket peat, Bellamy and Rieley (1967) found 1 mmol dm^{-3} in *Sphagnum fuscum* peat, and Waughman (Fig. 4.9) recorded about 2 mmol dm^{-3} in fen and bog peats. In round figures the ratio of water-soluble : extractable : total is about 1:10:100. In the Alaskan tundra, however, the "reserve" in exchangeable form seems to be much greater (Chapin et al., 1978) although the concentration of soluble inorganic phosphate is about the same. Similar differences exist for nitrogen. Sæbø (1970) recorded 0.1 to 0.3 μmol dm^{-3} NH_4 in suction water, but total nitrogen has been commonly recorded in the range 30 to 200 mmol dm^{-3} in peats from a wide range of temperate, boreal and arctic sites (for examples see Waksman and Stevens, 1929a, b; Malmer and Sjörs, 1955; Walsh and Barry, 1958; Gorham, 1961b; Malmer, 1962b; Stewart and Robertson, 1968; and Smith, undated) and from other regions, such as the Antarctic (Allen et al., 1967) and Borneo (Richards, 1963). Between 0.5 and 3% by weight seems to be the usual range. This is less than is found in most live plants, but not notably less than in *Sphagnum* (for example, see Malmer and Sjörs, 1955; and Malmer, 1962b). Much of the nitrogen seems to be associated with the "lignin" fraction (Fig. 4.8) and the relative constancy of binding capacity for non-exchangeable nitrogen in the "lignin" has already been mentioned. Mattson and Koutler-Andersson (1955) drew attention to the fact that the "lignin" in the top 50 cm is nearer to saturation than that in the rest of the profile (except the base, which might be expected to show minerogenic influences). They suggested that the surface anomaly is a result of direct absorption of ammonia from the air — ammonia produced in increasing amounts as a result of industrialization. The amounts they found correspond to about 11.6 kg ha^{-1} yr^{-1} since 1000 A.D., compared to precipitation measurements of about 3.5 kg ha^{-1} yr^{-1}. But there is good reason to suppose that the biological and physico-chemical processes in the top 50 cm of peat are markedly different from those lower down, and valid measurements of ammonia in precipitation are notoriously difficult to make. The surface anomaly *may* be a result of industrial activity, but it may be a normal feature of the surface.

Much the same can be said of the profiles of total phosphorus (Fig. 4.13), though the basal high values seem to be missing.

One factor which is always difficult to assess is the extent to which a particular vertical profile reflects very local changes — for example the difference between hummock and hollow. In some cases it can be demonstrated that what would appear in a single core of a few centimetres diameter as a "recurrence surface" (see Tallis, Ch. 9) extends at most a few metres (Walker and Walker, 1961; Casparie, 1972). The position is not clarified by the general lack of any reference to replication of analyses, let alone of samples, so that the precision

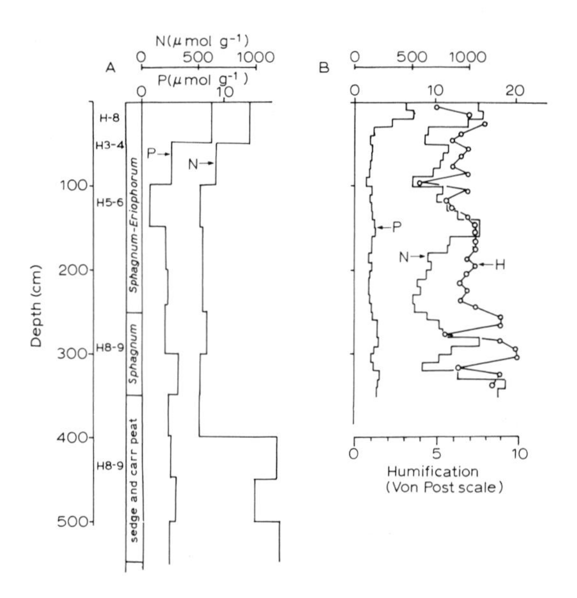

Fig. 4.13. Humification (Von Post scale) and concentration of N and P in two bog profiles. A. Raised bog at Cloncreen, Ireland. (B) Raised bog at Ramna, southern Sweden. Redrawn from Walsh and Barry (1958) and Mattson and Koutler-Andersson (1954).

and accuracy of published values can only be guessed. It is most unlikely that they justify the three or even four significant digits commonly reported. One exception to the lack of replication is the work of Stewart and Robertson (1968). They collected 110 samples from eighteen levels in a blanket peat site to a depth of just over 100 cm, spread over a lateral distance of about 410 cm. One vertical series of eighteen samples formed a profile, the other samples were taken at random. The peat in the top 40 cm was of a series of interdigitating patches of *Calluna*, *Sphagnum*, and *Eriophorum* in varying proportions. Below this the botanical composition was unclear but the peat was "fibrous" to a depth of about 90 cm, where it became "amorphous". Most of the measured substances showed the smallest range of concentration in the mid-depth zone between 40 and 80 cm deep. As a proportion of the whole range, the range in this zone was about 0.3 for nitrogen, 0.3 for phosphorus, 0.3 for sodium, 0.8 for potassium, 0.7 for calcium, 0.3 for magnesium and 0.3 for sulphur. For all but sulphur the range was proportionately greater in the "mixed" peat layers: about 0.4 for nitrogen, 0.5 for phosphorus, 1.0 for calcium, and so on. Only for sulphur, and perhaps magnesium, was the range about the same at all depths. The chosen profile also showed marked patterns in relation to the range of two elements: sodium was consistently high in the mid-layers, but low in the bottom layer; potassium was low in all but the bottom layers. The other elements showed no consistent pattern. Even in this work, however, it is not possible to separate sampling and analytical errors.

One may perhaps conclude that, in the absence of botanical information, erratic individual differences in concentration in adjacent samples by a factor of less than two may be uninterpretable, though a change of the same magnitude occurring gradually but steadily over several samples may be more meaningful.

Profiles of the surface layers

The surface layers (the top 50 cm or so) of a peat-forming community are specially important because within them rapid changes are in progress and much relocation of inorganic substances may occur. This upper layer has sometimes been called the "active" layer, overlying the "inert" layer. This is an unhappy choice of words, prejudicing consideration of what changes are going on in the lower layer. There is much to recommend the suggestion of Ingram (1978; see also Ch. 3) that the layers be called the acrotelm and catotelm, respectively.

Small differences in rate in the acrotelm may have a disproportionate effect on peat accumulation (Clymo, 1978; Jones and Gore, 1978). Perhaps the most useful indicator in this zone is the "redox potential". In practice this is measured as the electrical potential between a platinum probe and a calomel (or similar) reference junction. A KCl salt bridge, usually saturated, is used to connect the calomel to the peat (for the same reason that it is used in pH measurements) and its exact position is usually unimportant — anywhere within 1 to 100 cm will usually leave the potential unchanged. The potential may be changed however by the pretreatment of the platinum. Some workers (Urquhart and Gore, 1973) boil the platinum in concentrated oxidizing acids, and use one electrode for only one or at most a few measurements. Results which are *reproducible* may be had with no preparation of the platinum apart from a few immersions to 10 cm or so in the peat. This process seems to prepare the surface in some undefined way, so that on subsequent removal to sites of different redox potential the reading stabilizes within a few seconds. With platinum prepared by the long process, the reading usually drifts, and some arbitrary time must be chosen for the measurement. The less carefully prepared platinum may drift in oxidizing conditions, but very little in reducing conditions.

"Redox potential" is also affected by pH, and it is usual to apply a "correction" of about 58 mV per pH unit to an arbitrary standard, often to pH5 (E_5). Temperature affects the readings too, reducing the potential as temperature rises by about 0.8 mV per °C, and a correction to a standard temperature is usually made.

Potentials are usually reported relative to the calomel half cell. They may be changed to the standard hydrogen cell by adding about 242 mV. If the KCl bridge is not saturated, the shift is different (Conway, 1952).

It will be obvious that "redox potential", as measured in the field or on fresh peat newly returned to the laboratory, may be usefully compared within any one set of measurements but should not be compared too closely between sets, particularly if different methods have been used.

Three other features are commonly correlated with "redox potential" profiles: the position of the water table, the presence of sulphides, and the oxygen concentration. In acid peat and even in alkaline peat a sulphidic smell, usually attributed on no other evidence to hydrogen sulphide, is often conspicuous. The blackening of a freshly cleaned silver or copper wire or plate (Burgeff, 1961; Clymo, 1965; Urquhart, 1966) may be used in a semi-quantitative way. At low temperature particularly (as shown for example in Fig. 4.8) the sulphidic smell may be quite strong and yet there may be no detectable blackening of silver within an hour. The kinetics of blackening have not been investigated, but are obviously complex. Many profiles similar to that of Fig. 4.14 have been reported (for example, by Mörnsjö, 1969; and Urquhart and Gore, 1973). Most of these agree in showing the upper limit of S^{2-} to correspond to "E_5" about +200 mV. This is a figure similar to that found for the transition from SO_4^{2-} to S^{2-} in lake muds by Mortimer (1942). Urquhart and Gore (1973) reported that "E_4" showed a minimum at 12

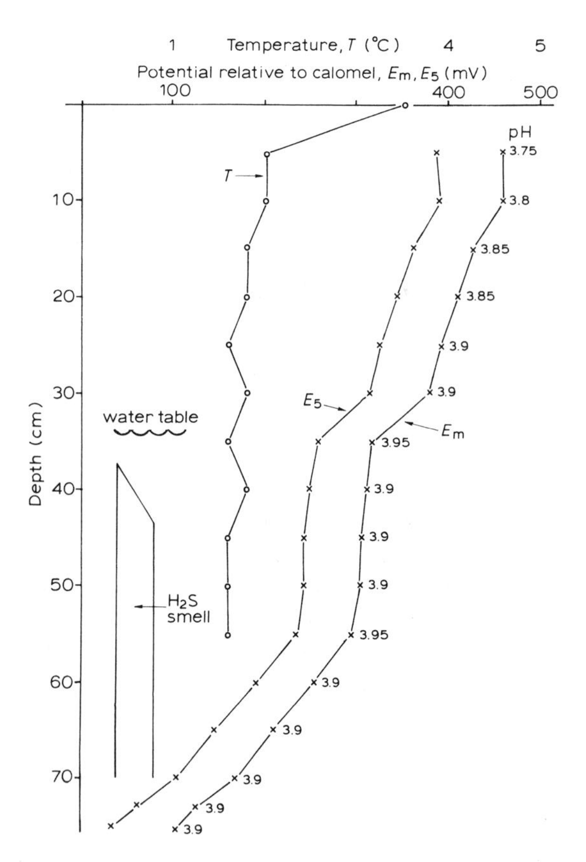

Fig. 4.14. "Redox potential" profile on September 17 in *Sphagnum fuscum* peat near Abisko, northern Sweden. E_m = as measured; E_5 = calculated at pH 5 using the temperature profile and pH value by each point. A platinum-tipped probe was used with a calomel reference junction. Results of Clymo (unpublished).

to 20 cm deep, but the evidence for this is much less strong than that for a large surface drop. Nevertheless, the profiles they reported are remarkably similar to those found in lakes in winter (Mortimer, 1942), and the parallel between mire surfaces and lake muds might be illuminating.

The sulphide zone seems to be of limited thickness (Mörnsjö, 1968; Urquhart and Gore, 1973). What happens to S^{2-} at the lower limit has not been investigated, but the seasonal variation in concentration of sulphide — lower in winter — perhaps indicates that sulphide is normally removed by chemical conversion to other forms, and that which is observed has been recently produced by microbiological means. The finding by Gorham (1956) of increased concentration of SO_4^{2-} in pools during dry weather in mid-May is consistent with this hypothesis.

If waterlogged peat is drained the oxidation of large amounts of sulphide may produce so much sulphate (as H_2SO_4) that the peat becomes extremely acid. Hart (1962) recorded that the pH of a drained mangrove swamp fell during six months from 6.2 to 3.8, and even greater acidity has been recorded in these circumstances. The seasonal variation of "redox potential" and the extent of blackening on a silver plate has been most thoroughly investigated by Urquhart and Gore (1973). They found that in wet sites (lowland, and 560 m altitude) E_4 fell, as temperature increased, by about 5 mV per °C between 12 and 20 cm depth, and by about 8 mV per °C between 22 and 30 cm depth. In drier sites the effects were erratic, and mostly of lower statistical significance. Blackening of a silver plate was clearly correlated with "wetness" — it was most intense (after 1 hr) in the wet sites, and in one case where the plate crossed the boundary between hummock and hollow the top of blackening was within 2 cm of the hollow surface, but 15 cm below the hummock surface and was actually lower (in altitude) below the hummock than below the hollow. Unfortunately the wet sites were examined in May and June and the dry sites in August, and common experience is that sulphidic smells on bogs are generally most noticeable in early summer, so this difference between sites may have been a result of measuring at different seasons of the year.

The interrelations of "redox potential" and sulphide content with water table in peat are obvious in general, but unclear in detail. One may deduce that during the measurements of Urquhart and Gore (1973) the top of the sulphide zone was usually just below the water table, and the data reported by Clymo (1965), Smith (undated) and those in Fig. 4.14 here show this too. On the other hand, Boggie (1972) found the mean water table was about 4 cm below the mean upper limit of blackening on silver-plated copper. His silvered plates were left in the peat for about a month however, compared to the period of one hour or so used by other workers, and it is to be expected that traces of sulphide not detectable in one hour would be conspicuous after 600 h. The problem is further complicated by the fact that the water table can, when near the surface, fluctuate by 5 to 10 cm over one to two days (Chapman, 1965); and the very idea of a water table (as such) in humified peat is as dubious as the existence of a water table in a domestic jelly, though there may be a clear and rapidly changing water table in a well in that jelly. Sulphide production is likely to be related to wetness of the peat and not, in general, to a free water table in an inspection pit, unless the hydraulic conductivity at the level is high. The following hypothesis seems to be consistent with most of the known facts. During winter the water level rises and covers some newly dead plant matter containing readily decomposable compounds, but temperature is low, so microbiological activity is low too. In late spring the temperature increases, and with it microbiological activity. The rate of oxygen diffusion (about 10^{-4} of the rate in air) is insufficient to support this activity, so the peat a few centimetres below the water table becomes anaerobic, as shown by a fall of "redox potential" by an average of about 150 mV (Urquhart and Gore, 1973). A different set of microbiological processes is then favoured, resulting in the relatively rapid production of sulphide. Later in the summer, as the water table falls and the readily metabolized material is exhausted, the rate of sulphide production declines, and more of the existing sulphide is reconverted to sulphate either chemically or by micro-organisms. It may be that production of methane is partly linked to this pattern.

There is a need for concerted studies of chemistry (both organic and inorganic) and of microbiology *in the field* at the same site. Tentative steps in this direction were taken during the International Biological Programme (for example, see the various accounts in Sonesson, 1973; Flower-Ellis, 1974; and Heal and Perkins, 1978), but far more needs to be done. The interrelations of "redox potential", sulphide concentration, and the concentration and movement of inorganic substances is equally unclear. The correlation between a particular chemical transition and redox potential is probably mediated by microbial activities. Table 4.11 gives four of the important transitions recorded in lake muds by Mortimer (1941, 1942). These are not the potentials for the chemical reactions, but the potentials which happen to be correlated with them. To what extent the bacteria cause the potentials or are favoured by them is obscure.

TABLE 4.11

Correlation of chemical transition with "redox potential" in mud from Lake Windermere [data from Mortimer (1942), adjusted to E_5 from E_7; relative to calomel]

Transition	"Redox potential", E_5 (mV)
$NO_3^- \rightarrow NO_2^-$	570–520
$NO_2^- \rightarrow NH_3$	520–470
$Fe^{3+} \rightarrow Fe^{2+}$	420–320
$SO_4^{2-} \rightarrow S^{2-}$	220–180

There seem to be no published measurements of valence states and few of sulphide or oxygen concentration in peat. Webster (1962) made measurements in ground water of a wet-heath community, but this ground water was probably moving through, and perhaps in, a mineral soil. That the concentration of oxygen and sulphide could be important is suggested by Fig. 4.15, which shows the maximum concentration of metals which may, in theory, remain in solution at various concentrations of sulphide and other anions. Ponnamperuma (1972) considered these and other equilibria. The order of precipitation from equimolar ionic solutions is: Ag^+ first, then Cu^+ (if it

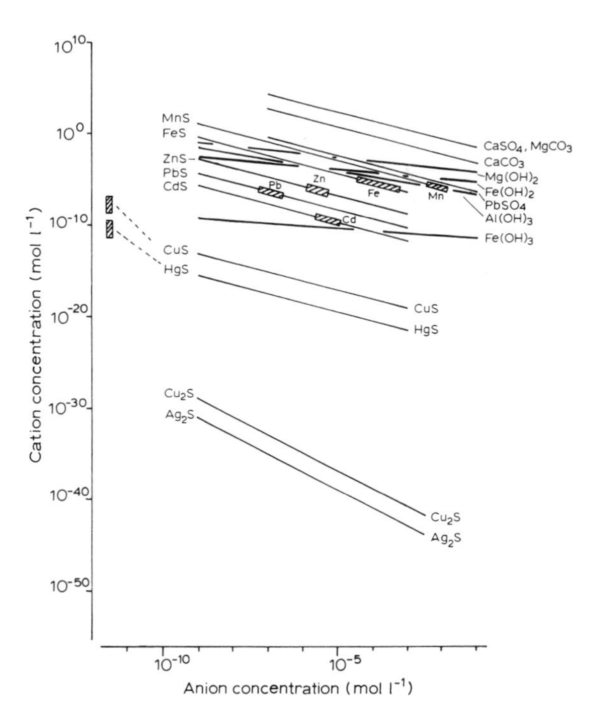

Fig. 4.15. Calculated relationship of cation and anion concentration for some sparingly soluble salts. Above the line the solution would be supersaturated. Hatched bars show the approximate peak volumetric concentration of the cation in bog peat. Calculated from solubility products in Weast (1965).

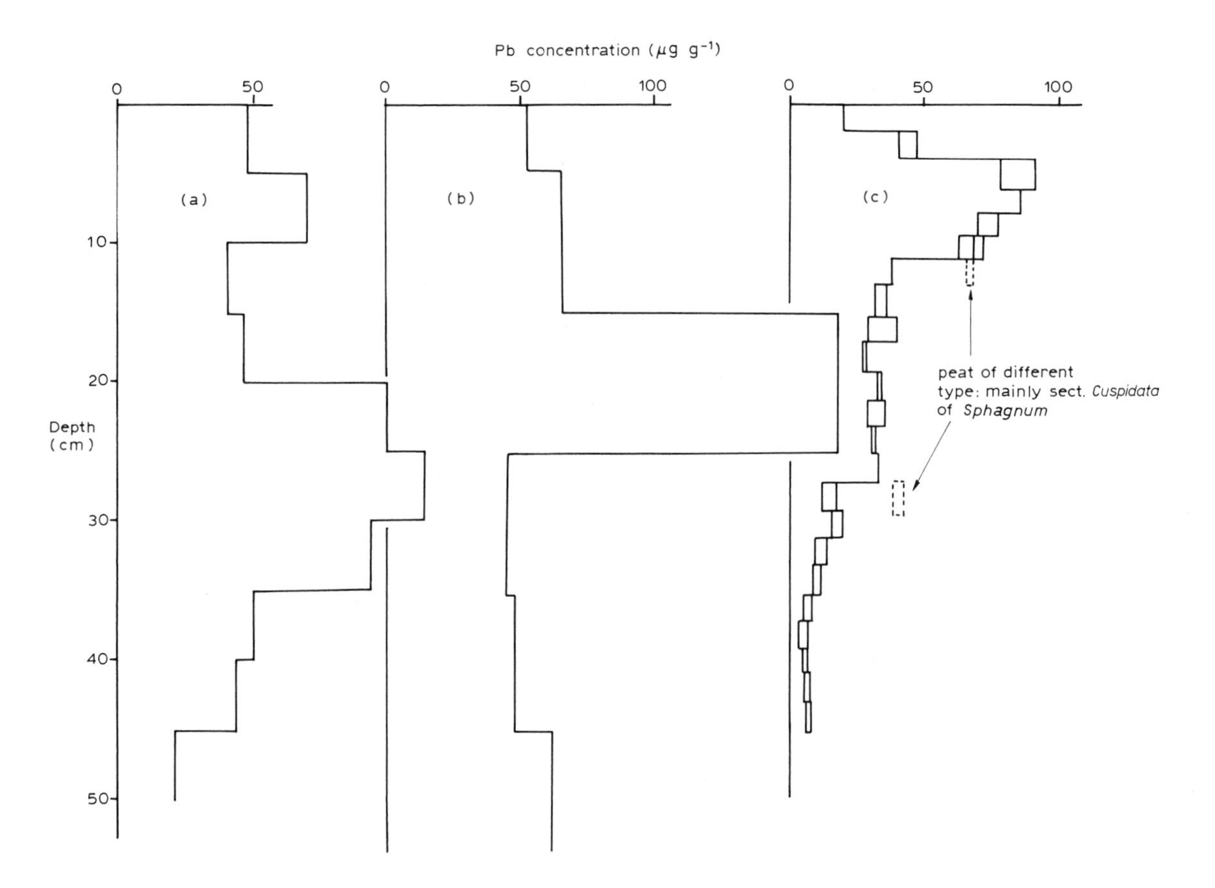

Fig. 4.16. Concentration of lead in three ombrotrophic peats from 60°N latitude: (a) Fjällmossen, southern Sweden; (b) Myras, southern Finland, *Sphagnum fuscum* peat; (c) Os, southern Sweden, *S. magellanicum* peat. Redrawn from Tyler (1972), Pakarinen and Tolonen (1977), and results of Clymo (unpublished).

forms), followed by Hg^{2+} and Cu^{2+}, then after a gap Cd^{2+}, Pb^{2+}, Zn^{2+}, Fe^{2+} and Mn^{2+} in that order. The actual formation of metal sulphides should be influenced by the solution concentration — strictly, activity — of the metals. As a first approximation the volumetric concentration in peat (Clymo, 1978, unpubl.; Damman, 1978) is shown, though this is probably higher than the solution concentration by a factor of perhaps 1000. All the sulphides shown are with divalent metals, so the slopes in Fig. 4.15 are all -1, and if the concentration of the metal is reduced by a factor $1/v$ then that of sulphide must be increased by a factor v. If the metal concentration in the peat water is a thousand-fold less, then the sulphide concentration must be a thousand-fold more. Applying this to Fig. 4.15, with a sulphide concentration of 1 mmol l^{-1} one would expect no MnS, but possibly some FeS, CdS and ZnS. The compounds PbS and, especially, CuS and HgS seem to have a high probability of forming. The blackening of silver and copper is some support for this prediction, but there are no reliable direct measurements of S^{2-} activity in peat, and H_2S is such a weak acid that one would expect only low S^{2-} concentration in acid conditions.

Detailed profiles of metal concentration in the top 20 to 50 cm are beginning to be available (largely as a result of the development of atomic absorption flame spectrophotometry). The resolution of such profiles has usually been 5 to 10 cm (Bellamy and Rieley, 1967; Sonesson, 1970; Sillanpää, 1972; Tyler, 1972; Pakarinen and Tolonen, 1977; Damman, 1978). It seems to be a common feature of such profiles that they show sporadic peaks — single samples with unusually high or low values; but there is almost never any replication of samples, so it is difficult to decide how to regard such one-sample peaks or troughs. Profiles sampled at 1-cm intervals (Clymo, 1978, unpublished; Aaby and Jacobsen, 1979; Aaby et al., 1979; see Figs. 4.16, 4.18) reduce these problems because the trends are usually clearly distinguishable against the sampling and analytical "noise". In the one published instance of replicate cores, sampled a year apart in time and analyzed by different workers with different equipment (Clymo, 1978), the mean difference in concentration of lead in 36 pairs of samples, each pair from the same depth down the profile, was 70 ppm, but the general trend was from a concentration of about 200 ppm in the top 19 cm, rising to 600 ppm at 29 cm, then falling to 80 ppm at 36 cm. This general point is illustrated in Fig. 4.16, which shows three lead profiles from peats at about the same latitude.

Even with these reservations it is possible to make some generalizations. The most interesting work of Damman (1978) reproduced in Fig. 4.17 may be used together with Fig. 4.18 to illustrate features which are found repeatedly in ombrotrophic mires. The element potassium has very high concentration in the top 1 to 2 cm of live *Sphagnum*, falling from 3500 ppm (0.1 mmol g^{-1}) to 150 to 200 ppm at 10 cm depth and to 25 ppm at 40 cm depth. The very high surface values are associated with the live moss; in dead patches the values are about 50 ppm (Clymo, unpublished). The element sodium, also monovalent, shows similar behaviour, though less extreme, with surface values of 400 ppm falling to 150 ppm at 20 cm depth. Granat (1975) reports the ratio of Na:K in precipitation about 100 km inland from the west coast of Sweden, and 200 km to the north of these peat sites, as about 1:1 (mass basis). The 6:1 ratio in peat at 40 cm is therefore rather surprising unless there has been selective loss of potassium by leaching.

The element manganese shows a characteristic pattern of very high surface concentration — 200 to 300 ppm — in the surface of hummocks, falling sharply from the surface to 5 to 10 cm above the water table. In peat formed by *Sphagnum magellanicum* growing in a "lawn" just above the water table, the surface concentration is about 50 ppm, and falls very steadily to about 1 to 2 ppm at 40 cm depth.

Other elements *increase* in concentration with depth and reach a peak at about the mean water table. In the case of iron and lead the peak is very marked. For zinc it may be less so, when the discrimination is fine. For yet other elements (aluminium, magnesium, calcium, phosphorus and nitrogen) the peak is even more obscure. For copper and cadmium there is no obvious pattern, though for copper (at least) replicates agree closely, so the small drifts in concentration may be of significance.

There are marked differences too in the prominence of peaks compared to the concentration at 40 cm depth. The peak is prominent for potassium, manganese, lead and zinc, but less so for sodium, and inconspicuous for aluminium, iron, magnesium and calcium.

The elements also differ in concentration in different peat types (Fig. 4.18). Conspicuously higher in concentration in *Eriophorum* peat than in *Sphagnum cuspidatum* peat are potassium, aluminium and insoluble ash. Slightly higher are lead, copper, cadmium, sodium, iron and calcium, but slightly lower are manganese, zinc and magnesium.

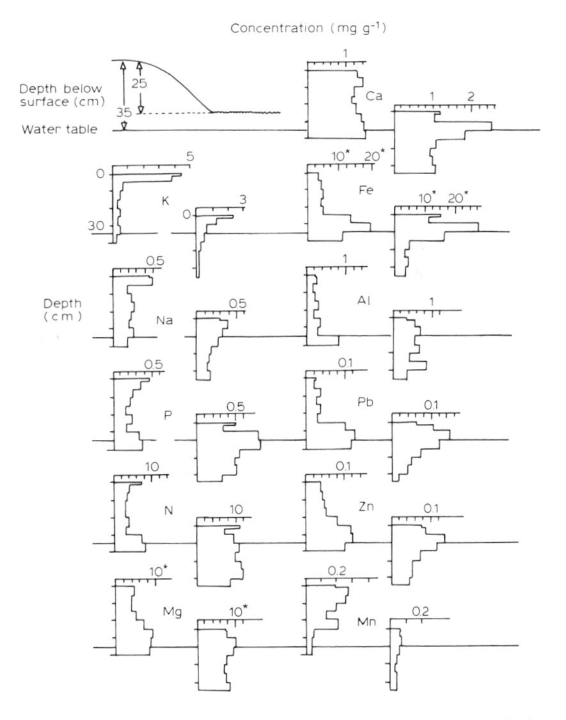

Fig. 4.17. Concentration, per unit dry mass of peat, of eleven elements in surface profiles of ombrotrophic peat from Storemosse, southern Sweden. Left columns: *Sphagnum fuscum*. Right columns: *S. magellanicum*. Roots of vascular plants were removed before analysis. Redrawn from Damman (1978).
(It seems likely, from other results in Damman's paper, that the Mg peak should be about 1 mg g^{-1}, not 10, and the Fe peak about 2 mg g^{-1}, not 20. If this is true, the scales should be reduced by a factor of ten.)

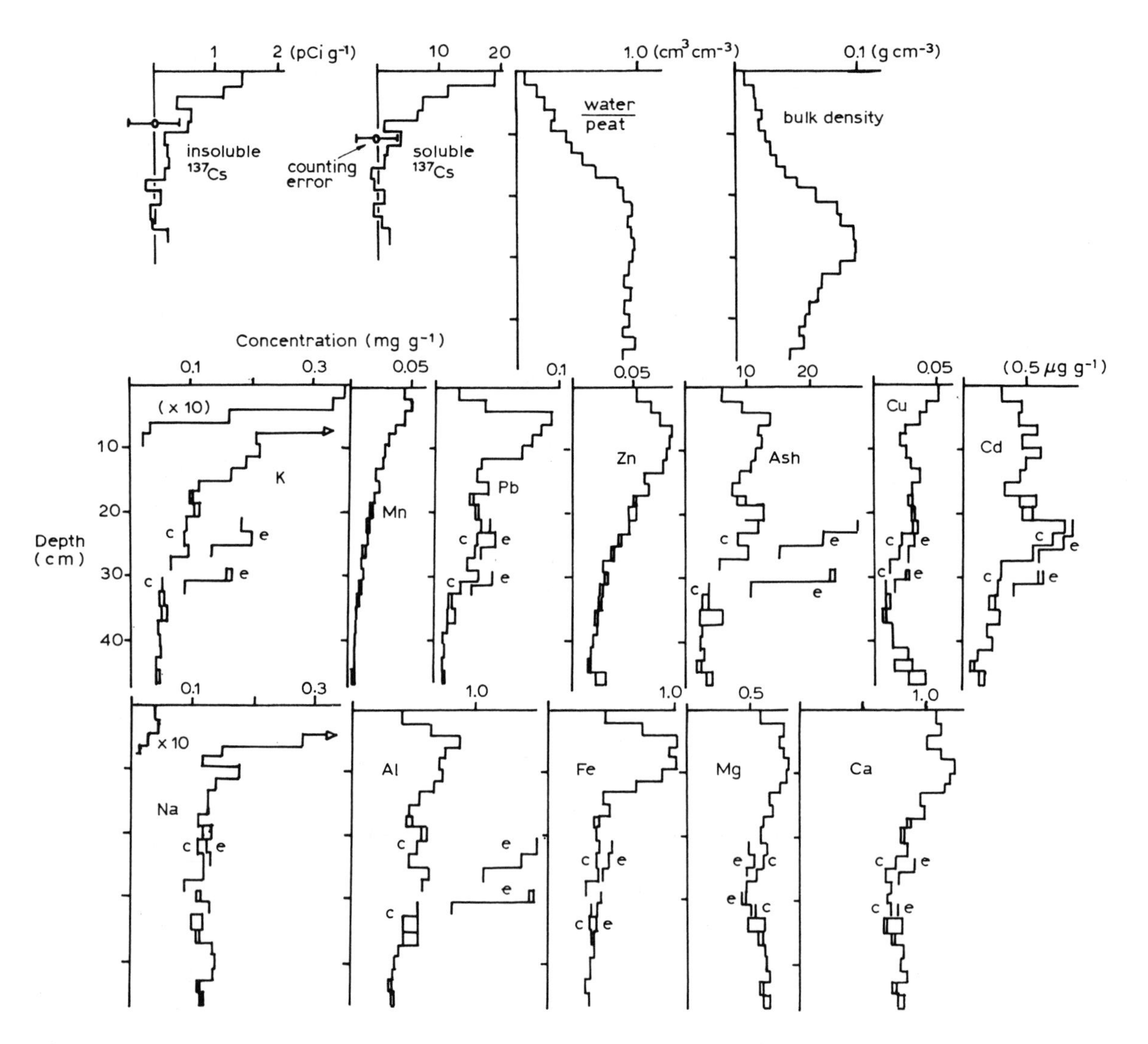

Fig. 4.18. Physical variables and concentration, per unit dry mass of peat, of twelve elements and of ash (%) in a surface profile of ombrotrophic peat from Os, Southern Sweden. Chemical units are mg g^{-1} except Cd and ^{137}Cs. The peat was predominantly *Sphagnum magellanicum* in the top 20 cm, but below that the proportion of *Eriophorum vaginatum* (*e*) and of the *Cuspidata* section of *Sphagnum* (*c*) increased and was analyzed separately. Roots were removed before analysis. Results of Clymo (unpublished).

There are as yet too few reports to be sure which of these patterns are common and which are particular to one or a few sites, but it does seem clear that there is a variety of patterns, and that some of these are correlated with the type of peat and with the oxidation-reduction state of the peat. The order of depth of the peaks of manganese, iron, zinc and copper is the reverse order to that of their sulphide solubility.

In several cases it has been assumed that bryophytes (Rühling and Tyler, 1971; Pakarinen and Tolonen, 1977) or peat profiles (Lee and Tallis, 1973; Livett et al., 1979) or both may record the history of deposition in the same way that ice accumulations do, but without the technical difficulties of that case (Murozumi et al., 1969) and with fewer geographic constraints. In one instance (Clymo, 1978, using information available in 1972) the cumulative amounts of magnesium, aluminium and titanium were used as indicators of age in the surface layers. It is important therefore to consider the extent and mechanism of movements of solutes in the surface layers, to a depth of perhaps 10 cm below the mean water table.

In a few cases it is possible to make crude estimates of the proportion of deposited matter which is retained. For a *Sphagnum magellanicum* profile in the Pennines (England) about 30 to 40% of magnesium and about 50% of caesium (^{137}Cs) were retained (Clymo, 1978). The calculations were possible because the profile for magnesium was relatively free of peaks, and most of the ^{137}Cs input is recent and fairly accurately known. For most other elements however the peaks imply relocation, so the best that can be done is to use the mean concentration in the lower "central" anaerobic section (Table 4.10) on the assumption that these approach "steady state" values and that in this section there is little relocation. These very crude calculations (Table 4.12) suggest that retention may be related to valence, ranging from values of 1 to 3% for sodium and potassium to 40% for aluminium. The valence state of manganese and iron is uncertain, but one might expect them to be in the reduced form.

The retention values in Table 4.12 are about 0.1 to 0.2 of those in Table 4.8. The calculations used in Table 4.8 were for the top 10 cm, and assumed about 1 cm of vertical growth per year, whilst those for Table 4.12 assumed that at greater depth only 1 mm of that is left.

The incomplete retention and the variety of profiles indicate that at least some elements are relocated. Equally clearly the processes of relocation differ in importance. In ombrotrophic peat-forming systems the following may be considered:

(1) Biological cycling of various kinds, particularly the movement from roots to shoots, and return either internally to rhizomes in winter, or externally by leaching and leaf-fall. These processes are considered in detail by Dickinson (Ch. 5), and are of obvious importance for elements such as potassium and caesium.

(2) Diffusion, which is probably of less importance because the available time is relatively short (perhaps 10 to 50 years) and the pathway is dissected and tortuous.

(3) Movement by mass flow, similar to the elution of a column of exchange resin or microporous gel. That such elution (or leaching) does occur seems nearly certain, and the relation of retention to valence is consistent with elution of cations from an exchange phase. If the surface layers have high hydraulic conductivity, as is often the case, then water probably flows down to the water table rapidly, and then laterally. The details are obviously complex; the profiles of Fig. 4.18 are not simply explicable by this mechanism, but have biological cycling and probably valence change as well as recent changes in deposition rate (of lead, for example) superimposed. One might expect Cs^+ to behave in a manner similar to K^+, and the pattern for ^{137}Cs (Figs. 4.10, 4.18), most of which

TABLE 4.12

Retention of deposited substances by the central section of peat deposits

	Na	K	Mg	Ca	Mn	Fe	Al	Precipitation (mm)
Concentration range								
in peat (Table 4.13)	4	1	30	20	0.05	2	10	
(μmol g^{-1})	50	5	80	50	0.2	20	30	
Deposition flux								
(μmol cm^{-2} yr^{-1})								
Chilton, 1973	5.2	0.8	0.9	1.9	0.02	0.48	1.2	574
Wraymires, 1973	18.3	0.9	2.3	1.4	0.01	0.23	0.6	1482
Approximate[1]								
retention (%)	1	3	30	20	10	30	40	

The peat concentrations are taken from Table 4.10. The deposition rates, selected from Cawse (1974), are from Chilton (central southern England, low precipitation, 90 km from sea, rural, predominantly arable) and Wraymires (English Lake District, high precipitation, 25 km from sea, rural, little arable, some industrial influence).

[1]Assuming peat growth of 1 mm yr^{-1}, density 0.1 g cm^{-3}.

fell in precipitation during 1963, does perhaps support this view. Much of the acid-soluble ^{137}Cs is still "suspended" biologically at the surface, with a small peak, probably representing the 1963 input, at about 10 cm in this profile. The acid-insoluble ^{137}Cs, perhaps glassy particles, peaks at about 8 cm depth, with a larger surface peak. The difference of 2 cm may represent elution of the soluble fraction in the ten years since it was deposited. The process cannot be as simple as this, however; Clymo (1978) reported that about 50% of the deposited ^{137}Cs was not recovered from peat profiles, and elution experiments in the field and laboratory (Mackay, unpublished, 1976) indicated that a substantial part of caesium and other cations added to *Sphagnum* is eluted. The same experiments indicate however that a fraction remains where it was applied. There may be one fraction which can be readily eluted and another which moves much more slowly.

The surface peak of acid-insoluble ^{137}Cs is rather surprising, but Mackay (unpublished, 1976) has shown that upward movement of particles and solutes occurs around *Sphagnum* during dry conditions. The top of the plants is the limit to movement in that direction, so particles may accumulate there. In the downward direction there is no limit, so particles moving in that direction may be dispersed and, eventually, lost laterally.

One may conclude that there is still much to be learned about the chemistry of peat.

ACCUMULATION OF PEAT

Peat accumulation results when the rate of addition of dry matter exceeds the rate of decay.

Matter is added mostly at the surface, and may be measured as productivity. For mosses — mainly *Sphagnum* in bogs or hypnoid mosses in fens — all the new matter is added at the surface, but vascular plants produce rhizomes and roots, and these add matter perhaps as much as 2 m below the surface in exceptional cases (Stavset, 1973). Most of the matter is added in the top 10 cm, however.

Decay on the other hand may occur not only at the surface but through the whole peat depth. The rate of decay seems to be highest in the surface layers, and to be much lower in waterlogged peat. In addition, some species — particularly some species of *Sphagnum* — seem to decay at a lower rate than others. These differences have several consequences: the depth of the water table is very important; a very low decay rate operating over 5 m of peat may be as important as a hundred-fold greater rate restricted to a 5 cm layer; selective decay of species may result in a peat dominated by one or a few only of the species which grew on the surface which formed it.

It is convenient to consider productivity first, then decay, and lastly models of the peat accumulation process.

Productivity

The productivity of vascular plants in peat-forming systems is reviewed by Bradbury and Grace (Ch. 8). Particularly valuable accounts of primary productivity of blanket-bog are given by Smith and Forrest (1978) and by Grace and Marks (1978), so only the contribution of bryophytes is considered here. In these acid habitats vascular plant productivity ranges from about 100 to about 800 g m^{-2} yr^{-1}. In fens, the contribution from non-vascular green plants — mostly hypnoid mosses — is probably small compared to that from vascular plants. In bogs, the contribution from non-vascular plants may be important and is usually dominated by *Sphagnum*. Exceptions include, for example, *Polytrichum commune* on the edge of some valley bogs in southern England. Individual plants up to 150 cm long may be recovered with annual increment of 12 cm and productivity up to 1400 g m^{-2} yr^{-1}. This productivity was maintained for a few years only during a transient rise in the water table, and over an area of about 1 ha only. Other cases where mosses are important are the "moss-banks" of the maritime Antacrtic. On Signy Island, Baker (1972) estimated the net productivity of the moss *Chorisodontium aciphyllum* at 440 g m^{-2} yr^{-1}, and Fenton (1980) estimated the net productivity of *C. aciphyllum* and *Polytrichum alpestre* to be from 160 to 350 g m^{-2} yr^{-1}. For systems with pH 4, a growing season of about five months, and a mean maximum temperature of 0°C, these seem at first to be surprisingly high values. They are probably explicable by adaptation to low temperature, but it may be that the mean monthly temperature of 0°C obscures the fact that the temperature is above 0°C for most of the growing season. If the mosses are able to respond to change in light-flux or in temperature as rapidly as aquatic subantarctic mosses can (Priddle, 1980a, b), then the productivity is less surprising.

These cases of relatively high productivity are all exceptional, and of interest only in pointing to what may be possible. On a world scale they are trivial; the only bryophytes whose production is important on this scale are a few species of *Sphagnum*. On blanket-bog in Britain their productivity in pure stands ranges from about 150 g m^{-2} yr^{-1} on hummocks to about 500 g m^{-2} yr^{-1} in lawns, but up to 800 g m^{-2} yr^{-1} in pools (Clymo, 1970). The methods of measurement have rather large inherent inaccuracy — perhaps $\pm 50\%$. As a component of shrub-dominated blanket bog *Sphagnum* contributes about 100 to 300 g m^{-2} yr^{-1} (summary in Smith and Forrest, 1978). The species of *Sphagnum* commonly grow in different micro-habitats, particularly in relation to the water table. Thus *S. capillifolium* (*S. rubellum*, *S. acutifolium*) commonly grows some way above the water table, *S. papillosum* and *S. magellanicum* grow just above the water table, and species of the section *Cuspidata* usually grow in the water or close to it. It is of some interest that species seem to grow in that habitat in which their productivity is greater than that of competing species of *Sphagnum* rather than that in which they grow best. For example, *S. capillifolium* grew best in pools, but better than other species tested on hummocks (Clymo and Reddaway, 1971; see also Fig. 4.19). Similar productivity — 100 to 600 g m^{-2} yr^{-1} — was reported by Sonesson (1973), who also found a similar difference between the drier hummocks and wetter depressions near Abisko in Arctic Sweden. Greater productivity by *Sphagnum* may occur in special habitats such as slow-flowing streams draining peatlands (where *S. subsecundum* shoots may grow more than 120 cm in a season) or, transiently, in the furrows made when preparing the ground for afforestation. Up to 1500 g m^{-2} yr^{-1} may be produced. Excepting these special cases, it is of interest to find the same range of values appearing for such diverse climates. If the value of 200 g m^{-2} yr^{-1} applied to all acid peatlands — say two-thirds of the 150×10^6 ha reported by Tibbetts (1968) — then the world annual production by *Sphagnum* would be about 200×10^6 t. Woodwell et al. (1978) estimated total productivity on the continents — that is, excluding that in seas and oceans — to be about 72×10^9 t yr^{-1}. *Sphagnum* may thus contribute about 0.3% of the total. It would be interesting to know how many other families contribute as much (though the family is a rather arbitrary division: the Sphagnales are a clearly isolated group). If the efficiency is calculated per unit of nitrogen or phosphorus employed, rather than per unit area, the importance of *Sphagnum* is even greater (Clymo, 1970).

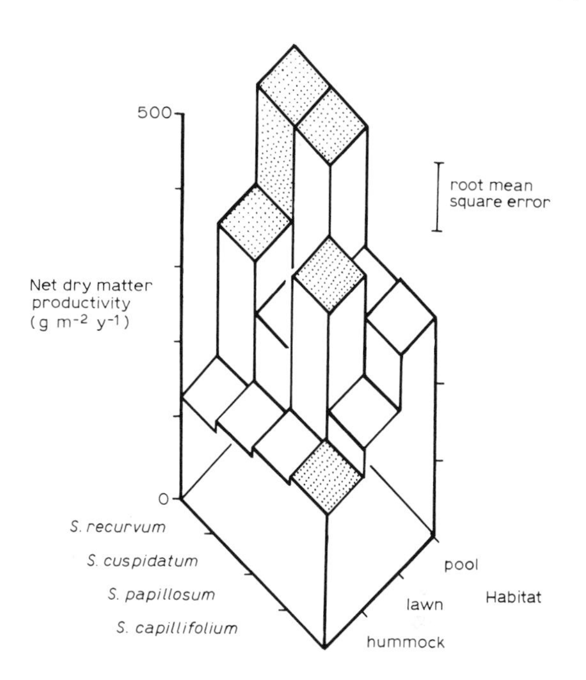

Fig. 4.19. Net productivity of four species of *Sphagnum* in three habitats on blanket bog at Moor House (England). The plants were put experimentally in all three habitats. Dotted area shows the usual natural habitat for each species. Redrawn from Clymo and Reddaway (1971).

Plant productivity, by vascular plants and bryophytes, in fens and bogs is clearly not exceptional for temperate regions. Yet peat accumulates, so the rate of decay must be exceptionally low. To this I now turn.

Decay rate

Some indication of the overall decay rate may be had by calculating how much of what was produced has remained as peat. Assuming productivity of 300 g m^{-2} yr^{-1}, no decay, and peat bulk density of 0.1 g cm^{-3}, then after 5000 years there would be 15 m depth of peat. Such a depth is uncommon — 2 to 5 m is perhaps more usual — so one may expect to find that half or more of the plant material decays. But it seems likely that there may be a rather delicate relationship between production and decay, so that a small change in either may have a big effect on how much remains in peat. The rate of decay is usually measured by one or more of five methods: by direct measurement of mass loss; by direct measurement of gas evolution; by indirect estimate from change in properties of a standard substance; by indirect estimate as a parameter of a model; or by indirect estimate as the result of simulation. These methods are considered in turn.

Direct measurement of mass loss

The commonest method is the "litter bag". The mesh size may be large (*c.* 1 cm), allowing animals ready access; medium (commonly *c.* 1 mm), restricting access by large soil animals and preventing the removal of whole chunks; or very small (*c.* 20 μm), preventing access by all but the smallest animals such as nematodes and protozoa. The nature of pretreatment of the plant material may have an important effect on results. In particular there may be quite a large fraction of the original plant mass which can be removed by simple leaching. For example, Coulson and Butterfield (1978) found that, if freshly gathered leaves were dried in vacuum at 50°C and put into water for seven days, then the loss in mass was 25% from *Rubus chamaemorus*, 14% from *Calluna vulgaris*, 5% from *Eriophorum vaginatum*, and 0% from *Sphagnum recurvum* (see Fig. 4.23). Another important technical point is the method of drying. *Sphagnum papillosum* dried at 105°C decayed less than half as fast as air-dried material (Clymo, 1965). Even air-drying may change the ease with which the plant is attacked.

A major problem with large-mesh litter bags is the possibility of loss of large fragments of litter. This danger becomes greater as time passes and plant structure breaks down. The amount remaining may therefore become increasingly biased on the low side. This effect is likely to be extreme in cases where individual leaves are tethered by a thread but not enclosed. Frankland (1966) used this technique with *Pteridium aquilinum* petioles on peat and found the rate of loss appeared to increase between the fourth and seventh years. The opposite effect — gain from material moving in — is occasionally found (e.g. Clymo, 1965).

A particularly elegant method can be applied in the special case of Antarctic moss peats (Baker, 1972; Fenton, 1978, 1980) in which the annual growth pattern of the moss is preserved for 200 years or more. Assuming that there has been no change in the average mass of a newly produced segment it is possible to calculate the loss after a known time. The plant material has been completely undisturbed. Fenton (1980) gives decay rate for ages as great as 170 years, whereas the best litter bag results (for example, Heal et al., 1978) rarely extend to more than five years. In all cases, however, difficulties develop by the time that less than 20% of the original mass remains.

Direct measurement of gas evolution

In this method it is the gaseous products of decay which are measured. There are many reports of the rate of evolution of carbon dioxide from mineral soils, forest litter, and so on, but few of the flux of carbon dioxide from peat. There are even fewer measurements of the flux of other carbon-containing gases such as methane. The fact that decaying organic matter may evolve methane has been known for 180 years (Dalton, 1802), and with the advent of gas chromatography it has become possible to show that small amounts of other paraffins (ethane, propane) may be evolved as well. Bog peat evolves easily measurable amounts of carbon dioxide and methane (Clymo and Reddaway, 1971; Svensson, 1974). The three main problems with this method are that enclosures used to collect the gas change the conditions (particularly temperature), the flux of gas varies during the day–night cycle, and the flux (especially of methane) is subject to erratic surges, probably associated with the escape of bubbles.

Indirect estimates on standard materials

The best-known method depends on the reduction in tensile strength of unbleached calico strips (Heal et al., 1974). The method can be calibrated, though not very precisely, against other measures of decay. It allows comparisons between sites to be made easily, but the almost pure cellulose is a "foreign" substrate and the results cannot be generalized.

Loss in mass from sheets of standard cellulose has been used in the same way (Rosswall, 1974).

Indirect estimates from models of peat growth

Suppose that the rate of addition of dry matter, p, is constant and that the rate of loss is a constant proportion, α, of the accumulated mass of peat, x. Then:

$$\frac{dx}{dt} = p - \alpha x \tag{7}$$

Here it is assumed that the time scale for the annual march of p and α is so short, compared to the time of interest, that the annual fluctuations can be treated as high-frequency components which have negligible effect on the low frequency trends, and p and α can be assumed constant. The solution to eq. 7 has been known since Newton's day and is:

$$x = \frac{p}{\alpha}(1 - e^{-\alpha t}) \tag{8}$$

This equation, illustrated in Fig. 4.26, has important consequences which are considered later. It can be shown (Clymo, *in lit.*) that where V is the peat mass measured from the surface *downwards* (i.e. backwards in time) and T is the age of the peat at a given depth, *relative to that at the surface*, then:

$$V = \frac{p}{\alpha}(1 - e^{-\alpha T}) \tag{9}$$

The variable V, the peat mass to a given depth below the surface, may be called the "depth in mass units". If the bulk density of dry peat were the same at all depths then V would be directly proportional to linear depth. In fact, of course, it rarely is so. Where mass of peat accumulated is being considered then V is a more useful variable than the linear depth.

The relationship (eq. 9) is formally the same as the relationship between the mass of peat accumulated and time elapsed (eq. 8). In effect one redefines zero time anywhere one wishes and looks backward. Given three depths (in mass units) and the corresponding ages, then the decay rate, α, is defined by:

$$\frac{V_2 - V_1}{V_3 - V_2} = \frac{1 - e^{-\alpha(T_2 - T_1)}}{1 - e^{-\alpha(T_3 - T_2)}} \tag{10}$$

This can be solved (by iteration) and, as a bonus, p may be obtained from eq. 9.

Published examples of dated profiles almost all relate to linear depth (measured in units of length) rather than depth in mass units. In such cases the calculation can be made for length in place of mass, but only if the bulk density profile is assumed constant — that is, there has been no consolidation. Jones and Gore (1978) and Clymo (1978) have given several profiles of bulk density, from which it is clear that there are variations within a profile and between profiles, so that the assumption of a uniform value could lead to non-trivial errors. To illustrate the principle of the method a very detailed profile from Draved Mose, Denmark (Aaby and Tauber, 1974) is shown in Fig. 4.20. Here eq. 9 has been fitted, by minimization of weighted squares of deviations, to all the points (not just to three) assuming no consolidation. The simple model gives a surprisingly complete description, with the decay parameter α estimated as 2×10^{-4} yr^{-1}.

A more complex model, including consolidation effects (Clymo, 1978), when applied to blanket-bog peat at Moor House gave most estimates between 10^{-3} and 10^{-5} yr^{-1}.

Indirect estimates from simulations

Simulations may be used to calculate the overall rate of decay from a knowledge of the rate and functional relationships of component processes. For example, Bunnell and Tait (1974) and Flanagan and Bunnell (1976) included the effects of temperature, moisture content and nature of plant substrate on microbial respiration rate. Given the march of temperature and moisture content, the model gives decay. Simulations of this kind have been used for litter of *Carex aquatilis* and *Eriophorum angustifolium*, but not for older peat.

Factors affecting the rate of decay

The rate of decay of plant material in peat-accumulating systems is affected, *inter alia*, by temperature, water supply, oxygen supply, nature of the plant material, and nature of the micro-organisms and invertebrates in the peat. These factors are not always independent, and all may be correlated with depth in the peat and with age. Their effects may interact, too. The rate of oxygen consumption by litter of *Eriophorum angustifolium*

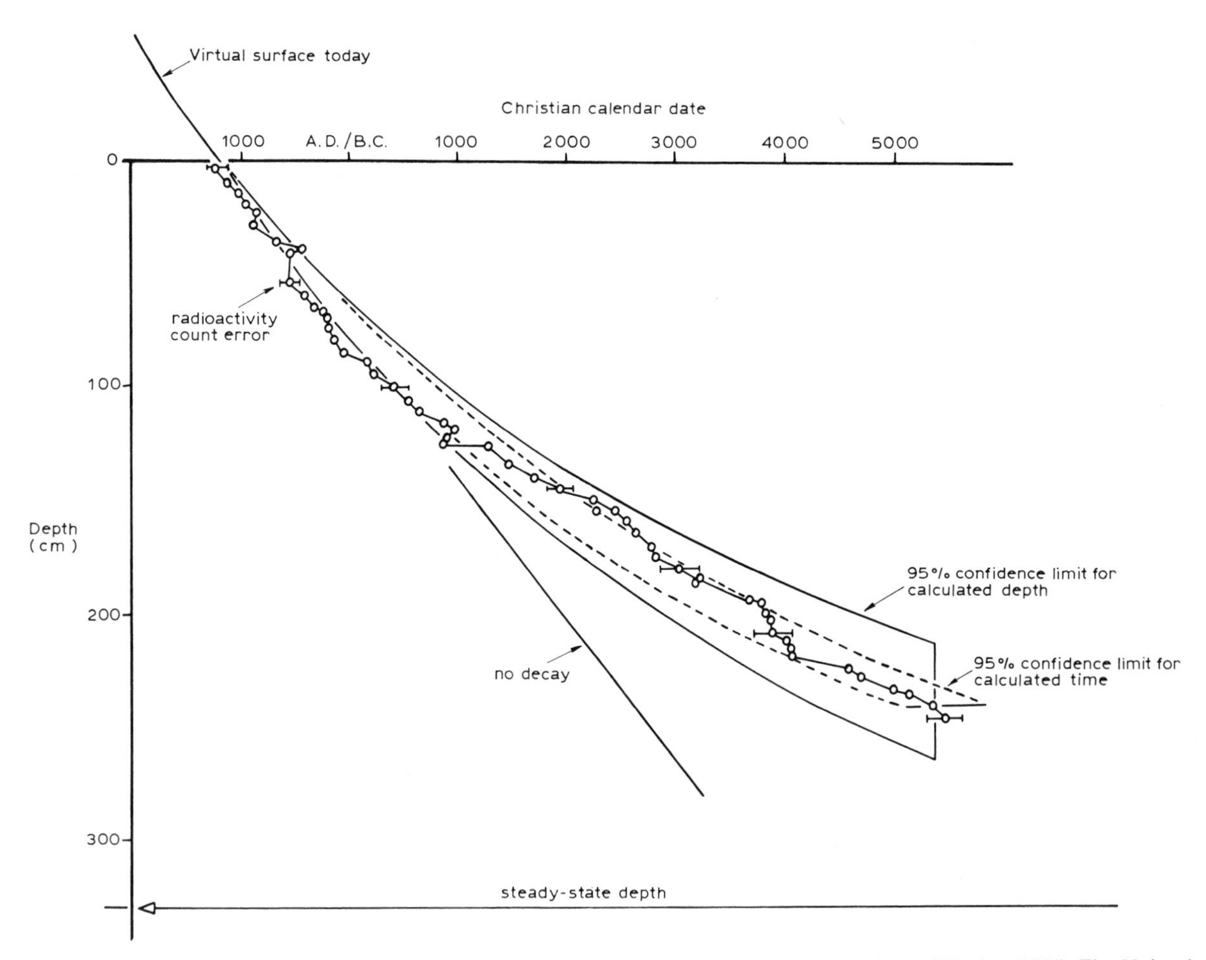

Fig. 4.20. Age (by ^{14}C counting) against depth curve for Draved Mose (Denmark). Results of Aaby and Tauber (1974). The 55 dated depths are shown by circles, seven of which have, for illustration, a horizontal bar giving the 95% confidence interval. Eq. 9 was fitted to these points (see text), assuming a uniform bulk density of 0.1 g cm^{-3}. The parameter estimates, ±standard error, are: productivity $(p) = 6.4 \times 10^{-3} \pm 2.8 \times 10^{-4}$ g cm^{-2} yr^{-1}; decay rate $(\alpha) = 1.9 \times 10^{-4} \pm 1.9 \times 10^{-5}$ yr^{-1}. The 95% confidence bands are shown assuming that error is all in the depth or all in the age. The diagonal straight line shows what would be expected for the same productivity, but no decay. The lower horizontal line at 330 cm is the steady-state depth (see text). The surface of this bog has been cut for fuel, so the whole bog may be considered haplotelmic (only the catotelm remaining; Ingram, 1978). The fitted line is extrapolated up to today, giving a "virtual surface" at about 55 cm above the actual surface. From Clymo (*in lit*).

at 20% water content is about 130 μl g^{-1} h^{-1} over the temperature range −4°C to 8°C (Flanagan and Bunnell, 1976; see also Fig. 4.21). At −4°C the rate is little changed over a range of "water" content from 20% to 320% at least. At a temperature above −2°C however, water content has a considerable effect. At 8°C and 320% water content the rate of oxygen consumption is double that at 20% water content. More usually, the relation between rate of oxygen consumption and temperature is a hollow curve. At 10% water content *Rubus chamaemorus* litter consumed oxygen at 5 to 8 μl g^{-1} h^{-1} at −5°C, increasing to 30 to 50 μl g^{-1} h^{-1} at 10°C and 240 to 360 μl g^{-1} h^{-1} at 30°C (Rosswall, 1974). *Carex aquatilis* litter gives similar results (Flanagan and Veum, 1974). These results show — not surprisingly — that temperature has an effect on the rate of oxygen consumption, and hence of decay rate, of litter. They also show that the process is significant at surprisingly low temperatures. Micro-organisms adapted to the general thermal environment seem to be the rule, so that the mean temperature is of less consequence than might be supposed. That psychrophilic micro-organisms occur in peat is known (see, for instance, Janota-Bassalik, 1963). What is not known is how fast the micro-organisms

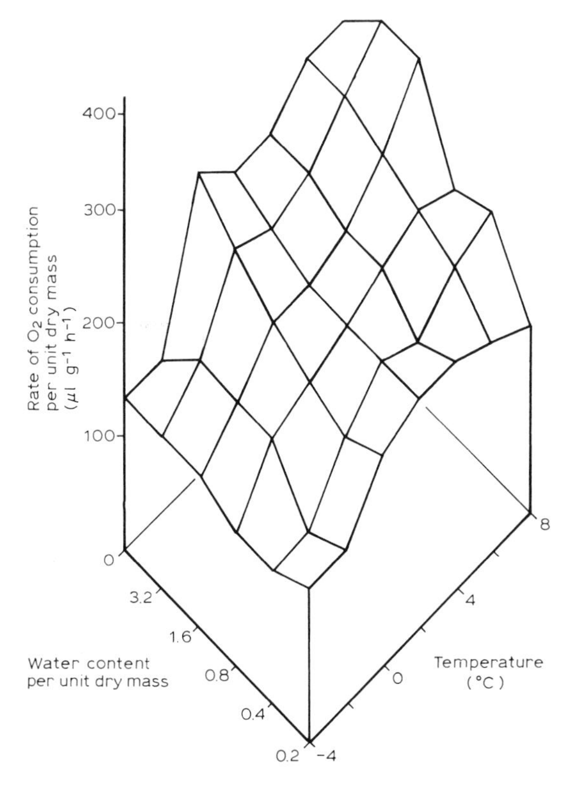

Fig. 4.21. "Respiration rate", measured in a respirometer, of dead leaves of *Eriophorum angustifolium* gathered at Barrow (Alaska, U.S.A.) in relation to water content and temperature. Redrawn from Flanagan and Bunnell (1976).

can adapt to temperature change. This may be important because peat, even water-saturated peat, is a fairly effective thermal insulator. The amplitude of temperature fluctuations is rapidly damped. The damping depth — that depth at which the amplitude of a sinusoidal oscillation has been reduced to $1/e = 0.37$ of that at the surface — is about 5 cm for the daily cycle and $5 \times \sqrt{365}$ cm ($\simeq 1$ m) for the annual cycle in peat (Monteith, 1973). At a depth three times the damping depth the amplitude is $e^{-3} = 0.05$ of that at the surface. At 15 cm depth, therefore, daily fluctuations in temperature are small, but at the surface they are twenty times larger, and the total amount of decay may be critically dependent on the speed with which decay organisms respond to change. This is particularly so if the relation between temperature and oxygen consumption rate is non-linear. It can be shown (Clymo, *in lit.*) that rapid response to a non-linear relation coupled with a smaller-amplitude thermal regime (though at the same mean temperature) should reduce the overall decay rate of *Rubus chamaemorus* at mean temperature 0°C from 0.35 yr^{-1} at the surface to 0.14 yr^{-1} at 100 cm depth, and at mean temperature 5°C from 0.62 yr^{-1} to 0.42 yr^{-1}. Of course *Rubus* litter is unlikely to survive in identifiable form to such a depth, but the same sort of effect should hold for material which does. It is this effect which makes the use of day-degree sums dubious.

Measurements of oxygen consumption by litter are necessarily made in artificial conditions; respirometers with shaken flasks are usual. The same material is often taken through a series of increases in temperature with no check of the extent of adaptation or damage. It is therefore of interest to find that measurements of decay of *Rubus chamaemorus* litter (in bags) in the field range from about 0.2 yr^{-1} (Rosswall, 1974) through 0.4 yr^{-1} (Heal et al., 1978) to 0.7 yr^{-1} (Coulson and Butterfield, 1978). These rates agree fairly well with those calculated from the respirometer measurements. Direct comparisons in the field at sites with different mean temperature have been made for standard cellulose material (Rosswall, 1974) and for uniform *Sphagnum* material of several species (Clymo, 1965). There are several reports of the rate of decay of plants of the same species, for example *Eriophorum vaginatum*, but these are not easy to compare because the plants were in all cases local to the site at which the measurements were made. In general the rate of decay of cellulose increases with temperature, or with time at a given temperature, and the decay of *Sphagnum* was about 0.05 yr^{-1} faster at a site with mean temperature about 4°C higher, but with the same amplitude.

The effects of oxygen supply on rate of decay have rarely been measured directly, but are deduced from the change in decay rate with depth. Waterlogged peat rapidly becomes deoxygenated because the rate of diffusion of oxygen in water is too slow (about 0.0001 of that in air) to replace that used in aerobic metabolism of micro-organisms. There are micro-anoxic patches in unwaterlogged peat too. The consequent change in the activities of micro-organisms is considered elsewhere in this volume. Here only the measured effects on decay are considered.

That decay rates can be low may be deduced in general from the fact that macroscopic and microscopic plant structures (not only pollen) are still identifiable after several thousand years as peat. This argues for a decay rate of less than 0.0001 yr^{-1}. In exceptional cases it may be even less. For example, between 2400 and 2600 years ago at Tregaron Bog (Wales) *Sphagnum magellanicum* was abundant (Turner, 1964). Even today there remains the equivalent of a productivity during that time of 3 t ha^{-1} yr^{-1}. This would be a respectable rate before decay had begun. Other instances are known too. There are records of about 700 discoveries of human bodies buried in peat (Dieck, quoted by Glob, 1969). In many cases the body was buried and fixed down in an old peat cutting. Usually the skin is well-preserved, and bones and other tissues may be too. Clearly, the processes of decay have been slower than in the aerobic conditions of the more conventional places for burial.

Direct measurements of plant loss by litter bag techniques have given results which differ in detail but show that there is a general reduction in decay rate of material in waterlogged peat when compared with the *same* material placed above the water table. *Sphagnum papillosum* decay rate fell abruptly from about 0.12 yr^{-1} at the surface of Thursley Bog to 0.07 yr^{-1} between 5 and 18 cm depth, and to 0.01 to 0.02 yr^{-1} below 18 cm depth — the mean depth of the water table (Clymo, 1965; see also Fig. 4.22). A similar surface decline and then abrupt drop of decay rate of *Calluna* stems at 18 cm depth in Moor House (England) blanket peat was found by Heal et al. (1978), and losses in tensile strength of cellulose strips at nine sites at Moor House, and in a wet site at Abisko (northern Sweden) all show the same abrupt drop (Heal et al., 1974), as do strips in an Antarctic *Rostkovia magellanica* peat bog (Anonymous, 1978). The only conspicuous exception is the decay of roots of *Eriophorum vaginatum*, which increases steadily from about 0.07 yr^{-1} at the surface of blanket bog at Moor House to about 0.19 yr^{-1} at 25 cm deep (Heal et al., 1978). The decay rate below the water table appears in most cases to be only 0.01 to 0.04 yr^{-1}. This is close to, or beyond, the accuracy of the litter bag techniques. An alternative is to measure the evolution of methane. At Moor House, on a wet blanket bog 3 m deep, the rate of evolution

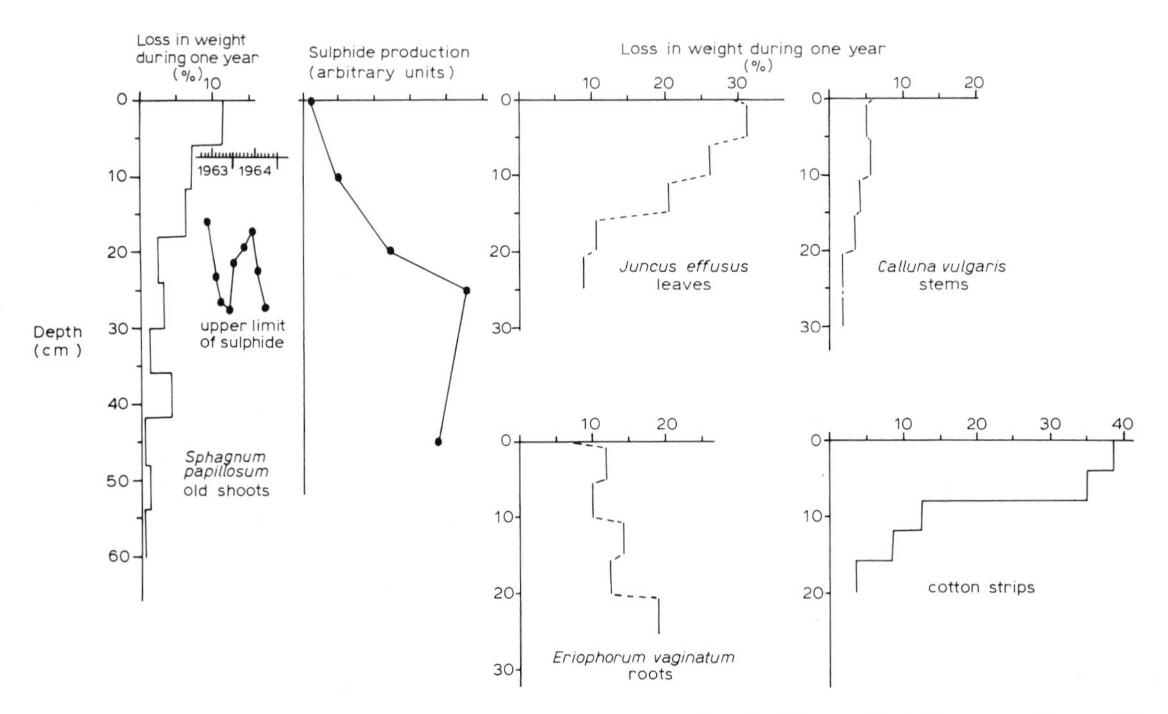

Fig. 4.22. Decay during the first year after death of a variety of materials at various depths in peat. For *Sphagnum papillosum* at Thursley, southern England, the annual cycle in the upper level of sulphide (detected by blackening of a silver wire) is shown, and the sulphide production, on a single occasion, of water samples added to an enrichment culture. The other four materials were measured at Moor House, northern England. Redrawn from Clymo (1965) and Heal et al. (1978).

of methane gas, calculated as carbon, was 1 to 7 g m^{-2} yr^{-1}, with least from hummocks and most from pools (Clymo and Reddaway, 1971). At Abisko, Svensson (1974) found similar rates of methane evolution. At Moor House the efflux of carbon dioxide, calculated as carbon, was 31 to 54 g m^{-2} yr^{-1}. If the methane is being produced at all depths, is the only product of decay, and reaches the surface by diffusion, then the decay rate is about 10^{-5} yr^{-1} (Clymo, 1978). This is certainly too simple a view: the erratic appearance of high methane concentration in traps suggests the evolution of gas bubbles; some of the methane may be oxidized and appear as carbon dioxide (Svensson, 1974); and there is no direct evidence that methane production is not localized near the current water table, with "fossil" methane in solution or even as trapped bubbles lower down. Svensson (1974) showed that a lot of methane is present in peat at a depth of 50 cm, but it is not clear how much of this is "fossil". There is much scope for experiment.

The other method of estimating decay at depth is from ^{14}C dates and a model of the peat accumulation process. The results from Draved Mose (Denmark) already quoted give a decay rate of about 2×10^{-4} yr^{-1}, which is (considering the inaccuracy of all existing methods) not inconsistent with 10^{-5} yr^{-1} from methane evolution at a different site.

These rates of decay in the anaerobic zone seem low, but they are not negligible, because they apply over perhaps 20 to 50 times the depth and 40 to 100 times the mass that the aerobic rates do, and they apply for a very long time. Just how important such rates are is shown later.

The effects of nutrition and of the spectrum of micro-organisms and invertebrates are closely connected with the nature, history and location in the peat of the plant material. The organic material itself is the main "nutrient" for the organisms of decay. They attack the different chemical constituents selectively, and may themselves produce new organic substances. Molecules which were attackable by aerobic organisms may be unavailable to anaerobic ones, and *vice versa*.

It has already been shown that the decay rate is lower in waterlogged peat whatever the plant material. In the surface layers, however, the rates of decay of different species and parts of species differ widely. One of the earliest demonstrations of this was made by Waksman and Tenney (1928). They measured carbon dioxide efflux at 26°C from air-dried and ground plant materials mixed with two or five times their own weight of water and inoculated with a "suspension of a good field soil". The rate of flux of carbon dioxide from maize stalks was about three times that from rye straw, pine needles, or *Sphagnum*. When small amounts of ammonium salts and phosphates were added, the efflux of carbon dioxide from all the materials except *Sphagnum* increased two- to five-fold, but the efflux from *Sphagnum* was unchanged unless the *Sphagnum* had been treated with acid, causing partial hydrolysis. These experiments were, of course, in very unnatural conditions and excluded the natural fauna, but they did show that "ordinary" soil micro-organisms could attack *Sphagnum* at much the same rate as they do several other plant materials, and that, uniquely among the materials tested, the rate of attack on *Sphagnum* could not be increased by adding inorganic nutrients alone. The rate of attack in this case seemed to be limited by organic nutrients alone. This may explain the observation (Strygin, 1968) that the rate of heating in piles of milled peat can be increased by the addition of hay or straw as sources of easily attacked organic nutrients, and that *Sphagnum papillosum* capitula, which presumably contain a relatively high concentration of easily attacked organic compounds as well as higher concentrations of nitrogen and phosphorus, decay about three times as rapidly as do newly dead stems and branches, whilst the addition of inorganic nitrogen and phosphorus has no significant effect on decay rate (Clymo, 1965). In a particularly elegant way, Coulson and Butterfield (1978) have confirmed these findings and have shown the importance of invertebrate grazers in decay in the aerobic zone. This is of particular interest because, whereas about 60% of the plant material on mineral soils adjacent to blanket bog at Moor House is assimilated, so that it is animal faeces which are the main food for the rest of the decomposers, only 5% of plant production on blanket bog is assimilated (Coulson and Whittaker, 1978). In particular, *Sphagnum* is scarcely ingested at all, except perhaps accidentally. Three species (the craneflies *Tipula subnodicornis* and *Molophilus ater*, and the enchytraeid worm *Cognettia sphagnetorum*) form about three-quarters of the biomass and account for about three-quarters of the assimilated plants. Coulson and Butterfield used litter bags and made four comparisons: of large-mesh (*c.* 1 mm) with fine-mesh (*c.* 20 μm) litter bags, thus allowing or preventing access by most of the soil fauna; of decay on mineral soil with that on peat; of a variety of species naturally occurring on the mineral soil or on the bog; and of plant material as it occurred naturally with that grown on plots fertilized with either nitrogen or phosphorus, and thus having larger concentrations of these elements. The first three of these comparisons are shown in Fig. 4.23. The species are arranged in rank of nitrogen concentration in the material used. Animals accounted for an important proportion of the weight loss of *Phleum pratense* on mineral soil, *Rubus chamaemorus* on mineral soil and on peat, *Calluna vulgaris* on both soils, *Festuca ovina* on peat, and *Sphagnum recurvum* on mineral soil. Animals had no effect on weight loss of *Eriophorum vaginatum* on either soil, or of *Sphagnum recurvum* on peat. Comparing soil types (with free access by animals) *Phleum* lost notably more weight on the mineral soil than on peat, whilst *Rubus* and to some extent *Calluna* showed a reverse effect. *Rubus* leaves recovered from peat after three months contained more en-

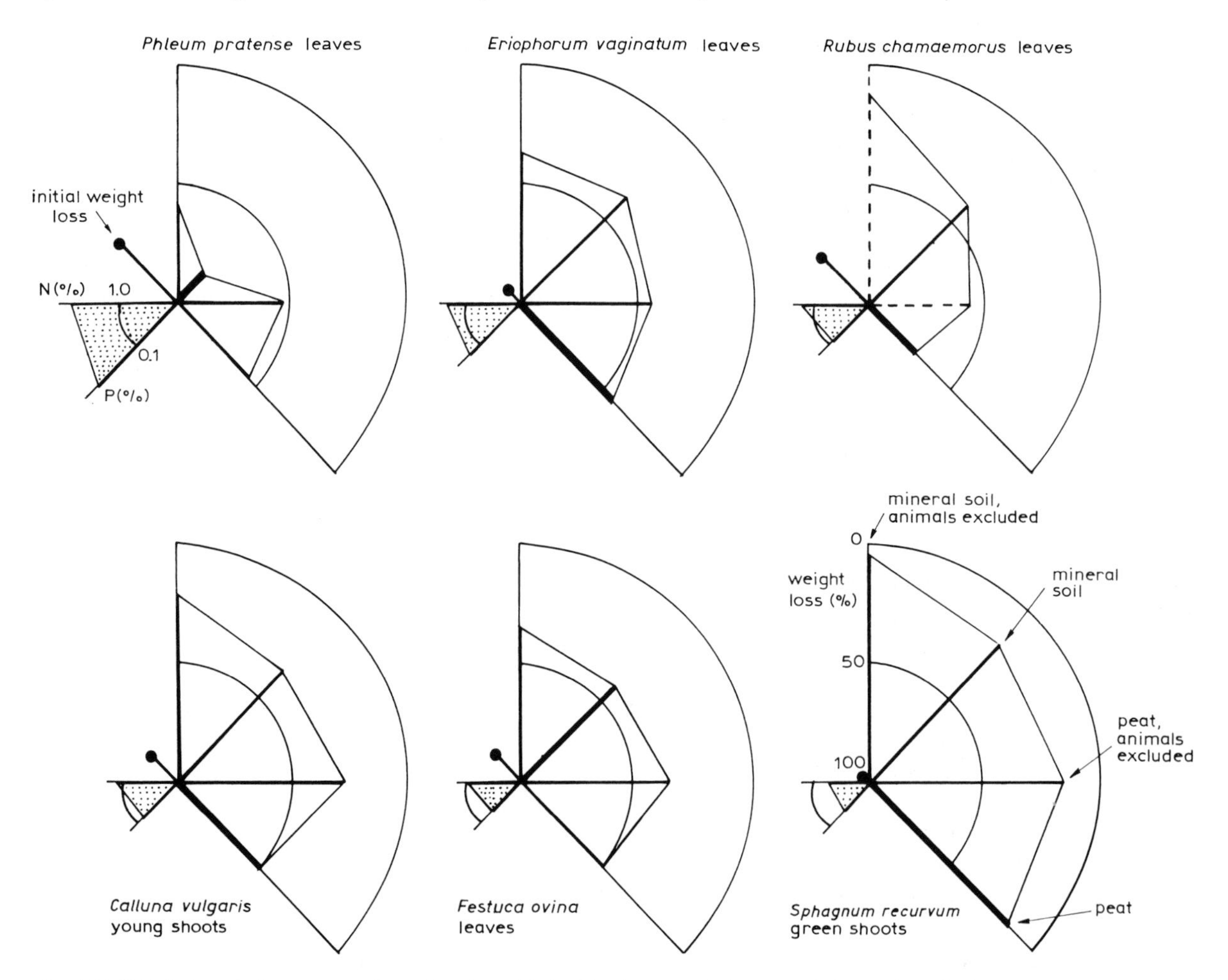

Fig. 4.23. Weight loss during one year from litter bags filled with plant material of one of six species. The bag mesh size was 20 μm (excludes most animals) or 0.8 mm, and bags were buried in mineral soil adjacent to blanket bog at Moor House (England) or in peat there. Arcs define 0% and 50% loss. The results for the four treatments are joined by straight lines, forming a polygon. *Rubus chamaemorus* in small-mesh bags was measured for 3 months only. The results have been calculated in the same proportion as that found for 3-month and 12-month measurements on large-mesh bags. The northwesterly "stick-and-knob" shows the leaching loss during the first seven days. All material was leached before the experiment began. To west and southwest are the initial concentrations of nitrogen and phosphorus. Thick bars show the natural habitat. Drawn from results of Coulson and Butterfield (1978).

chytraeids, dipteran larvae, mites and Collembola (by a factor of from 2 to 5) than did similar leaves on mineral soil.

Taken together, those results show a surprising range of rather specific effects. Invertebrates on peat are clearly important agents in the decay of *Rubus chamaemorus* leaves, but of little importance in the decay of *Eriophorum* leaves or *Sphagnum* shoots.

There are potentially important differences between species in the absolute rate of decay, measured over the first year, too (Table 4.13). The rank order of decay rate is much the same where it has been measured more than once, but the absolute values differ. There are uncontrolled differences between different sets of measurements, even for the same experimenters (e.g. Fig. 4.24). Water content can have a big effect. For example, Heal et al. (1978) showed that the relationship between loss in weight of *Rubus chamaemorus* leaves and water content is approximately hyperbolic with half-maximum rate ($0.3\ yr^{-1}$) at about 200% water content. In very wet conditions the rate was $0.6\ yr^{-1}$. Overall, however, it is clear that some parts of the species, such as leaves of *Rubus chamaemorus*, *Narthecium ossifragum* and *Calluna vulgaris*, decay rather rapidly, whilst stems and roots of *Calluna* and *Eriophorum*, and *Sphagnum*, decay more slowly. There is a tendency for materials with relatively high nitrogen and phosphorus concentration to lose weight more rapidly, whether they are available to animals or not (Figs. 4.23, 4.24). Coulson and Butterfield experimented with blanket bog which had been fertilized with ammonium nitrate or superphosphate (a mixture of calcium sulphate and calcium phosphate). The concentration of nitrogen and phosphorus in plants growing on these plots increased (by from 10 to 30%), and so did the abundance of many of the invertebrates. The decay rate of plants grown on unfertilized bog, but transferred to fertilized bog, was no different from similar material on fertilized bog, but plants grown on nitrogen-fertilized bog, and having a higher concentration of nitrogen, decayed more rapidly than did normal plants when put on unfertilized bog (Fig. 4.24). The decay rate of plants on phosphorus-fertilized bog was not more rapid, however. Indeed the loss of *S. recurvum* was only 10%, compared to 18% for unenriched plants.

The differences in the nitrogen experiment were rather small, but because the experiment was large and carefully designed the differences were significant at the 1 in 20 level. Earlier reports that nutrition was without effect probably mean that those experiments were not sufficiently sensitive to

TABLE 4.13

Rate of weight loss (yr^{-1}) from dead plants in large-mesh bags during one year on blanket bog at Moor House, England.

	Source		
	Heal et al. (1978)	Coulson and Butterfield (1978)	Clymo (1965)
Narthecium ossifragum leaves	0.45		
Rubus chamaemorus leaves	0.36	0.78[2]	
Eriophorum angustifolium leaves	0.24		
Eriophorum vaginatum leaves	0.22	0.46[2]	
Calluna vulgaris shoots	0.20	0.55[2]	
Sphagnum capillifolium shoots[1]			0.16
S. recurvum shoots		0.14[2]	
S. cuspidatum shoots[1]			0.15
S. papillosum shoots[1]			0.08
C. vulgaris above-ground stems	0.08		
C. vulgaris roots	0.05		
E. vaginatum roots	0.01		

[1] Excludes capitulum. [2] Pre-leached to remove readily soluble materials. Values recalculated to include these lost materials, to allow comparison with the first column.

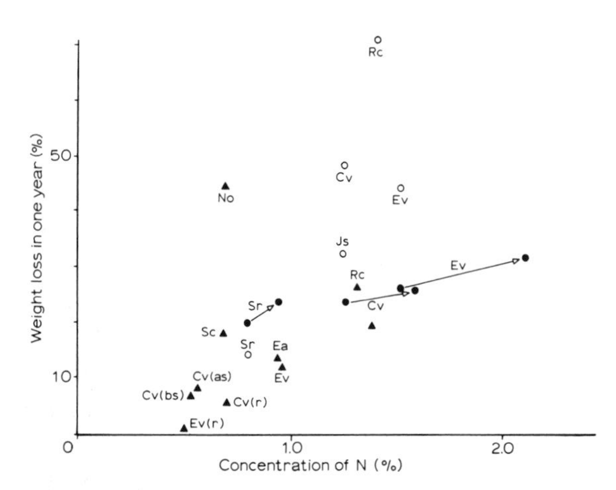

Fig. 4.24. Decay rate in relation to initial concentration of nitrogen in plant material in litter bags at Moor House (England):

Cv = *Calluna vulgaris* shoots
Cv(as) = *C. vulgaris* above-ground stems
Cv(bs) = *C. vulgaris* below-ground stems
Cv(r) = *C. vulgaris* roots
Ea = *Eriophorum angustifolium* leaves
Ev = *E. vaginatum* leaves
Ev(r) = *E. vaginatum* roots
Js = *Juncus squarrosus* leaves
No = *Narthecium ossifragum* leaves
Rc = *Rubus chamaemorus* leaves
Sc = *Scirpus cespitosus* leaves
Sr = *Sphagnum recurvum* shoots

Results from: ▲, Heal et al. (1978); ○, Coulson and Butterfield (1978), first experiment; ●, Coulson and Butterfield (1978), nutrition experiment. Arrow-heads point to the experimentally nitrogen-enriched material

detect the differences. The change in nitrogen concentration on fertilized bog was not large either, when considered as a proportion of the range shown in Fig. 4.24. Nevertheless, the results are important because there is high intercorrelation between the concentration of different elements in plants. Complex descriptive equations, such as that used by Heal et al. (1978) using (P+Ca) and (lignin × tannin) give no insight into mechanisms. Only by *experiment* is it possible to isolate the differences and to discover causal links.

Not only do different species and macroscopic parts of species disappear at different rates (Fig. 4.23), but so do different parts of cells. For example, the abaxial and adaxial surface of the cells in leaves of *Sphagnum* break down more rapidly than do the thickening bands and the enclosed cells (Dickinson and Maggs, 1974), and the attack is sporadic: some leaves are extensively damaged whilst others close by are virtually undamaged.

The time course of decay is difficult to measure and difficult to interpret. Broken litter may be lost from litter bags. The bags themselves become covered, and thus move into a different environment. Different chemical fractions decay at different rates. Equations for the time course of decay can therefore be nothing more than descriptions, but are essential if models of the peat accumulation process are to be made.

Two assumptions have commonly been made. The first (e.g. Baker, 1972) is that the rate of decay is constant, so that a constant proportion of the *original* mass disappears during each time interval. Thus, if the decay rate is 0.2 of the original mass per year, then the material will have disappeared in five years. A much commoner assumption, inspired by the observation that some peat is thousands of years old, and ignoring the fact that such peat if put in aerobic conditions will decay quite rapidly, is that the rate of decay is a constant proportion of what is left. Jenny et al. (1949) used this assumption, and have generally been followed, though with little or no evidence in support. The two assumptions may be expressed:

$$\frac{dm}{dt} = -\beta \qquad (11)$$

$$\frac{dm}{dt} = -\alpha m \qquad (12)$$

where m is the mass of peat, t is time, β a decay parameter (dimensions $M\,T^{-1}$) and α a decay parameter (dimension T^{-1}). These equations, when solved for m, give:

$$m_t = m_0 - \beta t \qquad (13)$$

$$m_t = m_0 e^{-\alpha t} \qquad (14)$$

Eq. 14 can be put in linear form as:

$$\ln(m_t/m_0) = -\alpha t \qquad (15)$$

The same results, plotted according to eqs. 13 and 15, are shown in Fig 4.25. There is little reason to claim that one set is closer to a straight line than the

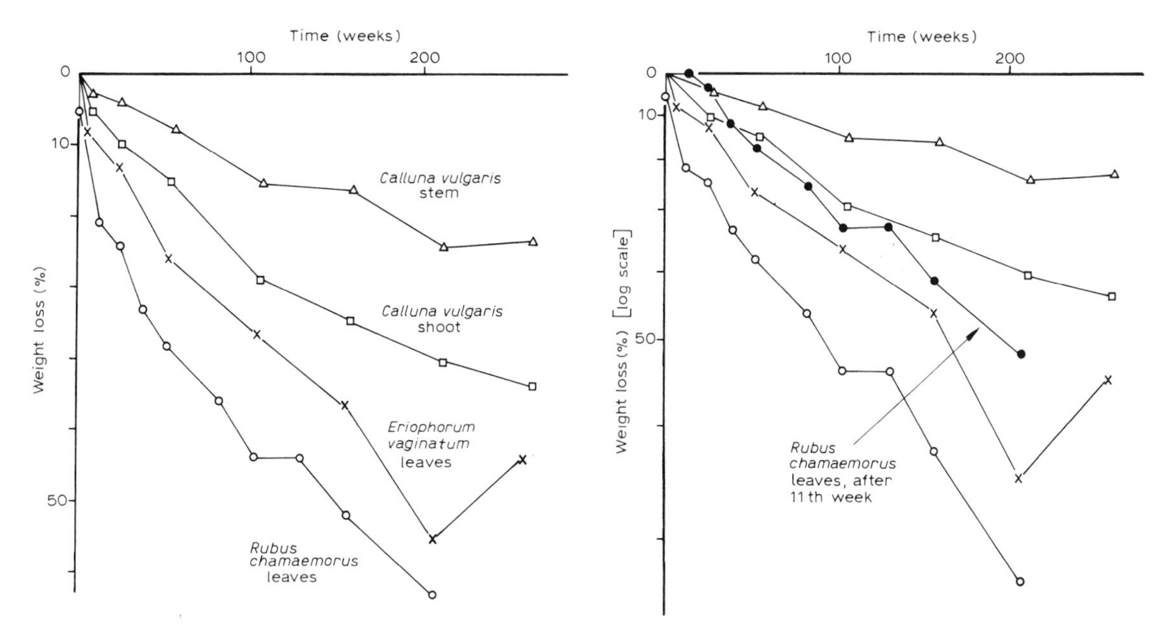

Fig. 4.25. Weight loss of various plant materials in litter bags during five years in peat at Moor House (England). On the left graph, a straight line would indicate that a constant proportion of the original weight was lost per unit time. On the right graph a straight line would indicate that the rate of loss is a constant proportion of the amount left at any time. *Rubus chamaemorus* is plotted again (filled circles) restarting at eleven weeks after the initial rapid loss in weight. Results of Heal et al. (1978), redrawn from Clymo (1978).

other, though there is some indication that there is a fraction — up to about 20% in *Rubus chamaemorus* — which decays within ten weeks. For the next 250 weeks, either assumption seems to fit equally well. Given a coefficient of variation of about 10 to 20%, which it is very difficult to reduce in experiments of this kind, it is unlikely that the evidence from litter bags will favour one assumption rather than the other until about 80% of the original mass has been lost. By that stage of course most of the structure has gone too, and litter bags become increasingly inaccurate. This is true also of Baker's (1972) method of measuring the loss from Antarctic *Chorisodontium aciphyllum* which had never been disturbed: the measurements fit either assumption fairly well, and become inaccurate beyond about 60% loss (Clymo, 1978, figure 1). Finally, one may perhaps have doubts about cases such as the last two points for *Eriophorum vaginatum* leaves in Fig. 4.25 where the *precision* is high but the accuracy is obviously much lower; errors of accuracy in long-term measurements are commonly unknown but are quite possibly much larger than errors in precision.

This uncertainty about which assumption is correct may appear to be unimportant, as it seems to matter only when the mass left is less than 30% of the original mass. That this is not true, and that the exact form of the decay curve is of vital importance, is best revealed by calculations on models, extended over hundreds or thousands of years. The existing evidence about decay is in some respects unsatisfactory and will probably have to remain so. The biggest single gap is the lack of reliable estimates of the actual rate of decay deep in the peat.

Before turning to models of peat accumulation it is worth summarizing the factors which are known to be important determinants of decay: (1) temperature; (2) water supply (at the surface of the peat-forming system); (3) oxygen concentration; (4) nature of the plant material; and (5) nature and activity of soil animals and micro-organisms. These interact with each other and with position in the profile, and they change with time.

Models of peat accumulation

The chief purposes of models of peat accumulation are: to test our understanding of the accumulation process; to reveal the consequences of assumed processes; and to estimate parameters, such as the rate of anaerobic decay over hundreds of years, which cannot be measured directly. There is much confusion about the reliability and value of model results. The apparent precision of symbols and numbers is often belied by grotesquely unlikely predictions. Some explanation is therefore necessary.

Models can be expressed in one or more (simultaneous) equations of the general form:

$$R_c = f(V_e, V_i, P) \qquad (16)$$

where R_c is the calculated result; V_e are external (driving) variables, such as temperature; V_i are internal (state) variables, such as the mass of some component; and P are parameters. These equations embody the assumptions about the *processes* and their interactions, if any. If there is evidence that a particular relationship does actually exist, and its functional form is known, then the ultimate test of understanding is likely to be more stringent because fewer arbitrary assumptions need be made; but lack of evidence has rarely inhibited the making of more or less credible assumptions. This stage of model making may start from assumptions about a process — for example, that decay rate is proportional to the amount of material — and derive equations algebraically. Alternatively it may be simply an *ad hoc* description, in symbols, of an observed geometric relationship (typically a graph) but enshrining no understanding of why the shape is as it is. Even in such cases there is room for choice: a hyperbola can be understood as some sort of saturating process, and is generally therefore a more illuminating assumption than a quadratic (or higher-degree polynomial) description which gives an equally good 'fit' to the observations. In choosing descriptions it is important to remember that in reality there are physical boundaries to the validity of any description; mass cannot be negative, rate cannot increase exponentially without limit. If these features are not included the model will, sooner or later, become unrealistic.

The fundamental strength of a model may, therefore, be tested by asking questions such as: Is there evidence that the processes occur over the range of variables to be used? Has the possibility of interaction between processes been considered? Do the descriptions arise from clear postulates, or are they *ad hoc*? Do the descriptions allow for discontinuity or non-linearity or both? When tested in this way, most of the models of peat accumulation are rather weak.

Models have been used in many ways. The simplest is to supply values for all the variables (V_e, V_i) and parameters (P) and to calculate the results R_c. This is a direct simulation. The results may be compared with measured values R_m. Such comparison may be of a crude "by eye" type, or may involve some numerical measure of badness of fit. There are difficult problems of weighting — the model may be a good fit in some places and bad in others, and this may or may not be expected. Essentially this is a direct test of the model. If agreement between R_c and R_m is poor then it may be that the *processes* are inaccurately described (the wrong shapes on graphs) or incompletely described, or it may be that the variable or parameter *values* are inaccurate (or both). The sensitivity of the model to errors in the values supplied may be examined analytically or by numerical trial. If agreement between R_c and R_m is close, then the model may be accepted as a useful working description, and as *confirmatory* evidence that the processes and values used are fairly accurate. The more complex the pattern of R_c, the greater the confirmatory value.

A more complex, indirect, way of using a model is to omit some or all the parameter values (P). Trial values are assumed and R_c calculated. Using known values of R_m, a measure of badness of fit is calculated. The values of P are then changed to arrive at a least bad fit between R_c and R_m. An example is the use already mentioned as a means for estimating the decay in anaerobic conditions over 5000 years at Draved Mose. This sort of use does test the description of processes to some extent; it might prove impossible to get an acceptable fit over the whole time span by any combination of p and α. But on the whole the test is not very strong. An equally close fit could be got by using an arbitrary quadratic (which has three parameters). This does not mean the model is "wrong" — only that the accuracy of the processes included in it cannot be well confirmed by this test.

The adjustment of parameter values is best done systematically. There are many ways of doing this, and choice of which is best is a matter for specialists. In general, there is no method of guaranteeing that *the* best combination of parameters has been found, but in practice it usually happens that much the same set of values is reached from different starting points, and this is taken to show that something near the best has been reached. Sometimes this minimization is done "by hand". A trial is made, then a new parameter value guessed and another trial made, and so on. This was the method used by Jones and Gore (1978) in estimating the decay rate that best fitted their model of the blanket bog at Moor House, England. They combined this approach with changes in the assumptions about processes too. It will be obvious that such "fitting" methods are likely to be inefficient. The parameters are usually estimated with different precision, and the estimates may be weakly or strongly correlated.

The earliest explicit model of organic matter accumulation was that of Jenny et al. (1949). They postulated an annual instantaneous injection of organic matter as litter fall from deciduous plants, and decay of a constant proportion of the accumulated organic matter during each year. This is mathematically awkward, so they assumed that both litter fall and decay were continuous, and arrived at the equivalent of eq. 8 on page 199:

$$x = \frac{p}{\alpha}(1 - e^{-\alpha t})$$

where p is net productivity ($M\ L^{-2}\ T^{-1}$) and x the mass accumulated at time t. This is shown in Fig. 4.26. If one assumes that litter addition is instantaneous but decay is continuous then one gets a graph similar to those in Fig. 4.26 but with annual negative exponential "teeth" superimposed.

Eq. 8 has interesting implications. One is that after a long time the whole system tends, asymptomically, towards a steady-state mass. More precisely, as $t \to \infty$, $x \to p/\alpha$. Suppose, for illustration, that present-day peat-forming systems are in a steady state. Forrest (1971) believed this to be true of the Moor House blanket bog, and one may deduce from curves relating peat depth to precipitation and bog diameter (Granlund, 1932; Wickman, 1951) that some Scandinavian peat-forming systems are approaching, or in, this state. If the peat is 5 m deep with bulk density 0.1 g cm^{-3} and if surface productivity is 200 g m^{-2} yr^{-1}, then α is 0.4×10^{-3} yr^{-1}. The model of eq. 8 makes no allowance for differences in rate of decay of different species or parts of species, or for position in relation to the water table, but a second implication is that a very low decay rate — 0.04% — is sufficient to change the accumulation curve from an indefinitely accumulating peat mass to one which has reached a steady state at a relatively modest depth of 5 m. A decay rate of 0.04% is very difficult to measure in the field. Litter bags are much too

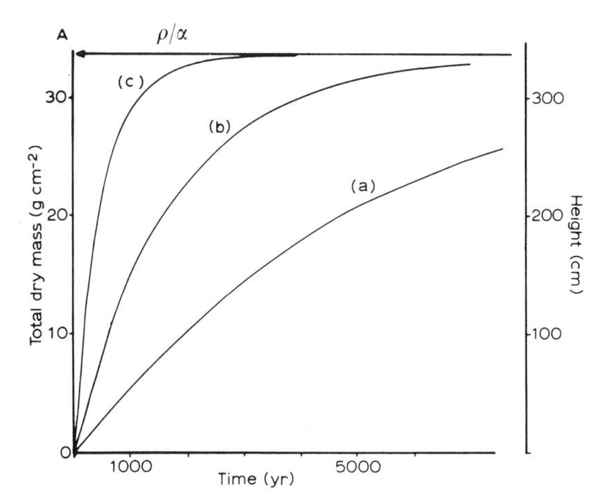

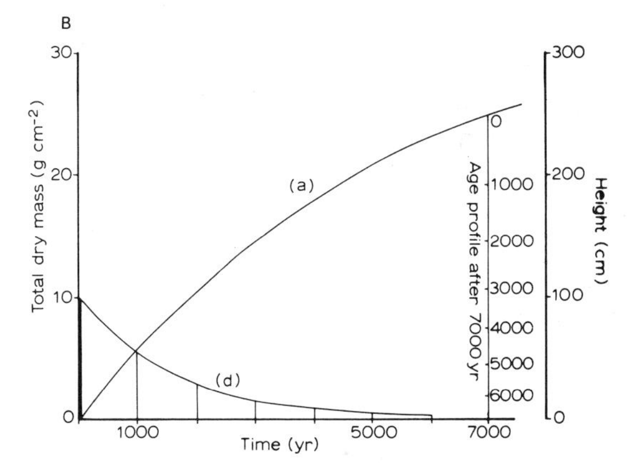

Fig. 4.26. A. Accumulation of a single-component peat according to eq. 8. The left axis shows mass, the right axis shows height assuming bulk density (ρ) to be 0.1 g cm^{-3}. The three curves (a), (b) and (c) have the same ratio of productivity (p, g cm^{-2} yr^{-1}) to decay rate (α, yr^{-1}). The steady state approached by all three is shown at p/α. The parameter values are:

	(a)	(b)	(c)
p(g cm^{-2} yr^{-1})	6.4×10^{-3}	1.9×10^{-2}	6.4×10^{-2}
α (yr^{-1})	1.9×10^{-4}	0.57×10^{-3}	1.9×10^{-3}

The values for (a) are those which best fit the Draved Mose data (Fig. 4.20).

B. Accumulation of a single-component peat [curve (a) repeated from Fig. 4.26A] and course of decay (d) of an initial mass of peat of 10 g cm^{-2} with the same decay rate as that of curve (a). The age profile of curve (a) at the arbitrary time of 7000 yr is shown.

imprecise and inaccurate. Gas evolution rates are better, but it is not easy to determine where the gas comes from. Indirect estimates, from models, may be sensitive enough but suffer from inaccuracy in their assumptions.

A third implication is now obvious: surface productivity, p, may be finite and perhaps large *and yet the rate of peat accumulation may be very small* or zero. It may even be negative, with a net loss of carbon to the atmosphere, if the rate of decay at depth has increased — perhaps because of an increase in temperature, or as a result of draining (e.g. Hutchinson, 1980) during the last century or two. The possible consequences of this were considered by Woodwell et al. (1978). The stable state may be seen formally in eq. 7 where, if $p=\alpha x$, then $dx/dt=0$, the rate of production, which is limited to the surface, is balanced by the rate of loss *summed over the whole depth* of peat.

The assumption that the peat-forming system has reached a steady state may be unimportant. It is easy to show that if Q is the proportion of the final asymptotic steady-state mass, x_∞, which has been reached by t_Q then:

$$t_Q = -\frac{\ln(1-Q)}{\alpha} \qquad (17)$$

For the example already used, the peat will have reached 90% ($Q=0.9$) of the steady state after about 5800 years. It is interesting to note that the time needed depends on the decay rate but not on the productivity.

The independence of rate of peat accumulation, dx/dt, and productivity, p, appears in another important context. It has become increasingly common to see layers of peat dated by ^{14}C age or by cross-reference of pollen zones to other ^{14}C-dated profiles. The distance between two such layers may be divided by the difference in dates, and a "peat accumulation rate" reported. There are two objections to this practice (apart from possible errors in the dating technique). First, the peat may have consolidated — a general term for elastic and plastic reduction in depth. That such consolidation may occur has been shown in experiments by, *inter alia*, Berry and Poskitt (1972) and Clymo (1978), and on a large scale in the English fens (partly as a result of drainage) by Hutchinson (1980). This problem could be avoided if peat accumulation rates were expressed on a mass basis rather than as depth. If this is to be done then a bulk density profile is needed, but those who take samples for ^{14}C measurements rarely measure bulk density too.

The second objection to such estimates of "peat accumulation rate" is that the calculation is valid only if the decay rate is zero. Suppose, for example, that ^{14}C dating establishes that a peat sample is 1000 years old, and another from 100 cm deeper is 2000 years old. Assume further that there is no consolidation. If there is no decay at all, then in another 5000 years this layer of peat will still be 100 cm thick. If the decay rate is as little as 0.04% per year, however, the peat layer will be only 13.5 cm thick (the general shape of the decay curve is shown in Fig. 4.26) and the peat accumulation rate would, wrongly, be reported as 13.5/5000 $=0.03$ mm yr^{-1}. Nor is the accumulation rate really $100/5000=0.2$ mm yr^{-1} because the youngest part of the layer is 1000 years old but the oldest is 2000 years old.

The solution to this problem is to use a minimum of three dated levels and apply eq. 10 and then eq. 9 to get first α and then p. More useful is to use the minimization technique already described to estimate p and α, as shown in Fig. 4.20. The values thus obtained ($p=64$ g m^{-2} yr^{-1}, $\alpha=2.8\times10^{-4}$ yr^{-1}) are small and refer to the zone of low decay — the anaerobic zone, or catotelm of Ingram (1978). In particular, p is the rate at which matter is added to the *anaerobic* zone — the material which has run the gauntlet of the aerobic acrotelm and reached the comparative safety of the anaerobic catotelm. Of course the material itself has not necessarily moved in space, but new plant production above it has gradually buried the original material, and the water table usually keeps at an approximately constant mean depth below the surface, so, as the water table rises, the material moves down *relative to the water table*. The rate at which material passes into the anaerobic catotelm is of much greater importance in determining the amount of peat accumulated than is the net primary productivity at the surface. For Draved Mose the rate of entry to the anaerobic zone is perhaps one-fifth to one-tenth of the net primary productivity. The value of 64 g m^{-2} yr^{-1} may be compared with 32 g m^{-2} yr^{-1} estimated by a different method for blanket bog in Ireland (Moore, 1972).

Differences in decay rate lead to changes in the

representation of different components too (components here mean chemical substances, tissues or whole plant parts). There is some experimental evidence for this in pine-needle litter (Minderman, 1968). It appears as a bend or as two lines joined by a curve on graphs similar to Fig. 4.25B. *Rubus chamaemorus* is a good example. The sort of effects to be expected are shown in Fig. 4.27. For two components with decay rates of 0.5 and 0.1 yr^{-1} and initially in equal proportions, then after ten years the proportions will be 0.2 and 0.8, and will change very little thereafter. If the decay rates are, say, ten times smaller, then it takes ten times as long to reach the same proportion. It seems that decay rate actually drops by a factor of about 100 when the catotelm is reached, so most of the selection probably happens in the first few years after death.

It is helpful, for illustration, to consider the results of a model which combines two layers and two components, giving four independent components of the mass variable m, and two independent decay parameters (α_a) in the aerobic acrotelm and two (α_c) in the anaerobic catotelm. There are two independent net primary productivity parameters, p_a and a fixed depth (in mass terms) for the acrotelm, giving seven parameters in all (Table 4.14). The output rate from the acrotelm is then the input p_c to the catotelm and is determined by the other p, α, and depth parameters. It is this p_c which can be estimated from the Draved Mose results. Starting from nothing, the acrotelm accumulates to the fixed depth, taking a characteristic time, t, determined by the p_a and α_a parameters,

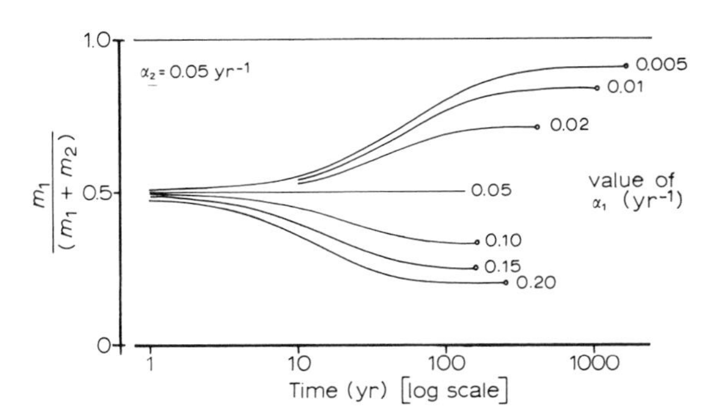

Fig. 4.27. Proportion that component 1 (mass $= m_1$) forms of the total mass ($m_1 + m_2$) in a two-component mixture in relation to time and decay rate (α). If both decay rates are multiplied by a factor, then the same graphs may be used, but with the time axis divided by the same factor. For example, if decay rates are multiplied by 10 then the time axis values must be divided by 10.

to do so. Thereafter the acrotelm is in a steady state and feeds matter at a constant rate to the catotelm. For matter added at the surface it takes the characteristic time for the catotelm to grow up and claim it, and at the end of that time whatever survives is in the catotelm. Some exploratory results of this model are shown in Table 4.14. The first case, which acts as a reference for all the others, gives the components equal productivity, but there is a four-fold difference in decay rates and a fifty-fold difference in decay rate between acrotelm and catotelm. The net productivity (1000 g m^{-2}) is between likely values for bog and fen. The acrotelm depth, assuming the approximately linear trend in bulk density from 0.01 g cm^{-3} at the surface to 0.1 g cm^{-3} lower down, corresponds to a depth of about 16 cm. It takes about sixteen years before the catotelm can claim newly produced organic matter. Only 24% survives so long, and of *this* only about 8% is of the faster decaying component 1, which originally formed 50% of the total. The final steady state corresponds to about 220 cm depth of peat; component 1 forms only 2% of all this, and much less than 1% in the lowest layers.

The second case gives both components the same decay rate in anaerobic conditions. This has little effect on the final steady-state depth (and the proportion of component 2 does not change in the catotelm).

In the third case, increasing the anaerobic decay rate of both components five-fold reduces the catotelm steady state in proportion to give a total depth of about 50 cm. It is very clear that the difference between a decay rate of 0.4% and 2% in the catotelm has a marked effect on the steady-state depth. It is the *proportional* change, not the absolute change, which matters.

The fourth case shows the effect of a four-fold decrease in the aerobic decay rate for the faster component. This results in a shorter time (ten years) for *both* components in the dangerous acrotelm. The steady-state depth approaches 4 m.

The same sort of effect is seen in the fifth case, in which the productivity of component 2 is doubled. The acrotelm is traversed in only seven years, so the input to the catotelm is great and the steady-state depth is over 7 m.

In the sixth case, the acrotelm has been deepened from about 16 cm to about 20 cm. This small difference results in a marked increase (from 16 to

TABLE 4.14

Parameter values and calculated results for a two-layer, two-component peat accumulating system (~ indicates the same value as in Case 1; an asterisk shows that the value is that of a parameter which is specified; other values are calculated results; bold type indicates the parameter value(s) in which each other case differs from Case 1)

Case	Component (Variable[1])	Aerobic zone (acrotelm) P_a (g cm^{-2} yr^{-1})	α_a (yr^{-1})	$m_a(t_c)$ (g cm^{-2})	$m_a(\infty)$ (g cm^{-2})†	t_c (yr)	Anaerobic zone (catotelm) p_c (g cm^{-2} yr^{-1})	α_c (yr^{-1})	$m_c(\infty)$ (g cm^{-2})†	Total Q_{2500} (%)	$m_a(t_c)+m_c(\infty)$ (g cm^{-2})†
1	1	*0.05	*0.20	0.24	0.25		0.002	*0.004	0.47		0.71
	2	*0.05	*0.05	0.56	1.00		0.022	*0.001	22.0		22.6
	Total	0.10		*0.80	1.25	16.4	0.024		22.5	92	23.3
2	1	*~	*~	~	~		~	*0.001	1.88		2.12
	2	*~	*~	~	~		~	*~	22.0		22.6
	Total	~		*~	~	~	~		23.9	92	24.7
3	1	*~	*~	~	~		~	*0.020	0.09		0.33
	2	*~	*~	~	~		~	*0.005	4.41		4.97
	Total	~		*~	~	~	~		4.5	100	5.3
4	1	*~	*0.05	0.40	1.00		0.030	*~	7.50		7.90
	2	*~	*~	0.40	~		0.030	*~	30.0		30.0
	Total	~		*~	2.00	10.2	0.060		37.5	84	37.9
5	1	*~	*~	0.19	~		0.012	*~	2.93		3.12
	2	*0.10	*~	0.61	2.00		0.070	*~	69.6		70.2
	Total	0.15		*~	2.25	7.3	0.082		72.5	92	73.3
6	1	*~	*~	0.25	~		<0.001	*~	0.05		0.30
	2	*~	*~	0.75	~		0.012	*~	12.5		13.2
	Total	~		*1.00	~	27.8	0.012		12.5	92	13.5

[1]Symbols: p_a = net productivity (must be specified); α_a = aerobic decay rate (must be specified); $m_a(t_c)$ = depth of aerobic zone in mass units (total must be specified); t_c = characteristic time that matter stays in the aerobic zone before being claimed by the rising anaerobic zone; $m_a(\infty)$ = steady-state mass which would have been reached by the aerobic zone alone; p_c = rate at which matter passes from the aerobic to the anaerobic zone; α_c = anaerobic decay rate (must be specified); $m_c(\infty)$ = steady-state mass in the anaerobic zone; Q_{2500} = proportion (as %) of the final steady state reached after 2500 years.

†An approximate value for depth (cm) may be obtained by multiplying the value for mass in g cm^{-2} by 10. This assumes that the bulk density is 0.1 g cm^{-3}. For the aerobic layer a factor of 20 would be more accurate because bulk density increases from 0.01 to about 0.10 g cm^{-3}.

28 years) in the time needed to traverse the acrotelm. In consequence, nearly all component 1 disappears. The steady-state depth is about 140 cm, but component 1 forms less than 0.4% of the whole mass. A 4 cm change in acrotelm depth has had a remarkable effect on the relative survival of the two components.

These few examples are sufficient to show that simple assumptions about the growth and decay processes can produce a great variety of effects.

Various modifications and extensions of this simple model have been used. The first (Gore and Olsen, 1967) was to increase the number of compartments (equivalent to zones) to include transfers of "net primary production" to "live plant" and thence to "dead plant" and "peat", with provision for transfer from live plant direct to peat (presumably as root and rhizome). The model was programmed for an analogue computer — at that time this was a sensible choice — but it is unclear from Gore and Olsen's account precisely what processes were supposed to be operating or where. The parameters of the model were adjusted by trial and error to fit the time course of matter in the different categories to that measured on a blanket bog recovering from cutting.

Subsequently, Jones and Gore (1978) devised another model which allowed several components to be put in at the surface at constant rate, but in which the rate of decay was a linear function of depth, decay decreasing at greater depth. This implies, of course, a potentially negative decay rate, and it is not clear what Jones and Gore did to prevent this happening. They showed that, if less dense leaves decay more slowly than denser wood does, then peat bulk density might increase for this reason alone. This effect has not been considered by other workers, and might be of some importance if the peat contained much wood. The simulated age–depth profiles did not agree closely with the observed ones down to 150 cm and 2500 years, and were particularly sensitive to changes in the depth–decay regression slope. Jones and Gore then tried *ad hoc* adjustment of parameters and of assumptions about the way in which decay rate changed with depth, but without much improvement in fit to the field results.

A different elaboration is to try to account for the effects of consolidation. At least two different effects much be considered. First is an "elastic" compression: a load, rapidly applied to and removed from peat which still retains macroscopic structure, produces elastic deformation. Over small ranges the stress–strain relation is linear (Hooke's Law) but over the range of interest, with compression over 20% or more, a negative exponential is a better description. The force involved is produced by the weight of the accumulated peat and of water above the water table. More important is long-term creep or plastic flow. Hanrahan (1954), Berry and Poskitt (1972) and Clymo (1978) have all shown experimentally that the amount of consolidation from this cause is proportional to the logarithm of time. These two processes can be combined with a two-zone (aerobic and anaerobic) single component model (Clymo, 1978). This model, which has seven parameters (p, α_a, α_c, rate of growth in length of plants, compression parameter, creep parameter, and depth of the acrotelm, which is equivalent to the transition from aerobic to anaerobic conditions) produces calculated profiles of bulk density and age. Initial tests of this model using measured parameter values were much less ambitious than those tried by Jones and Gore, and showed agreement within the 95% confidence bands for age and bulk density for about half the top 30 cm, corresponding to fifty years. It was obvious, however, that agreement varied systematically (Fig. 4.28). The model was particularly sensitive to the depth of the acrotelm, just as the one illustrated in Table 4.16 was, and just as Jones and Gore's model was to change in the regression slope. All these effects come from the same cause: the length of time spent in the zone of high decay rate is crucial to survival into the comparative safety of the anaerobic peat.

The consolidation model was also used to estimate the parameters for a fixed aerobic/anaerobic transition depth. The remaining six parameters were chosen at random on a logarithmic scale extending from 10^{-6} to 10^{6} times the measured value, and then adjusted to minimize a weighted sum of squares of deviations of observed bulk density and age from calculated bulk density and age. This process was repeated 28 times using different randomly chosen sets of parameter values to start the minimization. The frequency distribution of parameter values obtained after minimization is shown in Fig. 4.29. (The frequency distribution of *starting* values was, of course, much broader, even for anaerobic decay which is shown on a logarithmic scale.) In general the estimated biological parameters (except for anaerobic decay, which was ill-determined) agreed fairly well with the measured values, but the two physical ones did not agree.

Part of the explanation of this lack of agreement may be that the model takes no account of a process which certainly happens — the catastrophic collapse of mosses, with attendant increase in bulk density, when structure has been so weakened that it can no longer support the overlying weight of shoots and capillary water. Fenton (1978, 1980) worked with the very interesting and particularly favourable Antarctic "moss-banks". These contain no vascular plants and a negligible number of invertebrate grazers; they can be monospecific; the mosses retain their structure at all depths; and one species shows annual growth increments. Below 20

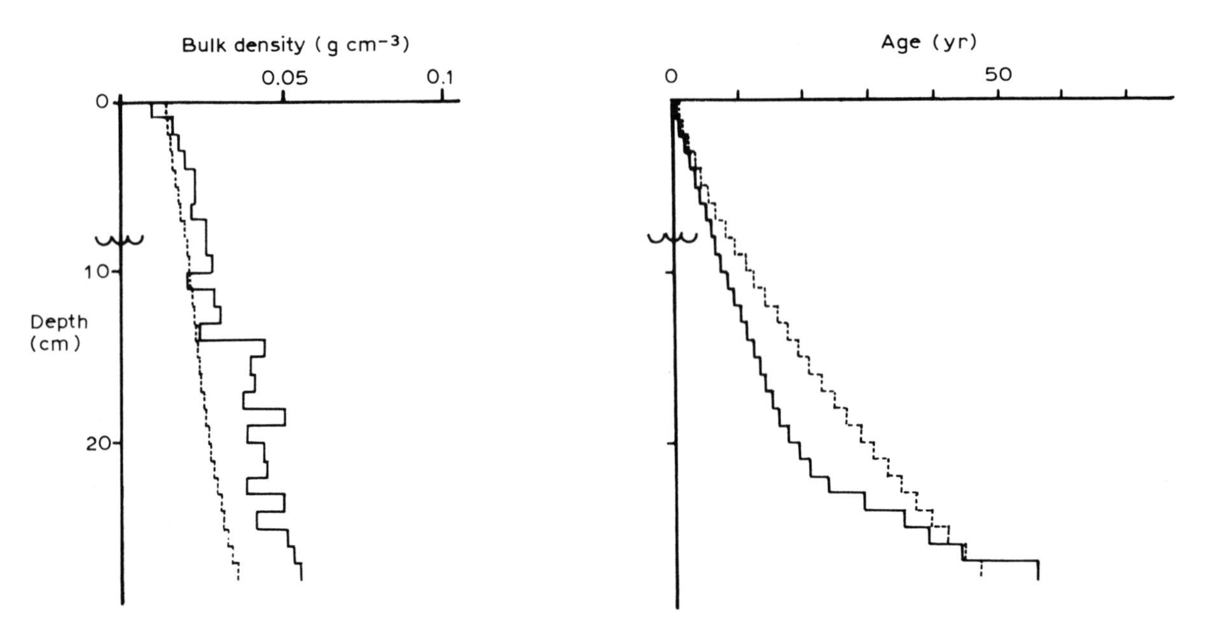

Fig. 4.28. Measured (full line) and calculated (dashed line) profiles of bulk density and age of the surface layer of peat at Moor House (England). The model included compression and creep. The transition from aerobic to anaerobic decay rate was at the water table at 8 cm depth. Measured values of the seven model parameters were used in the calculation. Redrawn from Clymo (1978).

to 30 cm the peat is permanently frozen, and, uniquely, the peat is nowhere anaerobic. Fenton assumed that the decay rate in the permafrost zone was zero and this zone might therefore be considered as the catotelm. The permafrost is at a depth below the surface which is approximately constant. It rises as the surface does, just as the anaerobic zone does in temperate peats. The special feature of the model which Fenton devised is that it allows for the observed fact that the shoots of mosses are parallel and upright at first, but at about 10 cm below the surface the combined effects of loss of strength and increasing weight of material above cause the shoots to bend. The bending may start as a catastrophic event, perhaps during the early autumn when there is a surface load of snow and ice, but during which the peat between the surface and the permafrost has not yet frozen.

The importance of this effect on bulk density can be quantified (Fig. 4.30). If the original depth of a layer is h_0, the depth after collapse is h, the shoot angle after collapse is θ, and the mean distance between shoots, normal to the shoot direction, is u_0 originally and u after collapse, then:

$$c = \frac{h}{h_0} = \frac{u}{u_0} = \sin\,\theta \qquad (18)$$

where c is the reciprocal of consolidation. Because the growth increments were preserved, all three measurements (thickness, intershoot distance, and angle) could be made. In other places it is unlikely that all three would be measurable, but in *Sphagnum fuscum* peat at Abisko (northern Sweden) h and h_0 were measurable (Clymo, unpublished). If this consolidation were all that were happening one would expect bulk density to be constant until collapse, and then to increase in proportion to consolidation. In fact it does not do this (Fig. 4.31) but decreases down to 10 cm depth and then increases. Fenton attributed the difference to decay, and pointed out that growth, collapse and decay might be combined in a model the parameters of which may be measured and the results checked. The measurements available are profiles of collapse, decay (assuming constant productivity), bulk density and age, and the annual growth in length of plants. Which is considered dependent is a matter of choice. The agreement between calculated and observed bulk density, for example, is quite close. Peat is reaching the permafrost zone (catotelm) at a rate of 90 to 160 g m^{-2} yr^{-1} — a rate similar to that at which it reaches the anaerobic zone in temperate bogs: 64 g m^{-2} yr^{-1} at Draved Mose, 32 at Glenamoy, Ireland (Moore, 1972), and 48 to 180 at

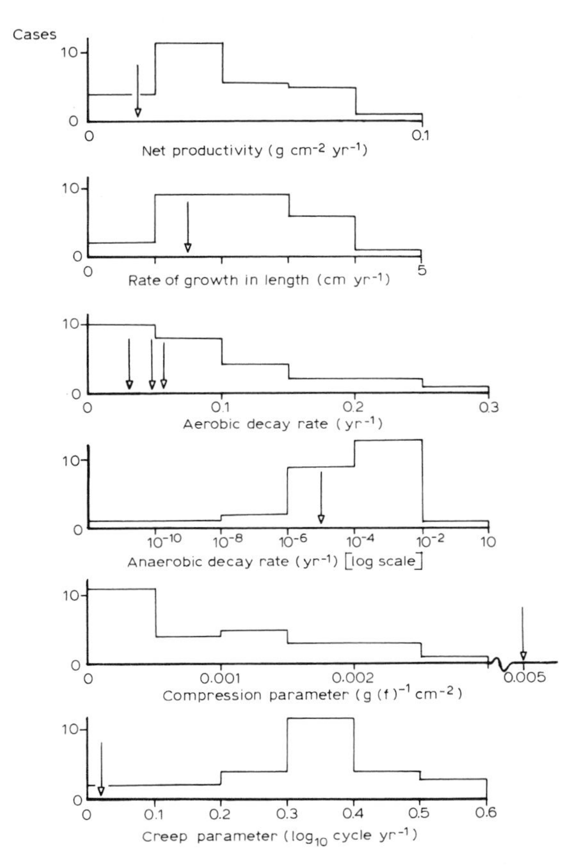

Fig. 4.29. Distribution of six parameter estimates minimizing a function of observed and calculated profiles of bulk density and of time in Fig. 4.28. The seventh parameter, water-table depth, was fixed at the measured depth of 8 cm. The distribution results from starting the minimization at different trial values. Vertical arrows show the measured values. None of the parameters is tightly constrained. The distribution of estimates of the top four (biological) parameters does centre at or near the measured value. For the bottom two (physical) parameters agreement is very poor. Redrawn from Clymo (1978).

Moor House (Clymo, 1978).

A major problem with models of complex systems — and models of peat growth are no exception — is that of how to assess the accuracy and reliable range of the model. In the simplest case considered here — that of a single component decay fitted to the Draved Mose age–depth profile — the estimated *values* of rate of transfer to the anaerobic zone and of decay rate are not incredible, and the general *shape* of the curve fits that postulated. But these tests are not stringent; the profile could be explained in part or whole by a steady increase of bulk density with depth (so that the value at 250 cm depth was about double that at the present-day

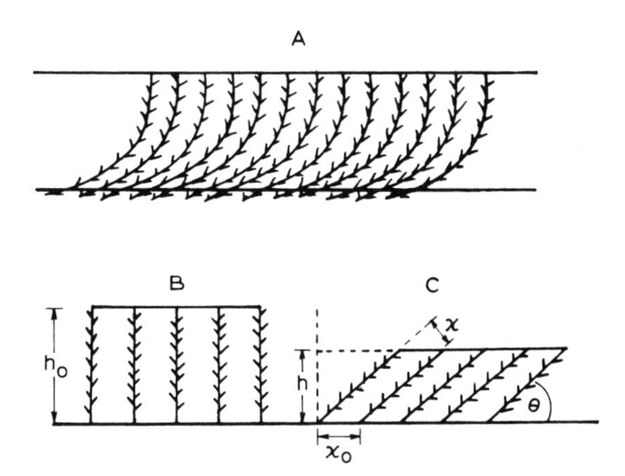

Fig. 4.30. A. Bending of moss shoots, with consequent increase in bulk density. B. Original upright shoots. C. After bending. See text for explanation and use of symbols. Redrawn from Fenton (1980).

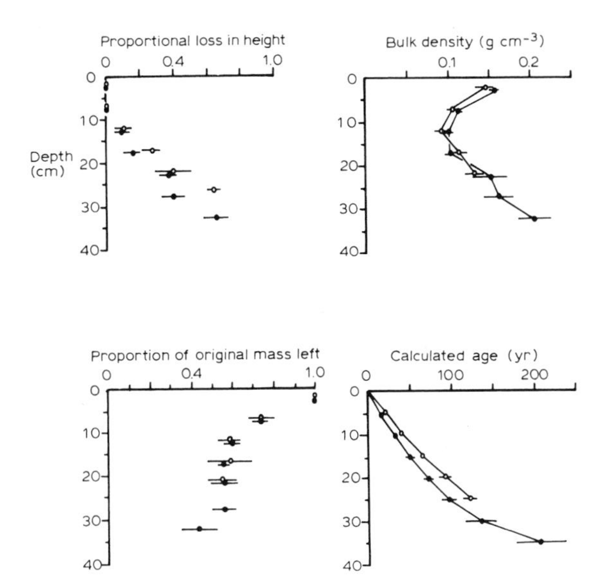

Fig. 4.31. Measured compression, bulk density and decay, and calculated age of *Polytrichum alpestre* moss-peat at two sites in the maritime Antarctic. Redrawn from Fenton (1980).

surface) and no decay at all. The addition of two zones and of consolidation (Clymo, 1978) or of several components and linear decrease of decay rate (Jones and Gore, 1978) allows tests of the profile of two variables (age and bulk density). But again, the expected shapes are simple curves, which could be explained by many different simple hypotheses. The way forward seems to be to include more variables, and to reach a point where much more complex curves are expected and may be measured. Forrester (1961) considered that the *pattern* of behaviour was all that was worth examining in complex models of industrial systems. These contained interlocking and nested feedback loops, however, whereas the existing models of peat forming systems have nothing more than simple feedback loops. Bunnell and Scoullar (1975) described a model of carbon flow in tundra which has 96 parameters. "Soil organic matter" and "soil humus" are only two of fourteen variables, however, and their account gives no results for these two variables. An ambitious attempt to make a complex model of a peat-forming system is that made by Wildi (1978). His model includes five variables (amount of peat, amount of water, amount of "nutrients", biomass of bog plants and biomass of fen plants) and twenty parameters. The variables are specified for each element of the model, and adjacent elements can be put at different height, so there is flow of water and "nutrients" between elements.

The complete model is so large that it is impracticable to use it for parameter minimization, but parameter values for a part, for example those controlling water-flow, were optimized. The whole model is able to simulate peat growth in relation to topography, and in a crude way to simulate succession of bog and fen plants (Fig. 4.32). There are clearly a large number of testable results. Unfortunately the tests which have actually been made are few and qualitative: "peat thickness ... comes very close to what can be observed in the field"; "the minimum water table is in the center of the bog"; "the high nutrient concentration at the upper limit on the slope which is occupied by a definite fen vegetation". Many of Wildi's assumptions about processes are necessarily simplistic, as he realized. The flow of water through peat, for example, does not always follow Darcy's Law (Rycroft et al., 1975b), the assumption that "the concentration of Ca^{++} may be the limiting factor in peat bogs" is less than certain, and the equation of "cations" with

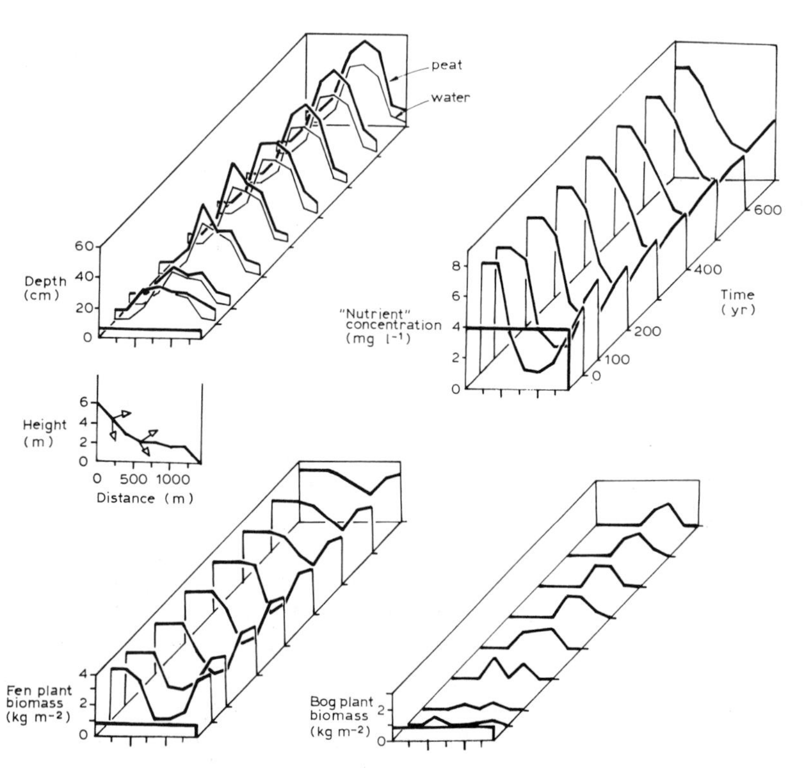

Fig. 4.32. Change in five variables during simulation of peat-bog growth on a sloping surface. The mineral soil surface level is shown at the centre left. Arrows indicate that at two places the water is allowed to flow laterally (or percolate into the ground). On all graphs, downhill is to the right. The peat and water are given as depth *above* the mineral soil surface defined in the small graph. Redrawn from Wildi (1978).

"nutrients" is even less so. Nevertheless such complex models have great promise because they give a large number of testable results, and, perhaps most important of all, they focus attention on those areas where ignorance is not bliss but a bane. The movement of water and "nutrients", and their effects on the growth and decay of bog and fen plants, are such areas.

ACKNOWLEDGMENTS

I thank Mrs. P. Ratnesar and T. Clymo for technical help, Professors Nils Malmer and Mats Sonesson for making possible the unpublished work reported here, and Professor D.W. Goodall for numerous helpful suggestions.

APPENDIX I: OUTLINE OF A COMMERCIAL CLASSIFICATION OF PEATS (Kivinen, 1977)

Three characters are used

1. Botanical composition
   (a) moss peat: >75% moss, <10% wood
   (b) herbaceous peat: >75% herbaceous plants, <10% wood
   (c) wood peat: >35% wood
   (d) mixed peat: any other type
2. Decomposition
   (a) little decomposed
   (b) medium
   (c) highly decomposed
3. Nutritional state
   (a) oligotrophic
   (b) mesotrophic
   (c) eutrophic

This scheme is the basis for the proposed International Classification.

APPENDIX II: TREATMENT OF ORGANIC SOILS IN THE U.S.D.A. 7th APPROXIMATION SOIL CLASSIFICATION (after Ragg and Clayden, 1973; Buol et al., 1973)

The ten orders of soils are:

(1) Entisols — recent soils
(2) Vertisols — clay soils with marked shrinking and swelling
(3) Inceptisols — embryonic soils
(4) Aridisols — arid region soils
(5) Mollisols — soils of steppe and prairie grassland
(6) Spodosols — soils with subsoil accumulations of iron and aluminium oxides and humus
(7) Alfisols — forest soils of high-base status
(8) Ulfisols — forest soils of low-base status
(9) Oxisols — highly weathered, sesquioxide-rich tropical soils
(10) Histosols — organic soils

During development a soil may pass from one order to another. The Histosols, in which peats fall, are perhaps the most outlying order, and have, on the whole, been neglected by soil surveyors. Histosols are defined as follows:

Soils with organic soil matter containing:
(1) no mineral layer 40 cm or more thick with upper boundary at or within 40 cm of the surface;
(2) less than 40 cm cumulative mineral layers in the top 80 cm.

The depth of organic matter must be one of:
(1) 60 cm or more if three-quarters or more (by volume) is fibric moss, or the bulk density is less than 0.1 g cm^{-3};
(2) 40 cm or more if: (a) the material is water-saturated for more than six months or is artificially drained; and (b) the organic material consists of sapric (H9–10) or hemic (H6–8) materials, or fibric (H1–5) materials if less than three-quarters by volume of fibric moss, and with a bulk density 0.1 g cm^{-3} or more;
(3) a layer more than twice the thickness of a mineral layer above and within 10 cm of the junction;
(4) any depth if the organic matter rests on fragmented mineral material with interstices partially or wholly filled by organic matter.

Sub-orders of Histosols

A depth of up to 130 cm (or 160 cm if the top 60 cm has three-quarters or more of fibric moss) is arbitrarily used for description ("control section"). Three layers ("tiers") are arbitrarily defined: upper 30 cm (or 60 cm if more than three-quarters of fibric moss), subsurface 60 cm, bottom 40 cm.

After separating Folists (see below), the main divisions are based on the state of decomposition. Within the suborders the main divisive criterion is the mean annual temperature, T_a.

(1) Folists — not water-saturated for more than a few days per year; less than three-quarters fibric *Sphagnum* and lithic material within 100 cm of surface
(2) Fibrists — humification about H1 to H5
  - Cryofibrists — frozen in some part two months after summer solstice
  - Sphagnofibrists — at least three-quarters fibric *Sphagnum* in top 90 cm.
  - Borofibrists — T_a less than 8°C
  - Tropofibrists — mean summer and mean winter temperatures differ by less than 5°C
  - Medifibrists — other fibrists with less than 2 cm humilluvic matter
  - Luvifibrists — others
(3) Hemists — humification about H6 to H8
  - Cryohemists, Borohemists, Tropohemists, Medihemists — as for corresponding Fibrists
  - Sulfihemists — sulphidic within 100 cm of surface
  - Sulfohemists — sulphidic within 50 cm of surface
  - Luvihemists — as for corresponding Fibrist

(4) Saprists — humification about H9 to H10

Cryosaprists Borosaprists Troposaprists Medisaprists Luvisaprists	as for corresponding Fibrists

APPENDIX III: OUTLINE OF A GEOLOGICAL/ECOLOGICAL CLASSIFICATION OF BIOGENIC SEDIMENTS (after West, 1968)

The classification is based on peat origin; symbols for use in stratigraphic diagrams are suggested by Troels-Smith (1955)

There are three main groups, of which only the second and third contain peats.

1. Sedimentary (allochthonous) deposits laid down in lakes, most with a large proportion of inorganic matter
2. Sedentary (autochthonous) deposits
   (i) Limnic peats, formed by plants growing in water, for example:
   (a) *Phragmites*
   (b) lake *Scirpus*
   (c) *Typha*
   (ii) Telmatic peats, formed by plants growing between low and high water

   Eutrophic types — examples:
   (a) *Cladium* peat
   (b) *Magnocaricetum* peat (formed from tall sedges)
   (c) fen moss peat (hypnoid mosses)
   Oligotrophic types — examples:
   (d) *Sphagnum cuspidatum* peat
   (e) *Eriophorum vaginatum* peat
3. Terrestrial peats, formed at or above high water mark

   Eutrophic types — examples:
   (a) *Parvocaricetum* peat (formed from small sedges)
   (b) fen wood peat (especially birch and alder)

   Oligotrophic types — examples:
   (a) *Sphagnum* peat formed by hummock species
   (b) *Sphagnum* + *Calluna* peat
   (c) bog shrub peat (*Myrica*, ericaceous shrubs other than *Calluna*)
   (d) bog wood peat (mainly *Betula*, but sometimes *Pinus*)
   (e) *Eriophorum vaginatum* peat [indistinguishable from 2(ii)(e)]
   (f) *Scirpus cespitosus* peat

REFERENCES

Aaby, B. and Jacobsen, J., 1979. Changes in biotic conditions and metal deposition in the last millennium as reflected in ombrotrophic peat in Draved Mose, Denmark. *Dan. Geol. Unders., Arb.*, 1978; 5–43.

Aaby, B., Jacobsen, J. and Jacobsen, O.S., 1979. Pb-210 dating and lead deposition in the ombrotrophic peat bog, Draved Mose, Denmark. *Dan. Geol. Unders., Arb.*, 1978: 45–68.

Aaby, B. and Tauber, H., 1974. Rates of peat formation in relation to degree of humification and local environment, as shown by studies of a raised bog in Denmark. *Boreas*, 4: 1–17.

Allen, S.E., Grimshaw, H.M. and Holdgate, M.W., 1967. Factors affecting the availability of plant nutrients on an Antarctic island. *J. Ecol.*, 55: 381–396.

Anderson, J.A.R., 1964. The structure and development of the peat swamps of Sarawak and Brunei. *J. Trop. Geogr.*, 18: 7–16.

Anonymous, 1978. *British Antarctic Survey, Annual Report 1977–78*, p. 78.

Bahnson, H., 1968. Kolorimetriske bestemmelser af humificeringstal i højmostorv fra Fuglsø mose på Djursland. *Medd. Dan. Geol. Foren.*, 18: 55–63.

Baker, J.H., 1972. The rate of production and decomposition of *Chorisodontium aciphyllum* (Hook. f. and Wils.) Broth. *Bull. Br. Antarct. Surv.*, 27: 123–129.

Basilier, K., 1973. Investigations on nitrogen fixation in moss communities. In: J.G.K. Flower-Ellis (Editor), *International Biological Programme, Swedish Tundra Biome Project, Tech. Rep.*, 16: 83–95.

Basilier, K., Granhall, V. and Stenström, T-A., 1978. Nitrogen fixation in wet minerotrophic moss communities of a subarctic mire. *Oikos*, 31: 236–246.

Belkevich, P.I. and Chistova, L.R., 1968. Exchange capacity of peat with respect to alkali and alkaline-earth metals. In: R.A. Robertson (Editor), *Proc. 2nd Int. Peat Congress, Leningrad, 1963*, pp. 909–918.

Bellamy, D.J. and Rieley, J., 1967. Some ecological statistics of a "miniature bog". *Oikos*, 18: 33–40.

Benda, I., 1957. Mikrobiologische Untersuchungen über das Auftreten von Schwefelwasserstoff in den anaeroben Zonen des Hochmoores. *Arch. Mikrobiol.*, 27: 337–374.

Berry, P.L. and Poskitt, T.J., 1972. The consolidation of peat. *Géotechnique*, 22: 27–52.

Bjarnason, O.B., 1968. Chemical investigation of Icelandic peat. In: R.A. Robertson (Editor), *Proc. 2nd Int. Peat Congress, Leningrad, 1963*, pp. 69–73.

Bland, D.E., Logan, A., Menshun, M. and Sternhell, S., 1968. The lignin of *Sphagnum*. *Phytochemistry*, 7: 1373–1377.

Boatman, D.J., 1961. Vegetation and peat characteristics of blanket bogs in County Kerry. *J. Ecol.*, 49: 507–517.

Boatman, D.J. and Roberts, J., 1963. The amounts of certain nutrients leached from peat by various extractants. *J. Ecol.*, 51: 187–189.

Boatman, D.J., Hulme, P.D. and Tomlinson, R.W., 1975. Monthly determinations of the concentrations of sodium, potassium, magnesium and calcium in the rain and in pools on the Silver Flowe National Nature Reserve. *J. Ecol.*, 63: 903–912.

Boelter, D.H., 1968. Important physical properties of peat materials. In: C. Lafleur and J. Butler (Editors), *Proc. 3rd Int. Peat Congress, Quebec, 1968*, pp. 150–154.

Boggie, R., 1972. Effect of water-table height on root development of *Pinus contorta* on deep peat in Scotland. *Oikos*, 23: 304–312.

Boggie, R., Hunter, R.F. and Knight, A.H., 1958. Studies of the root development of plants in the field using radioactive tracers. *J. Ecol.*, 46: 621–639.

Bonner, J., 1950. *Plant Biochemistry*. Academic Press, New York, N.Y., 537 pp.

Brackke, F.H. (Editor), 1976. *Impact of Acid Precipitation on Forests and Freshwater Ecosystems in Norway*. SNSF report 6, Oslo-Ås, 111 pp.

Bradis, E.M. and Andrienko, T.L., 1972. Bogs of the Ukrainian SSR. In: *Proc. 4th Int. Peat Congress, Otaniemi, 1972*, pp. 41–48.

Bunnell, F.L. and Scoullar, K.A., 1975. Abisko II. A computer simulation model of carbon flux in tundra ecosystems. In: T. Rosswall and O.W. Heal (Editors), *Structure and Function of Tundra Ecosystems. Ecol. Bull.*, No. 20. Swedish Natural Science Research Council, Stockholm, pp. 425–448.

Bunnell, F.L. and Tait, D.E.N., 1974. Mathematical simulation models of decomposition processes. In: A.J. Holding, O.W. Heal, S.F. Maclean Jr. and P.W. Flanagan (Editors), *Soil Organisms and Decomposition in Tundra*. Tundra Biome Steering Committee, Stockholm, pp. 207–225.

Buol, S.W., Hole, F.D. and McCrackern, R.J., 1973. *Soil Genesis and Classification*. Iowa State University Press, Ames, Iowa.

Burgeff, H., 1961. *Mikrobiologie des Hochmoores*. Fischer, Stuttgart.

Casparie, W.A., 1972. Bog development in south eastern Drenthe (The Netherlands). *Vegetatio*, 25: 1–271.

Cawse, P.A., 1974. *A Survey of Atmospheric Trace Elements in the U.K. (1972–73)*. A.E.R.E.-R 7669. H.M.S.O., London.

Chapin, F.S. III, Van Cleve, K. and Tieszen, L.L., 1975. Seasonal nutrient dynamics of tundra vegetation at Barrow, Alaska. *Arct. Alp. Res.*, 7: 209–226.

Chapin, F.S. III, Barsdate, R.J. and Barèl, D., 1978. Phosphorus cycling in Alaskan coastal tundra: a hypothesis for the regulation of nutrient cycling. *Oikos*, 31: 189–199.

Chapin, F.S. III, Van Cleve, K. and Chapin, M.C., 1979. Soil temperature and nutrient cycling in the tussock growth form of *Eriophorum vaginatum*. *J. Ecol.*, 67: 169–189.

Chapman, S.B., 1964a. The ecology of Coom Rigg Moss, Northumberland. I. Stratigraphy and present vegetation. *J. Ecol.*, 52: 299–313.

Chapman, S.B., 1964b. The ecology of Coom Rigg Moss, Northumberland. II. The chemistry of peat profiles and the development of the bog system. *J. Ecol.*, 52: 315–321.

Chapman, S.B., 1965. The ecology of Coom Rigg Moss, Northumberland. III. Some water relations of the bog system. *J. Ecol.*, 53: 371–384.

Clymo, R.S., 1963. Ion exchange in *Sphagnum* and its relation to bog ecology. *Ann. Bot. (Lond.) N.S.*, 27: 309–324.

Clymo, R.S., 1965. Experiments on breakdown of *Sphagnum* in two bogs. *J. Ecol.*, 53: 747–758.

Clymo, R.S., 1967. Control of cation concentrations, and in particular of pH, in *Sphagnum* dominated communities. In: H.L. Golterman and R.S. Clymo (Editors), *Chemical Environment in the Aquatic Habitat*. North-Holland, Amsterdam, pp. 273–284.

Clymo, R.S., 1970. The growth of *Sphagnum*: methods of measurement. *J. Ecol.*, 58: 13–49.

Clymo, R.S., 1978. A model of peat bog growth. In: O.W. Heal and D.F. Perkins, with W.M. Brown (Editors), *Production Ecology of British Moors and Montane Grasslands*. Springer-Verlag, Berlin, pp. 187–223.

Clymo, R.S. and Reddaway, E.J.F., 1971. Productivity of *Sphagnum* (bog-moss) and peat accumulation. *Hidrobiologia*, 12: 181–192. [A shorter version, without arbitrary cuts, appears as: A tentative dry matter balance sheet for the wet blanket bog on Burnt Hill, Moor House NNR. *Moor House Occas. Pap.*, No. 3 (1972): 15 pp.]

Collins, N.J., 1976. The development of moss-peat banks in relation to changing climate and ice cover on Signy Island in the maritime Antarctic. *Br. Antarct. Surv. Bull.*, 43: 85–102.

Collins, V.G., D'Sylva, B.T. and Latter, P.M., 1978. Microbial populations in peat. In: O.W. Heal and D.F. Perkins, with W.M. Brown (Editors), *Production Ecology of British Moors and Montane Grasslands*. Springer-Verlag, Berlin, pp. 94–112.

Conway, B.E., 1952. *Electrochemical Data*. Elsevier, Amsterdam, 374 pp.

Conway, V.M., 1937. Studies in the autecology of *Cladium mariscus* R. Br. III. The aeration of the subterranean parts of the plant. *New Phytol.*, 36: 64–96.

Conway, V.M., 1949. Ringinglow Bog near Sheffield. Part II. The present surface. *J. Ecol.*, 37: 148–170.

Coulson, J.C. and Butterfield, J.E., 1978. An investigation of the biotic factors determining the rates of plant decomposition on blanket bog. *J. Ecol.*, 66: 631–650.

Coulson, J.C. and Whittaker, J.B., 1978. Ecology of moorland animals. In: O.W. Heal and D.F. Perkins, with W.M. Brown (Editors), *Production Ecology of British Moors and Montane Grasslands*. Springer-Verlag, Berlin, pp. 52–93.

Coulter, J.K., 1950. Peat formations in Malaya. *Malay. Agric. J.*, 33: 63–81.

Coulter, J.K., 1957. Development of the peat soils of Malaya. *Malay. Agric. J.*, 40: 188–199.

Dalton, J., 1802. *Experimental Enquiry Into the Proportion of the Several Gases or Elastic Fluids Constituting the Atmosphere*. Manchester Philosophical Society.

Damman, A.W.H., 1978. Distribution and movement of elements in ombrotrophic peat bogs. *Oikos*, 30: 480–495.

Day, J.H., 1968. The classification of organic soils in Canada. In: C. Lafleur and J. Butler (Editors), *Proc. 3rd Int. Peat Congress, Quebec, 1968*, pp. 80–84.

Dickinson, C.H. and Maggs, G.H., 1974. Aspects of the decomposition of *Sphagnum* leaves in an ombrophilous mire. *New Phytol.*, 73: 1249–1257.

Donnan, F.G., 1911. Theorie der Membrangleichgewichte und Membranpotentiale bei Vorhandensein von nicht dialysierenden Elektrolyten. *Z. Elektrochem.*, 17: 572–581.

Duane, J., O'Brien, J.C. and Treacy, K., 1968. Recent work on milled peat in Ireland. In: C. Lafleur and J. Butler (Editors), *Proc. 3rd Int. Peat Congress, Quebec, 1968*, pp. 300–306.

Elina, G.A., 1972. Types of swamps in Northern Karelia of the USSR. In: *Proc. 4th Int. Peat Congress, Otaniemi, 1972*, pp. 59–74.

Eriksson, E., 1955. Current data on the chemical composition of air and precipitation. *Tellus*, 1: 134–139 [and in subsequent volumes].

Farnham, R.S., 1968. Classification of peat in the U.S.A. In: R.A. Robertson (Editor), *Proc. 2nd Int. Peat Congress, Leningrad, 1963*, pp. 115–132.

Farnham, R.S. and Finney, H.R., 1965. Classification and properties of organic soils. *Adv. Agron.*, 17: 115–162.

Fenton, J.H.C., 1978. *The Growth of Antarctic Moss Peat Banks.* Thesis, University of London, London, 162 pp.

Fenton, J.H.C., 1980. The rate of peat accumulation in antarctic moss banks. *J. Ecol.*, 68: 211–228.

Fieser, L.F. and Fieser, M., 1944. *Organic Chemistry.* Heath, Boston, Mass., 1091 pp.

Finney, H.R. and Farnham, R.S., 1968. Mineralogy of the inorganic fraction of peat from two raised bogs in Northern Minnesota. In: C. Lafleur and J. Butler (Editors), *Proc. 3rd Int. Peat Congress, Quebec, 1968*, pp. 102–108.

Flanagan, P.W. and Bunnell, F.L., 1976. Decomposition models based on climatic variables, microbial respiration and production. In: J.M. Anderson and A. Macfadyan (Editors), *The Role of Terrestrial and Aquatic Organisms in Decomposition Processes.* Blackwell, Oxford, pp. 437–457.

Flanagan, P.W. and Veum, A.K., 1974. Relationships between respiration, weight loss, temperature and moisture in organic residues in tundra. In: A.J. Holding, O.W. Heal, S.F. Maclean, Jr. and P.W. Flanagan (Editors), *Soil Organisms and Decomposition in Tundra.* Tundra Biome Steering Committee, Stockholm, pp. 249–277.

Florschütz, F., Menéndez Amor, J. and Wijmstra, T.A., 1971. Palynology of a thick Quaternary succession in Southern Spain. *Palaeogeogr., Palaeclimatol., Palaeoecol.*, 10: 233–264.

Flower-Ellis, J.G.K., 1974. Progress Report 1973. *International Biological Programme, Swedish Tundra Biome Project, Tech. Rep.*, 16: 212 pp.

Forrest, I., 1971. Structure and production of North Pennine blanket bog vegetation. *J. Ecol.*, 59: 453–479.

Forrester, J.W., 1961. *Industrial Dynamics.* M.I.T. Press, Cambridge, Mass., 464 pp.

Frankland, J.C., 1966. Succession of fungi on decaying petioles of *Pteridium aquilinum*. *J. Ecol.*, 54: 41–63.

Freistedt, E., 1968. Experiences in the design and construction of equipment for complete peat briquetting factories. In: C. Lafleur and J. Butler (Editors), *Proc. 3rd Int. Peat Congress, Quebec, 1968*, pp. 296–299.

Giles, B.R., 1977. *Root Function in* Eriophorum angustifolium. Thesis, University of London, London, 182 pp.

Gilliland, M.R. and Howard, A.J., 1968. Some constituents of peat wax separated by column chromatography. In: R.A. Robertson (Editor), *Proc. 2nd Int. Peat Congress, Leningrad, 1963*, pp. 877–886.

Given, P.H. and Dickinson, C.H., 1975. Biochemistry and microbiology of peats. In: E.A. Paul and D. Maclaren (Editors), *Soil Biochemistry, 3*, pp. 123–212.

Glob, P.V., 1969. *Mosefolket: Jernalderens mennesker bevaret i 2000 Ar.* Gyldendal, Copenhagen.

Godwin, H., 1956. *The History of the British Flora.* University Press, Cambridge, 384 pp.

Gore, A.J.P., 1968. The supply of six elements by rain to an upland peat area. *J. Ecol.*, 56: 483–495.

Gore, A.J.P. and Allen, S.E., 1956. Measurement of exchangeable and total cation content for H^+, Na^+, Mg^{++}, Ca^{++} and iron, in high level blanket peat. *Oikos*, 7: 48–55.

Gore, A.J.P. and Olsen, J.S., 1967. Preliminary models for accumulation of organic matter in an *Eriophorum/Calluna* ecosystem. *Aquilo, Ser. Bot.*, 6: 297–313.

Gorham, E., 1953. Some early ideas concerning the nature, origin and development of peat lands. *J. Ecol.*, 41: 257–274.

Gorham, E., 1956. On the chemical composition of some waters from the Moor House nature reserve. *J. Ecol.*, 44: 375–382.

Gorham, E., 1957. The development of peatlands. *Q. Rev. Biol.*, 32: 145–166.

Gorham, E., 1958. The influence and importance of daily weather conditions in the supply of chloride, sulphate and other ions to fresh waters from atmospheric precipitation. *Philos. Trans. R. Soc. Lond.*, B241: 147–178.

Gorham, E., 1961a. Chlorophyll derivatives, sulphur, and carbon in sediment cores from two English lakes. *Can. J. Bot.*, 39: 333–338.

Gorham, E., 1961b. Water, ash, nitrogen and acidity of some bog peats and other organic soils. *J. Ecol.*, 49: 103–106.

Gorham, E. and Sanger, J., 1967. Caloric values of organic matter in woodland, swamp, and lake soils. *Ecology*, 48: 492–493.

Göttlich, K-H. (Editor), 1976. *Moore- und Torfkunde.* E. Schwiezerbart'sche Verlagsbuchhandlung, Stuttgart, 269 pp.

Grace, J. and Marks, T.C., 1978. Physiological aspects of bog production at Moor House. In: O.W. Heal and D.F. Perkins, with W.M. Brown (Editors), *Production Ecology of British Moors and Montane Grasslands.* Springer-Verlag, Berlin, pp. 38–51.

Granat, L., 1972. *Deposition of Sulphate and Acid with Precipitation over Northern Europe.* Institute of Meteorology, University of Stockholm, Stockholm, Report AC-20, 30 + 19 pp.

Granat, L., 1975. *On the Variability of Rainwater Composition and Errors in Estimates of Areal Wet Deposition.* Department of Meteorology, University of Stockholm, Stockholm, Report AC-30, 34 pp.

Granlund, E., 1932. De svenska hogmossarnas geologi. *Sver. Geol. Unders. Afh.*, 26: 1–193.

Green, B.H. and Pearson, M.C., 1977. The ecology of Wybunbury Moss, Cheshire. II. Post-glacial history and the formation of the Cheshire Mere and Mire landscape. *J. Ecol.*, 65: 793–814.

Hanrahan, E.T., 1954. Factors affecting strength and deformation of peat. In: *Proc. 1st Int. Peat Congress, Dublin, 1954, Section b-3.2.*

Hart, M.G.R., 1962. Observations on the source of acid in empoldered mangrove soils. I. Formation of elemental sulphur. *Plant Soil*, 17: 87–98.

Heal, O.W. and Perkins, D.F., with Brown, W.M. (Editors), 1978. *Production Ecology of British Moors and Montane Grasslands.* Springer-Verlag, Berlin, 426 pp.

Heal, O.W., Howson, G., French, D.D. and Jeffers, J.N.R., 1974. Decomposition of cotton strips in tundra. In: A.J. Holding, O.W. Heal, S.F. Maclean Jr. and P.W. Flanagan (Editors), *Soil Organisms and Decomposition in Tundra.* Tundra Biome Steering Committee, Stockholm, pp. 341–362.

Heal, O.W., Latter, P.M. and Howson, G., 1978. A study of the rates of decomposition of organic matter. In: O.W. Heal and D.F. Perkins, with W.M. Brown, (Editors), *Production Ecology of British Moors and Montane Grasslands.* Springer-Verlag, Berlin, pp. 136–159.

Helfferich, F., 1962. *Ion Exchange.* McGraw-Hill, New York, N.Y., 472 pp.

Hill, M.O., 1973. Reciprocal averaging: an eigenvector method of ordination. *J. Ecol.*, 61: 237–249.

Howard, A.J. and Hamer, D., 1960. The extraction and constitution of peat wax: a review of peat wax chemistry. *J. Am. Oil Chem. Soc.*, 37: 478–481.

Hutchinson, J.N., 1980. The record of peat wastage in the East Anglian fenlands at Holme Post, 1848–1978 A.D. *J. Ecol.*, 68: 229–249.

Ingham, G., 1950. The mineral content of air and rain and its importance to agriculture. *J. Agric. Sci.*, 40: 55–61.

Ingram, H.A.P., 1978. Soil layers in mires: function and terminology. *J. Soil Sci.*, 29: 224–227.

Ingram, H.A.P., Rycroft, D.W. and Williams, D.J.A., 1974. Anomalous transmission of water through certain peats. *J. Hydrol.*, 22: 213–218.

Irwin, R.W., 1968. Soil water characteristics of some Ontario peats. In: C. Lafleur and J. Butler, (Editors), *Proc. 3rd Int. Peat Congress, Quebec, 1968*, pp. 219–223.

Janota-Bassalik, L., 1963. Psychrophiles in low-moor peat. *Acta Microbiol. Pol.*, 12: 25–40.

Jenny, H., Gessel, S.P. and Bingham, F.T., 1949. Comparative study of decomposition rates of organic matter in temperate and tropical regions. *Soil Sci.*, 68: 419–432.

Jones, H.E. and Gore, A.J.P., 1978. A simulation of production and decay in blanket bog. In: O.W. Heal and D.F. Perkins, with W.M. Brown (Editors), *Production Ecology of British Moors and Montane Grasslands.* Springer-Verlag, Berlin, pp. 160–186.

Kamula, A., 1968. Some observations on the autothermal gasification of peat dust on an industrial scale. In: R.A. Robertson (Editor), *Proc. 2nd Int. Peat Congr., Leningrad, 1963*, pp. 941–944.

Karesniemi, K., 1972. Dependence of humification degree on certain properties of peat. In: *Proc. 4th Int. Peat Congr., Otaniemi, 1972*, 2: 273–282.

Kivinen, E., 1977. Survey, classification, ecology and conservation of peatlands. *Bull. Int. Peat Soc.*, 8: 24–25.

Kivinen, E., Heikurainen, L. and Pakarinen, P., 1979. *Classification of Peat and Peatlands.* International Peat Society, V.V.T. Fuel Research Lab., Espo, 367 pp.

Knight, A.H., Crooke, W.M. and Inkson, R.H.E., 1961. Cation exchange capacities of tissues of higher and lower plants and their related uronic acid contents. *Nature (Lond.)*, 192: 142–143.

Knight, A.H., Boggie, R. and Shepherd, H., 1972. The effect of ground water level on water movement in peat: a study using tritiated water. *J. Appl. Ecol.*, 9: 633–642.

Korpijaakko, M. and Radforth, N.W., 1972. Studies of the hydraulic conductivity of peat. In: *Proc. 4th Int. Peat Congress, Otaniemi, 1972*, 3: 323–334.

Kulczyński, S., 1949. Peat bogs of Polesie. *Mem. Acad. Sci. Cracovie B*, 336 pp.

Küster, E. and Locci, R., 1964. Studies on peat and peat microorganisms. II. Occurrence of thermophilic fungi in peat. *Arch. Mikrobiol.*, 48: 319–324.

Kvét, J., 1955. *Biologicke Pusobeni Raseliny a Raseliniku.* Biologika Fakulta, University Karlovy, Prague, 105 pp.

Lee, J.A. and Tallis, J.H., 1973. Regional and historical aspects of lead pollution. *Nature (Lond.)*, 245: 216–218.

Livett, E.A., Lee, J.A. and Tallis, J.H., 1979. Lead, zinc and copper analyses of British blanket peats. *J. Ecol.*, 67: 865–891.

Luck, K.E., 1964. *Studies in the Autecology of* Calamagrostis epigeios *(L.) Roth and* C. canescens *(Weber) Roth.* Thesis, University of Cambridge, Cambridge, 165 pp.

MacFarlane, I.C. and Radforth, N.W., 1968. Structure as a base of peat classification. In: C. Lafleur and J. Butler (Editors), *Proc. 3rd Int. Peat Congress, Quebec, 1968*, pp. 91–97.

Malmer, N., 1962a. Studies on mire vegetation in the Archaean area of south western Götaland (south Sweden). I. Vegetation and habitat conditions on the Åkhult mire. *Opera Bot.*, 7(1): 1–322.

Malmer, N., 1962b. Studies on mire vegetation in the Archaean area of south western Götaland (south Sweden). II. Distribution and seasonal variation in elementary constituents on some mire sites. *Opera Bot.*, 7(2): 1–67.

Malmer, N. and Sjörs, H., 1955. Some determinations of elementary constituents in mire plants and peat. *Bot. Not.*, 108: 46–80.

Mattson, S. and Koutler-Andersson, E., 1954. Geochemistry of a raised bog. *K. Lantbrukhögsk. Ann.*, 21: 321–366.

Mattson, S. and Koutler-Andersson, E., 1955. Geochemistry of a raised bog. II. Some nitrogen relationships. *K. Lantbrukhögsk. Ann.*, 22: 219–224.

Mattson, S., Sandberg, G. and Terning, R.E., 1944. Electrochemistry of soil formation VI. Atmospheric salts in relation to soil and peat formation and plant composition. *K. Lantbrukhögsk. Ann.*, 12: 101–118.

Minderman, G., 1968. Addition, decomposition and accumulation of organic matter in forests. *J. Ecol.*, 56: 355–362.

Monteith, J.L., 1973. *Principles of Environmental Physics.* Arnold, London, 241 pp.

Moore, J.J., 1972. Report of the Glenamoy (Ireland) ecosystem study for 1971. In: F.E. Wielgolaski and T. Rosswall (Editors), *Proceedings IV. International Meeting on the Biological Productivity of Tundra.* Tundra Biome Steering Committee, Stockholm, pp. 281–282.

Moore, P.D. and Bellamy, D.J., 1974. *Peatlands.* Elek Science, London, 221 pp.

Morita, H., 1968. Polyphenols in the extractives of an organic soil. In: C. Lafleur and J. Butler (Editors), *Proc. 3rd Int. Peat Congress, Quebec, 1968*, pp. 28–31.

Mörnsjö, T., 1968. Stratigraphical and chemical studies on two peatlands in Scania, South Sweden. *Bot. Not.*, 121: 343–360.

Mörnsjö, T., 1969. Studies on vegetation and development of a peatland in Scania, South Sweden. *Opera Bot.*, 24: 1–187.

Mortimer, C.H., 1941. The exchange of dissolved substances between mud and water in lakes. I–II. *J. Ecol.*, 29: 280–329.

Mortimer, C.H., 1942. The exchange of dissolved substances between mud and water in lakes. III–IV. *J. Ecol.*, 30: 147–201.

Murozumi, M., Chow, T.J. and Patterson, C., 1969. Chemical concentrations of pollutant lead aerosols, terrestrial dusts and sea-salts in Greenland and Antarctic snow-strata. *Geochim. Cosmochim. Acta*, 33: 1247–1294.

Newbould, P.J., 1960. The ecology of Cranesmoor, a New Forest valley bog. *J. Ecol.*, 48: 361–383.

Nye, P.H. and Tinker, P.B., 1977. *Solute Movement in the Soil-Root System.* Blackwell, Oxford, 342 pp.

Ohira, Y., 1962. Some engineering researches on the experiments of the physical properties of the peat and on the sounding explorations of the peaty area in Hokkaido, Japan. *Mem. Defence Acad.* 11(2): 253–282.

Oldfield, F., Appleby, P.G., Cambray, R.S., Eakins, J.D., Barber, K.E., Battarbee, R.W., Pearson, G.R. and Williams, T.M., 1979. ^{210}Pb, ^{137}Cs and ^{239}Pu profile in ombrotrophic peat. *Oikos*, 33: 40–45.

Olenin, A.S., 1968. Peat resources of the USSR. In: R.A. Robertson (Editor), *Proc. 2nd Int. Peat Congress, Leningrad, 1963*, pp. 1–14.

Olsen, O.B., 1968. Peat and other substances. In: C. Lafleur and J. Butler (Editors), *Proc. 3rd Int. Peat Congress, Quebec, 1968*, pp. 264–267.

Osvald, H., 1923. Die Vegetation des Hochmoores Komosse. *Sven. Växtsociol. Sällsk. Handl.*, 1: 1–434.

Overbeck, F. and Schneider, S., 1940. Torfsetzung und Grenzhorizont, ein Beitrag zur Frage der Hochmoorentwicklung in Niedersachsen. *Angew. Bot.*, 22: 321–379.

Päivänen, J., 1969. The bulk density of peat and its determination. *Silva Fenn.*, 3: 11–12.

Pakarinen, P. and Tolonen, S., 1977. Distribution of lead in *Sphagnum fuscum* profiles in Finland. *Oikos*, 28: 69–73.

Paterson, M.P. and Scorer, R.S., 1973. Data quality and the European air chemistry network. *Atmos. Environ.*, 7: 1162–1171.

Pearsall, W.H., 1954. The pH of natural soils and its ecological significance. *J. Soil Sci.*, 3: 41–51.

Pearson, G.W., Pilcher, J.R., Baillie, M.G.L. and Hillam, J., 1977. Absolute radiocarbon dating using a low altitude European tree-ring calibration. *Nature (Lond.)*, 270: 25–28.

Peirson, D.H., Cawse, P.A., Salmon, L. and Cambray, R.S., 1973. Trace elements in the atmospheric environment. *Nature (Lond.)*, 241: 252–256.

Pennington, W., Cambray, R.S. and Fisher, E.M., 1973. Observations on lake sediments using ^{137}Cs as a tracer. *Nature (Lond.)*, 242: 324–326.

Pennington, W., Cambray, R.S., Eakins, J.D. and Harkness, D.D., 1976. Radionuclide dating of the recent sediments of Blelham Tarn. *Freshwater Biol.*, 6: 317–331.

Pigott, M.E. and Pigott, C.D., 1959. Stratigraphy and pollen analysis of Malham Tarn and Tarn Moss. *Field Stud.*, 1: 1–18.

Ponnamperuma, F.N., 1972. The chemistry of submerged soils. *Adv. Agron.*, 24: 29–96.

Potonié, R., 1908. Aufbau und Vegetation der Moore Norddeutschlands. *Englers Bot. Jahrb.*, 90.

Priddle, J., 1980a. The production ecology of benthic plants in some antarctic lakes. I. *In situ* production studies. *J. Ecol.*, 68: 141–153.

Priddle, J., 1980b. The production ecology of benthic plants in some antarctic lakes. II. Laboratory physiology studies. *J. Ecol.*, 68: 155–166.

Puustjärvi, V., 1956. On the cation exchange capacity of peats. *Acta Agric. Scand.*, 6: 410–449.

Puustjärvi, V., 1968. Standards for peat used in peat culture. *Peat Plant News*, 1: 19–26.

Puustjärvi, V., 1969. Fixing peat standards. *Peat Plant News*, 2: 3–8.

Puustjärvi, V., 1972. Standardization of peat products. In: *Proc. 4th Int. Peat Congress, Otaniemi, 1972*, 1: 415–420.

Ragg, J.M. and Clayden, B., 1973. The classification of some British soils according to the comprehensive system of the United States. *Soil Surv. Tech. Monogr.*, 3.

Ravera, O. and Premazzi, G., 1972. A method to study the history of any persistent pollution in a lake by the concentration of Cs-137 from fall-out in sediments. *Proc. Int. Symp. Radioecology Applied to the Protection of Man and his Environment*, 1: 703–722.

Richards, P.W., 1963. Soil conditions in some Bornean lowland plant communities. In: *Symposium on Ecological Research in Humid Tropics Vegetation.* Kuching, Sarawak, pp. 198–204.

Robertson, R.A., 1968. Scottish peat resources. In: R.A. Robertson (Editor), *Proc. 2nd Int. Peat Congress, Leningrad, 1963*, pp. 29–35.

Romanov, V.V., 1968. *Hydro-physics of Bogs.* Israel Programme for Scientific Translations, Jerusalem.

Rosswall, T., 1974. Cellulose decomposition studies on the Tundra. In: A.J. Holding, O.W. Heal, S.F. Maclean, Jr. and P. Flanagan (Editors), *Soil Organisms and Decomposition in Tundra.* Tundra Biome Steering Committee, Stockholm, pp. 325–340.

Rühling, Å. and Tyler, G., 1971. Regional differences in the deposition of heavy metals over Scandinavia. *J. Appl. Ecol.*, 8: 497–507.

Rühling, Å. and Tyler, G., 1973. Heavy metal deposition in Scandinavia. *Water, Air Soil Pollut.*, 2: 445–455.

Rycroft, D.W., Williams, D.J.A. and Ingram, H.A.P., 1975a. The transmission of water through peat. I. Review. *J. Ecol.*, 63: 535–556.

Rycroft, D.W., Williams, D.J.A. and Ingram, H.A.P., 1975b. The transmission of water through peat. II. Field experiments. *J. Ecol.*, 63: 557–568.

Sæbø, S., 1968. The autecology of *Rubus chamaemorus* L. I. Phosphorus economy of *Rubus chamaemorus* in an ombrotrophic mire. *Meld. Nor. Landbrukshøgsk.*, 47(1): 1–67.

Sæbø, S., 1969. On the mechanism behind the effect of freezing and thawing on dissolved phosphorus in *Sphagnum fuscum* peat. *Meld. Nor. Landbrukshøgsk.*, 48(14): 1–10.

Sæbø, S., 1970. The autecology of *Rubus chamaemorus* L. II. Nitrogen economy of *Rubus chamaemorus* in an ombrotrophic mire. *Meld. Nor. Landbrukshøgsk.*, 49(9): 1–37.

Sæbø, S., 1973. The autecology of *Rubus chamaemorus* L. III. Some aspects of calcium and magnesium nutrition of *Rubus chamaemorus* in an ombrotrophic mire. *Meld. Nor. Landbrukshøgsk.*, 52(5): 1–29.

Sæbø, S., 1977. The autecology of *Rubus chamaemorus*. L. IV. Potassium relations of *Rubus chamaemorus* in an ombrotrophic mire. *Meld. Nor. Landbrukshøgsk.*, 56(26): 1–20.

Salmi, M., 1954. Investigation of the calorific values of peats in Finland. In: *Proc. 1st Int. Peat Congress, Dublin, 1954*, B3: 1–9.

Schneider, S., 1968. Chemical and stratigraphical investigations of high-moor profiles in north-west Germany. In: R.A. Robertson (Editor), *Proc. 2nd Int. Peat Congress, Leningrad, 1953*, 1: 75–90.

Schnitzer, M., 1973. Chemical, spectroscopic, and thermal methods for the classification and characterization of humic substances. In: D. Povoledo and H.L. Golterman (Editors), *Humic Substances. Their Structure and Function in the Biosphere*. Centre for Agricultural Publishing and Documentation, Wageningen, pp. 293–310.

Schnitzer, M. and Kahn, S.U., 1972. *Humic Substances in the Environment*. Marcel Dekker, New York, N.Y.

Shaver, G.R., Chapin, F.S. III, and Billings, W.D., 1979. Ecotypic differentiation in *Carex aquatilis* on ice-wedge polygons in the Alaskan coastal tundra. *J. Ecol.*, 67: 1025–1046.

Sillanpää, M., 1972. Distribution of trace elements in peat profiles. In: *Proc. 4th Int. Peat Congress, Otaniemi, 1972*, 5: 185–191.

Sjörs, H., 1948. Myrvegetation i bergslagen. *Acta Phytogeog. Suec.*, 21: 1–299.

Skaven-Haug, S., 1972. Volumetric relations in soil materials. In: *Proc. 4th Int. Peat Congress, Otaniemi, 1972*, 2: 222–228.

Skene, M., 1915. The acidity of *Sphagnum* and its relation to chalk and mineral salts. *Ann. Bot. (Lond.)*, 29: 65–87.

Smith, R.A.H., undated. The environmental parameters of IBP experimental sites at Moor House. Aspects of the Ecology of the Northern Pennines *Moor House Occas. Pap.*, No. 4.

Smith, R.A.H. and Forrest, I., 1978. Field estimates of primary production. In: O.W. Heal and D.F. Perkins, with W.M. Brown (Editors), *Production Ecology of British Moors and Montane Grasslands*. Springer-Verlag, Berlin, pp. 17–37.

Sonesson, M., 1970. Studies on mire vegetation in the Torneträsk area, Northern Sweden. IV. Some habitat conditions of the poor mires. *Bot. Not.*, 123: 67–111.

Sonesson, M. (Editor), 1973. Progress Report 1972. *International Biological Programme, Swedish Tundra Biome Project, Tech. Rep.*, 14: 194.

Sparling, G.P. and Tinker, P.B., 1978a. Mycorrhizal infection in Pennine grassland. I. Levels of infection in the field. *J. Appl. Ecol.*, 15: 943–950.

Sparling, G.P. and Tinker, P.B., 1978b. Mycorrhizal infection in Pennine grassland. II. Effects of mycorrhizal infection on the growth of some upland grasses on γ-irradiated soils. *J. Appl. Ecol.*, 15: 951–958.

Sparling, G.P. and Tinker, P.B., 1978c. Mycorrhizal infection in Pennine grassland. III. Effects of mycorrhizal infection on the growth of white clover. *J. Appl. Ecol.*, 15: 959–964.

Spearing, A.M., 1972. Cation-exchange capacity and galacturonic acid content of several species of *Sphagnum* in Sandy Ridge Bog, Central New York State. *Bryologist*, 75: 154–158.

Sprent, J.I., Scott, R. and Perry, K.M., 1978. The nitrogen economy of *Myrica gale* in the field. *J. Ecol.*, 66: 657–668.

Stavset, K., 1973. Registering om molte i Andøy 1970–1972. *Medd. Nor. Myrselsk.*, 4: 153–156.

Stewart, J.M. and Robertson, R.A., 1968. The chemical status of an exposed peat face. In: C. Lafleur and J. Butler (Editors), *Proc. 3rd Int. Peat Congress, Quebec, 1968*, pp. 190–194.

Strygin, N.N., 1968. Research on spontaneous combustion processes in peat and their prevention. In: R.A. Robertson (Editor), *Proc. 2nd Int. Peat Congress, Leningrad, 1963*, pp. 509–513.

Svensson, B.H., 1974. Production of methane and carbon dioxide from a subarctic mire. In: J.G.K. Flower-Ellis (Editor), *International Biological Programme, Swedish Tundra Biome Project. Progress Report 1973*, pp. 123–143.

Swain, F.M., Blumentals, A. and Millers, R., 1959. Stratigraphic distribution of amino acids in peats from Cedar Creek Bog, Minnesota, and Dismal Swamp, Virginia. *Limnol. Oceanogr.*, 4: 119–127.

Tallis, J.H. and Switsur, V.R., 1973. Studies on Southern Pennine Peats. VI. A radiocarbon-dated pollen diagram from Featherbed Moss, Derbyshire. *J. Ecol.*, 61: 743–751.

Tamm, C.O. and Troedsson, T., 1955. An example of the amount of plant nutrients supplied to the ground in road dust. *Oikos*, 6: 61–70.

Tansley, A.G., 1939. *The British Islands and their Vegetation*. Cambridge University Press, London, 930 pp.

Theander, O., 1954. Studies on *Sphagnum* peat. III. A quantitative study on the carbohydrate constituents of *Sphagnum* mosses and *Sphagnum* peat. *Acta Chem. Scand.*, 8: 989–1000.

Tibbetts, T.E., 1968. Peat resources of the World — a review. In: C. Lafleur and J. Butler (Editors), *Proc. 3rd Int. Peat Congress, Quebec, 1968*, pp. 8–22.

Troels-Smith, J., 1955. Characterization of unconsolidated sediments. *Dan. Geol. Unders., IV Raekke*, 3 (10).

Turner, J., 1964. The anthropogenic factor in vegetational history. I. Tregaron and Whixall Mosses. *New Phytol.*, 63: 73–90.

Tyler, G., 1972. Heavy metals pollute Nature, may reduce productivity. *Ambio*, 1: 52–59.

Urquhart, C., 1966. An improved method for demonstrating the distribution of sulphide in peat soils. *Nature (Lond.)*, 211: 550.

Urquhart, C. and Gore, A.J.P., 1973. The redox characteristics of four peat profiles. *Soil Biol. Biochem.*, 5: 659–672.

Van Dijk, H. and Boekel, P., 1968. Effect of drying and freezing on certain physical properties of peat. In: R.A. Robertson (Editor), *Proc. 2nd Int. Peat Congress, Leningrad, 1963*, pp. 1051–1062.

Volarovich, M.P. and Churaev, N.V., 1968. Application of the methods of physics and physical chemistry to the study of peat. In: R.A. Robertson (Editor), *Proc. 2nd Int. Peat Congress, Leningrad, 1963*, pp. 819–831.

Von Naucke, W., 1976. Chemie von Moor und Torf. In: K.-H. Göttlich (Editor), *Moore- und Torfkunde*. E. Schweitzerbart'sche Verlagsbuchhandlung, Stuttgart, pp. 134–148.

Von Post, L., 1924. The genetic system of the organogen formations of Sweden. *Actes IVième Conf. Int. Pedologie*, p. 496.

Von Post, L. and Granlund, E., 1926. Södra Sveriges torvtillgångar I. *Sven. Geol. Unders.*, C: 335.

Waksman, S.A. and Stevens, K.R., 1928a. Contribution to the chemical composition of peat: I. Chemical nature of organic complexes in peat and methods of analysis. *Soil Sci.*, 26: 113–137.

Waksman, S.A. and Stevens, K.R., 1928b. Contribution to the chemical composition of peat: II. Chemical composition of various peat profiles. *Soil Sci.*, 26: 239–251.
Waksman, S.A. and Stevens, K.R., 1929a. Contribution to the chemical composition of peat: III. Chemical studies of two Florida peat profiles. *Soil Sci.*, 27: 271–281.
Waksman, S.A. and Stevens, K.R., 1929b. Contribution to the chemical composition of peat: IV. Chemical studies of highmoor peat profiles from Maine. *Soil Sci.*, 27: 389–398.
Waksman, S.A. and Stevens, K.R., 1929c. Contribution to the chemical composition of peat: V. The role of micro-organisms in peat formation and decomposition. *Soil Sci.*, 28: 315–340.
Waksman, S.A. and Tenney, F.G., 1928. Composition of natural organic materials and their decomposition in the soil: III. The influence of nature of plant upon the rapidity of its decomposition. *Soil Sci.*, 26: 155–171.
Walker, D. and Walker, P.M., 1961. Stratigraphic evidence of regeneration in some Irish bogs. *J. Ecol.*, 49: 169–185.
Walsh, T. and Barry, T.A., 1958. The chemical composition of some Irish peats. *Proc. R. Ir. Acad.*, 59B: 305–328.
Walter, H., 1977. The oligotrophic peatlands of Western Siberia — the largest peino-helobiome in the world. *Vegetatio*, 34: 167–178.
Watt, A.S., 1947. Pattern and process in the plant community. *J. Ecol.*, 35: 1–22.
Waughman, G.J., 1980. Chemical aspects of the ecology of some South German peatlands. *J. Ecol.*, 68: 1025–1046.
Weast, R.C. (Editor), 1965. *Handbook of Chemistry and Physics.* Chemical Rubber Co., Cleveland, Ohio, 46th ed.
Weber, C.A., 1908. Aufbau und Vegetation der Moore Norddeutschlands. *Englers Bot. Jahrb.*, p. 90.
Webster, J.R., 1962. The composition of wet-heath vegetation in relation to aeration of the ground-water and soil. I. Field studies of ground-water and soil aeration in several communities. *J. Ecol.*, 50: 619–637.
West, R.G., 1968. *Pleistocene Geology and Biology*. Longmans, London, 377 pp.
Wheatley, R.E., Greaves, M.P. and Russell, J.D., 1975. The occurrence of aphid wax in peat. *Soil Biol. Biochem.*, 7: 35–38.
Whitmore, T.C., 1975. *Tropical Rain Forests of the Far East.* Clarendon Press, Oxford, 282 pp.
Wickman, F.E., 1951. The maximum height of raised bogs. *Geol. för. Stockh. Förh.*, 73: 413–422.
Wildehain, W. and Henske, G., 1965. Organische Verbindungen aus Hochmoortorfextrakten. *Z. Chem.*, 5: 457–458.
Wildi, O., 1978. Simulating the development of peat bogs. *Vegetatio*, 37: 1–17.
Williams, B.L., 1974. Effect of water-table level on nitrogen mineralization in peat. *Forestry*, 47: 195–202.
Woodwell, G.M., Whittaker, R.H., Reiners, W.A., Likens, G.E., Delwiche, C.C. and Botkin, D.B., 1978. The biota and the world carbon budget. *Science*, 199: 141–146.

INTERNATIONAL PEAT CONGRESSES

The Proceedings of these Congresses are frequently referred to in these pages, and are an important source of information for the subject. They are bibliographically diverse, and not always easy of access, so it may be useful to give full details here:

Proceedings 1st International Peat Congress, Dublin, 1954. No Editor. No publication date.
Proceedings 2nd International Peat Congress, Leningrad, 1963. Edited by R.A. Robertson, H.M.S.O., London. Published 1968. 2 vols., 1090 pp.
Proceedings 3rd International Peat Congress, Quebec, 1968. Edited by C. Lafleur and J. Butler. No Publisher, but "Sponsored by Department of Energy, Mines and Resources, Ottawa, Canada and National Research Council of Canada". No publication date. 405 pp.
Proceedings 4th International Peat Congress, Otaniemi, Finland, 1972. No editor. No publisher. No publication date. 5 vols.: 484, 353, 569, 303, 284 pp.
Proceedings 5th International Peat Congress, Poznán, Poland, 1976. No editor. Wydawnictwa Czasopism Techniczynch Not, Warszawa. No publication date. 4 vols.: 482, 336, 335, 396 pp.
Proceeding 6th International Peat Congress, Duluth, Minn., 1980. No Editor. No Publisher. No publication date, 735 pp.

NOTE ADDED IN PROOF

A recent book summarizes a great deal of information about the organic chemistry and industrial uses of peat. It is:

Fuchsman, C.H., 1980. *Peat: Industrial Chemistry and Technology*. Academic Press, New York, N.Y., 279 pp.

Chapter 5

MICRO-ORGANISMS IN PEATLANDS[1]

COLIN H. DICKINSON

INTRODUCTION

Peat accumulations stand as impressive reminders of the high level of efficiency which is normally expected from decomposer micro-organisms. In most terrestrial habitats the activity of saprobic micro-organisms ensures that organic remains are quickly destroyed, thereby releasing the mineral nutrients that they contain for recirculation and plant nutrition. Even in ecosystems where litter is only slowly decomposed, such as temperate coniferous woodlands and dry heathlands, the activities of decomposers proceed considerably faster and further than in peat-forming habitats. Mineral nutrient cycling in the direct plant/soil sense is highly dependent upon the decomposer micro-organisms, among which the microflora and microfauna are both included. Features of these processes are also discussed in the chapters by Clymo (Ch. 4) and by Mason and Standen (Ch. 11). There are, however, other aspects of nutrient flow in which micro-organisms play an important role. Nitrogen input, for example, may be enhanced by free-living and symbiotic nitrogen-fixing organisms; and nutrient uptake by higher plants is often affected by soil micro-organisms, especially those in the rhizosphere. In some instances mycorrhizal associations supplement the roots' powers of nutrient gathering; and in others, mycorrhizas circumvent the normal decomposition processes by obtaining materials directly from newly-dead remains. Nutrient losses from green plants include materials leached from living roots and leaves. The extensive populations of micro-organisms which live epiphytically on higher plant organs almost certainly increase these losses by rapidly metabolizing those materials which are leaked out, thus ensuring a continual diffusion gradient from within the plant to its outer surface. Other nutrient losses occur through grazing by both large and small herbivores. Pathogenic micro-organisms are responsible for diverting nutrients into tissues which have rather different fates as far as their subsequent decomposition is concerned.

Following the senescence and death of primary producer tissues, the mineral nutrients contained therein are gradually released by autolysis and by the activity of decomposers. These nutrients are added to the cycling and reservoir pools in the substrate. Here micro-organisms are involved in a number of processes including transformations in the chemical form of the nutrients and their retention by exchange phenomena.

These general considerations provide a basis for examining the role of micro-organisms in peat ecosystems. There are, by definition, numerous limitations to nutrient cycling in mires. Bogs in particular are generally impoverished as regards essential plant nutrients, and their chemical and physical conditions cannot support extensive higher plant or microbial growth. One particular problem which assumes major importance in any consideration of nutrient cycling in mires is the occurrence of conditions with little or no oxygen. Such conditions can cause drastic alterations in qualitative aspects of nutrient transformations. Some of the special problems of anaerobic soil chemistry are dealt with separately in this volume by Sikora and Keeney (Ch. 6). Inevitably, however, there is a degree of overlap between the effects of aerobic and anaerobic microhabitats, especially as anaerobic conditions are not always restricted to well-defined, deeper layers of peat.

---

[1] Manuscript received January, 1979.

In general, information on this subject is still very sketchy because of the difficulties of relating the activities of communities of micro-organisms to *undisturbed* conditions in the peat.

NUTRIENT INPUT TO MIRE SYSTEMS

The importance of the distinction between minerotrophic and ombrotrophic mire systems, which is stressed elsewhere, is self-evident in this topic. Typically these two types of mire have vastly different scales of nutrient input. Many mires of the former type are characterized by luxuriant vegetation sustained by nutrient-rich water supplies flowing into them from elsewhere, whereas ombrotrophic mires depend upon the nutrients contained in atmospheric precipitation (Gore, 1968; Boatman et al., 1975; Kallio and Veum, 1975).

In both types of mire, especially under naturally waterlogged conditions, nitrogen is one of the more critical plant nutrients. Supplies of nitrogen can be supplemented by microbial activity. Alexander (1974) has reviewed the studies of tundra peats under the International Biological Programme (I.B.P.), and he noted that nitrogen fixation can contribute amounts which are equal to, or up to several times greater than, that in rainfall. Martin and Holding (1978) have provided data for ombrotrophic blanket-bog peat at Moor House (see Taylor, Part B, Ch. 1), which suggest that the ratio of fixation to precipitation ranges from 0.1 to 4.0. These authors also provided a useful summary of the probable magnitudes of the pools and main pathways of nitrogen in blanket peat (Fig. 5.1). Less is known concerning the amounts of nitrogen fixed in minerotrophic mires, but Moore and

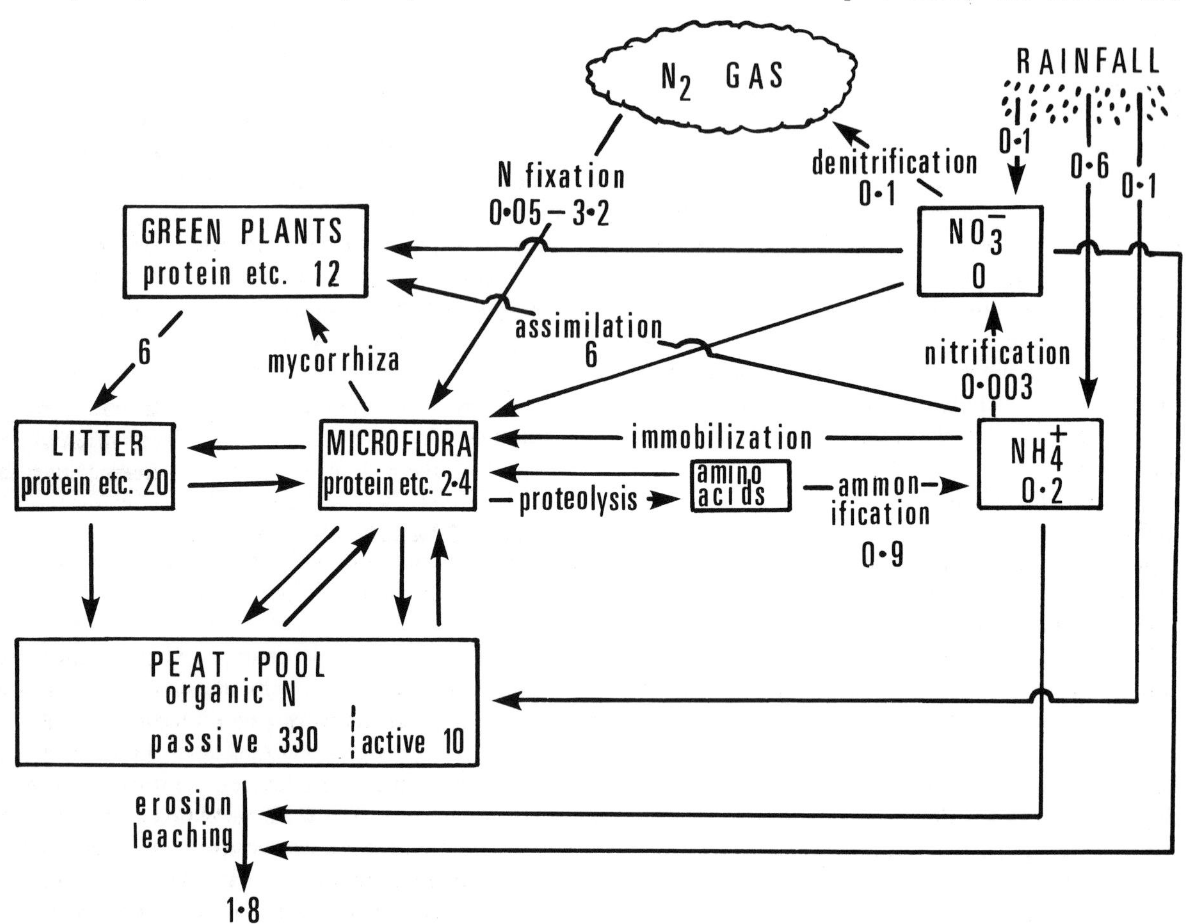

Fig. 5.1. A preliminary summary of the nitrogen cycle as it may operate in blanket peat (after Martin and Holding, 1978). Values in boxes represent the steady-state quantity (g m^{-2}) of nitrogen in that particular pool to a depth of 30 cm. Values beside arrows represent the calculated rate of transfer (g m^{-2} yr^{-1}) between pools. Where no value is given no estimate of the rate has been possible.

Bellamy (1974) suggested that they were far greater than are the additions to ombrotrophic systems.

As with mineral soils, nitrogen fixation may be carried out by free-living organisms, including the bacteria and blue-green algae, and by symbiotic organisms living in association with lichens, mosses and higher plants.

Free-living, nitrogen-fixing organisms

Bacteria

Gaseous nitrogen is utilized by aerobic bacteria, mainly members of the Azotobacteriaceae, facultative anaerobes, such as *Bacillus* and *Klebsiella*, and strict anaerobes, including *Clostridium*, *Desulfovibrio* and *Rhodospirillum*. These bacteria occur widely in soil, where their activity is enhanced in the rhizosphere which provides them with an enriched habitat, and some have also been shown to be able to grow and fix nitrogen on aerial plant surfaces (Ruinen, 1974).

Aerobic nitrogen-fixing species are generally most abundant in alkaline or neutral soils, but information regarding their occurrence in non-acid peats is scarce. The literature concerning the distribution of *Azotobacter* was discussed by Zimenko (1966) and Dunican and Rosswall (1974), and both concluded that these organisms are scarce or absent in most acid peatlands. Zimenko (1966) reinforced this conclusion with data indicating that the bacterium was absent from bog, fen and transitional peats in the Ptich' peat tract in central Belorussiya. Latter et al. (1967) reported that aerobic nitrogen-fixing bacteria were absent from blanket peats at Moor House in England, but Collins et al. (1978) were able to detect small populations of such organisms in neighbouring areas. Granhall and Lid-Torsvik (1975) have shown that significant contributions of fixed nitrogen are produced by the populations of aerobic, nitrogen-fixing bacteria in relatively acid mires at Stordalen, Sweden, and Hardangervidda, Norway.

Despite many indications to the contrary, peat is not inhabited by especially large populations of anaerobes (Latter et al., 1967; Stout, 1971; Clarholm et al., 1975; Collins et al., 1978). Anaerobic nitrogen-fixing bacteria were, however, found in the acid peats examined by Zimenko (1966), and Collins et al. (1978) demonstrated that there were anaerobic organisms with the potential to fix nitrogen in most samples of blanket-bog peat taken from Moor House. The numbers present at Moor House were similar to those of the aerobic nitrogen-fixers, and the populations were mostly concentrated in the litter horizons. Anaerobic organisms appeared to contribute somewhat less combined nitrogen than did aerobic bacteria in the Scandinavian sites described by Granhall and Lid-Torsvik (1975). However, not all the nitrogen fixed under anaerobic conditions is obtained by obligate anaerobes. Stout (1971) found that nitrogen was fixed in peat at Wicken Fen by the facultative anaerobe *Bacillus polymyxa* at a pH of 7.7 under anaerobic conditions.

It is important to reiterate that the presence of species with the potential to fix nitrogen, or any other micro-organisms, in peat does not imply that they are active in this substrate. The difficulties of demonstrating the quality and quantity of microbial processes under natural conditions are very considerable. Amongst the adverse factors affecting the growth of bacteria, the pH of many acid mires is well below that needed for optimum activity of the nitrogenase enzyme. This may account for the isolation from acid peats of aerobic *Azotobacter*-like bacteria which were unable to reduce acetylene in the standard test for nitrogen fixation (Dunican and Rosswall, 1974).

Blue-green algae (Cyanobacteria)

Members of the Cyanobacteria seem to thrive best in wet, mesotrophic, neutral to alkaline habitats (Stewart, 1974). This explains their abundance in open, wet prairie communities in the fresh-water part of the Florida Everglades (P.H. Given, pers. comm., 1972) and their relative rarity in acid ombrotrophic mires. Dooley and Houghton (1973) failed to isolate any blue-green algae from Irish blanket bog at Glenamoy, though they did obtain evidence for the occurrence of small numbers of light-sensitive organisms which were actively fixing nitrogen. Granhall and Lid-Torsvik (1975) reported that blue-green algae were rare in the more acid situations in Scandinavian mires, at pH 4.0 and below, though where the pH was slightly higher these algae were abundant in open pools and amongst moss carpets. Genera of blue-green algae found in acid peats include *Anabaena*, *Chlorogloea*, *Gloeocapsa* and *Scytonema*. *Calothrix*, *Fischerella*, *Hapalosiphon* and *Tolypothrix* thrive especially well as epiphytes in moss carpets, and they have been shown to fix as much as 12 g nitrogen m^{-2} yr^{-1} in these habitats. Broady (1977, 1979) has also demonstrated the occurrence of vigorous algal populations in moss-dominated communities in two maritime Antarctic mires on Signy Island, South Orkneys. The populations of blue-green algae were more varied and more abundant in a *Calliergon–Calliergidium–Drepanocladus* carpet, with a pH about 5.0, than in a *Polytrichum–Chorisodontium* turf, where the pH in the surface layers was about 4.0. *Nostoc muscorum* was especially common in the former habitat and it was regularly observed to form heterocysts, which are indicative of vigorous nitrogen fixation.

In addition to the reports of the presence of free-living blue-green algae in peat, several workers have described more intimate associations between these organisms and a number of mosses. Stewart (1966) described how filaments of a *Hapalosiphon* species appeared to be trapped within the hyaline cells of *Sphagnum*, and he obtained evidence that nitrogen was being fixed by this association. Alexander (1974) and Granhall and Lid-Torsvik (1975) reported the presence of intimate, intracellular associations between the mosses *Sphagnum* and *Drepanocladus* and filaments of the blue-green alga *Nostoc*. Where such associations exist they may represent a hitherto unrecognized form of loose symbiosis. They are known to occur in other members of the Bryophyta, and are presumably advantageous in that the algae will not be unduly limited by shading, which is an important factor restricting their activity elsewhere.

However, not all such mosses have algal associates. P.A. Broady (pers. comm., 1978) used a microscopic technique to examine moss leaves, but he failed to find any such associations in a number of Antarctic mosses, including *Drepanocladus*.

Carnivorous plants

It may be argued that *Drosera*, *Pinguicula*, *Sarracenia* and other carnivorous plants merely recycle nutrients within mire ecosystems rather than gathering additional minerals from elsewhere. However, to answer this point one needs to learn more about their insect diet than is known at present. Heslop-Harrison (1978) reviewed the biology of these remarkable plants and she noted that most species inhabit nutrient-poor habitats, which is what one would expect if the main purpose of the carnivorous habit is to supplement the minerals available in the substrate. Many of these plants are found in mires, although the 400 or so species are by no means confined to such ecosystems.

Some indications of the significance of the carnivorous habit were obtained by Chandler and Anderson (1976), who found that *Drosera* grew optimally when insect-fed plants were raised on a nitrogen-deficient medium. A positive growth response was also obtained on a sulphur-deficient medium, but growth was not stimulated by the provision of insects when the plants were raised on substrates lacking either phosphorus or plant micronutrients.

Symbiotic nitrogen-fixing organisms

Rhizobium

Rhizobium, although perhaps the best known symbiotic nitrogen-fixing organism, is not especially important in mire habitats owing to the scarcity of legumes. Some legumes, for example *Lathyrus palustris*, *Lotus pedunculatus* and *Genista tinctoria*, grow in fens, but the family is notably absent from more acid, ombrotrophic habitats.

Much of the information concerning the activity of *Rhizobium* in peatlands has been gained from peat reclamation schemes where clover has been used to improve the status of the newly established pasture. Jones (1966) found that, in an upland mire which had previously supported a *Calluna*, *Eriophorum* and *Sphagnum* community, the initial populations of *Rhizobium* in the cleared and cultivated peat, with a pH of 3.5, were of the order of 100 cells per gram of peat. By comparison there were 4.4×10^3 cells per gram in a nearby mineral soil with a pH of 4.5; and the highest population he recorded was 180×10^3 in a medium loam soil, with a pH of 5.7, under permanent pasture. Not only were the peat populations relatively small, but also a high proportion of the isolates were relatively ineffective in tests designed to determine their ability to induce nodules on clover roots.

It is tempting to relate the failure to produce nodules directly to the acidity of the environment. However, Masterson (1968) found no correlation between the pH of a number of reclaimed peats and the nodule-forming abilities of *Rhizobium* isolates from these peats. Holding and Lowe (1971) cited evidence suggesting that the activity of *Rhizobium* is governed by a complex of factors, including both pH and the availability of heavy-metal ions. In addition, both Jones et al. (1964) and O'Toole and Masterson (1968) found that high levels of calcium were required for successful nodule formation in reclaimed acid peats. It is therefore clear that, whilst *Rhizobium* can grow vigorously in reclaimed peatland, it is generally inactive in undisturbed mires.

Actinomycete-like organisms

Other types of symbiotic, nitrogen-fixing organisms inhabit root nodules in a wide variety of plants, including the genera *Alnus* and *Myrica* which grow in fen peat. The symbiotic micro-organisms involved in the *Alnus* associations are actinomycetes which have been classified in the genus *Frankia*. They have been grown in artificial culture but little is yet known concerning their taxonomy or biology. These organisms substantially increase the levels of combined nitrogen in their host plants, and this clearly benefits both the individual plants and eventually their respective ecosystems (Bond, 1974). The organisms inhabiting *Alnus* root nodules are especially effective in fixing very large quantities of nitrogen. In one experiment in which *A. glutinosa* was grown in nitrogen-free solutions, nodulated specimens were found to contain 492 mg nitrogen per plant, which was nearly 1000 times more than in the non-nodulated controls. This has led to this genus being employed in reclamation schemes for derelict and impoverished land. The *Alnus* nodule-forming organism was found by McVean (1959) to occur in shallow blanket bog, but establishment of alder from seed was more successful if an inoculum of crushed nodule was added to the mire surface.

Myrica gale nodule organisms also fix appreciable amounts of nitrogen. In a similar experiment to that mentioned above the nodulated plants contained 158 mg nitrogen as against 0.2 mg in the non-nodulated controls. Hence, the widespread occurrence of this species in fens means that it is an important contributor to the nutrient status of these types of mire.

Blue-green algae (Cyanobacteria) in lichens

Somewhat fewer than 10% of lichen species contain blue-green algae, either as their principal phycobionts or as the inhabitants of cephalodia, which are warty protuberances on thalli whose main algal partner is a member of the Chlorophyceae. Many of the blue-green algae inhabiting lichens have been shown to fix nitrogen within the thalli or cephalodia, and this enhances the nutritional status of lichen communities, especially in arctic and subarctic areas. Nitrogen-fixing lichens include species of *Collema*, *Leptogium*, *Lobaria*, *Nephroma*, *Peltigera*, *Solorina*, *Stereocaulon* and *Sticta* (Millbank, 1974). A number of these genera occur in tundra and related ecosystems, but their significance in peatlands is less than in drier areas elsewhere. Duncan (1970) listed some 75 British lichens which are characteristic of acid peatlands and of these only one or two have any potential for fixing nitrogen. Indeed, it is notable that most of the common peat-inhabiting genera, including *Bacidia*, *Cetraria* and *Cladonia*, possess the same green, non-fixing alga, *Trebouxia*. Lichens are not notable inhabitants of fens, but *Peltigera* can occasionally be found in these habitats and several species of this genus are noted for their nitrogen-fixing abilities.

The fate of nutrients which have been absorbed by plants

Nutrients employed in the metabolism of plant tissues may be lost from the plant whilst the tissue is still alive, they may be translocated from the dying tissues during senescence or they may remain in the dying tissues. Losses from the living plant include leaching from the aerial canopy, with the minerals being utilized *in situ* by epiphytic organisms or being washed to the ground by rain. Leaching from subterranean organs also occurs, but the limited activity of the microflora in peat may mean that root exudates are not as rapidly metabolized as in mineral soils, and hence leakage due to concentration gradients may be less than from aerial tissues. Translocation from dying tissues may substantially reduce the need for further uptake of particular elements, especially in long-lived perennials which have the ability to store food materials during dormancy. Excellent examples of this type of behaviour are provided by *Eriophorum vaginatum*, which has been shown to transfer phosphorus and potassium from dying leaves to perennating tissues in the tussocks (Goodman and Perkins, 1959), and *Molinia caerulea*, which removes up to 75% of the nitrogen and phosphorus from its leaves before their abscission (Morton, 1977). *Drosera* species bearing tubers store phosphorus in these organs during their dormant phase and then re-export it to the leaves when growth recommences (Chandler and Anderson, 1976). The facility of peat-inhabiting plants to conserve mineral nutrients has been shown by Small (1972) to be greater than that of species living elsewhere. He also demonstrated that evergreen species living in bogs are more economical in their use of nitrogen, as measured by the amount of photosynthate produced by unit of nitrogen, than deciduous species living either in bogs or elsewhere. However as Heal et al. (1978) pointed out this facility for conservation has the effect of rendering the resulting peat-bog plant litter a poorer substrate for micro-organisms.

The mineral nutrients which remain in the dead tissues become involved in nutrient cycling or become incorporated into the peat itself. Nutrient release from plant litter involves both physical and biological processes. Some nutrients are lost by simple leaching. Morton (1977) has shown that more than 90% of the calcium, magnesium and potassium in *Molinia* leaves is quickly removed after their death by the action of rainfall. The release of other elements, especially nitrogen, phosphorus and sulphur, is dependent upon decomposition which involves complex microbial food chains.

The importance of micro-organisms in the movement of elements from one pool to another has been highlighted by Chapin et al. (1978) in their studies of wet-meadow coastal tundra in northern Alaska, U.S.A. They found that there was a rapid and massive transfer of phosphorus from the dead organic matter to the soluble pool during the first ten days after snow-melt. This transfer was correlated with a significant decline in the microbial populations. Chapin et al. (1978) suggested that further decomposition was then limited by the rate at which the microbial population recovered.

Decomposition of aerial plant tissues in mire ecosystems is probably comparable with the processes on adjacent mineral soils, at least during the time the debris remains above the surface of the peat. Decay of buried debris, including particularly root litter formed *in situ*, is, however, substantially different to that of similar tissues in other soils. In mineral soils decay is actually enhanced following burial, because of the more favourable conditions which are then available for the growth of saprobic micro-organisms. In peat the situation is reversed, as conditions beneath the peat surface are generally inhospitable for decomposer organisms — which is an obvious corollary of the fact that peat has accumulated in the first place.

Nutrient release from litter can occur by a variety of processes (Dowding, 1974):

(a) Abiotic leaching of soluble materials from both the substrate and the decomposers.
(b) Incorporation into the decomposer biomass.
(c) Respiration and denitrification losses.
(d) Translocation, spore dispersal or migration of the microfauna.
(e) Transformation into faecal material.

The importance of each of these processes in mires depends on the abiotic environment, the types of plant tissue involved and microbial activity on and in the peat.

Many decomposition studies involving peatlands have not been accompanied by descriptions of associated nutrient changes. Some recent attempts to rectify this have resulted from the International Biological Programme (e.g. Martin and Holding, 1978). However, much of our knowledge of this aspect of nutrient cycling in mires must still be inferred from a knowledge of the decomposition processes on and in peat. The following account will be based on the vertical distribution of litter in mires.

Decomposition above the mire surface

In recent years the extensive use of litter bags, tagging techniques and simple sequential sampling procedures has provided a good deal of information concerning decomposition above the peat surface. These techniques have shown that there is considerable variation in the weight losses per year for different types of litter. Heal and French (1974) have described such variations as occur in tundra peatlands. On blanket bog in the English Pennines weight losses ranged in the first year from about 45% for *Narthecium* and *Rubus* leaves to less than 10% for *Calluna* stems. In nearly all instances there were smaller losses in the second year after abscission. Very much smaller weight losses were recorded elsewhere for several mosses, notably *Chorisodontium* on Signy Island, South Orkneys, *Sphagnum fuscum* in Stordalen mire, Sweden and

Drepanocladus at Hardangervidda, Norway.

During these periods on the surface of the mire the litter becomes incorporated into the superficial layers of peat. *Juncus squarrosus* leaves, for example, remained in the L layer[1] of a blanket-bog peat at Moor House for about twelve months and then were resident in the F_1 horizon for a further fifteen months (Latter and Cragg, 1967). During this time they lost *c.* 30% of their dry weight per year.

Heal et al. (1978), describing work done under the I.B.P. at Moor House, concluded that most of the observed weight losses resulted from microbial respiration. Leaching losses (for *Sphagnum* spp.) were estimated by Clymo and Reddaway (1971), working in the same area, to be about 15% of the total loss (gases plus leaching). Heal et al. (1978) stressed that total decomposition rates, estimated by gross changes in weight of litter, obscure the individual processes of microbes degrading a very large range of cellular and structural components at widely different rates.

Mason and Bryant (1975) showed that the leaves of *Phragmites* and *Typha* growing in fen peat in the Norfolk Broads (England) are completely decomposed in 1.4 and 2.1 years respectively, which emphasizes the slowness of these processes at the blanket-bog sites considered by Heal et al. (1978) (see also Mason and Standen, Ch. 11).

Few comparisons have been made between decomposition rates on mire surfaces and those on neighbouring mineral soils. Frankland (1966) has shown that *Pteridium* petioles lost 40% of their original dry matter during five years on the peat of a drained raised bog as compared with a maximum of 80% on a number of mull and moder mineral soils. By contrast, mineral nutrient losses from the petioles were greatest at the peat site. A more detailed analysis of this question has been presented by Coulson and Butterfield (1978). They exposed a range of plant tissues on peat and mineral soil; where the larger soil animals were excluded, they found that decomposition was either similar on both substrates or that it was greater on peat. With one exception, involving *Phleum* leaves, the same picture emerged when the soil animals were allowed access to the litter. They concluded that the nature of litter is of paramount importance in determining its rate of decomposition on peat, with the remains of important peat-forming species being most resistant to decomposition.

Nutrient losses from, or gains by, superficial litter have been discussed by Dowding (1974) for tundra peatlands and by Mason and Bryant (1975) for fens. In most instances the greatest changes in concentration occurred in soluble ions, such as potassium, sodium and magnesium. Nitrogen and phosphorus levels altered more slowly. In some instances a correlation could be determined between the rate of nutrient loss and the rate of weight loss from that tissue. Heal et al. (1978) determined the changes in the concentrations of carbon, nitrogen, phosphorus and potassium in the main litter components of blanket peat at Moor House over four or five years. There were no significant changes in carbon content as percent dry weight of litter, but the final concentration of nitrogen increased from 50 to 127% of the initial concentration and phosphorus increased by from 15 to 70%. Changes in potassium concentration, however, varied from a 5% increase in *Calluna vulgaris* shoots to a decrease of 84% in *Rubus chamaemorus* litter. The increases reflect the relatively greater loss of carbon during decomposition. Comparisons with concentrations of nutrients in similar materials within the peat suggested how rapidly equilibrium states could be expected to be reached. *Eriophorum vaginatum* leaves and *Calluna vulgaris* shoots were unlikely to change further after five years, but nutrient levels in *Calluna* stems were still below the concentrations in the peat, reflecting their low rate of decay. Dowding (1974) has attempted to co-ordinate such nutrient fluxes and the associated decomposition changes by considering a broader range of carbon/nutrient ratios than are usually considered in this respect. Ratios such as C/P, C/K, N/K and K/P are, however, at present of limited value because the processes affecting their fluctations are not necessarily linked together nor fully understood. Rates of decomposition of a variety of types of plant litter have been shown to be highly correlated with their

[1] Distinct horizons within the litter layer are often designated as the L or litter layer, where newly fallen debris accumulates but where microbial action is limited by extreme fluctuations in the environment, the F or fermentation layer, in which microbial activity is intense and which can often be divided into two sublayers based on the state of decay of the remains, and the H or humification layer, where the litter has lost most of its original form and where microbial activity has subsided to a steady but rather low rate.

nitrogen and phosphorus contents (Coulson and Butterfield, 1968). Enrichment of the litter with nitrogen enhanced decomposition, but extra phosphorus had no effect on its rate of decay. Nitrogen also produced an increase in the density of blanket-bog peat invertebrates, and this may in part account for the increased decomposition.

While litter is on the peat surface there may be alterations to its physical or chemical nature which will tend to inhibit further decomposition. It would seem likely, however, that such changes will be most pronounced within the surface layers of the mire.

Decomposition in the surface layers of mires

It may be argued that the failure of micro-organisms to complete the decomposition of tissues which have become incorporated into the mire surface, or to destroy materials formed *in situ* within the mire surface, is the principal cause of peat formation. Clearly several factors, such as temperature, pH, water content, the partial pressure of several gases, mineral nutrient levels, microbial and higher plant toxins and the anatomical characteristics of litter, play a part, individually or in combination, in limiting microbial activity. Whatever factors are involved, it is clear from a number of studies that there is an abrupt decline in microbial activity in the surface layers of mires.

Higher plant tissues.

Few experiments on litter decomposition have been designed to continue until the material has become incorporated into the surface layer of the mire. This may be because it becomes difficult to recover sufficient material for adequate measurement. Some attempts have, however, been made to determine weight losses for litter artificially buried at known depths (see p. 236 below for comments on *Sphagnum*). Heal et al. (1978) presented data for such an experiment carried out in a blanket bog in the Pennines. In this study they attempted to avoid the problems associated with use of living material or newly-dead litter in a study of weight losses at different depths in the peat. Weight loss from *Juncus effusus* (a species used because of its experimental convenience) and *Calluna vulgaris* stems decreased significantly with depth, but loss from *Eriophorum vaginatum* roots *increased* with depth. There was no obvious explanation for this unexpected result, but it was speculated that because the *Eriophorum* roots had been extracted from a depth of 10 to 20 cm they carried a microflora adapted to conditions at that depth. In most litter decomposition experiments the main method used for assessing microbial activity has been weight-loss determinations. It is, however, useful to supplement weighing with one or more assessments of the state of preservation of the tissues involved. The scanning electron microscope has considerable potential in this respect. Examples of the deterioration which has been observed at Hummel Knowe Moss, a raised bog in the northern Pennines, are shown in Fig. 5.2. More detailed studies using the light microscope and microtome sections have shown that, despite some loss in weight, the cellular anatomy of leaves and roots of *Eriophorum* and *Erica* remains intact for many years after the tissue dies.

Wilson (1977) has emphasized the importance of root tissues in peat formation. Many roots develop within microbiologically inert layers of peat and, unlike their counterparts in mineral soil (Waid, 1974), they are not invested with extensive epiphytic fungal populations. In studies at Hummell Knowe Moss a very high proportion of dead roots

Fig. 5.2. Scanning electron micrographs of living and dead tissues from Hummel Knowe Moss, Northumberland, England. Scale bar = 50 μm.

A. *Erica tetralix* green leaf showing the intricate epidermal cell wall corrugations and trichomes.

B. *E. tetralix* dead leaf from about 6 cm depth in peat; epidermal cell structure well preserved but with extensive superficial microbial colonization.

C. *E. tetralix* leaf dissected out from peat at about 6 cm depth, showing extensive damage to the outer walls of many of the epidermal cells but with little sign of microbial activity.

D. *Eriophorum vaginatum* dead, brown root, showing collapsed appearance of the epidermal cells, which is similar to that of the white living roots, but with no clear evidence of decomposition.

E. *E. vaginatum* newly-dead leaf, illustrating the typical appearance of this monocotyledon organ, no evidence of decomposition.

F. *E. vaginatum* leaf after 37 weeks in a litter bag on the peat surface in the field, showing extensive decomposition along the lines of epidermal cells.

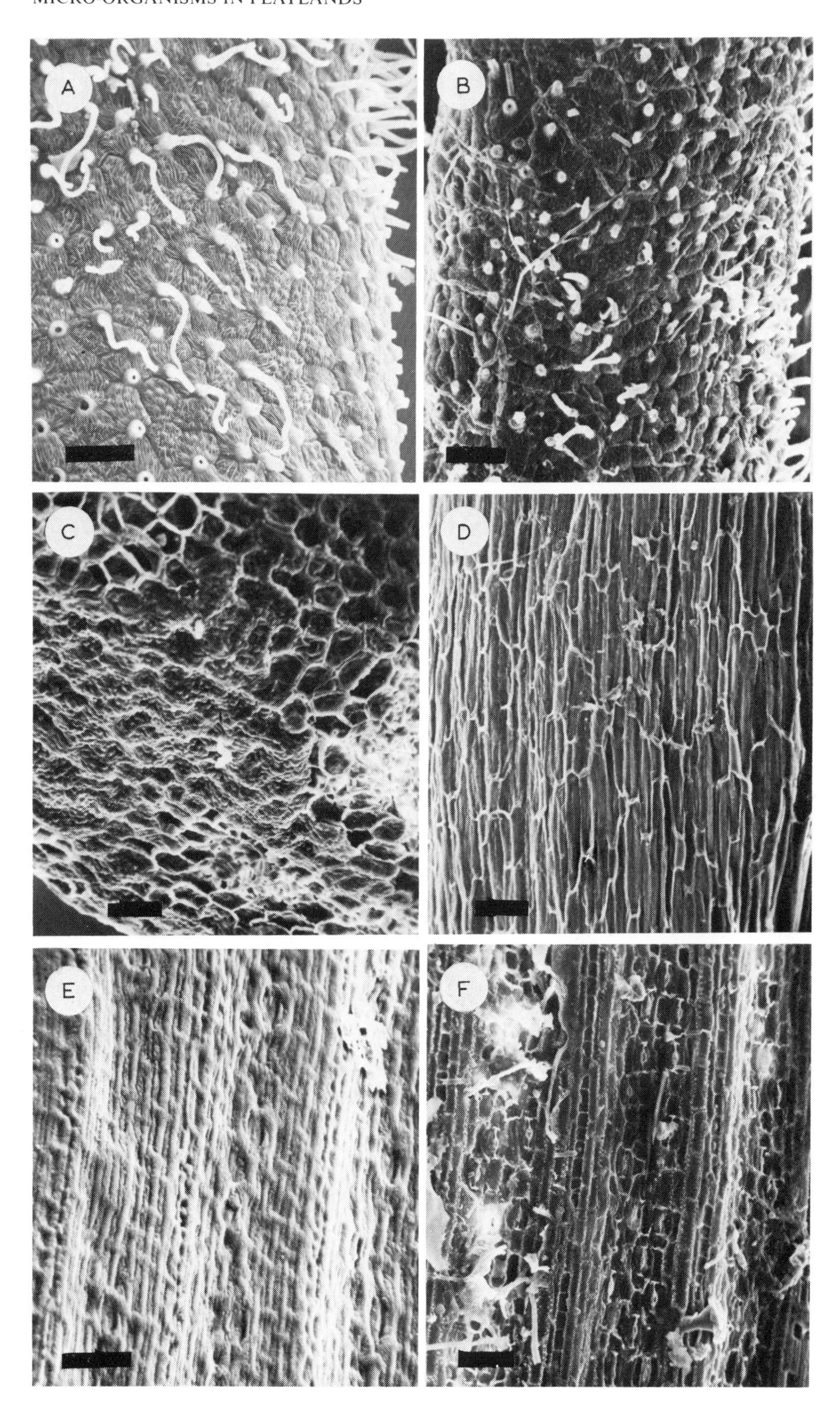
A
B
C
D
E
F

of *Erica tetralix* and *Eriophorum vaginatum* from several depths were found to be free from fungi (Wilson, 1977). Blakemore (1978) confirmed these results using the scanning electron microscope, and she has also shown that, despite the frequent flattening of many roots, their cellular anatomy is preserved extremely well. Sorting and weighing can also be used to illustrate the substantial contribution of roots to peat formation (Table 5.1).

These results contrast somewhat with the results for *Eriophorum vaginatum* roots discussed above (Heal et al., 1978), but they are supported by an earlier study (Heal and French, 1974) in which very small annual weight losses were reported for roots of *Eriophorum* and *Calluna* in an English blanket bog. Larger weight losses occurred from the roots of *Carex* and other plants at Hardangervidda.

The relative importance of roots as peat-building tissues elsewhere is largely unexplored. One exception concerns mangrove peats in Florida, which have been shown by Cohen (1968) to include large proportions of extremely well preserved roots. Tidal scour removes much of the aerial mangrove debris, and the peats would appear to build up by the continual intersusception of roots and rootlets into the substrate.

The significance of roots as peat-forming tissues will vary according to the rooting strategy of different plants (Table 5.2). Notable differences occur in this respect between plants such as *Erica* spp. and *Eriophorum* spp. Roots of the former inhabit superficial layers of mires whereas cotton-

TABLE 5.1

Composition of the surface layers of peat at Hummel Knowe Moss, Northumberland, England; eight cores from a *Sphagnum* lawn were frozen, sliced and sorted into five types of component

Depth (cm)	Volume sorted (ml)	Composition (mg dry weight ml^{-1})				
		Sphagnum	higher plants			amorphous material
			shoots	live roots	dead roots	
0–2	167	19.9	1.3	0.1	0.2	7.6
4–6	131	20.6	8.9	0.6	0.6	21.6
8–10	126	23.1	6.8	1.2	1.4	43.9
12–14	126	14.0	12.0	2.2	4.3	51.5
16–18	123	9.3	13.8	2.1	4.6	57.4
20–22	139	10.5	9.8	1.4	5.1	55.9
24–26	56	11.4	9.6	2.4	3.4	42.8

TABLE 5.2

Rooting characteristics of a number of peat-inhabiting plants

Species	Working depth (cm)	Maximum depth (cm)	Laterals	Hairs	Mycorrhizal infections
Erica tetralix[1,4]	0–15	18	+	–	+
Narthecium ossifragum[1,4]	2–20	25	+	–	+
Dupontia fisheri[2]	5–15	20	+	+	?
Carex aquatilis[2]	10–20	25	+	?	?
Juncus squarrosus[5]	0–15	30	+	+	–
Calluna vulgaris[1,4]	0–15	31	+	–	+
Molinia caerulea[1,4]	0–31	61	+	+	+
Scirpus cespitosus[1,4]	15–31	61	+	±	
Eriophorum angustifolium[2,4]	15–61	61	–	–	–
Eriophorum vaginatum[3,4]	15–61	100	–	–	–

[1]Data from Heath et al. (1938); [2]data from Shaver and Billings (1975); [3]data from Wein (1973); [4]data from Boggie et al. (1958); [5]data from Welch (1966).

grass roots penetrate to considerable depths, and thus there is a continuous alteration in the composition of peat layers originally formed many decades earlier. In Chapter 1 Gore (p. 18) refers to an attempt to estimate the effects of deep-rooting plants on peat accumulation. In general these effects were smaller than expected, but root penetration must be taken into account when considering nutrient cycling in that it extends the depth of the peat which is "active" in this respect.

Roots play another important role in peat formation in that they are important agents in the comminution of tissues. The leaf bases of tussock plants, such as *Eriophorum* and *Scirpus*, are partially fragmented by the formation of adventitious roots, and in sorting peat I have regularly seen *Sphagnum* leaves which have been pierced by roots.

Microbiological studies of naturally buried litter show that there is a marked fall-off of populations in the top layers of mires. Latter and Cragg (1967) have shown that fungal colonization of *Juncus squarrosus* leaves was most extensive at about 5 cm below the peat surface, but had declined markedly by 10 cm. Wilson (1977) showed that there was a similar marked decline in both the qualitative and quantitative aspects of the fungal populations on *Eriophorum* and *Erica* leaf litter in the top 10 cm of Hummel Knowe Moss. Dickinson and Maggs (1974) found that *Sphagnum* leaves were extensively colonized by fungi in the top 10 cm of Fozy Moss, Northumberland, England, but below this more than 90% of leaves failed to yield any fungal colonies (see p. 236).

There are to date insufficient data to make many generalizations concerning the exact pattern of decline in microbial activity in the top layers of mires, but Collins et al. (1978) have provided a detailed account of microbial distributions in the peat at Moor House. They used the colour of the upper peat horizons to distinguish fungal and bacterial habitats rather than standard units of depth. Rather larger numbers of both fungi and bacteria were found under *Calluna vulgaris* than under *Sphagnum*. In general fungal biomass exceeded that of bacteria in the litter zone (to 5 cm) but the situation was reversed with increasing depth, and although bacterial numbers decreased down the profile the decline was greater with aerobic than with facultative or anaerobic types. For example, the densities (in thousands per gram of dry matter) in the litter layer were 26 000 $\pm$ 6600 for aerobic heterotrophic bacteria, and 590 for anaerobic types, while in the green-brown layer (at about 13 cm) the numbers were respectively 7600 $\pm$ 1900 and 630. Over 2000 m g^{-1} of stained fungal mycelium was found in the litter layer, but only 750 m g^{-1} was recorded in the green-brown layer.

Sphagnum

Sphagnum is a pre-eminently successful peat-inhabiting and peat-forming plant. The growth habit of many species of *Sphagnum* is such that tissues live and die close to the water table and hence microbial breakdown of the remains is highly restricted. In addition this plant contains a number of phenolic compounds, including a substance known as sphagnol, and these materials have been shown to inhibit microbial activity (Given and Dickinson, 1975). The actual fate of *Sphagnum* remains has been investigated by at least four different methods.

(1) Perhaps the simplest method of determining the extent to which *Sphagnum* litter is decomposed involves examining peat profiles for evidence of tissue survival. Unfortunately, as with so many other aspects of peatland ecology, the major problem with this method is that one cannot know for certain what the status of living *Sphagnum* was in the mire community when any particular peat horizon was formed. Even where the peat is almost entirely composed of *Sphagnum* remains it does not mean that decomposition has been either negligible or constant. These difficulties of interpretation are illustrated by data from the surface layers of Hummel Knowe Moss (Table 5.1). Cores from *Sphagnum*-dominated lawns were frozen, sliced, measured, sorted, dried and weighed. The data show that the contribution of *Sphagnum* remains to peat accumulation decreases as one penetrates deeper into the peat, at least down to 18 cm depth. It is tempting to ascribe this to progressive decomposition; but the increased values for higher plant shoot material in these deeper layers may imply either that the higher plant remains have decomposed less rapidly than the *Sphagnum* litter, or that there has been a change in the composition of the community over the period involved in the formation of this layer of peat. Some of the effects of differential decomposition rates have been examined by Jones and Gore (1978) (see p. 17).

(2) An extension of this approach involves the production of a dry matter balance sheet for mire communities which are dominated by *Sphagnum*. Such a balance sheet includes estimates of productivity and the cumulative decomposition processes, determined by measuring gas evolution and solution losses. Clymo and Reddaway (1971) carried out such an exercise in a blanket bog at Moor House, and they calculated that there were net gains of 1.0, 1.8 and 2.7 g dry matter dm^{-2} yr^{-1} in *Sphagnum* hummocks, pools and lawns respectively. These imply that 42%, 52% and 71% of the annual production of hummocks, pools and lawns respectively is accumulating at present as peat at that site. Clymo and Reddaway concluded that "the productivity of these peatlands is unimpressively small. What is impressive is the accumulation of peat resulting from the even smaller rate of breakdown."

Clymo (1965) had already attempted to gain further insight into the problem of *Sphagnum* decomposition by an experiment involving litter bags buried in two mires, a lowland poor fen and upland blanket bog, for varying periods. Weight losses were greatest in the lowland site, and for *Sphagnum papillosum*, as compared with *S. cuspidatum* and *S. capillifolium*. Losses over a period of twelve months reached about 20% of the original weight in the surface layers, that is, down to 10 cm, but were less than 5% at 75 cm. However, as indicated by the work of Heal et al. (1978) described above, this approach may be erroneous as it short-circuits the microbial processing which normally occurs as green tissues senesce.

(3) Another method for studying the decomposition of *Sphagnum* was described by Dickinson and Maggs (1974). A combination of scanning electron microscopy and light microscopy was used to determine the state of preservation of *Sphagnum* leaves extracted from various horizons of Fozy Moss, an ombrotrophic raised mire in Northumberland, England. Some spectacular examples of decomposition were seen, but the quantitative data showed that there was considerable preservation of the leaves with little or no gross-structural damage in a high percentage of samples (Fig. 5.3). More damage was seen at depths of 16 cm and 30 cm than at 0.5 cm or 2.0 cm but very few leaves appeared to have been broken down into fragments. Stems were not examined in this study.

(4) Given the difficulties of trying to ascertain the decomposition of *Sphagnum* remains in the field, it is not surprising that a number of workers have attempted to analyse this problem in the laboratory. These studies, which have been reviewed by Given and Dickinson (1975), indicate that decomposition of this plant is carried out by a small and select group of peat micro-organisms. Many of the saprophytic fungi isolated from *Sphagnum* remains were shown to be ineffective in this respect.

All these studies would seem to suggest that *Sphagnum* debris is subject to some decomposition, but that this may only result in the loss of a relatively small fraction of the original weight. As already indicated, there are several possible reasons for the reduction in the rate of decomposition with increasing depth. Some evidence suggests that the leaf hyaline cells of *Sphagnum* may be especially resistant to decay. Cushions of the closely related moss *Leucobryum glaucum* have been shown to contain perfectly preserved leaves and stems, and the most acceptable explanation for this phenomenon involves toxins which inhibit microbial activity. These *Leucobryum* studies tend to suggest that differential rates of decay are unlikely to be important, though some earlier workers mentioned that *Sphagnum* stems appeared to be decomposed faster than its leaves, and this would account for the "collapse" shown by Clymo (1965).

Decomposition in deeper horizons of mires

Much of the information concerning microbial activity in the deeper strata of mires consists of

Fig. 5.3. Scanning electron micrographs of the outer surfaces of *Sphagnum* leaves from Fozy Moss, Northumberland, England. Scale bar = 50 μm.

A. *S. palustre* green leaf illustrating the appearance of intact, hyaline cells with their thickening bars and pores.
B. *S. palustre* leaf from 10 cm depth in peat, showing excellent preservation despite having been buried for many years.
C. *S. palustre* leaf from 6 cm depth showing an early stage in the breakdown of the hyaline cells.
D. *S. papillosum* leaf from 8 cm depth well preserved but with a fungal hypha entering a hyaline cell pore.
E. *S. papillosum* leaf from 10 cm depth showing fungal spores, hyphae and hyaline cell breakdown.
F. *S. papillosum* leaf from 12 cm depth showing the typical pattern of hyaline cell destruction with the walls having collapsed between the thickened bars.

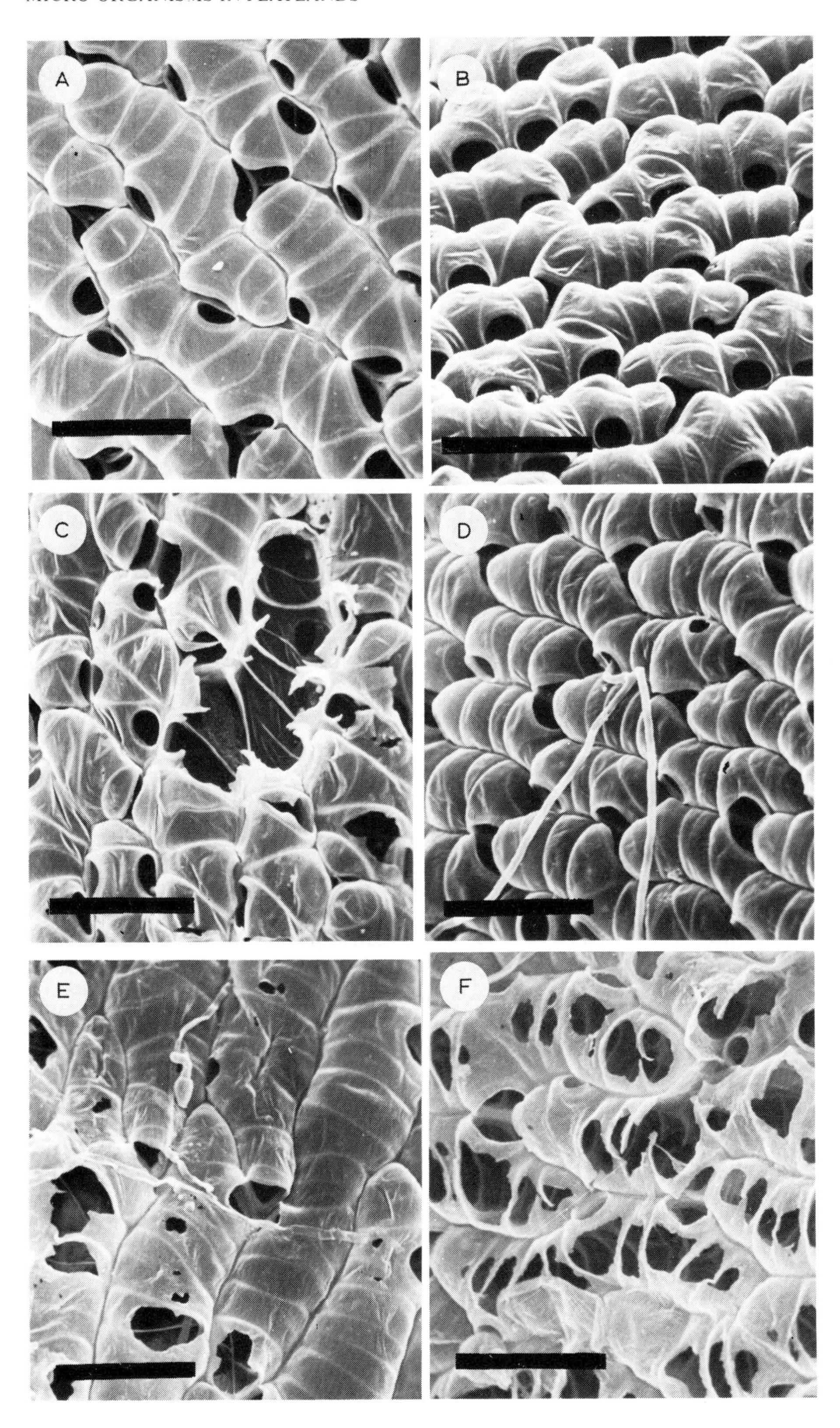
A
B
C
D
E
F

demonstrations that there are populations of viable organisms at considerable depths in peat (Waksman and Purvis, 1932; Dickinson et al., 1974). These populations are generally very small indeed by comparison with those in the surface layers of mires, but the time scale over which they can act to bring about changes in the peat is enormously long in microbial terms. The populations of organisms in deep layers of peat may continue decomposition processes begun in the surface layers, they may colonize newly formed plant roots and they may bring about subtle chemical changes in the substrate. Clymo (1965) found that filter paper lost weight at a rate of 3% yr^{-1} when placed at 75 cm in the blanket peat at Moor House but, as already noted, there are considerable shortcomings in experimental work of this kind. Stout (1971) examined the populations of bacteria and other micro-organisms in the fen peat at Wicken Fen near Cambridge (see Taylor, Part B, Ch. 1). The total bacterial numbers per gram wet weight (peat moisture content *c.* 75%) in three depths of undrained peat were 150×10^6 (0–7.5 cm), 23×10^6 (7.5–15 cm) and 0.8×10^6 (*c.* 1 m). These values are similar, when adjusted, to those reported above for observations by Collins et al. (1978) on bog peat. A garden soil on peat outside the water-retaining lode or dyke at Wicken Fen had a relatively low total number (6.5×10^6) of bacteria. Stout concluded from this and other evidence that the striking difference in the level of the Fen within the lode from that of the agricultural soil outside (2–3 m) was due to erosion, shrinkage and other physical factors, and not to microbial oxidation. However, Schothorst (1977) came to an opposite conclusion for similar fens southwest of Utrecht in The Netherlands. He considered that microbial oxidation following deep drainage was almost wholly responsible for the observed subsidence of 2 m over 1000 years.

Many roots dissected from deeper layers of oligotrophic peat were not colonized by fungi (Wilson, 1977). This may be explained by the very slow growth rate of organisms in deep layers of peat, which would mean that for long periods many pieces of debris would escape the attentions of micro-organisms. Subtle alterations to the chemistry of peat may be of minor significance to biologists studying mires but they are of interest to geochemists for whom such organic sediments may represent an important step in the genesis of coal (Given, 1972).

NUTRIENT UPTAKE BY HIGHER PLANTS

The uptake of ions by green plants growing in well-drained soils may be through the specialized absorbing surfaces associated with root apices, or, where the root is infected with a mycorrhizal associate, via fungal hyphae. Plants growing in peat are not likely to be exceptional in these respects, though the development of root hairs is suppressed in waterlogged, acidic and calcium-deficient environments. Most mycorrhizal associations appear to be beneficial in terms of the uptake of particular mineral ions, but even where they are absent rhizosphere micro-organisms, whose activity is stimulated by the developing root, may increase the availability of essential nutrients. Rhizosphere organisms play a number of roles in the biology of root systems, but little is known concerning the extent of their activity in peat.

Mycorrhizal associations

A substantial proportion of peat-inhabiting plants are known to be regularly infected by mycorrhizal fungi (Table 5.3). The associations are of five main types according to the classification proposed by Lewis (1975).

(1) Ectomycorrhizas

The ectomycorrhizal, or sheathing, type of association has proved to be relatively easy to study and a great deal is now known about the physiological relationships between the partners (Harley, 1969). These ectotrophic associations are only formed between species of Basidiomycotina and various trees, and as such they are likely to occur in relatively few types of mire. Their presence may be detected by the appearance of toadstool sporophores or by careful excavation and examination of the root system. Mycorrhizal roots are relatively broad, by comparison with uninfected roots, and the root system is dominated by clusters of numerous short branches.

Ectomycorrhizal associations facilitate the uptake of ammonium, phosphate and metal ions from newly-fallen litter. The fungal partners enable the

TABLE 5.3

Occurrence of mycorrhizal associations in a selection of peat-inhabiting plants

Type of mycorrhizal association	Higher plant partner examples	
	family	species
Ectomycorrhizal	Myrtaceae	*Eucalyptus* spp.
	Pinaceae	*Pinus* spp.
	Betulaceae	*Betula pubescens*
Ericoid	Empetraceae	*Empetrum nigrum*
	Ericaceae	*Andromeda polifolia*
		Calluna vulgaris
		Erica cinerea
		Erica mackaiana
		Erica tetralix
		Vaccinium myrtillus
		Vaccinium oxycoccos
		Vaccinium vitis-idaea
Arbutoid	Ericaceae	*Arctostaphylos uva-ursi*
Arbuscular	Cyperaceae	*Carex stans*
		Kobresia sp.
	Liliaceae	*Narthecium ossifragum*
	Poaceae	*Molinia caerulea*
		Nardus stricta
Orchidaceous	Orchidaceae	*Liparis loeselii*
		Listera cordata
		Malaxis paludosa
		Orchis fuchsii
		Spiranthes romanzoffiana
No associations found to date	Cyperaceae	*Cladium mariscus*
		Eriophorum vaginatum
		Schoenus nigricans
	Juncaceae	*Juncus squarrosus*
	Poaceae	*Phragmites australis*
	Menyanthaceae	*Menyanthes trifoliata*
	Nymphaeaceae	*Nuphar lutea*
		Nymphaea spp.
	Rosaceae	*Rubus chamaemorus*

higher plants to regain quickly the nutrients which they lose in deciduous debris and this may involve the fungi becoming active in the early stages of litter decomposition. It is likely that such ectomycorrhizal associations are as important in this respect in mire habitats as they are in mineral soils, but there is evidence that only a limited range of ectomycorrhizal fungi are able to function in moorland soils (Harley, 1969). For example, *Leccinum scabrum* and *Amanita muscaria* are the only mycorrhizal toadstools to infect birch in *Calluna* moorland, whereas a greater variety of fungi are associated with this host elsewhere (Dimbleby, 1953). Heikurainen (1955) reported that ectomycorrhizal roots only occurred near the surface of waterlogged soils, and those that were seen were relatively dark-coloured and slender, suggesting that they were in poor physiological condition. In drained mires there was an improvement in the appearance of infected roots; they were found deeper in the peat, and more fungal sporophores were found.

(2) Ericoid mycorrhizas

These infections are common in many members of the Ericaceae; they consist of intracellular hyphae in the outer cortical cells of the lateral roots. The only fungus so far identified as an ericoid associate is an ascomycete named *Pezizella ericae*, which is apparently widely distributed in many soils. Not surprisingly, in view of the ecological distribution of the host genera, these associations occur in peat with a similar frequency to that in mineral soils. Stribley and Read (1975) produced evidence to show that these associations are of special significance to plants growing in nutrient-poor soils. The fungal hyphae appear to act as auxiliary root extensions, absorbing both simple and complex forms of nitrogen and phosphorus from the peat and translocating them into the root. Here they are transferred to the root cells, possibly in exchange for carbohydrates from the host. The ability of the fungal enzymes to break down complex nitrogenous molecules could be especially important for peat-inhabiting plants, as the limited decomposition and peculiar chemistry of this habitat means that such complex nitrogen-containing molecules may be more common here than in mineral soils.

Burgeff (1961) made an extensive study of these ericoid mycorrhizal fungi, and he showed that they could decompose tannins and other compounds which inhibited the growth of other micro-organisms in the peat. They were also shown to stimulate the growth of other organisms through a change in the pH in the vicinity of their hyphae, and they enhanced the rate of bacterial proteolysis in the anaerobic layers of peat.

(3) Arbutoid mycorrhizas

The superficial appearance and developmental patterns of the mycorrhizal associations in *Arbutus* and *Arctostaphylos* roots resemble the ectotrophic system of *Pinus* and other trees. However, in these arbutoid plants there is a later stage which, unlike the *Pinus* pattern, involves the hyphae penetrating into, and subsequently being digested by, the host cells. In this latter respect these associations appear to have something in common with endotrophic mycorrhizas. Little is known concerning the identity of the fungal partners and the physiology of these systems. The possibility exists that the fungi involved supply not only minerals but also carbohydrates derived from organic matter in the substrate (Lewis, 1975).

(4) Arbuscular mycorrhizas

In mineral soils *Endogone* and other similar fungi infect the roots of a wide variety of hosts. They develop within the cortical cells of the root, forming coiled and swollen structures which are sufficiently characteristic for the fungi to be identified with certainty as arbuscular associates. In some soils this internal development is accompanied by the formation of spores in the surrounding soil, but these spores are not common in grassland and permanently wet soils (Sparling and Tinker, 1975). Arbuscular associations are, however, common in such habitats, and it is presumed that spread occurs by hyphal growth from root to root through the soil.

The importance of this type of mycorrhizal association centres on the supply of phosphorus and other ions which have a poor mobility in soil (Bowen et al., 1975). Arbuscular infections have been shown experimentally to enhance the availability of phosphorus, and this usually results in a pronounced growth response by the host plant.

Arbuscular infections have, however, been recorded in few peat-forming plants, and B. Mosse and D. Hayman (pers. comm., 1978) considered *Endogone* infections to be of little significance in wet blanket peat (see Martin and Holding, 1978). The members of the Poaceae listed in Table 5.3 constitute the best examples of peat-inhabiting species which may have such associations, as arbuscular infections are especially common in this family. Less is known concerning the details of the associations in other species listed.

(5) Orchidaceous mycorrhizas

A number of orchid species characteristically inhabit fens and bogs. In Europe about ten species occur in ombrotrophic bogs, though they are normally restricted to habitats where there is a substantial lateral water flow. These characteristic bog species include *Dactylorhiza maculata*, *D. lapponica*, *Malaxis paludosa*, *Listera cordata*, *Microstylis monophyllos* and *Platanthera hyperborea*. A more diverse orchid flora is found in calcareous fens, including species of *Dactylorhiza*, *Epipactis*, *Gymnadenia*, *Listera*, and *Spiranthes*. Orchids spread by seed, but for successful germination and establishment they must become infected with a *Rhizoctonia*-like basidiomycete. The resulting mycorrhizal association has an important role in providing carbohydrate during the early heterotrophic life of the chlorophyll-free seedling. Continued infection of the roots and rhizomes of mature plants, which is common but not obligatory, may merely signify the continuation of the supply of carbohydrates. Apart from the report by Smith (1966) that phosphorus is translocated into protocorms, there is little other evidence concerning the possible role played by these fungi in enhancing the uptake of mineral ions from the soil or other substrates (Purves and Hadley, 1975).

As with other aspects of microbial ecology in peat, the difficulties of relating the often low levels of activity to ecologically significant supplies of nutrients are considerable. This undoubtedly accounts for the paucity of knowledge concerning such mycorrhizal associations in mire-plant nutrition.

Species lacking mycorrhizal infections

Harley (1969) noted the scarcity of mycorrhizal infections in two families, the Cyperaceae and the Juncaceae, which are especially common in mires. Powell (1975) confirmed this generalization, and suggested that these families have avoided the need for mycorrhizal infections by the development of especially long, branched roots which bear numerous large root hairs. These root systems were shown by Powell to excel at obtaining phosphorus from soil. In one experiment *Juncus planifolius*, and mycorrhizal and non-mycorrhizal plants of *Poa colensoi*, were grown in three soils deficient in phosphorus. When the shoots of the three sets were analyzed the *Juncus* plants contained on average 0.13% phosphorus as against 0.19% for the mycorrhizal and 0.09% for the non-mycorrhizal *Poa* plants. Amongst peat-inhabiting genera it is notable that *Eriophorum* in particular has exceptionally long roots which have been shown to penetrate to depths greater than 1 m in ombrotrophic mires.

Other associations between fungi and green plants

Apart from the recognized mycorrhizal associations there are a number of other fungus/green plant relationships in mires whose status is less well understood. Several species of the basidiomycete genus *Omphalina* regularly fruit in *Sphagnum* lawns, and the zygomycete genus *Mortierella* is commonly associated with both *Erica* roots and *Sphagnum* and *Leucobryum* leaves (Burgeff, 1961; Dickinson and Maggs, 1974). There have been some suggestions that *Mortierella* species are mycorrhizal partners in the Ericaceae, but this has now been shown to be incorrect. *Omphalina* may be involved in the decomposition of *Sphagnum* remains; but *Mortierella* is less likely to be involved in this type of activity as it is a "sugar-fungus", whose abilities do not generally include the breakdown of cellulose or other polymers.

Rhizosphere formation in peats

There is very little evidence concerning the significance of the rhizosphere effect in peat. Martin (1971) was unable to detect an enriched rhizosphere-like population of bacteria on or around *Eriophorum vaginatum* roots, but Martin and Holding (1978) refer to large numbers of facultatively anaerobic, acid- and gas-producing bacteria being produced by the percolation of columns of peat with sodium glutamate. Such organisms were isolated in significant numbers only from the rhizosphere of *Eriophorum* species, and amino acids are known to be secreted by many plant roots. As far as fungi are concerned, Wilson (1977) has shown that the surfaces of living *Eriophorum vaginatum* roots are only sparsely colonized. On average 38% of 700 root pieces, each 1 cm long and taken from several depths between 4 cm and 32 cm at Hummel Knowe Moss, were free from fungi. The remainder were generally each colonized by only one type of fungus, and in the majority of instances this fungus was a sterile unidentifiable form. Similar results were also obtained for *Erica tetralix* roots.

In this respect it may be worth briefly considering the factors which would influence the development of active rhizosphere and root surface populations. The most obvious antagonistic factor which must be overcome is that resulting in a general inhibition of microbial growth within peat. Additionally the roots of some plants, such as *Eriophorum vaginatum*, grow through peat very rapidly and they are relatively short-lived (Forrest, 1971). Hence there may be too little time for extensive microbial colonization of either the rhizosphere or the root surface. On the positive side, however, root development will result in the release of oxygen, and of sugars, amino acids, and other microbial nutrients, which are otherwise lacking in deeper layers of peat (Geilser, 1955; Greenwood, 1971). Urquhart (1966) and Urquhart and Gore (1973) used silver-coated plates inserted into the surface layers of upland and lowland wet bog peats in northern England to illustrate the distribution of sulphide production. Their plates showed white streaks corresponding to the roots of *Eriophorum* species on the otherwise darkened silver surfaces, which suggested strongly that there were some root processes inhibiting the production of sulphides. These changes may not, however, ameliorate the environment sufficiently for micro-organisms to become active. Supra-optimal water levels and the occurrence of toxins may continue to exert an overriding inhibitory effect on microbial growth.

NUTRIENT TRANSFORMATIONS WITHIN PEAT

Where micro-organisms are active mineral nutrients are likely to undergo alterations in their chemical status. The best known examples of such transformations concern nitrogen, which can be changed from an organic form into ammonia, nitrite and nitrate, or may be converted to the elemental gaseous form with a consequent loss from the ecosystem. Sulphur is another element whose chemical form in soil is largely decided by the activity of micro-organisms, with the sulphate and sulphide ions playing a major part in soil chemistry.

Nitrogen transformations

Aerobic ammonia-forming bacteria are not uncommon in peat (Visser, 1964; Zimenko, 1966), but there is general agreement that nitrifying bacteria are scarce in most mires (e.g. Collins et al., 1978). Visser (1964) found tiny populations of these organisms in three Ugandan mires dominated by *Sphagnum*, *Cyperus papyrus* and *Phragmites* respectively and situated at altitudes of from 1900 to 3400 m above sea level with annual surface temperatures varying from 12 to 17°C. In arctic, wet-meadow tundra peat at Barrow, Alaska, the soil solution contains high levels of ammonia as compared to nitrate nitrogen, and this has been ascribed to the inhibitory effects of low pH, temperature and oxygen partial pressure on the nitrifying organisms (Flint and Gersper, 1974). Zimenko (1966) failed to find any nitrifying bacteria in bog and transitional mires at Ptich' and Obol' (U.S.S.R.), whose pH was around 3.1, but he recorded quite substantial populations in fens at Ptich' and Zhitkovichi, at a pH of 4.6. He noted that the absence of nitrifying bacteria resulted in the accumulation of ammonia, and he expected these bacteria would become more important, utilizing the ammonia, if the peat were brought into cultivation. This hypothesis has been confirmed by Dunican and Rosswall (1974) who only detected nitrifying bacteria in Glenamoy blanket-bog peat when it had been drained and seeded to produce grassland.

As was discussed earlier, Martin and Holding (1978) described in some detail the nitrogen transformations in blanket peat at Moor House (see Fig. 5.1). Their data show, for example, a steady-state pool of 0.2 g m^{-2} NH_4^+ in the top 30 cm of peat. Contributions to this pool were either from ammonification (0.9 g m^{-2} yr^{-1}) or from rainfall (0.6 g m^{-2} yr^{-1}). The part of the annual flux from the NH_4^+ pool due to nitrification was estimated to be at a rate of 0.003 g m^{-2} yr^{-1}. The resulting nitrate pool was so small as to be considered to disappear almost immediately due to denitrification (0.1 g m^{-2} yr^{-1}) and uptake by plants and microorganisms.

Denitrifying bacteria are more common in peat, though the low levels of nitrate in most habitats mean that these organisms occur in small numbers, except perhaps in the litter layer (Collins et al., 1978). The anaerobic conditions in many mires will favour the activity of several groups of denitrifying bacteria, and counts of up to 10^5 g^{-1} dry weight of peat have been reported for several ombrotrophic mires (Dunican and Rosswall, 1974). Visser (1964) recorded higher numbers of denitrifying bacteria in a *Sphagnum*-dominated mire, fewer in a *Phragmites* swamp, and least in a mire dominated by *Cyperus*. This result may be due in part to the occurrence of particular environmental conditions favouring the bacteria, as the levels of nitrate might have been expected to produce an opposite distribution pattern.

Sulphur transformations

Sulphur does not usually command the same attention as does nitrogen, but it is nevertheless an important nutrient for higher plants, and in some environments it may become limiting (Dunican and Rosswall, 1974). Perhaps the most important microbial transformations in this respect are those which occur under anaerobic conditions and which result in the formation of insoluble and possibly toxic sulphides. Given and Dickinson (1975) have summarized the chemistry involved and Sikora and Keeney (Ch. 6) have also commented on aspects of these transformations under anaerobic conditions.

Sulphate-reducing bacteria, mainly *Desulfovibrio* and *Desulfotomaculum*, occur both in acid peats (Paarlahti and Vartiovaara, 1958; Dunican and Rosswall, 1974; Collins et al., 1978) and in more alkaline fens (Visser, 1964). Collins et al. (1978) showed that the populations of these bacteria in peat at Moor House increased with depth from 5 to 20 cm. The larger populations occurred at those depths where the redox potentials indicated the presence of a very low oxygen tension. Stout (1971) did not find any sulphide formation as evidence of activity by these bacteria in sedge peat in England, but he did not attempt to isolate the organisms directly. Silver wires (see Urquhart, 1966) inserted by A.P. Gore (pers. comm., 1979) at Wicken Fen only showed traces of sulphide even after three months' exposure.

Thiobacillus and other thiosulphate-oxidizing species were uncommon in all three mires studied by Visser (1964), but substantial numbers were reported from blanket bog peats in Ireland (Dunican and Rosswall, 1974) and England (Collins et al., 1978) and from acid peat in Finland (Paarlahti and Vartiovaara, 1958). Collins et al. (1978) noted that some of their isolates grew in anaerobic conditions, when they also reduced nitrate to nitrogen. This activity thus provides a link between the nitrogen and sulphur cycles which may be especially important in nutrient-impoverished habitats, such as occur in most bog peats.

GENERAL SUMMARY

In many mire systems there is a low level of microbiological activity and the indications are that there is little mineral nutrient cycling, so that the flow of nutrients is essentially one way.

Nutrient supplies come from external sources, *viz.* rain (ombrotrophic) or the catchment area (minerotrophic), and, unlike many other terrestrial ecosystems, green plants obtain these nutrients without the intervention of micro-organisms. After the death of the primary producers, these nutrients within the litter are incorporated into peat *in situ*. Quantitative evidence in support of this generalization is, however, sparse due to the difficulties of measurement of the essential components of an appropriate model.

Knowledge of the microbiology of mires has recently improved considerably, especially as a result of the I.B.P. There are still, however, many gaps in the present understanding of these ecosystems. In particular the interactions between micro-organisms and green plants need further study. In this respect more attention should now be paid to the microbiology of root systems in peat, including the development and significance of rhizosphere and root surface populations and the possible roles played by mycorrhizal associations.

REFERENCES

Alexander, V., 1974. A synthesis of the I.B.P. Tundra Biome circumpolar study of nitrogen fixation. In: A.J. Holding, O.W. Heal, S.F. MacLean, Jr. and P.W. Flanagan (Editors), *Soil Organisms and Decomposition in Tundra.* Tundra Biome Steering Committee, Stockholm, pp. 109–121.

Blakemore, E.S.A., 1978. *Morphological and Chemical Aspects of the Formation of Peat.* Thesis, University of Newcastle upon Tyne, Newcastle upon Tyne, 76 pp.

Boatman, D.J., Hulme, P.D. and Tomlinson, R.W., 1975. Monthly determinations of the concentrations of sodium, potassium, magnesium and calcium in the rain and in pools of the Silver Flowe National Nature Reserve. *J. Ecol.*, 63: 903–912.

Boggie, R., Hunter, R.F. and Knight, A.H., 1958. Studies of the root development of plants in the field using radioactive tracers. *J. Ecol.*, 46: 621–639.

Bond, G., 1974. Root-nodule symbioses with actinomycete-like organisms. In: A. Quispel (Editor), *The Biology of Nitrogen Fixation.* North-Holland, Amsterdam, pp. 342–378.

Bowen, G.D., Bevege, D.I. and Mosse, B., 1975. Phosphate physiology of vesicular–arbuscular mycorrhizas. In: F.E. Saunders, B. Mosse and P.B. Tinker (Editors), *Endomycorrhizas.* Academic Press, London, pp. 241–260.

Broady, P.A., 1977. The Signy Island terrestrial reference sites. VII The ecology of the algae of site 1, a moss turf. *Br. Antarct. Surv. Bull.*, 45: 47–62.

Broady, P.A., 1979. The Signy Island terrestrial reference sites: IX. The ecology of the algae of site 2, a moss carpet. *Br. Antarct. Surv. Bull.*, 47: 13–30.

Burgeff, H., 1961. *Mikrobiologie des Hochmoores.* Fischer-Verlag, Stuttgart, 197 pp.

Chandler, G.E. and Anderson, J.W., 1976. Studies on the nutrition and growth of *Drosera* species with reference to the carnivorous habit. *New Phytol.*, 76: 129–141.

Chapin, III, F.S., Barsdate, R.J. and Barel, D., 1978. Phosphorus cycling in Alaskan coastal tundra: a hypothesis for the regulation of nutrient cycling. *Oikos*, 31: 189–199.

Clarholm, M., Lid-Torsvik, V. and Baker, J.H., 1975. Bacterial populations of some Fennoscandian tundra soils. In: F.E. Wielgolaski (Editor), *Fennoscandian Tundra Ecosystems. Part I. Plants and Microorganisms.* Springer-Verlag, Berlin, pp. 251–260.

Clymo, R.S., 1965. Experiments on breakdown of *Sphagnum* in two bogs. *J. Ecol.*, 53: 747–757.

Clymo, R.S. and Reddaway, E.J.F., 1971. Productivity of *Sphagnum* (Bog moss) and peat accumulation. *Hidrobiologia*, 12: 181–192. [Reproduced as: A tentative dry matter balance sheet for the wet blanket bog on Burnt Hill, Moor House NNR. *Moor House Occas. Pap.*, No. 3: 15 pp.]

Cohen, A.D., 1968. *The Petrography of Some Peats of Southern Florida (With Special Reference to the Origin of Coal).* Thesis, Pennsylvania State University, University Park, Pa.

Collins, V.G., D'Sylva, B.T. and Latter, P.M., 1978. Microbial populations in peat. In: O.W. Heal and D.F. Perkins (Editors), *Production Ecology of British Moors and Montane Grasslands.* Springer-Verlag, Berlin, pp. 94–112.

Coulson, J.C. and Butterfield, J.E.L., 1978. An investigation of the biotic factors determining the rates of plant decomposition on blanket bog. *J. Ecol.*, 66: 631–650.

Dickinson, C.H. and Maggs, G.H., 1974. Aspects of the decomposition of *Sphagnum* leaves in an ombrophilous mire. *New Phytol.*, 73: 1249–1257.

Dickinson, C.H., Wallace, B. and Given, P.H., 1974. Microbial activity in Florida Everglades peat. *New Phytol.*, 73: 107–113.

Dimbleby, G.W., 1953. Natural regeneration of pine and birch on the heather moors of north-east Yorkshire. *Forestry*, 26: 41–52.

Dooley, F. and Houghton, J.A., 1973. The nitrogen-fixing capacities and the occurrence of blue-green algae in peat soils. *Br. Phycol. J.*, 8: 289–293.

Dowding, P., 1974. Nutrient losses from litter on IBP tundra sites. In: A.J. Holding, O.W. Heal, S.F. MacLean, Jr. and P.W. Flanagan (Editors), *Soil Organisms and Decomposition in Tundra.* Tundra Biome Steering

Committee, Stockholm, pp. 363–373.

Duncan, U.K., 1970. *Introduction to British Lichens*. T. Buncle, Arbroath, 292 pp.

Dunican, L.K. and Rosswall, T., 1974. Taxonomy and physiology of tundra bacteria in relation to site characteristics. In: A.J. Holding, O.W. Heal, S.F. MacLean, Jr. and P.W. Flanagan (Editors), *Soil Organisms and Decomposition in Tundra*. Tundra Biome Steering Committee, Stockholm, pp. 79–92.

Flint, P.S. and Gersper, P.L., 1974. Nitrogen nutrient levels in arctic tundra soils. In: A.J. Holding, O.W. Heal, S.F. MacLean, Jr. and P.W. Flanagan (Editors), *Soil Organisms and Decomposition in Tundra*. Tundra Biome Steering Committee, Stockholm, pp. 375–387.

Forrest, G.I., 1971. Structure and production of north Pennine blanket bog vegetation. *J. Ecol.*, 54: 453–479.

Frankland, J.C., 1966. Succession of fungi on decaying petioles of *Pteridium aquilinum*. *J. Ecol.*, 54: 41–63.

Geilser, G., 1955. The morphogenetic effect of oxygen on roots. *Plant Physiol., Lancaster*, 40: 85–88.

Given, P.H., 1972. Biological aspects of the geochemistry of coal. *Adv. Organic Geochem.*, 1971: 69–92.

Given, P.H. and Dickinson, C.H., 1975. Biochemistry and microbiology of peats. In: E.A. Paul and D. McLaren (Editors), *Soil Biochemistry, 3*. Marcel Dekker, New York, N.Y., pp. 123–212.

Goodman, G.T. and Perkins, D.F., 1959. Mineral uptake and retention in cotton-grass (*Eriophorum vaginatum* L.). *Nature, Lond.*, 184: 467–468.

Gore, A.J.P., 1968. The supply of six elements by rain to an upland peat area. *J. Ecol.*, 56: 483–495.

Granhall, U. and Lid-Torsvik, V., 1975. Nitrogen fixation by bacteria and free-living blue-green algae in tundra areas. In: F.E. Wielgolaski (Editor), *Fennoscandian Tundra Ecosystems. Part I. Plants and Microorganisms*. Springer-Verlag, Berlin, pp. 305–315.

Greenwood, D.J., 1971. Studies on the distribution of oxygen around the roots of mustard seedlings (*Sinapsis alba* L.). *New Phytol.*, 70: 97–101.

Harley, J.L., 1969. *The Biology of Mycorrhiza*. Leonard Hill, London, 334 pp.

Heal, O.W. and French, D.D., 1974. Decomposition of organic matter in tundra. In: A.J. Holding, O.W. Heal, S.F. MacLean, Jr. and P.W. Flanagan (Editors), *Soil Organisms and Decomposition in Tundra*. Tundra Biome Steering Committee, Stockholm, pp. 279–309.

Heal, O.W., Latter, P.M. and Howson, G., 1978. A study of the rates of decomposition of organic matter. In: O.W. Heal and D.F. Perkins (Editors), *Production Ecology of British Moors and Montane Grasslands*. Springer-Verlag, Berlin, pp. 136–159.

Heath, G.H., Luckwill, L.C. and Pullen, O.J., 1938. The rooting systems of heath plants (with a section on the underground organs of heath bryophytes). *J. Ecol.*, 26: 331–352.

Heikurainen, L., 1955. Der Wurzelaufbau der Kiefernbestände auf Reisermoorböden und seine Beeinflussung durch die Entwässerung. *Acta Forest. Fenn.*, 65: 1–85.

Heslop-Harrison, Y., 1978. Carnivorous plants. *Sci. Am.*, 283: 104–116.

Holding, A.J. and Lowe, J.F., 1971. Some effects of acidity and heavy metals in the *Rhizobium*-leguminous plant association.*Plant Soil, Spec. Vol.*, 1971: 153–166.

Jones, D.G., 1966. The contribution of white clover to a mixed upland sward. II. Factors affecting the density and effectiveness of *Rhizobium trifolii*. *Plant Soil.*, 24: 250–260.

Jones, D.G., Munro, J.M.M., Hughes, R. and Davies, W.E., 1964. The contribution of white clover to a mixed upland sward. I. The effect of *Rhizobium* inoculation on the early development of white clover. *Plant Soil*, 21: 63–69.

Jones, H.E. and Gore, A.J.P., 1978. A simulation of production and decay in blanket bog. In: O.W. Heal and D.F. Perkins (Editors), *Production Ecology of British Moors and Montane Grasslands*. Springer-Verlag, Berlin, pp. 160–186.

Kallio, P. and Veum, A.K., 1975. Analysis of precipitation at Fennoscandian tundra sites. In: F.E. Wielgolaski (Editor), *Fennoscandian Tundra Ecosystems. Part I. Plants and Microorganisms*. Springer-Verlag, Berlin, pp. 333–338.

Latter, P.M. and Cragg, J.B., 1967. The decomposition of *Juncus squarrosus* leaves and microbiological changes in the profile of *Juncus* moor. *J. Ecol.*, 55: 465–482.

Latter, P.M., Cragg, J.B. and Heal, O.W., 1967. Comparative studies on the microbiology of four moorland soils in the northern Pennines. *J. Ecol.*, 55: 445–464.

Lewis, D.H., 1975. Comparative aspects of the carbon nutrition of mycorrhizas. In: F.E. Saunders, B. Mosse and P.B. Tinker (Editors), *Endomycorrhizas*. Academic Press, London, pp. 119–148.

McVean, D.N., 1959. Ecology of *Alnus glutinosa* (L.) Gaertn. VII. Establishment of alder by direct seeding of shallow blanket bog. *J. Ecol.*, 47: 615–618.

Martin, N.J., 1971. *Microbial Activity in Peat with Reference to the Availability and Cycling of Inorganic Ions*. Thesis, University of Edinburgh, Edinburgh.

Martin, N.J. and Holding, A.J., 1978. Nutrient availability and other factors limiting microbial activity in the blanket peat. In: O.W. Heal and D.F. Perkins (Editors), *Production Ecology of British Moors and Montane Grasslands*. Springer-Verlag, Heidelberg, pp. 113–135.

Mason, C.F. and Bryant, R.J., 1975. Production, nutrient content and decomposition of *Phragmites communis* Trin. and *Typha angustifolia* L. *J. Ecol.*, 63: 71–95.

Masterson, C.L., 1968. The effects of some soil factors on *Rhizobium trifolii*. *Trans. 9th Int. Congr. Soil Sci.*, II: 95–102.

Millbank, J.W., 1974. Associations with blue-green algae. In: A. Quispel (Editor), *The Biology of Nitrogen Fixation*. North-Holland, Amsterdam, pp. 238–264.

Moore, P.D. and Bellamy, D.J., 1974. *Peatlands*. Elek Science, London, 221 pp.

Morton, A.J., 1977. Mineral nutrient pathways in a Molinietum in autumn and winter. *J. Ecol.*, 65: 993–999.

O'Toole, M.A. and Masterson, C.L., 1968. Interaction of calcium and hydrogen ion concentration on nodulation of white clover in peat. *Ir. J. Agric. Res.*, 7: 129–131.

Paarlahti, K. and Vartiovaara, U., 1958. Observations concerning the microbial populations in virgin and drained bogs. *Commun. Inst. For. Fenn.*, 50: 1–38.

Powell, C.Ll., 1975. Rushes and sedges are non-mycotrophic. *Plant Soil*, 42: 481–484.

Purves, S. and Hadley, G., 1975. Movement of carbon com-

pounds between partners in orchid mycorrhiza. In: F.E. Saunders, B. Mosse and P.B. Tinker (Editors), *Endomycorrhizas*. Academic Press, London, pp. 175–194.
Ruinen, J., 1974. Nitrogen fixation in the phyllosphere. In: A. Quispel (Editor), *The Biology of Nitrogen Fixation*. North-Holland, Amsterdam, pp. 121–167.
Schothorst, C.J., 1977. Subsidence of low moor peat soils in the western Netherlands. *Geoderma*, 17: 265–291.
Shaver, G.R. and Billings, W.D., 1975. Root production and root turnover in a wet tundra ecosystem, Barrow, Alaska. *Ecology*, 56: 401–409.
Small, E., 1972. Photosynthetic rates in relation to nutrient recycling as an adaption to nutrient deficiency in peat bog plants. *Can. J. Bot.*, 50: 2227–2233.
Smith, S.E., 1966. Physiology and ecology of orchid mycorrhizal fungi with reference to seedling nutrition. *New Phytol.*, 65: 488–499.
Sparling, G.P. and Tinker, P.B., 1975. Mycorrhizas in Pennine grassland. In: F.E. Saunders, B. Mosse, and P.B. Tinker (Editors), *Endomycorrhizas*. Academic Press, London, pp. 545–560.
Stewart, W.D.P., 1966. *Nitrogen Fixation in Plants*. Athlone Press, London, 168 pp.
Stewart, W.D.P., 1974. Blue-green algae. In: A. Quispel (Editor), *The Biology of Nitrogen Fixation*. North-Holland, Amsterdam, pp. 202–237.
Stout, J.D., 1971. Aspects of the microbiology and oxidation of Wicken Fen soil. *Soil Biol. Biochem.*, 3: 9–25.
Stribley, D.P. and Read, D.J., 1975. Some nutritional aspects of the biology of ericaceous mycorrhizas. In: F.E. Saunders, B. Mosse and P.B. Tinker (Editors), *Endomycorrhizas*. Academic Press, London, pp. 195–207.
Urquhart, C., 1966. An improved method of demonstrating the distribution of sulphide in peat soils. *Nature*, 211: 550.
Urquhart, C. and Gore, A.J.P., 1973. The redox characteristics of four peat profiles. *Soil Biol. Biochem.*, 5: 659–672.
Visser, S.A., 1964. La présence et l'effet des microorganismes dans les tourbes tropicales. *Ann. Inst. Pasteur*, 107 (Suppl. No. 3): 303–319.
Waksman, S.A. and Purvis, E.R., 1932. The microbiological population of peat. *Soil Sci.*, 34: 95–109.
Waid, J.S., 1974. Decomposition of roots. In: C.H. Dickinson and G.J.F. Pugh (Editors), *Biology of Plant Litter Decomposition*. Academic Press, London, pp. 175–211.
Wein, R.W., 1973. Biological flora of the British Isles. *Eriophorum vaginatum* L. *J. Ecol.*, 61: 601–615.
Welch, D., 1966. Biological flora of the British Isles. *Juncus squarrosus* L. *J. Ecol.*, 54: 535–548.
Wilson, C.M., 1977. *Studies on the Formation of Peat in a Raised Mire*. Thesis, University of Newcastle upon Tyne, Newcastle upon Tyne, 110 pp.
Zimenko, T.G., 1966. The microflora of peat soils. In: *Microflora pochv severnoi i srednei chasti SSSR*, Nauka, Moscow, pp. 136–165 (in Russian).

Chapter 6

FURTHER ASPECTS OF SOIL CHEMISTRY UNDER ANAEROBIC CONDITIONS[1]

L.J. SIKORA and D.R. KEENEY

INTRODUCTION

The unaltered organic soil ecosystem has a profile generally consisting of an upper aerobic portion, a lower anaerobic portion, and a transitional layer separating the two. In the lower portion, the metabolism of organisms reduces the concentration of oxygen, resulting in anaerobic conditions with characteristic redox potentials (Eh), and reduced forms of carbon, nitrogen, phosphorus, sulfur, iron and manganese, and toxic substances. This review will discuss these distinct characteristics.

The heterogeneity of organic soils with respect to mineral content (resulting from erosion from the surrounding catchment) and layering (resulting from differing rates of decomposition and botanical makeup), necessitates that any conclusions on the physical and chemical environment of organic soils must be very general.

REDOX POTENTIAL

Even if an organic soil is waterlogged throughout the year, its profile will exhibit vertical gradients with respect to oxygen, redox potential (Eh), and temperature. This situation, coupled with the fact that the surface of the soil is in a relatively undecomposed condition, whereas decomposition has proceeded to varying degrees beneath the surface, indicates that the upper horizons of an organic soil are in a rather dynamic state chemically. Few measurements of the dynamic changes in peat profiles are available.

Urquhart and Gore (1973), in a systematic study of the Eh of four acid peat profiles (pH 2.7–3.17) in northern England over a one-year period, found that Eh decreased (the environment became more reducing) with depth[2]. Reducing conditions [i.e., Eh values below those at which dissolved oxygen (DO) becomes undetectable (Ponnamperuma, 1972)] prevailed below about 10 cm. Significant linear relationships between temperature and Eh existed, with Eh declining as temperature increased in three of the four profiles. The effects of temperature on the Eh values became more significant with depth, presumably because of the lessening of oxygen diffusion effects on the redox potential. Haavisto (1974) evaluated the redox changes in an acid (pH 3.2–3.8) floating *Sphagnum* peat mat after a heavy rainfall. The Eh dropped an average of 45 mV below values recorded after five days of dry weather. He concluded that these major and rapid changes in Eh could have marked effects on the chemistry of undisturbed natural peats.

While there are inherent problems in interpreting Eh values measured with the platinum electrode (Ponnamperuma, 1972; Bohn, 1971), used within their known limitations these values do provide a rapid, useful technique for evaluating the redox status of mires (see Clymo, Ch. 4, pp. 189–191). Table 6.1 summarizes the range of observed Eh values (platinum electrode) of a number of systems operating in flooded soils, and lists the predominant classes of micro-organisms involved. This sequence can be predicted by thermodynamics (Ponnamperuma, 1972; Stumm and Morgan, 1970).

---

[1] Manuscript received July, 1977.

[2] Redox potentials of peats measured using a bright platinum electrode are reproducible only under definitely reducing conditions; such conditions only applied to the lower horizons of two of the peats used in Urquhart and Gore's study (see Gore, Ch. 1, p. 18).

As is noted in Table 6.1, negative Eh values are associated with obligate anaerobic metabolism (also often referred to as fermentation). Negative Eh readings have often been observed in flooded paddy soil and lake sediment environments (Keeney et al., 1972; Ponnamperuma, 1972; Keeney, 1973; Yoshida, 1975). However, fewer measurements are available for organic soils. Given and Dickinson (1975) in a comprehensive literature review cite the minimum Eh readings in organic soils. The values ranged from +200 mV to 0 mV, with most often observed minimums being in the 0 to +100 mV range. In the study by Urquhart and Gore (1973), minimum Eh_4 value readings ranged from +15 to +310 mV, depending on the site and date of observation. Two of the sites had relatively low minimum Eh_4 readings (< 100 mV most of the year) and S^{2-} production (as shown by sulfide deposition on a silver-coated plate) was detected in these profiles. Similarly, Haavisto (1974) recorded Eh_3 minimum values of +230 to +250 mV, corresponding to Eh_7 values from 0 to slightly negative, and Wilde et al. (1950) reported that the Eh of a flooded moss peat with chlorotic sedges (pH 4.2) was −153 mV. These Eh values confirm that a waterlogged peat profile is generally anaerobic. Thus, denitrification can readily occur (although nitrate is often scarce in acid peats: see Dickinson, Ch. 5, p. 226), iron and manganese can become reduced and mobilized, S^{2-} ions can form, and ferrous sulfide can be precipitated. If the redox conditions are sufficiently low, methane and elemental hydrogen also can be formed and anaerobic nitrogen fixers may be operative.

HYDROGEN ION CONCENTRATION

A wide range of pH values from 2.8 to 7.4 has been reported for organic deposits (Given and Dickinson, 1975). In general, fen peats have pH values near neutral as a result of the effect of calcium and magnesium ions, which together with bicarbonate, raise the pH. Bog peats tend to be acid (i.e., pH *c.* 3). In Wisconsin, U.S.A., pH values of sedge peats ranged from 4.5 to 8.0, but the *Sphagnum* peats measured from 3.6 to 4.0 (Isirimah et al., 1970; Isirimah and Keeney, 1973a). Similar results were reported by Rigg and Gessel (1956) for Washington organic deposits. Marked differences in the chemistry and biochemistry of organic deposits will occur over such a wide range of pH values.

The factors responsible for such a low pH in an anaerobic system are of interest. It is well documented that soils and sediments, regardless of their initial pH, tend toward neutrality upon flooding. Ponnamperuma (1972) has reviewed in depth the physical chemistry of this phenomenon. The increasing pH of acid soils is largely a result of the consumption of hydrogen ions during reduction of iron through equilibrium reactions:

$$Fe(OH)_3 + 3H^+ + e = Fe^{2+} + 3H_2O \quad (1)$$

$$3\,Fe(OH)_3 + H^+ + e = Fe_3(OH)_8 + H_2O \quad (2)$$

$$1/2\,Fe_3(OH)_8 + 4H^+ + e = 3/2Fe^{2+} + 4H_2O \quad (3)$$

Precipitates of $FeCO_3$ and FeS may also be formed during long-term waterlogging. This system:

TABLE 6.1

Possible systems operating in flooded environments as related to redox potential[1]

System	Redox potential range (mV)[2]	Micro-organisms involved
Oxygen disappearance	+500 to +350	aerobes
Nitrate disappearance	+350 to +100	facultative anaerobes
Mn^{2+} formation	below +400	facultative anaerobes
Fe^{2+} formation	below +400	facultative anaerobes
Sulfide formation	0 to −150	
Hydrogen, methane formation	below −150	obligate anaerobes

[1]Takai and Kamura (1966); Connell and Patrick (1968, 1969); Turner and Patrick (1968); Bell (1969); Parr (1969); Meek et al. (1969); Ponnamperuma (1972); Graetz et al. (1973).
[2]Corrected to pH 7 [$Eh_7 = Eh - 0.059(pH - 7)$ V].

$$1/2Fe_3(OH)_8 + 3/2H_2S + H^+ + e = 3/2FeS + 4H_2O \quad (4)$$

should limit H_2S accumulation to negligible levels in neutral anaerobic systems if sufficient $Fe_3(OH)_8$ is present. Siderite ($FeCO_3$) has been found in marshes (Ponnamperuma, 1972). In flooded sodic and calcareous soils, the Na_2CO_3–H_2O–CO_2 and $CaCO_3$–H_2O–CO_2 systems, respectively, operate to control pH at about neutrality.

With the above discussion in mind, commonly held beliefs about the pH of peat soils need to be re-examined. Lowland peats, which commonly receive considerable amounts of minerals, would be expected to tend toward neutrality when flooded. The higher pH, besides promoting more rapid decomposition, would also permit more exchange sites to be available for bases from the ground water. Thus, even when drained, the pH of these soils would not decline dramatically. However, acid peats are definitely more abundant than those of neutral to alkaline pH. Apparently the iron in these soils is not participating in redox reactions. One possible cause of this situation is the low initial content of stable precipitates, particularly pyrite (FeS_2) and siderite ($FeCO_3$). Crystals of iron oxide and pyrite have been reported in acid (pH 3.5) peat (Davis and Lucas, 1959). An alternative possibility, however, also exists — namely that microbial activity is so low in these soils that microbially mediated redox transformations do not occur.

The source of acidity in organic soils is variously given as humic and fulvic acids, organic acids, or sulfuric acid accumulation (Gorham, 1961; Given and Dickinson, 1975). Organic acids of low molecular weight are major by-products of anaerobic metabolism, but are generally metabolized to methane and carbon dioxide by strict anaerobes (Ponnamperuma, 1972), at least in flooded mineral soils and in waste treatment processes.

ION EXCHANGE AND COMPLEX FORMATION

The cation exchange capacity (CEC) of organic soils is commonly in the range of 1 to 2 meq. g^{-1} when measured at pH 7 (Isirimah et al., 1970; Given and Dickinson, 1975). However, studies of the CEC at the pH of the sample have not been conducted. This value without doubt is much lower, especially for acid peats, since the exchange sites are of the pH-dependent type. Considerable indirect information exists to show that covalent bonding of divalent ions (Ca^{2+}, Mg^{2+}, Fe^{2+}, Cu^{2+}, Zn^{2+}, Mn^{2+}) occurs in peats and in isolated humic, fulvic, and hymatomelonic acids (Given and Dickinson, 1975). Monovalent ions are sorbed less strongly than divalent, and Mn^{2+} is more tightly held than Ca^{2+} (Black, 1968).

Buffer curves of acid peats typically have pK_a's of about 6 (Coleman and Thomas, 1967). This is weaker than predicted from consideration of the carboxyl groups present, presumably a result of the formation of covalent-bonded hydroxy complexes of iron and aluminum such that an increase of pH involves hydrolysis of these complexes rather than ionization of carboxyl groups (Coleman and Thomas, 1967). In this connection, it is noteworthy that Turner and Nichol (1962) found that pH of an aluminum-saturated peat in $0.01M$ $CaCl_2$ to be pH 3.1.

The ability of waterlogged organic soils to retain heavy metals is of considerable interest, because wetlands have been, and probably will continue to be, receptors of wastes. As was discussed by Keeney and Wildung (1977), it is highly probable that organometallic complexes exist, but unequivocal proof of these complexes still remains to be demonstrated. Further, for most donor groups, hydrogen is a highly efficient competitor with the metals, and this competition may limit the degree of metal complexation in extremely acid organic soils. Much work (reviewed by Ellis and Knezek, 1972; and Stevenson and Ardakani, 1972) shows that zinc and copper form strong complexes with organic matter. However, attention has been focused mainly on humic and fulvic acid isolates. Davies et al. (1969) noted that retention of copper by a sedge peat was greater than retention by the extracted humic acid. Added copper was not extracted from the peat by normal hydrochloric acid unless the pH was 2 or less. Treatment of the peat with hydrofluoric acid to remove iron and aluminum increased its capacity to retain copper. Peats drained and developed for agricultural purposes are often deficient in available copper (Davis and Lucas, 1959), probably in part as a result of the strong sorptive capacities of the organic matter for this element and also because organic soils tend to have relatively low concentrations of trace elements (Given and Dickinson, 1975; Maciak et al., 1963) and a low mass of solids in the root zone.

MICROBIAL ECOLOGY

Peat was once thought to be so low in microbial numbers as to be essentially sterile. However, Waksman and his co-workers (Waksman and Stevens, 1929; Waksman and Purvis, 1932) largely disproved this idea.

Changes in microbial populations occur in peats during transformations from aerobic to anaerobic conditions as well as when different substrates are present. The nature of peat was found to exert a marked influence upon the nature and abundance of the bacterial population. Bog peats have fewer bacteria than fen peats primarily because of the acidity of the bog types (Waksman and Stevens, 1929). Liming bog peat greatly stimulates the development of various groups of bacteria, especially those forming ammonia.

Fen peats contain actinomycetes at the surface and to some depth (Waksman and Stevens, 1929). These peats also have a vigorous flora of nitrifying and cellulose-decomposing bacteria. Bog peats do not contain certain of these types. Waksman and Stevens (1929) indicated that changes in actinomycetes, fungi, aerobic cellulose-decomposing bacteria, nitrifying bacteria and anaerobic bacteria occurred with depth in two fen and two bog peats. The fen peats were characterized by a decrease of total bacteria with depth, decreases in fungal and actinomycete numbers to zero at approximately 75 cm, decreases in nitrifying bacteria with depth, and increases in anaerobic bacteria. The proportion of anaerobic bacteria increased steadily down the profile and organisms forming butyric acid were quite abundant. The bog peats were characterized by complete absence of *Azotobacter* and nitrifying organisms, a general increase in total bacteria with depth, and increases of acid-resistant organisms with depth. The few fungi and actinomycetes present occurred at the surface only.

In undrained peats, Stout (1971) found bacterial spore-formers and anaerobes to increase proportionally with depth, but the total number of microorganisms decreased. In the surface layer, gram-negative bacteria dominated. Low numbers of Protozoa were recorded throughout the profile.

Characteristic groups of organisms normally associated with flooded organic soil must be mentioned. The transformation of SO_4^{2-} to S^{2-} under anaerobic conditions obviously occurs in marshes and bogs; therefore, *Desulfovibrio* spp. should be common in these areas. *Clostridium* spp. are probably also present, including the non-symbiotic nitrogen fixers, especially in marshes with high nutrient loading near urban areas. The methanogenic organisms are also obviously present under long-continuing anaerobic conditions, as well may be the hydrogen-oxidizing organisms. The denitrifiers are also present, but generally have little nitrate available for reduction.

CARBON TRANSFORMATIONS

A major characteristic of peats is their high percentage of organic matter. A representative peat contains 80% organic matter on a dry-matter basis (Black, 1968); this organic matter includes 50% of the carbon present (Davis and Lucas, 1959). The principal organic constituents of peat as described by Passer et al. (1963) include bitumens, water-soluble substances, hemicelluloses, cellulose, humic acid, and humins. They define these groupings according to their analytical fractionation. The end-products are a result of decomposition under both aerobic and anaerobic conditions and usually reflect the type of plant material found in the area. For instance, phenolic acids in *Sphagnum* peats arise from *Sphagnum* moss (Morita, 1968).

The degradation of soluble materials in an anaerobic zone of peats follows some general biochemical reactions. In the anaerobic degradation of carbohydrates, the array of substances produced includes: ethanol, butanol, 2-propanol, glycerol, 2,3-butanediol, and acetyl methyl carbinol; acetone and acetaldehyde; formic, acetic, butyric, valeric, caprioc, lactic, oxaloacetic, malonic, fumaric, and succinic acids; and the gases carbon dioxide, hydrogen, methane, and acetylene (Ponnamperuma, 1972; Yoshida, 1975).

Adamson et al. (1975) demonstrated the presence of several volatile organic products in glucose-amended soils under anaerobic conditions. These included carboxyl compounds, alcohols, acids, esters, and ketones. Only 10% of the added carbon, though, was volatilized.

Anaerobic breakdown of fatty acids, waxes, lipids, lignin, proteins, etc., also occurs, with a resulting series of end-products. These products are again characteristic of the plant material and the microbial populations of the peat.

The indirect role of carbon decomposition in peats is the effect of the generated electrons on acceptors such as NO_3^-, Fe^{3+}, Mn^{4+}, CO_2, and O_2. These acceptors are discussed in other parts of this review.

Methane is a gas that has been long recognized in flooded soils and sediment systems. Rennie (1810) indicated the presence of a gas containing a small number of carbon atoms per molecule in bogs. Waksman and Purvis (1932) described the evolution of a gas that took fire readily when taking samples of peats. Takai and Kamura (1966) showed that with a decrease in carbon dioxide production came an increase in methane.

Carbon dioxide, if not lost to the atmosphere or transformed to another product, exerts a major influence on pH and availability of certain mineral nutrients in flooded soils (Ponnamperuma, 1972). In mires with flowing water the predominant anion and cation are bicarbonate and calcium, respectively (Bellamy, 1968).

NITROGEN TRANSFORMATIONS

Most of the transformation of nitrogen probably occurs in the aerobic upper portion of the organic deposit and in the immediate subsurface anaerobic zones. Waksman and Purvis (1932) noted that the nitrogen fractions of *Sphagnum* peat were rapidly liberated as ammonium even though carbon decomposition was very slow. Thus, while the C/N ratio of the plant material was high, the (available C)/(available N) ratio was such that net mineralization occurred. This situation would explain the common observation that such materials are high in exchangeable ammonium nitrogen (Waksman and Stevens, 1929; Kivekas and Kivinen, 1959). Since the environmental conditions (pH and oxygen) conducive to nitrification are not present, this ammonium nitrogen will remain in an available form. On the other hand, peats formed from sedges and reeds show a high content of organic nitrogen as a result of the higher availability of organic carbon and subsequent immobilization by microbes.

While various workers have suggested that nitrogen limits the rate of organic-matter decomposition in peat, additions of nitrogen to peats do not increase the rate of decomposition (Waksman and Stevens, 1929; Kuster and Gardiner, 1968).

Denitrification (providing nitrate is present) should be rapid in waterlogged organic soils but will often occur at variable rates in different materials — presumably because of the different pH and availability of organic carbon. Kuster and Gardiner (1968) indicated that denitrification was rapid in a blanket-bog peat (pH 4.6), and that liming increased the rate of denitrification. Similarly, Avnimelech (1971) found a relatively rapid rate of denitrification (above 15°C) in a peat from Israel. This peat, which was drained for agricultural use, exhibited a high rate of nitrification under aerobic conditions; calculations indicated that the nitrification rate was so high that oxygen diffusion into the peat was the limiting factor. Avnimelech speculated that this caused the system to be in balance relative to nitrate accumulation since nitrate formed in the upper layer would be denitrified in the anaerobic zone. Indeed, anaerobic zones probably can develop in drained areas of organic soils, so that nitrification and denitrification can both occur and within a short distance (Greenwood, 1968). In flooded rice soils and in sediment/water systems, nitrification in the water, followed by denitrification in the soil or sediment, has been observed (Keeney, 1973; Engler and Patrick, 1975). Ishizuka et al. (1963) and Ishizuka and Ogata (1963) found, for example, that if the water table of cultivated peat was within 10 cm of the surface nitrogen deficiency symptoms occurred in plants grown in the glasshouse, even though nitrification was rapid at the surface. They hypothesized that this caused denitrification. Similar findings were reported by Ogata (1963). Isirimah and Keeney (1973b) found that from 4 to 36% of the nitrogen was lost from peat samples incubated under aerobic conditions for six months. While they did not speculate on the mechanism of loss, nitrification followed by denitrification is the only plausible explanation for their results.

Isirimah and Keeney (1973b) found that, as would be expected, denitrification was more rapid at higher temperatures (25°C as opposed to 10°C) and more rapid in a calcareous than an acid peat. They also used $^{15}N–NO_3^-$ and found that a significant portion of the nitrate nitrogen (7–15%) was immobilized in 21 days at 25°C.

In most cases, the rate of nitrification probably will limit the extent of denitrification that occurs in peats. However, organic soils could well act as a major sink for the nitrate added via ground-water seepage into and through a marsh area.

Nitrogen fixation by blue-green algae and by aerobic or anaerobic bacteria could occur in organic soils. Granhall and Selander (1973) used the acetylene reduction technique to investigate nitrogen fixation in an acid (pH 4) subarctic mire. They found that the rate of nitrogen fixation by anaerobic bacteria was very low but measurable. They concluded that biological nitrogen fixation could be of importance to the nitrogen economy of the mire. Nitrogen fixation rates would be expected to be even higher in warmer climates and at more favorable pH levels.

PHOSPHORUS

The proportions of total phosphorus in soils in inorganic and organic forms vary widely. While no definitive fractionation studies have been conducted, one would *a priori* expect organic phosphorus to dominate in organic soils. Thus, inorganic phosphorus availability in waterlogged organic soils would be largely a function of the net mineralization of organic phosphorus and the chemistry of inorganic phosphorus in the organic matrix. However, Paul (1954) reported that alkali-soluble forms (mainly phosphates of iron and aluminum) predominated in the peat soils he studied.

Cosgrove (1967) indicated that most of the organic phosphorus in mineral soils is of microbial origin, although its precise chemical nature is not known. A wide range of micro-organisms exists in soil which are capable of dephosphorylating known forms of plant phosphorus. Sommers (1971) fractionated the organic phosphorus in sediments from ten Wisconsin lakes. He found that the major part of it was apparently associated with the well-decomposed and presumably biologically stable organic matter. Much of the phosphorus was present in humic–fulvic complexes of relatively high molecular weight. Sagher et al. (1975) found that mineralization of organic phosphorus under anaerobic conditions in lake sediments was minimal.

The ability of organic soils to sorb inorganic phosphorus can be expected to vary widely, in accordance with the nature and amount of primary and secondary minerals and with pH. For Wisconsin lake sediments, most of the inorganic phosphorus is associated with the surfaces of oxides of iron and aluminum (Syers et al., 1973). Inorganic phosphorus associated with calcium appeared to be largely apatite derived from eroding soil materials. The role of calcium carbonate in sorbing phosphorus, while only poorly understood, seems to be much less than that of the iron oxides (Syers et al., 1973).

Since the labile iron-rich complex appears to be of importance in phosphorus sorption in the anaerobic environment of a lake sediment, one would predict that pH and Eh, which control the forms of iron, would play an important role in release of inorganic phosphorus. Release of phosphorus on flooding of soils has been well documented (Ponnamperuma, 1972), although the mechanisms responsible are complicated. Also, the amount released is dependent on the properties of the soils, being greatest for those low in iron. Further, since much of the water-soluble phosphorus may be in the organic form or present as soluble inorganic complexes, quantitative studies of phosphorus equilibria in submerged soils are difficult (Ponnamperuma, 1972).

Reduction of phosphate to phosphine in waterlogged soils is thermodynamically feasible (Tsubota, 1959; Ponnamperuma, 1972). However, the excellent work of Burford and Bremner (1972) indicated that phosphine evolution from waterlogged mineral soils does not occur. This may be because soils readily sorb phosphine. However, they warned that their findings were not unequivocal proof that phosphine production was not significant, since a condition could exist in which more phosphine was formed than could be sorbed. Such a situation would probably occur with organic soils.

The ability of organic soils to sorb phosphorus varies greatly. Isirimah and Keeney (1973b) found that a calcareous (sedge) peat (pH 8.0, 69% organic matter) sorbed ten times as much phosphorus from solution as did an acid *Sphagnum* peat (pH 3.6, 77% organic matter). Nicholls and MacCrimmon (1974) found inorganic phosphorus concentrations ranging from a trace to 0.026 mg l^{-1} in the subsurface water under an uncultivated marsh (pH 6.20) in Ontario, Canada. Isirimah and Keeney (1973a) found similar concentrations of inorganic phosphorus in a calcareous marsh in Madison, Wisconsin. Interestingly, in the latter study the concentration of inorganic phosphorus in the marsh drainage water was an order of magnitude higher than in the ground-water sample from the same marsh.

SULFUR

Odors long recognized in bogs and marshes are those of hydrogen sulfide and mercaptans. Sulfide is the main product of anaerobic transformations of sulfur and is derived from sulfate reduction (Ponnamperuma, 1972). In bogs, localized areas may have a high concentration of sulfur in the form of iron sulfide, especially where bog iron has formed and the drainage water is rich in sulfates. Sulfur content is generally lowest in bog peat, medium-high in fen peat, and highest in sedimentary peats (Davis and Lucas, 1959).

Although large amounts of hydrogen sulfide are produced in submerged soils, lakes, and anoxic waters, the concentration of water-soluble hydrogen sulfide may be small (Ponnamperuma, 1972). This is because it is removed as insoluble sulfides, chiefly FeS.

Fig. 6.1 depicts the sulfur cycle as diagrammed by Hallberg (1973). The reactions occurring in an anaerobic environment are those indicated in the lower half of the diamond. The formation of sulfides in sediments and flooded soils has considerable relation to heavy-metal availability. Metals are released from soils and bottom sediments as water-soluble humic complexes (Jackson, 1975). However, if conditions are sufficiently reducing, insoluble sulfides are formed. This occurrence is diagrammed by Hallberg (1973) in Fig. 6.2 where the redox cline, defined as a gradient correlated with the redox potential, governs the form of the metal.

Engler and Patrick (1975) demonstrated that sulfur from Na_2S, MnS, FeS, ZnS, and CuS in anaerobic, intensely reduced soils was apparently partially oxidized in the soil adjacent to the growing root of the rice plant and taken up. The degree of uptake was directly related to the solubility of the sulfides.

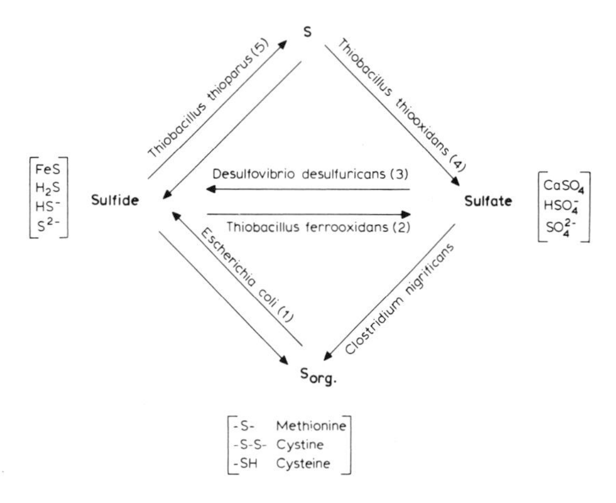

Fig. 6.1. The sulfur cycle in nature as described by Hallberg (1973).

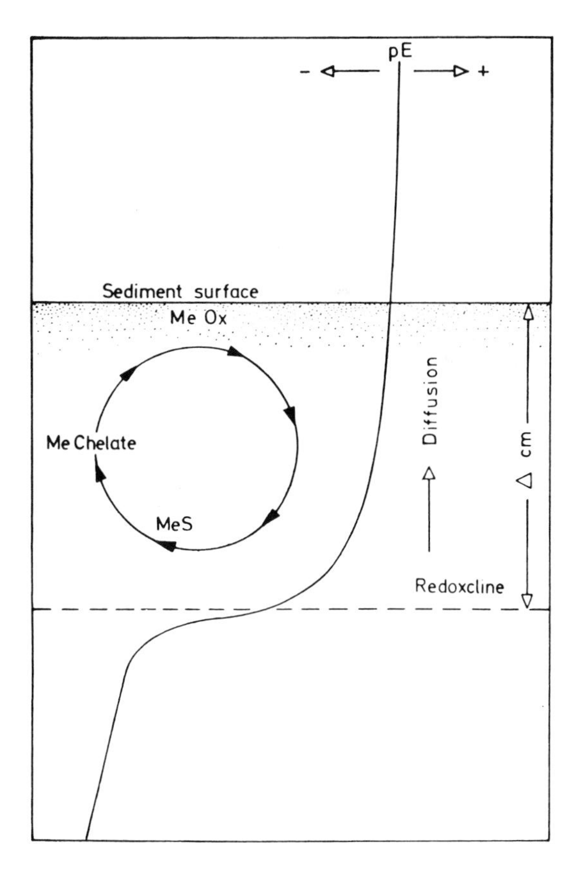

Fig. 6.2. The cycle of chelated heavy metals (Me chelate) above the redox cline in sediments. At the sediment surface, the oxidized form of the metal (Me Ox) is found, while near the redox cline the metal sulfide (MeS) is predominant (from Hallberg, 1973).

TOXIC SUBSTANCES PRODUCED IN WATERLOGGED SOILS

No reports are available to indicate that phytotoxic substances are produced in waterlogged organic soils under natural conditions. This is not surprising, since justification for such research would be difficult and since the species growing on such soils are ecologically adapted *a priori*. Toxicants identified in flooded rice paddy soils include S^{2-}, Fe^{2+}, volatile organic acids, and unidentified reducing substances (Ponnamperuma, 1972). The effects of Fe^{2+} and S^{2-} are interrelated, because soils high in Fe^{2+} are low in S^{2-}. The oxidizing activity of roots of marsh plants probably accounts for the adaptability of hydrophytes to anaerobic conditions in the rooting zone (Armstrong, 1967; Ponnamperuma, 1972).

Takijuma (1964, 1965) found that rice plants grown in peat paddy soils were injured by ferrous iron, sulfides, and unidentified reducing substances, while Lees (1972) identified sulfide as the agent responsible for toxicity to spruce seedings grown in waterlogged peat.

IRON AND MANGANESE

Under anaerobic conditions, the reduction of Fe and Mn occurs at a lower redox potential than the reduction of nitrate. Because of the greater amounts of iron occurring in soils than either manganates or nitrates, the resulting Eh is probably governed most by the Fe^{3+}–Fe^{2+} oxidation/reduction system (Takai and Kamura, 1966). However, as was discussed earlier, the iron in peats may be unavailable for transformations either because of insolubility or because of low microbial metabolism. Fig. 6.3, taken from Takai and Kamura (1966), shows the relationship between redox potential and reduction of manganese and iron.

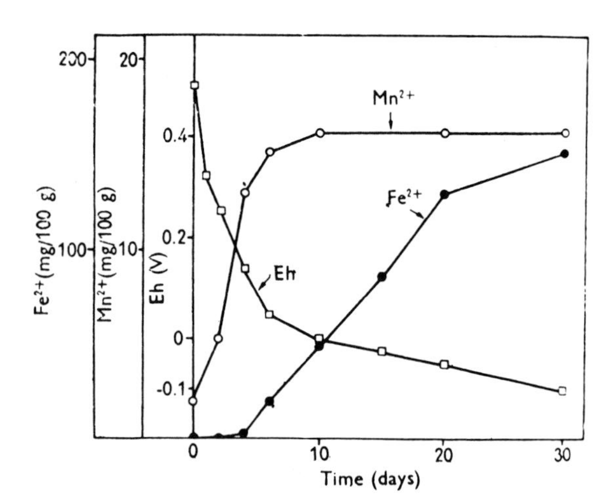

Fig. 6.3. The reduction of manganese and iron as related to Eh (from Takai and Kamura, 1966).

Most of the reduced iron in paddy soils is hydrated magnetite [$Fe_3O_4{\cdot}H_2O$ or $Fe_3(OH)_8$] along with some hydrotrolite ($FeS{\cdot}nH_2O$). If anaerobic conditions persist, these precipitates may age and produce the typical minerals of reduced sediments, magnetite (Fe_3O_4) and pyrite (FeS_2) (Ponnamperuma, 1972).

The reduction of iron has many important chemical consequences. They include: (a) increases in the concentration of water-soluble iron, which may, depending upon concentration, be toxic to plants; (b) increases in pH; (c) displacement of cations from exchange sites; (d) increases in solubility of phosphorus and silica (Ponnamperuma, 1972).

Robinson (1930) found higher manganese and iron concentrations in bog waters than ordinary drainage waters. If soils were submerged, he found increases in manganese solubility with time, and these were greatest in the presence of an abundance of organic matter.

The main transformations of manganese in submerged soils are the reduction of Mn^{4+} oxides to Mn^{2+}, an increase in the concentration of water-soluble Mn^{2+}, and precipitation of manganous carbonate. The manganese concentration in anoxic soil solutions represents a balance between the release of Mn^{2+} by reduction and its removal by cation exchange reactions (Ponnamperuma, 1972).

SUMMARY

The chemical state of nutrients in organic soils is greatly influenced by the concentration of oxygen. Under anaerobic conditions, characteristic forms of carbon, nitrogen, phosphorus, sulfur, iron and manganese are observed. The availability of these nutrients to plants, as well as their toxicities to plants, are affected by the concentration of oxygen.

A disturbance in the ecology of a flooded organic soil by draining or by deposition of waste materials in the area can result in changes in nutrient forms, micro-organisms and plant populations. Before such disturbances are initiated, careful consideration should be given to the final effects on the organic soil ecosystem.

REFERENCES

Adamson, J.A., Francis, M.J., Duxbury, J.M. and Alexander, M., 1975. Formation of volatile organic products in soils under anaerobiosis. I. Metabolism of glucose. *Soil Biol. Biochem.*, 7: 45–50.

Armstrong, W., 1967. The oxidizing activity of roots in waterlogged soils. *Physiol. Plant.*, 20: 920–926.

Avnimelech, Y., 1971. Nitrate transformation in peat. *Soil Sci.*, 111: 113–118.

Bell, R.G., 1969. Studies on the decomposition of organic matter in flooded soil. *Soil Biol. Biochem.*, 1: 105–116.

Bellamy, D.J., 1968. An ecological approach to the classification of European mires. In: C. Lafleur and J. Butler (Editors), *Proc. 3rd Int. Peat Congress, Quebec, 1968*, pp. 74–79.

Black, C.A., 1968. *Soil–Plant Relationships*. Wiley and Sons, New York, N.Y., 2nd ed.

Bohn, H.L., 1971. Redox potentials. *Soil Sci.*, 112: 39–45.

Burford, J.R. and Bremner, J.M., 1972. Is phosphate reduced to phosphine in waterlogged soils? *Soil Biol. Biochem.*, 4: 489–495.

Coleman, N.T. and Thomas, G.W., 1967. The basic chemistry of soil acidity. In: R.W. Pearson and F. Adams (Editor), *Soil Acidity and Liming. Am. Soc. Agron. Monogr.*, No. 12: 1–42.

Connell, W.E. and Patrick, Jr., W.H., 1968. Sulfate reduction in soil: Effects of redox potential and pH. *Science*, 159: 86–87.

Connell, W.E. and Patrick, Jr., W.H., 1969. Reduction of sulfate to sulfide in a waterlogged soil. *Soil Sci. Soc. Am. Proc.*, 33: 711–715.

Cosgrove, D.J., 1967. Metabolism of organic phosphates in soil. In: A.D. McLaren and G.H. Peterson (Editors), *Soil Biochemistry, 1.* Marcel Dekker, New York, N.Y., pp. 216–228.

Davies, R.I., Chesire, M.V. and Graham-Bryce, I.S., 1969. Retention of low levels of copper by humic acid. *J. Soil. Sci.*, 20: 65–71.

Davis, J.F. and Lucas, R.E., 1959. *Organic Soils, Their Formation, Distribution, Utilization and Management.* Special Bull. 425. Michigan State University, East Lansing, Mich., 156 pp.

Ellis, B.G. and Knezek, B.D., 1972. Adsorption reactions micronutrients in soils. In: J.J. Mortredt, P.M. Giodano and W.L. Linsay (Editors), *Micronutrients in Agriculture.* Soil Science Society of America, Madison, Wisc., pp. 59–78.

Engler, R.M. and Patrick, Jr., W.H., 1975. Stability of sulfides of manganese, iron, zinc, copper, and mercury in flooded and nonflooded soil. *Soil Sci.*, 119:217–221.

Engler, R.M., Antie, D.A. and Patrick, Jr., W.H., 1976. Effect of dissolved oxygen on redox potential and nitrate removal in flooded swamp and marsh soils. *J. Environ. Qual.*, 5: 230–235.

Given, P.H. and Dickinson, C.H., 1975. Biochemistry and microbiology of peats. In: E.A. Paul and A.D. McLaren (Editors), *Soil Biochemistry, 3.* Marcel Dekker, New York, N.Y., pp. 123–212.

Gorham, E., 1953. Some early ideas concerning the nature, origin, and development of peat lands. *J. Ecol.*, 41: 257–274.

Gorham, E., 1961. Water, ash, nitrogen, and acidity of some bog peats and other organic soils. *J. Ecol.*, 49: 103–106.

Graetz, D.A., Keeney, D.R. and Aspiras, R., 1973. Eh status of lake sediment–water systems in relations to nitrogen transformation. *Limnol. Oceanogr.*, 6: 908–917.

Granhall, U. and Selander, H., 1973. Nitrogen fixation in a subartic mire. *Oikos*, 24: 8–15.

Greenwood, D.J., 1968. Measurement of microbial metabolism in soil. In: T.R.G. Gray and D. Parkinson (Editors), *The Ecology of Soil Bacteria.* University of Toronto Press, Toronto, Ont., pp. 138–157.

Haavisto, V.F., 1974. Effects of a heavy rainfall on redox potential acidity of a waterlogged peat. *Can. J. Soil Sci.*, 54: 133–135.

Hallberg, R.O., 1973. The microbiological C–N–S cycles in sediments and their effect on the ecology of the sediment–water interface. *Oikos Suppl.*, 15: 51–62.

Ishizuka, Y. and Ogata, S., 1963. Depth of drainage of peat land for agricultural reclamation. 3. Effect of fluctuation of water table on the growth of forage crops and soil properties. *J. Sci. Soil Tokyo*, 34: 195–196 (in Japanese). [Cited in *Soils Fert.*, 27: 2702.]

Ishizuka, Y., Ogata, S. and Sekiya, S., 1963. The depth of drainage of peat land for agricultural reclamation. I. The relationships between depth of water table and growth of forage crops. *J. Sci. Soil Tokyo*, 33: 483–488 (in Japanese). [Cited in *Soils Fert.*, 27: 153.]

Isirimah, N.O. and Keeney, D.R., 1973a. *Contribution of Developed and Natural Marshland Soils to Surface and Subsurface Water Quality.* Technical Completion Report, Project No. OWRRA-049-Wis. Water Resources Center, University of Wisconsin, Madison, Wisc., 30 pp.

Isirimah, N.O. and Keeney, D.R., 1973b. Nitrogen transformations in aerobic and waterlogged histosols. *Soil Sci.*, 115: 123–129.

Isirimah, N.O., Keeney, D.R. and Lee, G.B., 1970. Chemical differentiation of selected Wisconsin histosols. *Soil Sci. Soc. Am. Proc.*, 34: 478–482.

Jackson, T.A., 1975. Humic matter in natural waters and sediments. *Soil Sci.*, 119: 56–64.

Keeney, D.R., 1973. The nitrogen cycle in sediment water systems. *J. Environ. Qual.*, 2: 15–29.

Keeney, D.R. and Wildung, R.E., 1977. Chemical properties of soils. In: F.J. Stevenson and L.F. Elliot (Editors), *Soils for the Management and Utilization of Organic Wastes and Waste Waters.* American Society of Agronomy, Madison, Wisc., pp. 75–100.

Keeney, D.R., Herbert, R.A. and Holding, A.J., 1972. Microbiological aspects of the pollution of freshwater with inorganic nutrients. In: G. Sykes and F.A. Skinner (Editors), *Microbial Aspects of Pollution.* The Society for

Applied Bacteriology, Symposium Series No. 1, 1971. Academic Press, London, pp. 181–200.

Kivekas, V. and Kivinen, E., 1959. Observations on the mobilization of peat nitrogen in incubation experiments. *Maataloust. Aikak*, 31: 268–281. [Cited in *Soils Fertil.*, 23: 499.]

Kuster, E. and Gardiner, J.J., 1968. Influence of fertilizers on microbial activity in peatland. In: C. Lafleur and J. Butler (Editors), *Proc. 3rd Int. Peat Congress, Quebec, 1968*, pp. 314–317.

Lees, J.C., 1972. Soil aeration and Sitka spruce seedling growth in peat. *J. Ecol.*, 60: 343–349.

Maciak, F., Makisimow, A. and Liwski, S., 1963. Chemical investigation of peat soils in Poland. In: R.A. Robertson (Editor), *Proc. 2nd Int. Peat Congress, Leningrad, 1963*, pp. 919–924.

Meek, B.D., Grass, L.B. and Mackenzie, A.J., 1969. Applied nitrogen losses in relation to oxygen status in soils. *Soil Sci. Soc. Am. Proc.*, 33: 575–578.

Morita, H., 1968. Polyphenols in the extractives of an organic soil. In: C. Lafleur and J. Butler (Editors), *Proc. 3rd Int. Peat Congress, Quebec, 1968*, pp. 28–30.

Nicholls, K.H. and MacCrimmon, H.R., 1974. Nutrients in subsurface and runoff waters of the Holland Marsh, Ontario. *J. Environ. Qual.*, 3: 31–35.

Ogata, S., 1963. The depth of drainage of peat land for its agricultural reclamation. 2. Relationships of the level of nitrogen fertilizing and water table level to the growth of forage crops. *J. Sci. Soil Tokyo*, 33: 489–495 (in Japanese). [Cited in *Soils Fert.*, 27: 155.]

Parr, J.F., 1969. Nature and significance of inorganic transformations in tile-drained soils. *Soils Fert.*, 32: 411–414.

Passer, M., Bratt, G.T., Elberling, J.A., Piret, E.L., Hartmann, L. and Madden, Jr., A.L., 1963. Proximate analysis of peat. In: R.A. Robertson (Editor), *Proc. 2nd Int. Peat Congress, Leningrad, 1963*, pp. 841–851.

Paul, H., 1954. Phosphorus status of peat soils. *Soil Sci.*, 77: 87–93.

Ponnamperuma, F.M., 1972. The chemistry of submerged soils. *Adv. Agron.*, 24: 29–96.

Rennie, R., 1810. *Essays on the Natural History and Origin of Peat Moss. III–IX.* Edinburgh. [Quoted from Gorham, 1953.]

Rigg, G.B. and Gessel, S.P., 1956. Peat deposits of the State of Washington. *Soil Sci. Soc. Am. Proc.*, 20: 566–570.

Robinson, W.O., 1930. Some chemical phases of submerged soil conditions. *Soil Sci.*, 30: 197–217.

Sagher, A., Harris, R.F. and Armstrong, D.E., 1975. *Availability of Sediment Phosphorus to Microorganisms.* Technical Report WIS WRC 75-01. Water Resources Center, University of Wisconsin, Madison, Wisc., 56 pp.

Sommers, L.E., 1971. *Organic Phosphorus in Lake Sediments.* Thesis, University of Wisconsin, Madison, Wisc., pp. 91.

Stevenson, F.S. and Ardakani, M.S., 1972. Organic matter reactions involving micronutrients in soils. In: J.J. Mortredt, P.M. Giordano and W.L. Linsay (Editors), *Micronutrients in Agriculture.* Soil Science Society of America, Madison, Wisc., pp. 79–114.

Stout, J.D., 1971. Aspects of the microbiology and oxidation of Wicken Fen soil. *Soil Biol. Biochem.*, 3: 9–25.

Stumm, W. and Morgan, J.J., 1970. *Aquatic Chemistry.* Wiley–Interscience, New York, N.Y.

Syers, J.K., Harris, R.F. and Armstrong, D.E., 1973. Phosphate chemistry in lake sediments. *J. Environ. Qual.*, 2: 2–14.

Takai, Y. and Kamura, T., 1966. The mechanism of reduction in waterlogged paddy soil. *Folia Microbiol.*, 11: 304–313.

Takijuma, Y., 1964. Studies on the mechanism of root damage of rice plants in the peat paddy fields. 1. Root damage and growth inhibitory substances found in the paddy and peat soil. *Soil Sci. Plant Nutr. (Tokyo)*, 10(6): 1–8.

Takijuma, Y., 1965. Studies on the mechanism of root damage of rice plants in the peat paddy fields. 2. Status of roots in the rhizosphere and the occurrence of root damage. *Soil Sci. Plant Nutr. (Tokyo)*, 11(5): 20–27.

Tsubota, G., 1959. Phosphate reduction in the paddy field. I. *Soil Plant Food Tokyo*, 5: 10–15.

Turner, R.C. and Nichol, W.E., 1962. A study of the lime potential: 2. Relation between the lime potential and percent base saturation of negatively charged clays in aqueous salt suspensions. *Soil Sci.*, 94: 58–63.

Turner, F.T. and Patrick, Jr., W.H., 1968. Chemical changes in waterlogged soils as a result of oxygen depletion. *Int. Congr. Soil Sci., Trans. (Adelaide, S.A.)*, 9(4): 53–65.

Urquhart, C. and Gore, A.J.P., 1973. The redox characteristics of four peat profiles. *Soil Biol. Biochem.*, 5: 659–672.

Waksman, S.A. and Purvis, E.R., 1932. The microbiological populations of peat. *Soil Sci.*, 34: 95–109.

Waksman, S.A. and Stevens, K.R., 1929. Contribution to the chemical composition of peat: V. The role of microorganisms in peat formation and decomposition. *Soil Sci.*, 28: 315–340.

Wilde, S.A., Youngberg, C.T. and David, J.H., 1950. Changes in composition of ground water soil fertility, and forest growth produced by the construction and removal of beaverdams. *J. Wildlife Manage.*, 14: 123–128.

Yoshida, T., 1975. Microbial metabolism of flooded soils. In: E.A. Paul and A.D. McLaren (Editors), *Soil Biochemistry, 3.* Marcel Dekker, New York, N.Y., pp. 83–123.

Chapter 7

ROOT SURVIVAL IN FLOODED SOILS[1]

R.M.M. CRAWFORD

INTRODUCTION

The inspiration behind this book, the unique nature of wetland ecosystems, suggests that the vegetation of mires is not only floristically different from that of well-drained land but also physiologically distinct. The greater part of the physiological experimentation on the effects of flooding on plants has unfortunately been carried out on those plants that are least able to withstand this stress, namely the crop plants of agriculture and horticulture. There is available an extensive literature on the various causes of death or loss in yield to crop plants as a result of flooding (Scott Russell, 1977; Cannell, 1977). Although these studies give an accurate indication of the negative aspects of inundated soils and the hazards these have for plants, they do not give any direct information on the positive attributes that mire plants may possess and which enable them to survive long and frequent periods of flooding (Hook and Crawford, 1978). In examining physiological tolerance in an ecological setting it is also necessary to set aside any assessment of the effects of flooding on plant yield. Under the environmentally suboptimal conditions which exist in all mixed-species communities, yield is irrelevant to the success of an individual and survival is the only attribute of ecological importance. As will be seen below, many bog plants, although restricted in nature to wetland sites, do in fact grow better when their soils are drained. Thus, in assessing flooding tolerance it is no measure of adaptation to note if a plant grows or not when flooded. *Larix decidua* produces a burst of growth of second-year needles when flooded, yet this growth is followed by the death of the tree within a few weeks of flooding (Figs. 7.1, 7.2). To assess the degree of flood damage it is a more apposite test to measure the ability of plants to resume normal growth after the period of flooding has passed, than to measure the ability to continue growth during the period of inundation. For these reasons this chapter will examine the survival of roots in flooded soils, and in particular the effects of oxygen deficiency on root physiology. Many plants alleviate the lack of oxygen in the soil atmosphere in flooded soils by means of an internal transport system whereby oxygen diffuses down to the root from the

Fig. 7.1. Breaking of bud dormancy caused by flooding. Root inundation has caused the second-year dwarf shoots to sprout on the first-year branches of *Larix decidua*. This is rapidly followed by complete defoliation as this species is intolerant of any flooding treatments that last longer than a week under glasshouse conditions at 20°C. (Photo: R.M.M. Crawford.)

[1]Manuscript received April, 1979.

Fig. 7.2. Breaking of bud dormancy by flooding in *Picea mariana*. The unflooded control plant is on the left. The flooded plant on the right remained healthy during a month of flooding under glasshouse conditions at 20°C. (Photo: R.M.M. Crawford.)

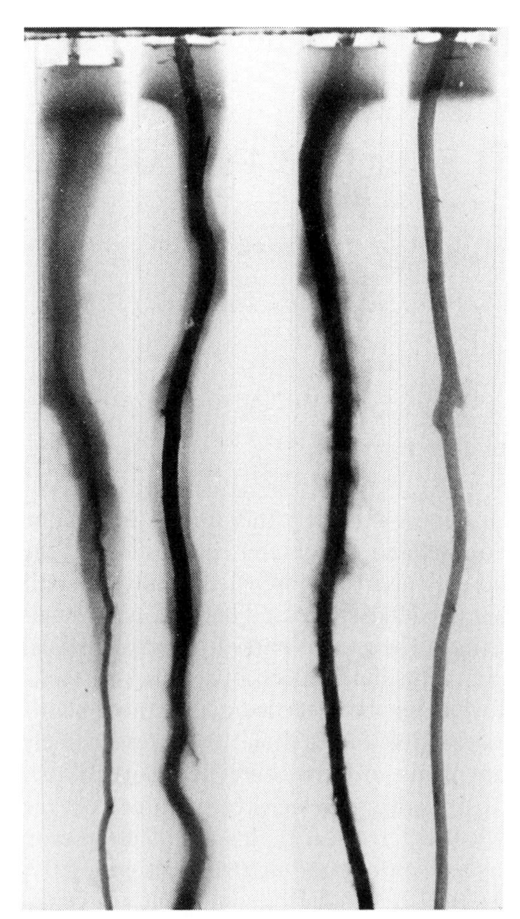

Fig. 7.3. Oxygen transport in tree roots of *Picea sitchensis* and *Pinus contorta* demonstrated by the oxidation of reduced indigo carmine solution to a coloured form by oxygen diffusing from the roots. From left to right the roots are: (a) *Pinus contorta*, with tissues external to the xylem removed; (b) *Pinus contorta*, bark intact; (c) *Picea sitchensis*, bark intact; and (d) *Picea sitchensis*, tissues external to the xylem removed. The roots are submerged to a depth of 40 cm in the dye. (Photograph of unpublished research supplied by courtesy of M.P. Coutts and J.J. Philipson, Forestry Commission, Northern Research Station, Roslin, Scotland.)

aerial shoot (Figs. 7.3, 7.4). In spite of this important adaptation, the perennial nature of much mire vegetation means that, at certain seasons of the year and at certain times in the life cycle such as at seed germination, most species will experience at least a state of *hypoxia* (partial oxygen deficiency) or even a state of *anoxia* (complete oxygen deficiency). For this reason the ability to compete in wetland sites is in many cases an endurance phenomenon which can be directly compared with the diving capacity of terrestrial animals. Just as beavers, whales and seals are all animals that have been able to exploit the aquatic habitat as an ecological refuge by holding their breath, so have our mire plants gained an entry to wetland habitat by being able to survive for varying periods with a restricted air supply.

Apart from the endurance of anoxia by the internal metabolism of the plant root, wetland plants are also exposed to the additional stress of the toxic ions and organic compounds that are present in many anaerobic soils. The need to take up ions through their roots (Buttery and Lambert, 1965), instead of through their leaves as in truly aquatic species, confronts the mire plants with the dangers of root poisoning from anaerobic soil toxins. The root in wetland plants has the important role of functioning as a detoxifying organ protecting the plant from the excessive accumulation of harmful ions. The capacity for detoxification will not be unlimited, and for this reason the ecology of wetland vegetation in relation to soil toxins will also be an endurance phenomenon, with different

Fig. 7.4. Free-hand transverse sections of *Pinus contorta* roots cut at a point 40 mm from the apex. A. Root growing in freely drained soil above the water table. No large cavities are present. B. Root growing in waterlogged soil, showing large cavities in the stele. [Photographs reproduced with permission from Coutts and Philipson (1978).]

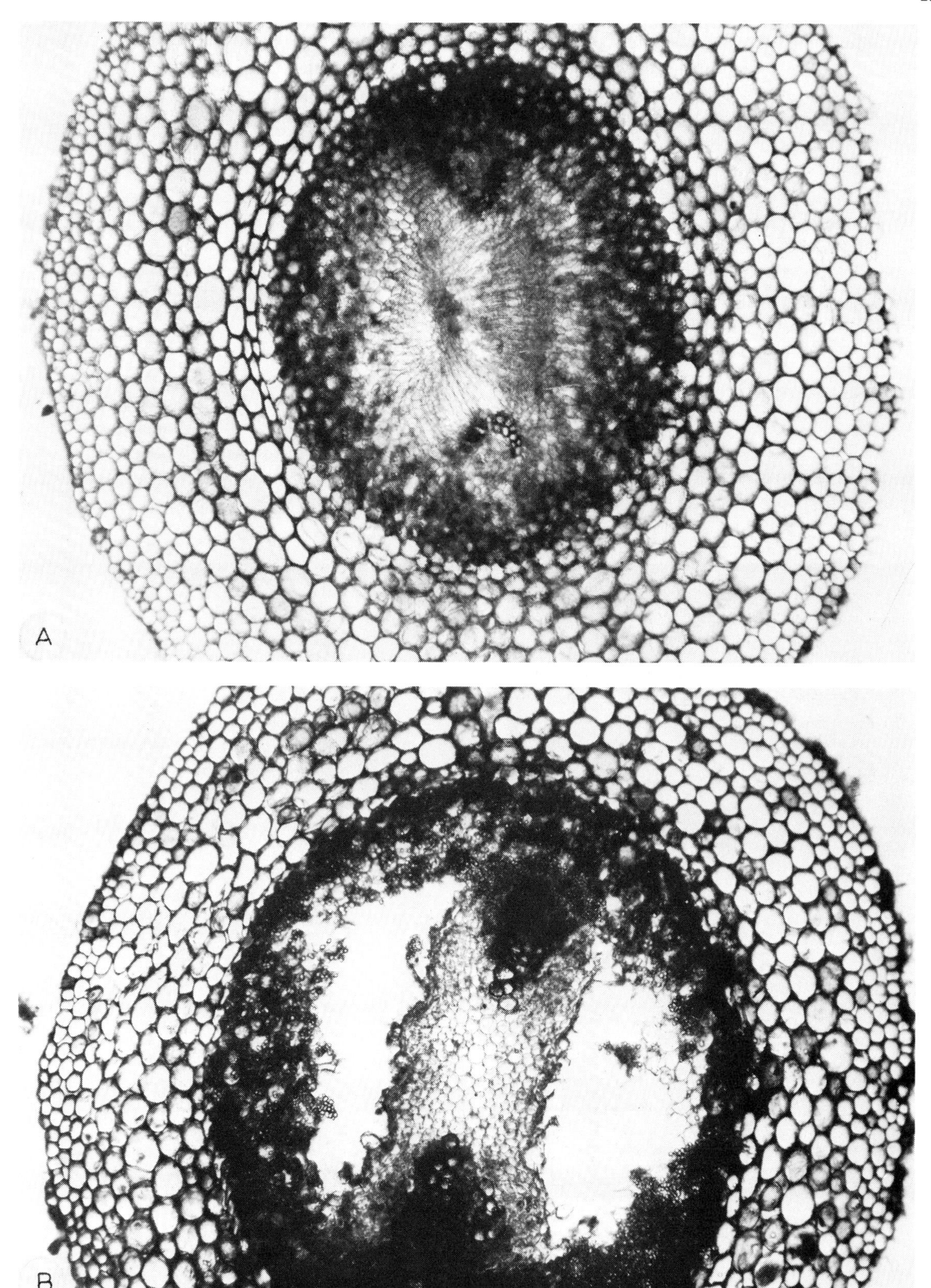
A
B

species showing varying degrees of susceptibility. It is therefore not surprising that wetland vegetation exhibits such a high degree of variation in its distribution. Every minor drainage channel or hummock in a bog distinguishes itself from its surroundings by contrasting vegetation. Similarly, minor fluctuations in the supply of nutrients or running water lead to dramatic changes in the flora of wetland sites. In a situation where all species are living under suboptimal conditions their individual limits of tolerance are readily displayed by minor intensifications or alleviations of the various stresses imposed by flooding.

SWIFTNESS OF DEATH

Flood-tolerant plants enjoy no special immortality under anoxia. Absence of oxygen is eventually fatal to all higher plant tissues (Figs. 7.5, 7.6). The habitat fitness of mire plants in relation to anoxia stems from their ability to take "an unconscionable time dying". The swiftness of death is therefore one of the principal discriminating factors in the ecology of mire plants.

Sudden death due to flooding is normally observed only in certain species of well-drained habitats where inundation is a rare event. The most intolerant of all plants to flooding are those that contain cyanogenic glycosides in their roots. The bark of apricot, plum and peach trees (*Prunus armeniaca*, *P. domestica* and *P. persica*), both stem and roots, is rich in cyanogenic glycosides and these species can be killed by as little as 24 h flooding (Rowe and Beardsell, 1973). Anaerobiosis causes hydrolysis of the cyanogenic glycoside in the roots, cyanide evolution takes place and tissue death rapidly ensues. In these species flooding sensitivity

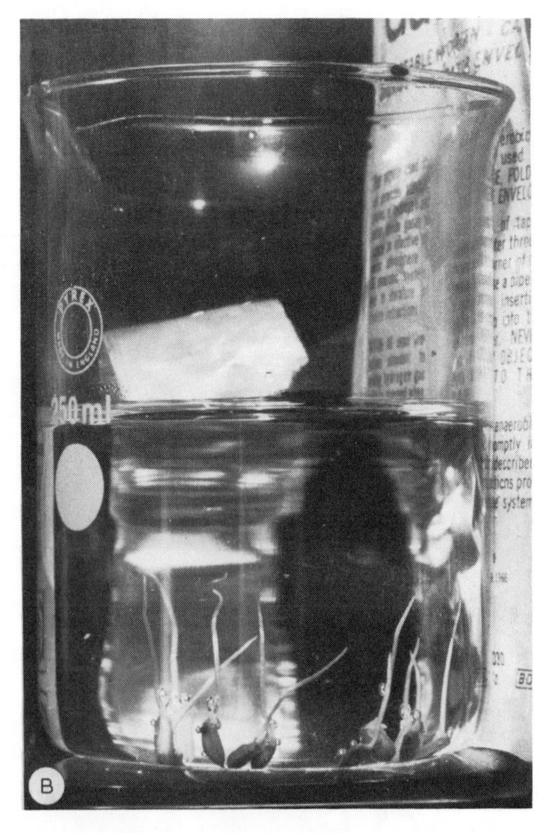

Fig. 7.5. The effects of anoxia on the germination and growth of rice seedlings under water. A. Rice germinating under water with the water surface exposed to air and showing normal development. B. Rice germinating under water with all oxygen excluded above the water surface by placing the beaker in an anaerobe jar. Note the arrested development and hypertrophy of the coleoptiles together with the failure of the roots to emerge. (Photo: R.M.M. Crawford.)

Fig. 7.6. Increase of flooding sensitivity with size in pot-grown plants of *Picea sitchensis*. All plants were flooded for one month under glasshouse conditions of 20°C and 16-h days. (Photo:R.M.M. Crawford.)

can be directly related to the glycoside content of roots. In species which have both cyanogenic and non-cyanogenic forms, moderately wet conditions usually favour cyanogenesis as this gives protection against grazing by the higher densities of snails that occur in such sites (Crawford-Sidebotham, 1972). However, in excessively wet conditions this predator-defence mechanism exposes the possessor to the risk of death by flooding. Thus *Lotus corniculatus*, which occurs on a variety of soils with free drainage conditions, has both cyanogenic and non-cyanogenic forms; but in flood-prone sites *L. corniculatus* is usually replaced in Britain by *L. uliginosus*, which is never cyanogenic.

There are many wetland species which, although they occur only in flooded sites, do so under severely suboptimal conditions, and their final elimination from regional bog floras can often be finally achieved by some additional stress such as grazing, burning or pollution (Sparling, 1967; see also Part B, Ch. 1). Thus, the rush *Schoenus nigricans* will grow sufficiently well to withstand grazing on flushed peat, but in blanket bogs it is only able to maintain itself in remote areas away from grazing pressure (Boatman, 1972). The disappearance of *Empetrum nigrum* from bogs adjacent to agricultural land has been attributed to aluminium toxicity from dust deposition (Bell and Tallis, 1974).

The inhospitality of mires in respect of wetness is frequently made apparent by the improved growth that is produced when either the degree of flooding (height of the water-table) or its duration is reduced. Increased growth of the bog asphodel (*Narthecium ossifragum*) is obtained when drainage is improved (Miles, 1976); similarly *Empetrum nigrum* also improves its performance if flooding is not excessive (Bell and Tallis, 1974). Accurate experimental confirmation of this beneficial effect of lowered water tables on the growth of wetland species has also been found with the two swamp trees *Taxodium distichum* and *Nyssa sylvatica*. In both these species growth is better in saturated soils than unsaturated, as might be expected from their distribution. However, the improved growth in the saturated soil over the unsaturated is only obtained when the saturated soil is allowed to dry out well below field capacity between successive waterings (Dickson and Broyer, 1972).

The most reliable data for the length of time that plants can survive continuous flooding is found with trees. Where water reservoirs have been completed without the trees being felled it is possible to collect accurate data on their growth and survival after the raising of the water level. In many cases this has been part of a deliberate study to seek out trees that can withstand prolonged inundation of roots, as they can do much to mask the unsightly draw-down zones around water reservoirs (Gill, 1977). Table 7.1 lists a selection of tree species and their reported endurance limits to flooding. Continuous flooding eventually kills all trees tested, although some species such as *Quercus lyrata* and *Q. nuttallii* can take up to four years to die (Broadfoot and Williston, 1973).

For most species the length of flooding period is critical only if it occurs in the growing season. The tulip tree (*Liriodendron tulipifera*) can withstand prolonged flooding in the dormant season but will die after four days' flooding in May and three days' flooding in June (McAlpine, 1961). This seasonal aspect of flooding tolerance has been known for over 2000 years. Cato (234–149 B.C.) in his work *De Agre Cultura* devoted a whole chapter to drainage, *Per heimem aquam de agro depellere*, in which he stressed the need for avoiding spring and autumn flooding of winter cereal crops. The need for a respite from flooding suggests the necessity of not accumulating an oxygen debt during the period of active growth. In a review of flooding tolerance in woody species, Gill (1970) pointed out that all-year-round flooding can only be tolerated in isolated

TABLE 7.1

Survival time under inundation of some flood-tolerant trees

Species	Flooding survival years	Reference
Quercus lyrata	3	1
Q. nuttalii	3	1
Q. phellos	2	1
Q. nigra	2	3
Q. palustris	2	2
Q. macrocarpa	2	2
Acer saccharinum	2	3
A. rubrum	2	3
Diospyros virginiana	2	3
Fraxinus pennsylvanica	2	3
Gleditsia triacanthos	2	3
Populus deltoides	2	3
Carya aquatica	2	3
Salix interior	2	3
Cephalanthus occidentalis	2	3
Nyssa aquatica	2	3
Taxodium distichum	2	3
Celtis laevigata	2	2
Quercus falcata	1	4
Acer negundo	0.5	4
Craetagus mollis	0.5	4
Platanus occidentalis	0.5	4
Pinus contorta	0.3	5

References: 1, Broadfoot and Williston (1973); 2, Yeager (1949); 3, Hall et al. (1946); 4, Bell and Johnson (1974); 5, Vester and Crawford (1978).

years, and in general even the most flood-tolerant trees need to remain unflooded for at least 55 to 60% of the growing season. Once a habitat is flooded for more than 40% of the growing season, woody species cannot colonize it, although once established they may survive. The upper limit of woody plants under continual inundation appears to be related to the need to regenerate their younger absorbing roots every two years. As this cannot take place under anaerobic conditions most tree species succumb by their third year of inundation, and only in exceptional cases do they manage to persist for four years.

In spite of this demonstration of the beneficial effects of a respite from flooding, there are plants which require continuous inundation for their survival. A typical example is the rush *Scheuchzeria palustris*. In this particular species it has been noted that its colonization of a bog depends on finding a habitat which is adjacent to permanently open water (Tallis and Birks, 1965). Another species noted for preferring bog sites adjacent to open water is the arctic–alpine bog cotton-grass, *Eriophorum scheuchzeri* (Raup, 1965; Fig. 7.7). In such habitats the adventitious roots will be bathed by free water, and should an oxygen debt accumulate with ethanol concentrations likely to prove dangerous to the plant, they will be rapidly dissipated by diffusion. In rice grown in flooded paddy conditions, as well as with higher plants of aquatic habitat such as *Phragmites australis*, a similar situation exists in that a mat of adventitious roots at or near the surface of the soil is able to rid itself of dangerous concentrations of ethanol into the adjacent bulk of free water. In germinating rice seedlings, after the emergence of the radicle over 97% of the ethanol produced by the submerged plant passes into the water (Fig. 7.8). In conclusion, it would appear to be a general phenomenon in higher plants that, unless the oxygen debt can be alleviated by removal of toxic by-products into free water, constant inundation of the root system either inhibits or reduces growth.

PLANT MORPHOLOGY IN RELATION TO FLOODING TOLERANCE

Morphological adaptations in roots

In many mire systems monocotyledenous species constitute a high proportion of the vegetation. With regard to their roots many monocotyledons show a seasonal dimorphism which can be directly compared with the cessation of leaf production that is found during the stress period in many desert shrubs and temperate deciduous trees. In the monocotyledons the almost complete renewal of the adventitious root system with each successive growing season obviates the problem of preserving these organs during the winter flooding period. (see Clymo, Ch. 4). In spring the renewed growth of the adventitious root system takes place as the water tables descend and the soil profile and bog active-layer become aerated again.

The lack of lignification in adventitious root systems and their branched and often shallow spreading form also does much to facilitate aeration and permit the dissipation of the oxygen debt by diffusion of ethanol. The nearer the roots are to

Fig. 7.7. *Eriophorum scheuchzerii* growing beside open water at Mesters Vig, northeast Greenland (72°N). The proximity of open water is a regular habitat characteristic for this species. For other species with similar ecological preferences, see p. 262. (Photo: R.M.M. Crawford.)

the surface of a wetland soil, the greater will be their contact with the movement of aerated water. It is therefore not surprising that moving water is so often cited as an important factor in bog ecology (Ingram, 1967). The flushing of soils is usually considered to benefit the plants by ensuring the continued supply of adequate mineral nutrients (see Gore, Ch. 1). However the removal of metabolic toxins must also be considered, as it would inevitably be facilitated by this same process.

The ability of certain tree species to withstand flooding has been linked by several authors with their ability to mimic the monocotyledons in the production of adventitious roots at the soil surface (Gill, 1977). Although these roots do little to provide anchorage and support for the plant, they do serve as efficient exchange organs for absorbing oxygen into the plant and allowing ethanol, acetaldehyde and ethylene to diffuse out of the plant (Chirkova, 1968).

Table 7.2 lists a number of flood-tolerant trees in relation to their morphological response to flooding. In the majority adventitious root formation is the characteristic response. However, there is a significant number of exceptions where other morphological changes take place. Thus, in *Taxodium distichum*, which does not form adventitious roots, there is instead a form of stem hypertrophy with the formation of various types of buttress root: shallow, cone, bottle and bell, each of which corresponds to different flooding regimes (Kurz and Demaree, 1934). Similar to the buttress roots are the knee roots which also form on *Taxodium distichum* and other species in relation to flooding. Opinion is divided on whether these knee roots serve to aerate submerged tissues, as it appears that their cambial layers may use all the oxygen they absorb (Kramer et al., 1952). Doubt also exists in relation to the function of pneumatophores that are found in certain tropical salt-tolerant mangrove species such as *Xylopia staudtii* and *Anthocleista nobilis*. In the pneumatophores of the former there are no air spaces while those of the latter contain chloroplasts and are positively phototropic. Nevertheless, Scholander et al. (1955) were able to demonstrate that in at least two species, *Rhizophora*

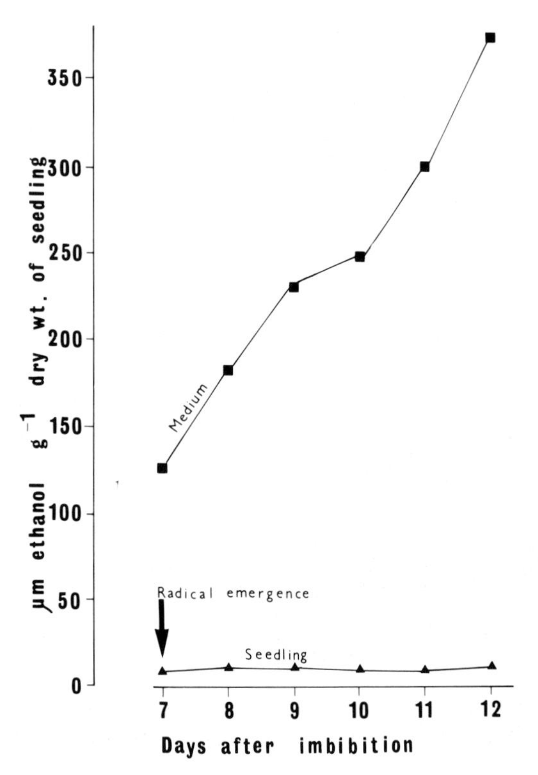

Fig. 7.8. Ethanol production by germinating rice seedlings immersed in sterile water under air as in Fig. 7.5A. Note that the ethanol content of the seedling tissues remains low as most of the ethanol diffuses out into the bathing medium (Crawford, unpublished).

mangle and *Avicennia nitida*, the pneumatophores are capable of aerating submerged portions of the root. If the lenticels in the aerial portion of the root in these species are blocked, then the oxygen concentration in the submerged portion falls from between 15 and 18% to 2% or less (Fig. 7.9). Most significantly, these authors note that there is a pressure drop in pneumatophores while submerged at high tide which would draw air in when the roots are re-exposed at low tide. It must be emphasized that mangrove vegetation is a very special case of flooding-tolerance in trees. Not all the trees in a mangrove swamp develop pneumatophores, though most show some form of specialized root development, stilt-root formation being one of the commonest. Mangrove swamps are sensitive to wave action and tidal currents, and mainly form in sheltered tropical seas, particularly where off-shore coral reefs reduce the amount of exposure (Walter, 1962). The stilt-root formation will confer greater stability on the tree. It is also possible that the

TABLE 7.2

Morphological adaptations to flooding in higher plants

Adaptation	Species	Reference
Buttress and knee roots	*Taxodium distichum*	Hook and Scholtens (1978)
Pneumatophores	*Rhizophora* spp. *Avicennia* spp.	Scholander et al. (1955)
Adventitious roots and root branching	*Salix* spp.	Gill (1970)
	Alnus glutinosa	Gill (1970)
	Cephalanthus occidentalis	Gill (1970)
	Pinus contorta	Gill (1970)
	Thuja plicata	Gill (1970)
	Tsuga heterophylla	Gill (1970)
	Ulmus americana	Gill (1970)
Lenticel proliferation and hypertrophy	*Nyssa aquatica*	Hook and Scholtens (1978)
	Salix spp.	Chirkova and Gutman (1972)
Reduction in suberisation of epidermis and casparian strip	*Nyssa aquatica*	Hook and Scholtens (1978)
Stelar embolism with oxygen transport	*Pinus contorta*	Coutts and Philipson (1978)
Positive phototropism	*Trifolium subterraneum*	Bendixen and Peterson (1962)
	Mentha aquatica	Crawford (see text)
Leaf xeromorphy	*Erica tetralix*	Jones (1971)

greater surface area that this affords the roots increases the exchange of toxic substances and improves aeration. With all mangrove formations, flooding is of short duration, with regular periods of relief between high tides. This forest therefore contrasts markedly with the tropical swamp forest of the upper Amazon (Rio Negro and *igapo* lagoons) where for three months during the wet season the trunks of the trees are flooded to a depth of several metres (Gessner, 1968) (see also Junk, Part B, Ch. 9). In these forests there is no record of any pneumatophore formation, nor are there the stilt roots that are so characteristic of mangrove forests. As temperatures are high during the flooding period and flooding is deep, with no special aerating mechanisms, it can only be assumed that in these forests the roots are well adapted to withstand long periods of anaerobiosis.

Fig. 7.9. Pneumatophores of *Avicennia nitida* showing pronounced lenticel formation. (Photo: R.M.M. Crawford.)

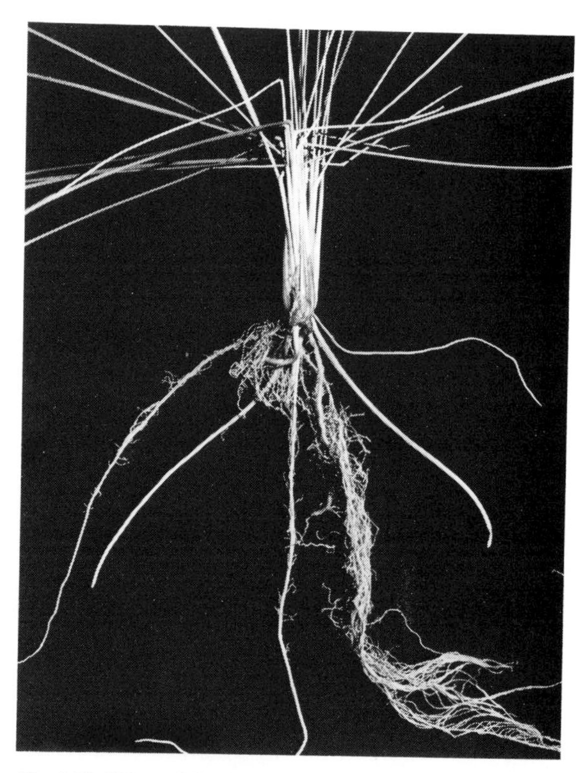

Fig. 7.10. Effect of flooding on root growth in *Nardus stricta*. This plant was grown in sand culture which was subsequently flooded. The two thickened unbranched roots were produced after the plant was flooded. (Photo: R.M.M. Crawford.)

A frequent characteristic of many wetland species is their ability to change their root form when flooded. Roots that are grown in water culture or in a flooded soil are frequently thicker than those grown in aerated soil or sand culture (Figs. 7.10, 7.11). The increase in root diameter is largely taken up by an increase in root air space with the development of aerenchyma tissues. Table 7.3 compares the varying percentages of root space that is air-filled in roots grown under varying conditions. In many herbaceous species this development of aerenchyma appears to be brought about by loose packing of the cortical parenchyma, whereby the root contains much larger air spaces or lacunae. In such cases the root air spaces are due to cell separation (schizogeny) during the maturation of the root. In other species cell breakdown (lysigeny) is the principal cause of space formation (Arber, 1920; Sculthorpe, 1967). In some species such as grasses and sedges, cell collapse can accompany schizogeny.

Gas transport in roots is not necessarily restricted to the cortex. In *Pinus contorta* the ability of the root to transport oxygen is much greater if the root tip grows down into the inundated region of the soil than when the water table rises to flood the root (Philipson and Coutts, 1978). When *Pinus contorta* roots are able to adapt in this way to flooding, gas transport takes place via the stele. This path of gas transport was confirmed by anatomical studies which showed that roots which had adapted to flooding by growing down into the water table developed gas-filled cavities in the pericycle (Coutts and Philipson, 1978; Fig. 7.4).

In spite of these convincing demonstrations of the development of aerenchyma in response to flooding, it must be noted that this does not take place in all flood-tolerant plants. Notable exceptions where aerenchyma is not found include *Phalaris arundinacea* and *Filipendula ulmaria* (Table 7.3). Further, the possession of aerenchyma and the ability to allow oxygen to diffuse to the flooded root does not appear to prevent roots from suffering the metabolic consequences of anaerobiosis. When *Pinus contorta* roots are flooded

Fig. 7.11. A view of Rannoch Moor, Scotland, in early spring showing the white dead shoots of last season's growth of *Nardus* in its preferred habitat of wet hollows and flushes. (Photo: R.M.M. Crawford.)

there is a small increase in ethanol accumulation, which indicates that flooding, in spite of the ability to permit the downward diffusion, does not prevent an increase in the state of hypoxia in the submerged root (Crawford and Baines, 1977). Equally significant are the findings of Raymond et al. (1978) that, when roots of maize and rice are placed in an atmosphere of nitrogen with their shoots in air, there is a significant drop in energy charge as compared with the controls when both roots and shoots are exposed to normal air (Fig. 7.12).

Energy charge, which gives an indication of the amount of nucleotide material in the high-energy form as opposed to the low-energy form, is represented by the ratio:

$$\text{energy charge} = \frac{[\text{ATP}]+\frac{1}{2}[\text{ADP}]}{[\text{ATP}]+[\text{ADP}]+[\text{AMP}]}$$

where [ATP], [ADP] and [AMP] represent the concentrations of the tri-, di-, and mono-adenosine phosphates, respectively (Atkinson, 1968). The concept of energy charge was devised as a parameter of metabolic control in energy-requiring and -producing reactions. Irrespective of its merits or faults in this field, it is a useful monitor of the effect of hypoxia on the energy balance within the root. It is therefore important to note that, even although rice roots are well known for their ability to obtain oxygen by diffusion, they also suffer a drop in energy charge within 30 min if the atmosphere surrounding their roots is deprived of oxygen.

This inability of the root to maintain its normal metabolic pattern even when oxygen is available from the shoot supports the findings of Vartapetian and Nuritdinov (1976), where, with the use of inhibitors for cytochrome oxidase activity and the

TABLE 7.3

Percent air space in roots of various wetland species of higher plants determined by weighing intact and disintegrated roots in pycnometer bottles (unpublished data of Crawford and Smirnoff); mean values based on observations on whole root system unless otherwise stated

Species	Percentage air space	Standard error
Nardus stricta (primary root)	45	±1.2
Eriophorum angustifolium	42	±1.9
Glyceria maxima (primary root)	35	±0.3
Eriophorum vaginatum	31	±1.8
Scirpus cespitosus	26	±3.1
Juncus effusus	25	±3.5
Spartina townsendii	22	±1.4
Narthecium ossifragum (primary root)	19	±1.5
Deschampsia cespitosa	19	±2.7
Ranunculus lingua (primary root)	19	±0.4
Ranunculus lingua (lateral root)	12	±1.7
Lythrum salicaria	12	±1.8
Glyceria maxima (lateral root)	11	±0.8
Potentilla palustris	10	–
Nardus stricta	8	±2.0
Mentha aquatica	7	±0.6
Phalaris arundinacea	6	±0.6
Myosotis scorpioides	4.9	±0.7
Narthecium ossifragum (lateral root)	3.3	±1.4
Filipendula ulmaria	2.0	±0.5

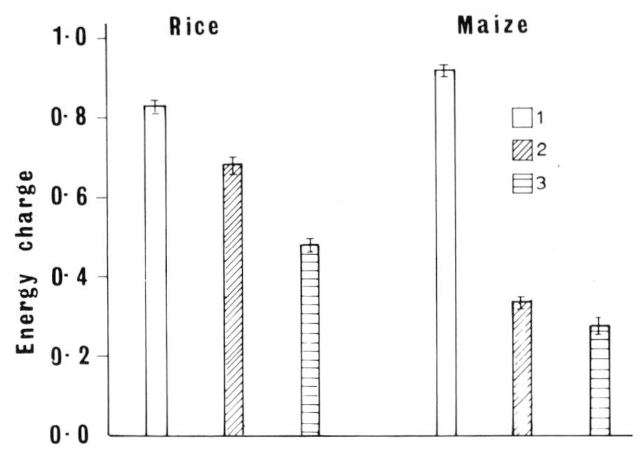

Fig. 7.12. Change in energy charge of intact roots of rice and maize plants as they are subjected to varying degrees of anoxia. Legend: *1* = roots and shoots in air; *2* = shoots in air while roots are in nitrogen; *3* = roots and shoots in nitrogen. (From Raymond et al., 1978.)

alternative oxidation pathway it was shown that diffusion could account for only 8% of the oxygen needs of cotton roots. Although oxygen may diffuse out of roots it cannot therefore be assumed that it has satisfied all the respiratory needs of the roots. Oxygen diffusion is purely a physical process, and as such must take the pathway of least resistance. It is possible that, although oxygen diffusion may be of adaptive significance to flood-tolerant roots, this may be for reasons which are not connected with the respiratory needs of the root.

The apparent inability of gaseous diffusion to satisfy the oxygen demands of the root prompts a re-examination of the functional significance of aerenchyma. Previous discussions have attributed the following roles to aerenchyma in alleviating the hazards of inundation: (1) the provision of a reservoir of oxygen (Williams and Barber, 1961); (2) the facilitation of root aeration from the shoot (Armstrong, 1972); and (3) the provision of an increased volume of root for a given amount of living tissue (Williams and Barber, 1961).

An assessment of the oxygen reservoir capacity of aerenchyma tissue for a number of species is made in Table 7.4. By making measurements of root air space with pycnometer bottles (Jensen et al., 1969), and having first obtained the Q_{O_2} values for these same roots, it can be seen that the oxygen reserves of the aerenchyma vary between 2 and 126 min. The high figure for *Eriophorum angustifolium* agrees with the findings of Armstrong (1975) that the internal supply of oxygen for this species will support root respiration for only one hour's submergence when the plants are kept in the dark. The aerenchyma therefore, if it does serve the plant as an oxygen reserve, can provide oxygen only for very short term needs. In the case of *Spartina townsendii*, where flooding is of short duration during high tide, this adaptation may be sufficient to avoid acute anaerobiosis.

The most striking observation to emerge from these data is that air-space tissue in roots occurs mainly in the monocotyledons. An examination of a number of dicotyledon species characteristic of wetland sites revealed a very low proportion of air space in their roots. When this was compared with the metabolic needs of these roots (Fig. 7.13), then the fact that the air reserves in a species such as *Filipendula ulmaria* are sufficient for only 2 min of aerobic respiration at 18°C raises a serious doubt as to whether root aeration for the purposes of maintaining aerobic respiration is a necessary feature of adaptation to wetland habitats.

These doubts concerning the generally accepted

TABLE 7.4

The capacity of root air space to serve as an oxygen reservoir, calculated on the length of time that the air volume of the root could sustain full activity of aerobic dark respiration at 18°C without replenishment (Smirnoff and Crawford, unpublished data) assuming that the oxygen content of the root atmosphere is 20%

Species	Oxygen reservoir capacity (min)	Oxygen consumption [μl O_2 (mg dry wt.)$^{-1}$ h^{-1}]	Air space [ml (g dry wt.)$^{-1}$]
Eriophorum vaginatum	126	0.362	3.81
Eriophorum angustifolium	102	0.781	6.62
Scirpus cespitosus	67	0.168	0.94
Glyceria maxima	37	0.520	1.61
Ranunculus lingua	32	0.670	1.76
Juncus effusus	34	0.645	1.84
Carex curta	22	0.633	1.15
Lythrum salicaria	22	0.515	0.94
Narthecium ossifragum	14	0.388	0.46
Mentha aquatica	13	0.650	0.71
Lycopus europaeus	12	0.755	0.76
Potentilla palustris	9	0.520	0.39
Filipendula ulmaria	2	0.375	0.07

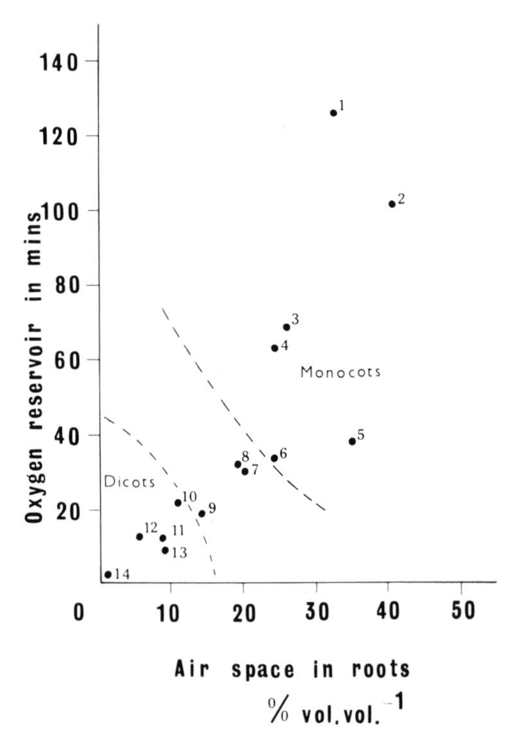

Fig. 7.13. Relationship between the capacity of the oxygen reservoir in the roots to maintain aerobic respiration at the normal rate observed in air and the percentage air space in the roots of a number of species (for method see Jensen et al., 1969). Legend: 1 = *Eriophorum vaginatum*; 2 = *Eriophorum angustifolium*; 3 = *Scirpus cespitosus*; 4 = *Spartina townsendii*; 5 = *Glyceria maxima*; 6 = *Juncus effusus*; 7 = *Ranunculus lingua*; 8 = *Carex curta*; 9 = *Narthecium ossifragum*; 10 = *Lythrum salicaria*; 11 = *Lycopus europaeus*; 12 = *Mentha aquatica*; 13 = *Potentilla palustris*; 14 = *Filipendula ulmaria*. (Crawford and Smirnoff, unpublished.)

view of the function of aerenchyma having been raised, it is necessary to examine what other function of an adaptive nature can be attributed to this tissue and the diffusion of oxygen from submerged roots (see Clymo, Ch. 4). The oxygen that diffuses out of roots will oxidize the reduced ions which approach the root with the mass movement of the soil solution. It is particularly important that ferrous iron, manganous ions and sulphide be oxidized as all these ions are highly phytotoxic. As the concentration of reduced ions increases in the soil they are no longer oxidized externally to the root, but are deposited on the root surface and in the cell walls of the cortex. As the root matures these outer cortical tissues are sloughed off as a result of microbial attack (Greaves and Darbyshire, 1975) and the root thus rids itself of its precipitated load of iron, manganese and other potentially toxic ions. This process of detoxification by oxidation, precipitation and eventual removal will protect the root from poisoning until the precipitate-carrying capacity of the root is saturated, or the passage of water is prevented by the precipitation of oxidized ions such as iron. Fig. 7.14 illustrates the effect of increasing ferrous iron concentration on the growth of *Nardus stricta*. There is no decrease in shoot growth until the capacity of the roots to absorb ferric iron is saturated. In this situation any mechanism which increases the amount of wall tissue, without increasing the respiratory needs of the root by increasing cytoplasm content, will increase the capacity of the root to function as a detoxifying organ. Furthermore, if the increased cell wall area surrounds an aerated chamber, then this oxidation and sequestration mechanism will be ideally fitted for the removal of reduced ions from the plants incoming soil solution. Aerenchyma can therefore be considered as the ideal tissue form for increasing the detoxifying power of roots. In addition the

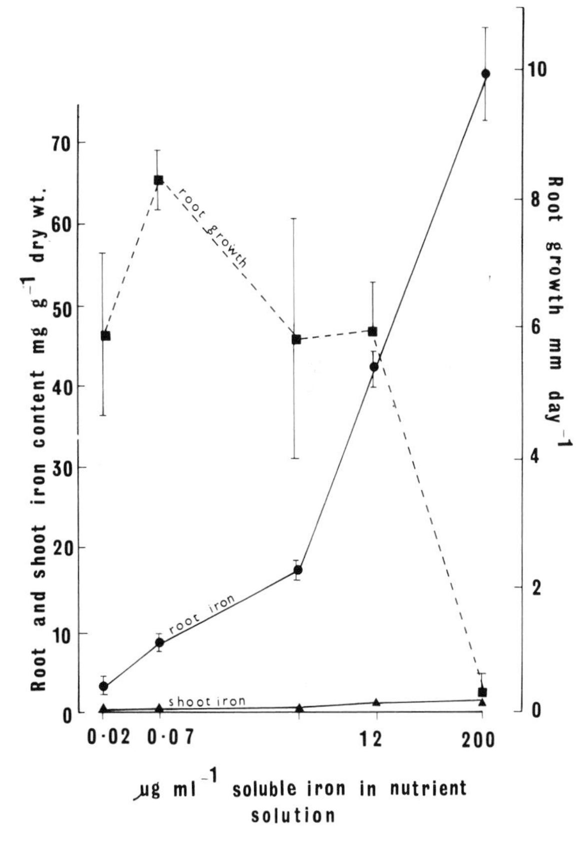

Fig. 7.14. Absorption of iron by roots and shoots of *Nardus stricta* grown in culture solutions with varying concentrations of soluble iron supplied as ferrous sulphate (Crawford and Smirnoff, unpublished).

increased surface area for ion absorption due to aerenchyma development should also prolong the functional capacity of the root as a water-absorbing organ. Although only a few species have been examined, it would appear that there is a case for reassessing the functional significance of aerenchyma and the radial loss of oxygen from flooded roots in terms of environmental detoxification, and not as a means of supplying the respiratory needs of submerged roots.

In woody plants gas has long been observed in the xylem. The intercellular spaces in the medullary ray tissue which are also gas-filled have been thought to permit the radial movement of gas. The cambium, however, has no visible spaces and will therefore present a barrier to gaseous diffusion in woody species. It has been demonstrated that the resistance of the radial cambium to the passage of gas varies between trees and this resistance is least in the trees that are the most tolerant of flooding (Hook et al., 1972). In pines and certain conifers the late wood tracheids lose their water-filling soon after differentiation. These tracheids, which are then gas-filled, are lined with bordered pits which do not close the pit aperture as they do in early wood (Petty, 1970), and thus allow a path of both vertical and horizontal gaseous diffusion. It has been speculated (Coutts, 1976) that the products of microbial infection in certain trees induce a degree of embolism in the xylem which leads to the formation of the improved pathways of gas transport similar to those observed in trees such as the flood-tolerant *Pinus contorta* (Fig. 7.4).

Stem adaptations to flooding

Wetland vegetation does not restrict its morphological adaptations to roots. The growth and form of the aerial portions of plants are also altered by the flooding of the roots. One of the most striking and immediate responses to shoot growth as a result of root flooding is the change in phototrophic response that can often be noted in the stem. *Mentha aquatica* assumes a lax semi-prostrate form when grown in freely drained sand culture but as soon as the sand is flooded the stems grow in a directly upright position. When the stolons of *Trifolium fragiferum* are grown along a gas-filled tube they remain diaphototropic as long as the tube is filled with air containing 20% oxygen. However, when this is replaced by nitrogen, the stem apex becomes positively phototropic (Bendixen and Peterson, 1962). The adaptive significance of upright shoot growth in flood-prone habitats needs no comment.

Common, but not commented on in the literature, is the breaking in dormancy of stem buds that is caused by flooding. Fig. 7.15 shows the stem of an alder (*Alnus incana*) after a few weeks' flooding with the renewed growth from the base of the original stem, which would account for the typical form of alders growing in flooded-carr conditions (Figs. 7.15–7.17). This bush form contrasts with the pole form which alder assumes when grown in well-drained sites. The bush form has greater surface area of bark per unit volume of wood, and this will automatically improve the degree of ventilation immediately above the flood level. The importance of stem ventilation has been investigated by a

Fig. 7.15. Effect of flooding for one month under glasshouse conditions of 20°C and 16-h days on *Alnus incana*. Note the breaking in dormancy of the basal buds. The leaves that wilted after the initial flooding (see Fig. 7.16) have now been shed and are being replaced by foliage that remains healthy throughout several months of continuous flooding. (Photo: R.M.M. Crawford.)

Fig. 7.16. The initial effects of flooding on alder. Pot-grown plants of *Alnus incana* show leaf discolouration and wilting within four days of flooding. (Photo: R.M.M. Crawford.)

Fig. 7.17. Natural form of alder (*Alnus glutinosa*) with branching from the base of the trunk as usually observed in wetland sites. Pole-form alders are restricted to better-drained sites. The production of this form is presumably due to the flood-induced breaking of dormancy of the basal stem buds as shown in Fig. 7.15. (Photo: R.M.M. Crawford.)

number of authors. If the lenticels at the base of willow stems are flooded, the radial loss of oxygen ceases in the adventitious roots (Armstrong, 1968). Similarly the marked development of lenticels that occurs around the base of alder and willow stems has also been shown to serve for the removal of acetaldehyde, ethanol and ethylene from flooded trees (Chirkova and Gutman, 1972).

A common feature of many wetland plants is their possession of hollow stems. The hollow stem of the common reed (*Phragmites australis*) is found in many other rushes and sedges. This hollow stem remains filled with air even when the stem is flooded. It has been suggested (Billings and Godfrey, 1967) that this may have a double adaptive significance in that not only will it facilitate aeration to submerged portions of the plant but, as this air accumulates carbon dioxide, it will provide an enriched atmosphere in the aerial parts of the stem for the photosynthetic needs of the plant. Any increase in photosynthetic activity will not only improve growth but will also increase the oxygenation of the gas supply diffusing downwards from shoot to root.

Leaf morphology and flooding tolerance

The xeromorphic features in the leaves of many bog species have been discussed by numerous authors (Armstrong, 1975; Small, 1973). Some controversy exists however as to their adaptational significance. One hypothesis suggests that oligotrophic bog vegetation is very easily poisoned by soluble reduced ions, and any means by which their uptake can be reduced will increase the flooding tolerance of the plant. *Erica cinerea* is readily poisoned by ferrous iron when transpiring freely. Spraying the leaves of this species with an anti-transpirant both reduces the iron content of the foliage and increases its ability to withstand flooding (Jones, 1971). The use of an anti-transpirant can make *E. cinerea* as flood-tolerant as *E. tetralix*.

An alternative explanation of the frequent occurrence of xeromorphy in oligotrophic bog vegetation emphasizes the poor nutrient status of the bogs rather than the hazards of flooding. In the leaves of *Ledum groenlandicum* lysigenous air spaces similar to those that are found in aquatic plants appear as the foliage develops. This phenomenon has been attributed to the lack of sufficient nutrient material to maintain the cell structure of the developing leaf (Sifton, 1940).

The low-nutrient status of bogs has also been suggested as the basic cause of the xeromorphic foliage that is characteristic of many species of oligotrophic sites. The best economy in leaf production in any area of low growth potential is effected by having evergreen foliage that will serve for more than one season (see Dickinson, Ch. 5). Many plants of oligotrophic bogs are evergreen, and this could be an adaptation which economizes in the use of mineral nutrients. The evergreen habit although it can compensate for low growth rates will require to be resistant to winter desiccation injury, and this may account for the xeromorphic nature of many bog plants (Small, 1973). As with many theories of adaptation it is possible that more than one purpose may be served by the same device, and in bog plants xeromorphy may protect foliage against mineral poisoning and winter desiccation.

METABOLIC ADAPTATIONS TO FLOODING

The varying speed of death in different species following flooding, already discussed on p. 260, clearly indicates a need for a comparative study of the differential effect of hypoxia on flood-tolerant and non-flood-tolerant species. When roots fail to receive an oxygen supply that is adequate for their metabolic needs then it is only to be expected that this abnormal stress (if the plant is non-flood-tolerant) will cause a number of imbalances in the proper functioning of the root. Not only will growth cease but the production of growth substances and absorption of water and minerals will all suffer. Thus epinasty, the breaking of bud dormancy, wilting and leaf chlorosis (Fig. 7.16) are all symptoms which often precede death in non-flood-tolerant plants (Scott Russell, 1977). Important as it is in agriculture to recognize these symptoms, it is possible that they are far removed from the original physiological stress that flooding imposes on the unadapted root. Thus, death from microbial attack or hormone imbalance can be but the *coup de grâce* in a series of physiological maladjustments caused by the inability of the plants metabolism to survive on a reduced oxygen supply. In an ecological enquiry into flooding tolerance it is essential to ascertain as closely as possible the features which allow certain species to survive periods of hypoxia. For any adaptation to be considered as of fundamental importance to the survival of plants in wetland habitats it should be present in a number of different species of this habitat and absent in dry-land plants. This section on the metabolism of plants will therefore examine the differences in wetland and dryland vegetation on the basis of *group-behaviour*. For convenience the discussion is divided into two sections, one on *metabolic rate*, and the other on *metabolic pathways*. It must be remembered, however, that in some cases the pathway that is most utilized by a tissue will depend on its metabolic rate, and the two phenomena cannot be viewed in isolation.

Metabolic rate and anoxia

Due to the low K_m[1] values for cytochrome oxidase which range from 4.0 to 0.024 μmol dm^{-3}

[1] K_m is the substrate concentration necessary for half-maximum reaction velocity.

(Hayaishi, 1962), the concentration of oxygen within the plant tissue must be less than 1% before availability alone determines its rate of consumption. Below 1% oxygen the aerobic metabolism of the plant cannot match the activity of the glycolytic pathway, and the products of anaerobic metabolism accumulate (the Pasteur Point). In spite of the low saturating value for oxygen there are many plant tissues where anaerobic metabolism predominates even under well-aerated conditions. In such cases it has to be concluded that the oxygen level within the tissues (provided the aerobic respiration mechanism is intact) is below the 1% level that is necessary to saturate the cytochrome oxidase system. This situation is found most commonly in germinating seeds and in the meristematic regions of roots (Crawford, 1977, 1978; see also Fig. 7.18). In these tissues any reduction in the external oxygen supply below that normally found in air will cause glycolytic products of respiration to accumulate. In many plant tissues the position is aggravated by the operation of the Pasteur effect, whereby a reduction in citrate and ATP levels removes the allosteric regulation of the glycolytic pathway, resulting in an acceleration of anaerobic respiration. This phenomenon can readily be observed in the roots and seeds of flood-intolerant plants. In addition the longer the tissue is kept under anaerobic conditions the greater is the increase in anaerobic activity. Fig. 7.19 shows the effect of flooding roots for one month on the rate of ethanol production. In the flood-intolerant species there is an increase in ethanol production which is not observed in the flood-tolerant plants. Associated with this increased production of ethanol there is always a marked increase in alcohol dehydrogenase (ADH) activity (Fig. 7.20). ADH is an inducible enzyme and can have its activity increased in non-flood-tolerant plants growing in aerated water culture merely by introducing its substrate acetaldehyde (Crawford and McManmon, 1968). This induction of ADH in flood-intolerant plants has been observed by several investigators and has even been extended to the behaviour of flood-tolerant and flood-intolerant varieties of a single species. Separation of the isoenzymes of ADH in *Zea mays* (Marshall et al., 1973) has shown that the possession of alternative forms of ADH can affect the fitness of individual plants to withstand flooding. In

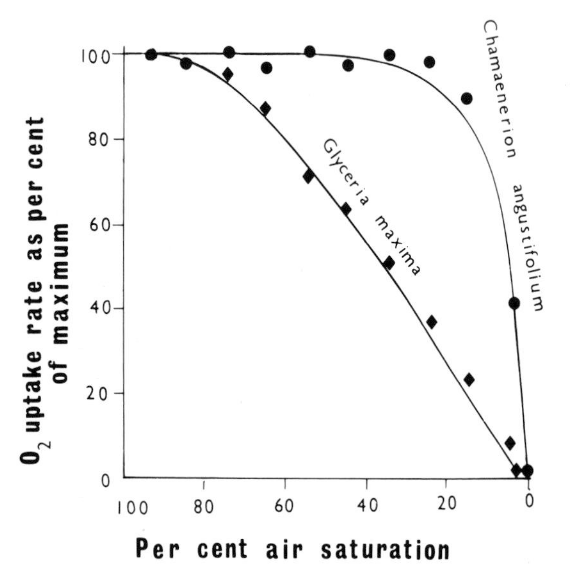

Fig. 7.18. Root oxygen uptake as a function of the air saturation of the bathing medium. Measurements were made using a Clark oxygen-electrode containing six root tips, 1 cm each in length and immersed in phosphate buffer pH 6.0 (0.1*M*) with 2% sucrose and kept at 20°C (Crawford and McCreath, unpublished).

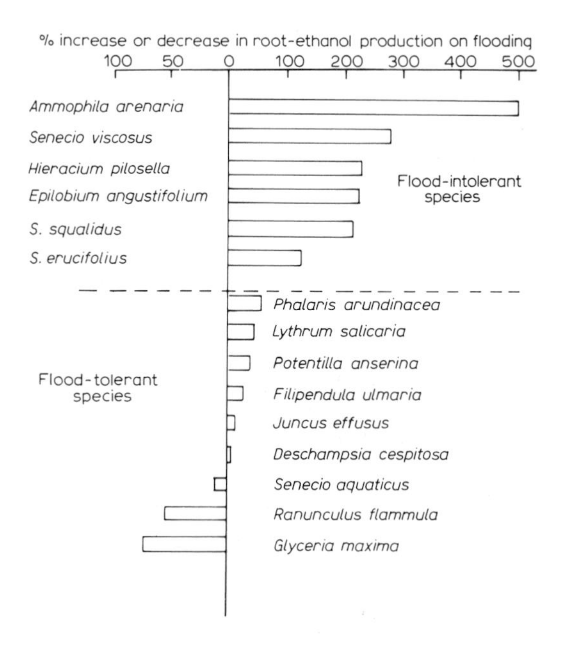

Fig. 7.19. Change in rate of ethanol production in roots of fifteen species of higher plants of varying flooding tolerance after flooding in sand culture for one month as compared with unflooded controls. Ethanol production was determined in excised roots incubated with 2.5% glucose under nitrogen (Crawford, 1978).

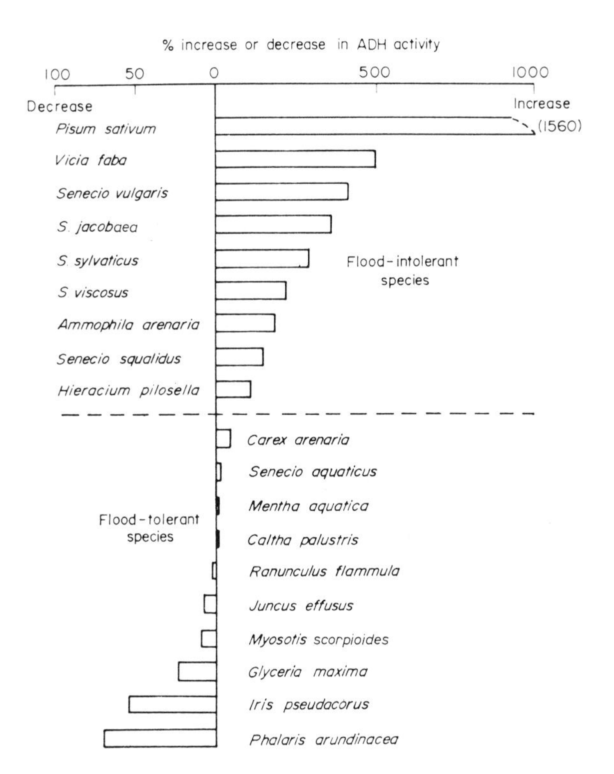

Fig. 7.20. Changes in alcohol dehydrogenase (ADH) activities of nineteen species after flooding in sand culture for one month as compared with unflooded controls. Enzyme activity is calculated on a soluble protein basis (Crawford and McManmon, 1968).

strains of maize possessing a fast-migrating, active and heat-labile form of ADH, tolerance to flooding is less than in those strains possessing the less active slower migrating form of the enzyme. This pattern of behaviour has also been observed in *Trifolium subterraneum*, where it has been suggested that the differential behaviour of ADH is sufficiently regular, that, provided standard portions of roots are taken, the activity of this enzyme might be used in the selection of flood-tolerant provenances of this species (Francis et al., 1974). In *Lupinus angustifolius*, another species with varying tolerance of flooding, the tolerant strain contains only one electrophoretic band of ADH isoenzyme, whereas the increase in ADH activity that is found on flooding intolerant plants is associated with the possession of an additional and distinct ADH band (Marshall et al., 1974). A similar increase in ethanol production associated with an increase in ADH activity is also found in seeds that are unable to withstand soaking (Crawford, 1977).

The only plants which are exceptions to the rule that flooding intolerance is associated with increased glycolytic activity and ethanol production are those plants where ethanol can readily be removed from the tissues by diffusion. Thus, in the adventitious roots of willow (Chirkova, 1973), in rice (John and Greenway, 1976), and in *Nyssa sylvatica* var. *biflora* (Hook et al., 1971) flooding causes an increase in ethanol production. However, in all these cases the ethanol is readily lost to the surrounding medium by the adventitious root system. As already mentioned (p. 262), over 97% of the ethanol produced in rice roots passes immediately into the culture solution (Fig. 7.8). This behaviour should be considered with the observation that only when roots have access to free water do plants endure constant inundation better than intermittent flooding with periods of respite during which water tables are lowered. This would suggest that glycolytic activity is only capable of sustaining root growth in flushed sites where toxic products can be readily removed from the roots.

In these cases it is necessary to consider if there is any adaptive significance in the increased ADH activity that is found in so many non-flood-tolerant plants. When flooding is of short duration (24–48 h), or where ethanol is prevented from reaching toxic concentrations by diffusion into open or moving water, it is possible for the plant to increase its glycolytic rate without exposing itself to the dangers of self-poisoning from the products of anaerobiosis. In such a situation there does remain the danger however of acetaldehyde poisoning. This intermediate in the production of ethanol is very much more toxic than ethanol itself. However, if the K_m values for ADH with respect to acetaldehyde are reduced in the ADH that is synthesized *de novo* during anaerobiosis, then this risk is minimized. This explanation could account for the reduction in K_m values which is always associated with ADH synthesis in plants that are unable to withstand prolonged flooding (Table 7.5).

Where flooding is prolonged (greater than 48 h), or where there is no means of ethanol removal from the root, then any increase in glycolytic activity will increase the problems associated with the accumulation of an excessive oxygen debt which cannot be either repaid or removed. In seeds during the period of anaerobiosis that precedes testa rupture, many species accumulate their oxygen debt in the form of

TABLE 7.5

Effect of flooding on the apparent K_m of alcohol dehydrogenase with respect to acetaldehyde in the roots of eight species (McManmon and Crawford, 1971); the figures are in terms of an acetaldehyde concentration of 1 mmol dm^{-3}

	Unflooded	Flooded
Flood-tolerant		
Senecio aquaticus	21	16
Caltha palustris	7.6	18
Mentha aquatica	*c.* 55	*c.* 55
Ranunculus flammula	33	26
Flood-intolerant		
Senecio jacobaea	3.5	1.2
Hieracium pilosella	13.7	3.0
Senecio viscosus	12.0	1.0
Pisum sativum	27	1.1

lactate. If the seed lies in flooded soil this process cannot continue indefinitely. In the unvacuolated seed lactate accumulation will reduce cytoplasmic pH and inhibit lactate production, in favour of the decarboxylation reaction which leads to ethanol formation (Davies et al., 1974). Low metabolic rates thus prolong the period over which lactate is able to accumulate the oxygen debt. A comparison of metabolic rate under anoxia ($Q_{CO_2}{}^{N_2}$) in relation to the ability of seeds to withstand soaking injury (Crawford, 1977) has shown that it is the seeds that are resistant to soaking injury that have the lowest metabolic rates under anoxia, show a minimal Pasteur effect and accumulate lactate. By contrast the intolerant seeds have higher metabolic rates, show a pronounced Pasteur effect and accumulate the oxygen debt exclusively in the form of ethanol (Fig. 7.21). The relationship between anaerobic metabolic rate and survival under anoxia can be seen by examining the effect of storage temperature on seeds kept in the imbibed state under anoxia (Fig. 7.22). The lower the temperature the greater is the ability of the seed to survive the period of low-oxygen stress. Thus, one might expect that the seed bank that lies buried in wetland sites will be greatest in areas of low temperatures (Table 7.6). The record for length of viability of a buried seed is held by *Nuphar nelumbo* (Fig 7.23) which has been reported to have survived 250 years of burial in peat in Manchuria (Turrill, 1957). Only in certain quick-germinating crop species have exceptions been

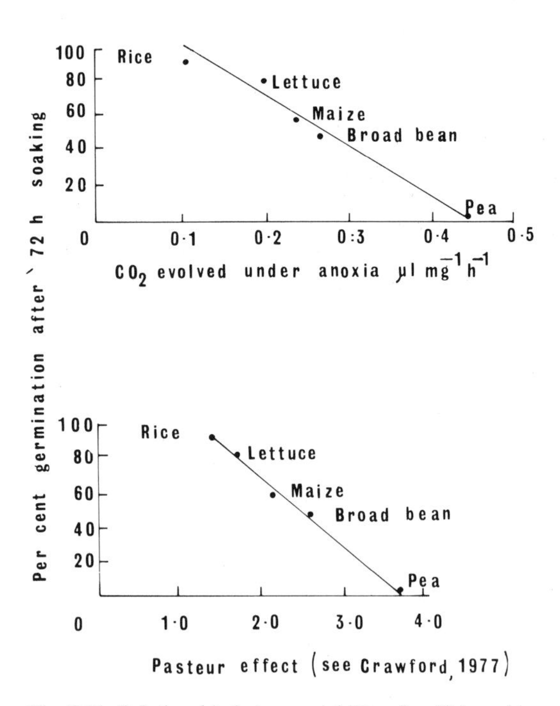

Fig. 7.21. Relationship between viability after 72 h soaking and (above) anaerobic respiration rate $Q_{CO_2}{}^{N_2}$ [μl CO_2 (mg dry wt.)$^{-1}$ h^{-1}] and (below) Pasteur effect as determined from carbon dioxide evolution (Crawford, 1977).

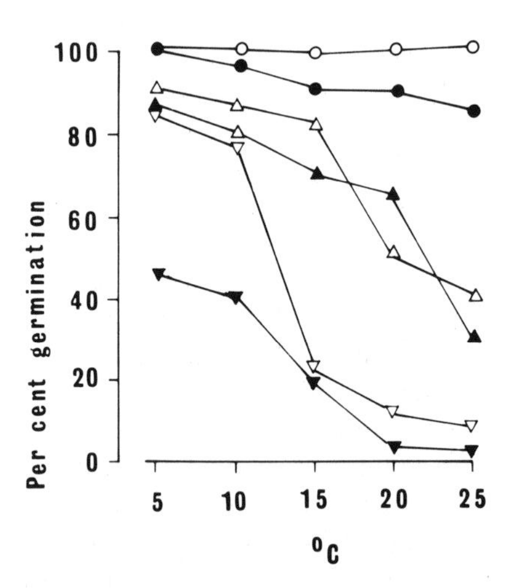

Fig. 7.22. Deterioration of seed viability of six graminaceous species during storage in imbibed condition for six weeks under complete anoxia in anaerobe jars. Legend: ○ = *Oryza sativa*; ● = *Festuca rubra*; △ = *Lolium perenne*; ▲ = *Dactylis glomerata*; ∇ = *Agrostis tenuis*; ▼ = *Zea mays*.

reported of the advantage of low temperatures to the survival during anoxia. Thus in pea seeds (Perry and Harrison, 1970) low temperature aggravates the effects of soaking — an effect which may be due

TABLE 7.6

Number of seeds germinated from sub-surface samples of wetland soils

Area	Vegetation	Seedlings obtained (m^{-2})	Reference
Coastal dune slack; Fife, Scotland with prolonged flooding	*Juncus effusus* *Filipendula ulmaria* *Galium palustre*	1134	Crawford and Taylor (unpublished data)
Coastal dune slack; Fife, Scotland with short-term flooding	*Salix repens*	24 400	Crawford and Taylor (unpublished data)
Wet dune-heath; Fife, Scotland with occasional flooding	*Erica tetralix* *Calluna vulgaris* *Empetrum nigrum* *Carex flacca*	1010	Crawford and Taylor (unpublished data)
Raised bog; Fife, Scotland	*Erica tetralix* *Calluna vulgaris* *Eriophorum vaginatum* *Sphagnum* spp.	9575	Crawford and Taylor (unpublished data)
Nutrient-rich fen; Fife, Scotland	*Juncus effusus* *Filipendula ulmaria*	8568	Crawford and Taylor (unpublished data)
Acadia Forest, New Brunswick Canada	*Sphagnum* bog	0	Moore and Wein (1977)

Note: The present author's experience in Scotland does not agree with the negative report of seed viability in Canadian bogs of Moore and Wein (1977).

to the inability of these seeds to reconstitute their membranes sufficiently rapidly at low temperatures, with the result that the seed tissues are flooded by the soak water.

Apart from the exceptions noted above it appears that flooding tolerance is frequently associated with a non-acceleration of glycolysis under anoxia (i.e., the absence of a Pasteur effect) whereas in the flood-tolerant species there is a marked increase in glycolysis. This is due in part to the operation of the Pasteur effect, and in part to an inductive increase in glycolytic rate with any prolongation of the condition of hypoxia. This activity rapidly leads to considerable rises in ethanol content of the tissues, and is followed by membrane leakage and organelle damage, and finally microbial infection and death.

Metabolic pathways and anoxia

The study of anaerobiosis in relation to the survival of plants has received systematic study only in the past 10 to 15 years. Many textbooks of plant physiology therefore limit their consideration of the subject to the production of ethanol, together with notes on the interesting occurrences of lactate in potato and *Equisetum* spp. (James, 1953). In certain flood-tolerant herbaceous species the malic acid content of the root has been shown to rise as a result of flooding (Crawford and Tyler, 1969). The ecological significance of malate accumulation in flood-tolerant plants has been elegantly demonstrated by Linhart and Baker (1973) in a study of a population of *Veronica peregrina* growing in and around an ephemeral pond in California. Populations taken from the centre of the pond were derived from plants that were able to establish themselves and grow in flooded soils while those taken from the periphery were plants that began growth only when the flood levels of the pond subsided. Experimental flooding of the populations taken from the centre of the pond was always accompanied by an increase in the malic acid content of the roots. The populations taken from the periphery of the pond showed no regularity of behaviour in relation to malic acid content of the roots and experimental flooding (Fig. 7.24). It is a striking example of the localized effects of ecotypic

Fig. 7.23. The Indian water lotus *Nuphar nelumbo*. Seeds of this species have been reported to have survived burial in anaerobic mud for 250 years (Turrill, 1957). (Photo: R.M.M. Crawford.)

differentiation that, in one species over a distance of only 1 to 2 m, there can exist a physiological distinction both in the ability to survive flooding and in the physiological reaction to this stress.

In micro-organisms it has long been known that anaerobiosis usually results in the formation of a number of products. These mixed fermentations (Krebs, 1972) also occur in animals, and the list of products that serve to accumulate the oxygen debt in parasites of the gut and blood stream, bivalve molluscs, cephalopods and insects grows steadily longer (Crawford, 1978). In higher plants, due to the practice of working with non-flood-tolerant crop plants for most physiological research, the biochemical diversity of anaerobic adaptation has remained largely unexplored. It is now clear, however, there are many parallels between plants and animals in their metabolic reactions to hypoxia. Table 7.7 lists those substances which have been reported to accumulate under anaerobic conditions in higher plants and animals. All these compounds are capable of accumulating the oxygen debt generated under anoxia as they allow the regeneration of nicotineamide adenine dinucleotide (NAD) from its reduced form NADH. The system of carbon flow which would allow this regeneration is shown in Fig. 7.25. As can be seen there are several striking similarities between plants and animals in metabolite accumulation under anoxia. Glycerol for example is produced by blood-inhabiting African trypanosomes (Gilmour, 1965), where it can account for 50% of the glucose utilized. It is also accumulated at diapause by silkworms and in the adipose tissue of hibernating hedgehogs. In plants it can be found to increase in alder roots after eight days' flooding (Crawford, 1972). Alanine also oc-

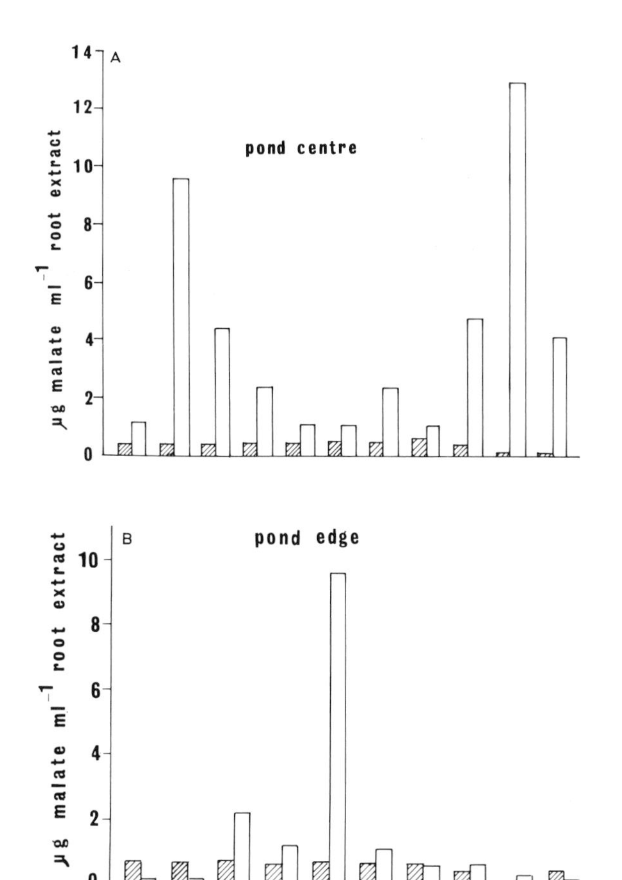

Fig. 7.24. Malic acid content of root extracts made from pot-grown populations of *Veronica peregrina*; hatched: soils kept drained; blank: soils kept wet. A. Populations collected from the pond centre. B. Populations collected from the pond edge. (Data from Linhart and Baker, 1973.)

curs in insects and molluscs during periods of anaerobiosis, as well as in the roots of flooded trees (Dubinina, 1961). If a comparison is made of the flood-tolerant *Senecio aquaticus* with the intolerant *S. jacobaea*, it is the former flood-tolerant species that accumulates malic acid (Crawford and Tyler, 1969) as well as proline and glutamic acid and probably also homoserine (Lambers, 1976). In both plants and animals this multiplicity of end-products in anaerobic respiration avoids the dangers that are imposed by the exclusive use of one product to accumulate the oxygen debt. In plants, the avoidance of ethanol accumulation in situations where it cannot easily be removed from the root increases the probability of root survival after periods of prolonged flooding. In animals, apart from increasing the efficiency of proton disposal during anaerobiosis, the coupling of amino-acid and carbohydrate metabolism allows an increase in the number of molecules of ATP that are produced per mole of glucose respired. As yet, no comparable increase in respiratory efficiency has been demonstrated in plants.

Little information is available for plants on the control mechanisms which determine the nature of the end-products of glycolysis under anoxia. The change from ethanol to glycerol production in alder takes about eight days and this time delay suggests the induction of an adaptive enzyme system. The production of lactate by flood-tolerant seeds is favoured by low metabolic rates and is influenced by pyruvate concentration and cell pH (Davies et al., 1974). Enzyme deletion was originally proposed to account for the accumulation of malate instead of ethanol in flood-tolerant marsh plants (McManmon and Crawford, 1971). The absence of malic enzyme in carboxylation and decarboxylation by-pass of pyruvate kinase which was thought to operate in most plant tissues (Mazelis and Vennesland, 1957) would allow malate to accumulate at the expense of ethanol. It now appears that malic enzyme is not absent from these marsh species (Davies et al., 1975), but that it is inhibited by flooding in flood-tolerant species (Chirkova, 1978). The overall effect is similar to deletion, and the pathway of malate accumulation in flood-tolerant plants can still be expected to follow the original suggestion of McManmon and Crawford (1971).

Recently other metabolic systems have been shown to be inhibited differentially in flood-tolerant and non-flood-tolerant plants. In *Senecio aquaticus* oxygen uptake is reduced by 70% when the air supply is restricted, and it has been suggested that this is due to the inhibition of the alternative oxidase system (Lambers and Steingrover, 1978). In general the half-saturation values for oxygen consumption are lower in flood-tolerant than in flood-intolerant plants (Lambers and Smakman, 1978; see also Fig. 7.18).

THE EVOLUTION OF FLOODING TOLERANCE

It is a curious paradox of evolutionary advance that, after the millions of years that were necessary for plant and animal life to make the initial transfer

TABLE 7.7

Substances reported to accumulate under anaerobic conditions in higher plants and animals (references are intended only as a source of further information and do not indicate priority of discovery or complete range of occurrence)

Substance	Animal occurrence	Plant occurrence
Lactic acid	vertebrate skeletal muscle[1]	germinating seeds[2], tubers[3]
Pyruvic acid	vertebrate skeletal muscle	willow roots[4]
Formic acid	parasitic helminths[5]	–
Acetic acid	bivalve molluscs, cestodes[6]	–
Acetoin	nematodes[5]	–
Propionic acid	molluscs, cestodes[5]	–
Butyric acid	parasitic protozoa[5]	–
Succinic acid	bivalve molluscs[6]	seeds[2]
Malic acid	–	roots of marsh plants[8,9]
Shikimic acid	–	iris and water-lily roots[10]
Glycolic acid	–	willow roots[4]
Ethanol	parasitic protozoa[5], helminths[5]	flood-intolerant roots and seeds[10]
Sorbitol	insects[16]	–
Glycerol	insects	alder roots[10]
Alanine	sea turtles[7], molluscs[19]	flood-tolerant roots[4]
Aspartic acid	marine annelids[11]	flood-tolerant roots[4]
Glutamic acid	marine annelids[11]	flood-tolerant roots[4]
Serine	–	flood-tolerant roots[4]
Proline	–	flood-tolerant roots[4]
Octopine	cephalopods[12]	crown-gall tissues[13]
γ-Amino butyric acid	–	tomato roots[14]
		radish leaves[17]
Methyl butyrate	parasitic nematodes[15]	–
Methyl valerate	parasitic nematodes[15]	–
Glycerophosphate	insects[16]	–
Hydrogen	parasitic protozoa[5]	–
Ethylene	–	roots and fruits[18]

References: 1, Hochachka and Storey (1975); 2, Wager (1961); 3, Davies et al. (1974); 4, Dubinina (1961); 5, Von Brand (1966); 6, Gäde et al. (1975); 7, Hochachka et al. (1975); 8, Crawford and Tyler (1969); 9, Linhart and Baker (1973); 10, Crawford (1972); 11, Zebe (1975); 12, Hochachka et al. (1977); 13, Menage and Morel (1964); 14, Fulton et al. (1964); 15, Bryant (1971); 16, Gilmour (1965); 17, Streeter and Thompson (1972); 18, Pratt and Goeschl (1969); 19, De Zwaan and Wijsman (1976).

from an aquatic to a terrestrial environment, so many present-day aquatic species have apparently evolved from terrestrial ancestors. For species that are able to survive the hazards of inundation and submersion, water provides an ecological refuge safe from the interference and predation of dryland competitors. In animals the advantages of the aquatic habitat are clearly seen in turtles, crocodiles and other marine reptiles, for without the possibility of a sub-aquatic retreat it is unlikely that any of these species would be alive today. Even birds exploit the aquatic refuge, as is evident from the large numbers of diving ducks, grebes, cormorants and auks.

Among plants, the fundamental division into monocotyledons and dicotyledons first made by John Ray (1628–1705) has long been thought to be due to the re-entry into the aquatic environment of terrestrial species. It is over 150 years since De Candolle (1827) proposed an aquatic origin for the monocotyledons. Although taxonomic opinion differs in detail, there is general agreement that the modern monocotyledons have evolved from an earlier dicotyledonous group not unlike the modern Nymphaeales (Takhtajan, 1969). In aquatic species many of the vegetative structures essential for life in terrestrial habitats either have degenerated or else are much simplified. These reductions include: loss of secondary thickening, simplification of leaf structure, suppression of main root development and the reduction of the conducting system. The same simplifications are also typical of monocotyle-

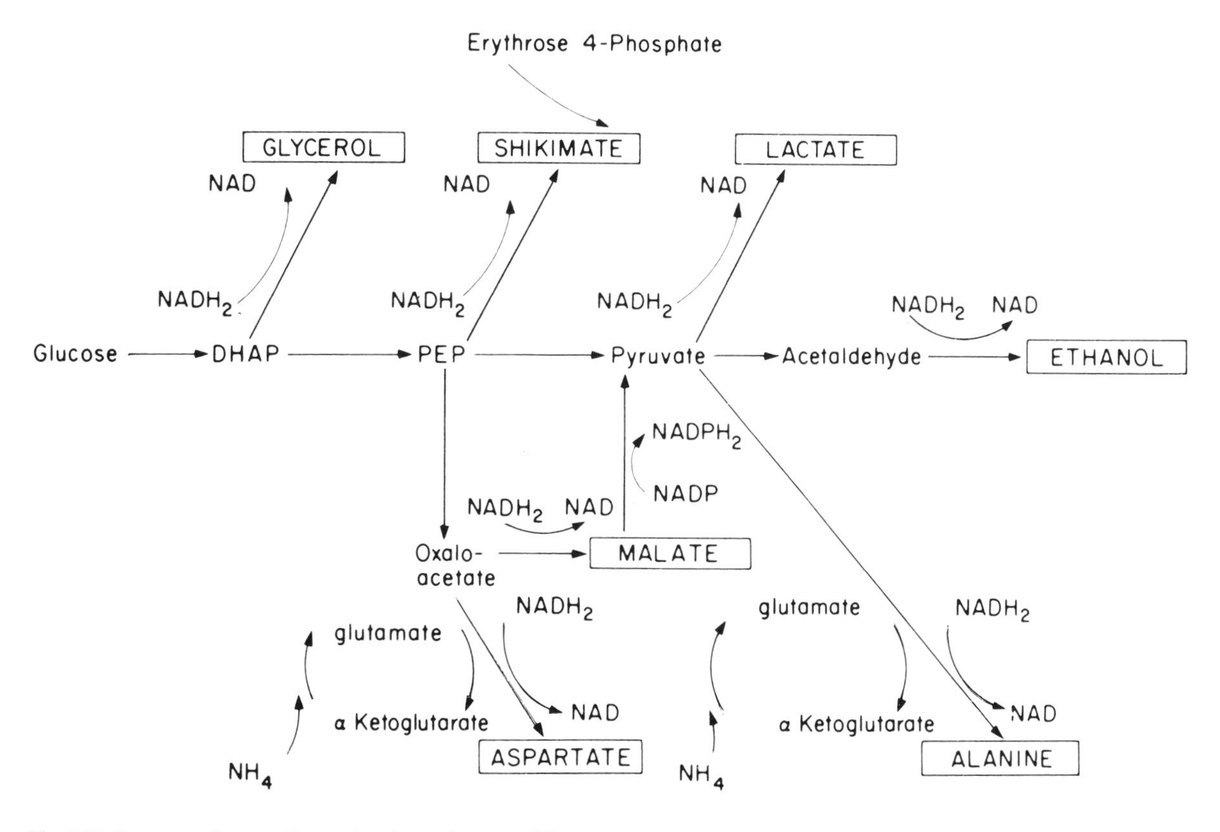

Fig. 7.25. Summary diagram illustrating the various possible means of proton disposal and the range of end-products which may be associated in plants with the capacity to endure prolonged periods of anaerobiosis (see Table 7.7 for references).

donous structures. The loss of the cambium is a fundamental deprivation as it precludes the formation of any vessels that are not found in primary tissue. Only after returning to land again has the monocotyledonous shoot developed an effective vascular system (Cronquist, 1968). The evolutionary progress of the monocotyledons has suffered a number of convolutions, as some terrestrial species which were originally derived from aquatic ancestors have evolved to produce further groups which have returned to a hydrophilous habit. Thus, in the modern subclass Alismatidae there is a progressive adaptation from a terrestrial to an aquatic and even marine habitat. A submerging gradation takes place beginning with the Scheuchzeriaceae with rush-like bog-inhabiting plants and progressing to the marsh-dwelling Juncaginaceae and the rooted aquatics of the Potamogetonaceae and finally to the completely submerged marine eel grasses of the Zosteraceae. Similarly, in the subclass Commelinidae, families such as the Mayacaceae, Sparganiaceae and Typhaceae appear to have been derived from terrestrial ancestors (Cronquist, 1968).

The lack of cambium in the monocotyledons is also responsible for their reliance on a fibrous adventitious root system. In mire species this has the advantage of being adapted to give an adventitious root system that is renewed each year after the period of maximum flooding stress.

Being without a main root system, monocotyledons have to rely on other structures for their perennation. In wetland sites this presents the particular difficulty of renewing metabolic activity and growth in tissues which are either submerged or buried in anaerobic soils and without any connection with the atmosphere. The renewal of growth in such situations must depend on glycolytic activity, and, as this is costly in terms of carbohydrate supplies, well-developed perennating organs are essential. It has been suggested that the primitive monocotyledons were possibly marsh plants with stout rhizomes rather than true aquatic species (Parkin, 1923). From these marsh plants some descendents would have become hydrophytes and others geophytes. Irrespective of whether the monocotyledons had an aquatic or amphibious origin, wet and flooded soils appear to have played a significant role in their origin. It is therefore not surprising that so large a portion of the world's wetland flora belongs to the monocotyledons.

The simplified structure of aquatic angiosperms, in addition to promoting a more flexible morphology for life in lakes and marshes, also aids the physiological fitness that is necessary for survival in such sites. Hollow stems, adventitious roots and absence of woody tissues all aid the downward diffusion of oxygen to the submerged root as well as assisting the removal of the more volatile products of anaerobiosis, such as acetaldehyde, ethanol and carbon dioxide. As they require experimentation for their observation the physiological adaptations associated with flooding tolerance are less obvious than the anatomical conformities of wetland vegetation. Nevertheless, the existence in diverse tissues, such as seeds and roots, as well as in different life forms from herbs to trees, of similar properties of metabolic control and end-product diversity (Table 7.7) gives a striking example of the evolutionary moulding of physiological adaptation to wetland conditions. Many groups have had both their form and function shaped to a common pattern to meet the constraints of life in anaerobic environments.

Parallel evolution by diverse species under the influence of a common stress manifests itself in many different ecological situations. The desert plants of the Old and New Worlds resemble one another in form, even although they are taxonomically distinct. The stress of anaerobiosis acts in varying ways on both plants and animals, but for both adaptation has resulted in the evolution of certain common metabolic features. Diving in marine mammals and aquatic birds induces bradycardia (a slowing of the heart beat) and the absorption of the oxygen debt in a system that is buffered against rapid changes in pH due to accumulating lactic acid (Hochachka and Storey, 1975). Similarly in plants, adaptation to anoxia is widely accompanied by a control of anaerobic respiration rates (Fig. 7.19) and the possible diversification of the end-products of glycolysis in a manner very similar to animals (Table 7.7). The accumulation of glycerol in silkworms at diapause and in alders on flooding, as well as the use of amino acids to accumulate the oxygen debt in marsh plants, marine annelids, turtles and molluscs, together with the many other examples shown in Table 7.7 would suggest that similar metabolic pathways are serving the same ends in a very diverse range of species (Crawford, 1978). The adaptations that occur in plants and animals in relation to anoxia give a striking example of parallel evolution functioning at a molecular level. The remarkable fact that evolution has selected the same mechanisms in such widely different groups of plants and animals suggests that there can be only a few efficient metabolic solutions to this particular environmental stress. Thus, one can view the metabolic features that are common to higher plants, mammals and reptiles that have returned to the aquatic habitat in the same light: namely, the selection of the only solutions that are available. For both plants and animals the entry of terrestrial species into the anaerobic habitat, even if it is only for refuge during periods of temporary stress, is not the most efficient use of their energy resources. However, in both cases the ability to exploit this niche, free from the disturbing effects of unadapted dryland species, demonstrates that the expenditure is ecologically well justified.

REFERENCES

Arber, A., 1920. *Water Plants*. Cambridge University Press, Cambridge, 436 pp.

Armstrong, W., 1968. Oxygen diffusion from the roots of woody species. *Physiol. Plant.*, 21: 539–543.

Armstrong, W., 1972. A re-examination of the functional significance of aerenchyma. *Physiol. Plant.*, 27: 173–77.

Armstrong, W., 1975. Waterlogged soils. In: J.R. Etherington (Editor), *Environment and Plant Ecology*. Wiley and Sons, New York, N.Y., pp. 181–218.

Atkinson, D.E., 1968. The energy charge of the adenylate pool as a regulatory parameter. Interaction with feedback modifiers. *Biochemistry*, 7: 4030.

Bell, D.T. and Johnson, F.L., 1974. Flood-caused tree mortality around Illinois reservoirs. *Trans. Ill. State Acad. Sci.*, 67: 28–37.

Bell, J.N.B. and Tallis, J.H., 1974. The response of *Empetrum nigrum* L. to different mire water regimes, with special reference to Wynbury Moss, Cheshire and Featherbed Moss, Derbyshire. *J. Ecol.*, 62: 75–95.

Bendixen, L.E. and Peterson, M.L., 1962. Tropism as a basis for flooding tolerance of strawberry clover to flooding conditions. *Crop Sci.*, 2: 223.

Billings, W.D. and Godfrey, P.J., 1967. Photosynthetic utilization of internal carbon dioxide by hollow-stemmed plants. *Science*, 158: 121–123.

Boatman, D.J., 1972. The growth of *Schoenus nigricans* L. on blanket bog peats. II Growth on Irish and Scottish peats. *J. Ecol.*, 60: 469–477.

Broadfoot, W.M. and Williston, H.L., 1973. Flooding effects on southern forests. *U.S. J. For.*, 71: 584–587.

Bryant, C., 1971. *The Biology of Respiration*. Edward Arnold, London, 59 pp.

Buttery, B.R. and Lambert, J.M., 1965. Competition between *Glyceria maxima* and *Phragmites communis* in the region of Surlingham Broad. II. The competition mechanism. *J. Ecol.*, 53: 163–181.

Cannell, R.Q., 1977. Soil aeration and compaction in relation to root growth and soil management. In: T.H. Coaker, (Editor), *Applied Biology, 2*. Academic Press, New York, N.Y., pp. 1–86.

Chirkova, T.V., 1968. Oxygen supply to roots of certain woody plants kept under anaerobic conditions. *Sov. Plant Physiol.*, 15: 475–477.

Chirkova, T.V., 1973. Rol'nitratnogo dykhaniya korneĭ v zhiznedeyatel'nosti nekotorykh drevesnykh rästentĭ v usloviyakh vremennogo anaerobioza. [The role of anaerobic respiration in the adaptation of some woody plants to temporary anaerobiosis.] *Vestn. Leningr. Univ., Biol.*, 3: 88.

Chirkova, T.V., 1978. Some regulatory mechanisms of plant adaptation to temporal anaerobiosis. In: D.D. Hook and R.M.M. Crawford (Editors), *Plant Life in Anaerobic Environments*, Ann Arbor, Mich., pp. 137–154.

Chirkova, T.V. and Gutman, T.C., 1972. O fiziologicheskoĭ roli chechevichek vetveĭ ivi i topolya v usloviyakh kornevogo anaerobioza. [The physiological role of branch lenticels of willow and poplar under conditions of root anaerobiosis.] *Fiziol. Rast.*, 19: 352–359.

Coutts, M.P., 1976. The formation of dry zones in the sapwood of conifers. I. Induction of drying in standing trees and logs by *Fomes annosus* and extracts of infected wood. *Eur. J. For. Pathol.*, 6: 362–381.

Coutts, M.P. and Philipson, J.J., 1978. The tolerance of tree roots to waterlogging. II Adaptation of sitka spruce and lodgepole pine to waterlogged soil. *New Phytol.*, 80: 71–77.

Crawford, R.M.M., 1972. Physiologische Ökologie: Ein Vergleich der Anpassung von Pflanzen und Tieren an sauerstoffarme Umgebung. *Flora*, 161: 209–223.

Crawford, R.M.M., 1977. Tolerance of anoxia and ethanol metabolism in germinating seeds. *New Phytol.*, 79: 511–517.

Crawford, R.M.M., 1978. Biochemical and ecological similarities in marsh plants and diving animals. *Naturwissenschaften.*, 65: 194–201.

Crawford, R.M.M. and Baines, M., 1977. Tolerance of anoxia and ethanol metabolism in tree roots. *New Phytol.*, 79: 519–526.

Crawford, R.M.M. and McManmon, M., 1968. Inductive responses of alcohol and malic dehydrogenases in relation to flooding tolerance in roots. *J. Exp. Bot.*, 19: 435–441.

Crawford, R.M.M. and Tyler, P., 1969. Organic acid metabolism in relation to flooding tolerance in roots. *J. Ecol.*, 57: 235–244.

Crawford-Sidebotham, T.J., 1972. The role of slugs and snails in the maintenance of cyanogenesis polymorphisms of *Lotus corniculatus* and *Trifolium repens*. *Heredity*, 28: 405–411.

Cronquist, A., 1968. *The Evolution and Classification of Flowering Plants*. Nelson, London and Edinburgh, 396 pp.

Davies, D.D., Grego, S. and Kenworthy, P., 1974. The control of production of lactate and ethanol by higher plants. *Planta*, 118: 297–310.

Davies, D.D., Nascimiento, K.H. and Patil, K.D., 1975. The distribution and properties of malic enzyme in flowering plants. *Phytochemistry*, 13: 2417–2425.

De Candolle, A.P., 1827. *Organographie végétale*. Paris, 2 vols.

De Zwaan, A. and Wijsman, T.C.M., 1976. Anaerobic metabolism in Bivalvia (Mollusca); characteristics of anaerobic metabolism. *Comp. Biochem. Physiol.*, 54B: 313–324.

Dickson, R.E. and Broyer, T.C., 1972. Effect of aeration water supply and nitrogen source on growth and development of tupelo gum and bald cypress. *Ecology*, 53: 626–634.

Dubinina, I.M., 1961. Metabolism of roots under various levels of aeration. *Sov. Plant Physiol.*, 8: 395–406.

Francis, C.M., Devitt, A.C. and Steele, P., 1974. Influence of flooding on the alcohol dehydrogenase activity of roots of *Trifolium subterraneum* L. *Aust. J. Plant Physiol.*, 1: 9–13.

Fulton, J.M., Erickson, A.E. and Tolbert, N.E., 1964. Distribution of C^{14} among metabolites of flooded aerobically grown tomato plants. *Agron. J.*, 56: 527–529.

Gäde, G., Wilps, H., Kluytmans, J.H. and De Zwaan, A., 1975. Glycogen degradation and end products of anaerobic metabolism in the fresh water bivalve (*Anodonta cygnea*). *J. Comp. Physiol.*, 104: 79–85.

Gessner, F., 1968. Zur ökologischen Problematik der Überschwemmungs-wälder des Amazonas. *Int. Rev. Ges. Hydrobiol.*, 53: 525–547.

Gill, C.J., 1970. The flooding tolerance of woody species — a review. *For. Abstr.*, 31: 671–678.

Gill, C.J., 1977. Some aspects of the design of reservoir and management of reservoir margins for multiple use. In: T.H. Coaker (Editor), *Applied Ecology, 2*. Academic Press, New York, N.Y., pp. 129–182.

Gilmour, D., 1965. *The Metabolism of Insects*. Edinburgh, 195 pp.

Greaves, M.P. and Darbyshire, J.E., 1975. Microbial decomposition of plant roots. In: G. Kilbertus, D. Reisinger, A. Mourey and J.A. Consela de Fonseca (Editors), *Proc. 1st Int. Colloquium on Biodegradation and Humification, University of Nancy, 1974*, pp. 108–111.

Hall, T.F., Penfound, W.T. and Hess, A.D., 1946. Water level relationships of plants in the Tennessee Valley with particular reference to malaria control. *J. Tenn. Acad. Sci.*, 21: 18–59.

Hayaishi, O., 1962. Biological oxidations. *Annu. Rev. Biochem.*, 31: 25–46.

Hochachka, P.W. and Storey, K.B., 1975. Metabolic consequences of diving in animals and man. *Science*, 187: 613–621.

Hochachka, P.W., Owen, T.G., Allen, J.F. and Whittow, G.G., 1975. Multiple end products of anaerobic metabolism in diving vertebrates. *Comp. Biochem. Physiol.*, 50B: 17.

Hochachka, P.W., Hartline, P.H. and Fields, H.H.A., 1977. Octopine as an end product of glycolysis in the chambered *Nautilus*. *Science*, 195: 72–74.

Hook, D.D. and Crawford, R.M.M., 1978. *Plant Life in Anaerobic Environments*. Ann Arbor Science, Ann Arbor, Mich., 564 pp.

Hook, D.D. and Scholtens, J.R., 1978. Adaptations and flood-tolerance of tree species. In: D.D. Hook and R.M.M. Crawford (Editors), *Plant Life in Anaerobic Environments*. Ann Arbor Science, Ann Arbor, Mich., pp. 299–331.

Hook, D.D., Brown, C.L. and Kormanik, P.P., 1971. Inductive flood-tolerance in swamp tupelo (*Nyssa sylvatica* var. *biflora* Walt. Sarg.). *J. Exp. Bot.*, 20: 78.

Hook, D.D., Brown, C.L. and Wetmore, R.H., 1972. Aeration in trees. *Bot. Gaz.*, 133: 443–454.

Ingram, H.A.P., 1967. Problems of hydrology and plant distribution in mires. *J. Ecol.*, 55: 711–724.

James, W.O., 1953. *Plant Respiration*. Clarendon Press, Oxford.

Jensen, C.R., Luxmoore, R.J., Van Grundy, S.D. and Stolzy, L.H., 1969. Root air space measurements by a pycnometer method. *Agron. J.*, 61: 474–475.

John, C.D. and Greenway, H., 1976. Alcoholic fermentation and activity of some enzymes under anaerobiosis. *Aust. J. Plant Physiol.*, 3: 325–336.

Jones, H.E., 1971. Comparative studies of plant growth and distribution in relation to waterlogging. II An experimental study of the relationship between transpiration and the uptake of iron in *Erica cinerea L.* and *E. tetralix*. *J. Ecol.*, 59: 167–178.

Kramer, P.J., Riley, W.S. and Bannister, T.T., 1952. Gas exchange of Cypress (*Taxodium distichum*) knees. *Ecology*, 33: 117–121.

Krebs, H., 1972. The Pasteur effect and the relations between fermentation and respiration. In: P.N. Campbell and F. Dickens (Editors), *Essays in Biochemistry, 8*. Academic Press, pp. 1–34.

Kurz, H. and Demaree, D., 1934. Cypress buttresses and knees in relation to water and air. *Ecology*, 15: 36–41.

Lambers, H., 1976. Respiration and NADH-oxidation of the roots of flood-tolerant and flood-intolerant *Senecio* species as affected by anaerobiosis. *Physiol. Plant.*, 37: 117–122.

Lambers, H. and Smakman, G., 1978. Respiration of the roots of flood-tolerant and flood-intolerant *Senecio* species: affinity for oxygen and resistance to cyanide. *Physiol. Plant.*, 42: 163–166.

Lambers, H. and Steingrover, E., 1978. Efficiency of root respiration of a flood-tolerant and a flood-intolerant *Senecio* species as affected by low oxygen tension. *Physiol. Plant.*, 42: 179–184.

Linhart, Y.B. and Baker, G., 1973. Intra-population differentiation of physiological response to flooding in a population of *Veronica peregrina* L. *Nature (Lond.)*, 242: 275.

McAlpine, R.G., 1961. Yellow poplar seedlings intolerant to flooding. *J. For.*, 59: 566–568.

McManmon, M. and Crawford, R.M.M., 1971. A metabolic theory of flooding-tolerance: the significance of enzyme distribution and behaviour. *New Phytol.*, 70: 299–306.

Marshall, D.R., Broué, P. and Pryor, A.J., 1973. Adaptive significance of alcohol dehydrogenase isoenzymes in maize. *Nature (Lond.) New Biol.*, 244: 16–17.

Marshall, D.R., Broué, P. and Oram, R.N., 1974. Genetic control of alcohol dehydrogenase in narrow leaved lupins. *J. Hered.*, 65: 198–203.

Mazelis, M. and Vennesland, B., 1957. Carbon dioxide fixation into oxaloacetate in higher plants. *Plant Physiol.*, 32: 591–600.

Menage, A. and Morel, G., 1964. Sur la présence d'octopine dans les tissus de crown gall. *C.R. Acad. Sci. Paris*, 259: 4795–4796.

Miles, J., 1976. The growth of *Narthecium ossifragum* in some southern English mires. *J. Ecol.*, 64: 849–858.

Moore, J.M. and Wein, R.W., 1977. Viable seed populations by soil depth and potential site recolonisation after disturbance. *Can. J. Bot.*, 55: 2408–2412.

Parkin, J., 1923. The strobilus theory of angiospermous descent. *Proc. Linn. Soc. Lond. Bot.*, 153: 51–64.

Perry, D.A. and Harrison, J.G., 1970. The deleterious effect of water and low temperature on germination of pea seed. *J. Exp. Bot.*, 21: 504–512.

Petty, J.A., 1970. Permeability and structure of the wood of Sitka spruce. *Proc. R. Soc. Lond., B* 175: 149–166.

Philipson, J.J. and Coutts, M.P., 1978. The tolerance of tree roots to waterlogging. III Oxygen transport in Lodgepole pine and Sitka spruce roots of primary structure. *New Phytol.*, 80: 341–349.

Pratt, H.K. and Goeschl, J.D., 1969. Physiological roles of ethylene in plants. *Annu. Rev. Plant Physiol.*, 20: 541–584.

Raup, H.M., 1965. The flowering plants and ferns of the Mesters Vig district of Northeast Greenland. *Medd. Grønl.*, 166: 1–119.

Raymond, P., Bruzau, F. and Pradet, A., 1978. Étude du transport d'oxygène des parties aeriennes aux racines à l'aide d'un paramètre du métabolisme: la charge énergétique. *C.R. Acad. Sci. Paris*, 286: 1061–1063.

Rowe, R.N. and Beardsell, D.V., 1973. Waterlogging of fruit trees. *Hortic. Abstr.*, 43: 534–548.

Scott Russell, R., 1977. *Plant Root Systems*. McGraw-Hill, New York, N.Y., 298 pp.

Scholander, P.F., Van Dam, L. and Scholander, S.I., 1955. Gas exchange in the roots of mangroves. *Am. J. Bot.*, 42: 92–98.

Sculthorpe, C.D., 1967. *The Biology of Aquatic Vascular Plants*. Edward Arnold, London, 610 pp.

Sifton, H.B., 1940. Lysigenous air spaces in the leaf of Labrador Tea, *Ledum groenlandicum* Oder. *New Phytol.*, 39: 75–79.

Small, E., 1973. Water relations of plants raised in sphagnum peat bogs. *Ecology*, 53: 726–728.

Sparling, J.H., 1967. The occurrence of *Schoenus nigricans* L. in blanket bogs. II Experiments on the growth of *S. nigricans* under controlled conditions. *J. Ecol.*, 55: 15–31.

Streeter, J.G. and Thompson, J.F., 1972. Anaerobic accumulation of γ-aminobutyric acid in radish leaves (*Raphanus sativus* L.). *Plant Physiol.*, 49: 572–578.

Takhtajan, A., 1969. *Flowering Plants: Origin and Dispersal*. Oliver and Boyd, Edinburgh, 310 pp.

Tallis, J.H. and Birks, H.J.B., 1965. The past and present distribution of *Scheuchzeria palustris* in Europe. *J. Ecol.*, 53: 287–298.

Turrill, W.B., 1957. Germination of seeds. 5. The vitality and longevity of seeds. *Gard. Chron.*, 142: 37.

Vartapetian, B.B. and Nuritdinov, N., 1976. Molecular oxygen transport in plants. *Naturwissenschaften*, 63: 246.

Vester, G. and Crawford, R.M.M., 1978. Verschiedene Provenienzen von *Pinus contorta* Loudon und Überflutungsstress:

Klassifikation auf Grund morphologischer und metabolischer Kriterien. *Flora*, 167: 433–444.
Von Brand, T., 1966. *Biochemistry of Parasites*. New York, 429 pp.
Wager, H.G., 1961. The effect of anaerobiosis on acids of the tricarboxylic cycle in peas. *J. Exp. Bot.*, 12: 34–46.
Walter, H., 1962. *Die Vegetation der Erde in ökologischer Betrachtung*. Fischer-Verlag, Jena, 538 pp.
Williams, W.T. and Barber, D.A., 1961. The functional significance of aerenchyma in plants. *Symp. Soc. Exp. Biol.*, 15: 132–154.
Yeager, L.E., 1949. Effect of flooding on a river-bottom timber area. *Bull. Ill Nat. Hist. Surv.*, 25: 33–65.
Zebe, E., 1975. *In Vivo* Untersuchungen über den Glucose Abbau bei *Arenicola marina* (Annelida, Polychaeta). *J. Comp. Physiol.*, 101: 133–148.

Chapter 8

PRIMARY PRODUCTION IN WETLANDS[1]

I.K. BRADBURY and J. GRACE

INTRODUCTION

The purpose of this chapter is to review the published data on the primary production of wetlands, to identify the principal limiting factors and to examine some of the processes involved in plant growth in these communities. It is difficult to formulate and apply a precise definition of wetland ecosystems, largely because there is rarely a clearly defined boundary between open water and wetland on the one hand and wetland and terrestrial communities on the other. In addition wet–dry continua are usually subject to seasonal shifts. Despite these difficulties, however, wetlands are conventionally characterized by a water table which, for a significant part of the year, lies above or close to the surface of the substrate in which the vegetation is rooted. This characteristic has important implications for decomposition processes and it is the rate of organic matter decomposition which determines the threshold level of net primary production above which organic matter will accumulate.

The balance between decomposition and net production determines whether, and at what rate, organic matter accumulates in the wetland ecosystem and provides a convenient basis for distinguishing wetlands in which there is an appreciable build-up of organic matter (bogs and fens) from those in which organic matter accumulation is negligible (marshes and swamps).

For the purposes of this chapter three broad categories of wetland are recognized. The first includes emergent aquatic vegetation such as swamp, marsh and wet meadow. The second includes vegetation developed on peat — referred to in the literature as bog, mire and fen. The third includes the very cold wetlands — those tundra vegetation types which lie wet for much of the growing season and which contain acknowledged wetland genera such as *Carex* and *Eriophorum*. This division is somewhat arbitrary, being influenced by the community types in which investigations have been carried out as much as by biological, geographical or physicochemical considerations. Salt-marsh communities are excluded: their productivity has been reviewed by Keefe (1972; see also Chapman, 1977).

METHODS FOR ESTIMATING PRODUCTIVITY

Before presenting published data on net primary productivity it is necessary to consider briefly the methods that have been used to derive the figures. It will be shown that these figures should be regarded as *estimates* of the true values, to an extent which depends on the characteristics of the vegetation concerned.

Sequential harvests of vegetation

Most published results are derived from sequential harvest of vegetation which undergoes an annual growth cycle. Two harvests are normally taken, one at the beginning and the other at the end of the growth cycle of the dominant species; production is then assumed to equal the difference in the two values. The errors associated with this procedure are as follows:

(1) Statistical error. A large number of measurements must be made if the estimate of the mean is to be reliable. This source of error is more acute for "natural" vegetation than for crops because of greater spatial heterogeneity.

[1]Revised manuscript received June, 1979.

(2) Below-ground error. Many authors have failed to measure the below-ground component of productivity, because of the practical difficulty of excavation or the unacceptably large statistical errors which have arisen when excavation has been achieved. This is apparent in the data of Forrest (1971), where the below-ground estimates were so variable that it was impossible to detect any statistically significant changes during a year. Since many of the wetland communities considered here contain long-lived perennials, with considerable below-ground components, this source of error may be large. As an expedient, Forrest (1971) assumed that the ratio of production to biomass below-ground was equal to that observed above ground, although most of the below-ground material was woody tissue, in which the distinction between live and dead was not clear, and the annual radial increment was small. In herbaceous perennial species generally, the ratio of annual root to shoot production appears to decline with age (Weaver and Zink, 1946; Bradbury and Hofstra, 1976a).

(3) Errors due to mortality. The loss of material from the stand, resulting from abscission and grazing, should be measured, as it represents production which would escape detection by the harvest method. This correction factor has sometimes been overlooked.

Mortality of individual plants or shoots may introduce errors into the estimate of production (Wiegert and Evans, 1964; Bradbury and Hofstra, 1976b). At one extreme all individual shoots may appear simultaneously and develop synchronously, and many production studies make this assumption. At the other extreme shoots may appear and die asynchronously throughout the year. The latter situation results in a standing crop which fluctuates little, and for which the harvest method would indicate a productivity of zero. The level of mortality can be assessed by marking individual shoots initially so that their performance and longevity can be followed. Mathews and Westlake (1969) found that only 70% of *Phragmites australis* shoots present in the early spring survive the growing season in southern England. Boyd (1971b) showed that mortality was low in a population of *Juncus effusus* in South Carolina, but Bernard and MacDonald (1974) reported that annual shoot production of *Carex lacustris* estimated from the maximum and minimum standing crop values was only 54% of the value obtained when shoot mortality was taken into account. The importance of the mortality factor as a source of error in production studies probably varies considerably according to vegetation composition. It would appear, however, that some knowledge of population dynamics, at least of the dominant species, is a prerequisite for obtaining accurate production values.

Methods based on observations of plant density or height

Mathews and Westlake (1969) adopted a technique involving the integration of shoot number per area of ground and the mean weight of surviving individuals. This method, previously used in the demographic analysis of fish populations (Allen, 1951), involves a graphical representation of the relationship between the density of individuals (shoots or plants) and the mean dry weight of individuals during the growth cycle. The product of these factors for any point on the resulting curve gives the standing crop at that time, whilst the area under the curve represents population net production. If little mortality occurs, terminal standing crop will be similar to population net production, but if a large number of individuals has been lost the latter value can exceed the former by a considerable margin. Mathews and Westlake (1969) estimated that annual net production of a *Glyceria maxima* shoot population exceeded terminal standing crop by a factor of 3.4. This method may however overestimate net production because it assumes that shoots which die are equal in weight to those which survive, whereas mortality is normally highest in the smallest size classes of a population (Harper and White, 1974). Again, the method is difficult to apply where individuals within the community develop asynchronously, and not in recognizable cohorts.

In mosses productivity has been estimated by measuring extension growth. Clymo (1970) described the use of a cranked wire, which is inserted into carpets of *Sphagnum* to define a reference position against which subsequent increments may be measured. If the weight of a unit of extension is known, then the observed rates of extension may be converted into rates of dry matter production.

Sources of error in this method are those due to decay, and to compaction of the moss carpet, especially when covered by snow. These errors were relatively unimportant however in the *Sphagnum* community studied by Clymo (1970) and Clymo and Reddaway (1974).

Methods based on the measured rate of photosynthesis

Several workers have estimated the productivity of wetlands from measurements of photosynthesis, conducted in the field or laboratory.

One approach is to determine the flux of carbon dioxide to the vegetation using an appropriate micrometeorological method [see Unsworth (1981) for a discussion of these techniques]. Coyne and Kelley (1972) measured vertical profiles of wind speed, temperature and the concentration of carbon dioxide over wet tundra. The profile of wind speed was used to calculate K, the coefficient of turbulent diffusion. They assumed that this value of K, which strictly applies to the transfer of momentum, could equally be applied to the transfer of carbon dioxide, and used it to calculate the downward flux of carbon dioxide from the CO_2 profile (see p. 304). Measurements like these must be made throughout the season to obtain a complete carbon budget of the vegetation. Moreover, the site requirements for such work are rather rigorous — very large areas of more-or-less homogenous vegetation are required.

An alternative method, which also involves the measurement of carbon dioxide exchange, is that in which an area of ground, an individual plant, or a single leaf is enclosed in a chamber, so that the apparent rate of photosynthesis may be found. This involves considerable sampling error, and in general is more useful as a technique for exploring the sensitivity of photosynthesis to natural changes in the environment. This technique has been applied to cold wetlands by Hadley and Bliss (1964).

Photosynthesis of shoots or areas of leaf may also be measured in the laboratory under controlled conditions. The extrapolation of such results to the field situation may not be straightforward, as discussed by Grace and Woolhouse (1970, 1973, 1974) in their study of a *Calluna–Sphagnum* community. Nevertheless, the agreement between productivity estimated by this method, and that obtained by sequential harvesting at the same site by Forrest (1971) was encouraging.

When using photosynthetic measurements (CO_2 flux) to determine primary productivity (dry matter) a conversion factor is required. A review of the literature (Scott and Billings, 1964) showed that widely different conversion factors have been used for this purpose, even though the carbon content of plant material does not vary much.

PRODUCTIVITY ESTIMATES IN WETLANDS

Estimates of primary productivity from a number of studies in marsh and swamp communities are shown in Table 8.1. Most of our knowledge of the productivity of swamps and marshes has been obtained from relatively limited areas in the northern temperate zone, especially the southeastern sector of the United States, central Europe and the British Isles. The tropics and temperate areas in the Southern Hemisphere are not represented at all. Reliable productivity estimates are, of course, much easier to obtain in temperate areas, in which there is a clearly defined growing season, than in tropical or subtropical zones where standing crop usually varies little throughout the year. As an example, it was found that the sedge *Cladium jamaicense* (saw grass) in the Florida Everglades (U.S.A.) maintains a more or less constant standing crop of 3000 g m^{-2} throughout the year, although estimates of shoot production following a fire suggest that productivity is about 1400 g m^{-2} yr^{-1} (Steward and Ornes, 1975). It would be of considerable biological interest to obtain good productivity data for tropical wetlands such as those occupied by *Cyperus papyrus* in Africa (Thompson et al., 1979).

Species diversity in swamps tends to be low and frequently one species forms an almost pure stand. Certain families such as the Typhaceae (e.g. *Typha*), the Juncaceae (e.g. *Juncus*), the Cyperaceae (e.g. *Carex* and *Scirpus*) and the Poaceae (e.g. *Glyceria* and *Phragmites*), all of which are monocotyledonous, are well represented in this type of wetland.

Many of the productivity estimates shown in Table 8.1 are based on maximum standing crop values which, as discussed earlier, normally underestimate actual production because they fail to account for shoot mortality and abscission. Actual shoot productivity values are therefore likely to be

TABLE 8.1

Primary productivity of marsh and swamp communities

Dominant species	Location	Productivity[1] ($g\ m^{-2}\ yr^{-1}$)		Reference
		shoots	below ground	
Alnus rugosa/Fraxinus nigra	Michigan (U.S.A.)	570–640		Parker and Schneider (1975)
Alisma plantago-aquatica	central Alberta (Canada)	444	–	Van der Valk and Bliss (1971)
Carex aquatilis	western Alberta (Canada) (1455 m)[2]	550	–	Gorham and Somers (1973)
Carex elata	southern Sweden	337–548	–	Mörnsjö (1969)
Carex lacustris	New York State (U.S.A.)	1580	–	Bernard and MacDonald (1974)
		965	208	Bernard and Solsky (1977)
	Wisconsin (U.S.A.)	1034	147	Klopatek and Stearns (1978)
Carex rostrata	Minnesota (U.S.A.)	780	197	Bernard (1974)
	western Alberta (Canada)	740	–	Gorham and Somers (1973)
Carex spp.	northern England	420–630	–	Pearsall and Gorham (1956)
	New Jersey (U.S.A.)	1904	–	Jervis (1969)
	Quebec (Canada)	820	205	Auclair et al. (1976a)
	southern Poland	560		Baradziej (1974)
Eleocharis palustris	central Alberta (Canada)	341–447	–	Van der Valk and Bliss (1971)
Eleocharis quadrangulata	South Carolina (U.S.A.)	725	–	Polisini and Boyd (1972)
		881	–	Boyd and Vickers (1971)
Equisetum fluviatale	central Alberta (Canada)	491–707	–	Van der Valk and Bliss (1971)
Glyceria maxima	southern England	660	–	Westlake (1966)
Glyceria maxima/ Solanum dulcamara	southern England	1237	–	Buttery and Lambert (1965)
Glyceria maxima/ Phragmites communis	southern England	959	–	Buttery and Lambert (1965)
Iris pseudoacorus/ Equisetum fluviatile	southern Poland	795	–	Baradziej (1974)
Juncus effusus	South Carolina (U.S.A.)	1592	267	Boyd (1971b)
Juncus spp.	northern England	690–800	–	Pearsall and Gorham (1956)
Molinia caerulea	southern Sweden	125	–	Mörnsjö (1969)
	northern England	400	–	Pearsall and Gorham (1956)
Oryza sativa	The Philippines	1625	–	Tanaka et al. (1966)
Panicum hemitonium	South Carolina (U.S.A.)	1075	–	Polisini and Boyd (1972)
Phalaris arundinacea	Wisconsin (U.S.A.)	1353	675	Klopatek and Stearns (1978)
Phalaris arundinacea/ Phragmites australis	southern England	800	–	Buttery and Lambert (1965)
Phragmites australis	Denmark	781–829	620	Anderson (1976)
	northern Britain	1140–2500	–	Gorham and Pearsall (1956)
	northern England	1300	–	Pearsall and Gorham (1956)
	eastern England	551–1080	–	Mason and Bryant (1975)
	Czechoslovakia	1614	–	Květ (1971)
Pontederia cordata	South Carolina (U.S.A.)	716	–	Polisini and Boyd (1972)
Saururus cernuus	southeastern U.S.A.	184–1199	–	Boyd and Walley (1972)
Scirpus americanus	South Carolina (U.S.A.)	140	–	Boyd (1970)
		410	–	Polisini and Boyd (1972)
Scirpus fluviatilis	Wisconsin (U.S.A.)	1116	417	Klopatek and Stearns (1978)
Scirpus lacustris	Denmark	420–1325	1224	Anderson (1976)
Scirpus validus	South Carolina (U.S.A.)	1381	–	Polisini and Boyd (1972)
Scirpus spp./ Equisetum fluviatile	southern Quebec (Canada)	914		Auclair et al. (1976b)
Taxodium distichum	Georgia (U.S.A.)	692[3]	–	Schlesinger (1978)
Thuja occidentalis/ Betula papyrifera	Minnesota (U.S.A.)	1032	–	Reiners (1972)

TABLE 8.1 (*continued*)

Dominant species	Location	Productivity[1] (g m^{-2} yr^{-1})		Reference
		shoots	below ground	
Typha angustifolia	Denmark	807	1800	Anderson (1976)
	Czechoslovakia	2592	–	Květ (1971)
	eastern England	1515	–	Mason and Bryant (1975)
Typha australis	South Carolina (U.S.A.)	1483	–	Polisini and Boyd (1972)
Typha latifolia	northern England	400	–	Pearsall and Gorham (1956)
	South Carolina (U.S.A.)	684	–	Boyd (1970)
		530–1132	–	Boyd (1971a)
		574	–	Polisini and Boyd (1972)
	Wisconsin (U.S.A.)	1643	1550	Klopatek and Stearns (1978)
	New Jersey (U.S.A.)	1904	–	Jervis (1969)
	Oregon (U.S.A.) (60 m)[2]	418	–	McNaughton (1966)
	Oregon (U.S.A.) (610 m)[2]	330	–	McNaughton (1966)
	North Dakota (U.S.A.)	404	–	McNaughton (1966)
	South Dakota (U.S.A.)	378	–	McNaughton (1966)
	Nebraska (U.S.A.)	416	–	McNaughton (1966)
	Oklahoma (U.S.A.)	730	–	McNaughton (1966)
	Texas (U.S.A.)	1336	–	McNaughton (1966)
	Idaho (U.S.A.)	1707	–	Pearson (1965)
	Oklahoma (U.S.A.)	1527	–	Penfound (1956)
	southeastern U.S.A.	428–2252	–	Boyd and Hess (1970)
	central Alberta (Canada)	322	–	Van der Valk and Bliss (1971)
Typha hybrid swarm	Minnesota (U.S.A.)	1440–1680	–	Bray et al. (1959)
Various deciduous trees/shrubs/herbs	Minnesota (U.S.A.)	707	–	Reiners (1972)

[1]A range of values denotes that at least two similar sites were investigated within the same locality. [2]Altitude above sea level. [3]Above water.

in excess of those shown. There are few published data for below-ground productivity in swamps and marshes which can be considered reliable, and those presented must be considered approximate. Overall, the data shown support claims that swamps and marshes are amongst the most productive types of community (Westlake, 1963; Whittaker, 1975). However, even allowing for high below-ground productivity and considerable shoot mortality it appears that productivity in swamps and marshes rarely exceeds 3000 g m^{-2} yr^{-1} and is more generally around 1500 to 2000 g m^{-2} yr^{-1} (Table 8.1).

The relatively high productivity of swamp and marsh vegetation is best illustrated by comparing values with those for other community types in the same vicinity. In Minnesota (U.S.A.), shoot productivity of two *Typha* stands was 1440 and 1680 g m^{-2} yr^{-1} compared with 120 to 160 g m^{-2} yr^{-1} for nearby old fields and 1390 g m^{-2} yr^{-1} for above- and below-ground components of fertilized maize (Bray et al., 1959). Similarly, during oxbow-lake succession in central Alberta (Canada), emergent aquatic stages exhibited higher shoot standing crop (465 g m^{-2}) than dry meadow stages (325 g m^{-2}) (Van der Valk and Bliss, 1971), whilst Bernard (1974) found that annual production in a *Carex*-dominated marsh in Minnesota was about 30% greater than in an adjacent grass-dominated old field.

The relatively high productivity of swamp and marsh vegetation is generally attributed to a plentiful supply of nutrients, due to flushing with nutrient-rich water, and low water stress for most of the year.

Some attempts have been made to relate the productivity of swamp and marsh communities to edaphic characteristics. In Britain Gorham and Pearsall (1956) reported that shoot production by

Phragmites australis tended to be higher as the mineral status and depth of water increased, whilst Allen and Pearsall (1963) demonstrated a positive relationship between *Phragmites* standing crop and nitrogen concentration in the leaves.

The productivity of *Typha latifolia* and *Saururus cernuus* in the southern United States has been shown to be strongly correlated with phosphorus availability (Boyd and Hess, 1970; Boyd and Walley, 1972). The comparatively high productivity of a *Carex lacustris* stand in New York State (U.S.A.) has been attributed to nitrogen enrichment by actinomycetes in the root nodules of *Alnus rugosa* situated in an adjacent thicket (Bernard and MacDonald, 1974).

Primary productivity data from a number of bog and wet tundra locations are presented in Table 8.2. Wet tundra sites frequently contain cyperaceous species whilst peat bog or mire communities of lower latitudes are characterized by combinations of ericaceous shrubs, sedges (e.g. *Carex* and *Eriophorum*) and bryophytes, particularly *Sphagnum* spp.

Primary productivity of peat bogs or mires has been found to be between 300 and 1000 g m^{-2} yr^{-1}. Peat is formed in situations where the water table is high enough to inhibit normal decomposition processes and therefore occurs on sites where precipitation and other water inputs exceed evaporation, run-off and drainage capacity. The plants which are characteristic of peat bogs must tolerate relatively cool, wet conditions and a rooting medium which may be extremely low in available nutrients, of poor aeration and high acidity. In view

TABLE 8.2

Primary productivity of peat-bog and wet tundra sites

Vegetation	Location	Productivity (g m^{-2} yr^{-1})		Reference
		shoots	below ground	
Alpine grass/sedge meadow	Wyoming (U.S.A.)	112	–	Bliss (1962)
Caltha leptosepala	Alaska (U.S.A.)	108	–	Webber (1972)
Calluna vulgaris	northern England	177	–	Bellamy and Holland (1966)
	northern England	272	–	Bellamy et al. (1969)
Calluna vulgaris/	northern England	407	228	Forrest (1971)
Eriophorum vaginatum	northern England	364–768	100–210	Forrest and Smith (1975)
Conifers/bryophytes	southern Manitoba (Canada)	523	187	Reader and Stewart (1972)
Deschampsia flexuosa	Alaska (U.S.A.)	156	–	Webber (1972)
Ericaceous shrubs/ bryophytes	southern Manitoba (Canada)	482	1461	Reader and Stewart (1972)
Ericaceous shrubs/ bryophytes/conifers	southern Manitoba (Canada)	399	594	Reader and Stewart (1972)
Ericaceous shrubs/sedges	Ireland	338	–	Doyle (1973)
Ericaceous shrubs/	central England	638	–	Summerfield (1973)
sedges/bryophytes	north Wales	857	–	Summerfield (1973)
Salix planifolia	Alaska (U.S.A.)	459	–	Webber (1972)
Salix spp./*Carex* spp./	Norway	359	474	Wielgolaski (1975)
bryophytes	southern Manitoba (Canada)	1118	513	Reader and Stewart (1972)
Sedges/bryophytes	Alaska (U.S.A.)	90	–	Anderson (1972)
Sedge meadow	northern Sweden and northern Norway	136–163	–	Pearsall and Newbould (1957)
	Alaska (U.S.A.)	156	–	Brown and West (1970)
	Alaska (U.S.A.)	102	–	Dennis and Tieszen (1972)
	Alaska (U.S.A.)	73	–	Haag (1974)
	Devon Island (northern Canada)	42–115	70–191	Muc (1973)
	northern U.S.S.R.	356	–	Gorchakovsky and Andreyashkina (1972)

of this it is perhaps surprising that productivity values are as high as those reported.

Below-ground productivity values given by Forrest (1971) and Reader and Stewart (1972) should be treated with caution because in each study it was assumed that the productivity/standing crop ratio was the same for above- and below-ground components. No data are available, however, which support this assumption.

The principal peat-forming taxon on a world scale is the moss genus *Sphagnum*. Some data concerning the productivity of *Sphagnum* and other mosses are shown in Table 8.3. Clearly, the contribution made by mosses to primary productivity on bogs sites may be considerable; Clymo (1970), for example, found values of around 300 g m^{-2} yr^{-1} in an upland bog in northern England and 400 g m^{-2} yr^{-1} in a valley bog in southern England, whilst Overbeck and Happach (1956) reported an anomalous value of 1660 g m^{-2} yr^{-1} for *Sphagnum recurvum* in northern Germany. The generally low productivity of peat-bog communities is illustrated by the effects of improvement on a site in Ireland (Doyle, 1973). Production by the native ericaceous-dominated vegetation was less than 350 g m^{-2} yr^{-1}, but on a nearby site which had been drained, fertilized, and re-seeded with a grass/clover mixture production exceeded 1650 g m^{-2} yr^{-1}.

On bog sites in northern Europe phosphorus availability is often considered to be the main factor limiting productivity (see, for instance, Sæbo, 1968), and some workers have reported large increases in productivity after applying phosphatic fertilizers (Tamm, 1954; McVean, 1959). Gore (1961), however, found that neither *Eriophorum vaginatum* nor *Molinia caerulea* on bog sites responded to phosphorus or calcium application, even though the phosphorus status of the treated shoots was considerably increased by fertilization.

On wet tundra sites primary productivity tends to be low (Table 8.2). The available evidence suggests that shoot production is normally less, often considerably less, than 350 g m^{-2} yr^{-1}. Climatically, of course, arctic tundra sites experience a very short growing season during which temperatures are frequently low, although the shortness of growing season is compensated to some extent by long photoperiods during mid-summer (see p. 294). There may also be edaphic limitations to plant production in the arctic tundra. On wet

TABLE 8.3

Productivity of bryophytes on some wetland sites

Species	Location	Productivity (g m^{-2} yr^{-1})	Reference
Aulacomnium palustre	southern Manitoba (Canada)	5–36	Reader and Stewart (1972)
Homalothecium nitens	central Alberta (Canada)	190	Busby et al. (1978)
Hypnum pratense	southern Manitoba (Canada)	31	Reader and Stewart (1972)
Pleurozium schreberi	southern Manitoba (Canada)	108	Reader and Stewart (1972)
Polytrichum alpestre	Antarctica	342–507	Longton (1970)
Polytrichum juniperinum	southern Manitoba (Canada)	35	Reader and Stewart (1972)
Sphagnum fuscum	northern England	269	Bellamy and Rieley (1967)
Sphagnum magellanicum	northern England	68	Forrest and Smith (1975)
Sphagnum papillosum/ *Sphagnum magellanicum*	northern England	77–105	Chapman (1965)
Sphagnum recurvum	northern Germany	1660	Overbeck and Happach (1956)
Sphagnum capillifolium	northern England	80–135	Clymo and Reddaway (1974)
Sphagnum spp.[1]	northern England[2]	300	Clymo (1970)
	southern England[3]	400	
Sphagnum spp. (several)	Ireland	50	Doyle (1973)
Tortula robusta	Antarctica	300	J. Lawson (pers. comm.)
Various spp.	northern U.S.S.R.	10–70	Gorchakovsky and Andreyashkina (1972)
Various spp.	Devon Island (northern Canada)	10–293	Pakarinen and Vitt (1973)

[1] Approximate average values for four species in three habitats. [2] 575 m above sea level. [3] 30 m above sea level.

sedge-meadow sites in the Canadian tundra considerable increases in productivity resulted from nitrogen and, to a lesser extent, phosphorus fertilization (Haag, 1974).

The magnitude of the below-ground standing crop, particularly in relation to the above-ground component, is of considerable ecological interest because of its importance in nutrient cycling and energy transfer processes. Some values obtained from various types of wetland community are shown in Table 8.4. The high proportion of total standing crop which is below ground on wet arctic tundra sites was suggested by Rodin and Bazilevich (1967) to represent an adaptation to the severe aerial environment of these latitudes. The very high below-ground/above-ground ratio reported by Tieszen (1972), however, is partly due to an accumulation of dead organic matter below ground: the ratio for live tissue is considerably less. Below-ground/above-ground biomass ratios, although lower in marsh, swamp and bog communities than on tundra sites, are still generally greater than unity, and since the dominant plant types are perennial, considerable below-ground standing crops are maintained throughout the year.

Large below-ground/above-ground biomass ratios may have important implications for the functioning of wetland ecosystems. For example, a significant proportion of the available nutrients are likely to be contained in this component, whilst a small change in the concentration of non-structural carbohydrates in below-ground tissue may represent a considerable transfer of energy resources within the system.

MAXIMUM PRODUCTIVITY

Estimates of potential productivity of vegetation have been made from knowledge of the photosynthetic performance of leaves or cells in laboratory conditions. These estimates provide a useful standard against which we may judge the observed productivity of wetland communities.

Estimates

Under the best possible conditions in the laboratory, algal cells require about ten quanta of photosynthetically active radiation (PAR) to fix one molecule of carbon dioxide; this figure is referred to as the quantum requirement of photosynthesis. Now in the photosynthetically active part of the spectrum (380–720 nm) the average energy content of a quantum is 3.6×10^{-19}J. The heat of combustion of one molecule of (CH_2O) is 7.7×10^{-19}J, and hence the efficiency of photosynthesis, based on the known quantum requirement is:

$$e = \frac{7.7 \times 10^{-19}}{10 \times 3.6 \times 10^{-19}} = 0.215$$

In natural light, only about half of the total solar radiation is in the required wave-band (380–720 nm); so if the efficiency is re-expressed in relation to total solar radiation, the resulting value is only half the figure above. Thus, the quantum requirement can be used to find an upper limit of productivity, expressed as grams of dry matter produced for every kJ of solar energy received. The energy content of dry vegetation varies a little according to its chemical composition, most published results falling in the range 16–20 kJ g^{-1} (see Golley, 1961). Taking a representative figure of 17 kJ g^{-1} for non-woody vegetation, an estimate of the upper limit of productivity is $(0.215 \times 0.5/17) = 6.32$ mg per kJ of solar energy.

Assuming a high daily total of solar radiation as 25 MJ m^{-2} day^{-1}, the factor of 6.32 mg kJ^{-1} implies a productivity of 158 g m^{-2} day^{-1}. This is unnaturally high, having never been recorded nor even approached.

One main reason why such productivity is not achieved is that the quantum requirement for photosynthesis has been measured at low light intensities, when the rate of photosynthesis is not limited by the activity of carboxylating enzymes or the supply of carbon dioxide. In nature, this will apply only to shaded leaves, while sunlit leaves will normally be operating at "light saturated" conditions with considerable quantum wastage. Realistic estimates of maximum productivity thus require a knowledge of the relationship between irradiance and rate of photosynthesis.

Monteith (1972) calculated the rate of dry matter production of crop canopies composed of leaves with a known light response curve for photosynthesis. The light response curve was characterized by two parameters, one describing the initial

TABLE 8.4

Underground standing crop and below-ground/above-ground standing crop ratio in (a) swamp and marsh, (b) peat bog and (c) wet tundra sites

Vegetation	Locality	Standing crop[1]		Reference
		below ground ($g\ m^{-2}$)	below ground/ above ground	
(a) Swamp and marsh				
Acorus calamus	Czechoslovakia	995–1180	2.0–2.7	Fiala et al. (1968)
Carex elata	southern Sweden	1491–1855	3.4–4.4	Mörnsjö (1969)
Carex lacustris	New York State (U.S.A.)	433	0.3	Bernard and MacDonald (1974)
Carex rostrata	Minnesota (U.S.A.)	328	0.2	Bernard (1974)
Glyceria maxima	southern England	844	1.3	Westlake (1966)
Molinia caerulea	southern Sweden	571	4.6	Mörnsjö (1969)
Phragmites australis	Czechoslovakia	2300–4418	1.8–2.9	Fiala et al. (1968)
	Czechoslovakia	3170	2.0	Květ (1971)
	Denmark	2480	3.2	Anderson (1976)
Saururus cernuus	southern U.S.A. (various locations)	not given	0.4–1.7	Boyd and Walley (1972)
Scirpus lacustris	Czechoslovakia	2965–3517	2.0–2.9	Fiala et al. (1968)
	Denmark	8570	6.5	Anderson (1976)
Sparganium emersum	Czechoslovakia	520–1300	0.6–1.1	Fiala et al. (1968)
Typha angustifolia	Czechoslovakia	3570	1.8	Fiala et al. (1968)
	Czechoslovakia	2314	1.0	Květ (1971)
Typha latifolia	U.S.A. (various locations)	556–2646	1.1–3.1	McNaughton (1966)
	Denmark	3600	4.5	Anderson (1976)
Typha spp.	Wisconsin (U.S.A.)	3200	1.6	Klopatek and Stearns (1978)
(b) Peat bog				
Calluna vulgaris/ *Eriophorum vaginatum*	northern England	940	0.6	Forrest (1971)
Conifers/bryophytes	southern Manitoba (Canada)	2280	0.5	Reader and Stewart (1972)
Ericaceous shrubs/ bryophytes	southern Manitoba (Canada)	1956	4.6	Reader and Stewart (1972)
Ericaceous shrubs/ bryophytes/conifers	southern Manitoba (Canada)	1644	1.8	Reader and Stewart (1972)
Salix spp./*Carex* spp./ bryophytes	southern Manitoba (Canada)	1304	0.5	Reader and Stewart (1972)
(c) Wet tundra sites				
Sedge meadow	Alaska (U.S.A.)	1910	22.9	Dennis and Tieszen (1972)
	northern U.S.S.R.	2840	5.6	Gorchakovsky and Andreyashkina (1972)
	Devon Island (northern Canada)	633–2020	1.4–2.5	Muc (1973)

[1]A range of values denotes that at least two similar sites were investigated within the same locality.

slope and the other defining the maximum rate of photosynthesis per unit of leaf area. The crops under consideration formed three groups: the first contained tropical grasses such as *Zea mays* which have a very high rate of net photosynthesis (see p. 304), the second contained other graminaceous crops such as *Oryza sativa* (rice) which have a lower rate of net photosynthesis, while the third contained certain dicotyledonous crops such as *Glycine max* (soybean) which has still lower maximum rates of net photosynthesis. Fig. 8.1A presents the result of the calculations by Monteith (1972), made on the assumption that the leaves making up the canopy are so numerous that they intercept all the incident radiation. The calculated maximum productivity, per unit area of land, was 50 g m^{-2} day^{-1}: this may be regarded as the maximum daily productivity likely to be achieved anywhere. The curves are steepest at low irradiance — here all the leaves in the canopy would be working below the point of light saturation. At high irradiance the curves are less steep because the leaves near the top of the canopy would be light-saturated for much of the day.

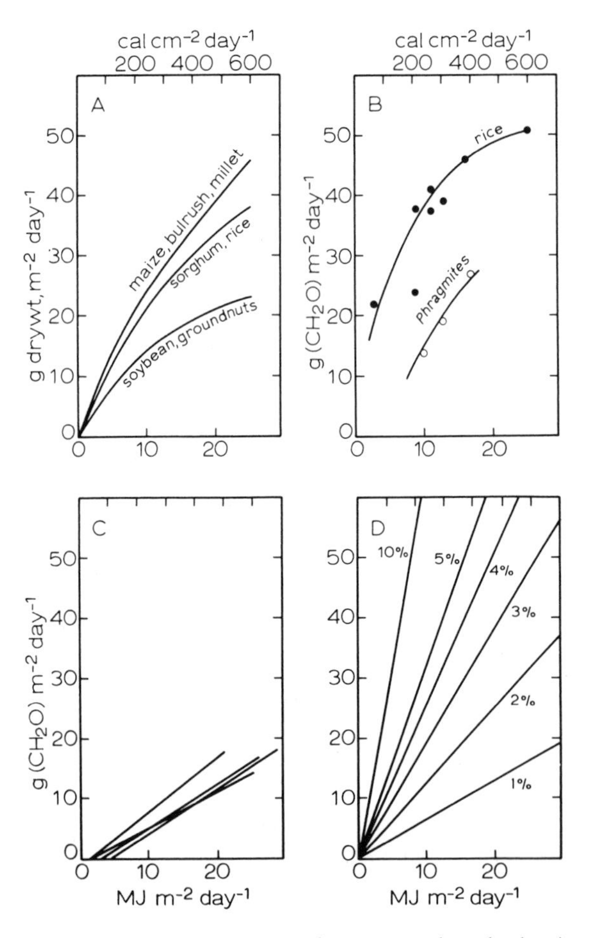

Fig. 8.1. Solar energy capture by crops and wetlands. A. Tropical crops according to Monteith (1972). B. Rice in the Philippines (Tanaka et al., 1966) and *Phragmites australis* in Czechoslovakia (Dykyjová et al., 1970). C. Wet tundra, Alaska (Tieszen, 1972). D. Solar energy conversion efficiency (see text).

Observations

Figs. 8.1B and 8.1C show actual observations made on wetland communities, displayed in the same way so as to be comparable with Monteith's calculations. Tanaka et al. (1966) collected eight days' records for a population of "Peta" rice in which the leaf area index (LAI) was approximately 6. The plot was in the Philippines and had received 100 kg ha^{-1} of nitrogen. Productivity was calculated from observations of community photosynthesis during the daylight hours only, no allowance having been made for respiratory losses at night. In a separate study on a similar stand of rice, Tanaka et al (1966) found that respiration at night was 11 g (CH_2O) m^{-2} day^{-1}. This result is generally like Monteith's calculation for rice (Fig. 8.1A), but the rates are somewhat higher. In the same figure are shown data from Dykyjová et al. (1970) for *Phragmites*; here the production data were obtained from harvests over a period of weeks during which the LAI increased from 0.7 to 2.6. Later in the year, when the LAI had risen to 4 or more, the productivity declined to less than 11 g m^{-2} day^{-1}. The *Phragmites* data included only the shoot component and it is likely that the early-season estimates are too high due to translocation of organic materials from subterranean parts which are known to store assimilates fixed in the previous year. In plotting the *Phragmites* data we have interpolated solar irradiance from Figure 5 of Dykyjová et al. (1970), multiplying by 2 to bring PAR to total solar irradiance.

The data of Tieszen (1972) were derived from observations of community photosynthesis at wet tundra sites near Barrow, Alaska. The lines in Fig. 8.1C show regressions based on about ten observations at each of four sites. Although the communities had a low LAI ($\simeq 1.0$), they display maximum rates of carbon assimilation (17 g m^{-2} day^{-1}) which approach the corresponding figure reported by Gaastra (1963) for a sugar beet crop in The Netherlands (23 g m^{-2} day^{-1}). Unlike the data for rice, the relationship between productivity and solar irradiance for Tieszen's data is a linear one. This may be the result of a strong correlation between irradiance and temperature; high insolation might be expected to result in high leaf temperatures, which (in an environment where temperatures in general are low) might result in higher rates of photosynthesis than might otherwise be expected.

In Tieszen's data (Fig. 8.1C), the intercepts on the abscissa give estimates of the energy required daily to exactly balance the energy expended in respiration.

Fig. 8.1D gives the relationship between assimilation rate and irradiance corresponding to various efficiencies of energy capture as defined above. As the abscissa in Fig. 8.1D is *total* irradiance, a ten-quantum requirement implies an efficiency of 0.11. Comparison between Fig. 8.1D and Fig. 8.1B shows that, on the day with least irradiance, the rice crop apparently achieved this maximum efficiency. On the day with most irradiance, only 0.03 of the energy was captured. For *Phragmites* the figure was less than 0.03, while the most productive tundra site was about 0.015 efficient.

It should be borne in mind that the figures assembled here refer to the *maximum* productivity observed for that part of the season when the plants are actively growing, and LAI is high. In some respects this is a more useful means of comparing the productivity of different vegetation types, because the *average* productivity over a whole year is influenced so much by the length of the growing season. The comparisons shown here, based on rather limited data, suggest that *Phragmites* has a maximum productivity similar to that of cultivated rice [a correction to the rice data of Tanaka et al. (1966) must be made to allow for night respiration], whereas plants of the cold wetlands of Alaska appear to perform at an appreciably lower rate.

Respiration losses

There are insufficient data available to compare the proportion of the day's assimilation which is lost as carbon dioxide at night. In tropical rice Tanaka et al. (1966) found that this proportion increased from 0.35 to 0.55 as the crop aged, amounting in absolute terms to between 11 and 26 g (CH_2O) m^{-2} day^{-1}. Proportions as high as this are more likely to be a characteristic of tropical conditions, because respiration is markedly temperature-sensitive ($Q_{10} = 2$). In temperate crops, the corresponding proportion appears to be between 0.17 and 0.37, according to Gaastra (1963). Complete carbon budgets for various vegetation types are now available (Monteith, 1976), but no figures are published for wetlands.

PHOTOSYNTHETIC PERFORMANCE

Absolute rates of photosynthesis

On the basis of carbon fixation pathways plant species may be simply classified as C_4 or C_3 plants. In the first group the first stable products of carbon fixation are 4-carbon molecules of dicarboxylic acid, and rates of photosynthesis from 50 to 90 mg CO_2 dm^{-2} h^{-1} are typical. In this group, light saturation normally does not occur even at the highest irradiances encountered naturally. In the second group the first stable product of photosynthesis is the 3-carbon molecule phosphoglyceric acid and the maximum rate of photosynthesis is usually less than 40 mg CO_2 dm^{-2} h^{-1}, while light saturation occurs at irradiances equal to or less than that of full sunlight. Within this group, woody plants normally display rates from 5 to 30 mg dm^{-2} h^{-1}, but slow-growing perennials may only have rates ranging from 1 to 10 mg dm^{-2} h^{-1}.

The great majority of plant species belong to the C_3 group, although within the angiosperms the C_4 pathway occurs in at least 13 families and 117 genera (Downton, 1975). Relatively few of these contain wetland species.

However, C_4 species occur widely in the Cyperaceae: 19 C_4 species of *Cyperus*, seven C_4 species of *Fimbristylis*, three of *Kyllinga* and one each of *Scirpus* and *Scleria* have been reported. Other wetland C_4 species are halophytes, in the genera *Atriplex*, *Suaeda* and *Spartina* (Downton, 1975; Raghavendra and Das, 1978). Those species which are the main constituents of the highly productive swamp and marsh vegetation of temperate latitudes are not C_4 plants as far as is known.

The maximum rate of photosynthesis recorded for *Phragmites australis*, in central California, is about 30 mg CO_2 dm^{-2} h^{-1} (Pearcy et al., 1972) which is typical of C_3 species. For the same species elsewhere average values of 6.1 mg CO_2 dm^{-2} h^{-1} have been recorded in Canada (Walker and Waygood, 1967), while in a Czechoslovakian population values ranged between 16 and 22 mg CO_2 dm^{-2} h^{-1} (Glosser, 1977). Exceptionally high rates between 43 and 68 mg CO_2 dm^{-2} h^{-1} have been reported for *Typha latifolia* by McNaughton and Fulkem (1970) but these values require confirmation.

In the tropical sedge *Cyperus papyrus*, Jones and

Milburn (1978) report a maximum photosynthetic rate of 25 mg CO_2 dm^{-2} h^{-1}. Since this species exhibits a *Kranz* anatomy and a low carbon dioxide compensation point, both characteristic features of C_4 species, it is surprising that assimilation rates are not higher in this species.

Generally, there have been very few detailed studies on the photosynthesis of wetland species. Published light-response curves for cultivated rice are shown in Fig. 8.2A. On the same figure the maximum rate for other crop plants is shown.

In Fig. 8.2B the photosynthetic curves for plants growing in *Sphagnum*-rich communities are given. Where the rates were determined on a single flat leaf illuminated from above, as in *Rubus chamaemorus*, the initial slope provides information on the photochemical efficiency of photosynthesis. For this species the initial slope is equivalent to an efficiency of conversion of PAR to carbohydrate of about 0.065, or 33 quanta per molecule of carbon dioxide fixed.

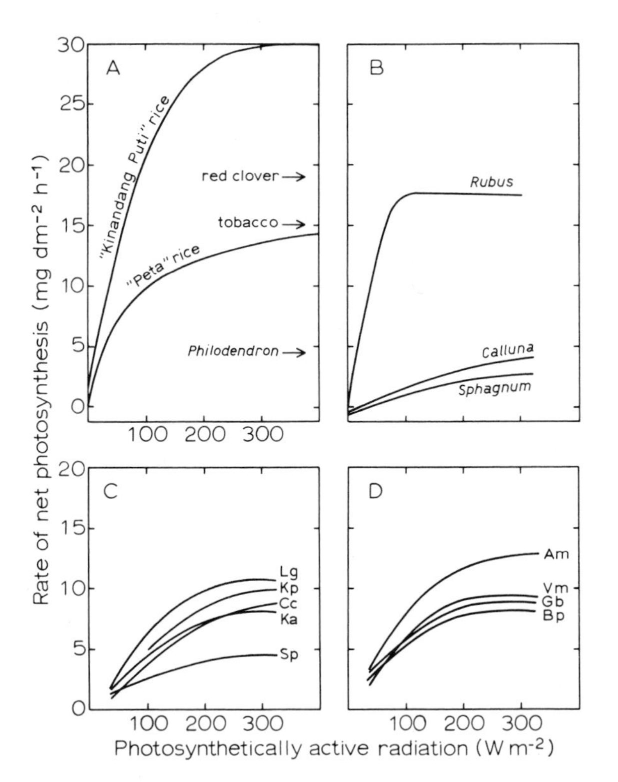

Fig. 8.2. Light response curves for net photosynthetic rates in wetland species. A. Rice in the Philippines (Tanaka et al., 1966). B. *Rubus chamaemorus*, *Calluna vulgaris* and *Sphagnum rubellum* from wet moorland in England (Grace and Marks, 1978). C and D. Bog species at Ottawa, Canada (Small, 1972) (data converted to area basis and smooth curves fitted by eye). Species key: *Am* = *Aronia melanocarpa*; *Bp* = *Betula populifolia*; *Cc* = *Chamaedaphne calyculata*; *Gb* = *Gaylussacia baccata*; *Ka* = *Kalmia angustifolia*; *Lg* = *Ledum groenlandicum*; *Sp* = *Sarracenia purpurea*; *Vm* = *Vaccinium myrtilloides*.

The other photosynthetic curves in Fig. 8.2B were obtained for cut shoots. Here the geometry of the shoot and the existence of mutual shading of the leaves preclude the calculation of efficiency (see Grace and Woolhouse, 1973).

Observed rates of photosynthesis for woody perennials in a *Sphagnum* bog are in the range expected for woody species (Small, 1972; see Figs. 8.2C,D). A comparison was made between these data and those obtained from woody species growing in a nearby field; Small concluded that the non-bog species had a mean rate of photosynthesis at high irradiance that was about equal to the rate observed for the bog species.

Presentation of such photosynthetic curves is not meant to suggest that each is a fixed characteristic of the species. Grace and Woolhouse (1970) investigated the photosynthetic rates of *Calluna vulgaris* over the course of a year and found that the maximum rate was variable, depending on the previous temperature regime to which the plants had been exposed, the age of the shoots, and whether or not the plant bore flowers. In winter the rate of photosynthesis was always low. In these respects, the photosynthetic behaviour of *Calluna* resembles that of woody perennials in general.

Rates of photosynthesis for plants of the cold wetlands have been reported by Hadley and Bliss (1964), Johansson and Linder (1975), Skre (1975) and Kjelvik et al. (1975). In the growing season, quite high rates are sometimes achieved, despite the rather low temperatures prevailing; on a subarctic mire Johansson and Linder (1975) found that *Rubus chamaemorus* and *Betula michauxii* achieved rates of from 7 to 10 mg CO_2 dm^{-2} h^{-1}.

Mosses are very often a conspicuous component of the vegetation of wetlands. It is not practical to investigate the photosynthesis of single moss leaves, so it is usual to measure the rate of photosynthesis per unit area of ground. Observed rates vary from 1 to 5 mg CO_2 dm^{-2} h^{-1} (Oechel and Collins, 1973; Kallio and Heinonen, 1975; Kallio and Vallane, 1975; Grace and Marks, 1978). In *Sphagnum capillifolium* the maximum rate of photosynthesis was found to be a function of water content; when the capitulum was very wet the rate was reduced, presumably because the water film constituted an extra diffusion resistance for carbon dioxide. In their wettest state the capitula held 22 times their own dry weight of water and maximum rates of photosynthesis were obtained when the water content had been allowed to fall to 1200%. Thereafter the rate fell abruptly as the water content declined to 400% (Grace and Marks, 1978).

Fig. 8.3 shows the daily course of photosynthesis in contrasting wet habitats. The first habitat is a wet spring at Death Valley, California (Pearcy et al., 1972), where mid-day air temperatures exceeded 40°C. High rates of photosynthesis were obtained and the stomata remained open. The rapid loss of water from the leaves by evaporation reduced the leaf temperature below that of the ambient air. The second habitat is wet tundra at Barrow, Alaska (Tieszen, 1972). Despite the low temperatures and low irradiances, quite high rates of photosynthesis were achieved. Added to this is the benefit of a long photoperiod, so that the daily total of carbon assimilation was very high, although not as high as in the *Phragmites* stand previously referred to. The final graph (C) in Fig. 8.3, also for Barrow, Alaska, concerns the moss *Polytrichum alpinum*, on a day when the air temperature hardly exceeded 0°C (Oechel and Collins, 1973). The data are expressed on a ground area basis. Although the daily sum of photosynthesis was less than for the previous two examples, it is nevertheless appreciable, and perhaps remarkable in view of the low prevailing temperatures.

Photosynthesis in relation to nutrients

Small (1972) has drawn attention to the nutrient deficiency of peat-bog plants in relation to their capacity to photosynthesize. The peat-bog substrate is exceptionally poor in available nutrients,

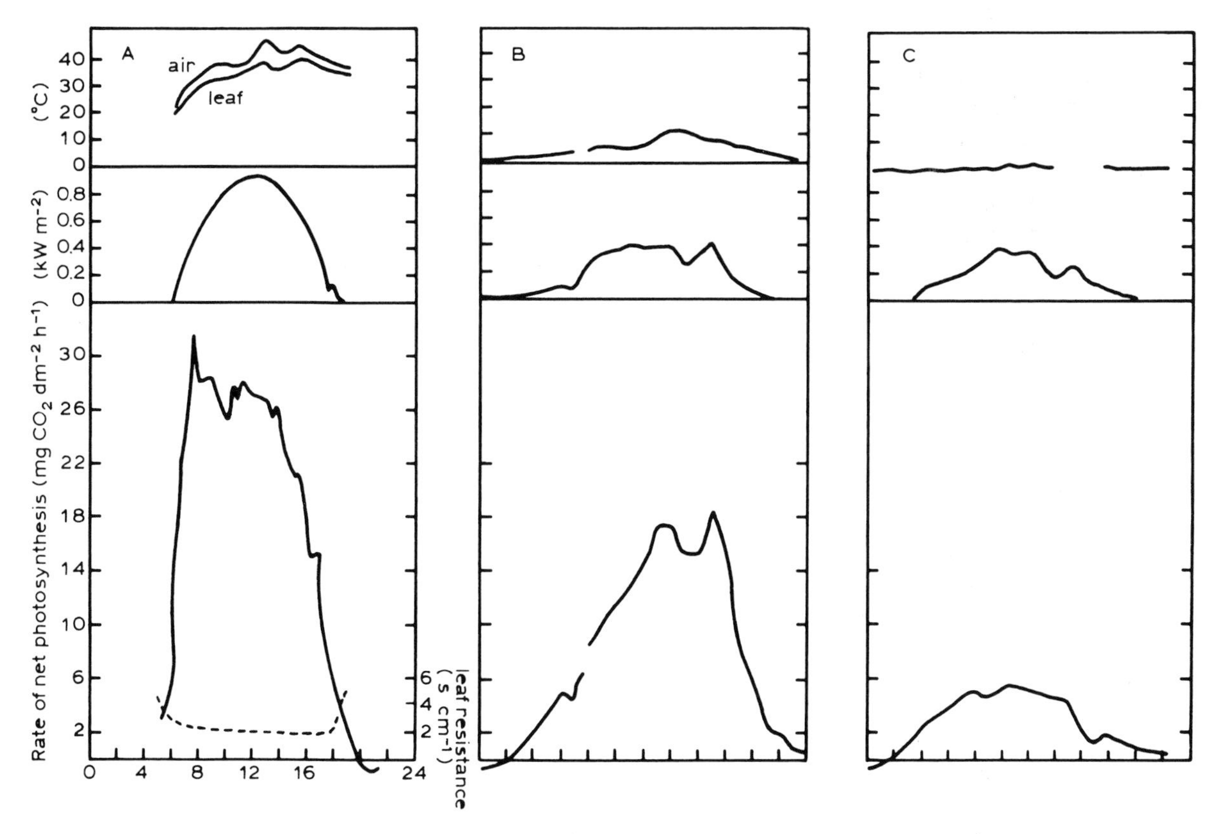

Fig. 8.3. Photosynthesis in contrasting wetland situations. A. *Phragmites australis* in a wet spring at Death Valley, California (Pearcy et al., 1972). B. *Carex aquatilis* at Barrow, Alaska (Tieszen, 1972). C. *Polytrichum alpinum* at Barrow, Alaska (Oechel and Collins, 1973).

especially nitrogen and phosphorus. Crisp (1966) attempted to draw up a nutrient budget for an ombrogenous blanket bog in the Pennines of England. Although he may have overlooked certain inputs, especially that due to nitrogen fixation (Waughman and Bellamy, 1972; Kallio and Kallio, 1975; Granhall and Lid-Torsvik, 1975), his figures suggest that the outputs of most nutrients from the catchment area greatly exceeded the input from rain. The inference to be drawn from this is that, unless substantial inputs have been overlooked, the system must be deteriorating, at least over the period of the observations. In these circumstances the efficiency with which the plants handle scarce nutrient resources is critical. Small (1972) compared photosynthetic rates and foliar nutrient levels of fifteen bog species with thirteen non-bog species. The bog species were found to re-absorb significantly more nitrogen from their foliage preceding litter fall. He calculated the potential photosynthate which the species could manufacture during the time a given unit of nitrogen remained in the plant before being lost through leaf fall. In bog evergreens this figure was 235% higher than for bog deciduous plants, and about 60% higher in bog deciduous species than in the non-bog deciduous species. He advanced the hypothesis that the increased time available to use nitrogen photosynthetically before it is recycled is of adaptive significance in bog plants, especially evergreens.

WATER RELATIONS

On a global scale shortage of water is a severe limitation to the growth of higher plants, and in agricultural practice there are few crops which do not benefit from irrigation. The high productivity of certain wetland communities of the world has been attributed to their more favourable supply of water. On the other hand, it is worth remembering that many bog plants possess xeromorphic features, such as rolled leaves, sclerenchyma and well-developed cuticular wax (Small, 1972). These characteristics suggest that the water is in some way unavailable, resulting in an evolutionary response to tissue water shortage. The purpose of this section is to assess whether plant growth in wetland sites is indeed sometimes limited by tissue water stress.

Origins of leaf water stress

When water evaporates from the leaf tissues it is replaced by water from the vessels. Because of the cohesive nature of water molecules columns of water are thus pulled through the plant/soil continuum in response to the evaporative demand of the atmosphere. The exact pathway of this transport, particularly the extent to which water moves through leaf and root tissues exclusively via the cellulose micelles within the cell walls (see Weatherley, 1970; and Newman, 1973), is still a matter of conjecture. In stems and petioles water certainly moves by viscous flow through xylem vessels. In the soil around the roots, water is drawn radially towards the root surface through the pore spaces of the soil.

Leaf water stress develops almost entirely because the water columns experience frictional resistance as they are drawn through the plant/soil continuum. In the simplest possible model the relationship between the flow, q, and the water potential, ψ, at any place may be written in a form which is analogous to Ohm's Law:

$$q = \frac{1}{R}\int_{z_0}^{z_1} \frac{d\psi}{dz}\,dz = \frac{\psi_1 - \psi_2}{R}\ ,$$

where z represents distance along the water conducting pathway (usually thought of as vertical distance), and R is the frictional resistance.

According to this simple model, tissue water stress ($\psi < 0$) is a necessary consequence of a finite transpiration rate and proportional to the water flux through the plant (when $q = 0$, values of ψ will approach zero, though in practice they do not achieve zero because of the force of gravity, the effect of dissolved minerals in the xylem water, and a demand for water due to growth). The equation can be rewritten in a form which shows the influence of the water potential of the soil on leaf water potential

$$\psi_{leaf} = \psi_{soil} - q(R_{soil} + R_{plant})$$

Thus, when ψ_{soil} approaches zero, as it often must in wetland communities, the magnitude of ψ_{leaf} depends on the magnitude of R_{soil} and R_{plant}. For wet soils, and especially if a proportion of the roots have developed in free water above the rooting substrate, R_{soil} may be very small. The dependence of soil hydraulic resistance on the soil water content and the size of the soil particles has been shown by Gardner (1960). When ψ_{soil} falls from -0.01 to -0.1 MPa[1] the resistance to water flow increases very suddenly. Particle size influences resistance appreciably only in very wet soils.

Experimental evidence that R_{soil} is large even in wet soils was obtained by Tinklin and Weatherley (1968), working with *Ricinus*. Whereas ψ_{leaf} was near zero for transpiring plants in water culture, the presence of soil — albeit wet — led to considerably negative values of ψ at high transpiration rates.

Several attempts have been made to detect a zone of dry soil around the roots of plants growing in soil (Newman, 1969a; Dunham and Nye, 1973); such a zone would indicate that the "perirhizal" resistance is large. However, such a steep gradient of ψ has not been detected, and Newman (1969b) suggested that the soil resistance is that involved in water transport from more distant parts of the soil. Recently, Weatherley (1976) has described an observation which seems pertinent: when plants were grown in soft plastic pots a manual squeezing of the pot relieved the tissue water stress. He discusses the possible existence of an air space around roots which might result from the root's contraction in diameter at high rates of transpiration.

The conclusion of this part of the discussion is that leaf water stress may be expected to develop whenever plants are rooted in the soil and transpiring rapidly even though the soil is wet.

In many wetland plants the possibility of uptake directly from the free water through the epidermis of submerged stems should not be overlooked.

Observed water potentials

Scholander et al. (1965) investigated xylem water potentials of many different sorts of vegetation, using the pressure chamber. Wetland communities (discounting the sea-shore) displayed water potentials of -0.5 to -2.0 MPa during the daytime (Fig. 8.4). Although this is less negative than the values recorded in other habitats it is well within the range suggested by Hsiao (1973) as having significant deleterious effects on plant growth; considerable effects on cell division and expansion are to be expected at -0.5 MPa.

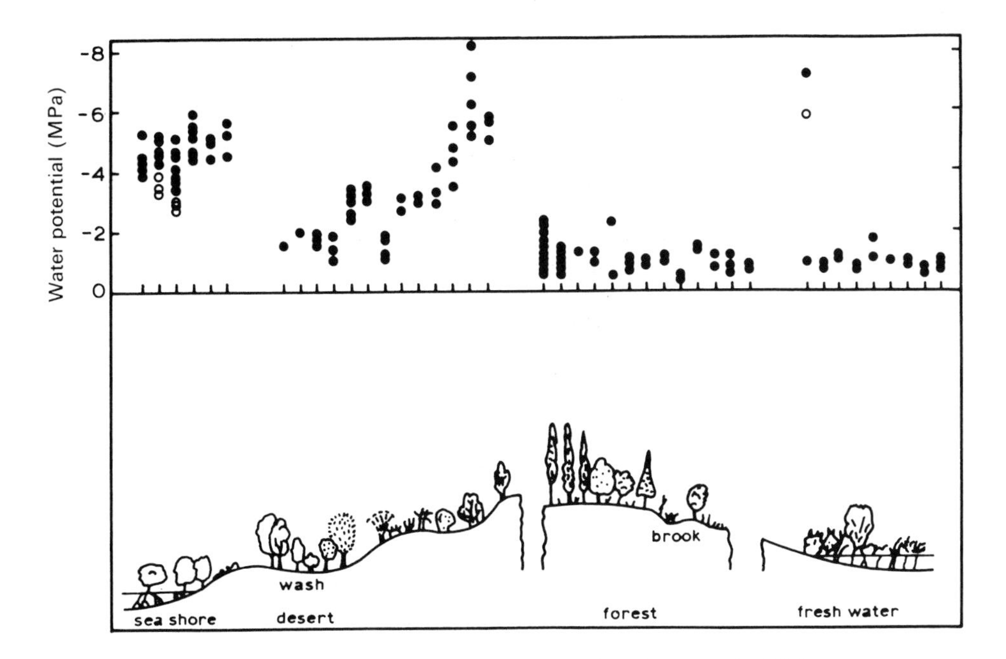

Fig. 8.4. Xylem pressure potentials in different habitats. From Scholander et al. (1965).

[1] 1 MPa = 10 bar.

Other measurements of xylem potential made in wetland vegetation since the pioneering work of Scholander et al. (1965) are tabulated (Table 8.5). In the work of Scholander and Perez (1968) the tree roots were covered by several metres of standing water yet a potential of −2.0 MPa was developed in the leaves. Small (1972) compared water potentials measured in twelve woody species rooted in *Sphagnum* with those recorded for another twelve woody species in a nearby field. The differences were surprisingly small, even though the *Sphagnum* bog was by far the wetter place. Ashmore (1975), working in a *Sphagnum* bog in the north of England, found that on sunny days midday stomatal closure occurred at about −0.8 MPa in *Calluna vulgaris*. He noted that *Calluna* has a relatively low root/shoot ratio when growing in this wet habitat, and suggested that the hydraulic resistance due to the root system was consequently high.

Water uptake by plant roots is markedly sensitive to temperature (more so than can be explained by viscosity effects) and to anaerobiosis (Slatyer, 1967). From this it can be deduced that somewhere in the plant water transport involves the passage of a living barrier. This is usually supposed to be located in the endodermis where the cell wall pathway is effectively blocked by the development of the Casparian strip. The presence in many wetland species of aerenchyma, which facilitates oxygen diffusion, is usually regarded as an adaptation to anaerobic conditions (see Crawford, Ch. 7). In the cold wetlands low temperatures may further contribute to the plant resistance to water flow, as the viscosity of water increases markedly at low temperatures.

The data available show that plants in wetland communities do indeed suffer water stress. In some cases this stress is accompanied by stomatal closure, and in all the data the stress is sufficiently large to suspect that it may influence cell division and expansion. On the other hand the xylem potential in wetland species has usually been less negative than that recorded in most other types of vegetation, and certainly much less severe than in arid zones and on the sea shores, which collectively represent the classic environments for the evolutionary response towards xeromorphy. Within the wetland vegetation types the distribution of xeromorphy remains a largely unresolved question. However, since xeromorphy in wetlands is associated primarily with woody vegetation on nutrient-poor bogs in seasonally cold climates, it is

TABLE 8.5

Observed shoot water potentials in some wetland species

Species	Location	Maximum ψ (MPa)	Conditions	Minimum ψ (MPa)	Conditions	Reference
Calluna vulgaris on *Sphagnum*	northern England	≃0	early morning with dew	−1.5	sunny	Ashmore (1975)
Taxodium distichum	glasshouse, California (U.S.A.)	−1.09	saturated soil	−1.79	unsaturated soil	Dickson and Broyer (1972)
Nyssa aquatica	glasshouse, California (U.S.A.)	−0.72	saturated soil	−1.28	unsaturated soil	Dickson and Broyer (1972)
Phragmites australis	Death Valley, California (U.S.A.)	−1.0	morning, sunny	−1.7	mid-day	Pearcy et al. (1972)
Clethra alnifolia	Okefenokee swamp, Georgia (U.S.A.)	−0.4 / −0.3	early morning and late evening	−1.3 / −1.0	mid-day	Schlesinger and Chabot (1977)
Symmeria paniculata and *Ruprechtia* spp.	Rio Negro, Amazonia (Brazil)	−1.0	overcast	−2.0	sunny	Scholander and Perez (1968)
Twelve woody species on *Sphagnum*	Ottawa (Canada)	−1.2	warm and sunny	−1.8	warm and sunny	Small (1972)
Carex aquatilis	Barrow, Alaska (U.S.A.)	−0.1	early morning	−0.75	1400 h	Stoner and Miller (1975)
Dupontia fischeri	Barrow, Alaska (U.S.A.)	−0.05	early morning	−1.0	0900 h	Stoner and Miller (1975)
Arctophila fulva	Barrow, Alaska (U.S.A.)	−0.1	early morning	−0.9	1400 h	Stoner and Miller (1975)
Eriophorum angustifolium	Barrow, Alaska (U.S.A.)	−0.05	early morning	−0.7	1400 h	Stoner and Miller (1975)

tempting to regard this feature as an adaptation to conditions when the peat is frozen and the plants are very vulnerable to leaf water loss.

CHARACTERISTICS OF STANDS

The rate of community photosynthesis depends on the physical conditions that develop within the vegetation in relation to the position of the photosynthetic organs. The development of a microclimatic profile depends on the structure of the leaf canopy, and is the result of transfers of energy, mass and momentum between the vegetation and the atmosphere.

Radiation balance

The local environment of the vegetation is determined largely by the partitioning of solar radiation between sensible and latent heat. Wetlands are characterized by a large availability of water, either from the substrate or from the leaves. Consequently, more of the available energy is dissipated as latent heat and less is manifest as an increase in surface temperature. The surface temperatures of wetland communities are thus likely to be lower than those of dryer habitats — a feature which may be expected to be advantageous in hot climates as already noted for *Phragmites* (Fig. 8.3), but disadvantageous in cold climates where the growth of plants is limited by low temperatures and the shortness of the growing season.

The utilization of available energy by rice and wet tundra is summarized in Table 8.6. In rice, nearly 80% of the energy is used to evaporate water, while the corresponding figure for the tundra immediately after snowmelt is 72%. Comparisons between vegetation in widely differing climates should be made with care, because the proportion is likely to be much influenced by the absolute level of net radiation, as well as the ambient humidity. However, the figures for very dry graminaceous vegetation have been included in Table 8.6; they show that at the end of the season, when the soil is very dry, only 50% of the available energy is used to evaporate water from prairie.

The proportion of the energy reaching the ground depends on the stage of development of the canopy. The temperature of rice fields and of irrigation water has long been a subject of interest in Japan, and is perhaps relevant to reed bed and other situations where the substrate is flooded by a thin layer of water. In the early part of the season, when the LAI is less than 1, the water temperature is found to be as much as 4°C above ambient, though this temperature excess falls gradually as the LAI of the paddy field increases to about 2.5. Thereafter, the mean water temperature is lower than the air temperature (Uchijima, 1976). During this time the diurnal amplitude of water temperature declines (Uchijima, 1976). A similar effect was reported for *Sphagnum* surface temperatures in a *Sphagnum–Calluna* community (Grace and Marks, 1978).

Canopy development and light interception

From the viewpoint of photosynthesis, the annual cycle of canopy development is of crucial

TABLE 8.6

Partition of available energy between evaporation and sensible heat

Vegetation and location	Season	Proportion of energy in:			Reference
		evaporation	sensible heat		
			air and leaves	soil and roots	
Tundra; Barrow, Alaska (U.S.A.)	post-melting	0.72	0.18	0.09	Weller and Cubley (1972)
	mid-summer	0.66	0.32	0.02	Weller and Cubley (1972)
Rice; Japan	July–September	0.79	0.21	0	Uchijima (1976)
Prairie; southern Saskatchewan (Canada)	July	0.65	0.22	0.11	Ripley and Redman (1976)
	early September	0.50	0.50	0	Ripley and Redman (1976)

importance because it determines the capture of solar energy. Also, it is of interest to know the structure of the canopy at the period of most rapid growth, to enable comparison with the most productive canopies that are known from agricultural work.

The annual cycle of LAI in five wetland communities is compared in Fig. 8.5. The rice crop in a tropical environment not only has the highest maximum LAI, but it is developed at a faster rate. The *Phragmites* community develops a LAI of 4.5, and values as high as those for rice have not been reported. In the colder wetlands the LAI is always low. It should be realized that a LAI less than 1 will always fail to intercept incoming radiation completely, the extent depending on the leaf angles and the degree to which the leaves are clustered together.

The capture of light energy and the distribution of photon flux in the canopy may be regarded as a problem of geometry (De Wit, 1966; Duncan et al., 1967). Using photosynthetic and optical properties appropriate for a leaf of *Zea mays*, Duncan et al.

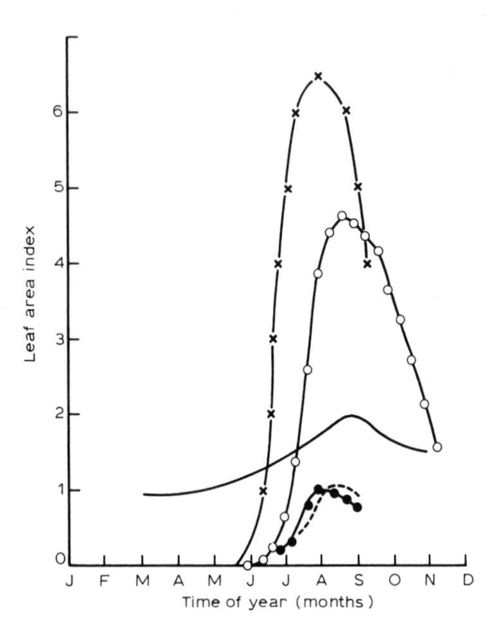

Fig. 8.5. Leaf area index of wetland communities: × = "Peta" rice in the Philippines (Tanaka et al., 1966); ○ = *Phragmites australis* in Czechoslovakia (Dykyjová et al., 1970); — = *Calluna vulgaris* in northern England (Forrest, 1971); ---- = wet meadow in Norway (Berg et al., 1975); ● = wet tundra, Alaska (Tieszen, 1972). The data of Forrest (1971) have been converted to an area basis assuming 60 cm^2 g^{-1}.

(1967) showed that maximum productivity would be achieved with a LAI of 8 and a leaf angle near vertical. If, on the other hand, leaves were horizontal no benefit would be derived from having a LAI that exceeded 3, and the calculated productivity would be only 70% of that found in the previous situation. The reason for this is that the distribution of light is much more even in canopies of erect leaves, whereas for horizontal leaves the uppermost members are light-saturated while the lower members are in deep shade. Although these calculations were performed for the conditions prevailing in California, it was shown that latitude does not greatly influence the result. A similar conclusion was reached independently by De Wit (1966).

The most productive wetlands, such as rice, *Phragmites* and *Typha* communities, all approach the ideal, having peak LAI values that are high ($\simeq 4$) and a leaf posture that approaches vertical (Ondok, 1973a; Uchijima, 1976), although this state is sustained for a short period only. In the colder wetlands, where productivity is much lower, this state is never attained, and leaf angles vary widely (Berg et al., 1975).

Towards the end of the season, ungrazed stands of vegetation frequently contain much dead material which contributes to the extinction of light but not to photosynthesis: this has been demonstrated for rice (Tanaka et al., 1966), *Phragmites* (Dykyjová et al., 1970), and for tundra wetlands (Caldwell et al., 1972).

The only detailed study of canopy development in relation to light penetration for wetland vegetation is concerned with *Phragmites* (Ondok, 1973a,b,c, 1975). At the start of the season, when the LAI is low, a large proportion of the leaves are sunlit (Fig. 8.6). Later, only the topmost layers of leaves are sunlit, the proportion depending on the solar angle (Fig. 8.6C). Towards the end of the season, the LAI is reduced because of death and abscission, but the light distribution is hardly changed because any tendency for radiation to penetrate further is offset by the accumulation of standing dead material and by the trend towards a more horizontal leaf posture. This general pattern of events is similar to that described for rice (Uchijima, 1976).

Many wetlands are made up of patchy vegetation, often occupying the margins of large bodies of water. Here, it is not appropriate to measure the

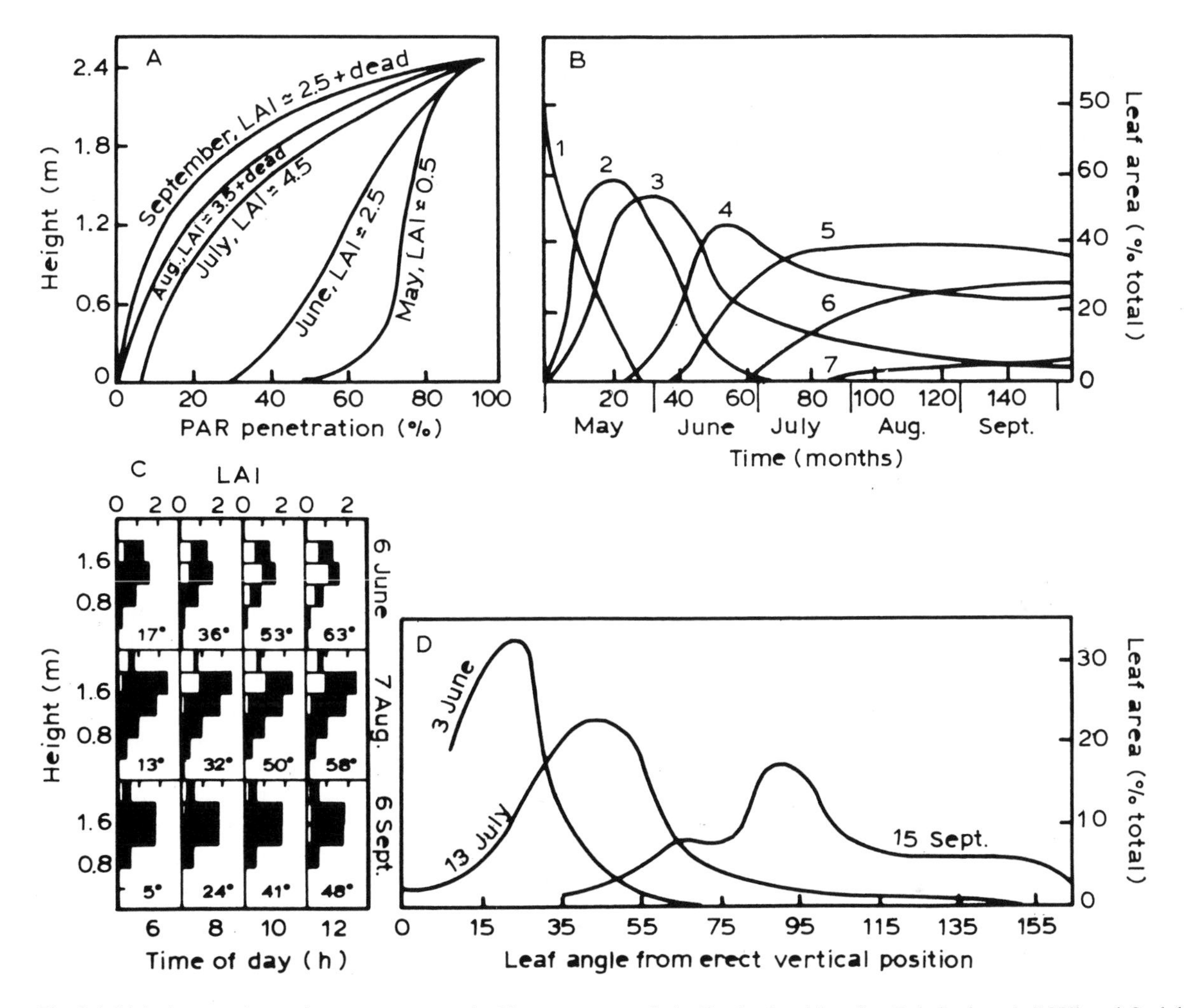

Fig. 8.6. Light interception and canopy structure in *Phragmites australis* in Czechoslovakia, after Dykyjová et al. (1970) and Ondok (1973a,b,c). A. Penetration of photosynthetically active radiation into the stand at different times of the year. B. Percentage of the total leaf area to be found in successive 400 mm strata as the stand develops and grows taller. C. Sunlit (open histogram segments) and shaded (dark histogram segments) foliage area at different times of day and at three different times of year (The number in each cell represents the solar angle). D. The change in leaf angles within the canopy as the season progresses.

penetration of light into the horizontal layers of vegetation, because there is considerable horizontal heterogeneity. The extreme condition is where the vegetation forms discrete clusters of leaves such that the leaves at the edge of the cluster which faces the sun are brightly illuminated while those on the opposite edge are in deep shade. This situation is discussed by Grace and Woolhouse (1973) in relation to *Calluna* plants growing in a *Sphagnum* bog.

Turbulent transfer and microclimate

Turbulent transfer, in which momentum, mass and heat are transported by turbulent eddies, is the principal mechanism whereby a stand of vegetation obtains its supply of carbon dioxide from the atmosphere, dissipates sensible heat from the leaves to the atmosphere, and loses water by evaporation. The extent of turbulent transfer depends on the prevailing atmospheric conditions, especially the wind speed, and on the quantity and configuration of the roughness elements that comprise the vegetation. For a full discussion the reader is referred to Thom (1975).

Turbulent transfer is essentially a diffusive process, the flux being proportional to the gradient of each entity being considered. The coefficient of turbulent transfer, also known as the eddy diffusivity K, is defined in relation to the fluxes as follows:

$$E = -K_V(\delta\chi/\delta z)$$

$$C = -K_H(\delta(\rho c_p T)/\delta z)$$

$$\tau = -K_M(\delta(\rho u)/\delta z)$$

where E is the flux of water vapour in evaporation (kg m^{-2} s^{-1}); C is the sensible heat flux (W m^{-2}); τ is the momentum flux (N m^{-2}); K_V, K_H and K_M are the eddy diffusivities for, respectively, water vapour, heat and momentum (m^2 s^{-1}); χ is the water vapour concentration (kg m^{-3}); z is the vertical distance (m); ρ is the density of air (kg m^{-3}); c_p is the specific heat of air at constant pressure (J kg^{-1} K^{-1}); T is the temperature (°K); and u is the windspeed (m s^{-1}).

Since these three entities are transferred by the same turbulent mixing, the similarity principle is often called upon, viz. $K_V = K_H = K_M$. This is useful when interpreting microclimatological profiles, as K_M may be found from suitable measurements of wind over the vegetation, and then used to calculate fluxes of CO_2 and H_2O from measured profiles of concentration. It should be realized however that the similarity principle is by no means exact in the circumstances of most vegetation (see Thom, 1975).

In some kinds of vegetation, where LAI is high and the leaves are acting as strong sinks for carbon dioxide, the rate of turbulent transfer may limit the assimilation rate. This is almost certainly the case with *Zea mays*, in which independent groups of workers have reported a midday depletion of carbon dioxide by about 50% (Uchijima, 1970; Allen, 1971). Under such conditions, assimilation rate would be a function of the wind speed. At the other extreme, tall crops with needle-like leaves and a low unit rate of leaf photosynthesis, such as coniferous forest, deplete the local carbon dioxide concentration by about 5 $\mu l\ l^{-1}$ only (Baumgartner, 1968; Jarvis et al., 1976). Similarly, such structures are good at dissipating sensible heat by turbulent transfer; the temperatures of leaves in coniferous forests rarely depart more than 1°C from those in the surrounding air.

The most productive wetland vegetation, such as rice crops and *Phragmites*, may be expected to fall midway between these two extreme cases. Microclimatological profiles within rice and calculated values of K are shown in Fig. 8.7A, taken from Uchijima (1976). The concentration of water vapour declined from the ground to above the canopy in daytime, indicating that the ground, not the leaves, was the main source of water loss (Fig. 8.7B). Air temperatures were low near this evaporating surface, and reached a maximum at a height of 0.4 h (h is the height of the stand). Below this level there was a downward flux of sensible heat, as energy was being used for evaporation (this state of affairs contrasts with a crop on dry soil where midday temperatures are often maximal at ground level). As many wetlands are dominated by graminaceous plants, in which the vegetative apex is near the ground, and for which the rate of growth is essentially a function of apical temperature (Watts, 1972; Peacock, 1975), this observation may be of considerable significance.

As found over all crops, the turbulent transfer coefficient, K, increased with height above the ground in the paddy field (Fig. 8.7A). The concentration of carbon dioxide is also shown in Fig. 8.7C, although these measurements were made on a different occasion to those of Fig. 8.7B. The lowest concentration is found to coincide with the greatest leaf amount, but the depletion is here only about 10 $\mu l\ l^{-1}$ which is much less than that reported in *Zea mays*, and not enough to reduce the assimilation rate appreciably. For "Peta" rice in the Philippines on the other hand, a carbon dioxide depletion of 50 $\mu l\ l^{-1}$ was reported by Tanaka et al. (1966) as being typical for an "ordinary" day whereas on a "windy" day the depletion was only 20 $\mu l\ l^{-1}$. There is an appreciable upward flux of carbon dioxide from the ground, as a result of below-ground respiration. In Fig. 8.7C this is equivalent to a carbon dioxide evolution of between 0.002 and 0.2 g m^{-2} h^{-1}, compared to the downward flux from the atmosphere of about 6 g m^{-2} h^{-1} at noon (Uchijima, 1976). This agrees with previous estimates for paddy fields (see Ishibashi, 1969; and Ohtaki et al, 1969).

For wetland communities which do not form a large homogenous area of vegetation, such as reeds

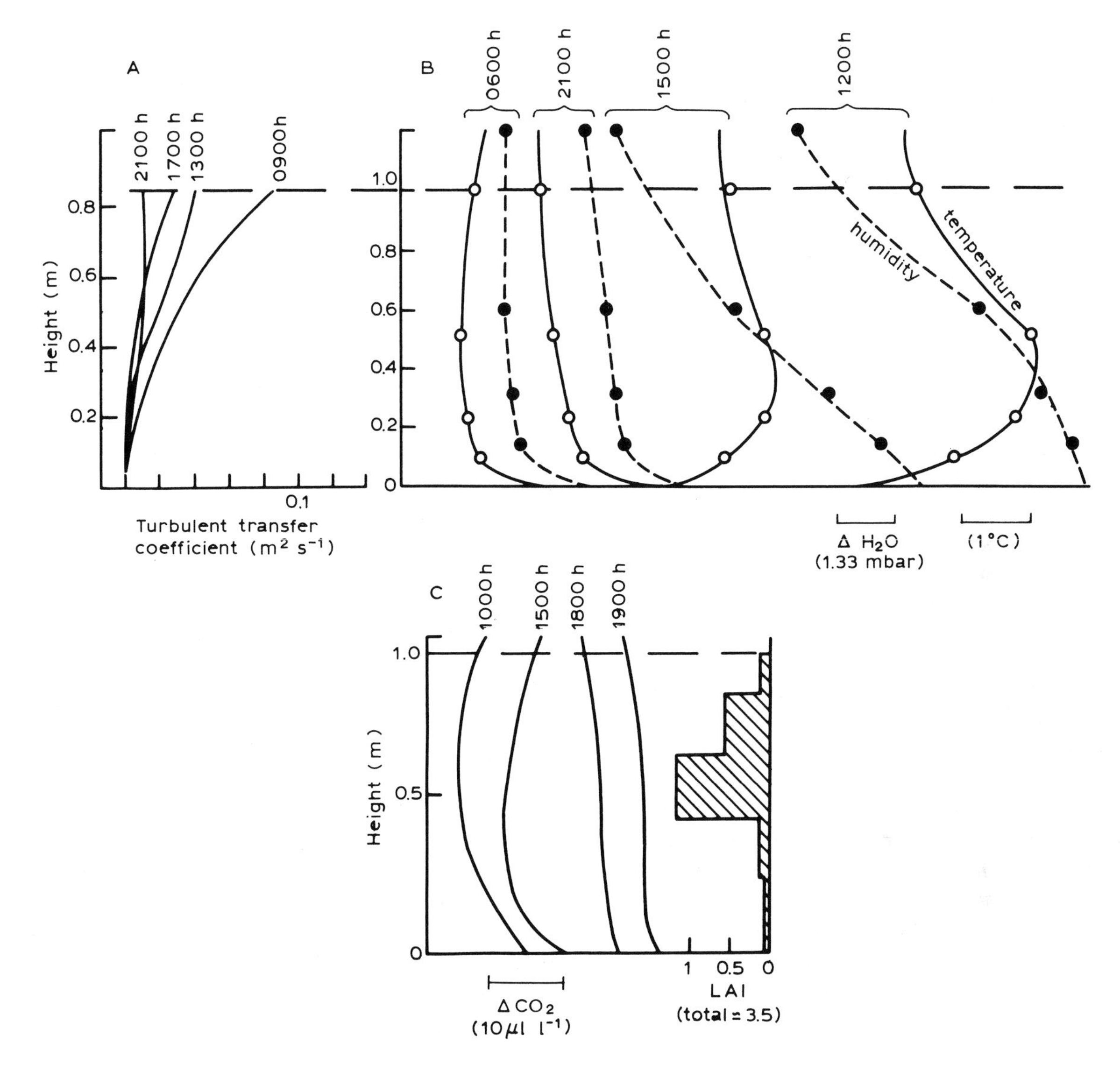

Fig. 8.7. Heat and mass transfer in rice crops. A. The turbulent transfer coefficient *K*. B. Humidity and air temperature. C. Carbon dioxide. A and B were obtained on the same day, C is a different year. From Uchijima (1976).

at the margin of lakes, ventilation may be expected to be particularly efficient, due to the flow of air into the edge of the stand. Here, large depletions of carbon dioxide are unlikely to occur. This "oasis effect" or the "clothes-line effect" has been cited as the explanation for the extremely high evaporation rates observed from swamp vegetation planted in isolated experimental tanks.

CONCLUSIONS

(1) Although there is a considerable literature on the productivity of wetlands, it is still not possible to state precisely the normal range of primary productivity in these systems.

(2) Nevertheless, from the available data we suggest that a figure of 3500 g m^{-2} yr^{-1} may be a realistic upper limit for swamps and marshes in temperate zones (this value is high compared with other natural communities and even in relation to

intensive agricultural systems). In cold wetlands and temperate peat bogs an upper limit of 1000 g m^{-2} yr^{-1} seems appropriate.

(3) Many wetland communities have a considerable underground biomass, but the productivity of this component has rarely been measured with any accuracy.

(4) We have found little available data on the productivity of tropical swamps and marshes.

(5) On a daily basis the productivity of swamps and marshes may be comparable with that of rice, being equivalent to an efficiency of solar energy capture of from 3 to 5%. Even wet tundra vegetation, during the short period of growth, may achieve a productivity of 20 g m^{-2} day^{-1}, equivalent to a 1% efficiency of solar energy capture.

(6) The photosynthetic rate on an area basis varies too much for generalization to be possible. There is a shortage of data for the major species of swamps and marshes, such as *Phragmites*, *Typha* and *Cyperus*.

(7) Leaf water stress in wetland plants is less acute than that observed in other habitats, but leaf water potentials are nevertheless sufficiently negative to reduce growth.

(8) The canopy structure in swamp and marsh vegetation is usually characterized by high values of leaf area index and near-vertical leaf posture. Such a structure approaches the calculated "ideal" for agricultural productivity. In contrast, the leaf area index in the cold wetlands is very low.

(9) In almost all conditions of wind speed, turbulent transfer is likely to prevent the local depletion of carbon dioxide within the vegetation, particularly where stands occupy margins of water and so are ventilated by a "clothes-line effect".

REFERENCES

Allen, K.R., 1951. The Horokiwi Stream: a study of a trout population. *Fish. Bull., Wellington, N.Z.*, 10: 1–238.

Allen, L.H., 1971. Variations in carbon dioxide concentration over an agricultural field. *Agric. Meteorol.*, 8: 5–24.

Allen, S.E. and Pearsall, W.H., 1963. Leaf analysis and shoot production in *Phragmites. Oikos*, 14: 176–189.

Anderson, F.O., 1976. Primary production in a shallow water lake with special reference to reed swamp. *Oikos*, 27: 243–250.

Anderson, J.H., 1972. Phytocenology and primary production at Eagle Summit, Alaska. In: S. Bowen (Editor), *Proceedings of IBP Tundra Biome Symposium, University of Washington, Seattle, Wash.*, pp. 61–70.

Ashmore, M.R., 1975. *The Ecophysiology of* Calluna vulgaris *(L.) Hull in a Moorland Habitat.* Thesis, University of Leeds, Leeds, 145 pp.

Auclair, A.W.D., Bouchard, A. and Pajaczkowski, J., 1976a. Productivity relations in a *Carex*-dominated ecosystem. *Oecologia*, 26: 9–31.

Auclair, A.W.D., Bouchard, A. and Pajaczkowski, J., 1976b. Plant standing crop and productivity relations in a *Scirpus–Equisetum* wetland. *Ecology*, 57: 941–952.

Baradziej, E., 1974. Net primary production of two marsh communities near Ispinia in the Niepolomice Forest (southern Poland). *Ekol. Pol.*, 22: 145–172.

Baumgartner, A., 1968. Ecological significance of the vertical energy distribution in plant stands. In: F.E. Eckardt (Editor), *Functioning of Terrestrial Ecosystems at the Primary Production Level.* UNESCO, Paris, pp. 367–374.

Bellamy, D.J. and Holland, P.J., 1966. Determination of the net annual aerial production of *Calluna vulgaris* (L.) Hull, in Northern England. *Oikos*, 17: 272–275.

Bellamy, D.J. and Rieley, J., 1967. Some ecological statistics of a "miniature bog". *Oikos*, 18: 33–40.

Bellamy, D.J., Bridgewater, P., Marshall, C. and Tickle, W.M., 1969. Status of the Teesdale rarities. *Nature (Lond.)*, 222: 238–243.

Berg, A., Kjelvik, S. and Wielgolaski, F.E., 1975. Measurement of leaf angles and leaf areas of plants at Hardangervidda, Norway. In: F.E. Wielgolaski (Editor), *Fennoscandian Tundra Ecosystems, Part 1, Plants and Microorganisms*, Springer-Verlag, Berlin, pp. 103–110.

Bernard, J.M., 1974. Seasonal changes in standing crop and primary production in a sedge wetland and an adjacent dry old-field in central Minnesota. *Ecology*, 55: 350–359.

Bernard, J.M. and MacDonald, J.G., 1974. Primary production and life-history of *Carex lacustris. Can. J. Bot.*, 52: 117–123.

Bernard, J.M. and Solsky, B.A., 1977. Nutrient cycling in a *Carex lacustris* wetland. *Can. J. Bot.*, 55: 630–638.

Bliss, L.C., 1962. Net primary production in tundra ecosystems. In: H. Lieth (Editor), *Die Stoffproduktion der Pflanzendecke.* Fischer-Verlag, Stuttgart, pp. 35–46.

Boyd, C.E., 1970. Production, mineral accumulation and pigment concentrations in *Typha latifolia* and *Scirpus americanus. Ecology*, 51: 285–290.

Boyd, C.E., 1971a. Further studies on productivity, nutrient and pigment relationships in *Typha latifolia* populations. *Bull. Torrey Bot. Club*, 98: 144–150.

Boyd, C.E., 1971b. The dynamics of dry matter and chemical substances in a *Juncus effusus* population. *Am. Midl. Nat.*, 86: 28–45.

Boyd, C.E. and Hess, L.W., 1970. Factors influencing shoot production and mineral nutrient levels in *Typha latifolia. Ecology*, 51: 276–300.

Boyd, C.E. and Vickers, D.H., 1971. Relationships between production, nutrient accumulation and chlorophyll synthesis in an *Eleocharis quadrangulata* population *Can. J. Bot.*, 49: 883–888.

Boyd, C.E. and Walley, W.W., 1972. Production and chemical composition of *Saururus cernuus* L. at sites of different fertility. *Ecology*, 53: 927–932.

Bradbury, I.K. and Hofstra, G., 1976a. The partitioning of net energy resources in two populations of *Solidago canadensis* during a single developmental cycle in southern Ontario. *Can. J. Bot.*, 54: 2449–2456.

Bradbury, I.K. and Hofstra, G., 1976b. Vegetation death and its importance in primary production measurements. *Ecology*, 57: 209–211.

Bray, J.R., Lawrence, D.B. and Pearson, L.C., 1959. Primary production in some Minnesota terrestrial communities. *Oikos*, 10: 38–40.

Brown, J. and West, G.C., 1970. Tundra biome research in Alaska. In: *The Structure and Function of Cold-Dominated Ecosystems. U.S. IBP Tundra Biome Report.* University of Alaska, Fairbanks, Alaska, pp. 70–71.

Busby, J.R., Bliss, L.C. and Hamilton, C.D., 1978. Microclimate control of growth rates and habitats of the Boreal forest mosses *Tomenthypnum nitens* and *Hylocomium splendens*. *Ecol. Monogr.*, 48: 95–110.

Buttery, B.R. and Lambert, J.M., 1965. Competition between *Glyceria maxima* and *Phragmites communis* in the region of Surlingham Broad. *J. Ecol.*, 53: 163–181.

Caldwell, M.M., Tieszen, L.L. and Fareed, M., 1972. Comparative Barrow and Niwot Ridge canopy structure. In: S. Bowen (Editor), *Proceedings of IBP Tundra Biome Symposium, University of Washington, Seattle, Wash.*, pp. 22–28.

Chapman, S.B., 1965. The ecology of Coom Rigg Moss, Northumberland III. Some water relations of the bog system. *J. Ecol.*, 53: 371–384.

Chapman, V.J. (Editor), 1977. *Wet Coastal Ecosystems.* Ecosystems of the World, 1. Elsevier, Amsterdam, 428 pp.

Clymo, R.S., 1970. The growth of *Sphagnum*: methods of measurement. *J. Ecol.*, 58: 13–49.

Clymo, R.S. and Reddaway, E.J.F., 1974. Growth rate of *Sphagnum rubellum* Wils. on Pennine blanket bog. *J. Ecol.*, 62: 191–196.

Coyne, P.I. and Kelley, J.J., 1972. CO_2 exchange in the Alaska arctic tundra: meteorological assessment by the aerodynamic method. In: S. Bowen (Editor), *Proceedings of IBP Tundra Biome Symposium, University of Washington, Seattle, Wash.*, pp. 36–39.

Crisp, D.T., 1966. Input and output of minerals for an area of Pennine moorland: the importance of precipitation, drainage, peat erosion and animals. *J. Appl. Ecol.*, 3: 327–348.

Dennis, V.G. and Tieszen, L.L., 1972. Seasonal course of dry matter and chlorophyll by species at Barrow Alaska. In: S. Bowen (Editor), *Proceedings of IBP Tundra Biome Symposium, University of Washington, Seattle, Wash.*, pp. 16–21.

De Wit, C.T., 1966. Photosynthesis of leaf canopies. *Agric. Res. Rep. (Wageningen)*, 663: 1–57.

Dickson, R.E. and Broyer, T.C., 1972. Effects of aeration, water supply and nitrogen source on growth and development of tupelo gum and bald cypress. *Ecology*, 53: 626–634.

Downton, W.J.S., 1975. The occurrence of C_4 photosynthesis among plants. *Photosynthetica*, 9: 96–105.

Doyle, G.J., 1973. Primary production estimates of native blanket bog and meadow vegetation growing on reclaimed peat at Glenamoy, Ireland. In: L.C. Bliss and F.E. Wielgolaski (Editors), *Proceedings of IBP Tundra Biome Symposium, Production and Production Processes, Dublin*, pp. 141–151.

Duncan, W.G., Loomis, R.S., Williams, W.A. and Hanau, R., 1967. A model for simulating photosynthesis in plant communities. *Hilgardia*, 38: 181–205.

Dunham, R.J. and Nye, P.H., 1973. The influence of soil water content on the uptake of ions by roots 1. Soil water content gradients near a plane of onion roots. *J. Appl. Ecol.*, 10: 585–598.

Dykyjová, D., Ondok, J.P. and Přibáň, K., 1970. Seasonal changes in productivity and vertical structure of reed-stands (*Phragmites communis* Trin.). *Photosynthetica*, 4: 280–287.

Fiala, K., Dykyjová, D., Květ, J. and Svoboda, J., 1968. Methods of assessing rhizome and root production in reed-bed stands. In: M.S. Ghilarov, V.A. Kovda, L.N. Novichkova-Ivanova, L.E. Rodin and V.M. Sveshnikova (Editors), *Methods of Productivity Studies in Root Systems and Rhizosphere Organisms. IBP Symposium, Leningrad*, pp. 36–47.

Forrest, G.I., 1971. Structure and production of North Pennine blanket bog vegetation. *J. Ecol.*, 59: 453–479.

Forrest, G.I. and Smith, R.A.H., 1975. The productivity of a range of blanket bog vegetation types in the Northern Pennines. *J. Ecol.*, 63: 173–202.

Gaastra, P., 1963. Climatic control of photosynthesis and respiration. In: L.T. Evans (Editor), *Environmental Control of Plant Growth.* Academic Press, New York, N.Y., pp. 113–140.

Gardner, W.R., 1960. Soil water relations in arid and semi-arid conditions. In: *Plant-Water Relationships in Arid and Semi-arid Conditions — Reviews of Research.* Arid Zone Research, 15. UNESCO, Paris, pp. 37–61.

Glosser, J., 1977. Characteristics of CO_2 exchange in *Phragmites communis* Trin. derived from measurements in situ. *Photosynthetica*, 11: 139–147.

Golley, F.B., 1961. Energy values of ecological materials. *Ecology*, 42: 581–584.

Gorchakovsky, P.L. and Andreyashkina, N.I., 1972. Productivity of some shrub, dwarf shrub and herbaceous communities of forest tundra. In: F.E. Wielgolaski and Th. Roswall (Editors), *Proceedings of IBP Tundra Biome Meeting, Biological Productivity of Tundra, Leningrad*, pp. 113–116.

Gore, A.J.P., 1961. Factors limiting plant growth on high level blanket peat 1. Calcium and phosphate. *J. Ecol.*, 49: 399–402.

Gorham, E. and Pearsall, W.H., 1956. Production ecology III. Shoot production in *Phragmites* in relation to habitat. *Oikos*, 7: 206–214.

Gorham, E. and Somers, M.G., 1973. Seasonal changes in the standing crop of montane sedges. *Can. J. Bot.*, 51: 1097–1108.

Grace, J. and Marks, T.C., 1978. Physiological aspects of bog production at Moor House. In: O.W. Heal and D.F. Perkins (Editors), *The Production Ecology of British Moors and Montane Grasslands.* Springer-Verlag, Berlin, pp. 38–51.

Grace, J. and Woolhouse, H.W., 1970. A physiological and

mathematical study of the growth and productivity of a *Calluna–Sphagnum* community 1. Net photosynthesis of *Calluna vulgaris* (L.) Hull. *J. Appl. Ecol.*, 7: 363–381.

Grace, J. and Woolhouse, H.W., 1973. A physiological and mathematical study of the growth and productivity of a *Calluna–Sphagnum* community II. Light interception and photosynthesis in *Calluna. J. Appl. Ecol.*, 10: 63–76.

Grace, J. and Woolhouse, H.W., 1974. A physiological and mathematical study of the growth and productivity of a *Calluna–Sphagnum* community IV. A model of growing *Calluna. J. Appl. Ecol.*, 11: 281–295.

Granhall, U. and Lid-Torsvik, V., 1975. Nitrogen fixation by bacteria and free-living blue-green algae in tundra areas. In: F.E. Wielgolaski (Editor), *Fennoscandian Tundra Ecosystems, Part I, Plants and Microorganisms.* Springer-Verlag, Berlin, pp. 305–315.

Haag, R., 1974. Nutrient limitations to plant production in two tundra communities. *Can. J. Bot.*, 52: 103–116.

Hadley, E.B. and Bliss, L.C., 1964. Energy relationships of alpine plants on Mount Washington, New Hampshire. *Ecol. Monogr.*, 34: 331–370.

Harper, J.L. and White, J., 1974. The demography of plants. *Annu. Rev. Ecol. Syst.*, 5: 419–463.

Hsiao, T.C., 1973. Plant responses to water stress. *Annu. Rev. Plant Physiol.*, 24: 519–570.

Ishibashi, A., 1969. Carbon dioxide evolution from standing water in a paddy field. In: M. Monsi (Editor), *Photosynthesis and Utilization of Solar Energy — Level 3 Experiments.* Japanese National Subcommittee for IBP, PP, Tokyo, pp. 27–29.

Jarvis, P.G., James, G.B. and Landsberg, J.J., 1976. Coniferous forest. In: J.L. Monteith (Editor), *Vegetation and the Atmosphere, 2.* Academic Press, London, pp. 171–240.

Jervis, R.A., 1969. Primary production in the freshwater marsh ecosystem of Troy meadows, New Jersey. *Bull. Torrey Bot. Club*, 96: 209–231.

Johansson, L.G. and Linder, S., 1975. The seasonal pattern of photosynthesis of some vascular plants on a subarctic mire. In: F.E. Wielgolaski (Editor), *Fennoscandian Tundra Ecosystems, Part 1, Plants and Microorganisms.* Springer-Verlag, Berlin, pp. 194–200.

Jones, M.B. and Milburn, T.R., 1978. Photosynthesis in papyrus (*Cyperus papyrus* L.). *Photosynthetica*, 12: 197–199.

Kallio, P. and Heinonen, S., 1975. CO_2 exchange and growth of *Rhacomitrium lanuginosum* and *Dicranum elongatum.* In: F.E. Wielgolaski (Editor), *Fennoscandian Tundra Ecosystems, Part 1, Plants and Microorganisms.* Springer-Verlag, Berlin, pp. 138–148.

Kallio, S. and Kallio, P., 1975. Nitrogen fixation in lichens at Kevo, north-Finland. In: F.E. Wielgolaski, (Editor), *Fennoscandian Tundra Ecosystems, Part 1, Plants and Microorganisms.* Springer-Verlag, Berlin, pp. 292–304.

Kallio, P. and Vallane, N., 1975. On the effect of continuous light on photosynthesis in mosses. In: F.E. Wielgolaski (Editor), *Fennoscandian Tundra Ecosystems, Part 1, Plants and Microorganisms.* Springer-Verlag, Berlin, pp. 149–162.

Keefe, C.W., 1972. Marsh production; a summary of the literature. *Contrib. Mar. Sci.*, 16: 163–181.

Kjelvik, S., Wielgolaski, F.E. and Jahren, A., 1975. Photosynthesis and respiration of plants studied by field technique at Hardangervidda, Norway. In: F.E. Wielgolaski (Editor), *Fennoscandian Tundra Ecosystems, Part 1, Plants and Microorganisms.* Springer-Verlag, Berlin, pp. 184–193.

Klopatek, J.M. and Stearns, F.W., 1978. Primary productivity of emergent macrophytes in a Wisconsin freshwater marsh ecosystem. *Am. Midl. Nat.*, 100: 320–332.

Květ, J., 1971. Growth analysis approach to the production ecology of reedswamp communities. *Hidrobiologia*, 12: 15–40.

Longton, R.E., 1970. Growth and productivity of the moss *Polytrichum alpestre* Hoppe in Antarctic regions. In: M.W. Holdgate (Editor), *Antarctic Ecology, 2.* Academic Press, London, pp. 818–837.

McNaughton, S.J., 1966. Ecotype function in the *Typha* community type. *Ecol. Monogr.*, 36: 297–326.

McNaughton, S.J. and Fulkem, L.W., 1970. Photosynthesis and photorespiration in *Typha latifolia. Plant Physiol.*, 45: 703–707.

McVean, D.N., 1959. Ecology of *Alnus glutinosa* (L.) Gaertn. VII. Establishment of alder by direct seeding of shallow blanket bog. *J. Ecol.*, 47: 615–618.

Mason, C.F. and Bryant, R.J., 1975. Production, nutrient content and decomposition of *Phragmites communis* Trin. and *Typha angustifolia* L. *J. Ecol.*, 63: 71–95.

Mathews, C.P. and Westlake, D.F., 1969. Estimation of production by populations of higher plants subject to high mortality. *Oikos*, 20: 156–160.

Monteith, J.L., 1972. Solar radiation and productivity in tropical ecosystems. *J. Appl. Ecol.*, 9: 747–766.

Monteith, J.L., 1976. *Vegetation and the Atmosphere, 2.* Academic Press, London, 439 pp.

Mörnsjö, T., 1969. Studies on vegetation and development of a peatland in Scania, south Sweden. *Opera Bot.*, 24: 187 pp.

Muc, M., 1973. Primary production of plant communities of the Truelove lowland, Devon Island, Canada — sedge meadows. In: L.C. Bliss and F.E. Wielgolaski (Editors), *Proceedings of IBP Tundra Biome Symposium, Primary Production and Production Processes, Dublin*, pp. 3–14.

Newman, E.I., 1969a. Resistances to water flow in soil and plant 1. Soil resistances in relation to amounts of root: theoretical estimates. *J. Appl. Ecol.*, 6: 1–12.

Newman, E.I., 1969b. Resistances to water flow in soil and plant II. A review of experimental evidence on the rhizosphere resistance. *J. Appl. Ecol.*, 6: 261–272.

Newman, E.I., 1973. Root and soil water relations. In: E.W. Carson, (Editor), *The Plant Root and Its Environment.* University of Virginia Press, Charlottesville, Va., pp. 363–440.

Oechel, W.C. and Collins, N.J., 1973. Seasonal patterns of CO_2 exchange in bryophytes at Barrow, Alaska. In: L.C. Bliss and F.E. Wielgolaski (Editors), *Proceedings of IBP Tundra Biome Symposium, Primary Production and Production Processes, Dublin*, pp. 197–203.

Ohtaki, E., Seo, T. and Takasu, K., 1969. Flux of carbon dioxide over a paddy field. In: M. Monsi (Editor), *Photosynthesis and Utilization of Solar Energy — Level 3 Experiments.* Japanese National Subcommittee for IBP, Tokyo, pp. 30–32.

Ondok, J.P., 1973a. Photosynthetically active radiation in a stand of *Phragmites communis* Trin. 1. Distribution of irradiance and foliage structure. *Photosynthetica*, 7: 8–17.

Ondok, J.P., 1973b. Photosynthetically active radiation in a stand of *Phragmites communis* Trin. 2. Model of light extinction in the stand. *Photosynthetica*, 7: 50–57.

Ondok, J.P., 1973c. Photosynthetically active radiation in a stand of *Phragmites communis* Trin. 3. Distribution of irradiance on sunlit foliage area. *Photosynthetica*, 7: 311–319.

Ondok, J.P., 1975. Photosynthetically active radiation in a stand of *Phragmites communis* Trin. 4. Stochastic model. *Photosynthetica*, 9: 201–210.

Overbeck, F. and Happach, H., 1956. Über das Wachstum und den Wasserhaushalt einiger Hochmoor-Sphagnen. *Flora*, 144: 335–402.

Pakarinen, P. and Vitt, D.H., 1973. Primary production of plant communities of the Truelove lowland, Devon Island, Canada — moss communities. In: L.C. Bliss and F.E. Wielgolaski (Editors), *Proceedings of the IBP Tundra Biome Symposium, Primary Production and Production Processes, Dublin.* pp. 37–46.

Parker, G.R. and Schneider, G., 1975. Biomass and productivity of an alder swamp in northern Michigan. *Can. J. For. Res.*, 5: 403–409.

Peacock, J.M., 1975. Temperature and leaf growth in *Lolium perenne* II. The site of temperature perception. *J. Appl. Ecol.*, 12: 115–124.

Pearcy, R.W., Berry, J.A. and Bartholemew, B., 1972. Field measurements of the gas exchange capacities of *Phragmites communis* under summer conditions in Death Valley. *Carnegie Inst. Year Book*, 71: 161–164.

Pearsall, W.H. and Gorham, E., 1956. Production ecology 1: Standing crops of natural vegetation. *Oikos*, 7: 193–201.

Pearsall, W.H. and Newbould, P.J., 1957. Production ecology IV. Standing crops of natural vegetation in the sub-arctic. *J. Ecol.*, 45: 593–599.

Pearson, L.C., 1965. Primary productivity in a northern desert area. *Oikos*, 15: 211–228.

Penfound, W.T., 1956. Primary production of vascular aquatic plants. *Limnol. Oceanogr.*, 1: 92–101.

Polisini, J.M. and Boyd, C.E., 1972. Relationships between cell-wall fractions, nitrogen and standing crop in aquatic macrophytes. *Ecology*, 53: 484–488.

Raghavendra, A.S. and Das, V.S.R., 1978. The occurrence of C4 photosynthesis: a supplementary list of C4 plants reported during late 1974 — mid 1977. *Photosynthetica*, 12: 200–208.

Reader, R.J. and Stewart, J.M., 1972. The relationship between net primary production and accumulation for a peatland in south-eastern Manitoba. *Ecology*, 53: 1024–1037.

Reiners, W.A., 1972. Structure and energetics of three Minnesota forests. *Ecol. Monogr.*, 42: 71–94.

Ripley, E.A. and Redman, R.E., 1976. Grasslands. In: J.L. Monteith (Editor), *Vegetation and the Atmosphere, 2.* Academic Press, London, pp. 349–398.

Rodin, L.E. and Bazilevich, N.I., 1967. *Production and Mineral Cycling in Terrestrial Vegetation.* Oliver and Boyd, London, 288 pp.

Sæbo, S., 1968. The autecology of *Rubus chamaemorus* L. 1. Phosphorus economy of *R. chamaemorus* in an ombrotrophic mire. *Sci. Rep. Agric. Coll. Norway*, 49: 1–36.

Schlesinger, W.H., 1978. Community structure, dynamics and nutrient cycling in the Okefenokee cypress swamp forest. *Ecol. Monogr.*, 48: 43–65.

Schlesinger, W.H. and Chabot, B.F., 1977. The use of water and minerals by evergreen and deciduous shrubs in Okefenokee swamp. *Bot. Gaz.*, 138: 490–497.

Scholander, P.F. and Perez, H., 1968. Sap tension in flooded trees and bushes of the Amazon. *Plant Physiol.*, 43: 1870–1873.

Scholander, P.F., Hammel, H.T., Bradstreet, E.D. and Hemmingsen, E.A., 1965. Sap pressure in vascular plants. *Science*, 148: 339–346.

Scott, D. and Billings, W.D., 1964. Effect of environmental factors on the standing crop and productivity of an alpine tundra. *Ecol. Monogr.*, 34: 243–270.

Skre, O., 1975. CO_2 exchange in Norwegian tundra plants studied by infra-red gas analyser technique. In: F.E. Wielgolaski (Editor), *Fennoscandian Tundra Ecosystems, Part 1, Plants and Microorganisms.* Springer-Verlag, Berlin, pp. 168–183.

Slatyer, R.O., 1967. *Plant–Water Relationships*, Academic Press, London, 366 pp.

Small, E., 1972. Photosynthetic rates in relation to nitrogen recycling as an adaptation to nutrient deficiency in peat bog plants. *Can. J. Bot.*, 50: 2227–2233.

Steward, K.K. and Ornes, W.H., 1975. The autecology of sawgrass in the Florida Everglades. *Ecology*, 56: 162–171.

Stoner, W.A. and Miller, P.C., 1975. Water relations of plant species in a wet coastal tundra at Barrow, Alaska. *Arct. Alp. Res.*, 7: 109–124.

Summerfield, R.J., 1973. The growth and productivity of *Narthecium ossifragum* on British mires. *J. Ecol.*, 61: 717 727.

Tamm, C.O., 1954. Some observations on the nutrient turnover in a bog community dominated by *Eriophorum vaginatum* L. *Oikos*, 5: 189–194.

Tanaka, A., Kawano, K. and Yamaguchi, J., 1966. *Photosynthesis, Respiration, and Plant Type of the Tropical Rice Plant.* The International Rice Research Institute, Manilla, Technical Bulletin 7.

Thom, A.S., 1975. Momentum, mass and heat exchange in plant communities. In: J.L. Monteith (Editor), *Vegetation and the Atmosphere, 1.* Academic Press, London, pp. 57–109.

Thompson, K., Shewry, P.R. and Woolhouse, H.W., 1979. Papyrus swamp development in the Upemba Basin, Zaire: studies of population structure in *Cyperus papyrus* stands. *Bot. J. Linn., Soc.*, 78: 299–316.

Tieszen, L.L., 1972. CO_2 exchange in the Alaskan arctic tundra: measured course of photosynthesis. In: S. Bowen (Editor), *Proceedings of IBP Tundra Biome Symposium, University of Washington, Seattle, Wash.*, pp. 29–35.

Tinklin, R. and Weatherley, P.E., 1968. On the relationship between transpiration rate and leaf water potential. *New Phytol.*, 65: 509–517.

Uchijima, Z., 1970. Carbon dioxide environment and flux within a corn crop canopy. In: I. Setlik (Editor), *Prediction and Measurement of Photosynthetic Productivity.* Centre for Agricultural Publishing and Documentation, Wageningen, pp. 179–196.

Uchijima, Z., 1976. Maize and rice. In: J.L. Monteith (Editor), *Vegetation and the Atmosphere, 2.* Academic Press, London, pp. 33–64.

Unsworth, M.H., 1981. The exchange of carbon dioxide and air pollutants between vegetation and the atmosphere. In: J. Grace, E.D. Ford and P.G. Jarvis (Editors), *Plants and Their Atmospheric Environment*. Blackwell, Oxford, pp. 111–138.

Van der Valk, A.G. and Bliss, L.C., 1971. Hydrarch succession and net primary production of oxbow lakes in central Alberta. *Can. J. Bot.*, 49: 1177–1199.

Walker, J.M. and Waygood, E.R., 1967. Ecology of *Phragmites communis* I. Photosynthesis of a single shoot *in situ*. *Can. J. Bot.*, 46: 549–555.

Watts, W.R., 1972. Leaf extension in *Zea mays*. II Leaf extension in response to independent variation of the temperature of the apical meristem, of the air around the leaves, and of the root zone. *J. Exp. Bot.*, 76: 713–721.

Waughman, G.J. and Bellamy, D.J., 1972. Acetylene reduction in surface peat. *Oikos*, 23: 353–358.

Weatherley, P.E., 1970. Some aspects of water relations. In: R.D. Preston (Editor), *Advances in Botanical Research, 3*. Academic Press, New York, N.Y., pp. 171–206.

Weatherley, P.E., 1976. Introduction: water movement through plants. *Philos. Trans. R. Soc. Lond. B.*, 273: 435–444.

Weaver, J.E. and Zink, E., 1946. Annual increase of underground materials in three range grasses. *Ecology*, 27: 115–127.

Webber, P.J., 1972. Comparative ordination and productivity of tundra vegetation. In: S. Bowen (Editor), *Proceedings of IBP Tundra Biome Symposium, University of Washington, Seattle, Wash.*, pp. 55–60.

Weller, G. and Cubley, S., 1972. The microclimates of the arctic tundra. In: S. Bowen (Editor), *Proceedings of 1972 IBP Tundra Biome Symposium, University of Washington, Seattle, Wash.*, pp. 5–12.

Westlake, D.F., 1963. Comparisons of plant productivity. *Biol. Rev.*, 38: 385–425.

Westlake, D.F., 1966. The biomass and productivity of *Glyceria maxima*. 1. Seasonal changes in biomass. *J. Ecol.*, 54: 745–753.

Whittaker, R.H., 1975. *Communities and Ecosystems*. MacMillan, New York, N.Y., 385 pp.

Wiegert, R.G. and Evans, F.C., 1964. Primary production and the disappearance of dead vegetation on an old field in south-eastern Michigan. *Ecology*, 45: 1–49.

Wielgolaski, F.E., 1975. Primary productivity of alpine meadow communities. In: F.E. Wielgolaski (Editor), *Fennoscandian Tundra Ecosystems Part 1, Plants and Microorganisms*. Springer-Verlag, Berlin, pp. 121–128.

Chapter 9

CHANGES IN WETLAND COMMUNITIES[1]

J.H. TALLIS

INTRODUCTION

Swamp, bog, fen and mire communities in general, by definition, are located in situations of excess water: in low-lying depressions, or on flat or very gently sloping surfaces, where water is constantly at or near the soil surface. In these situations the normal processes of organic decomposition are slowed down or even halted by anaerobiosis, so that the dead macro- and micro-remains of the constituent plants and animals gradually accumulate, layer by layer, over long periods of time, as peat. In this peat there is thus preserved a direct record of the antecedents of the present-day flora and fauna. It has been known for centuries that these antecedent communities often differ markedly from the present-day ones in the plants and animals they contain, and that in many instances the basal peat layers are formed from aquatic plant remains. The work of Weber (1908) and Clements (1916), in particular, focused attention on these sequences of aquatic and wetland plant communities preserved in the peat deposits as representing successive stages in the conversion of open water to dry land — the so-called process of **terrestrialization** (German: *Verlandung*) brought about by **hydroseral succession**. Later work, while substantially modifying the basic premises of Weber and Clements, has consistently emphasised both the changeability of wetland communities and the opportunities they afford for studying changes over long periods of time through the preserved plant remains. The following account accordingly examines documented patterns of change in wetland communities, and some of the possible reasons for these changes.

THE PARAMETERS OF WETLAND COMMUNITIES

This inherent changeability of wetland communities results in large part from their occurrence in environments where a single extremely variable habitat factor, that of water supply, is predominant. The following simple model, based on one proposed by Kulczyński (1949), suggests a mode of operation for this water-supply factor, and outlines some of the ways in which its operation can vary.

The situations where wetland communities occur in a given landscape are determined by the underlying topographic and hydrologic patterns of that landscape. All topographic complexities, the model assumes, can be resolved into two components: discrete catchment areas of sloping ground, funnelling water towards the permanent streams and rivers; and intervening watershed[2] or water-parting areas of flat elevated ground. Within that system the ground-water mass exists as a continuous and more or less convex body of water saturating the permeable subsurface layers, and subject to discharge (via streams and springs) and recharge (via precipitation). The upper surface of this ground-water mass forms the water table, which on watershed areas, where the water mass is static, experiences only small *vertical* oscillations, dependent on the balance between discharge and recharge. Elsewhere within the catchment areas, there is *lateral* flow of ground water towards points of discharge, with gradual concentration into defined flow lines and water courses.

[1]Revised manuscript received June, 1975.

[2]Throughout this chapter, the word "watershed" is used in the British sense of a water-parting, in contradistinction to the American sense of a catchment.

Wetland habitats are located where water comes to lie close to or at the soil surface, but where any water flow is diminished or absent, so that the accumulating peat is not swept away. Wetland communities of differing composition develop in part in response to varying mean levels of the water table either above or below the soil surface, and to varying seasonal patterns of fluctuation of level (which, it is assumed, operate by determining the day-to-day aeration pattern of the upper soil or peat layers). In addition to these effects, the water supply also generates the supply of ions to the habitat. This ionic supply may accordingly be derived from one or more of three principal sources: precipitation (rain water), immobile ground water, and mobile ground water. Rain water contains only minimal quantities of ions, whereas the ground-water mass typically contains larger quantities of ions dissolved out from the rocks with which it is in contact; the precise quantities will depend on the chemical composition of the rocks. Wetland habitats sustained wholly or in part by ground water (*minerotrophic*) must, therefore, receive a greater ionic supply than those sustained wholly or in part by precipitation (*ombrotrophic*); and, moreover, those habitats influenced by mobile ground water will be more strongly minerotrophic than those sustained solely by an immobile ground-water mass, since the flowing waters will ensure continual replenishment of ions.

Differences between wetland communities can, according to this model, be expressed in terms of two major environmental variables, water supply and ionic supply. Any given wetland community has only a limited tolerance range to variations in these variables, as data collected by Spence (1964) makes clear. Spence examined the distributions of twenty-one different wetland communities around more than one hundred lakes and swamps in Scotland, and measured the range of water-table levels over which each community occurred. Each community had a characteristic mean water-table level and a limited tolerance range (Fig. 9.1). This tolerance range was less for fen communities (i.e., communities rooted *above* mean water level) than for swamp communities (i.e., communities rooted *below* mean water level). If long-term changes in water level take place which exceed the tolerance range of a particular wetland community, then changes will occur leading to its replacement by another community — and, from Spence's data, fen communities would appear to be more susceptible to change than swamp communities. In Fig. 9.1, moreover, there is clustering of points, implying that each wetland community has not a unique niche in terms of water supply alone. Two or more communities often have rather similar mean water-table levels and tolerance ranges. It is assumed that, where alternative wetland communities with similar water-level requirements exist, then these communities will differ in terms of their ionic requirements.

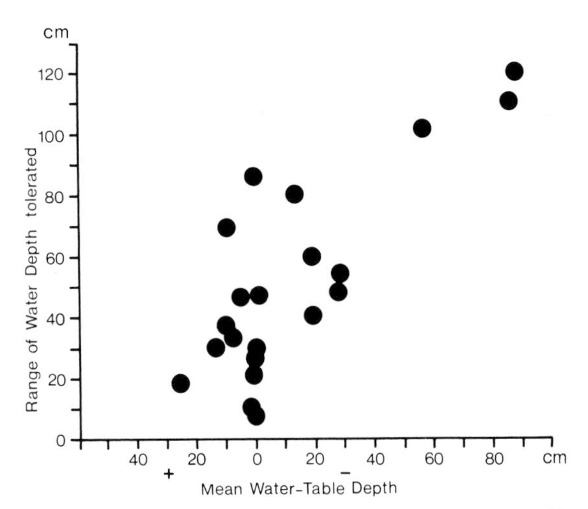

Fig. 9.1. The tolerance ranges of different wetland communities in Scotland to water-table depth, plotted against the mean height above (+) or below (−) the water table at which these communities occur (data of Spence, 1964).

Changes in water level and water movement within a wetland habitat can be brought about in a variety of ways; these changes basically result from modifications of the surface topography or of the hydrological pattern. Some of the modifications may result directly from the growth of the wetland plants themselves (*autogenic* changes), as, for example, where the gradual accrual of peat displaces or diverts the water supply to a particular community. Other modifications may result from the action of external factors (*allogenic* changes), as, for example, changes brought about by long-term trends in climate or by artificial drainage of a habitat. It is customary, following Clements (1916), to regard true vegetational succession as resulting from autogenic processes alone, but in the broader context of vegetation change in wetland communities both autogenic and allogenic processes are necessarily involved. Accordingly, in the following pages documented instances of both types of process will be considered.

FACTORS BRINGING ABOUT CHANGE IN WETLAND COMMUNITIES

Autogenic consequences of peat accumulation

Where a low-lying depression is sunk below the water table, so that a permanent body of open water is present, then the progressive accumulation of organic debris underwater results in a gradual shallowing of the water mass, and ultimately perhaps leads to its total obliteration. In the 1890's Weber examined a large number of such basin sites in northern Germany, and formulated a "typical" sequence for the infilling deposits, the various stages of which could be related, through the preserved macroremains, to present-day wetland communities. These present-day communities characteristically occurred in different, but defined, relationships to the water table: (1) with the constituent plants either free-floating or rooted in deep water (**limnetic** communities); (2) rooted in shallow water and capable of forming peat at or just below mean water level (**telmatic** communities); (3) rooted above the water table, but seasonally flooded (and hence forming peat above water level, **semi-terrestric** communities); or (4) adapted to lowered water tables, intolerant of flooding, and accumulating peat only slowly if at all (**terrestric** communities). The typical sequence of basin deposits formulated by Weber (1908) was then as follows (from below upwards):

lake mud→peat mud (limnetic)→*Phragmites* peat (telmatic)→*Alnus* wood peat (semi-terrestric)→*Pinus–Betula* wood peat (terrestric)→*Scheuchzeria* or *Carex–Sphagnum* peat (telmatic)→*Sphagnum* peat (semi-terrestric).

Weber envisaged that, at some stage in the accrual of peat, the surface layers would emerge above the water, and be totally waterlogged only during periods of high water levels; the surface peat layers would then be exposed in part to an ombrotrophic rather than a minerotrophic water supply. Ultimately the peat mass might build up to such a height that the surface layers were no longer influenced by minerotrophic waters at all — as was possible in the relatively humid climate of northern Germany, where rainfall excess over evaporation combined with the capillary action of the peat mass was sufficient to counterbalance the downward, gravitational movement of water, and produce a locally convex water table sustained by precipitation. The apparent anomaly at the northern German sites examined by Weber of a second telmatic peat layer overlying terrestric peat was explained by him as follows: the slowly accumulating wood peat (terrestric) formed an irregular and largely impermeable surface, so that local flooding occurred after heavy rain; in the small more or less ombrotrophic pools so formed, *Sphagnum cuspidatum* became established, often together with *Scheuchzeria palustris* and *Carex lasiocarpa*. Gradually these local foci of *Sphagnum* growth built up and spread laterally over the woodland, eventually killing the trees and forming a continuous *Sphagnum*-dominated carpet (semi-terrestric). In drier climates than that of northern Germany (e.g. eastern Europe; Kulczyński, 1949), this *Sphagnum* build-up was not possible, and autogenic growth ceased at the terrestric peat stage. In all cases, however, a conversion of an open-water habitat to dry land occurred, solely as a result of the growth of the component plants.

Such was the so-called *Verlandungshypothese* of Weber, which has influenced all subsequent thinking on hydroseral change. In some ways this hypothesis has been too pervasive and, as mentioned elsewhere (pp. 319, 324, 327), the formation of peat directly on to "dry land" (**paludification**) is a very widespread phenomenon. In this present context of change however, the *Verlandungshypothese* is clearly very relevant.

Weber also recognized that allogenic processes might be important during peat accumulation. He showed that the ombrotrophic *Sphagnum* peats in northern Germany were clearly divisible into two regions, with an upper weakly humified peat overlying more highly humified lower peat. The point of contact between the two regions (the *Grenzhorizont*) was sharply defined, with a thin layer of telmatic *Scheuchzeria–Sphagnum cuspidatum* peat abruptly succeeding the intensely humified layers of the lower semi-terrestric peat; tree stumps often occurred in the uppermost layers of this lower peat. Weber interpreted this stratigraphic sequence as resulting from a climatic change towards increased wetness after a prolonged dry period. Frenzel (Ch. 2) critically discusses this and other aspects of the *Grenzhorizont* in the general context of **recurrence surfaces**. Such surfaces show sharp changes from well-humified to relatively undecomposed peat within the profiles of widely separated mires.

At about the same time, workers in Sweden, whilst confirming Weber's general findings (Von Post, 1912), were, however, recording a more complex patterning of the ombrotrophic *Sphagnum* peats there, with overlapping lens-shaped masses of weakly humified peat separated by thin highly humified layers. Von Post and Sernander (1910) suggested that this "lenticular structure" resulted from *autogenic* processes of bog growth (and not from climatic change), which took the form of cyclical changes, with each complete local vegetation cycle represented by a single peat lens. Von Post and Sernander identified appropriate stages in the cycle in the present-day vegetation mosaic at two bog sites in central Sweden. In particular, they described local patches in the vegetation where peat accumulation was slowed down, so that the patches became encircled by actively growing bog vegetation at a higher level, and converted into muddy depressions or open-water pools. These foci of retarded peat growth were typically associated with luxuriant growth of lichens (especially *Cladonia* spp.), mosses, hepatics, or algae. The bare hollows and pools so formed were gradually recolonized by peat-forming plants, and once a continuous *Sphagnum* cover had been established rapid upward growth was resumed. Ultimately peat accumulation slowed down again, and a new **regeneration cycle** was entered. The thin highly humified layers in the peat profiles corresponded to the phases of retarded peat growth, whilst the weakly humified peat comprising the bulk of each peat lens was formed during the periods of rapid upward *Sphagnum* growth. Subsequently Osvald (1923), on the basis of detailed studies at Komosse (a raised-bog complex in southwestern Sweden), formalized these ideas into a system involving the obligatory alternation in time of hummock- and hollow-type communities, which were represented by a corresponding spatial mosaic of hummocks and hollows on the present-day bog surface. The form of this present-day **hummock/hollow complex** varied somewhat on different parts of Komosse and Osvald was able to quantify these differences in terms of varying proportions of a limited number of major plant associations (Table 9.1). The different forms of the hummock/hollow complex appeared to represent different degrees of activity of the current regeneration cycle, and Osvald termed these different forms the *Regenerationskomplex* (the typical facies), the *Teichkomplex* (with a predominance of open-water pools), and the *Stillstandkomplex* (with a predominance of hummock communities).

Osvald's concept of the regeneration cycle can be summarized as follows. Rapid upward growth during hummock-building (the *Calluna–Sphagnum magellanicum* association) results in a progressively increasing dryness of the surface layers, and the ultimate establishment of communities with negligible peat accumulation (*Calluna–Cladonia* association). Cessation of hummock growth is accompanied by rapid upward growth in intervening hollow areas (*Eriophorum vaginatum–S. magellanicum* association), and as the water table is gradually raised in this accumulating *Sphagnum* mass, so the degenerating hummocks become waterlogged and converted into hollows (*Zygogonium–Sphagnum* association). The regeneration cycle idea thus added on to Weber's original conception of successional changes in wetland habitats (i.e. *progressive* change in floristic compostion) a system of *cyclical*

TABLE 9.1

The relative proportions (%) of the major hummock and hollow associations on the northeast part of Komosse, Sweden (data from Osvald, 1923)

	Regenerations-komplex	*Teichkomplex*	*Stillstand-komplex*
Calluna–Cladonia	10.6	19.1	34.8
Calluna–Sphagnum magellanicum	48.6	39.6	18.4
Eriophorum vaginatum–S. magellanicum	15.7	12.6	1.0
Zygogonium hollows	24.2	13.7	45.8
Open-water pools	0.9	15.0	

changes (also autogenic in character), which was capable of maintaining a constant floristic composition in the climax bog vegetation. The idea was enthusiastically taken up by many workers, and claims for peat showing lenticular structure (and thus presumably representing a series of successive regeneration cycles of the vegetation) were made, for instance, for Finland (Eurola, 1962; Tolonen, 1966), Germany (Rudolph and Firbas, 1927; Ernst, 1934), Poland (Kulczyński, 1949), Ireland (Tansley, 1949; Mitchell, 1956), and England (Godwin and Clifford, 1939; Clapham and Godwin, 1948). In only relatively few cases, however, was the evidence based on examination of exposed peat faces rather than on isolated borings, and often subsequent re-examinations of bogs in the same general area have resulted in different interpretations (Walker and Walker, 1961; Tolonen, 1971; Casparie, 1972; see p. 339).

Both the *Verlandungshypothese* of Weber and the regeneration cycle idea of Von Post and Sernander invoke mechanisms of autogenic growth involving progressive upward growth of peat in relation to a water table experiencing small *vertical* fluctuations of level only; but the accumulation of a peat mass does not act solely by displacing water in a vertical direction. Most wetland habitats, at least in the early stages of development, are affected by laterally flowing ground waters. As the peat mass builds up, so these patterns of water flow are diverted or modified — the exact form of the modification depending upon the rate of lateral water movement (Kulczyński, 1949). Under conditions of relatively *rapid* water movement, appreciable quantities of silt are deposited around the bases of the reed-swamp plants, whilst at the same time considerable decomposition of the accumulating plant remains occurs because of the oxygen-charged waters flowing over the peat surface. As a result the peat builds up only slowly and is of high density. Although the water flow is initially directed over the surface of this high-density peat, the flow gradually becomes canalized into definite water courses, between which are areas of peat affected by the mobile waters only during periods of excessive inflow. Accordingly, there is a slow progression towards localized ombrotrophic conditions. Where the rate of water flow within the wetland habitat is initially *low*, on the other hand, peat accumulates rapidly in largely deoxygenated water, and being of low density forms a floating mat. The floating mat acts as a physical impedance to continued water flow, so that the ground waters gradually "back up" and are diverted laterally around the peat dam (Millington, 1954). The peat surface accordingly becomes raised above ground-water influence again, whilst the peat mass comes to form a reservoir above the actual water table, with water held within it against gravity by capillary action (a so-called "perched" water table). Thus in both cases, whether the water flow be fast or slow initially, the continued upward growth of peat ultimately removes large areas of the wetland habitat from the direct influence of moving water, so that they are subjected to inundation by ground water only when the water table within the habitat rises following periods of heavy rainfall. Lateral water movements within the peat mass are thus replaced by vertical oscillations, and largely ombrotrophic conditions come to predominate at the surface.

Pronounced vertical oscillations of ground water can be compensated for in part by the capacity of a peat mass to change in volume according to its water supply, contracting in volume when the supply is lowered and swelling like a sponge when flooded. Peat deposits of appreciable thickness may avoid surface flooding altogether by this process, whereas shallower marginal peats will show periodic submergence and desiccation. The capacity for compensation is particularly well developed in floating mats of low-density peat, where the mat rises and falls with changes in water level. The superficial layers of the mat, therefore, never dry out nor become flooded, but remain uniformly wet throughout the year. Buell and Buell (1941) found that the floating sedge mat in a Minnesota (U.S.A.) bog moved up and down a distance of 40 cm during the course of a year, to compensate for changes in water level. As has already been shown (Fig. 9.1), many wetland communities cannot tolerate fluctuations in water level of this magnitude. Where rooted in the lake bottom or in high-density peat, these communities would thus be eliminated, but where rooted in a floating mat they would be able to survive, because of the compensatory movements in level of the mat. Only when the raft becomes "grounded" by the continued accumulation of peat does this mechanism cease to function.

Climatic change

Upward growth of a peat mass cannot continue indefinitely. Physical factors diminishing the water supply to the surface layers become increasingly important once the peat mass begins to rise above the water table. Continued saturation of the surface layers is then dependent upon water supplied by precipitation and by capillary rise within the peat offsetting water losses by evaporation, by surface run-off, and by gravity — either in relation to a vertically oscillating water table or to lateral flow within the peat as a result of the peat reservoir rising beyond the physical confines of the basin. As soon as the water supply by rainfall and by capillary action becomes insufficient, then the surface layers dry out and the rate of peat accumulation slows down. Ultimately a state of equilibrium between accrual and decomposition of plant residues is reached. Stratigraphically this state of equilibrium is manifested by a layer of highly humified peat or forest peat.

Climatic change towards increased wetness can upset this state of equilibrium by enhancing the water supply to the surface layers, and thus lead to renewed peat accumulation. Stratigraphic studies by Blytt and Sernander in Scandinavia towards the end of the nineteenth century (Sernander, 1908) suggested that at least two such climatic changes towards increased wetness had occurred in the Postglacial period, which could accordingly be subdivided into a number of alternating drier and wetter climatic phases. The drier phases were characterized stratigraphically by terrestric peat accumulation, the wetter phases by telmatic or semi-terrestric peat accumulation. Wetter conditions probably developed around 7000 B.P. and 2500 B.P., the latter date being that suggested by Weber for the formation of the *Grenzhorizont* in northern German bogs (see also Frenzel, Ch. 2). The influence of these climatic changes on the *overall* pattern of hydroseral change during the Postglacial period has recently been demonstrated by Rybníček (1973), in an extensive survey of past and present wetland communities in central Europe. In the early part of the Postglacial period, when meltwaters from the ice sheets provided abundant fresh-water habitats for colonization by aquatic plants, there was a predominance of limnetic and telmatic communities (Fig. 9.2). Sudden and radical changes in the hydroseral processes, which were probably the result of climatic change, are indicated in Fig. 9.2 by the expansion of telmatic and semi-terrestric communities around 9000 B.P., and of bog communities in particular after 7000 B.P.

The very precise dependence of peat growth on climate was first expressed in quantitative terms by Granlund in 1932. Granlund studied the distribution of nearly 700 peat sites in southern Sweden, and showed that their final form could be related to annual rainfall. Ombrotrophic peat never developed in areas receiving less than 460 mm pre-

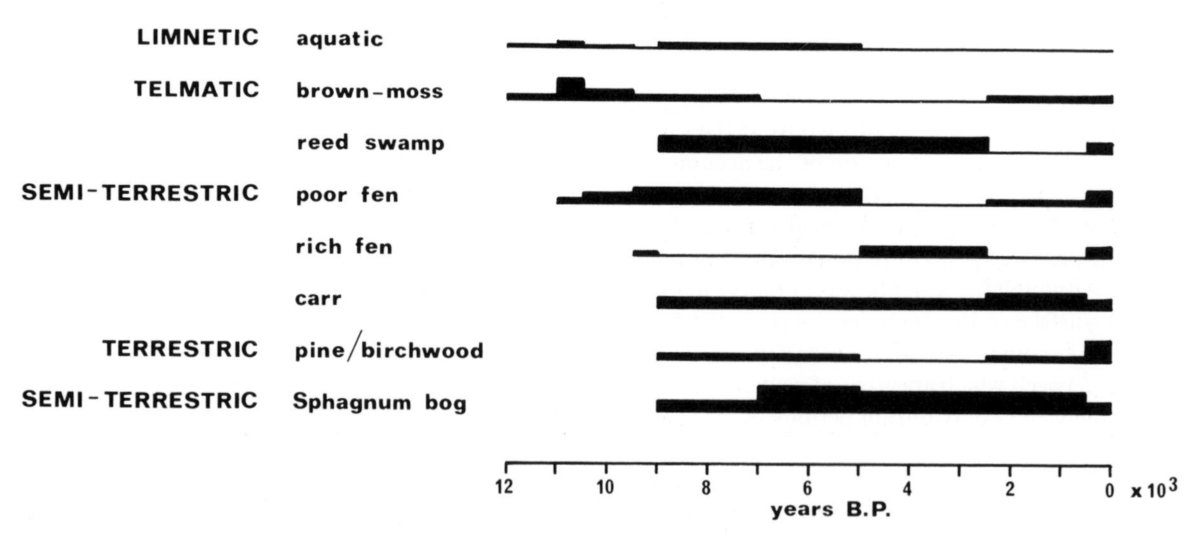

Fig. 9.2. The relative frequencies (arbitrary scale) of different wetland communities in central Europe during the Late- and Postglacial periods (adapted from Rybníček, 1973).

cipitation each year; in areas of higher rainfall, convexly domed peat masses (**cupolas**) were built up above the water table in basin sites to form raised bogs or *Hochmoore*. The degree of convexity of these raised bogs was apparently determined by the annual precipitation, with the flattest cupolas characteristic of the lower-rainfall areas in the southeast of Sweden, while in the higher-rainfall areas farther west the raised bogs were more pronouncedly domed. Thus two bogs of similar diameter, one in a low- and one in a high-rainfall area, would differ not only in the absolute height of the cupola (measured as height above the surrounding mineral ground), but also in the steepness of slope of the bog margins. In a detailed analysis of over 300 raised bogs in the province of Småland, Granlund grouped the sites according to annual rainfall values and, by plotting length of bog in profile against height of cupola, was able to show that for any given rainfall value there was a maximum **limiting height** for the cupola, which was almost never exceeded even in the largest bogs (Fig. 9.3). Granlund postulated that, as a raised bog increased in height, so water losses from the peat surface by evaporation and by run-off downslope increasingly counteracted water supplied by precipitation and by capillary rise from below (which he showed rarely to exceed 50 cm). The surface layers thus became drier, and peat formation slowed down. Over sixty of the sites investigated in detail by Granlund had highly humified surface layers, with negligible peat accumulation, and thus

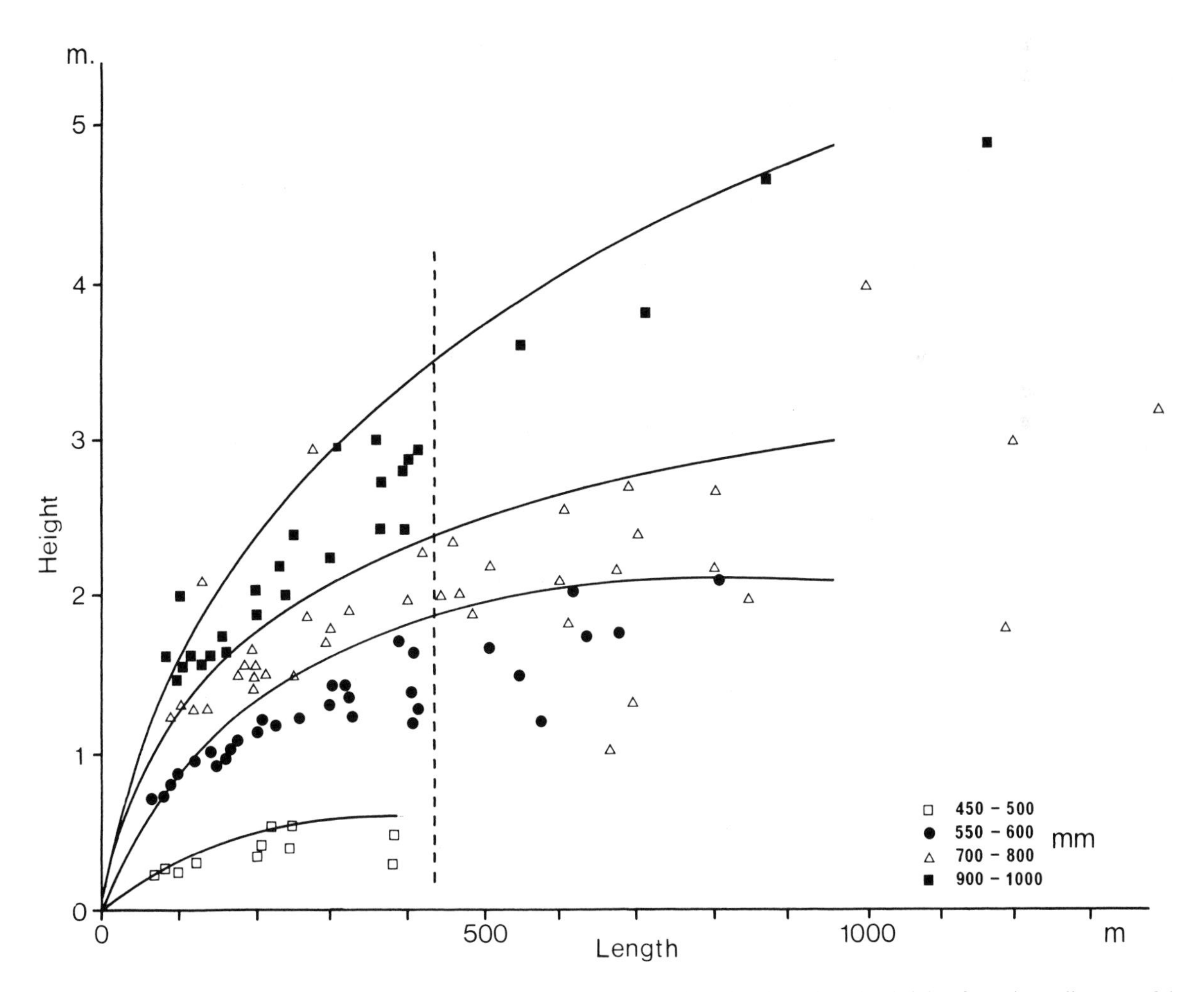

Fig. 9.3. The relationship between degree of convexity of raised bogs in southern Sweden (plotted as height of cupola vs. diameter of the bog basin) and the mean annual rainfall. The data are re-plotted from Granlund (1932), and many values to the left of the broken line have been omitted for the sake of clarity; the continuous lines define so-called "limiting heights" for a given rainfall range.

appeared to have reached "limiting height"; whilst nearly all sites had slower rates of accumulation of peat towards the surface than lower down (see Frenzel, Ch. 2, p. 44).

If this relationship between degree of convexity of a raised bog and the annual rainfall were a true one, then it was to be expected that the equilibrium could be disturbed by climatic change. Granlund found evidence of such disturbances in many Swedish bogs in the form of **recurrence surfaces** — abrupt transitions from highly humified peat to overlying weakly humified peat. He suggested that each recurrence surface represented a rejuvenation of growth in a bog previously stagnating at "limiting height", the rejuvenation being triggered off by a change to a wetter climate. At one site near Stockholm, Snöromsmosse, Granlund found five successive recurrence surfaces, but all the other investigated profiles had four or fewer. These successive recurrence surfaces marked the positions of a series of "limiting heights" for a particular bog at different times in the past, and under different, lower, rainfall conditions (see Frenzel, Ch. 2, p. 44, etc.).

Subsequent workers have found similar stratigraphic sequences in other parts of northern Europe (see references in Dickinson, 1975), but no consistent picture, either of numbers of recurrence surfaces or of approximate datings, has emerged. Thus Overbeck et al. (1957), on the basis of 25 radiocarbon dates, recorded three major recurrence surfaces in seven profiles from northern German bogs, whilst Nilsson (1964) found seven major recurrence surfaces at Ageröds Mosse in southern Sweden. Nilsson brought together more than fifty radiocarbon dates for recurrence surfaces in Sweden, central Europe, and the British Isles, and showed that, although the distribution of dates was far from random, any clustering of dates was rather diffuse and occurred over quite long time intervals (Fig. 9.4). In many of the Poles'ye bogs (U.S.S.R.) studied by Kulczyński (1949), similar recurrence surfaces were also found, with rejuvenated peat growth succeeding on wood peat layers. Kulczyński, however, interpreted these sequences as resulting not from climatic change but from regional rises in the water table caused by other processes (see below). Kulczyński also pointed out that climatic change need not affect all wetland habitats in the same way. A change towards in-

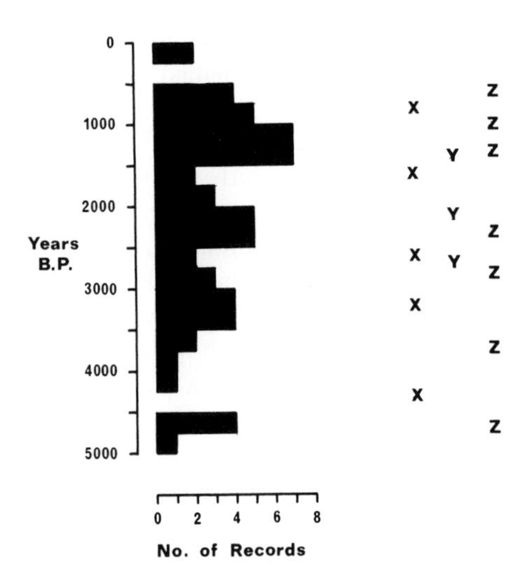

Fig. 9.4. Numbers of records of recurrence surfaces in northern Europe within 250-year radiocarbon-dated intervals (data of Nilsson, 1964), together with dates of recurrence surfaces according to Granlund (1932) (*X*), Overbeck et al. (1957) (*Y*), and Nilsson (1964) (*Z*).

creased wetness of climate which produced renewed *Sphagnum* growth in southern Swedish bogs might, in the higher rainfall regions of western Britain (where precipitation consistently exceeds evaporation throughout the year), result in peat erosion as a consequence of increased surface run-off of waters downslope. Likewise in wetland habitats affected by mobile ground waters, increased inflow resulting from a change to a wetter climate could create conditions of more rapid flow unfavourable for continued peat accumulation.

Large-scale topographic changes

Much of the earlier work on change in wetland communities was based on the assumption that each wetland habitat was a largely isolated entity, unaffected by changes occurring elsewhere. Kulczyński's work in Poles'ye (1949) probably represented the first attempt to consider wetland habitats as part of an integrated and evolving landscape, in which long-term topographic and hydrologic changes played a major role in determining local vegetation changes. The landscape of Poles'ye (the Pripet Marshes of some geographers) is one of subdued relief, based on a wide flat basin formed of extremely permeable glacial sands; some 20% of the land is peat-covered. The water table follows the surface configurations of the landscape, being gently domed below watershed regions, though the height of the dome never exceeds a few metres (Fig. 9.5). Because of the uniformly permeable substrate enclosing the ground waters, any local changes in water table lead to readjustments of the whole regional water table. Since wetland habitats in Poles'ye occur only where the water table lies at or above the mineral soil surface, regional changes in water-table levels can lead to correlated changes in all the individual wetland habitats. These regional changes can involve either a rising or a falling water table.

Kulczyński showed that regional rises in the water table could occur as a result of the accumulation of peat in the lower valleys, and the gradual impedance of water flow and the "backing up" of ground waters. As the water table rose in the valleys, so the water table in adjacent watershed areas would rise also, and affect wetland habitats there (Fig. 9.5). According as this rise was progressive or step-wise, so peat growth in the watershed areas varied also. Cessation of peat growth in the valleys, either from autogenic or allogenic causes, resulted in a corresponding cessation of growth in the watershed systems, and the formation of "limiting horizons". These were analogous to Granlund's recurrence surfaces, but originated in an entirely different way. In the Lake Agassiz Peatlands Natural Area, Minnesota, U.S.A., Heinselman (1970) found regional rises of the water table of up to 6.7 m since peat formation started 8000 years ago.

Pearsall (1950) has shown how, in upland Britain, the high rainfall leads inevitably towards soil podzolization and peat accumulation. On flat or gently sloping ground the formation of an impermeable pan layer in the soil results in impeded drainage, seasonal or permanent waterlogging, and the initiation of peat accumulation. Once bog conditions become established at any point on a hillside, then there is an inexorable tendency for bog seepage to extend downwards and to produce widespread degeneration and waterlogging in the soils below. Tree remains in these marginal peats show that this extension of bog downslope is often at the expense of woodland, and, while not all tree remains in peat can be interpreted in this way (Tallis, 1975), the process is essentially one of **paludification** referred to above.

Over the long periods of time involved in peat accumulation, geomorphological changes in the landscape may have noticeable effects. Headward erosion of streams, in particular, may lead to large-scale changes in the drainage patterns. As a result of stream capture, for example, directions of water flow may be reversed or checked altogether. Downcutting of streams, by lowering the regional water table, may lead to cessation of growth in valley bogs — and ultimately, too, in watershed bogs. Kulczyński (1949) cited a number of instances of geomorphological change in the Poles'ye landscape, including one where river capture appears to be actually in progress at the present day. The headwaters of the River Zehulanka flow southwards and then eastwards in a flat trough filled with a well-developed bog complex (Fig. 9.6). The two most westerly affluents discharge into the north side

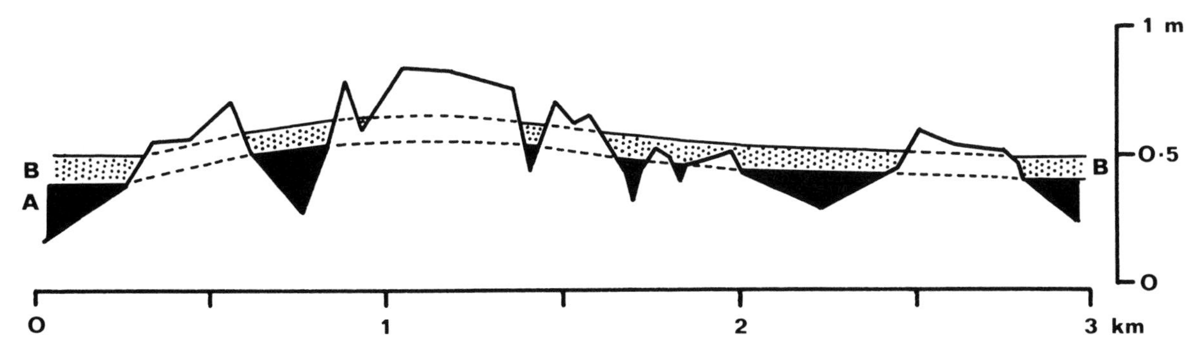

Fig. 9.5. The effects of regional rises in the water table in Poles'ye (redrawn from Kulczyński, 1949). The broken lines represent two successive positions, *A* and *B*, of the regional water table; peat deposits formed in relation to water table *A* are shown in black. Continued peat accumulation in the valleys (stippled) results in a rise in the regional water table (to *B*), and leads to further peat formation in the water-parting area also.

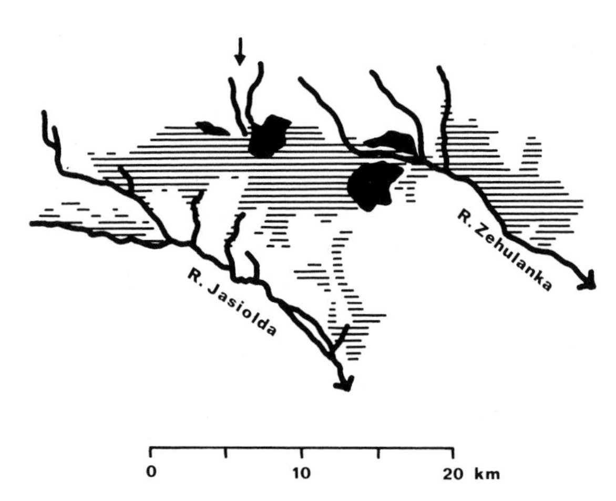

Fig. 9.6. River capture in Poles'ye (redrawn from Kulczyński, 1949). The two tributary streams at top centre (arrowed) formerly flowed into the River Zehulanka, but now flow southwest through the peat mass into the River Jasiolda (Yasel'da). Ombrotrophic peat areas are shown black, and the major minerotrophic peat areas by horizontal lines.

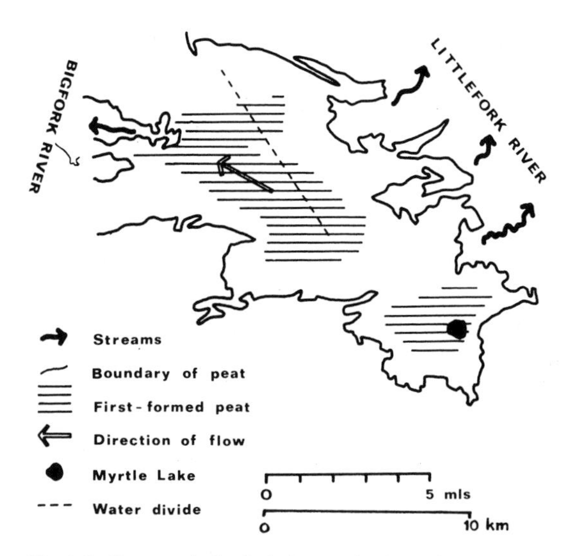

Fig. 9.7. Geomorphological changes in the Lake Agassiz Peatlands (redrawn from Heinselman, 1970). For explanation see text.

of this peat mass, but no longer drain out eastwards to the Zehulanka. Instead the waters flow southwestwards through the peat towards the River Jasiolda (Yasel'da), which is here more deeply incised than the Zehulanka. The central parts of the bog complex, formerly developed under a west–east water flow, are now becoming deprived of mobile ground waters, and consequently ombrotrophic conditions are beginning to develop at the surface.

In Minnesota, Heinselman (1970) demonstrated similar geomorphological changes affecting the Lake Agassiz Peatlands. The Peatlands are formed on a more or less flat plateau which is today drained by two rivers — the Bigfork River to the west and the Littlefork River to the northeast (Fig. 9.7). The plateau slopes unidirectionally northwestwards towards the Bigfork River, and when widespread peat formation commenced *c.* 8000 years ago this was largely confined to the more or less flat trough sloping in that direction. Here minerotrophic peats developed, under the influence of mobile ground waters. About 3000 years ago affluent streams of the Littlefork River eroded headwards to the northeast peat margin and created a two-way drainage system of the Peatlands. A northwest/southeast water divide was thus brought into being which was no longer affected by mobile ground waters, and oligotrophic peat accordingly accumulated here. Genuine ombrogenous bog is unlikely in this region due to the narrow excess of precipitation over evaporation.

Geomorphological changes of a more catastrophic nature have been invoked in a few instances to explain stratigraphical peculiarities. Thus, Kulczyński (1949) has explained anomalies in the peat profiles of the Degerö Stormyr bog complex in northern Sweden as resulting from a limited period of erosion following tectonic upheaval. The peculiar morphology and stratigraphy of certain small basin sites in the West Midlands of England, it has been suggested, are the result of local land subsidence caused by ground-water solution of underlying saliferous strata (Green and Pearson, 1968, 1977; Tallis, 1973). Changes in the relative levels of land and sea, finally, are known to have led to the submergence of some coastal habitats below the sea, and to the gradual emergence of others above the sea (Godwin, 1943). Taylor (Vol. B, Ch. 1) gives further details of the results of this type of influence.

Anthropogenic changes

Over the last one hundred years anthropogenically induced changes in wetland ecosystems have been widespread. In densely populated countries few wetland ecosystems have escaped disturbance in one form or another, and this disturbance often superimposes a new dimension of change on to those considered above, since secondary wetland communities may be formed which have no exact parallels in the stratigraphic record. The concept of hydroseral development as formulated by Clements (1916) is particularly valuable in such cases, because it is often desirable to predict what the future outcome of a given anthropogenic disturbance may be. A knowledge of the major pathways of change in wetland systems, and how these are related to environmental changes, may enable such a prediction to be made.

Anthropogenic disturbance can operate in a variety of ways, but it is possible to distinguish two basically different methods of operation:-

(1) Disturbances which bring about changes in the *existing* parameters of the system (water supply and water chemistry). In this category one can include drainage, peat cutting, sporadic burning, lumbering, and eutrophication.

(2) Disturbances which impose an entirely *new* parameter of change on to the system, for example, grazing, mowing, and regular burning.

Drainage of a wetland ecosystem by ditching is usually undertaken with the aim of lowering the water table and drying out the surface peat layers; colonization by trees and shrubs almost invariably follows drainage, and may merely represent an acceleration of the normal pathways of hydroseral change. Similarly, artificially induced changes in the water levels of a lake may affect the marginal wetland communities solely by advancing or retarding the hydroseral pathway. Following drainage, however, the peat mass may shrink in volume, and develop a more or less impermeable dried-out surface skin, which restricts water uptake and modifies the autogenic responses of the peat mass to changes in water supply. In extreme cases erosion of the peat mass may set in, as a result of increased surface run-off. Inefficient maintenance of drainage ditches may lead to unforeseen consequences. Thus, Green and Pearson (1968) described how, at Wybunbury Moss, Cheshire (England), shrinkage of the peat following drainage, together with blockage of the main outflow ditch, resulted in drainage waters accumulating in the peripheral ring ditch and becoming ponded on to the mire surface along the radial drainage ditches. As a result of run-off from the surrounding agricultural land, strongly minerotrophic waters were thus introduced secondarily on to an oligotrophic *Sphagnum* carpet.

Drainage may also accompany **peat cutting**. At its simplest, peat cutting results in an accentuation of the microrelief of a wetland ecosystem, producing water-filled hollows and elevated drier baulks or ridges. When peat cutting ceases, normal bog growth may be resumed. If, however, the peat surfaces exposed at lower levels by the cutting differ in chemical composition from the original surface (which they often do), then new secondary communities will develop. At Woodwalton Fen, Huntingdonshire (England), Poore (1956) showed how peat cutting in the past had produced an irregular peat surface composed of a mosaic of minerotrophic and oligotrophic areas. These different areas had been recolonized by contrasting vegetation complexes. On the minerotrophic peats, *Calamagrostis epigejos*, *Filipendula ulmaria*, *Galium palustre*, *Salix atrocinerea* and *Urtica dioica* were the major species; on the oligotrophic areas the major species were *Betula* spp., *Calluna vulgaris*, *Cladium mariscus*, *Molinia caerulea* and *Myrica gale*.

Lumbering has been particularly characteristic of North American wetland communities, where a "climax" vegetation of commercially utilizable tree species is widespread. Tree removal involves more than just the setting-back of the succession by one stage, and the permitting of light-demanding species to return, since by reducing losses of water by evapotranspiration tree removal can increase the wetness of the surface peat layers. Thus Kulczyński (1949) described how deforestation of the large bog Mak in Poles'ye fifty years ago resulted in renewed *Sphagnum* growth on a surface where previously active peat accumulation had ceased. Tree felling can result in local soil erosion, and lead to increased quantities of nutrients entering the drainage waters (and hence to local eutrophication). Lumbering too is often accompanied by burning, to dispose of unwanted brushwood, etc., and if so the whole wetland ecosystem may be considerably altered.

The term "eutrophication", in its initial usage, meant "lake enrichment owing to any and all nutrient substances", and as such could refer to natural long-term ageing of lakes as well as to anthropogenically induced changes. Mechanisms of eutrophication are discussed by Lund (1971). Increased ionic supplies, resulting from eutrophication, must necessarily modify patterns of change in wetland communities (see later, p. 343).

The effects of **burning** on wetland ecosystems depend on its severity. If burning occurs when the surface peat layers are wet, then only superficial removal of the above-ground vegetation may result, and natural regeneration from the undamaged rooting parts quickly follows. Under dry conditions, however, or when the fire is particularly fierce, oxidation of the surface peat layers can occur, with consequent total death of many plants. Burning seldom affects a wetland area uniformly, because of local variations in the wetness of the surface peat layers. Local drier areas may undergo deep burning, continuing to smoulder for several weeks, so that a final very irregular surface is produced, with frequent depressions which quickly fill with water. An impermeable skin to the peat may also result, as with drainage. In addition, the resulting surface deposit of plant ash gives rise to temporarily increased nutrient levels, particularly of phosphate. Consequently, after severe burning there may be an initial recolonizing stage dominated by "nitrophilous" species (e.g. *Urtica dioica*) and by "weedy" plants such as *Chamaenerion angustifolium* and (in North America) *Aster* spp. and *Solidago* spp. (Vogl, 1969). Periodic controlled burning is widely practised in upland pastureland in Britain to promote new plant growth for grazing animals, and may also affect adjacent peat communities. Such regular burning may lead to temporary stability of wetland communities, but with the probability of long-term degradation of the habitat, through nutrient losses and soil and peat erosion (Gimingham, 1972).

The imposition of long-continued **mowing** or **grazing** practices on to a wetland community, by contrast to the anthropogenic influences previously described, introduces a new major environmental factor to the system. Wetland communities in many countries are regularly utilized for the provision of hay or thatching materials; this mowing is most widely carried out on telmatic or semi-terrestric communities, and can lead to the suppression or even elimination of the normal dominant plants of these hydroseral stages, and their replacement by other species better adapted to regular mowing (Godwin, 1929; Frolik, 1941). Later hydroseral stages may similarly be regularly grazed, with the consequent spread of graminoid species (Pearsall, 1950). Regular mowing and grazing both tend to override inherent processes of change produced by the operation of the hydrological factors, and to promote temporary vegetation stability. Normally, however, the parameters of water supply and ionic supply are not affected to any extent, so that discontinuance of grazing or mowing allows the normal pathways of change to be resumed (Godwin et al., 1974).

DOCUMENTED PATTERNS OF CHANGE IN WETLAND COMMUNITIES

The investigation of stratigraphic sequences

The possibility of investigating successional trends in wetland communities by the systematic examination of a large number of sediment profiles (as opposed to the previous practice of examining a few, perhaps atypical, profiles) was first developed by D. Walker in 1970. Walker examined published accounts of stratigraphic sequences at over forty sites in the British Isles, and recorded the frequencies of occurrence of changes from one particular type of sediment to another. The sediments were classified into twelve types, each corresponding to a different present-day wetland community. Walker found that certain pathways of change occurred more frequently than others, and that these changes could be interpreted as representing progressive hydroseral development. Walker's original data are summarized in Fig. 9.8 where the number of different sediment types has, for simplicity, been reduced to five:

(1) Lake mud (limnetic) deposits, accumulating in deep water in which floating-leaved and submerged macrophytes are typically growing.

(2) Swamp mud (telmatic) deposits, accumulating in shallow water below reed-swamp communities.

(3) Fen peat (semi-terrestric), accumulating below sedge–tussock and herbaceous fen communities.

(4) Wood peat (semi-terrestric), formed below swamp and fen carr.

(5) Bog peat, largely composed of *Sphagnum*.

These five sediment types correspond to successive stages in Weber's terrestrialization scheme, so that if hydroseral changes really have occurred, then the predominant pathways of stratigraphic change should be in the one, "forward" direction (i.e. from lake mud towards bog peat), and not in the reverse direction (i.e. from bog peat towards lake mud). Fig. 9.8 shows that this is indeed the case.

There is, as Walker pointed out, no single preferred pathway of terrestrialization, though the majority of sequences involve a reed-swamp stage, and most, if not all, are directed towards a final ombrotrophic bog stage (as Weber surmised). Walker showed that the full sequence of stages proposed by Weber, with relatively late invasion by *Sphagnum* spp., was most likely to occur in large inland lake basins; in small kettle-type basins, on the other hand, invasion by *Sphagnum* typically occurred at an early stage, following directly on either the open water or reed-swamp stages, and giving rise to a floating raft of vegetation.

In North America most suggested terrestrialization schemes also involve a floating raft of vegetation (though composed principally of *Carex* spp. in the early stages), and the available stratigraphic data, though more limited in quantity than for the British Isles, support this contention. Tables 9.2 and 9.3 and Fig. 9.9 summarize the stratigraphic sequences recorded from 36 profiles distributed equally between the New England States (Connecticut, Maine, Massachusetts, Vermont), the Lake States (Michigan, Minnesota, Wisconsin), and the north

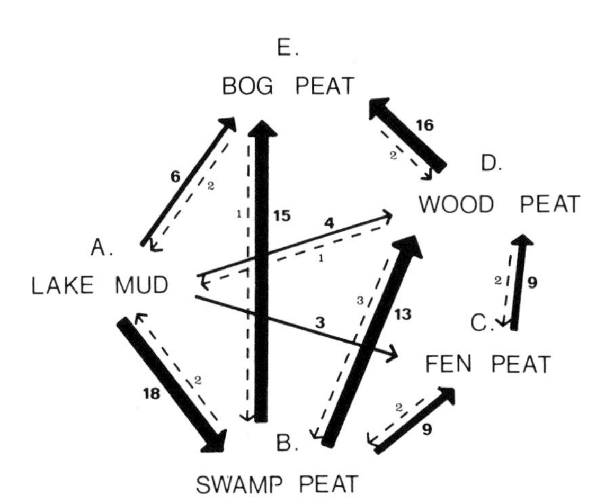

Fig. 9.8. Stratigraphic sequences in British peat profiles (data from Walker, 1970). The arrows connect pairs of superposed strata in published profile descriptions; interrupted arrows indicate reversals of the normal hydroseral sequence. The number against each arrow gives the number of recorded instances of that particular transition.

Pacific coast (British Columbia, Washington). At all these sites, in contradistinction to some of the sites examined by Walker, a supposedly climax vegetation had developed — either bog forest or *Hochmoor* (Table 9.2). The classification of peat types necessarily differs somewhat from that adopted above for the British Isles, and is as follows:

(1) Lake mud (limnetic) deposits.

(2) Sedimentary peat — finely comminuted organic debris accumulating below a floating raft of vegetation.

(3) *Phragmites* peat (telmatic).

(4) Moss peat (telmatic), with predominant remains of hypnoid mosses.

(5) Sedge peat (semi-terrestric), typically produced by a floating raft of vegetation.

(6) Forest peat (semi-terrestric) — typically formed of *Sphagnum* or Ericaceae, with abundant woody remains.

(7) *Sphagnum* peat (semi-terrestric), without woody remains.

Table 9.3 and Fig. 9.9 (which includes both "forward" and "reverse" successions) show that in North America, as in the British Isles, there is no

TABLE 9.2

The present-day vegetation of the 36 North American sites from which stratigraphic sequences were examined (numbers of sites)

	Bog forest	*Sphagnum* bog
New England States	5	7
Lake States	6	6
North Pacific coast	0	12

The data are derived from the following sources: Dachnowski (1925, 1926); Dachnowski-Stokes (1927); Osvald (1935, 1955); Rigg and Richardson (1938); Rigg (1940); Conway (1949); Davis (1958); Janssen (1967); and Heinselman (1970).

TABLE 9.3

The numbers of North American sites at which different initial hydroseral stages have been recorded

	Lake mud	Forest peat	Sedge peat	Moss peat
New England States	7	2	1	2
Lake States	3	4	4	1
North Pacific coast	11	–	1	–

The sources for the data are as in Table 9.2.

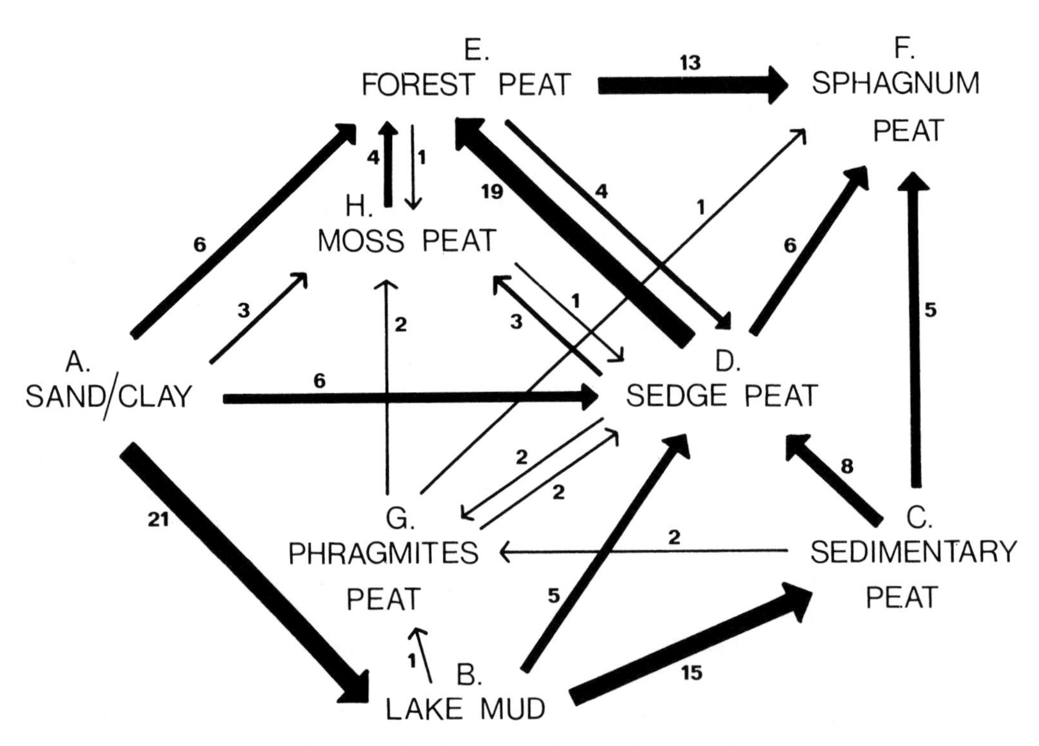

Fig. 9.9. Stratigraphic sequences in peat profiles from North America. The arrows connect pairs of superposed strata in published profile descriptions; the number against each arrow gives the number of recorded instances of that particular transition.

single preferred course of development. Development is again almost exclusively progressive, and involves both **terrestrialization** and **paludification**. It is clear, however, that a reed-swamp stage has never figured prominently in the course of terrestrialization in North America, being replaced largely by a floating sedge mat, whilst the typical carr genera in Britain, *Alnus*, *Betula*, and *Salix*, are largely replaced by species of *Larix*, *Picea*, and *Thuja*.

The very detailed stratigraphical records published by Tolonen (1967) for a wide variety of peat sites in southeastern Finland reveal very different patterns of change again, which are almost certainly a function of the more northerly latitudes. In southeastern Finland some 30% of the land surface is peat-covered, but only exceptionally has there been an initial open-water stage (only five of Tolonen's 35 profiles have basal limnetic deposits); similarly only four profiles show tree remains at the base (pine or birch). In the vast majority of cases minerotrophic wetland communities, apparently similar in composition to those occurring today, developed directly above the glacial sands — another example of paludification. Table 9.4 shows the macro-remains recorded at the bases of Tolonen's profiles. Only two species in this Table (*Carex limosa* and *Scheuchzeria palustris*) occur today in southeastern Finland in ombrotrophic commu-

TABLE 9.4

Numbers of records of different species from basal peat deposits in Finland (thirty sites examined by Tolonen)

Species	Records
Carex spp. (*C. chordorrhiza*, *C. lasiocarpa*, *C. limosa*, *C. rostrata*)	24
Equisetum fluviatile	22
Menyanthes trifoliata	11
Scorpidium scorpioides	10
Sphagnum teres	9
Drepanocladus spp.	7
Acrocladium stramineum	7
Sphagnum subsecundum	6
Bryales	5
Phragmites australis	4
Sphagnum, Series *Cuspidata*	3
Sphagnum papillosum	2
Scheuchzeria palustris	2
Paludella squarrosa	2

nities; the majority are characteristic of "poor fen" (weakly minerotrophic) conditions today. Two species (*Scorpidium scorpioides* and *Sphagnum teres*) occur in eutrophic situations. Although *Carex lasiocarpa* and *C. chordorrhiza* figure prominently in both the Finnish and the North American profiles, their behaviour is very different in the two regions; there is no evidence that in southeastern Finland a floating sedge mat composed of these species was ever formed. Rather, they were direct colonizers of mineral soil influenced by mobile ground waters.

Most of the upland blanket peat in northern Britain similarly develops directly on mineral soil, but there are few detailed records of the "pioneer" species. Normally the peat overlies strongly oligotrophic communities dominated by *Calluna vulgaris*; tree remains (principally birch and pine) may be prominent in or below the peat on steeper slopes (Tallis, 1975). Mrs. J. Bostock (unpublished data, 1975) has recorded a variety of macro-remains from the basal peat layers on flatter areas of the Berwyn Mountains, Wales, including the following: *Aulacomnium palustre*, *Calluna vulgaris*, *Carex* spp., *Drepanocladus fluitans*, *D. revolvens*, *Eriophorum* spp., *Juncus effusus*, *J. squarrosus*, *Polytrichum commune*, *Racomitrium* spp., *Rhytidiadelphus squarrosus*, *Sphagnum* spp. (*S. capillifolium*, *S. cuspidatum*, *S. fuscum*, *S. palustre* and *S. recurvum*). Charcoal was widespread in the basal peat layers, as in the southern Pennines (Tallis, 1975), indicating that regular burning could have been a factor influencing the development of these communities. The possibility that blanket peat formation in upland areas of Britain was not an entirely natural process, but was triggered off by woodland clearance, has recently been suggested by Moore (1973, 1975) and others including Taylor (Part B, Ch. 1).

Kulczyński's (1949) investigations in eastern Poland revealed great stratigraphic diversity in the bogs there, but as mentioned earlier the patterns of hydroseral development were largely governed by regional variations in the water table, and not by autogenic processes within a single basin. For tropical and subtropical countries stratigraphical data on wetland sequences are few at present (see Anderson, Part B, Ch. 6) and generalizations are not yet possible.

The present-day spatial relationships of wetland communities

The pathways of apparent hydroseral change observable in peat stratigraphic profiles can often be paralleled fairly precisely by the concentric zoning of present-day wetland communities around bodies of open water. The sequence of communities in space is accordingly inferred to reflect, or recapitulate, the sequence of communities in time. This inference may subsequently be confirmed by long-term observations of vegetation change at the site, or by the investigation of the antecedent communities from peat borings. In the following paragraphs an attempt is accordingly made to interpret some of the major stratigraphic sequences shown in Figs. 9.8 and 9.9 in terms of described vegetation zonations. Only examples of fairly complete vegetation zonations are quoted; numerous cases of partial zonations are described in the literature, but the piecing together of such fragments to form supposedly complete zonations must necessarily be a somewhat speculative process.

A → B → C → D (Fig. 9.8)

The "typical" zonation of wetland communities around open water in Europe, as elaborated by Warming (1896) and many subsequent workers, involved a sequence of stages as follows:

(1) Submerged aquatics, such as *Myriophyllum* and *Chara* spp.

(2) Floating-leaved aquatics, such as *Nuphar* and *Potamogeton* spp.

(3) Reed-swamp, dominated initially by *Phragmites australis*, *Typha* spp., or *Scirpus lacustris*, and peripherally by *Iris*, *Equisetum*, *Menyanthes*, etc.

(4) Fen communities dominated either by large tussock-forming sedges such as *Carex paniculata*, or by smaller *Carex* spp. and emergent herbaceous plants.

(5) Fen woodland or carr, dominated by *Alnus glutinosa* or *Salix* spp.

(6) Mesophytic deciduous woodland.

Present-day zonations of vegetation in Great Britain approximating to Warming's scheme have been described from, for instance, Esthwaite Water (Pearsall, 1918) in the Lake District, Calthorpe Broad in East Anglia (Godwin and Turner, 1933), and Sweat Mere in Shropshire (Sinker, 1962).

Fig. 9.8 shows that a corresponding sequence of peat types (A→B→C→D) forms one of the major pathways of stratigraphic change in peat profiles from the British Isles; but Tansley (1949) also recognized a number of other alternative pathways of apparent hydroseral succession, based largely on present-day vegetation zonations and determined by regional climatic differences. These are summarized in Fig. 9.10.

A→B→D (Fig. 9.8)

Comparison of Figs. 9.8 and 9.10 reveals one major discrepancy: the absence in Tansley's scheme of the sequence A→B→D, which is so widespread in the peat profiles. The diagrams of vegetation zonations around Scottish lakes published by Spence (1964) do indeed show, in some instances, a zone of fen carr, or even bog vegetation, succeeding directly on the reed-swamp zone without any intervening fen zone (e.g. Loch Marlee). In lowland Britain it is possible that some sort of fen zone was present, but was largely unrepresented (or not recognized) in the peat profiles. Thus in many of the Norfolk Broads *Phragmites* remains a prominent constituent of the fen vegetation, and the trees in the surrounding carr zone are rooted in a floating and semi-liquid swamp peat (Lambert, 1965). At Sweat Mere the alder carr succeeds a zone of swamp carr, here of *Carex paniculata* tussocks bearing *Salix cinerea* bushes, and developed on a floating raft of *Typha angustifolia*.

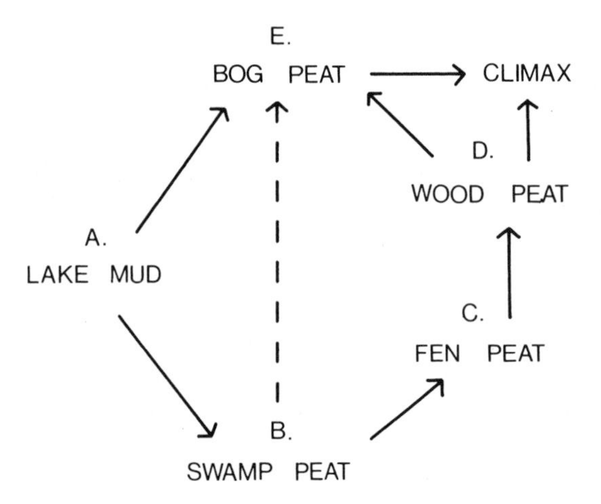

Fig. 9.10. The inferred relationships of peat types in Great Britain, based on present-day vegetation evidence (from Tansley, 1949).

A→B→(C)→D→E (Fig. 9.9)

In some Scottish lochs the reed-swamp stage is replaced by a floating sedge mat, dominated by *Carex rostrata* or *C. lasiocarpa*; this sedge mat passes outwards into either a *Sphagnum–Myrica* bog zone, or a fen zone with prominent *Deschampsia cespitosa* and *Filipendula ulmaria* (Spence, 1964). A similar floating sedge mat, often dominated also by *C. lasiocarpa*, and abutting on open water, occurs widely in eastern North America. Below the floating sedge mat there is an accumulation of finely particulate or colloidal organic debris (often forming a loosely flocculated "false bottom"), but towards the margin of the basin the mat is "grounded" under its own weight. Dansereau and Segadas-Vianna (1952) described a generalized sequence of vegetation zones as follows:

(1) A pioneer stage, consisting of a floating graminoid mat dominated by *Carex* spp., *Eriophorum* spp., or *Menyanthes*, over open water, with subsequent invasion by *Andromeda* and *Myrica*.

(2) The consolidation of this floating mat by the formation of a continuous carpet of *Sphagnum*, in which first ericaceous shrubs (*Chamaedaphne*, *Kalmia*, and *Ledum*), and then, later, taller woody plants (*Alnus rugosa*, *Nemopanthus*) become established.

(3) The gradual development of a forest cover, starting with scattered individuals of *Larix laricina* or *Picea mariana* throughout the *Sphagnum* carpet, and passing outwards into dense forest dominated by *Abies balsamea*, *Acer saccharum*, or *Betula* spp.

Conway (1949) gave *Carex lasiocarpa*, *C. prairea*, *Decodon verticillatus* and *Typha latifolia* as characteristic of the pioneer stage in Minnesota, but pointed out that a bog forest zone might follow on directly from this pioneer mat without the intervention of *Sphagnum* or *Chamaedaphne*. *Betula glandulosa* var. *glandulifera*, when present, replaced *Alnus* and *Nemopanthus* in the tall scrub stage.

A→B→C→D→F (Fig. 9.9)

Rigg (1925) has described a similar sequence of stages from the Pacific coastlands of North America, but with somewhat different participating species. *Carex* spp., *Menyanthes*, *Potentilla palustris*, or *Typha latifolia* are characteristic of the pioneer floating mat, whilst *Kalmia*, *Ledum*, and

Vaccinium oxycoccos are the major shrubby species. *Sphagnum* bog appears to culminate the sequence at the present day but there is some evidence that ultimately trees may invade the bog, with *Tsuga heterophylla*, *Pinus* spp., and *Thuja plicata* succeeding each other towards a climax forest dominated by *Pseudotsuga menziesii*.

A→B→C→D→E (Fig. 9.9)
Frolik (1941) described a comparable "bog sere" from Wisconsin, but he also described an alternative "hydrosere", characterized by successive zones of aquatic plants, reed-swamp (with *Typha latifolia*, *Scirpus validus*, *S. acutus*, or *Phragmites australis*), sedge meadow (dominated by *Carex* and *Calamagrostis* spp.), scrub (with *Salix* spp., *Cornus stolonifera*, and *Betula pumila*), and finally *Populus tremuloides* forest. This sequence, comparable to the classic European sequence described earlier, figures less prominently in American literature than the so-called "bog sere".

A→B→E (Fig. 9.8)
Direct overgrowth of open water by a floating mat of *Sphagnum* spp., in a fashion analogous to that of the American sedge mat, is suggested by the vegetation zonation recorded round several small basins in Britain. At Black Lake, Cheshire (Tallis, 1973), for example, the following zones can be recognized:
(1) Open water.
(2) A partially submerged floating lawn of *Sphagnum cuspidatum*, *S. recurvum*, and *Eriophorum angustifolium*.
(3) A floating lawn of *S. recurvum*, *E. angustifolium*, *Carex rostrata*, *Drosera rotundifolia*, and *Vaccinium oxycoccos*.
(4) A marginal hummocky carpet with *S. recurvum*, *Polytrichum commune*, *V. oxycoccos*, and *Erica tetralix*. Occasional *Pinus sylvestris* seedlings occur on the hummocks, so that some sort of bog forest might ultimately develop.
This floating *Sphagnum* mat is probably commoner in Britain than the scanty references in the literature suggest.

A→B→H→D/E (Fig. 9.9)
Vegetation zonations involving a quaking bryophyte mat dominated by hypnoid mosses, and abutting on open water, have been described from the muskeg country of Alberta by Lewis et al. (1928) and by Moss (1953). The mat is dominated by species of *Drepanocladus* (e.g. *D. aduncus*, *D. revolvens*, *D. sendtneri*, *D. vernicosus*), either alone or in association with small sedges (*Carex aquatilis*, *C. chordorrhiza*, *C. lasiocarpa*, *C. paupercula*, *C. rostrata*), and typically passes outwards into a *Sphagnum–Ledum* zone or a bog forest zone dominated by *Betula glandulosa* and *Larix laricina*.

D→E (Fig. 9.8), **A→E→F** (Fig. 9.9)
The degeneration of an established forest community and its supercession by *Sphagnum* (so-called **paludification**) is a process recorded from many northerly and upland situations. Thus Pearsall (1950) inferred the extension of blanket peat communities downslope in northern Britain and the gradual death of adjoining woodland as a result of waterlogging, partly from observations on the present-day spatial relationships and partly from observations of dead tree remains engulfed below peat. Heilman (1966) records in close proximity in Alaska:
(1) Mature *Betula papyrifera–Alnus crispa* forest.
(2) Degenerating *Picea mariana* forest on north-facing slopes, with a well-developed moss layer of *Hylocomium*, *Pleurozium*, *Polytrichum*, and *Dicranum* spp.
(3) *Sphagnum* bog with stunted *Picea* trees.
Heilman inferred a successional relationship between these communities, partly from the spatial relationships and partly from the stratigraphic evidence of birch remains below the moss layer.

Direct observations of hydroseral change

More concrete evidence of the successional relationships of particular wetland communities can be obtained from direct observations of vegetation change at a given site over long periods of time; but since rates of change relative to the human life span are generally slow, one can usually hope to observe directly only two, or at best three, stages in any sere. The synthesis of many such fragments of evidence to build up a complete seral sequence for a given site may then sometimes be possible, but it should not be forgotten that such a sequence is hypothetical (particularly when involving evidence from changes brought about by artificial disturbance) and needs to be checked against other lines of evidence.

Some of the most detailed documentary records of hydroseral change in the British Isles are for the small fen at the north end of Esthwaite Water, in the Lake District. The vegetation here was mapped and described in detail by W.H. Pearsall in 1914 and 1929 (Pearsall, 1918; Tansley, 1949), and further surveys were carried out in 1960 by G.I. Forrest and G. Tennant (unpublished), and in 1967–69 by C.D. Pigott and J. Wilson (Pigott and Wilson, 1978). Taken in conjunction with the first edition of the 1:10 560 Ordnance Survey map (1848), and with aerial photography coverage since 1944, these maps and surveys enable one to form a fairly clear picture of the major hydroseral changes occurring at Esthwaite North Fen during a period of over a century.

Esthwaite North Fen is developed around the mouth of the inflow stream, Black Beck, which discharges large quantities of inorganic silt into the shallow waters at the north end of the lake. Water depths here are less than 2 m for a distance of more than 200 m out from the mouth of the Beck. The wetland vegetation of the North Fen has developed in relation to a gradient of silting, from heavy silting at the western side adjacent to Black Beck to minimal silting at the eastern side. A variety of communities along this gradient has been described by Tansley (1949); in the maps in Fig. 9.11 these have been grouped into four major categories:

(1) Reed-swamp communities, dominated by *Phragmites australis*, *Scirpus lacustris* or *Typha latifolia*.

(2) Fen communities, dominated by *Calamagrostis epigejos*, *Carex* spp., *Filipendula ulmaria*, *Molinia caerulea* or *Phalaris arundinacea*.

(3) Open carr, with invading bushes of *Salix cinerea* and *S. purpurea*.

(4) Closed carr, with predominant *S. cinerea*.

The maps in Fig. 9.11 are compiled from the vegetation surveys of 1914, 1929, 1960 and 1967–69, and from the 1944 aerial photograph. Certain inaccuracies are to be expected, because of the difficulties associated with surveying along the lake margin, with the reduction of the maps to a common scale, with the interpretation of vegetation boundaries from the aerial photograph, and with the interpretation of the different vegetation types recognized by different surveyors. The most recent survey, 1967–69, covered only the southern part of the North Fen (Pigott and Wilson, 1978), and did not recognize vegetation categories directly comparable with the closed carr and open carr of earlier surveys. For purposes of comparison, therefore, the areas occupied at different dates by the four major communities have been expressed on a percentage basis (of the total area mapped), and these values are shown graphically in Fig. 9.12. The expansion of carr communities between 1914 and 1967 is clearly shown, as is the encroachment of the fen communities on to reed-swamp. The major changes occurring since 1848 can be summarized as follows:

(1) An extension to the south and southwest of the reed-swamp margin into open water — by 25 m between 1848 and 1914, and by at least 19 m between 1914 and 1929; relatively little change has occurred since 1929 (Pigott and Wilson, 1978).

(2) A massive replacement of *Phragmites* reed-swamp by sedge communities (dominated by *Carex elata*, *C. rostrata* or *C. vesicaria*) in the eastern part of the Fen.

(3) An expansion of open- and closed-carr communities. A photograph of the North Fen taken in 1888 shows almost no bushes on the surface, and tree ring counts of the oldest willows confirm that they are only 60 to 70 years old (Pigott and Wilson, 1978). Between 1929 and 1969 *Alnus glutinosa*, *Betula pubescens* and *Fraxinus excelsior* have increased greatly in the carr, whilst a few saplings of *Quercus petraea* and *Corylus avellana* are now established in the oldest parts. The major expansion of *Salix cinerea* occurred in the period from 1914 to 1929; *Alnus*, now co-dominant throughout nearly all the closed carr, spread more gradually over the period from 1914 to 1953. The *Salix* is no longer regenerating, and some individual trees are now moribund (Pigott and Wilson, 1978).

(4) A spread of *Molinia* and *Sphagnum palustre* within the closed carr, at the expense of *Carex elata* and *Filipendula ulmaria*.

It is still uncertain from the present trends, however, whether the vegetation is developing towards a mesophytic woodland climax, or whether the expansion of *Sphagnum* will continue and bog vegetation will ultimately replace the carr communities. The observed hydroseral changes at Esthwaite North Fen may not, moreover, represent an undisturbed succession. There is some evidence of changing lake levels in Esthwaite Water during

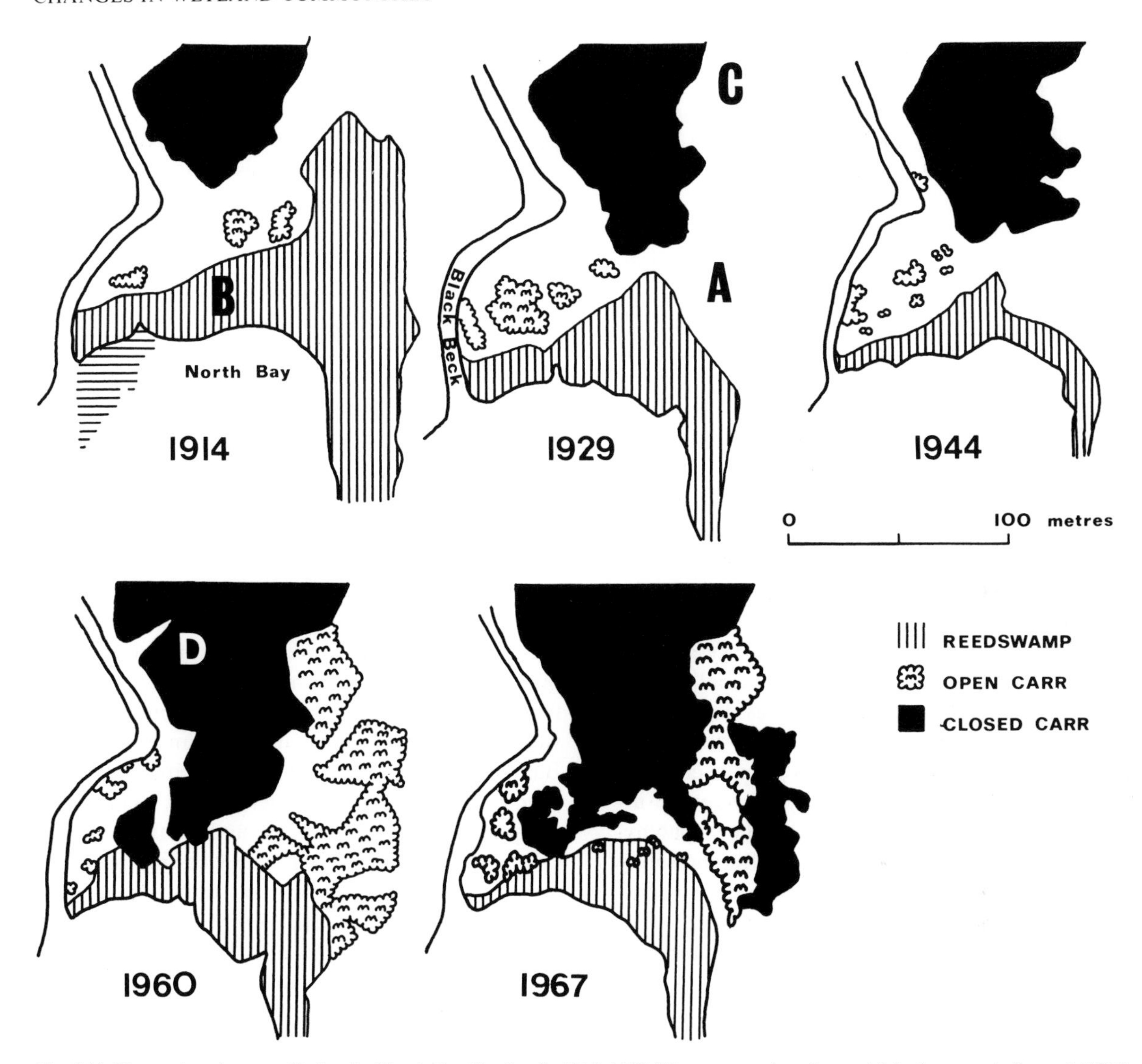

Fig. 9.11. Vegetation change at Esthwaite North Fen (England), 1914–1969. The maps are based on published surveys in Pearsall (1918), Tansley (1949) and Pigott and Wilson (1978), an unpublished survey by Forrest and Tennant (1960), and an aerial photograph of the Fen taken in 1944. *A*, *B*, *C* and *D* refer to specific areas mentioned in the text.

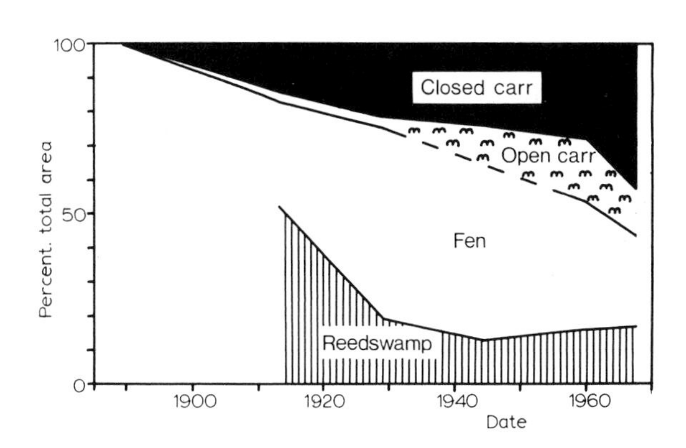

Fig. 9.12. Vegetation change at Esthwaite North Fen. The areas occupied by the four major vegetation types at different dates are shown graphically as a percentage of the total area mapped.

the last century, which may have been artificially induced to ameliorate flooding around the margins. The 1:2500 Ordnance Survey map of 1904 shows the mean lake level as 214.3 feet (65.32 m); in 1958 it is shown as 213.5 feet (65.07 m) — a drop of 25 cm. A fall of this magnitude could have profoundly affected the wetland communities of the North Fen, removing large areas from regular inundation, since monthly records of lake levels taken at the Freshwater Biological Association's gauge, at National Grid Reference 34/359975, show that only exceptionally do annual fluctuations exceed 30 cm. The maps indicate marked vegetation changes between 1914 and 1929, especially in relation to the reed-swamp and fen communities, so that it is possible that the fall in lake levels (assuming the Ordnance Survey maps to be correct) took place principally during this period.

Rates of hydroseral change have not been the same in all parts of the North Fen. Thus in area *A* on Fig. 9.11 the reed-swamp community of 1914 had progressed to open carr by 1960; a small area of reed-swamp at *B* had similarly progressed to *closed* carr by 1960. By contrast, the *Molinia*-dominated areas immediately to the north at *C* have changed little during the present century. *Sphagnum* invasion of the closed carr areas has been patchy, with the greatest spread at *D*. These differences may merely reflect differences in degree of silt accumulation, but Pigott mentions the possibility that hydroseral changes in the North Fen are currently being affected by eutrophication, since phosphate levels in the reed-swamp peat are now three times what they were half a century ago (Pigott and Wilson, 1978).

Additional, but less precise evidence of the rapidity of hydroseral change in Britain is provided by the work of Spence (1964). Spence examined ten lochs in Scotland which had been surveyed and photographed by West (1905, 1910) over fifty years earlier. In six of these lochs there had been no detectable change in the marginal vegetation (either reed-swamp or a floating *Carex lasiocarpa* mat). At one site there were small local changes. At the other three sites there had been rapid advance of reed-swamp into open water (though this advance was not quantified), and spread of scrub. In all three cases the changes were associated with rapid deposition of silt by inflow streams in sheltered bays. In the absence of silting no perceptible vegetation changes had occurred during the fifty-year period.

By contrast, there are for the United States a number of references in the literature to rapid encroachment of a floating sedge mat on to open water, suggesting that in this situation at least hydroseral change proceeds rapidly. Thus Gates (1942) reported an advance of 46 m by the sedge mat margin at Mud Lake Bog, Michigan, between 1916 and 1940. Jewell and Brown (1929) stated that at Cecil Bog, Michigan, an area of open water 92 m by 18 m was completely overgrown between 1923 and 1926. They also reported a rise in the loosely flocculated "false bottom" of the lake by up to 3 m over a 25-year period. Gates (1926) reported a similarly rapid rise of the false bottom at Bryant's Bog, Michigan, of 2.3 m during the period from 1912 to 1918. On the other hand, Buell et al. (1968) found almost no change in the position of the sedge mat margin at Cedar Creek Bog, Minnesota, between 1934 and 1967; periodic observations along a permanent transect showed instead a massive colonization of the sedge mat by *Larix laricina* between 1948 and 1967, resulting in a contraction in width of the open sedge-mat zone from 23 m to 3 m. Buell et al. concluded that the concentric zones of vegetation round open water do not maintain a constant width, but expand and contract under changing conditions (such as a succession of dry or wet summers). Their view is supported by the observations of Isaak et al. (1959), who recorded nearly 50% mortality of *Larix laricina* trees on the sedge mat at Floating Bog, Minnesota, in the period 1946 to 1957. This mortality was correlated with the above-normal precipitation values from 1941 to 1951 in northern Minnesota. Conway (1949) deduced that hydroseral changes in Minnesota bogs do not proceed smoothly, but in a step-wise fashion, and she cited as evidence for this view the abrupt transitions between one vegetation zone and another, the apparently uniform age of individuals of a given species within a particular zone, and the occurrence of sedge mat overgrown by *Chamaedaphne* and abutting on open water at some sites. All the examples quoted come from sites towards the southern limits of the area of maximum abundance of bog plants in North America designated by Transeau (1903), so that short-term fluctuations of climate here may have proportionately greater effects on the pattern of succession than further north.

There is no doubt that hydroseral change can proceed rapidly following artificially induced changes. Thus, Swan and Gill (1970) recorded the development of a floating mat of *Chamaedaphne*, *c.* 400 × 400 m in area, over shallow water at the north end of Harvard Pond, Massachusetts, between *c.* 1890 and 1968. The pond was created by the flooding of a forested valley following the construction of dams at either end; open-water conditions were present shortly before 1890. Photographic evidence indicated only a 2-m advance of the mat margin between 1938 and 1967, but rapid mat development had occurred in the initial stages of colonization by the coalescence of pioneer *Chamaedaphne* plants established on the numerous submerged tree stumps left behind from the former forest cover. Tallis (1973) mentions a shallow pond in Delamere Forest, Cheshire (England), *c.* 100 m in diameter, which had become completely overgrown with a floating lawn of *Sphagnum recurvum* following clearance of the outflow drainage ditch around 1930.

At Wicken Fen, Cambridgeshire, records of vegetation changes in an area of *Cladium–Molinia* fen, formerly cut regularly for thatch and litter, have been kept since 1923 when the area was fenced off (Godwin, 1936; Godwin et al., 1974). In 1923 there was a single central clump of old *Rhamnus catharticus* bushes, occupying rather less than 8% of the total area, and various other scattered bushes, principally of *Frangula alnus* and *Salix cinerea*. By 1934 *Rhamnus* occupied 22% of the area, and by 1972 56%. *Frangula* increased proportionately, to occupy much of the remaining area in 1972. During the period from 1923 to 1972 a number of species increased greatly in amount (e.g. *Crataegus monogyna*, *Phalaris arundinacea* and *Urtica dioica*); others decreased equally spectacularly (e.g. *Phragmites australis* and *Salix cinerea*). Many disappeared completely, including the original dominants *Cladium* and *Molinia*; and a whole new category of carr species appeared (e.g. *Epilobium montanum*, *Sambucus nigra* and *Solanum nigrum*). Since 1944 there has been increasing colonization of the *Rhamnus* carr by deciduous trees, notably *Alnus glutinosa*, *Populus tremula*, and *Betula* sp., but also locally *Quercus* and *Fraxinus*. The rapid spread of carr at Wicken Fen may not, however, represent an undisturbed succession, since water levels in the Fen have fallen over the last twenty years as a result of increased control of flood waters in winter by the locks.

LeBarron and Neetzel (1942) described a similarly rapid colonization of peat in Michigan by trees following drainage in 1928. The vegetation then was largely open bog dominated by *Chamaedaphne* and *Sphagnum*, but with some trees around the margins. By 1930 *Ledum*, *Kalmia*, *Alnus incana*, and *Vaccinium canadense* had become important components of the vegetation, and scattered tree seedlings (of *Abies balsamea*, *Picea mariana*, *Sorbus* sp., and *Thuja occidentalis*) were present. *Acer rubrum* was first recorded in 1932. By 1941 the bog had a well-developed tree cover, consisting of *Acer*, *Abies*, *Alnus*, *Picea*, *Sorbus*, and *Thuja*.

RATES OF CHANGE DEDUCED FROM STRATIGRAPHIC DATA

In addition to measurements of change obtained by direct long-term observations, natural rates of hydroseral change can also, in theory, be worked out from stratigraphic sequences, provided that reliable datings for particular levels in a profile are known, and that the sediment between dated levels is more or less uniform in composition. Walker (1970) compiled data for rates of accumulation of a number of different types of organic sediment in the British Isles by this method, and was unable to find any significant differences between sediment types. However, data assembled in this way must necessarily be heterogeneous, pooling rates of accumulation achieved under very different topographic and climatic conditions, so that perhaps all that can be hoped to emerge is a general picture of the range of recorded variation for rates of peat accumulation, and of the most frequently recorded rates. Fig. 9.13 shows rates of accumulation of 133 samples, derived partly from Walker's data and partly from recent radiocarbon datings of blanket peat profiles. Certain of Walker's data have been omitted — notably those from sediments accumulated during pollen zones III and IV, when climatic conditions were very different from present-day ones (see Taylor, Part B, Ch. 1), and also from certain lake sediments accumulated in deeper water. The data in Fig. 9.13 show a peaked but asymmetric distribution, with a mean value of almost exactly

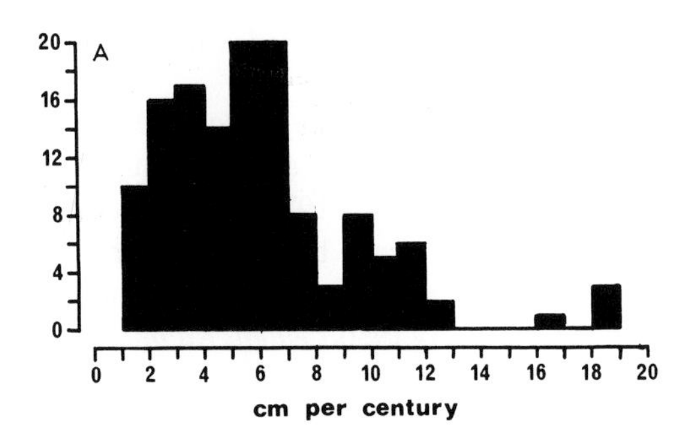

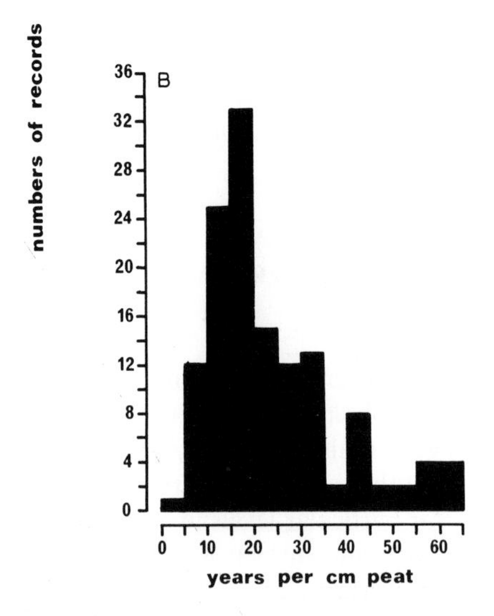

Fig. 9.13. Histogram of rates of accumulation of 133 dated samples of raised and blanket peat in Great Britain. Data from Walker (1970), Hicks (1971), Pennington et al. (1972), Turner and Chambers (1972), Tallis and Switsur (1973), and Turner et al. (1973).

6.0 cm per century; 65% of the values lie between 2.0 and 7.0 cm per century. On the other hand, the total range of recorded variation is from 1.3 cm per century to 55.3 cm per century (Hicks, 1971).

Comparable data for North America are more difficult to compile, principally because of the small numbers of radiocarbon-dated profiles. Table 9.5 summarizes some of the available data. The mean rate of accumulation for 32 samples is 5.8 cm per 100 years.

Thus it may perhaps be concluded that a mean rate of peat accumulation of *c*. 6.0 cm per century has some general validity, both for different sediment types and for different countries.

Peat accumulation necessarily represents an excess of *input* (from dead plant and animal remains) over *output* (losses by decomposition) in the system, and it is possible to make some crude estimates of the magnitude of this imbalance in general terms, using the relationship:

$$\%\text{preservation} = \frac{\text{rate of accumulation} \times \text{peat bulk density} \times 100}{\text{annual input} \times \text{time}}$$

Recorded peat bulk densities range from 0.02 to 0.26 g dry wt. cm^{-3} saturated volume (Boelter, 1966, 1968). Published values for annual input (primary productivity) range from *c*. 300 g m^{-2} (Clymo and Reddaway, 1974) to *c*. 2500 g m^{-2} (Westlake, 1963). If a mean rate of peat accumulation of 6 cm per century is assumed, then the extent of preservation of the annual input required to give the recorded extremes of peat density can be worked out (Table 9.6). This ranges from less than 1% to 52%. In very general terms it can thus be concluded that only exceptionally is more than half of the annual input preserved as peat. The mean rate of accumulation of 6 cm per century is subject to at least the variation noted in Fig. 9.13A. Under some circumstances with very high rates of accumulation, due to low decomposition and low compaction, changes of habitat can be brought about very quickly. Consequences of these and other interactions are dealt with in more detail elsewhere, however (see Gore, Ch. 1; and Clymo, Ch. 4).

THE ECOLOGICAL REQUIREMENTS OF HYDROSERAL COMMUNITIES

Telmatic communities

Probably the major role in the process of terrestrialization of an open-water habitat is carried out by the telmatic communities; these are responsible for the build-up of organic residues to more or less the level of the ground-water table. Kulczyński (1949) recognized two differing types of telmatic communities:

(1) Communities composed of large **immersive** perennials, which are rooted below water in mineral or peat soils of high bulk density, and which accumulate peat by apposition.

(2) Communities composed of small **emersive** perennials, which are rooted in peat of low bulk

TABLE 9.5

Rates of peat accumulation at North American sites

Locality	Reference	Peat accumulation [cm (100 yr)$^{-1}$]
A. Lake mud deposits		
British Columbia	Mathewes (1973)	2.4, 2.6, 7.0, 14.6
Colorado	Pennak (1963)	1.8, 2.7, 3.4, 4.2
Connecticut	Davis (1969)	5.0, 12.0
Indiana	Wetzel (1970)	3.9, 11.3
Manitoba	Ritchie (1969)	3.9
Minnesota	Fries (1962)	10.0, 10.0
	Janssen (1968)	4.3, 6.3, 6.9
Pennsylvania	Walker and Hartman (1960)	5.0
Washington	Rigg and Gould (1957)	4.2
Mean of 20 values		*6.1 cm (100 yr)$^{-1}$*
B. Fibrous peat or *Sphagnum* peat		
Colorado	Pennak (1963)	1.8, 1.9, 3.2, 9.6
Minnesota	Heinselman (1963, 1970)	4.6, 5.0, 13.6
Northwest Territories	Nichols (1967)	1.7, 2.9, 3.0, 3.6
Washington	Rigg and Gould (1957)	3.7
Mean of 12 values		*4.5 cm (100 yr)$^{-1}$*

TABLE 9.6

Percentage preservation of peat at two bulk densities and two levels of annual input, given an accumulation rate of 6 cm 100 yr^{-1}

Input (g m^{-2} yr^{-1})	Bulk density (g dry wt. cm^{-3})	
	0.02	0.26
2500	0.48%	6.24%
300	4.0%	52.0%

density and thus typically form a floating raft of vegetation, accumulating peat underwater by deposition from the base of the raft.

Kulczyński cited as characteristic immersive plants *Phragmites*, *Scirpus*, *Typha*, larger species of *Carex*, *Iris pseudacorus*, and *Ranunculus lingua*; and as characteristic emersive plants, several smaller species of *Carex*, *Menyanthes*, *Sphagnum cuspidatum*, and *Drepanocladus* spp. However, although the recognition of immersive and emersive community types is a useful and valid distinction, the categorization of specific plants as immersive or emersive is probably not. Thus, *Phragmites australis* (an immersive species according to Kulczyński) frequently forms floating raft communities in Europe (Pallis, 1915; Lambert, 1965), whilst *Carex lasiocarpa* (supposedly emersive) is a common paludifying species on wet mineral soil in Scandinavia (Tolonen, 1967). According to Kulczyński, telmatic communities of immersive perennials are found in situations with a pronounced water flow, and mesotrophic to eutrophic waters; emersive communities develop under conditions of much reduced water flow, and mesotrophic to oligotrophic waters. Both types of communities occur in situations with pronounced vertical oscillations of the water table. Immersive communities survive by virtue of their resistance to conditions of alternating flooding and desiccation, emersive communities by virtue of compensatory movements of the floating mat. Kulczyński pointed out that emersive communities possess great capacities for accumulating peat and at the same time for raising the water table locally; immersive communities, by contrast, form a high bulk density peat with little powers of expansion and contraction and, since the surface layers are periodically exposed to aerobic decomposition, have low rates of peat accrual. Kulczyński doubted whether immersive communities could bring about terrestrialization without additional processes whereby the water table is raised independently.

It is clear that floating rafts of vegetation can originate in a number of different ways, dependent in part at least upon the morphometry of the site. In enclosed basins with fairly steeply sloping sides, a raft community may form by marginally rooted species with buoyant rhizomes gradually extending out across the surface of the open water (as in North America). In shallow waters there may be direct colonization by floating stems of *Sphagnum* spp. or other bryophytes, followed by gradual interlacing with rhizomes of *Carex* or *Eriophorum* spp. Alternatively in shallow water a community of immersive perennials (e.g. *Phragmites*, *Typha*) rooted below water may initially develop, but as the rhizome mat consolidates so the increasing buoyancy results in uplift to the water surface to form a floating raft.

Over much of Europe the principal dominant of telmatic communities is *Phragmites australis*, although in northern Scandinavia, for example, hypnoid mosses are prominent (Lohammar, 1965). In the northern United States, by contrast, floating sedge mat communities dominated by *Carex lasiocarpa* are the major type (see p. 326); farther north in Canada, as in Scandinavia, bryophytes become an increasingly important component (Moss, 1953). The contrasting roles played by *Phragmites* and *Carex lasiocarpa* in Europe and North America are particularly noteworthy, though no systematic attempt has ever been made to analyse the reasons for the different patterns of terrestrialization in the two regions. A number of interacting factors are almost certainly involved:

Topographical factors

Floating mat communities are capable of centripetal encroachment upon a body of open water, provided there is no pronounced water flow, and can thus grow out over quite steeply shelving lake bottoms. *Phragmites*, by contrast, is most characteristically found on very gently shelving areas where there is a definite water flow over the surface. Although reed-swamp frequently forms a fringe of vegetation around the shores of many lakes today, Kulczyński (1949) doubted whether this vegetation can advance progressively on to open water, as can the floating mat communities. Under conditions of more rapid water flow, exceeding 0.15 to 0.20 m s^{-1}, *Phragmites* may be replaced by *Typha* or *Scirpus* spp. (Rodewald-Rudescu, 1974), or by *Glyceria maxima* (Lambert, 1965).

Kulczyński characterized situations of reed-swamp dominance as having an obligatory alternation of minerotrophic and ombrotrophic water supplies, achieved by canalization of water flow over the very gently sloping surfaces (river terraces, margins of large lakes). At periods of high water levels the whole surface is inundated (typically with deposition of silt); at periods of low water levels, flow is confined to definite channels, and the intervening areas of reed-swamp become raised above the water table and influenced by rain water.

Seedling establishment of *Phragmites* appears not to take place if the soil surface is permanently flooded (Spence, 1964; Haslam, 1971), and this may be an important factor limiting *Phragmites* to such situations as those described above.

Chemical factors

The apparent differences in patterns of water supply suggest that *Carex lasiocarpa* communities might be more oligotrophic than *Phragmites* reed-swamp, but this is not borne out by the available water analyses (Table 9.7). Vitt and Slack (1975) and Schwintzer (1978) have stated that *Carex lasiocarpa* sedge mats are characteristic of mesotrophic ("fen") conditions in Minnesota and Michigan, and in oligotrophic conditions their place is taken by *Sphagnum*-dominated communities. *Phragmites* reed-swamp is also clearly a "fen" community, and Daniels (1975) found a positive correlation between shoot growth of

TABLE 9.7

Water analyses from two communities compared

	Community dominated by	
	Phragmites australis	*Carex lasiocarpa*
pH	4.7–8.1	3.7–7.0
Ca (mg l^{-1})	3.6–306.0	0.8–75.0
Mg (mg l^{-1})	0.8–194.0	2.8–17.8
K (mg l^{-1})	0.7–8.1	0.2–3.9
References	Rodewald-Rudescu (1974)	Gorham (1950) Vitt and Slack (1975) Schwintzer (1978)

Phragmites and calcium concentration in the water at Roydon Common, Norfolk. Haslam (1965) showed that the supply of phosphate appeared to be a major factor determining dominance patterns in the Breck fens of East Anglia: at high phosphate concentrations *Phragmites* was the predominant species, but at lower concentrations it was replaced by *Schoenus nigricans* or *Cladium mariscus*. Boyd and Hess (1970) similarly found that the major factor controlling the productivity of *Typha latifolia* in the southeastern United States was the supply of phosphate. Spence (1964) found a positive correlation between shoot growth of *Phragmites* and nitrogen concentration in Scottish lakes, whilst Haslam (1972) stated that nitrogen supply was a major factor influencing early seedling growth in *Phragmites*.

Climatic factors

The productivities of both *Phragmites*- and *Carex*-dominated communities appear to be positively correlated with summer warmth (Spence, 1964; Gorham, 1974), but *Phragmites* is more sensitive to cold. Thus towards the northern part of its range fruit is rarely formed, whilst the young shoots are increasingly damaged by late spring frosts (Haslam, 1972). There is only a single period each year of new shoot emergence from subterranean buds, and in colder climates emergence is delayed and the growing season correspondingly curtailed; thus in Malta new shoots emerge in late January, in The Netherlands in mid-February, and in Connecticut in April or May (Haslam, 1972). Other species which show growth earlier in the spring may be able accordingly to suppress *Phragmites* (Buttery and Lambert, 1965). Severe cold in winter (as in continental-type climates) accompanied by low lake levels (as in northern Sweden; Lohammar, 1965), enabling frost to penetrate deeply into the underlying soil, may prevent *Phragmites* growth in many situations, especially as *Phragmites* does not appear to be able to grow in as deep water under oligotrophic conditions as under mesotrophic or eutrophic conditions (Haslam, 1972).

Invasion by trees and shrubs

A wide variety of tree and shrub species are capable of growing on peat, forming some sort of closed-canopy woodland (carr, bog forest, etc.). Among the larger trees, conifers are predominant: *Pinus sylvestris* in Europe, and *Larix laricina*, *Picea mariana*, *Pinus contorta*, *Thuja occidentalis*, and *Chamaecyparis thyoides*, for example, in North America. Invasion and establishment of tree seedlings is liable to occur once the surface peat layers are raised above the level of the water table for a substantial part of the year. Godwin and Bharucha (1932) maintained continuous records of water-table levels at a series of observation pits on Wicken Fen, Cambridgeshire (England) over a period of two years, and found that establishment of shrubs (principally *Frangula alnus* and *Rhamnus catharticus*) occurred at sites where the peat surface was flooded for less than 175 days each year on average. Uninterrupted submergence of the peat surface for several weeks in winter prevented shrub establishment. Consequently it is not surprising that a number of workers have found experimentally that the young seedlings of many typical bog trees are intolerant of flooding. Thus, *Larix laricina* seedlings will not survive more than one week's continuous submergence (Duncan, 1954); *Betula* seedlings will only survive up to three weeks of continuous submergence, and *Alnus glutinosa* up to five weeks (McVean, 1956). The developing roots of many bog trees quickly become horizontal, remaining confined to the better aerated surface-peat layers (Rigg and Harrar, 1931); as a result, death may ensue during periods of excessive drought (Duncan, 1954; see also Crawford, Ch. 7). Particularly when establishing on ombrotrophic peats, tree and shrub seedlings may make poor growth because of nutrient deficiencies: nitrogen and phosphorus deficiencies in *Pinus sylvestris* (McVean, 1963) and *Acer rubrum* (Moizuk and Livingston, 1966), and phosphorus deficiency in *Alnus* and *Betula* (McVean, 1956).

The type of pioneer woody community that develops is influenced very strongly by the preceding hydroseral stages and in particular by the processes by which the peat surface is raised above the water table. Thus, Lambert (1951, 1965) found that different patterns of carr establishment in the Norfolk Broads could be explained in terms of differing types of fen vegetation formed over the initial spongy, semi-floating raft of *Phragmites*. Where the large tussock-forming *Carex paniculata* was the dominant fen species, in situations of fairly free water movement, then precocious invasion of seedlings of *Salix atrocinerea* and *Alnus glutinosa* on the drier tops of the tussocks was possible. As the tussocks built up and the weight of the trees increased, so the tussocks became more and more unstable, either rolling over or sinking gradually into the underlying unconsolidated raft. Thus the mature **swamp carr** exhibited a characteristic pool-and-tussock structure, with many dead or dying trees on the depressed waterlogged tussocks. In situations of more restricted water movement, where instead the rhizomatous *Carex acutiformis* dominated the fen stage, invasion of woody plants was delayed until the overall level of the peat had been raised uniformly; even so, invasion occurred before the raft was fully stabilized, and the weight of the maturing trees again led to local subsidence, and the formation of an unstable **semi-swamp carr**. With a fen stage dominated by *Cladium mariscus*, establishment of tree seedlings was almost entirely prevented by the thick surface mat of dead leaves (and also probably by regular cutting for thatch in the past); thus a prolonged fen stage resulted, with sufficient time elapsing for the raft to become grounded and fully stabilized. A necessary prelude to the establishment of tree seedlings appeared to be invasion and opening-up of the dense *Cladietum* by *Myrica gale*, *Salix repens*, or even *Phragmites*. The mature **fen carr** that developed was thus a much more stable system than the swamp carr, with appreciable representation of *Quercus robur*, *Fraxinus excelsior*, and *Betula pubescens*. In North America there is often a similar invasion of an unstable floating raft by tree seedlings (typically *Larix laricina*). High seedling mortality is characteristic of this pioneer bog forest stage (Duncan, 1954), with the pattern of invasion very much dependent on short-term climatic fluctuations (see p. 330). Although in many instances the invasion of *Larix* seedlings is delayed until the raft is grounded and *Chamaedaphne* established, Conway (1949) showed that in Minnesota direct invasion of the sedge mat could also occur, with the consequent bypassing of the *Sphagnum–Chamaedaphne* stage. Pioneer forest could thus be found on peats ranging in pH from 4.3 to 7.3. Campbell (Part B, Ch. 5) describes similar but more extreme consequences of trees growing on floating islands in Tasmania. In Europe, *Pinus sylvestris* is a common invader of ombrotrophic *Sphagnum*-dominated communities, but seedling establishment is almost entirely confined to the drier raised hummocks. Gradual subsidence of the hummocks under the weight of the maturing pine trees, and the ensuing death of the trees by waterlogging, has been described from lowland mires in Cheshire (England) by Green and Pearson (1968) and Tallis (1973). In eastern Fennoscandia and Poland *P. sylvestris* characteristically dominates the drier hummocks of the "regeneration complex" (see pp. 314, 315); the trees here never attain any great age, since upward growth of the intervening hollows leads to a general raising of the water table, and consequent waterlogging of the hummocks and death of the pines (Kulczyński, 1949).

It is probable that most individual trees in the pioneer forest stages have a relatively short life-span, as a result of unnaturally early death from a variety of causes. In addition to death by waterlogging, as described above, many trees may apparently be killed by fungal or insect attack. Thus Godwin (1936) described massive "die-back" of *Frangula alnus* at Wicken Fen between 1923 and 1934, caused by *Nectria cinnabarina* and *Fusarium* sp. In a Wisconsin bog nearly half the *Populus tremuloides* population died between 1940 and 1966, probably as a result of fungal attack by *Ganoderma applanatum* (Vogl, 1969). The disappearance of *Viburnum opulus* from Wicken Fen between 1936 and 1972 was ascribed by Godwin et al. (1974) to defoliation by the beetle *Galerucella viburni*. Both Gates (1942) and Duncan (1954) noted that populations of *Larix laricina* in North America can be devastated by the larvae of the sawfly *Pristiphora erichsonii*. In some cases at least it has been shown that seedlings of the characteristic pioneer trees and shrubs are intolerant of shading, and hence cannot grow beneath the parent plants [*Larix laricina* (Conway, 1949; Duncan, 1954), and *Chamaecyparis thyoides* (Buell and Cain, 1943)]; consequently these trees tend to be replaced after the first generation by other species whose seedlings are shade-tolerant — *Picea mariana* or *Thuja occidentalis* in North America, for example (Conway, 1949).

A mature forest cover developed over peat might be expected to cause the surface peat layers to dry out, as a result of interception of precipitation by the tree canopy and increased losses of water by evapotranspiration. The ensuing increased rate of decomposition of plant remains might then ultimately create conditions favourable for the establishment of a mesophytic woodland community. Evidence relating to this hypothesis is, however, scanty. Opinion is divided as to whether present-day vegetation zonations, with coniferous or deciduous woodland adjacent to bog forest or carr, reflect a true natural succession, as often this woodland is developed on mineral soil or thin peat. Direct measurements of the effects of a tree canopy on the water balance of a peat mass are almost non-existent, though there are limited data for mineral-soil areas. Rutter (1963) recorded a 32% reduction of annual precipitation reaching the ground below a 20-year-old *Pinus sylvestris* plantation, and Rutter and Fourt (1965) estimated that evapotranspiration from this plantation was 60% higher than from adjoining grass plots. Moore (1975) cited increases in stream discharge ranging from 8 to 40% following deforestation, and the increase presumably represents the water intercepted or transpired formerly by the tree canopy.

In wet or cold climates, moreover, establishment of *Sphagnum* on the forest floor is a common occurrence. Watt and Heinselman (1965) noted a marked reduction in the growth of *Picea mariana* in "muskeg" areas of northern Minnesota, associated with thick deposits of *Sphagnum* peat. Foliar analyses showed that this poor growth was probably correlated with deficiency concentrations of nitrogen and phosphorus in the surface peat layers. Heilman (1966, 1968) investigated similar situations in Alaska, and also found deficiency concentrations of nitrogen and phosphorus, and probably potassium as well. He showed that soil deterioration was initiated not by *Sphagnum* but by a dense carpet of hypnoid mosses. Natural ion deficiencies were accentuated under *Sphagnum*, since *Sphagnum* dominance on the forest floor resulted in a rise in the height of the soil permafrost layer, and the main concentration of most ions occurred at a lower depth in the *Sphagnum* peat than in the mineral soil.

Invasion by *Sphagnum* spp.

The advent of *Sphagnum* into a wetland community represents a critical stage in the pattern of change, since all species of *Sphagnum* appear to be capable of rapidly and profoundly modifying the chemical properties of the habitat. Ever since the pioneer experiments of Skene in 1915 it has been known that *Sphagnum* spp. can bring about acidification of solutions in which they are growing, by the uptake of cations from solution and the release of equivalent numbers of hydrogen ions into solution. Cations of higher valencies are selectively taken up. Clymo (1963) showed that the sites of cation exchange were located in the cell walls, and were probably unesterified polyuronic acid molecules. *Sphagnum* possessed a greater percentage by weight of such cation exchange sites than any other known plant. By contrast, the capacity for anion uptake was only poorly developed.

Conway (1949) examined changes in acidity in sedge-mat communities in Minnesota bogs brought about by the invasion of different *Sphagnum* species. She measured pH values within the living *Sphagnum* clumps, and also at different depths in the underlying *Sphagnum* peat. Her results are summarized in Table 9.8. They show the ability of both *S. recurvum* and *S. magellanicum* to become established in near-neutral conditions and to convert them into ones of high acidity. *S. capillifolium* had much more limited capacities for acidification.

Several workers have also shown the ability of other bryophytes to bring about acidification. Kurz (1928) showed a reduction of pH from 8.0 to 6.0 in *Drepanocladus* peat in a Michigan bog, and from 7.3 to 4.4 in living *Aulacomnium palustre* mats. Conway (1949) similarly mentioned *Aulacomnium* as a potentially important acidifying species in Minnesota bogs. Chouard and Prat (1929) recorded a pH reduction from 8.0 to 5.8 by a hypnoid mat in the Pyrenees, and suggested that these mosses played an important role in lowering the pH of alkaline waters to levels at which *Sphagnum* spp.

TABLE 9.8

pH values in living *Sphagnum* clumps and in the underlying peat (data of Conway, 1949)

	Living *Sphagnum*	15 cm depth	37–45 cm depth
Sphagnum recurvum	4.75 (10)	6.28 (8)	
Sphagnum magellanicum	4.70 (10)	4.71 (5)	6.49 (4)
Sphagnum capillifolium	4.29 (7)	4.48 (4)	

The pH values are mean values; the numbers of readings are given in brackets.

could invade. The initial acidification of soils by hypnoid mosses in Alaska was mentioned above (p. 337). Wilde et al. (1954) examined similar sites in Ontario, and showed the ability of a moss carpet composed of *Hypnum*, *Calliergon*, *Hylocomium*, and *Dicranum* spp. to bring about degeneration of lime-enriched soils and permit the entry of *Sphagnum* spp. Under the moss carpet pH levels were reduced to 5.7 and dissolved oxygen levels to 0.9 p.p.m. Bellamy and Rieley (1967) studied the chemical changes from top to bottom of a large *Sphagnum fuscum* hummock, 56 cm in height, growing in fen vegetation in northern England, which was irrigated by calcareous spring waters. A conversion to ombrotrophic conditions (for instance, a lowering of pH from 7.3 to 4.1) was achieved within the basal 8 cm of this *S. fuscum* hummock, and above this the chemical changes were negligible. At the base of the hummock the predominant macro-remains were of *Campylium stellatum* and *Cratoneuron commutatum* var. *falcatum*, and these appeared to have carried out the initial acidification. On the living fen surface these mosses formed swelling tufts which tended to divert the flow of ground water. Bellamy and Rieley calculated that, at the present-day rate of peat accumulation below *S. fuscum*, the complete transformation from minerotrophic to ombrotrophic conditions could be accomplished within seven years.

Comparable chemical changes, documenting the change from minerotrophic to ombrotrophic conditions, can usually be detected in peat profiles (see, for instance, Mörnsjö, 1968). The Ca/Mg ratio (based either on total or exchangeable quantities), in particular, has been widely used in this connection. Wherever possible, it is desirable to compare the chemical characteristics of particular levels in the peat profile with similar values from organic debris immediately below a range of present-day wetland communities growing in the same area, since the chemical composition of the water supply to the bog surface need not be reproduced in the chemical composition of the accumulating plant residues (Tallis, 1973). Selective uptake of certain ions, and their continued recycling within the plant, is known to occur in living plants; and the considerable decomposition occurring in many peats suggests substantial humification or even mineralization of organic material, so that repeated recycling of ions in short supply (e.g. phosphate) cannot be ruled out. Such a process would modify the chemical composition of the derived peats. Removal or redistribution of ions or colloidal material by water flowing laterally through more easily permeable peat layers may even occur. Mörnsjö (1968), in addition, has deduced secondary effects from percolating ground waters in two peat profiles from southern Sweden, based on discrepancies between the chemical and botanical compositions of the peat.

The development of surface patterning

Characteristic of almost all wetland communities raised above the water table is their heterogeneity. This heterogeneity, at its simplest, may merely be a reflection of the growth habits of the component species; thus predominance of a tussock-forming species such as *Eriophorum vaginatum* will necessarily result in a more non-uniform mire surface than one dominated by *Sphagnum recurvum*. Even a continuous *Sphagnum* carpet, however, is seldom uniform; more typically, some differentiation into incipient hummocks and lower-lying hollows is present, and on most mature raised-bogs there is a well-defined hummock/hollow complex, with marked differentiation of distribution of the component species between hummocks and hollows. The most pronounced surface heterogeneity is displayed by the so-called "boreal mires" of northerly latitudes (Sjörs, 1961), with strikingly regular hummock and pool systems.

On mires with an almost level surface, surface patterning, when present, takes the form of an irregular mosaic of hummocks and hollows, or of an extensive but irregular pool complex. Where a distinct cupola is present the hummocks and hollows become differentiated into elongated ridges and hollows aligned parallel to the surface contours so as to form a concentric pattern. Where such a cupola has developed over gently sloping terrain, so that the apex of the cupola is displaced upslope, then the surface patterning of ridges and hollows is correspondingly asymmetric. In the so-called "eccentric raised mires", where the slope of the mire surface is unidirectional, the patterning consists of a regular alternation of narrow ridges and broad hollows aligned parallel with the surface contours. In all these types both pools and hollows are ombrotrophic in character, but in the more northerly "aapa mires", where the patterning is similar to that of the eccentric raised mires, the hollows are minerotrophic in character due to through-flow of water.

In the case of these aapa mires it seems likely that differentiation into ridges and hollows arose some 2500 to 4000 years ago (Björkbäck, 1965; Kivinen and Tolonen, 1972), and that the configuration of the ridges and hollows has persisted relatively unchanged ever since. In the Red Lake Peatland of northern Minnesota, surface patterning apparently developed over a uniform sedge-mat community about 2000 years ago (Griffin, 1977). At Claish Moss in northwestern Scotland, Moore (1977) found evidence of stable patterning extending back over some 5000 years. No such uniformity of origin is apparent for other patterned mires. In some it is claimed that several successive generations of alternating hummocks and hollows extend downwards through the peat for depths of more than 3 m (Tansley, 1949); in other mires it is clear that the present-day surface patterning has developed only recently, over a much more uniform mire surface (Godwin and Conway, 1939; Hansen, 1966); at yet other sites the peat stratigraphy reveals an ancient hummock/hollow complex buried below relatively undifferentiated *Sphagnum* peat (Casparie, 1972).

The classical "regeneration complex" ideas propounded by Von Post and Sernander and by Osvald (see p. 314) would require a direct continuity between the surface patterning and the underlying peat stratigraphy. In some mires this seems indeed to be the case, but in other mires the peat stratigraphy suggests very different growth mechanisms by which the bulk of the *Sphagnum* peat was laid down. Continued cyclical regeneration of a hummock/hollow system should result in a lenticular-structured peat; but, whilst this type of peat has been widely reported (p. 315), there is little uniformity of opinion as to the size of the individual peat lenses or of the duration of each regeneration cycle. Neither Von Post and Sernander nor Osvald attempted to define these parameters. Both Kulczyński (1949) and Overbeck (1963), largely on theoretical grounds, envisaged a rather short cycle duration (perhaps only 50 years). Rudolph and Firbas (1927) found two cycles within a depth of 23 cm, whilst the diagrammatic profile from Willis's Piece in Somerset (Clapham and Godwin, 1948) indicates five or six cycles within a 30 cm depth. This 30 cm interval probably represents a duration of some 200 years (from 50 A.D. to 250 A.D.), and thus a cycle length of less than 50 years. Such a short cycle length is unlikely to cause confusion with the recurrence surfaces discussed earlier (p. 318), which are typically separated by much larger time intervals. On the other hand, Tansley (1949) described two profiles from Ireland, each with six successive cycles, with a mean cycle interval of 0.6 m; it is improbable that the earliest cycle post-dates 3500 B.P. (on the basis of comparisons with the profiles and pollen diagrams of Mitchell, 1956). That is to say, the minimum length of each cycle must be of the order of 500 to 600 years. Backéus (1972), moreover, re-mapping parts of Skagershultmosse sixty years after Von Post's original survey (Von Post and Sernander, 1910), could find no appreciable changes in the vegetation of the surface hummock/hollow complex.

Detailed examination of exposed peat faces by Walker and Walker (1961) and by Casparie (1972) suggests that at certain sites, or under certain conditions, a hummock/hollow system once established might be able to maintain itself unchanged over long periods of time without any cyclical alternation (as apparently in the northern aapa mires, see above). Thus at Bourtanger Moor Casparie found evidence of hummocks persisting locally for up to 1000 years (profiles Emmen 17 and 18). Walker and Walker concluded that in Irish bogs the principal growth mechanism involved periodic rejuvenation of the bog surface under conditions of increased wetness (cf. recurrence surfaces), followed by progress towards a more mature vegetation under constant or drying conditions; periods of alternation of dominance by *Sphagnum* and by higher plants thus produced a stratified peat profile, with alternating layers of less humified and more humified peat (which could superficially resemble regeneration-complex peat). Hummocks, once established, could persist through several rejuvenation cycles, expanding laterally over the bog surface during dry periods and contracting again during wet periods. Rather similar conclusions have recently been reached by Moore (1977) and by Boatman and Tomlinson (1977). The intervals between successive layers of more highly humified peat in Walker and Walker's profiles were typically between 15 and 20 cm. Since Mitchell (1956) had earlier deduced a mean depth of 275 cm of peat formed in thirteen Irish bogs since *c.* 3500 B.P., a depth of 15 to 20 cm in Walker and Walker's profiles would probably represent a time interval of some 200 to 250 years.

From exposed peat faces at Klaukkalan Isosuo, in southern Finland, Tolonen (1971) described a rather different stratigraphy, which he suggested was of widespread occurrence in northern European bogs. Weakly humified *Sphagnum fuscum* peat constituted the bulk of the upper peat layers, but included within it were numerous dark streaks, the streaks typically being less than 1 cm thick and often pursuing a wavy course laterally through the peat for more than 10 m. The streaks consisted of highly humified peat with predominant *Calluna* or *Eriophorum vaginatum* remains, and differed from the encompassing *S. fuscum* peat in rhizopod and pollen composition and in calorific value. Thirty such streaks occurred in the three profiles analysed, with a mean vertical interval between adjacent streaks of 7.5 cm. This interval probably represented a time duration of less than 100 years. Tolonen suggested that the streaks were formed when *Sphagnum* growth was interrupted during short-term drier periods. At Bourtanger Moor (Casparie, 1972), in profiles Emmen 9 and 34, similar highly humified layers were separated by a vertical interval of from 10 to 20 cm.

Dominance of a single *Sphagnum* species (typically either *S. fuscum* or *S. imbricatum*) over considerable depths of weakly humified peat is a common phenomenon in many European peat profiles (Overbeck, 1963). If, as Tolonen, and Walker and Walker, suggested, intercalated bands of more humified peat are also widespread, then probably all profiles so far described from northern Europe show recurrent indications of retarded and rejuvenated peat growth ("rejuvenation cycles"). The growth mechanisms which have been suggested, however, differ as to whether these cycles are interpreted as resulting from autogenic or allogenic processes. Kulczyński (1949) suggested that different patterns of growth would result from continuous as opposed to intermittent rises of the water table. If the rise were gradual and continuous, then a stable surface topography (whether plain or patterned) might be maintained; if, on the other hand, the rise in water table were halted periodically, allowing drying out of hummocks present and colonization by lichens, then a profile of alternating hummock and hollow phases could result. Such a mechanism would imply that allogenic processes are of more significance than autogenic processes.

Present-day hummock/hollow complexes show considerable diversity as to size, shape, and configuration of pools and hummocks, and no satisfactory mechanism has yet been proposed for their origin. Implicitly or explicitly it has been suggested by many workers from Weber's time onwards that a hummock/hollow mosaic can arise following the flooding of a desiccated peat surface, with the hummocks and hollows perpetuating the irregular character of this underlying surface — an irregularity resulting from the differential distributions of tussock-forming and woody species. Whilst there are many stratigraphic instances of hummock/hollow complexes arising in this way, there are also present-day bog surfaces where hummocks and hollows appear to be differentiating *de novo* from a continuous *Sphagnum* carpet (Tallis, 1973). Pearsall (1956) suggested an alternative mechanism, based on detailed studies of two northern Scottish mires, whereby on gently sloping ground mass movement of semi-liquid peat results in tearing and pressure-ridging of the mire surface. Such a mechanism presupposes a correlation between surface patterning and the underlying mineral ground contours, and in other mires where detailed surveying has been carried out this relationship does not exist — the surface patterning is aligned according to the present-day surface contours (Boatman and Tomlinson, 1973). However, mass movement and tearing *has* been demonstrated by Pearson at Muckle Moss in Northumberland (Pearson, 1979), where marker stakes at the margins of pools have moved downslope for distances of up to 3.5 m over an eight-year period.

The recognition that the topmost 5 to 10 cm of uncompressed and largely unhumified peat in a peat mass forms an "active layer" for water movement, with hydraulic conductivities several thousand times higher than in the underlying more or less impermeable peat (Boelter, 1965), provides a possible explanation for the conformity of surface patterning with the surface contours. Hummock/hollow complexes typically only occur under conditions of permanently high water table, so that on sloping ground there would be considerable slow lateral flow of water in this active layer. Since a high water table reduces the extent of decomposition of plant residues (Clymo, 1970), it is possible that local differences in primary productivity over the mire surface (Forrest, 1971) could lead to small-scale surface heterogeneity; in addition, differences in resistance to decay (resulting from, for instance, differences in lignin content) could lead to local differences in hydraulic conductivity within the active layer, which might produce ponding of water. Once an incipient pool system is formed it is easy to envisage the development of concentric patterning by local fusion of contiguous pools during periods of high water level, as suggested by Boatman and Armstrong (1968) and by Vitt et al. (1975).

Although mechanisms for the initial development of surface patterning are still obscure, there are many indications as to how a surface micro-relief, once established, can be perpetuated. Conditions in ombrotrophic pools are apparently suboptimal for the growth of *Sphagnum cuspidatum* (Boatman, 1977), so that its establishment in a pool may lead to a reduced rate of peat accumulation (Boatman and Tomlinson, 1977). In deeper pools, colonization by *S. cuspidatum* may be completely inhibited (Boatman and Armstrong, 1968), whereas in shallow bare pools initial colonization by algae may lead to the development of a rapidly photosynthesizing scum which retards the accumulation of peat (Moore and Bellamy, 1973). In periods of prolonged drought, hummocks may remain moist for a longer time than adjacent hollows, on account of the greater water-retaining abilities of the hummock-forming *Sphagnum* spp. (Overbeck, 1963). Differential growth rates of the component species have been shown to occur in hummock/hollow situations (Clymo, 1970), and in some situations at least hummock species may show faster growth rates than hollow species (Boatman and Tomlinson, 1977). In the driest situations, on the higher hummocks, peat accumulation may be negligible.

DISCUSSION

The evidence presented in the preceding pages demonstrates convincingly the inherent changeability of the wetland habitat. Whether it be from direct observations of a given habitat over a long period of years, or from examination of the plant and animal remains preserved in the underlying peat deposits, there is ample evidence of sequential changes in the vegetation, and these can be paralleled with greater or lesser precision at other similar sites. That is to say, there are some common patterns in the vegetation changes. However, instead of one or two major pathways of vegetation change in the past, there appear to have been a number of variations. Some of these variations may have been the result of regional climatic differences, which undoubtedly operate at the present day to give clear-cut geographical patterns in the distribution of the major peatland types (Moore and Bellamy, 1973). Further analysis of the diversity of vegetation changes in the past is probably not possible in the present state of our knowledge, though it is clear that explanations based solely on the classic *Verlandungshypothese* are inadequate. Some sort of *Verlandung* sequence can, it is true, be recognized in the stratigraphic sequences from all regions, since the sequence represents a particular set of plant responses to a local situation of abundant free water. In regions of low rainfall these situations (open-water lakes and ponds) may be the only places where peat accumulation is possible. In regions of high rainfall situations suitable for the development of wetland communities are far more widespread; instead of considering just the vegetation changes occurring in a single basin, it may be necessary to view changes in wetland communities in the context of the evolution of the total landscape. Allogenic processes may accordingly be of more consequence than autogenic processes. It may be possible to predict the course of some vegetation changes, but the total pattern of change is inevitably more complex. Heinselman (1963) summarized his findings for the Lake Agassiz Peatlands in Minnesota as follows: "The various peatland types cannot be regarded as merely stages in the orderly development of vegetation towards mesophytic communities. Instead, each is only the net result of the interplay of a multitude of physical and biological events in the history of the landscape. ... *Raising* of the bog surface above the water-table is said to be the essential process of bog succession. ... Yet peat accumulation has actually *caused* a rise in water-tables over thousands of square miles."

Even within a single basin, a far more complex interplay of autogenic and allogenic processes than

Weber envisaged may be involved, as is shown by the work of Casparie (1972) on Bourtanger Moor, on the northern border between western Germany and The Netherlands (i.e., within the same general region as that in which Weber carried out his classic studies). Casparie studied a large number of exposed peat faces over a 20 km^2 area in the southwest of Bourtanger Moor, occupying a more or less flat-bottomed ice-margin valley, the Hunze Depression. The basic sequence of deposits he found was similar to that formulated by Weber (p. 313), but the overall developmental history was far more complex. The earliest organic deposits were formed along the deeper, western, side of the Hunze Depression, under the influence of gently northwestward-flowing ground water. Extensive desiccation of the fen peat surface occurred around 7300 B.P., as a result of the impedance of this water flow by peat growth immediately to the south. Colonization by pines ensued, but on the eastern side of the valley further fen peat started to accumulate under the influence of iron-rich waters welling up to the surface through underlying permeable sands. Radiocarbon dating showed not one but three distinct phases of pine growth: 7300–6900 B.P., 6500–6000 B.P. and 5050–4900 B.P. The progressive development of ombrotrophic peat after *c.* 6500 B.P. led to the gradual obliteration of the primary water sources, and the formation of a new drainage system flowing northward between six contiguous domes of ombrotrophic peat. An increase in precipitation around 4000 B.P. led to the development of persistent hummock/hollow complexes over these areas, of four distinct types. Casparie found evidence of two phases of peat erosion in the Hunze Depression. The first, around 5050 B.P., of a rather local character, probably resulted from the sudden drainage of a lake to the west of the Moor, which formed the major site of infiltration for the iron-rich seepage waters. The second erosion phase, around 2500 B.P., was much more widespread, and resulted from the catastrophic discharge of a superficial bog lake situated between the ombrotrophic peat domes. This bog-burst effectively drained off the surface waters from the pools of the hummock/hollow complexes, and led to the formation of a more uniform *Sphagnum* carpet. Casparie's work in the Hunze Depression is further discussed by Frenzel (Ch. 2).

The *Verlandungshypothese* relies on two basic assumptions: (1) that the accrual of peat leads inevitably to autogenic and predictable changes in the supply of water and ions to the habitat; and (2) that changes in the vegetation lead invariably towards a stable climax state. Clementsian ideas of plant succession would identify this climax state with mesophytic forest in nearly all regions where peat accumulation is possible. However, it is still uncertain whether mesophytic forest communities can ever become established on *deep* peat, whilst it seems probable that mire types such as blanket-bog and muskeg can represent true climatic climaxes in certain topographic situations. In northern Minnesota, where bog forest communities are widespread both at the present day and in the stratigraphic record, Heinselman (1963) has demonstrated a wide variety of processes of change associated with wetland communities. These include:

(1) Lake filling (terrestrialization), typically via a floating sedge-mat stage.

(2) Downslope and lateral bog expansion (paludification), often but not necessarily preceded by gleying or podzolization of the soils.

(3) Upslope peat advance (paludification), by obstruction of the water flow and consequent ponding-up of the water table.

(4) The local isolation of sites from mineral-influenced waters as a result of peat growth.

(5) Correlated rises in the peat surface and the local water table, associated with the development of ombrogenous *Sphagnum* peat.

(6) The development of patterned wetland communities, related to some sort of ridge-and-hollow surface topography, and often involving an expansion rather than a contraction of open water (Vitt et al., 1975).

Whilst change in wetland communities has undoubtedly occurred in the past and is occurring today, it is important to remember the limitations of the evidence. The direct evidence rarely relies solely on a single person's observations; because of the long periods of time involved, the observations are usually based on comparisons with earlier data or maps compiled by other people. These data or maps may not be in an immediately comparable form; in the case of maps, the delimitation of vegetation boundaries may be a somewhat subjective process (particularly when they are derived from aerial photographs), whilst for telmatic communities there are often considerable technical difficulties involved in accurate surveying at the edge of open water. The stratigraphic evidence relies on the accurate identification of plant and animal tissues; in addition to the problems associated with the precise recognition of fragments of plant or animal material, the extent of decomposition in many peats can be considerable. How far differential preservation may be a factor affecting the reconstruction of past wetland communities is still largely unknown. Moreover, most present-day wetland communities, once the peat surface rises above the level of the water table, are characterized by an irregular surface topography; in extreme cases (Kulczyński, 1949) the community may consist of distinct layers of immersive and emersive perennials (e.g. large species of *Carex* and hypnoid mosses), which behave independently. In the stratigraphic record these layers would be preserved as superposed strata, and hence could be interpreted erroneously as representing successive communities in time.

Whilst the evidence for vegetation change is based largely on *first-hand* information (direct observations or preserved remains), the evidence for associated habitat changes is largely indirect and inferential. Much of the evidence relies on the analogy between vegetation changes in the past and the present-day spatial relationships of wetland communities: the habitat differences existing today in space are inferred to have taken place in time in the past. As always in ecological investigations, one can rarely be sure that the measurable parameters of the habitat are in fact the immediately significant ones for the plant. Where chemical analyses of peat deposits are carried out, inferences about habitat changes are particularly difficult to make, as explained above (p. 338). Where long-term observations of a single habitat reveal appreciable vegetation changes, then some assessment of accompanying habitat changes may be possible, but, as the American examples quoted earlier show, over the periods of time covered by direct observations relatively short-term climatic fluctuations may override or obscure the major trends of hydroseral change. In addition, small-scale allogenic disturbances, such as slow land subsidence or minor anthropogenic modifications of lake levels, which could be significant in determining patterns of vegetation change, may not be recognized if observations are made only periodically at sites where local knowledge is not readily available.

West (1968) gives values of 1 to 3 cm per century for rates of land subsidence due to isostatic movements in parts of northwestern Europe, and since, as shown earlier, the majority of peat deposits accumulate at rates of from 2 to 8 cm per century, in certain situations the two processes may more or less counterbalance each other. The gradual and progressive rise in water table associated with the land subsidence is thus offset by peat accrual. Such a mechanism might account for the occurrence in some peat profiles of thick uniform layers formed from wetland communities which at the present day have only a limited tolerance range to changes in the water table (Kulczyński, 1949). More typically at the present day, changes in water levels are sudden and not progressive, as a result of anthropogenic disturbance, and the existing wetland communities do not have time to adapt. Changes in water level probably have a greater effect on semiterrestric and terrestric communities, since these apparently have the smallest ranges of tolerance (Fig. 9.1). However, where changes in water level are sufficient to upset the balance between the supply of minerotrophic and ombrotrophic waters to a particular peat surface, then the visible vegetation changes may be most marked.

A factor of great significance to patterns of vegetation change in wetland communities is eutrophication, and this is potentially the most important of contemporary anthropogenic disturbances. Eutrophication may involve the additional input of certain minerals into the ground-water supplies, as a result of the application of fertilizers to neighbouring agricultural land, or of increased leaching and soil erosion associated with modern methods of crop husbandry. The widespread use of detergents today is a major source of increased phosphate levels in many drainage waters, and as has already been shown the supply of phosphorus may be limiting in many wetland communities. In addition, eutrophication may involve increases in the mineral content of natural rain water, as a result of increased particulate matter in the atmosphere from agriculture (fertilizer dust) or industry (soot, etc.). Sulphur dioxide pollution can also lead to increased acidity in rain water. The patterns of change in wetland communities are accompanied typically by trends of increasing oligotrophy and mineral deficiencies. Since eutrophication necessarily acts by upsetting the balance of ions supplied to the peat surface, it may be creating situations at the present day where entirely new pathways of change and entirely new climax communities have to be envisaged.

REFERENCES

Backéus, I., 1972. Bog vegetation re-mapped after sixty years. *Oikos*, 23: 384–393.

Bellamy, D.J. and Rieley, J., 1967. Some ecological statistics of a "miniature bog". *Oikos*, 18: 33–40.

Björkbäck, F., 1965. Älgfloarna, a mixed mire complex in Jämtland. *Acta Phyt. Suec.*, 50: 188–192.

Boatman, D.J., 1977. Observations on the growth of *Sphagnum cuspidatum* in a bog pool on the Silver Flowe National Nature Reserve. *J. Ecol.*, 65: 119–126.

Boatman, D.J. and Armstrong, W., 1968. A bog type in northwest Sutherland. *J. Ecol.*, 56: 129–141.

Boatman, D.J. and Tomlinson, R.W., 1973. The Silver Flowe. I. Some structural and hydrological features of Brishie Bog and their bearing on pool formation. *J. Ecol.*, 61: 653–666.

Boatman, D.J. and Tomlinson, R.W., 1977. The Silver Flowe. II. Features of the vegetation and stratigraphy of Brishie Bog, and their bearing on pool formation. *J. Ecol.*, 65: 531–546.

Boelter, D.H., 1965. Hydraulic conductivity of peat. *Soil Sci.*, 100: 227–231.

Boelter, D.H., 1966. Hydrological characteristics of organic soils in Lake States watersheds. *J. Soil Water Conserv.*, 21: 50–53.

Boelter, D.H., 1968. Important physical properties of peat materials. In: C. Lafleur and J. Butler (Editors), *Proc. 3rd Int. Peat Congress, Quebec*, pp. 150–154.

Boyd, C.E. and Hess, L.W., 1970. Factors influencing shoot production and mineral nutrient levels in *Typha latifolia*. *Ecology*, 51: 296–300.

Buell, M.F. and Buell, H.F., 1941. Surface level fluctuation in Cedar Creek Bog, Minnesota. *Ecology*, 22: 317–321.

Buell, M.F. and Cain, R.L., 1943. The successional role of southern white cedar, *Chamaecyparis thyoides*, in southeastern North Carolina. *Ecology*, 24: 85–91.

Buell, M.F., Buell, H.F. and Reiners, W.A., 1968. Radial mat growth on Cedar Creek Bog, Minnesota. *Ecology*, 49: 1198–1199.

Buttery, B.R. and Lambert, J.M., 1965. Competition between *Glyceria maxima* and *Phragmites communis* in the region of Surlingham Broad. I. The competition mechanism. *J. Ecol.*, 53: 163–181.

Casparie, W.A., 1972. Bog development in southeastern Drenthe (The Netherlands). *Vegetatio*, 25: 1–271.

Chouard, P. and Prat, H., 1929. Note sur les tourbières du Massif de Néouvielle (Hautes-Pyrénées). *Bull. Soc. Bot. Fr.*, 76: 113–130.

Clapham, A.R. and Godwin, H., 1948. Studies in the post-glacial history of British vegetation. VIII. Swamping surfaces in peats of the Somerset Levels. *Philos. Trans. R. Soc. B*, 233: 233–249.

Clements, F.E., 1916. *Plant Succession*. Carnegie Institution of Washington, Washington, D.C., 512 pp.

Clymo, R.S., 1963. Ion exchange in *Sphagnum* and its relation to bog ecology. *Ann. Bot. Lond., N.S.*, 27: 309–324.

Clymo, R.S., 1970. The growth of *Sphagnum*: methods of measurement. *J. Ecol.*, 58: 13–49.

Clymo, R.S. and Reddaway, E.J.F., 1974. Growth rate of *Sphagnum rubellum* Wils. on Pennine blanket bog. *J. Ecol.*, 62: 191–196.

Conway, V.M., 1949. The bogs of Central Minnesota. *Ecol. Monogr.*, 19: 173–206.

Dachnowski, A.P., 1925. Profiles of peatlands within limits of extinct glacial lakes Agassiz and Wisconsin. *Bot. Gaz.*, 80: 345–366.

Dachnowski, A.P., 1926. Profiles of peat deposits in New England. *Ecology*, 7: 120–135.

Dachnowski-Stokes, A.P., 1927. The botanical composition and morphological features of "highmoor" peat profiles in Maine. *Soil Sci.*, 27: 379–388.

Daniels, R.E., 1975. Observations on the performance of *Narthecium ossifragum* (L.) Huds. and *Phragmites communis* Trin. *J. Ecol.*, 63: 965–977.

Dansereau, P. and Segadas-Vianna, F., 1952. Ecological study of the peat bogs of eastern North America. *Can. J. Bot.*, 30: 490–520.

Davis, M.B., 1958. Three pollen diagrams from central Massachusetts. *Am. J. Sci.*, 256: 540–570.

Davis, M.B., 1969. Climatic changes in southern Connecticut recorded by pollen deposition at Rogers Lake. *Ecology*, 50: 409–422.

Dickinson, W., 1975. Recurrence surfaces in Rusland Moss, Cumbria (formerly North Lancashire). *J. Ecol.*, 63: 913–935.

Duncan, D.P., 1954. A study of some of the factors affecting the natural regeneration of tamarack (*Larix laricina*) in Minnesota. *Ecology*, 35: 498–521.

Ernst, O., 1934. Zur Geschichte der Moore, Marschen und Wälder Nordwestdeutschlands IV. Untersuchungen in Nordfriesland. *Schr. Naturwiss. Ver. Schleswig-Holstein*, 20: 209–334.

Eurola, S., 1962. Über die regionale Einteilung südfinnischen Moore. *Ann. Bot. Soc. "Vanamo"*, 33: 1–243.

Forrest, G.I., 1971. Structure and productivity of North Pennine blanket bog vegetation. *J. Ecol.*, 59: 453–479.

Fries, M., 1962. Pollen profiles of late Pleistocene and recent sediments from Weber Lake, northeastern Minnesota. *Ecology*, 43: 295–308.

Frolik, A.L., 1941. Vegetation on the peat lands of Dane County, Wisconsin. *Ecol. Monogr.*, 11: 117–140.

Gates, F.C., 1926. Plant successions about Douglas Lake, Cheboygan County, Michigan. *Bot. Gaz.*, 82: 170–182.

Gates, F.C., 1942. The bogs of northern Lower Michigan. *Ecol. Monogr.*, 33: 327–374.

Gimingham, C.H., 1972. *Ecology of Heathlands*. Chapman and Hall, London, 266 pp.

Godwin, H., 1929. The "sedge" and "litter" of Wicken Fen. *J. Ecol.*, 17: 148–160.

Godwin, H., 1936. Studies in the ecology of Wicken Fen. III.

The establishment and development of fen carr (scrub). *J. Ecol.*, 24: 82–116.

Godwin, H., 1943. Coastal peat beds of the British Isles and the North Sea. *J. Ecol.*, 31: 199–247.

Godwin, H. and Bharucha, F.R., 1932. Studies in the ecology of Wicken Fen. II. The fen water table and its control of plant communities. *J. Ecol.*, 20: 157–191.

Godwin, H. and Clifford, M.H., 1939. Studies of the post-glacial history of British vegetation. I. Origin and stratigraphy of Fenland deposits near Woodwalton, Hunts. *Philos. Trans. R. Soc. B*, 229: 323–363.

Godwin, H. and Conway, V.M., 1939. The ecology of a raised bog near Tregaron, Cardiganshire. *J. Ecol.*, 27: 313–363.

Godwin, H. and Turner, J.S., 1933. Soil acidity in relation to vegetational succession in Calthorpe Broad, Norfolk. *J. Ecol.*, 21: 235–262.

Godwin, H., Clowes, D.R. and Huntley, B., 1974. Studies in the ecology of Wicken Fen. V. Development of fen carr. *J. Ecol.*, 62: 197–214.

Gorham, E., 1950. Variation in some chemical conditions along the borders of a *Carex lasiocarpa* fen community. *Oikos*, 2: 217–240.

Gorham, E., 1974. The relationship between standing crop in sedge meadows and summer temperature. *J. Ecol.*, 62: 487–491.

Granlund, E., 1932. De svenska högmossarnas geologi. *Sver. Geol. Unders. Ser. C*, 373: 1–193.

Green, B.H. and Pearson, M.C., 1968. The ecology of Wybunbury Moss, Cheshire. I. The present vegetation and some physical, chemical and historical factors controlling its nature and distribution. *J. Ecol.*, 65: 245–267.

Green, B.H. and Pearson, M.C., 1977. The ecology of Wybunbury Moss, Cheshire. II. Post-glacial history and the formation of the Cheshire mere and mire landscape. *J. Ecol.*, 65: 793–814.

Griffin, K.O., 1977. Palaeoecological aspects of the Red Lake Peatland, northern Minnesota. *Can. J. Bot.*, 55: 172–192.

Hansen, B., 1966. The raised bog Draved Kongsmose. *Bot. Tidsskr.*, 62: 146–185.

Haslam, S.M., 1965. Ecological studies in the Breck Fens. I. Vegetation in relation to habitat. *J. Ecol.*, 53: 599–619.

Haslam, S.M., 1971. The development and establishment of young plants of *Phragmites communis* Trin. *Ann. Bot., Lond., N.S.*, 35: 1059–1072.

Haslam, S.M., 1972. Biological Flora of the British Isles. *Phragmites communis* Trin. *J. Ecol.*, 60: 585–610.

Heilman, P.E., 1966. Change in distribution and availability of nitrogen with forest succession on north slopes in interior Alaska. *Ecology*, 47: 825–831.

Heilman, P.E., 1968. Relationships of availability of phosphorus and cations to forest succession and bog formation in interior Alaska. *Ecology*, 49: 331–336.

Heinselman, M.L., 1963. Forest sites, bog processes, and peatland types in the glacial Lake Agassiz region, Minnesota. *Ecol. Monogr.*, 33: 327–374.

Heinselman, M.L., 1970. Landscape evolution, peatland types, and the environment in the Lake Agassiz Peatlands Natural Area, Minnesota. *Ecol. Monogr.*, 40: 235–261.

Hicks, S.P., 1971. Pollen-analytical evidence for the effect of prehistoric agriculture on the vegetation of north Derbyshire. *New Phytol.*, 70: 647–667.

Isaak, D., Marshall, W.H. and Buell, M.F., 1959. A record of reverse plant succession in a tamarack bog. *Ecology*, 40: 317–320.

Janssen, C.R., 1967. Stevens Pond: a postglacial pollen diagram from a small *Typha* swamp in northwestern Minnesota, interpreted from pollen indicators and surface samples. *Ecol. Monogr.*, 37: 145–172.

Janssen, C.R., 1968. Myrtle Lake: a Late- and Postglacial pollen diagram from northern Minnesota. *Can. J. Bot.*, 46: 1397–1408.

Jewell, M.E. and Brown, H.W., 1929. Studies on northern Michigan bog lakes. *Ecology*, 10: 427–475.

Kivinen, E. and Tolonen, K., 1972. Excursion guide. F. Joensuu. Virgin peatlands, forestry, cultivation. In: *Proc. 4th Int. Peat Congress, Otaniemi, 1972.*

Kulczyński, S., 1949. Peat bogs of Polesie. *Mém. Acad. Sci. Cracovie B*, pp. 1–356.

Kurz, H., 1928. Influence of *Sphagnum* and other mosses on bog reactions. *Ecology*, 9: 56–69.

Lambert, J.M., 1951. Alluvial stratigraphy and vegetational succession in the region of the Bure Valley Broads. III. Classification, status and distribution of communities. *J. Ecol.*, 39: 149–170.

Lambert, J.M., 1965. The vegetation of Broadland. In: E.A. Ellis (Editor), *The Broads*. Collins, London, pp. 69–92.

LeBarron, R.K. and Neetzel, J.R., 1942. Drainage of forested swamps. *Ecology*, 23: 457–465.

Lewis, F.J., Dowding, E.S. and Moss, E.H., 1928. The vegetation of Alberta. II. The swamp, moor, and bog forest vegetation of central Alberta. *J. Ecol.*, 16: 19–70.

Lohammar, G., 1965. The vegetation of Swedish lakes. *Acta Phyt. Suec.*, 50: 28–47.

Lund, J.W.G., 1971. Eutrophication. In: E. Duffey and A.S. Watt (Editors), *The Scientific Management of Animal and Plant Communities for Conservation.* Blackwell, Oxford, pp. 225–240.

McVean, D.N., 1956. Ecology of *Alnus glutinosa* (L.) Gaertn. III. Seedling establishment. *J. Ecol.*, 44: 194–218.

McVean, D.N., 1963. Growth and mineral nutrition of Scots pine seedlings on some common peat types. *J. Ecol.*, 51: 657–670.

Mathewes, R.W., 1973. A palynological study of postglacial vegetation changes in the University Research Forest, southeastern British Columbia. *Can. J. Bot.*, 51: 2085–2103.

Millington, R.J., 1954. *Sphagnum* bogs of the New England plateau, New South Wales. *J. Ecol.*, 42: 328–344.

Mitchell, G.F., 1956. Post-Boreal pollen diagrams from Irish raised bogs. *Proc. R. Ir. Acad.*, 57 B: 185–251.

Moizuk, G.A. and Livingston, R.B., 1966. Ecology of red maple (*Acer rubrum* L.) in a Massachusetts upland bog. *Ecology*, 47: 942–950.

Moore, P.D., 1973. The influence of prehistoric cultures upon the initiation and spread of blanket bog in upland Wales. *Nature, Lond.*, 241: 350–353.

Moore, P.D., 1975. Origin of blanket mires. *Nature, Lond.*, 256: 267–269.

Moore, P.D., 1977. Stratigraphy and pollen analysis of Claish Moss, northwest Scotland: significance for the origin of surface-pools and forest history. *J. Ecol.*, 65: 375–397.

Moore, P.D. and Bellamy, D.J., 1973. *Peatlands*. Elek, London, 221 pp.

Mörnsjö, T., 1968. Stratigraphical and chemical studies on two peatlands in Scania, South Sweden. *Bot. Not.*, 121: 343–360.

Moss, E.H., 1953. Marsh and bog vegetation in northwestern Alberta. *Can. J. Bot.*, 31: 448–470.

Nichols, H., 1967. Pollen diagrams from sub-arctic Central Canada. *Science, N.S.*, 155: 1665–1668.

Nilsson, T., 1964. Entwicklungsgeschichtliche Studien im Ageröds Mosse, Schonen. *Lunds Univ. Åsskr., N.F. 2*, 59: 5–34.

Osvald, H., 1923. Die Vegetation des Hochmoores Komosse. *Sven. Växtsociol. Sälls. Handl.*, 1: 1–436.

Osvald, H., 1935. A bog at Hartford, Michigan. *Ecology*, 16: 520–528.

Osvald, H., 1955. The vegetation of two raised bogs in north-eastern Maine. *Sven. Bot. Tidskr.*, 49: 110–118.

Overbeck, F., 1963. Aufgaben botanisch-geologischer Moorforschung in Nordwestdeutschland. *Ber. Dtsch. Bot. Ges.*, 76 B: 2–12.

Overbeck, F., Münnich, K.O., Aletsee, L. and Averdieck, F.R., 1957. Das Alter des "Grenzhorizonts" norddeutscher Hochmoore nach Radiocarbon-Datierungen. *Flora*, 145: 37–71.

Pallis, M., 1915. The structure and history of Plav: the floating fen of the Delta of the Danube. *J. Linn. Soc. Bot.*, 43: 233–290.

Pearsall, W.H., 1918. The aquatic and marsh vegetation of Esthwaite Water. *J. Ecol.*, 6: 53–74.

Pearsall, W.H., 1956. Two blanket bogs in Sutherland. *J. Ecol.*, 44: 493–516.

Pearsall, W.H. and Pennington, W., 1973. *The Lake District. A Landscape History*. Collins, London, 320 pp.

Pearson, M.C., 1979. Patterns of pools in peatlands (with particular reference to a valley head mire in northern England). *Acta Univ. Oulu., Ser. A*, 82 Geol., 3: 65–72.

Pennak, R.W., 1963. Ecological and radiocarbon correlations in some Colorado mountain lake and bog deposits. *Ecology*, 44: 1–15.

Pennington, W., Haworth, E.Y., Bonny, A.P. and Lishman, J.P., 1972. Lake sediments in northern Scotland. *Philos. Trans. R. Soc.*, B 264: 191–294.

Pigott, C.D. and Wilson, J.F., 1978. The vegetation of North Fen at Esthwaite in 1967–9. *Proc. R. Soc. Lond.*, B 200: 331–351.

Poore, M.E.D., 1956. The ecology of Woodwalton Fen. *J. Ecol.*, 44: 455–492.

Rigg, G.B., 1925. Some *Sphagnum* bogs of the North Pacific coast of America. *Ecology*, 6: 260–278.

Rigg, G.B., 1940. Comparisons of the development of some *Sphagnum* bogs of the Atlantic coast, the interior, and the Pacific coast. *Am. J. Bot.*, 27: 1–14.

Rigg, G.B. and Gould, H.R., 1957. Age of Glacier Peak eruption and chronology of post-glacial peat deposits in Washington and the surrounding areas. *Am. J. Sci.*, 255: 341–363.

Rigg, G.B. and Harrar, E.S., 1931. The root systems of trees growing in *Sphagnum*. *Am. J. Bot.*, 18: 391–397.

Rigg, G.B. and Richardson, C.T., 1938. Profiles of some *Sphagnum* bogs of the Pacific Coast of North America. *Ecology*, 19: 408–434.

Ritchie, J.C., 1969. Absolute pollen frequencies and carbon-14 age of a section of Holocene lake sediment from the Riding Mountain area of Manitoba. *Can. J. Bot.*, 47: 1345–1349.

Rodewald-Rudescu, L., 1974. *Das Schilfrohr* Phragmites communis *Trinius*. Die Binnengewässer, 27. Schweizerbart'sche, Stuttgart, 302 pp.

Rudolph, K. and Firbas, F., 1927. Paläofloristische und stratigraphische Untersuchungen böhmischer Moore. III. Die Moore des Riesengebirges. *Beih. Bot. Zentralbl., Abt. 2*, 43: 69–144.

Rutter, A.J., 1963. Studies in water relations of *Pinus sylvestris* in plantation conditions. I. Measurements of rainfall and interception. *J. Ecol.*, 51: 191–203.

Rutter, A.J. and Fourt, D.F., 1965. Studies in water relations of *Pinus sylvestris* in plantation conditions. III. A comparison of soil water changes and estimates of total evaporation on four afforested sites and one grass-covered site. *J. Appl. Ecol.*, 2: 197–209.

Rybníček, K., 1973. A comparison of the present and past mire communities of Central Europe. In: H.J.B. Birks and R.G. West (Editors), *Quaternary Plant Ecology*. Blackwell, Oxford, pp. 237–261.

Schwintzer, C.R., 1978. Vegetation and nutrient status of northern Michigan fens. *Can. J. Bot.*, 56: 3044–3051.

Sernander, R., 1908. On the evidence of Postglacial changes of climate furnished by the peat-mosses of northern Europe. *Geol. Fören. Stockholm Förh.*, 30: 465–473.

Sinker, C.A., 1962. The North Shropshire meres and mosses: a background for ecologists. *Field Stud.*, 1: 101–138.

Sjörs, H., 1961. Surface patterns in Boreal peatland. *Endeavour*, 20: 217–224.

Skene, M., 1915. The acidity of *Sphagnum*, and its relation to chalk and mineral salts. *Ann. Bot. Lond.*, 29: 65–87.

Spence, D.H.N., 1964. The macrophytic vegetation of freshwater lochs, swamps and associated fens. In: J.H. Burnett (Editor), *The Vegetation of Scotland*. Oliver and Boyd, Edinburgh, pp. 306–425.

Swan, J.M.A. and Gill, A.M., 1970. The origins, spread, and consolidation of a floating bog in Harvard Pond, Petersham, Massachusetts. *Ecology*, 51: 829–840.

Tallis, J.H., 1973. The terrestrialisation of lake basins in north Cheshire, with special reference to the development of a 'schwingmoor' structure. *J. Ecol.*, 61: 537–567.

Tallis, J.H., 1975. Tree remains in southern Pennine peats. *Nature, Lond.*, 256: 482–484.

Tallis, J.H. and Switsur, V.R., 1973. Studies on southern Pennine peats. VI. A radiocarbon-dated pollen diagram from Featherbed Moss, Derbyshire. *J. Ecol.*, 61: 743–751.

Tansley, A.G., 1949. *The British Islands and their Vegetation, 2*. Cambridge University Press, Cambridge, 443 pp.

Tolonen, K., 1966. Stratigraphic and rhizopod analyses on an old raised bog, Varrassuo, in Hollola, South Finland. *Ann. Bot. Fenn.*, 3: 147–166.

Tolonen, K., 1967. Über die Entwicklung der Moore im finnischen Nordkarelien. *Ann. Bot. Fenn.*, 4: 219–416.

Tolonen, K., 1971. On the regeneration of Northeuropean bogs. I. Klaukkalan Isosuo in S. Finland. *Acta Agralia Fenn.*, 123: 143–166.

Transeau, E.N., 1903. On the geographic distribution and

ecological relationships of the bog plant societies of northern North America. *Bot. Gaz.*, 36: 401–420.

Turner, J. and Chambers, K., 1972. Quaternary studies of Valley Bog. *Moor House Annu. Rep.*, 13: 16–23.

Turner, J., Hewetson, V.P., Hibbert, F.A., Lowry, K.H. and Chambers, C., 1973. The history of the vegetation and flora of Widdybank Fell and the Cow Green reservoir basin, Upper Teesdale. *Philos. Trans. R. Soc.* B 265: 327–408.

Vitt, D.H. and Slack, N.G., 1975. An analysis of the vegetation of *Sphagnum*-dominated kettle-hole bogs in relation to environmental gradients. *Can. J. Bot.*, 53: 332–359.

Vitt, D.H., Achuff, P. and Andruss, R.E., 1975. The vegetation and chemical properties of patterned fens in the Swan Hills, north central Alberta. *Can. J. Bot.*, 53: 2776–2795.

Vogl, R.J., 1969. One hundred and thirty years of plant succession in a southeastern Wisconsin lowland. *Ecology*, 50: 248–255.

Von Post, L., 1912. Über die stratigraphische Zweigliederung schwedischer Hochmoore. *Sver. Geol. Unders., Ser. C*, 248: 1–52.

Von Post, L. and Sernander, R., 1910. Pflanzenphysiognomische Studien auf Torfmooren in Närke. *Livretguide des excursions en Suède du XIe Congr. Géol. Int. 14 Stockholm*, 48 pp.

Walker, D., 1970. Direction and rate in some British Post-glacial hydroseres. In: D. Walker and R.G. West (Editors), *Studies in the Vegetational History of the British Isles*. University Press of Cambridge, Cambridge, pp. 117–139.

Walker, D. and Walker, P.M., 1961. Stratigraphic evidence of regeneration in some Irish bogs. *J. Ecol.*, 49: 169–185.

Walker, P.C. and Hartman, R.T., 1960. The forest sequence of the Hartstown Bog area in western Pennsylvania. *Ecology*, 41: 461–474.

Warming, E., 1896. *Lehrbuch der ökologischen Pflanzengeographie*. Gebrüder Borntraeger, Berlin, 412 pp.

Watt, R.F. and Heinselman, M.L., 1965. Foliar nitrogen and phosphorus level related to site quality in a northern Minnesota spruce bog. *Ecology*, 46: 357–361.

Weber, C.A., 1908. Aufbau und Vegetation der Moore Norddeutschlands. *Engler's Bot. Jahrb.*, 40(Suppl.): 19–34.

West, G., 1905. A comparative study of the dominant phanerogamic and higher cryptogamic flora of aquatic habit in three lake areas of Scotland. *Proc. R. Soc. Edinb.*, 25: 967–1023.

West, G., 1910. A further contribution to a comparative study of the dominant phanerogamic and higher cryptogamic flora of aquatic habit in Scottish lakes. *Proc. R. Soc. Edinb.*, 30: 65–181.

West, R.G., 1968. *Pleistocene Geology and Biology*. Longmans, London, 377 pp.

Westlake, D.F., 1963. Comparisons of plant productivity. *Biol. Rev.*, 38: 385–425.

Wetzel, R.G., 1970. Recent and postglacial production rates of a marl lake. *Limnol. Oceanogr.*, 15: 491–503.

Wilde, S.A., Voigt, G.K. and Pierce, R.S., 1954. The relationship of soils and forest growth in the Algoma District of Ontario, Canada. *J. Soil Sci.*, 5: 22–38.

Chapter 10

THE ANIMALS[1]

MARTIN C.D. SPEIGHT and ROBERT E. BLACKITH

INTRODUCTION

More than three-quarters of the world's biological species are animals, and a very high proportion of these animals may find their way into accumulative wetlands at least on a temporary basis, but only characteristically wetland species are considered in any detail in this chapter. This is not to say that more generally occurring species are of no significance in wetlands; locally they may be of considerable significance — for a small swamp in Sri Lanka, the daily visit of a family of elephants on its rounds probably modifies the vegetation more than do the activities of its more permanent inhabitants.

Unless ideas and facts are selected with great care, attempts to describe wetland animals as a group become attempts to compress the greater part of animal ecology into a framework which is neither appropriate nor adequate. This chapter seeks to give some broad principles (with which some may disagree), illustrating these by reference to particular groups of organisms. Animal groups about which we do not have adequate information have been simply ignored; this cavalier treatment is applied partly from necessity and partly from defects peculiar to the authors.

In keeping with the other chapters of this volume most emphasis is given to peaty or accumulative wetlands. While references to other types of wetland are not excluded, no attempt has been made to distinguish animals in this sense.

SURVIVAL IN WETLANDS

For animals, the physical character of bog, fen, swamp and mires in general undergoes a most uncomfortable oscillation of conditions; for part of the time, even the major part, the general ground surface is covered in water to a depth of anything from a few centimetres to a metre or so, for the rest of the time it is dry or perhaps soggy. The inhospitable nature of these accumulative wetlands is most clearly manifest in the fact that they *are* accumulative; were detritivore and herbivore able to take full advantage of the annual primary production, rendering it all vulnerable to microbiological attack, and were the micro-organisms responsible for decay processes able to function effectively, peat would not accumulate. The exacting nature of these wetlands is also shown by the range of adaptations exhibited by wetland animals. Many of these adaptations tend to render wetland species inviable outside wetlands, so wetland faunas contain many species which do not also occur elsewhere. Of course, other species do occur both in and out of wetlands, particularly in different parts of their geographical range. Thus, at the southern end of its range in continental Europe the lizard *Lacerta vivipara* is restricted to reed-beds along the Danube (Turcek, 1976), while in the oceanic climate of Great Britain it is ubiquitous. The ground beetle *Carabus granulatus* exhibits a similar tendency, being confined to wetlands in Great Britain but ubiquitous in the extremely oceanic climate of Ireland.

Adaptations promoting survival in water are inappropriate during the dry phases of wetlands, and adaptations for dry land equally out of place during wet phases. Inevitably, structures like the webbed feet possessed by ducks and their allies, which can be used both in water and on dry land,

---

[1]Manuscript received July, 1977.

are at a premium among wetland animals. But some functions, notable among them being respiration, cannot so easily be performed by such dual-purpose apparatus. Seasonal oscillation between wet and dry conditions involves change not only in the character of the physical environment in which the wetlands animal is attempting to survive, but also in the character of the biotic environment: wetlands have marked cycles of biological activity related to the local hydrologic cycle. As a consequence wetland animals also need mechanisms enabling them to "keep in step" with the bursts of production, timing their life cycles or their migrations accordingly.

Survival *in situ*

Animals remaining on-site throughout the year in a wetland environment require methods of coping with the oscillating conditions to which they are then subject, and major modifications of structure and biology exhibited by wetland animals have evidently been developed in response to this particular need. Where the wet/dry oscillations are of brief duration there is a tendency for mechanical devices to be employed that enable the organisms concerned to remain active throughout. At the other extreme, some form of suspended animation is characteristic. Modifications enabling species to cope with other attributes of wetlands are evident too — for instance the elongate, spatulate toes of the lily-trotter *Jacana* spp. (lotus bird) and their analogue in the modified pasterns of the sitatunga (*Tragelaphus spekii*), a swampland antelope, which enable these species to move freely over floating vegetation that would otherwise not support their weight.

Physical gills

Terrestrial animals which have come to occupy wetland habitats have respiratory systems based upon the use of atmospheric oxygen, so that in times of flooding survival *in situ* is, for them, dependent upon mechanisms providing contact with a gaseous supply of oxygen. Pulmonate molluscs of swamp and fen, such as *Lymnaea truncatula*, a species characteristic of temporarily flooded areas in Europe, use the simple expedient of climbing up to the water surface to replenish their oxygen supply, poking through the surface film a periscopic device tipped by the pneumostome. These snails store their air supply in the mantle cavity, replenishing it each time they return to the water surface. Other pulmonates of marshy places, such as *Cochlicopa lubrica*, are primarily terrestrial and, though they can withstand temporary submersion, they do not possess modifications permitting prolonged immersion. A variant on the periscope theme has been evolved in the annelid genus *Alma*. These worms inhabit anaerobic tropical swamp muds, and when submerged extend their tail ends up to the water surface, exposing a lung-like structure everted from within a dorsal groove in the body wall. This "lung" can then absorb atmospheric oxygen while most of the animal remains in the bottom mud. According to Bouché (1972), the temperate-region wetland lumbricid *Scherotheca savignyi* obtains its oxygen in a similar fashion, though it is not equipped with such an elaborate device for the purpose.

Wetland arthropods belonging to groups terrestrial in origin, such as spiders, beetles or flies, have developed various forms of so-called "physical gill", some of which are quite complex. In these structures a supply of air is trapped on some part of the body surface from which there is direct access to both the respiratory system and the surrounding water. In this way the animal can extract oxygen from the trapped air bubble and have its supply constantly renewed by diffusion into the bubble of dissolved oxygen from the water. A simple form of this plastron respiration is seen in the large fenland spider *Dolomedes*, which is covered in hydrofuge hairs, so that when it moves into water it becomes covered in a silvery film of trapped air adequate to provide enough oxygen for some hours. *Dolomedes* actively runs down into the water as a defence reaction (Bristowe, 1971). More specific arrangements of recurved surface hairs developed to hold a plastron of air are found in various insect groups, for example subaquatic and aquatic bugs of the families Saldidae and Notonectidae. In the form of a ramifying network of cavities within the chorion, a plastron is also incorporated into the eggs of many subaquatic flies such as Sphaeroceridae and Stratiomyiidae (Hinton, 1953, 1961, 1962). Another approach, enabling an organism to deal with both atmospheric and dissolved oxygen, is to evolve both lung and gill. This combination is found in the notorious African snail *Biomphalaria sudanica*, an inhabitant of seasonal swampland and vector of *Schistosoma* (*Bilharzia*).

Suspended animation

Where there is an annual dry phase in the wetland cycle, many aquatic wetland animals use some form of suspended animation as a means of surviving *in situ* what would otherwise be conditions lethal for them. Dryland species frequently use the same technique as a means of out-lasting periods when the ground surface is flooded. Outside the tropics, the period of lowest productivity is the winter. Within the tropics, the equivalent is the dry periods, so for many organisms there is an added advantage in suspending activity during these times, beyond simply avoiding an unfavourable physical environment: avoidance of food shortage.

For organisms like terrestrial invertebrate predators, reduced activity is forced during winter by low temperature and poor food supply in most types of ecosystem, so the only additional problem facing them in wetlands is to find appropriate hibernation sites above the winter high-water levels. Emergent vegetation is much used for this purpose, large numbers of insects and spiders accumulating in late autumn in seed-heads of large aquatic grasses such as reeds. Where stem-boring insects have left hollow stems, the borings are also used as hibernation sites, as are old birds' nests. Trees and bushes such as occur in carr and fen woodland provide a range of high-level sites under bark, in rotten wood and in the burrowings of hole-nesting birds and wood-boring insects.

Aquatic species that survive dry periods by means of suspended animation tend to be small organisms that outlive the dry as drought-resistant eggs. The notostracan Crustacea, the fairy shrimps and their allies, which characteristically inhabit nutrient-rich, temporary pools and puddles, do just this. The fish genus *Nothobranchius*, an inhabitant of seasonal swamps in East Africa, adopts the same stratagem. An alternative is encystment, employed by many Protozoa. For Protozoa the problem may be more subtle than simply surviving desiccation; on its return, the water may possess the wrong pH, a phenomenon which can recur for a number of years. In these circumstances Protozoa may remain encysted, re-emerging only when the water chemistry is tolerable. Encystment can also be triggered by changes in water temperature or chemistry — of the 98 species found living on *Sphagnum* in Great Britain, the majority have been found to die or encyst during the winter (Heal, 1964). Of larger animals, lungfish are among the few aquatic groups that go into suspended animation. They burrow deep into the bottom ooze of their ponds (these burrows may be as much as 4 m in length), where they make chambers in which they aestivate, surrounded by a mucus which they secrete themselves and which retards moisture loss. A similar practice is adopted by many Amphibia, in some cases to avoid times of cold, low productivity, or both.

Metamorphism

The capacity to undergo change from one form to another during the course of the life history is peculiarly suited to a seasonal wetland, if one phase of the life history is aquatic and another terrestrial and these phases can be adjusted to coincide seasonally with the wet and dry phases, respectively, of the wetland. Anuran Amphibia, among vertebrates, and stratiomyiid Diptera, among invertebrates, are demonstrably effective in achieving this adjustment throughout the world. In both cases, the larvae are aquatic and the adults terrestrial. The frog *Rana temporaria* provides an excellent example, its tadpoles occurring in vast numbers in many a temporary pool or pond which dries up during the summer, by which time they have developed into frogs and are able to survive out of water. Further, these frogs may then live through a number of dry years during which the ponds do not reappear, thus surviving to breed when standing water is once more present.

In urodele Amphibia, adaptations of the process of metamorphosis to permit survival in wetlands with seasonal and irregular water supply can become most complex, involving not only a larval and an adult morph but polymorphism of both, together with paedogenesis. A recent study of the tiger salamander (*Ambystoma tigrinum*) demonstrates how the various morphs of this species are adapted to different environmental conditions (Rose and Armentrout, 1976). Essentially, smaller morphs and paedogenetic larvae are in this species characteristic of the most temporary water bodies. Because individuals can mate prior to metamorphosis, and metamorphosis can occur very rapidly (ten days) if the pond begins to dry out, even though the larvae be then only 6 cm in length (large morphs average 14–15 cm), this salamander is able to complete its life cycle in water bodies colonized by few other amphibia. Equally, when water persists, the species is able to produce large individuals which are longer-lived and more fecund.

Among insects, the resting (pupal) phase in the life cycle can be used to great effect in wetlands as a means of surviving seasons of unfavourable environmental conditions. In the tropics the dry seasons are passed in this way by many aquatic and subaquatic species. The adults then hatch with the onset of the first rains, and one or more generations are passed through rapidly before the wetlands again dry out. Certain Lepidoptera of temperate fenlands live through the winter months as dormant pupae, the swallowtail (*Papilio machaon*) being one prominent species with this habit in the British Isles (Dempster et al., 1976). However, the pupal resting phase is not necessarily used in this way; another fenland butterfly, the large copper (*Lycaena dispar*), overwinters as a diapausing larva (Duffey, 1968), while the stem-feeding wainscot moths (*Leucania* spp., etc.) overwinter as eggs.

Environmental manipulation

Few organisms are able to "manufacture" appropriate living conditions for themselves, but an outstanding example of this phenomenon among wetlands animals is provided by the beaver (*Castor fiber*). Beavers live in burrows excavated in high banks, with the burrow entrance below water level. To this end they require a reasonably constant water level, so where this does not occur naturally they construct dams and even canals in order to regulate water flow. The main structural elements in their dams are logs, to which are added a matrix of inter-woven branches and eventually mud and debris to fill the remaining holes and cracks. The resultant structure frequently causes an upstream build-up of water to such an extent that a large pond is produced. Once instituted, these dams are semi-permanent structures, added to and repaired by generation after generation of beavers, so the beaver ponds then develop their own hinterland of fen.

For the beaver, the winter is a time of both unfavourable environmental conditions and poor food supply, but the animal neither migrates nor hibernates. To avoid being flooded out by winter high-water levels, their burrows are extended upwards and into large, hollow mounds of twig and mud built above, on the bank top. These beaver lodges incorporate an aeration system in the roof, so that the occupants do not suffocate. To provide food for the winter, branches are cut in the autumn, dragged under-water and wedged there, somewhere within easy reach of the lodge. An extended account of the beaver's constructional activities is given by Wilsson (1971). In the past, when the numbers of beaver in northern Eurasia and North America had not been greatly reduced by man, the influence of this species, as a wetland-creating and maintaining agency, must have been considerable in wooded regions.

Migration and dispersal

An alternative stratagem to attempting to cope with unfavourable conditions *in situ* is to migrate elsewhere till favourable conditions return. Migrations of wetland animals occur in their most dramatic form to and from wetlands located where there is marked seasonal variation in climate and productivity (Dorst, 1974). Thus the northern Temperate Zone sees an annual movement of enormous numbers of waterfowl to and from their breeding grounds in the tundra and muskeg of the far north, the birds remaining there long enough to breed while food supplies are available during the short northern summer (during which the almost perpetual daylight allows well-nigh continuous feeding activity and thus very rapid chick growth), then returning south to feeding grounds in low-lying areas flooded by the autumn rains in Western Europe, West Africa, etc. In this journey south the birds are moving from one seasonally available food supply to another, for the winter-flooded pastures of Western Europe and the inundation lands of the African Sahel region have a high productivity when flooded by the autumn rains. Those European waterfowl that characteristically use the Sahel [e.g., the garganey (*Anas querquedula*), the ruff (*Philomachus pugnax*), and the black-tailed godwit (*Limosa limosa*) (Moreau, 1972; Roux, 1976)] have to overfly the Sahara *en route*, a journey which has evidently become significantly longer during the last 3000 years; seasonal water bodies which would have been usable by these species would have existed in southern

Algeria some 3000 years ago (Lhote, 1976), shortening the migration distance by perhaps 1500 km. It is tempting to postulate that, as it is today, this migration is the dwindling remnant of a much larger-scale phenomenon established at a time when, in terms of energetics, it made better sense. Such a hypothesis would not explain the migration of Palaearctic waders to the southern tip of Africa, where during the northern winter Palaearctic migrants make up over 80% of the wader populations along the coast in estuaries, *vleis* (seasonally flooded areas) and lagoons (Cooper et al., 1976). Of more general occurrence are the local movements of wetland animals, occasioned by the advance or retreat of water. Again these may be associated with periods of seasonal production. Thus, with the flooding of massive areas of swampland during the rainy seasons in the tropics, various fish species characteristically invade the temporarily flooded areas and use them as spawning grounds. Terrestrial swampland animals, conversely, retreat in the face of the rising water. Where there is an abundance of emergent vegetation, this retreat may take the form of an ascent among the plants, a stratagem employed by many small animals like insects and spiders.

Aquatic insects inhabiting seasonal water bodies are normally fully flighted, so that should their flooded area dry out they can move off to water bodies elsewhere. However, it has been demonstrated that their flights are random rather than directional. Many such species do not use temporary water bodies as breeding sites, so they do not encounter the problem of unflighted larvae or nymphs caught in a drying pool; among those which do, corixid bugs have been found to possess a flight-muscle polymorphism related to the water temperature they experience as nymphs (Young, 1965). Thus nymphs reared at higher temperatures, such as would be experienced in a shallow pool liable to dry out, develop into adults with functional wing muscles. Conversely, nymphs reared at the lower temperatures they would experience in deep pools do not produce flighted adults.

Problems of dispersal are overcome by non-flighted wetland organisms by a variety of methods. One energy-saving tactic is phoresy. Fryer (1974) has summarized the phoretic travels of two bivalve Mollusca of the genus *Pisidium*, which attach themselves to the legs of corixid water bugs and are thus transported with them in the bugs' flight to other wetlands. Only one mollusc is attached to each leg of the bug, and in one striking case Fryer observed that 20% of the bugs carried at least one mollusc. High rates of phoresy by mites on grasshoppers and phytophagous bugs in wetlands are commonplace. Tiny organisms can be carried to one place from another by wind, this providing a means of transport for encysted Protozoa, or drought-resistant Crustacean eggs. The adaptive cost of providing for means of dispersion from sites which became unfavourable is discussed later (pp. 358, 359).

TROPHIC CHARACTERISTICS OF WETLAND FAUNAS

There is a strikingly obvious difference between the larger vertebrate fauna of those nutrient-rich wetlands into which ground water drains, and that of nutrient-deficient ones into which only rain water falls. The former wetlands, particularly in the tropics, support a varied and often spectacular fauna of water birds and other animals such as water buffalo and crocodiles. The latter, by comparison, are near-deserts, unable to support more than a few vertebrates on any regular basis. In some tropical swamps the productivity is spectacular, flocks of ibis (*Ibis*), flamingo (*Phoenicopterus*) and pelicans (*Pelecanus* spp.) feeding or flying among water buffalo (*Bubalus*), elephants (*Elephas*), etc., as happens during the wet season in the partly artificial swamp systems of Sri Lanka. The biomass of frogs (Ranidae) alone can be remarkable, for example at Wilpattu in northwestern Sri Lanka the rest-house furniture in the National Park is apt to be covered with an attractive greenish motif consisting of living frogs, which also neatly frame the pictures on the wall.

Since pools are an integral part of wetlands, their fauna is an integral part of the wetlands fauna, but it is distinct in many of its components from that of the "terrestrial" parts of the wetlands. This is a problem that the ecologist faces whenever a mosaic site is discussed; the fauna of open pools may be more appropriately considered in a discussion of lakes or ponds, but it is still a feature of swamps, bogs and mires, of which it may be a substantial element. The assemblages of fresh-water animals are not just a feature of the type of pool found, but also vary with pool size. Tucker (1958) noted that erratic fluctuations occurred in small fresh-water ponds, perhaps associated with the larger fluctuations in organic matter content found there, as compared with the larger water masses. In many respects such pools function as self-contained ecological entities.

In the shallow waters of accumulative wetlands the humic acids form an important constituent. Buckney and Tyler (1973) considered that these acids probably determine the type of biotic community which inhabits them. In general, intensely coloured and acid water may preclude food chains based on phytoplankton, in favour of those based on bacteria. The entire food chain is thus a reflexion of the basic chemistry of the wetland (Bayly et al., 1972), and particularly of its water masses, the component most likely to be changed by human interference. So the biological nature of a wetland springs as much from the chemistry of its water as from the physics; Schroevers (1967) has summarized this situation clearly. The role of humic acids has not been investigated in detail for the mesofauna of peaty soils, but there seems to be a tendency for humates to precipitate on any animals which penetrate into the normally anaerobic layers, as do certain Collembola of the genera *Onychiurus*, *Folsomia* and *Tullbergia* in Ireland. These animals are only found at such levels in the autumn when partial drainage aids in aeration, but they emerge coated with a thick film of brown material which may influence their survival (Blackith, 1974).

Plant feeders

The most noticeable plant feeders of accumulative wetlands are rodents and waterfowl, though the activities of the latter are largely restricted to swamp and fen, in locations where open water is also present. The largest wetland rodent is the capybara (*Hydrochoerus hydrochoeris*), found in riverine swamps along the Amazon (see Junk, Part B, Ch. 9). More familiar species are the muskrat (*Ondatra zibethicus*) and coypu (*Myocastor coypus*), introduced to various parts of Europe from the Americas. The devastating effect which the coypu can have upon fen vegetation, due to its grazing activities, was demonstrated when its numbers built up in the Broadland area of eastern England: within six years, 10 ha (*c*. 24 acres) of reed and sedge were converted to open water round four small lakes studied (Nature Conservancy, 1966). Reed grass (*Glyceria maxima*) and lesser reed mace (*Typha angustifolia*) were almost eliminated locally. This selective feeding caused different plant assemblages to develop when vegetation returned. The biology of the muskrat has been studied in detail in North America by Errington (1963), but it is in Europe, where it is now widely distributed, that its excesses as a consumer of wetland vegetation have become noticeable. In European lake-side fen and swamp communities the muskrat's feeding activities can become the predominant factor influencing the relative abundance of the plant species present (Fiala and Květ, 1971). The muskrat's preferred food there is reed mace (*Typha latifolia*) — shoots in the summer, rhizomes during the winter — which it may eradicate over wide areas to the advantage of the reed *Phragmites australis*, which then progressively colonizes the areas previously containing *Typha*. The succession from the one vegetation type to the other is rapid, Fiala and Květ (1971) noting that, although muskrat lodges are normally built in *Typha* stands, the older lodges while still occupied become surrounded by *Phragmites*, and then have trails leading from them to the nearest *Typha* stand.

A major risk for muskrat is the inundation of fresh-water marshes by saline water after hurricanes or other natural disasters, which may be followed by spectacular mortality. In North America, muskrats are predated by the mink (*Mustela vison*) and to a lesser extent by the horned owl (*Bubo virginianus*), though probably not with any effects of consequence on muskrat populations. The muskrat suffers from epizootics of haemorrhagic disease and tularaemia, and undergoes an apparent ten-year cycle of abundance for the causes of which a variety of environmental factors have been studied. The lack of success in identifying environmental controls of natural cycles stimulates thought on the possibility that such cycles are endogenous.

Herbivorous rodents tend to ingest unexpectedly vast quantities of vegetable matter because of their relatively inefficient digestive system, but many of them, including the beaver, increase the effectiveness of their digestion by the process of coecotrophy[1] (Harder, 1949; Wilsson, 1971). In the

[1] The re-ingestion of partly digested food from the anus.

beaver, according to Wilsson, coecotrophy is a daily occurrence throughout the year, the special excrement being voided as a porridge-like mass which is consumed immediately. In contrast to the other wetland rodents, the beaver feeds mainly on tree species, using in particular deciduous trees found by water (e.g., *Alnus glutinosa*, *Populus tremula*, *Salix* spp.).

The food preferences of plant-eating waterfowl vary considerably with the species. Some waterfowl graze waterside vegetation and grassland further afield, others crop emergent, floating or submerged aquatic plants, so that when present in numbers waterfowl can exert a controlling influence over the amount of both aquatic and water-edge vegetation present. Indeed, in the North Temperate Zone, numbers of plant-feeding ducks can be so high that wetlands used by them as stop-over areas when on migration are frequently unable to provide adequate food sources, to the extent that wild-fowling interests plant areas of appropriate "duck food" (e.g. millet) on the wetlands while they are not flooded during the summer (Linde, 1969). Among the commonest and most ubiquitous groups are the surface or dabbling ducks of the genus *Anas*. These include the synanthropic, cosmopolitan mallard *A. platyrhynchus*. To some extent these species are omnivores, taking aquatic insect life, worms, molluscs, etc., as well as plant food, so that their food varies significantly with season and location; but up to 90% of the food of the mallard can be plant matter (Dorst, 1974). These ducks gather a sizeable proportion of their food while in the water, skimming it from the surface and then filtering it through horny filter plates inside their bills. For the rest they graze waterside vegetation. Other common waterfowl, like some of the diving ducks, coot (*Fulica atra*) and swans, crop subsurface aquatic plants preferentially (e.g. *Chara*, *Myriophyllum*, *Potamogeton*), together with water-edge plants. A further notable group of plant-feeding waterfowl are the geese, which, although today more local than the dabbling ducks, because of their large size exert a considerable influence upon wetland vegetation where they do occur. For instance, in European fens the food preferences of the greylag (*Anser anser*) have been found to produce vegetation changes reminiscent of those caused by muskrat grazing, mentioned earlier. But in the case of the greylag the preferred foods are the reeds *Phalaris arundinacea* and *Phragmites australis*, so that mixed stands of *Phragmites* and *Typha* develop from pure *Phragmites* stands under the influence of goose grazing (Fiala and Květ, 1971).

The largest herbivore of the northern temperate region to make particular use of wetland vegetation is the moose (*Alces alces*), which feeds on aquatic and subaquatic vegetation during the spring and summer, but moves to higher and drier ground for the rest of the year. Its tropical counterpart, the hippopotamus, feeds primarily in the water on subaquatic swamp vegetation, but also makes forays onto dry land to graze, though in doing so it rarely travels far from water. Both species serve the function of trail-breakers for other wetland animals, a phenomenon of particular significance in thick, tall vegetation like that of papyrus swamp. These trails and tracks made through wetland vegetation by the larger herbivores are also important as habitats for many invertebrate species, which would not find appropriate conditions in the uniform reed-bed environment. This is well illustrated by the Tetrigidae, or ground hoppers, a group of orthopteroid insects with many wetland species. These insects are microphagous, grazing algal mats, and are characteristic of peaty animal tracks in swamp and fen. An example is *Tetrix subulata*, which occurs in the British Isles. The grazing activities of larger animals are also important to invertebrate plant feeders in that, especially where plants with sub-apical meristems are involved (e.g. *Phragmites*), their grazing effectively prolongs the growing season, causing a burst of regrowth following loss of the first set of new shoots. This enables animals such as aphids to go through extra generations *in situ*, without having to migrate elsewhere in search of new growth.

It is noticeable that all of the vertebrate herbivores mentioned in the preceding pages are animals of swamp and fen. The plant-feeding vertebrates of temperate bogs are exceedingly few, and those which do occur, like the blue hare (*Lepus timidus*), willow grouse (*Lagopus lagopus*) and red grouse (*L. lagopus* var. *scoticus*), are more typical of other ecotopes, surviving in boglands only in very low numbers. To a substantial extent this reflects the much lower utilizable productivity of acid peatlands. Work by Lance and Mahon (1975; see also Lance, 1978) demonstrates the mechanism whereby this factor acts to limit red grouse populations on

Irish bogs. They found that the males required to maintain significantly larger territories than are found on moorland, and that the low nutrient status of the available vegetation frequently left the birds in poor condition to carry out territorial behaviour effectively.

The plant-feeding invertebrate faunas of temperate wetlands are unusual in that they comprise a high proportion of stem-suckers, stem-borers and gall-formers, leaf-eaters being correspondingly few (Imhof and Burian, 1972). The aquatic forms are largely microphages, using algae and diatoms as a food source. Because of the scarcity of triturating plant feeders, detritivores of bog and fen receive a smaller proportion of the annual plant production already comminuted, and derived from living plant tissue, than do their counterparts in other ecosystems. In boglands, plant-feeding invertebrates of all types are few, and apart from aquatic microphages tend to be species associated with bogland plants that also occur in other habitats, such as the emperor moth (*Saturnia pavonia*) and the fox moth (*Macrothylacia rubi*), whose larvae feed on a wide range of plants in various heath and moor environments. Similarly widespread are the cuckoo-spit bugs *Philaenus spumarius* and *Neophilaenus lineatus*. An exception is the large grasshopper *Stethophyma grossum*, which at the northern end of its range in Europe appears to occur only in *Molinia* bogs. Another group which can be abundant in bogs are the slugs, whose distribution is not so restricted in these calcium-poor wetlands as is that of shelled molluscs. The large black slug, *Arion ater* agg., is common even in parts of the oceanic blanket bogs of the west of Ireland, though of the two *Deroceras* species which can be abundant in fenland (Brindley, 1925), *D. laeve* is rare and *D. agreste* is unrecorded for Ireland (Kerney, 1976).

Detritivores

Whatever the precise ecological interrelationships between given wetland animal species may be, of more general significance and less subject to variation is the sort of role they each perform in the dynamics of an ecosystem they inhabit. This is amply illustrated in temperate wetlands by the build-up of peat which characterizes sites where detritivore species are unable adequately to comminute the annual production of plant debris so that it can all be broken down by micro-organisms. This inability of detritivores to function effectively is probably at its extreme in acid bogs. Chemical deficiencies almost entirely exclude two groups, lumbricids and isopods, from both wet and dry phases of the bog, while waterlogging and anaerobiosis perennially prevent access to the lower layers of the peat for other groups like mites (Acari) and springtails (Collembola). The low temperatures, low pH and anaerobiosis also inhibit the activity of both microfungi and bacteria, which are in many cases the actual food of detritivores, for many of which comminution of vegetable debris is primarily an unavoidable "side-effect" of grazing the micro-organisms living on plant remains. Finally, because of the scarcity of chewing plant feeders, the plant remains reaching the ground surface are largely uncomminuted, and of lower nutrient content than in many other environments, arriving senescent as they do when shed by the bog plants, rather than as new growth, comminuted and part-processed (as they would be had they passed through the guts of herbivores). This again can only retard micro-organism activity.

The relative scarcity of micro-organisms in peat bog appears likely to act as an additional factor limiting numbers of organisms like springtails there. Indeed, Blackith (1974) has provided evidence to suggest that Collembola are common in Irish blanket bogs only where the roots of living plants provide foci of bacterial activity. The same data demonstrated that each plant species had its own characteristic community of associated collembolan species, a phenomenon indicating the specialized nature of bogland microhabitats and the difficulties inherent in attempts to generalize about the "animal communities" of entire ecosystems. For instance, the species *Tetracanthella britannica* occurred almost exclusively in association with purple moor grass (*Molinia caerulea*) and black bog-rush (*Schoenus nigricans*), while *T. wahlgreni* was found with lichens (*Cladonia* spp.). Heather (*Calluna vulgaris*) plants carrying encrusted lichens on their stems supported a mixed assemblage of species associated with heather and lichens. Unpublished experiments (1975) by Miss B. Dickey have extended this work to the raised bogs of eastern Ireland, which must differ ecologically from the blanket bogs of the west if only because *Schoenus nigricans* is not an element of the flora in the east. This investigation showed that Collembola were about as closely associated with the particular plant species in the east as in the west, but that the species complexes were appreciably different in the two areas. Part of this difference no doubt arises from the different suites of plants and animals to be found in the two sets of bogs, and Neill (1974) has suggested that the interactions between some elements of an assemblage of plants and animals will be strongly influenced by the other elements which happen to be present. Several competition coefficients which he calculated from populations of micro-crustaceans in equilibrium microcosm communities changed significantly with changes induced in the species composition of the remainder of the community. Miss Dickey's results seem strongly to confirm Neill's findings, for there is no other very obvious reason why the relative abundances of the different Collembola, and their association with individual plant species, should differ so strikingly. A multivariate analysis of the combined data on collembolan distributions in eastern and western bogs has been undertaken, which showed that the faunal assemblages for any given plant species tended to cluster together on the major axes of variation, as one would expect, save for one axis which clearly represented the contrast between eastern bogs as a whole and western bogs as a whole. This result may give some encouragement to those who hope that the understanding of plant–animal communities may be enhanced by the construction and interpretation of a community matrix (Levins, 1968), as against the strong pessimism with which Neill views such attempts at analysis, for virtually no consideration has as yet been given to the possibility that, although there are indeed high-order interactions between the elements of a community, these interactions may be more orderly and readily interpretable than Neill supposed.

A major faunal distinction between fen and bog peat is the presence in acid bog of enchytraeid worms, which take over from the lumbricid worms whenever the pH falls much below 4.0 (Bouché, 1972) and can represent a substantial proportion of the biomass of bog invertebrates. In fen peat itself perhaps the most characteristic earthworm is *Allolobophora chlorotica*, whose typical green form is even more tolerant of the fen conditions than is the white form. Other species move in as soon as the drainage is improved, so that species lists for earthworms in a fenland may include several species that occur there almost only in such slightly drier sites. The same is true of bog peat in which earthworms are very local and uncommon, although slight improvements of drainage or nutrient status allow species of the genus *Dendrobaena* to thrive. Bouché (1972) considers that the redox potential of flooded soil is the limiting factor in determining the distribution of earthworms, only one species, *Allolobophora oculata*, thriving in frankly putrid soils.

It is apparent that in the reed-bed (*Phragmites*) environment a significant proportion of the annual increment of plant litter can be processed by aquatic organisms during the time that it is flooded. Snails, aquatic isopods (*Asellus* spp.) and caddis-fly (Trichoptera) larvae are prominent among the comminuting animals involved (Imhof and Burian, 1972).

In the sense that they are all forms of winter-flooded grassland, heavily grazed by livestock, or cut for hay when the waters have receded, many turloughs[1], dolines[2], water meadows, bottom lands and callows of the temperate regions have much in common. It has been suggested that the amount of peat build-up that occurs on such areas depends primarily on the proportion of the annual production of plant growth that is cropped by the grazing animals or removed by cutting; without some form of cropping, peat build-up increases (or begins), demonstrating the inability of the decomposer organisms to cope with the quantities of plant growth these sites can produce. For a clearer idea of the role of grazing in this respect reference must be made to some form of dynamic model discussed elsewhere, for example by Gore (Ch. 1).

Predators

In seasonal wetlands, the food supply available to predators can vary enormously in both volume and character; in the tropics seasonal swamplands have little to offer at the end of the dry season, yet

[1]**Turlough:** a seasonal water body, found in karst limestone areas, flooding with ground water from swallow or sink holes in its basin.

[2]**Doline:** a basin-shaped feature, in karst limestone regions, caused by subsurface solution of the rock and subsequent sinking of the overlying material.

they teem with fish fry, etc., shortly after flooding, while in tundra regions vast swarms of chironomid midges hatch during the brief summer, into an environment where active life of any sort virtually ceases during the winter. Where these pronounced short-term oscillations in the abundance of potential food supplies occur, the populations of resident predators are frequently greatly outnumbered by those of seasonal immigrants, which remain only during seasons of high productivity. Waterfowl invade the flooded areas of seasonal swamps in great numbers in the tropics, and breed during this period of abundant food. The flooded ground then serves an ancillary function of providing safer nest sites, many breeding colonies of birds such as ciconiids (storks), plataleids (spoonbills and ibis) and ardeids (herons and egrets) being established in scrub and trees standing within the flooded areas.

There are indications that smaller predators which are present in temperate bog and fen, such as ground beetles (Carabidae), are dependent primarily upon the detritivores as a food source (Dawson, 1965), and are themselves the main prey of larger predators such as frogs (Blackith and Speight, 1974), though in blanket bog frogs (*Rana temporaria*) will evidently also consume the slug *Arion ater*, where it is available. With their limited food resources, the larger resident predators of bogland are correspondingly few; in Irish blanket bog only the pygmy shrew (*Sorex minutus*) and a species of frog (*Rana temporaria*) occur in any numbers [the lack of the grass snake (*Natrix natrix*) and the water shrew (*Neomys fodiens*) may be due to zoogeographic factors — there is no evidence that these animals ever reached Ireland during the Postglacial]. The frog does not feed in the water, so for its food it is dependent upon the terrestrial inhabitants of the wetlands in which it lives. Some wetland predators like the grass snake hunt both in water and out of it, so a much wider range of prey is available to them. The European pond tortoise (*Emys orbicularis*), a characteristic denizen of swamp and fen, hunts food only in the water. If it finds food on land it returns to water to consume it. Arthropod predators tend to be similarly restricted, in that those species which capture prey out of water do not capture prey in water, and vice versa, so that different suites of predators can be found in the same area in its wet and dry phases. However, it would be misleading to imply that the aquatic and terrestrial organisms of wetlands function as completely independent groups. For instance, the predatory insects and spiders inhabiting fen vegetation derive a major part of their food from the aquatic insect fauna, in the form of the non-aquatic adults of insect species aquatic as larvae or nymphs (e.g. may flies, stone flies, caddis flies). Similarly, as water levels drop, non-aquatic predators like ground beetles advance along the receding water line, consuming dead and dying aquatic animals that have been left stranded.

Rising water levels can also be used to advantage by terrestrial predators. For instance, organisms such as springtails appear to be subject to intense predation during the winter months in bogland, when they are forced up into the surface vegetation layers by the rising ground water. In Irish blanket bog, collembolan numbers suffer a 66% reduction at this time.

Many families of flies (Diptera) have predatory, subaquatic larvae, which live in the surface debris layers of the waterlogged soils of fens and marshes, preying upon other invertebrates present. The Sciomyzidae are notable in that they predate water snails, some species attacking the snails under water, others consuming them out of it. These flies are not predaceous as adults. The larvae of the horse flies and clegs (Tabanidae), another typically wetland group, are found in similar habitats. Some of them specialize in feeding on the larvae of other Diptera (Chvala et al., 1972), but others are much less discriminating and will even attack larvae of their own species (Oldroyd, 1964).

Where shallow standing water is present, planktonic invertebrates are abundant, many of them predatory upon microphagous plankton. Apart from providing food for fish-fry, etc., this planktonic life represents a sufficiently reliable food source to have supported the evolution of bird species specialized for its consumption, namely the tropical skimmers (*Rhynchops* spp.) and the spoonbill (*Platalea leucorodia*). Similarly, the large number of invertebrates living in reed and sedge-bed habitats support a varied assemblage of warblers, buntings and rails.

THE COST OF DISPERSION FROM WETLANDS

There is a sense in which all ecosystems, looked at *sub specie aeternitas*, are temporary habitats, but, as the previous pages suggest, wetlands are perhaps more temporary than most. Because all wetland animals are liable to find themselves homeless sooner or later, for such species to persist they require not only adaptations ensuring survival within the wetland but also effective dispersal techniques. But adaptations for these two purposes are obviously potentially antagonistic in character; the ideal stratagem for staying in a wetland is likely to be far removed from that required for dispersal. This concept of the cost of dispersal out of a habitat seems not to have been given the attention it deserves since Wolfenbarger's pioneer studies (Wolfenbarger, 1946); it is a part of the total load on a species somewhat analogous to the genetic load, and in this context could be defined as the mean number of "surplus" individuals per generation that need to be produced in order to secure the colonization of areas of wetland separated from that in which the population is established, given some broad parameters concerning the frequency with which suitable wetlands occur within the dispersal range of the animals. That this necessary surplus is likely to be larger for wetland animals than for woodland ones stems from the fact that, in general, forests, at least until quite recently in the earth's history, have formed large, more or less continuous systems, whereas wetlands tend to have been discrete, unstable units within such areas, of very uneven distribution in both time and space.

Although the smallest denizens of wetlands may be characteristically dispersed by wind action, larger animals are in a more difficult position, for although they may make use of air currents to aid dispersal, positive action on their part is normally required for them to become airborne.

It is probably no accident that two of the most important species of plague locust pass their solitary phase in wetlands — the migratory locust (*Locusta migratoria*) in the Niger basin, and the red locust (*Nomadacris septemfasciata*) in the Rukwa Valley of East Africa (Albrecht, 1967) — only passing into the migratory phase (which is well adapted to arid lands) when a sufficient degree of crowding within the wetland has taken place. It is ironic that wetlands should be an essential, if intermittent, habitat of some locusts, associated as locust plagues are with periods of aridity, as in the Spanish (South American) saying "Primero la sequia, ahora la lengosta! [First the drought, then the locust!]" but in fact locusts are closely adapted to both wet and dry habitats, and the evolutionary cost of maintaining both sets of adaptations can be considered part of their dispersive load.

One extreme dispersion stratagem is to migrate only when the habitat deteriorates; but that is also a time when reproduction is hardest. Another is to migrate during each generation — again not ideal because the generation time may be short compared with changes in the ecosystem, and there is waste involved in sending colonists to already occupied localities. A compromise adopted by many wetland animals is to migrate when exceptionally favourable conditions allow a substantial surplus of individuals to be produced: a peculiarly suitable stratagem in view of the seasonally productive nature of accumulative wetlands, and promoting the annual phases of dispersion among polyvoltine species (those with several generations a year).

Organisms inhabiting the riverine swamps of enormous river systems such as the Nile must carry a lower dispersive load than their counterparts in other wetlands, since these swamps have evidently been continuously in existence over periods of geological time sufficient for evolution to occur *in situ*, to judge from the large numbers of endemics found there; without the counteracting need for dispersal, adaptation to specific swamp conditions has been possible.

ZOOGEOGRAPHICAL CONSIDERATIONS

There is no reason for supposing that the distribution patterns of wetland animals differ significantly from those of other organisms, in relation to the major biogeographic regions of the world. But within any biogeographic region the especially transitory nature of wetlands themselves can influence the constitution of their faunas. Most accumulative wetlands represent the intermediate phase (or phases) in processes of both geomorphological and vegetational succession. The fens which succeed the open-water phase in shallow lake basins are good examples of this phenomenon, normally evolving within a few thousand years or less into fen woodland. In the case of Wicken Fen, in England, the drying-out process has been accelerated by drainage operations in the surrounding countryside, and the consequent faunal changes have been recorded by Gardiner (1921–28). Omer-Cooper et al. (1928) described in detail the changes that occurred in the beetle fauna, demonstrating clearly a replacement within the last hundred years of typically fenland species by more ubiquitous animals of somewhat drier habitats. By contrast, wetlands on the flood plains of large rivers can be regarded as a reasonably permanent component of the landscape, since the frequent changes of course exhibited by these meandering waters repeatedly produce new oxbows to fill in with fen, or sweep away old reed beds in the development of new channels. In this way the swamps of the Danube or Nile have been constantly rejuvenated, and it could be argued that the faunal diversity of these deltaic wetlands is due not only to their undoubted productivity but also to their longevity and continuity of existence. Organisms associated with the more transitory forms of wetland, such as the valley bog, are dependent for the continued existence of their habitat upon recurrent phases of wetland formation. There are indications that the relative frequencies of such phases has increased in recent geological time and, in the form of the Ice Ages, they are highly characteristic of the present geological period, the Pleistocene (*sensu* West, 1968, chapter 1). Indeed, it could be argued that, for wetland animals in the Holarctic, the major determinant of present distribution patterns has been the Ice Ages, until man began to exert an influence.

Prior to the Pliocene, some fifteen million years ago, the number and distribution of wetlands in the world had evidently decreased progressively from the beginning of the Tertiary epoch onwards. Whether or not this led to the extinction of a wider range of wetland animals than of animals associated with other environments has not been ascertained, but contractions in range suffered by wetland animals during the Early Tertiary were, at least in Eurasia, difficult to recover from, since this was also a period of mountain-building; the Alps, the Atlas Mountains and the Himalayas, all of them formidable zoogeographic barriers, were formed at this time. With the establishment during the Pliocene of the climatic oscillations which spawned the glacial periods, came cycles of wetland formation due to the scouring and earth-moving activities of ice and melt water. Since wetland animals are by definition found close to water there is an excellent record of wetland faunas provided by deposits dating from various interglacials throughout the Pleistocene — there is a better record of the faunas from fen, bog, pond and river margin than there is from any other non-marine habitat. Inevitably, remains of smaller animals are more frequently preserved. Assemblages of beetle fragments are particularly abundant. From the fact that the species represented among these remains appear identical to present-day species it can be concluded that the Ice Ages did not promote a tremendous burst of evolutionary activity (Coope, 1965, 1969) among wetland insects, so from them it is possible to consider some of the zoogeographical events of the Pleistocene directly. Most of these species assemblages have been extracted from interglacial and interstadial detritus peat from Western European sites, and so relate primarily to the zoogeography of Palaearctic organisms. It is clear from these assemblages that, during the warm phases represented by the interglacials, wetland animals rapidly re-invaded those parts of northwestern Europe that were ice-covered during the glacial phases. This is perhaps not surprising, since each river margin would represent a linear continuum of wetland environments, providing convenient stepping stones for immigrant wetland organisms. Indeed, the relative ease with which wetland animals could have dispersed under such conditions may well partly explain their disproportionately strong representation in assemblages from early in interglacial periods (rather than this abundance being due entirely to their close proximity to the preserving medium, as mentioned above). Given that the spread of wetland animals may well have been particularly easy as the climate warmed in Europe during each interglacial, it is hardly necessary to contrive "glacial refugia" for them near to where they occur in interglacial assemblages, in order to explain their interglacial distribution; the problem of "refugia" has bedevilled general discussions of interglacial floral and faunal distribution for some time, especially when islands like the British Isles are under consideration. Whether or not warm-climate species found "refugia" in the southwest of the British Isles during full glacial phases, there is little doubt that tundra species were able to survive there then. Today it is these cold-climate species which would need "refugia" in the British Isles, and it is notable that some of the more obvious species are not found there now (wetland species recorded in Great Britain earlier in the Pleistocene, but absent there today, are listed in Table 10.1). It is tempting to suggest that *Diachila* species, for example, are absent in the British Isles today simply because appropriate "refugia" do not exist (*D. arctica* was present 13 500 years ago: Pearson, 1962); but Table 10.1 shows that many wetland insects are known to have become extinct in Great Britain during the last 200 years due to man's drainage operations, so that man's activities earlier in the Postglacial could possibly also account for the disappearance of species like *D. arctica*.

Whatever the date and cause of their disappearance, it is evident that various wetland invertebrates now absent in Great Britain were there earlier in the Pleistocene. Some of these, like the ground beetle *Diachila*, are repeatedly found in the record. The same is true for at least one outstanding aquatic vertebrate, the hippopotamus, which managed to penetrate into northwestern Europe as far as Great Britain during two interglacials, the later of the two being about 90 000 years ago.

In continental Europe there are dramatic examples of cold-climate wetland species occurring in isolated populations well south of the rest of their present range; the most plausible explanation for these is that they represent relict populations surviving today only in appropriate refugia. Thus, in Western Europe the vole *Microtus oeconomus* is found outside Scandinavia and Poland only in coastal marshes in Holland (Wijngaarden et al., 1964) and inland salt marshes along the Danube (Turcek, 1976).

In general, it could be said that warm-climate forms have progressively failed to recolonize Europe following each successive glaciation, and it is difficult to escape the conclusion that the east–west mountain barriers in southern Europe have been instrumental in this decline. Though not strictly from wetlands in that they were seemingly primarily inhabitants of permafrost tundra, two cold-climate animals which demand mention are the woolly rhinoceros (*Coelodonta antiquitatis*) and hairy mammoth (*Mammuthus primigenius*). These large mammals both evolved and became extinct during the Pleistocene. Although the reasons for the disappearance of the woolly rhinoceros are more obscure, there is now good evidence for believing that the mammoth was eradicated by early

TABLE 10.1

Accumulative-wetland invertebrates which were present in Great Britain earlier in the Pleistocene, but which are unknown in Great Britain today

Interglacial/interstadial species		Postglacial species exterminated by loss of habitat
Cold-climate species	warm-climate species	
Chlaenius costulatus (Coope, 1965)	*Oodes gracilis* (West, 1957)	*Lycaena dispar* (Heath, 1974)
Diachila arctica (Coope, 1969)		*Chlaenius tristis* (Lindroth, 1974)
Diachila polita (Coope, 1969)		*Agabus striolatus* (Hammond, 1974)
Hydrobius arcticus (Coope, 1969)		*Rantus aberratus* (Hammond, 1974)
Agabus vittiger (Coope, 1969)		*Hydaticus stagnalis* (Hammond, 1974)
Vertigo genesii (Evans, 1972)		*Laelia coenosa* (Heath, 1974)
		Trachea atriplicis (Heath, 1974)
		Costaconvexa polygrammata (Heath, 1974)
		Tenthredo rossi (Benson, 1932)

man, together with more strictly wetland Pleistocene mammals like the giant beavers (*Castoroides*) (Martin, 1967).

The present-day distribution of wetland animals in the Palaearctic is so extensively influenced by man that attempts to explain the absence of particular species from particular places in terms of zoogeographical barriers or climatic change must inevitably be cautious. Nonetheless, it is also apparent that the last major phase of wetland production petered out some 10 000 years ago, and that the progressive decrease in wetland area which has occurred since has now reached zoogeographically significant proportions: man's activities have served (through drainage, in particular) greatly to accelerate the rate of wetland loss, producing a situation which for many species must now be critical (Dauphin, 1972). This is reflected in the disjunct distributions exhibited by wetland animals today, the number of local extinctions known to have occurred during the historic period (for instance, the species recently lost in Great Britain, listed in Table 10.1) and the incipient speciation evidenced in now-isolated wetlands (for instance, *Papilio machaon* in its remaining British Isles haunt in eastern England). It is arguable that a zoogeographical phenomenon analogous to, and at least as meaningful as, the "island effect" (Macarthur and Wilson, 1967) has been produced, the remaining wetlands representing for wetland animals "islands" of appropriate habitat in a sea of dry land.

When considering zoogeographical phenomena on a larger scale one can make a broad distinction between the distribution of the larger animals and that of those animals small enough to be dispersed on wind-borne soil or vegetation. For the larger animals, their occurrence in wetlands will be determined by the position of the site on the earth's surface, and one can usefully speak of their zoogeography. For the smaller animals, as Bonnet (1964) has argued, there is no real zoogeography, as any species may be found in any part of the world, if the site at which it happens to be deposited is favourable. Included in this group are testaceous protozoa, tardigrades, rotifers and some free-living soil nematodes, all of which are capable of occupying a wide variety of habitats because of their small size and capacity to pass inclement periods in a state of suspended animation, in some cases lasting for many years.

IS THERE A GENERALIZED ANIMAL ASSEMBLAGE OF ACCUMULATIVE WETLANDS?

Analyses of wetland zoology involve several high-risk interpretative situations. Fashions in models succeed one another at a rate which suggests that several more will be needed before an asymptotic approach to a workable model is found. The once fashionable energy-flow approach now looks inappropriate for attempts at a synthesis of the principles governing animal life in wetlands, partly because it starts from assumptions which seem likely to be substantially false, such as the proposition that wetlands are limited by energy rather than by nutrients. An outstanding characteristic of most types of wetland discussed in this chapter is that peat is currently accumulating in the waterlogged levels; the large, usually predominant, contribution to this energy sink must be the outstanding feature of an energy-flow model of an accumulative wetland. In this respect bogs have much in common with deserts — to regard them as wet deserts would not be entirely fanciful; hot dry deserts share with bogs an abundant latent energy in the form of accumulated organic matter, but in each instance the availability of that energy is limited by the excess or deficiency of moisture, or nutrient imbalance. Once rainfall, or irrigation, have made the organic matter available in deserts, or drainage or drought have performed the same service for fens and bogs, the productivity can be quite high if other climatic conditions permit the full exploitation of the energy flow, unless nutrient scarcity intervenes. Where so large a proportion of the annual energy input is shunted into a sink in this way, it seems perverse to maintain that energy could be the limiting factor, except in the way in which a compulsive gambler may claim to be short of money. A modeller who shuts his eyes so firmly to such an outstanding feature of the system he is modelling cannot be expected to be much more sensitive to the disadvantages of averaging behaviour of the elements of the fauna whose very *raison d'être* is that they behave in contrasting ways. This point has been most cogently argued by Goodall (1974), who notes that an ecosystem model which ignores species differences runs the risk of neglecting important elements in its dynamics. He also comments that spatial heterogeneity, like biological diversity, may contribute greatly to the dynamics of an ecosystem.

Systems ecology seems to have reached the same stage of awareness that physiology attained at the end of the last century, when Claude Bernard put forward the classic aphorism "L'homme moyen n'existe pas [There is no average man]" on finding that the average of the physiological parameters of a series of human beings could nowhere be matched by any actual living person, and was presumably an inviable compromise. It is easy to show that some modellers' compromises are lethal, by neglecting the differences between the rules of the road in Britain and France, modelling the system as a whole to show that on an average traffic flows down the centre of the road in both directions, or, averaging out traffic flows, that the road system is effectively a storage unit for industrial output between production and decomposition!

Some energy-flow models are successful in predicting broad, but important, properties of the ecosystem, as Nixon and Oviatt (1973) have done for a New England salt marsh, but the compromises necessary to make such models tractable can wreck their fine structure. In wetlands their product is likely to be a misleading average of a series of nested sub-systems. These nested sub-systems are essentially of two kinds: those associated with the different niches occupied by related species of animal (which would generally be considered as a single entity for the purposes of energy-flow modelling); and those associated with the mosaic elements of the wetlands, such as open water, emergent vegetation, algal mats and so on, each of which may be supposed to contain an intrinsic fauna together with elements from the faunas of adjacent parts of the mosaic. So the product of such modelling is misleading essentially because it involves the implicit assumption that at some very general level, such as "the salt marsh", or "the valley bog" an ecologically meaningful animal community exists. Perhaps, in relation to their responses to the physical environment, wetland animals can be assigned to broad categories, as implied by the preceding pages. However, the extent to which such groupings represent meaningful ecological units is highly debatable. While it may be true that blanket-bog faunas are species-poor in comparison with fenland faunas, and that the general pattern of activity in both wetland types is governed primarily by the climate in general and the hydrologic cycle in particular, such information is unhelpful when the comparative dynamics of the faunal assemblages associated with *Calluna* plants and *Molinia* tussocks come under scrutiny, because there is every indication that such assemblages function independently. If so, then in such cases perhaps the most meaningful level at which animal communities could be discussed would be in relation to the individual plant, or bog pool, or ditch. The proven existence of recurrent species assemblages like those recorded by Blackith (1974) and Morris (1969) would lend support to this view. But even then, the same criticism can be levelled at such an approach as has been levelled at the more all-embracing community concept — it is a far cry from demonstrating the existence of recurrent species associations to establishing the ecological relationships which exist between their constituent species. Probably in response to the developing awareness of this problem, there is today a resurgence of interest in autecological studies; but it will be many years before an adequate corpus of data can be amassed from such work for more realistic models of wetland animal communities to be constructed.

REFERENCES

Albrecht, F.O., 1967. *Polymorphisme Phasaire et Biologie des Acridiens Migrateurs.* Masson, Paris, 194 pp.

Bayly, I.A.E., Lake, P.S., Swain, R. and Tyler, P.A., 1972. Lake Pedder: its importance to biological science. In: *Pedder Papers. Anatomy of a Decision.* Australian Conservation Foundation, Parkville, Vic.

Benson, R.B., 1932. Additions and corrections to the preliminary list of the saw-flies of Wicken Fen. In: J.S. Gardiner (Editor), *The Natural History of Wicken Fen.* Bowes and Bowes, Cambridge, pp. 544–547.

Blackith, R.E., 1974. The ecology of Collembola in Irish blanket bogs. *Proc. R. Irish Acad. B*, 74: 203–226.

Blackith, R.M. and Speight, M.C.D., 1974. Food and feeding habits of the frog *Rana temporaria* in bogland habitats in the West of Ireland. *J. Zool., Lond.*, 172: 67–79.

Bonnet, L., 1964. Le peuplement thécamoebien des sols. *Rév. Ecol. Biol. Sol.*, 1: 123–408.

Bouché, M.B., 1972. *Lombriciens de France: écologie et systématique.* Institut National de la Recherche Agronomique, Paris, 671 pp.

Brindley, H.H., 1925. The mollusca of Wicken Fen. In: J.S. Gardiner (Editor), *The Natural History of Wicken Fen.* Bowes and Bowes, Cambridge, pp. 154–161.

Bristowe, W.S., 1971. *The World of Spiders.* New Naturalist Series, No. 38, revised edition. Collins, London, 304 pp.

Buckney, R.T. and Tyler, P.A., 1973. Chemistry of some sedgeland waters: Lake Pedder, South-west Tasmania. *Aust. J. Mar. Freshwater Res.*, 24: 267–273.

Chvala, M., Lyneborg, L. and Moucha, J., 1972. *The Horse-flies of Europe (Diptera, Tabanidae)*. Entomological Society of Copenhagen, Copenhagen, 498 pp.

Coope, G.R., 1965. Fossil insect fauna from late Quaternary deposits in Britain. *Adv. Sci.*, 21: 564–575.

Coope, G.R., 1969. The response of Coleoptera to gross thermal change. *Mitt. Int. Ver. Limnol.*, 17: 173–183.

Cooper, J., Summers, R.W. and Pringle, J.S., 1976. Conservation of coastal habitats of waders in the South-Western Cape, South Africa. *Biol. Conserv.*, 10: 239–247.

Dauphin, C., 1972. La mort des marais des Ardennes. *Entomologiste*, 28: 104–105.

Dawson, N., 1965. A comparative study of the ecology of eight species of fenland Carabidae (Coleoptera). *J. Anim. Ecol.*, 34: 299–314.

Dempster, J.P., King, M.L. and Lakhani, K.H., 1976. The status of the Swallowtail butterfly in Britain. *Ecol. Entomol.*, 1: 71–84.

Dorst, J., 1974. *The Life of Birds*. Weidenfeld and Nicolson, London, 2 vols., 717 pp.

Duffey, E., 1968. Ecological studies on the large copper butterfly, *Lycaena dispar* Harv. *batavus* Obth. at Woodwalton Fen National Nature Reserve, Huntingdonshire. *J. Appl. Ecol.*, 5: 69–96.

Errington, P.L., 1963. *Muskrat Populations*. Iowa State University Press, Ames, Iowa, 665 pp.

Evans, J.G., 1972. *Land Snails in Archaeology*. Seminar Press, London, 436 pp.

Fiala, K. and Květ, J., 1971. Dynamic balance between plant species in South Moravian reed-swamps. In: E. Duffey and A.S. Watt (Editors), *The Scientific Management of Animal and Plant Communities for Conservation. Proc. 11th Symp. Brit. Ecol. Soc.* Blackwell, Oxford, pp. 241–269.

Fryer, G., 1974. Attachment of bivalve molluscs to corixid bugs. *Naturalist*, No. 928: 18.

Gardiner, J.S. (Editor), 1921–1928. *The Natural History of Wicken Fen*. Bowes and Bowes, Cambridge, 652 pp.

Goodall, D.W., 1974. Problems of scale and detail in ecological modelling. *J. Environ. Manage.*, 2: 149–157.

Hammond, P.M., 1974. Changes in the British coleopterous fauna. In: D.L. Hawksworth (Editor), *The Changing Fauna and Flora of Britain*. Academic Press, London, pp. 323–370.

Harder, W., 1949. Zur Morphologie und Physiologie des Blinddarmes der Nagetiere. *Verh. Dtsch. Zool. Ges., Mainz*, pp. 95–109.

Heal, O.W., 1964. Observations on the seasonal and spatial distribution of Testacea (Protozoa: Rhizopoda) in *Sphagnum*. *J. Anim. Ecol.*, 33: 395–412.

Heath, J., 1974. A century of change in the Lepidoptera. In: D.L. Hawksworth (Editor), *The Changing Fauna and Flora of Britain*. Academic Press, London, 461 pp.

Hinton, H.E., 1953. Some adaptations of insects to environments that are alternately dry and flooded, with some notes on the habits of the Stratiomyidae. *Trans. Soc. Br. Entomol.*, 11: 209–227.

Hinton, H.E., 1961. How some insects, especially the egg stages, avoid drowning when it rains. *Proc. S. Lond. Entomol. Nat. Hist. Soc.*, 1960: 138–154.

Hinton, H.E., 1962. Respiratory systems of insect egg-shells. *Sci. Prog. (Lond.)*, 50: 96–113.

Imhof, G. and Burian, K., 1972. *Energy-flow Studies in a Wetland Ecosystem (the Reed Belt of the Lake Neusiedler See)*. Austrian Academy of Sciences, Special Publication for I.B.P., Vienna, 15 pp.

Kerney, M.P. (Editor), 1976. *Atlas of the Non-marine Mollusca of the British Isles*. Conchol. Soc. G.B. and Ireland.

Lance, A.N., 1978. Territories and the food-plant of individual red grouse. II. Territory size compared with an index of nutrient supply in heather. *J. Anim. Ecol.*, 47: 307–314.

Lance, A.N. and Mahon, G., 1975. Foods of a marginal red grouse population in western Ireland. *J. Wildl. Manage.*, 39: 183–187.

Levins, R., 1968. *Evolution in Changing Environments*. Princeton University Press, Princeton, N.J., 120 pp.

Lhote, H., 1976. *Vers d'autres Tassilis: nouvelles découvertes au Sahara*. Arthaud, Paris, 258 pp.

Linde, A.F., 1969. *Techniques for Wetland Management*. Res. Pap., 45. Dep. Nat. Resour., Madison, Wis., 156 pp.

Lindroth, C.H., 1974. *Carabidae, Coleoptera*. Handbooks for the Identification of British Insects, 4(2). Royal Entomological Society, London, 148 pp.

Macarthur, R.H. and Wilson, E.O., 1967. *The Theory of Island Biogeography*. Monographs in Population Biology, No. 1. Princeton University, Princeton, N.J., 203 pp.

Martin, P.S., 1967. Prehistoric overkill. In: *Proc. VII Congr. Int. Assoc. Quaternary Res.*, pp. 75–120.

Moreau, R.E., 1972. *The Palaearctic—African Bird Migration Systems*. Academic Press, New York, N.Y., 384 pp.

Morris, M.G., 1969. Associations of aquatic Heteroptera at Woodwalton Fen, Huntingdonshire, and their use in characterising artificial aquatic biotopes. *J. Appl. Ecol.*, 6: 359–373.

Nature Conservancy, 1966. *The 1965 Report on Broadland*. The Nature Conservancy, London, 98 pp.

Neill, W.E., 1974. The community matrix and interdependence of the competition coefficients. *Am. Nat.*, 108: 399–408.

Nixon, S.W. and Oviatt, C.A., 1973. Ecology of a New England salt marsh. *Ecology*, 43: 463–498.

Oldroyd, H., 1964. *The Natural History of Flies*. Weidenfeld and Nicolson, London, 324 pp.

Omer-Cooper, J., Perkins, G.L. and Tottenham, C.E., 1928. The Coleoptera of Wicken Fen. In: J.S. Gardiner (Editor), *The Natural History of Wicken Fen*. Bowes and Bowes, Cambridge, pp. 267–297.

Pearson, R.G., 1962. Coleoptera from a detritus mud deposit of full glacial age at Colney Heath, near St. Albans. *Proc. Linn. Soc., Lond.*, 173: 37–55.

Rose, F.L. and Armentrout, D., 1976. Adaptive strategies of *Ambystoma tigrinum* Green inhabiting the Uano Estacado of West Texas. *J. Anim. Ecol.*, 45: 713–730.

Roux, F., 1976. The status of wetlands in the West African Sahel: their value for waterfowl and their future. In: M. Smart (Editor), *Conservation of Wetlands and Waterfowl. Proc. Int. Congr. Heiligenhafen, 1974*. IWRB, Slimbridge, pp. 272–287.

Schroevers, P.J., 1967. Is water H_2O? In: M.F. Morzer Bruyns (Editor), *10 Jaren Rivon*. Rivon — State Institute for Nature Conservation Research, Zeist, pp. 75–89.

Tucker, D.S., 1958. The distribution of some fresh-water invertebrates in ponds in relation to annual fluctuations in the chemical composition of the water. *J. Anim. Ecol.*, 27: 105–123.

Turcek, F.J., 1976. Birds and mammals of salt marshes and salt steppes in southern Slovakia (CSSR). *Biol. Conserv.*, 9: 29–36.

West, R.G., 1957. Interglacial deposits at Bobbitshole, Ipswich. *Philos. Trans. R. Soc., Lond.*, B 241: 1–31.

West, R.G., 1968. *Pleistocene Geology and Biology*. Longmans, London, 377 pp.

Wijngaarden, A.V., Rooth, J. and Leffef, W.J.B., 1964. Animal species in the Netherlands peculiar to wetland habitats. In: *Proc. MAR Conf. 1962. IUCN Publ., N. S., No. 3(1)*: 116–121.

Wilsson, L., 1971. Observations and experiments on the ethology of the European Beaver (*Castor fiber* L.). *Swed. Wildl.*, 8: 115–266.

Wolfenbarger, D.O., 1946. Dispersal of small organisms. Distance dispersion rates of bacteria, spores, seeds, pollen and insects: incidence rates of diseases and injuries. *Am. Midl. Nat.*, 35: 1–152.

Young, E.C., 1965. Flight muscle polymorphism in British Corixidae; ecological observations. *J. Anim. Ecol.*, 34: 347–389.

Chapter 11

ASPECTS OF SECONDARY PRODUCTION[1]

C.F. MASON and VALERIE STANDEN

INTRODUCTION

Swamps and fens represent rather disparate habitats from those of bogs and oligotrophic mires in terms of nutrient regimes and plant communities. Correspondingly the animal communities of swamps and fens will be treated separately from those of bogs and oligotrophic mires, so following the broad division based on trophic status used elsewhere in this book. Here swamps are considered as transitional habitats, with few plant species but many animals, while fens are rich in both plant and animal species. Bogs are generally poor in both. The animal communities are markedly influenced by the structure and palatability of the vegetation, the water level and the pH. The sizes and shapes of the habitat are also likely to be important in the development of the animal communities. Swamps and fens usually occupy relatively small areas and abut on to markedly different terrain. Bogs, on the other hand, are often very extensive, forming a landscape of their own. This difference has a profound influence on the distribution of animals so that, although quantitative information about animal populations and production is inevitably patchy and disconnected, an attempt has been made to set the available material into a spatial context at the local level of plant structure and at the wider level of landscape.

The majority of secondary production studies so far have been either on reed-swamps or on bogs of upland temperate or higher-latitude regions, with fens and similar more eutrophic mires being largely neglected. These latter habitats, therefore, receive only cursory treatment in the ensuing pages.

[1]Manuscript received August, 1977.

SWAMP AND FEN

In this chapter, swamps are defined as those communities which are dominated by emergent, non-woody plants which colonize the edges of waters, usually in nutrient-rich conditions. Swamps may vary in size from a narrow band of vegetation, less than a metre wide, along river edges, to many hundreds of square kilometres in the deltas of major rivers. Common dominants in Europe are *Phragmites australis*, *Typha latifolia* and *Scirpus lacustris*. Much litter is produced by these communities, which under some conditions (see Tallis, Ch. 9) results gradually in raised, drier conditions, a lower pH and the development of a typical fen vegetation, often dominated by *Cladium mariscus* or *Juncus subnodulosus*. The vegetational succession is very variable and in the case of the English Norfolk Broads, where many of the results described here were obtained, the succession from swamp to fen may proceed in several ways, depending on the water level and flow (Lambert, 1965; Mason, 1976a). Dense growths of submerged or floating macrophytes, such as *Stratiotes aloides*, occurring offshore of emergent swamps, are not considered, being primarily plants of fresh-water habitats. Swamp and fen woodlands are not discussed either.

Swamps, although poor in plant species, are highly productive. Some species such as *Phragmites australis* and *Typha latifolia* may form pure stands, particularly when carefully managed. Fens are usually rich in plant species. Faunal diversity is high in both habitats, though few quantitative studies have been undertaken, partly because of taxonomic problems, but also because the often treacherous nature of the ground makes sampling for quantitative work difficult. Of particular importance to these transitional habitats is the water regime. Some swamps are almost permanently inundated, others have the water table almost permanently below the surface. The majority show a gradation of conditions from waterside to dry land, with a seasonal variation in the extent of inundation. Where there is little water flow, anoxic conditions frequently develop for long periods. These marked variations and fluctuations in water levels and oxygenation influence the relative importance of the aquatic and terrestrial elements of the fauna of swamps. This topic is dealt with in more detail later.

Many animals are permanent members of the swamp community and diversity is greatest in those aquatic groups which breathe atmospheric air, such as Coleoptera and some dipterous larvae (see later, p. 369). In permanently flooded swamps fish may also be resident, if they can tolerate the low oxygen conditions during the summer. Under the more extreme conditions found in the tropics, most of the permanent swamp fishes of the River Zaïre, such as the lung fishes *Polypterus*, *Protopterus* and *Papyrocranus*, have accessory respiratory organs (Beadle, 1974; Lowe-McConnel, 1975). Those swamps, on the other hand, whose water tables are almost permanently below ground contain resident populations of animals which are typically terrestrial, as the house mouse *Mus musculus* reported in South Australia by Newsome (1969).

Swamp faunas include a large temporary element — that is, those animals which spend only the early developmental stages there or which return only to breed, examples being many insects, fishes and birds. Many other species use swamps only for feeding or protection during periods of inactivity. Swamps, being frequently linear in form, attract temporary fauna from both neighbouring aquatic and terrestrial habitats.

The fauna of swamps can be divided into four groups:

(1) The bottom fauna: species which live in the bottom sediment and detritus, chiefly detritivores and their predators.

(2) The epiphytic fauna: those animals occurring on the stems of plants below the water level. These are mainly micro-herbivores, feeding on attached periphyton, and their predators. Some species are detritivores, often burrowing into dead plant stems, such as the may-fly *Povilla adusta*, and larvae of the midge *Chironomus tentans*.

(3) Planktonic species.

(4) Those animals which inhabit the aerial part of swamps, consisting largely of herbivores.

Group 4 is distinct, but groups 1 to 3 are intimately related — for instance, most planktonic crustaceans are also abundant in the periphyton.

(1) **The bottom fauna**

Dvořák (1970a, b) and Mason and Bryant (1974) have studied the structure of the animal communities living in the bottom sediments of swamps. Seasonal changes in water level have a marked effect on the distribution of the bottom fauna. Both the oxygen concentration and the pH fall steadily at increasing distances from open water. Dvořák (1970b) recorded a drop in oxygen concentration in a swamp from 4 mg l^{-1} at 0.5 m from open water to almost zero at 5.5 m. In the same study the pH dropped from 7.0 to 6.1 over a distance of 20 m. Similarly a drop from pH 8.4 to 6.4 over 20 m was recorded by Mason and Bryant (1974) in Norfolk.

The animals present at increasing distances from open water in a Czechoslovakian swamp dominated by *Glyceria maxima* are shown in Fig. 11.1 (Dvořák, 1970b). At stations *A* to *C*, from 20.5 to 5.5 m from open water, molluscs and culicid larvae were most abundant. Those species respiring atmospheric oxygen (the snails *Aplexa hypnorum* and *Segmentina nitida*, and the flies *Theobaldia annulata* and *Culex pipiens*), and hence largely free of the constraints imposed by a de-oxygenated aquatic environment, were especially abundant at *A* and *B* within the body of the swamp. Leeches, requiring oxygen from the water, were most abundant at Station *D*, 0.5 m from open water. Dvořák (1970a) found that total biomass decreased at greater distances from open water, from 66 g m^{-2} at 2 m from open water to 34.5 g m^{-2} at 9 m, presumably again a reflection of the harsher environmental conditions within the swamp. The most abundant species in a *Phragmites australis* swamp (area 1.6 ha, width 30 m) at Alderfen Broad, Norfolk, are shown in Table 11.1. Hemiptera (*Sigara* spp.), ostracods and the caddis *Cyrnus flavidus* were especially abundant near the water's edge (T1), where there was also a large aquatic element in the fauna. At the greatest distance from the water's edge (T3) Diptera were most abundant and there was a terrestrial

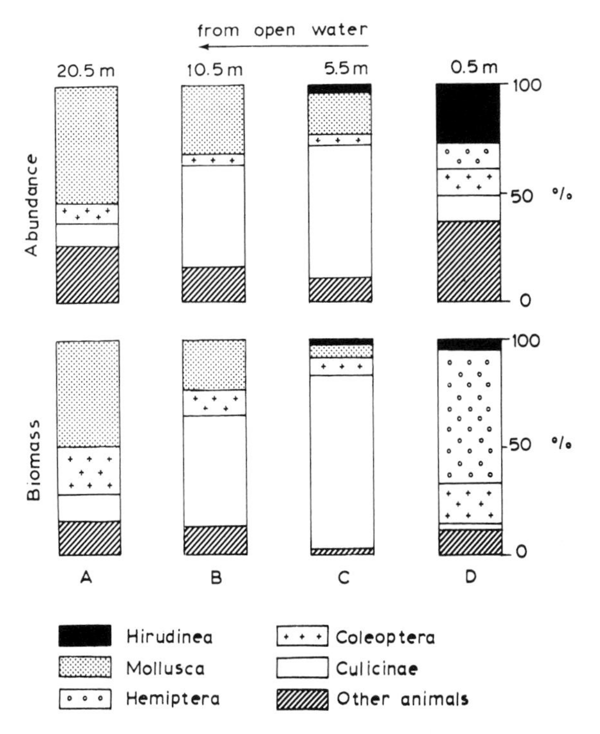

Fig. 11.1 The percentage share of the most important groups of animals in total macrofauna at four stations in a Bohemian reed-swamp (from Dvořák, 1970b with permission).

element in the fauna (e.g. Carabidae). The central area of the swamp (T2) was dominated by beetles, of which 28 species were present. Beetle assemblages seem characteristic of swamps. Obrtel (1973) recorded 116 beetle species in a Czechoslovakian swamp, of which 61 species (and most individuals) were carnivores, and 16 species were detritivores. The remaining species were described as scavengers, necrophages, herbivores or accidentals.

Species diversity in the Alderfen reed-swamp was measured as:

$$H' \approx \frac{c}{N} \left(N \log_{10} N - \sum_i n_i \log_{10} n_i\right)$$

where c is a scale factor for conversion from base 10 to base 2 logs, N is the total number of individuals and n_i is the number of individuals in the ith species. Evenness was measured as H'/H'_{max} where H'_{max} was the maximum diversity obtainable if all species in the samples were equally abundant. Species diversity and evenness were highest in the central part of the Alderfen reed-swamp (T2, see Table 11.1). This area of mud and detritus, mosses and pools of water (during the summer) was the most extensive habitat in the swamp. Since it was a broad habitat band, each site within it persisted for longer before being altered to the next successional stage (dry marsh), allowing time for greater niche differentiation. The narrow (2 m) permanently aquatic habitat represented by T1 was continually moving forward into the water (at 0.5 m per year) and being filled in

TABLE 11.1

The most abundant animal species occurring at three depths within a *Phragmites* swamp (Mason and Bryant, 1974)

Transect number: T1	T2	T3
Distance from water's edge: 1 m	10 m	20 m
pH: 8.4	7.3	6.4
Planorbis albus	*Rhynchelmis limosella*	*Rhynchelmis limosella*
Ostracoda	*Lumbriculus variegatus*	*Lumbriculus variegatus*
Eurycercus lamellatus	*Planorbis albus*	*Cyphon* sp.
Asellus aquaticus	*Asellus aquaticus*	*Prionocera turcica*
Asellus meridianus	*Asellus meridianus*	*Ptychoptera albimana*
Caenis horaria	*Gammarus pulex*	*Pericoma nubila*
Sigara dorsalis	*Cyphon* sp.	
Sigara falleni	*Hydroporus angustatus*	
Agrypnia pagetana	*Hydroporus lineatus*	
Cyrnus flavidus	Chironomidae	
Chironomidae	*Ptychoptera albimana*	
Chaoborus crystallinus		
Total number of species: 60	74	51
Species diversity (H'): 4.2	4.5	3.4
Evenness (H'/H'_{max}): 0.7	0.7	0.6

behind. However, there was less fluctuation in water, oxygen and nutrient conditions. The low-diversity habitat represented by T3 was less structured than the other habitats in the swamp, having no surface water. It was also a more extreme habitat, being liable to dry out in summer and flood in winter, making colonization by either primarily aquatic or terrestrial species difficult.

In the extensive swamps of the Danube Delta (up to 2000 m wide), Zinevici (1971) found the richest fauna to occur among reed stems with a medium density of 37 m^{-2}, there being smaller numbers of animals and few species when the stem density was higher or lower than this. Also in the Danube Delta Rudescu and Popescu-Marinescu (1970) compared the fauna attached to the stems of emergents with the benthic fauna. They recorded 323 species, of which 134 species were attached to plants and 189 species lived mainly on the bottom amongst detritus and mosses. A plankton fauna of 127 rotifer species and 23 gastrotrich species was also found. Overall faunal standing crops of up to 575 kg ha^{-1} were recorded in stands of *Phragmites australis* close to open water, this total including both benthic and epiphytic fauna. Faunal biomass in *Phragmites* was much greater than in *Typha angustifolia* (maximum 22 kg ha^{-1}) or *Scirpus lacustris* (maximum 15 kg ha^{-1}). The benthic fauna tended to decrease away from the edges of swamps and the periphytic fauna became more important, presumably because of decreased oxygen levels in the floor of the swamp when water flow was more restricted. Animals which were primarily benthic on the edge of the swamps became periphytic further in, and the total biomass was lower deep in the reed-swamps than it was along the edge.

Rudescu and Popescu-Marinescu also found that the invertebrate biomass was greater in those reed beds which were harvested by man than in unmanaged reed beds. This may be due to higher oxygen levels in the benthos, as much of the plant material was removed before it fell into the water and began to decompose; consequently the microfloral oxygen demand was much lower. Fish production was also found to be two to three times higher in harvested reed beds.

In a reed-swamp in Tjeukemeer, The Netherlands, Chambers (1971) recorded a production of 780 kg ha^{-1} of just one species, the detritivorous amphipod *Gammarus tigrinus*.

Erman and Erman (1975) studied the composition and production of fauna inhabiting calcareous fens in northern California, and have provided one of the few quantitative studies of fenland animals. The fens were dominated by the mosses *Cratoneuron filicinum* and *Drepanocladus aduncus*, with some forty other species of plant present. Detritivorous aquatic oligochaetes were the most abundant members of the fauna, comprising some 89% of the total. *Kincaidiana freidris* (Lumbriculidae) and *Mesenchytraeus* sp. (Enchytraeidae) were the most abundant worms. Diptera, especially Ceratopogonidae, were also common. The life zone for invertebrates was limited by oxygen, which extended down only a few centimetres. The fauna resembled more that of a bog than a swamp. Production is discussed later.

(2, 3) **The epiphytic and planktonic fauna**

It is very difficult to distinguish those animals which live on the stems of emergent plants from those living in the bottom detritus and there is much movement between the two. Similarly most planktonic animals are also epiphytic at times.

The periphyton is an intimate community of bacteria, fungi, algae, protozoa, nematodes and rotifers, which are grazed upon by larger animals. Meschkat (1934) in Germany found the most numerous animals in the periphyton to be nematodes, the most abundant of eight species being *Chromadora ratzeburgensis* and *Chromadora bioculata*. All fed on periphytic algae. The nematodes were most abundant in the first 30 m of the swamp from the open water, with the numbers falling towards the middle of the swamp. Seasonal changes in their depth distribution were also found.

Hirai (1970) in Japan compared the crustaceans and rotifers attached to reeds and floating macrophytes with those of the plankton. All attached species were also planktonic, though most individuals were epiphytic, at least during the day. Densities on the surface of reeds were often high (over 2000 cm^{-2}).

In Lake Chilwa in Malawi, 32 species of animals were living on *Typha australis*, a permanent member of the reed-swamp flora, forming extensive stands (McLachlan, 1975). The mean biomass of animals per unit area of *Typha* was 255 ± 33 mg m^{-2}, 67% of which consisted of one species,

Chironomus brevipalpus. Much of this biomass, however, was on dead floating swamp material. Two other plant species, *Diplacne fusca* and *Aeschynomene pfundii*, colonized the mud of the lake during a dry phase and died a few years after the lake had refilled. During these years *Diplacne* supported a much greater biomass of animals (1912 ± 370 mg m^{-2}) than *Typha*, and *Aeschynomene* a much smaller biomass ($1 \pm$ mg m^{-2}), though the latter figure increased markedly on the death of the plant.

Animals may exert a considerable grazing effect on the periphytic algae (Mason and Bryant, 1975b). The standing crops of periphyton were measured from February to November in Alderfen Broad, Norfolk, on dead stems of *Typha angustifolia* and on glass rods supported from a polystyrene float in the water amongst the *Typha* stems. The periphytic algae (diatoms and filamentous green algae) were similar on the two substrates. Chironomid larvae, which made up most of the fauna, were counted on the *Typha* stems. They were very scarce on the glass rods due to difficulties in attachment. The periphyton standing crop on *Typha* fell from 1.8 mg cm^{-2} in early April to nearly zero by November (see Fig. 11.2), whereas the standing crop on the glass rods reached a peak in May (1.93 mg cm^{-2}) and July (1.94 mg cm^{-2}) before falling to a steady level of 1.6 mg cm^{-2}. The chironomid population was highest in May on the *Typha* stems and then gradually declined. The guts of the dominant chironomids, *Glyptotendipes glaucus* and *Cricotopus dizonias*, were full of diatoms and filamentous algae. It appeared that the decline in standing crops of periphyton on *Typha* was largely due to grazing by chironomids, and was at a daily rate of 74 mg dry weight per square metre of stem surface. Chironomids moved from the mud to the *Typha* stems during April, probably because periphyton was a better food, being richer in nitrogen, phosphorus and essential amino acids. Successive generations of chironomids occurred on the *Typha* stems through

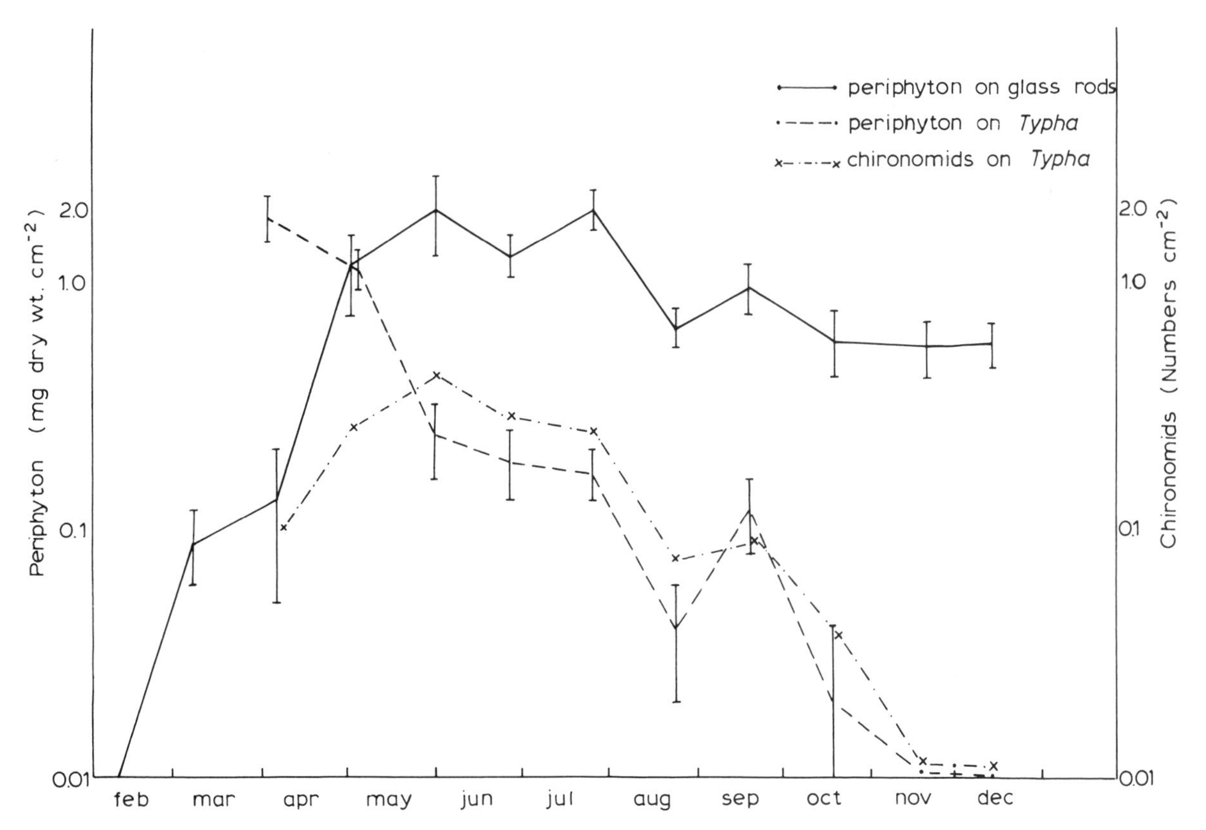

Fig. 11.2. The standing crop of periphyton on *Typha* stems and glass rods (mg dry wt. cm^{-2} $\pm$2SE) and the density of chironomid larvae on *Typha* stems (from Mason and Bryant, 1975b with permission).

the summer, but those present in the autumn went down to the mud when the periphyton supply was exhausted, the temperature fell and wave action made the stems inhospitable. A thick felt of periphyton developed on the *Typha* stems through the winter in the absence of chironomids.

(4) **Herbivore communities**

The quality and quantity of food are likely to be of more importance in determining the numbers and diversity of herbivores than of other groups, since, living in an aerial environment, they largely avoid the limitations associated with water level and oxygen concentration.

Some 39 species of insects are known to be closely associated with *Phragmites australis* (Haslam, 1972) and the herbivorous fauna of this plant has been studied by Waitzbauer (1971) in the swamps of the lake Neusiedlersee, Austria. He recognized four basic groups:

(1) Those animals which form galls on the surface or inside the plant, such as the dipterous larvae of *Lipara lucens* (Chloropidae) and *Perrisia inclusa* (Cecidomyiidae).

(2) Stem-miners, such as the Lepidoptera *Phragmatoecia castaneae* (Cossidae) and *Schoenobius gigantellus* (Pyralidae).

(3) Shoot suckers, such as *Chaetococcus phragmitidis* (Hemiptera, Coccidae).

(4) Leaf suckers, such as the aphid *Hyalopterus pruni*.

Phragmatoecia castaneae is a relatively large insect which prefers thick, vigorous stems and was not found in the marginal zone of the reed-swamp. The species had a consumption of 58.3 kJ m^{-2} yr^{-1} as larvae, with a production of 12.5 kJ m^{-2} yr^{-1}. However, these stem miners had little influence on the growth of *Phragmites* (Pruscha, 1973). The gall-forming *Lipara lucens* tended to be restricted to the landward zone of the swamp, because of the later development of the reeds, and consumed only 0.7 kJ m^{-2} yr^{-1}, with a production of 0.46 kJ m^{-2} yr^{-1}. However, by destroying the vegetative tips of *Phragmites*, *Lipara* stunted the growth of reed and prevented flowering. In Czechoslovakia *Phragmites* stems infected by *Lipara* were only 63% of the weight of healthy stems (Rychnovsky, 1973). The energy intake of the phytophagous insects of the Neusiedlersee was estimated at from 168 to 252 kJ m^{-2} yr^{-1}, which was less than 1% of the net primary production (Imhof, 1973).

The predators of phytophagous insects in swamps and fens have received little attention. Observations by Green and Bibby (1973) and Bibby et al. (1976) showed that reed aphids (*Hyalopterus pruni*) were abundant in some reed beds in southern England in late summer. The aphids were patchily distributed, and the timing of their peak abundance was broadly correlated with the passage of sedge warblers (*Acrocephalus schoenobaenus*). Aphids formed a significant component of the warblers' food, the feeding rate of the birds being some 0.24 attempts per second. Sedge warblers stayed longer and gained weight faster in years when aphids were abundant. Aphids were considered important in enabling these birds to lay down fat reserves prior to migration.

Studying an English breeding population of reed warblers (*Acrocephalus scirpaceus*), Catchpole (1974) found that 80% of nests were in reed-swamp,but that 64% of foraging occurred in adjacent *Salix* scrub, compared with 23% in the swamp. This study again emphasizes the importance of juxtaposition of reed-swamp with other habitats, as well as suggesting that the swamp did not contain sufficient food, in the form of herbivorous arthropods and their predators, during the spring to support broods of warblers.

Mammalian grazers may make a greater impact on primary production in swamps than insects. For instance, muskrats (*Ondatra zibethicus*) may be important in eastern Europe (Pelikan et al., 1970) and introduced coypus (*Myocastor coypus*) in eastern England. Muskrats and coypus eat the rhizomes and newly emerging shoots. A 40% reduction (from 942 g dry wt. m^{-2} to 524 g dry wt. m^{-2}) in the standing crop of a *Phragmites* swamp in East Anglia between two successive years was considered to be largely due to grazing by coypus on newly emerged shoots in the latter year (Mason and Bryant, 1975a). The grazed shoots were not replaced, and there was no compensation by increased growth of the surviving shoots. Gosling (1972) showed how coypus utilized the basal meristems of *Carex elata* in spring, quickly destroying the young plants, though replacement shoots were not utilized after May and grew to maturity. Gosling showed a number of seasonal patterns in feeding by coypus which were related to food quality. For instance, *Phragmites* leaf blades were mainly eaten in late summer, while rhizomes were excavated in the winter. The seed-pods of *Nuphar lutea* were utilized in August and September. Ellis (1963) has shown how selective feeding by coypus at high densities has altered the structure of plant communities of the Norfolk Broads. *Rumex hydrolapathum* and *Cicuta virosa*, for instance, have been markedly reduced in numbers.

Decomposition

Most of the animals in swamps are detritus feeders. Carnivores are important in the benthos in the central areas of swamps, where they may compose 20% of the total number of animals. The net production of the most important detritivores in the reed beds of the Neusiedlersee has been reported by Imhof and Burian (1972) and Imhof (1973). The production of both *Asellus aquaticus* (at densities of 190 to 400 m^{-2}) and chironomids (at a density of 700 m^{-2}) was estimated at 42 kJ m^{-2} yr^{-1}. Four species of pulmonate snail had annual production varying from 20 to 33 kJ m^{-2} yr^{-1} (total for the four species 105 kJ m^{-2} yr^{-1}), with *P:B* (production to biomass) ratios ranging from 2.1 to 4.4.

Mason and Bryant (1975a) have examined the role of animals in the decomposition of leaf litter of *Phragmites australis* and *Typha angustifolia*. A known weight of leaf litter was placed in large-mesh (aperture 4600 μm) and small-mesh (aperture 250 μm) nylon bags and placed in water along the edge of the reed-swamp at Alderfen Broad. Groups of bags were removed at regular intervals and the weight loss was determined. The small-mesh bags excluded all but micro-organisms, whereas the large-mesh bags developed large populations of detritivores and their predators. The litter of *Typha* and *Phragmites* took from 7 to 14% longer to decompose in small-mesh than in large-mesh bags, though this difference was not significant. Similar rates of breakdown were suggested by laboratory experiments on the effect of detritivores on the particle size of detritus — from 3 to 11% of the total annually.

The majority of the decomposition of swamp vegetation is, therefore, due to the micro-organisms, though animals play a role in making the substrate available to the micro-organisms. *Typha* litter grazed by the snail *Lymnaea peregra* had a higher respiration rate than control litter, probably because the action of the snail's radula increased the surface area of litter available for colonization by micro-organisms (Mason and Bryant, 1975a). A similar effect was not recorded, however, in the feeding of *Gammarus pulex* and *Asellus aquaticus*, although the faeces and non-ingested food produced by these animals resulted in a suspension of particles in the medium which respired at a significantly higher rate than the control medium.

Further to these experiments selective antibiotics have been used to determine the relative importance of fungi and bacteria in the decomposition of *Phragmites* leaves (Mason, 1976b). Fungi and bacteria were equally responsible for the respiration and weight loss of litter in the early stages of decomposition, but fungi became much less important later.

It must be emphasized that in swamps a proportion of the litter remains undecomposed, though there are no direct estimates of this available. Osborne and Moss (1977) have shown that the rate of sediment deposition in a shallow lake in Norfolk, surrounded by reed-swamp, varied from 1.2 mm to 3.1 mm per annum between 1720 and 1950 A.D. The deposition rate then increased to about 5 mm in the 1950's, to 10 mm in the 1960's and 12 mm per year during the 1970's, during a period of increasing eutrophication. However, the reed-swamp was by no means the only source of sediment, material from sewage treatment works, phytoplankton and bank erosion also being involved.

Erman and Erman (1975) have estimated the production of the fauna of minerotrophic fens in California. The fauna was dominated by detritivorous oligochaetes (see p. 370), with annual production ranging from 7.4 to 58 g wet wt. m^{-2} over seven sites and *P:B* ratios ranging from 5.5 to 9.3. The oligochaete production increased with increasing peat depth over the range of peat depths studied (17.3–87.4 cm). It was suggested that this was due to the greater buffering capacity afforded by deeper peats, which reduce the daily changes in oxygen, temperature and water level; the survival of oligochaetes was probably less in shallow fens. Overall annual production of invertebrates at these seven Californian sites ranged from 8.8 to 59.4 g wet wt. m^{-2}.

The importance of swamps to aquatic ecosystems

Swamps have an important role to play in the energy and nutrient dynamics of the lakes and rivers with which they are often intimately associated. Thus, large amounts of detritus and nutrients are washed into aquatic ecosystems, playing an important part in the primary and secondary production of fresh waters (Banoub, 1975; Howard-Williams and Lenton, 1975; Mason and Bryant, 1975a; Schröder, 1975). Schröder (1975) has calculated that in the Bodensee–Untersee (Lake of Constance) the littoral zone contributes annually 340 mg m^{-2} dissolved phosphorus and 620 mg m^{-2} nitrogen. (The littoral occupies 14 km^2 of a total lake area of 66 km^{-2}, and one-third of the littoral is covered with *Phragmites*, the remainder having summer growth of submerged and floating macrophytes.) This nutrient loading is relatively small compared with that received from human sources. However, the input is greatest in summer when the lake level rises and coincides with a low nutrient concentration in lake water. It may thus lead to local eutrophicating effects. Mason and Bryant (1975a) found that flushing of the reed-swamp occurred in winter in Alderfen Broad, Norfolk, when water levels rose, potentially releasing 230 kg nitrogen and 16 kg phosphorus in 1972. They suggested that the reed-swamp may function as a nutrient regulator, taking nutrients up in the spring and summer for growth and releasing them during the winter, when phytoplankton growth was limited by light and temperature. Sharp increases in nitrogen and phosphorus occurred in the lake water in December, but were not followed by an increase in algae. Much of these nutrients would be lost from the lake with increased winter outflows of water.

Swamp vegetation appears much more resistant to the effects of excessive eutrophication than are floating and submerged macrophytes. Where these latter disappear the emergent vegetation becomes the sole site for oviposition of many invertebrates and fishes. These often become restricted to the littoral reed zones, which form a reservoir of animals from which recolonization can take place if the conditions in the open water improve sufficiently for the re-establishment of macrophytes (Mason and Bryant, 1975c). Wortley (1974) has shown how artificial plants (nylon rope attached to plastic bases) in the rather barren benthos of Alderfen Broad quickly become colonized by invertebrates normally restricted to the reed-swamp in the littoral zone of the lake.

BOGS AND OLIGOTROPHIC MIRES

Those habitats classed as bogs include a wide variety of plant communities which are frequently intricately interspersed one with another. Changes in the water status of bogs — induced naturally and by man — have resulted in a variety of successions and the production of complex mosaics of vegetation. Upland blanket bogs in the British Isles frequently contain localized areas of deep wet peat and these are subject to natural drainage via erosion. Where hydrostatic stability has been achieved, as in the convex parts of blanket bogs and managed moorlands, more uniform areas may occur. Even so, in their fine detail such areas are still extremely heterogeneous and are intricately patterned with plants having a wide range of forms. The fauna associated with these habitats are also varied; but, for the purpose of this survey of secondary production, it has been necessary to treat the fauna as a whole and to neglect many of the fine differences of scale present in the habitat.

Cragg's (1961) survey of the fauna of some moorland sites including various types of bog on the Moor House National Nature Reserve, Cumbria (England), was one of the first attempts to analyze information on the fluctuations in numbers, distribution and dispersion of a wide range of species. Cragg chose to study moorland as being "a relatively simple natural habitat", but preliminary studies showed it to be a "highly complex system, the result of long term natural changes combined with the effects brought about by man in managing the area for game and sheep". A series of ecological investigations by individual research workers has been in progress since the reserve was acquired by the Nature Conservancy in 1952, and the results of these studies have been used extensively in the present analysis. The bog and mire systems at Moor House may have been altered from their natural state due to man-made effects but the intensive studies on these systems are useful because there is a scarcity of information on secondary production in unaltered bogs elsewhere in the world. Coulson and

Whittaker (1978) discuss many aspects of the Moor House studies, and include tables of population and biomass estimates of individual groups.

When considering one particular aspect of a large number of species it is convenient to put them into groups simply because it allows one to consider fewer species at any given time. It is helpful to do this according to the extent to which they interact with and influence one another. One attribute of animals on which to base such a division when considering the energy flow through a community would be according to the type of food eaten by species at the beginning of the food chain — that is, whether they are herbivores, saprovores or coprophages. However, the food of many species is not known, while many other species are known to feed on several different types of food and the extent to which they utilize the different groups is not known. For this reason no attempt has been made to assign animals to different food-based communities, and the animals of bogs are considered here as belonging to either an above-ground or a below-ground group. The above-ground food chain is based largely but not entirely on the consumption of live plant tissue by herbivores, while the below-ground animals derive their food mainly from dead plant and animal material. This simple spatial division is followed here as being the most practical. A similar approach has been made in the earlier part of this chapter dealing with swamps, and, as noted there, there are many respects in which they overlap. In the case of bogs some species feed below ground on dead plant material but also utilize live root tissue, and conversely some animals such as Tipulidae (craneflies) live and feed below ground for most of their life cycle, but later they emerge and are then included in the dynamics of life above ground.

One other aspect of the transference of food through a particular group of animals should be mentioned here, which is that the activity, as measured by biomass, consumption and production, of the below-ground animals may be greatly affected by the activity of the herbivore populations above the ground. Net primary production, which is neither eaten by a herbivore nor retained by the plant as increased biomass, dies and falls to the ground and is then potential food for the below-ground detritivore (saprovore) populations. Food consumed by herbivores and not assimilated by them also goes back to the ground as faeces and may be utilized by below-ground populations. Thus, where the herbivore population is large, a large proportion of primary production is converted to above-ground secondary production or respired; and also a large proportion, after being consumed, is not assimilated but is deposited as faeces — more readily decomposable, perhaps, than the plant material which has only died and fallen to the ground. Many invertebrates feed on plant material which has been partially decomposed after passage through the gut of a herbivore. Coulson and Whittaker (1978) suggest that the scarcity of herbivores and the consequent low input of faeces contributes to the reduced activity (i.e. secondary production) of below-ground animals, and is thus associated with the accumulation of dead plant matter in bogs and oligotrophic mires. Svendsen (1957) found two species of earthworms associated with sheep dung and absent elsewhere on blanket bog.

The animal community above the ground

Although above-ground primary production of blanket bog (*Calluneto-Eriophoretum*) at Moor House in northern England is high, averaging 426 g m^{-2} yr^{-1} (8874 kJ m^{-2} yr^{-1}) from a summer standing crop of 1564 g m^{-2} oven-dry weight (Forrest, 1971), as noted above, there are very few vertebrate herbivores. Rawes and Welch (1969), in their study of production by sheep grazing on a variety of vegetation types over the whole area, considered that this type of vegetation was used very little by sheep. In arctic and subarctic regions of North America and northern Europe large vertebrate herbivores include caribou (*Rangifer* spp.) and moose (*Alces* spp.) which browse over large areas containing muskeg bogs. These animals too are highly selective in their feeding preferences, and the extent to which they utilize bog vegetation as opposed to other types within the same area is difficult to determine in the field, and may be very small. Published data give biomass and production figures for wild populations ranging over a great variety of vegetation types. For example the American moose (*Alces alces*) feeds mainly on aquatic plants in summer and taiga trees in winter (Longhurst, 1960), and Pimlott (1960) gives density estimates up to one per five square miles (approximately 170 kg km^{-2}) with net production up to 22% in Newfoundland moose populations.

Other mammalian herbivores include bog lemmings (*Synaptomys* sp.) and species of *Microtus* and *Pedomys*, but populations of these microtines are low compared to those of the lemmings (*Lemmus* sp.) in tundra which according to Batzli (1975) may be from 10 to 25 km^{-2}. At Moor House there are few mammalian herbivores on the bog sites, but the red grouse (*Lagopus lagopus scoticus*) does feed on heather on blanket bog. These animals feed selectively from the heather plant (Moss, 1972) to the extent of choosing shoots with a higher nitrogen content, and thus the quantity of material actually available to them is far less than the total primary production. Butterfield and Coulson (1975) have shown that adult red grouse supplement their food by eating insects — mainly tipulids. Grouse populations in English and Scottish moors generally fluctuate widely from year to year, but may reach one pair per two hectares (65 kg km^{-2}) (Jenkins et al., 1963). Production is usually about twice the standing crop (P:B=2:1) (Leopold and Ball, 1931).

Thus, although primary production on blanket bog (*Calluneto-Eriophoretum*) is high, the few vertebrate herbivores which do feed there are able to utilize only a fraction of the plant material.

Invertebrate herbivores of bogs are also few in species with a low total biomass, although exceptionally a few species occur in large numbers. At Moor House the northern eggar (*Lasiocampa quercus* var. *callunae*) was virtually unknown before 1972, but then occurred in large numbers which have since been maintained (J. Butterfield, pers. comm., 1976). In areas of bog in England at lower altitude (e.g. Malham Tarn bog, Yorkshire) and in Scotland, the heather beetle (*Lochmaea suturalis*) occasionally reaches outbreak proportions. At these times large areas of heather are killed by the feeding activity of the beetle, which eats around the base of the stems and thus indirectly kills off the whole stem. Actual secondary production by the beetle is probably relatively small compared with its effect on production of its host plant. Grimshaw (1911) extracted approximately 360 beetles per square meter from soil samples in winter. At Moor House one of the most numerous invertebrate herbivores is the small plant-sucking hemipteran *Strophingia ericae* which feeds on *Calluna*. Hodkinson (1973) found this species in densities up to 11 000 m^{-2} with an average of 2000 m^{-2} and biomass 2.9 kJ m^{-2}; but, as its life cycle takes two years, production is low and he estimated that the *S. ericae* population consumed only between 0.1 and 1% of the shoot production of heather. This is another example of selective feeding — the Hemiptera suck the plant sap with a high amino acid content. Even so their effect on the primary production of the host plant is probably not very great. Hodkinson (1973) could detect no difference in photosynthesis between plants heavily infested with *S. ericae* and plants without this animal. Other Hemiptera have been studied on bog vegetation at Glenamoy, Eire. *Neophilaenus lineatus* occurs at a maximum density of late larvae of 5 m^{-2} and the late larvae of *Philaenus spumarius* were found at a maximum density of 10 m^{-2} (Nixon et al., 1975). *N. lineatus* is also numerous on *Juncus squarrosus* at Moor House. Production of up to 80 mg m^{-2} yr^{-1} oven-dry weight was recorded for this species by Whittaker (1967) from a population with a density of late larvae of about 20 m^{-2}.

Carnivores preying on above-ground herbivores are scarce. Cherrett (1964) examined the spider fauna at Moor House and found up to 155 m^{-2} on *Calluneto - Eriophoretum* blanket bog and 470 m^{-2} on moor dominated by *Juncus squarrosus*, a transition type of vegetation between blanket bog and grassland on the Moor House Reserve. The majority of the species were liniphyids. Very few lycosids or orb-web spinning spiders were found on blanket bog, and Cherrett accounted for their scarcity as being due to the architecture of the site and the inability of some species to withstand desiccation, rather than to a lack of food. In other areas lycosid species may be abundant on bogs. Norgaard (1951) has described the distribution of *Pirata piratica* and *Lycosa pullata* from a Danish *Sphagnum* bog, but there are few quantitative data on standing crop or production.

Probably the main above-ground predator/prey interaction at Moor House is that between the meadow pipit (*Anthus pratensis*) and adult crane-flies (Tipulidae). The prey species feed as larvae below ground in the peat and within the *Sphagnum* mat. Meadow pipits raise their first brood with adult *Tipula subnodicornis* as 83% of their food, but even so Coulson (1962) calculated that this was less than 1% of those available. In this example the close synchronization of emergence of the adult tipulids from the bog sites meant that they were an abundant source of food for the predator for a limited time only. Food brought to the second brood was composed mainly of animals emerging from streams. Coulson and Whittaker (1978) suggest that the necessity to hunt for food at streams accounts for the observed distribution of meadow pipit nests, which are limited to the edges of blanket bog.

Frogs are a common vertebrate carnivore at Moor House (Houston, 1973) and at Glenamoy (Blackith and Speight, 1974), but examination of the gut contents of frogs collected at these localities suggests that they feed principally on animals from nearby waterside, woodland and grassland habitats rather than on the bog itself.

In general the picture presented by bogs, at least those upland bogs referred to here and for which quantitative information exists, is one of little above-ground activity. Numbers of both individuals and species of herbivores are low, and this is reflected by the scarcity of carnivores.

Below-ground animals

The most abundant soil mesofauna of many blanket bogs are Enchytraeidae (Oligochaeta) and Tipulidae (Diptera). The majority of individuals are found in the upper 6 cm of peat soils at Moor House, although the zone of activity seems to extend to 10 cm depth in the case of enchytraeids in drier peat. The Lumbricidae which are so important in the processes of decomposition and soil formation in mineral soils are virtually absent from temperate bogs; only Hedberg (1954) reported the presence of unspecified earthworms in *Carex runssoroensis–Sphagnum* bogs in alpine East Africa, which according to him eat the *Sphagnum* and give rise to a substance unlike normal peat.

Primary production of plant tissue below ground in the form of roots is approximately one-third of the total production on blanket bog at Moor House, although probably less than one-third in many other types of bog. Forrest (1971) estimated 209 g m^{-2} yr^{-1} (4492 kJ m^{-2} yr^{-1}) below-ground production from a standing crop of 886 g m^{-2} (oven-dry weight).

Tipula subnodicornis feeds on living and dead plant material, and may reach very high densities on *Juncus squarrosus* (transitional) moor. At Moor House populations of fourth-instar larvae of over 200 m^{-2} have been recorded (Coulson, 1962). In one year, annual production was calculated as 5.2 g m^{-2} (107 kJ m^{-2}) from an average biomass for all stages of 1.25 g m^{-2}, giving an annual $P{:}B$ of 4.2. *T. subnodicornis* occurs in *Calluneto-Eriophoretum* bog at Moor House at approximately one-third of the density found under *Juncus squarrosus* (J.C. Coulson, pers. comm., 1977).

One other species of tipulid occurs at Moor House in large numbers in *Juncus squarrosus*, and in lower densities on *Calluneto-Eriophoretum*. The larvae of the small wingless cranefly *Molophilus ater* feeds probably on a mixture of decomposing organic material and decomposer micro-organisms. Hadley (1971) recorded from 150 to 2000 final instar larvae of this species per square metre, and commented that mortality after the first instar was low. Smith (1973) calculated production on *Juncus squarrosus* as 0.5 g dry wt. m^{-2} yr^{-1} (13.2 kJ m^{-2} yr^{-1}) from an average biomass for all stages of 0.2 g m^{-2}. Annual $P{:}B$ for this tipulid is thus 2.5:1.

Production of enchytraeids in blanket bog at Moor House reached 2.3 g dry wt. m^{-2} yr^{-1} (53.9 kJ m^{-2}) from an average biomass of 2 g m^{-2} for the dominant species *Cognettia sphagnetorum*. There was an additional 0.06 g m^{-2} yr^{-1} production by four other species on the same site (Standen, 1973). *C. sphagnetorum* is probably a saprovore, and is widespread and abundant in many bogs in the west of Ireland and in northwestern Scotland; it is also the dominant enchytraeid in peat soils in Denmark (Nielsen, 1961) and in coniferous forest soils in Norway (Abrahamsen, 1972). This enchytraeid reproduces by fragmentation, and grows slowly, and consequently has a low annual $P{:}B$ ratio (approximately 1:1). It is not acclimatized to low temperatures, and as a result population metabolism over the long winter period is relatively low. Material utilized directly by a population of *C. sphagnetorum* for tissue production and respiration amounted to only 2% of the annual input of litter to the bog. However, given that most detritivores have a low assimilation efficiency, the amount of material consumed but not assimilated is likely to be correspondingly much greater. Experiments using litter of *Calluna*, *Eriophorum* and *Sphagnum* enclosed in fine nylon mesh bags showed that when *C. sphagnetorum* was included with the litter the rate of weight loss by micro-organisms over three months was increased by a factor of 1.35 compared with litter without these animals (Standen, 1978). It is also suggested that fragmentation is a method of reproduction which conserves minerals and energy, and is an adaptation enabling *C. sphagnetorum* to live in nutrient-poor bogs and oligotrophic soils generally.

Mites, Collembola and nematodes are numerous in bogs and oligotrophic mires. Some of these feed as detritivores, but others are specialized as carnivores or feed selectively on fungi. Blackith (1974) recorded high numbers of Collembola associated with particular plant species in bogs in Eire, particularly on *Scirpus*, *Molinia*, *Schoenus* and *Sphagnum*, but not on *Calluna* or *Eriophorum*. One species, *Friesia mirabilis*, fed on other species of live or moribund Collembola and was locally abundant (up to 16 000 m^{-2}). Hale (1966) found that populations of Collembola in blanket bog were fairly stable, with numbers up to 80 000 m^{-2} and an average biomass of 0.1 g m^{-2} in heather litter. Nine species were involved, one of which, *Tetracanthella wahlgreni*, was restricted to blanket bog. Blackith (1974) agreed with Hale (1966) that Collembola are of minor importance in promoting energy flow through the system, but referred to Macfadyen's (1963) suggestion that, by grazing the microflora, they may be of great indirect importance in the process of litter decomposition.

In a study of the mites at Moor House, Block (1966) found that they were more abundant and varied in peat soils than in nearby mineral soils. In peat soils the Cryptostigmata made up 63% of the biomass which, on peat soils, ranged from 0.36 to 2.48 g m^{-2} live weight between March and June.

Nematodes in peat soils are probably microbial feeders. At Moor House Banage (1963) extracted from 1.9 to 3.1×10^6 m^{-2} nematodes from *Juncus squarrosus* moor (478–751 mg m^{-2} wet weight), and lower numbers than this from blanket bog. The testate amoebae are another group of microfauna abundant in peat soils which have been studied quantitatively. According to Heal (1962) they feed on detritus and, taking several species together, he found up to 16×10^6 m^{-2} (approximately 1 g m^{-2}) in *Sphagnum recurvum*.

Few carnivores are associated with these populations of below-ground detritivores, although tipulids on emerging as adults are taken by spiders and by insectivorous birds. Larval tipulids are probably included in the diet of lycosid spiders which are sometimes abundant on the surface of *Sphagnum* bogs, and Collembola and mites are probably preyed upon by the adults and larvae of Carabidae which have been trapped on the surface of bogs. Houston (1971), using pitfall traps, found eight species of beetle occurring commonly on wet blanket bog, but overall density was low at less than 10 m^{-2}.

DISCUSSION

Bogs and oligotrophic mires support few species of animals, but those that do occur characteristically have high population densities. Individuals tend to be small and feed selectively. Few animals have become adapted to utilize primary production above the ground as herbivores, and this in turn has reduced the diversity of material entering the pool of dead organic matter and consequently the number of saprovore species able to feed on it.

Climate is probably not directly responsible for reduced diversity or production. Coulson and Whittaker (1978) compared bog sites with limestone grassland sites at Moor House National Nature Reserve. Primary production is high on blanket bog but herbivore production is very low and there is a net accumulation of peat in bogs, while on neighbouring mineral grasslands herbivore activity is high and there is no accumulation of organic material.

Possibly many species of bog plants are unpalatable to herbivores. Certainly many bog plants are conservative in their use of nutrients, and nutrient supply may be the key to low secondary productivity where relatively few species are adapted to feed selectively (e.g. *Strophingia*).

The biomass of soil animals in peat is about one-quarter of that in limestone soils, and most of the difference is due to the absence of earthworms from peat. The standing crop of tipulids in peat is reduced but that of the enchytraeids is broadly similar.

Many species of earthworms do not occur in soils with low pH value, but the material available to the saprovores in bogs is mainly unmodified leaf litter due to the lack of herbivores, and this may be unpalatable to many species. Where plants like *Calluna* withdraw nutrients from leaves before they

TABLE 11.2

Population characteristics of some invertebrates of mires

Species or group	Locality	Community type	Average number (m^{-2})	Biomass[1]		Annual production[1]		Annual $P{:}B$	Reference
				kJ m^{-2}	g m^{-2}	kJ m^{-2}	g m^{-2}		
BOGS									
Neophilinus lineatus	Glenamoy, Eire	*Schoenus* bog	0.1–5.0	–	–	–	–	–	Nixon et al. (1975)
Neophilinus lineatus	Moor House, England	*Juncus squarrosus*	35	–	0.02	–	0.08	4.0:1	Whittaker (1967)
Strophingia ericae	Moor House, England	*Calluneto-Eriophoretum*	11 000	–	0.1	–	0.14	1.4:1	Hodkinson (1973)
Lochmaea suturalis	various moors	Heather moor	360	–	–	–	–	–	Grimshaw (1911)
Tipula subnodicornis	Moor House, England	*Juncus squarrosus*	600	–	1.25	107	5.2	4.2:1	J.C. Coulson (pers. comm., 1971)
Molophilus ater	Moor House, England	*Juncus squarrosus*	4000	–	0.2	13.2	0.5	2.5:1	Smith (1973)
Molophilus ater	Moor House, England	*Calluneto-Eriophoretum*	2000	–	0.05	3.9	0.15	3.0:1	Smith (1973)
Cognettia sphagnetorum	Moor House, England	*Calluneto-Eriophoretum*	30 000	–	2.0	53.9	2.3	1.0:1	Standen (1973)
Collembola	Moor House, England	*Calluneto-Eriophoretum*	80 000	–	0.1	–	–	–	Hale (1966)
Mites	Moor House, England	*Calluneto-Eriophoretum*	50 000		1.0 f.w.	–	–	–	Block (1966)
Nematodes	Moor House, England	*Juncus squarrosus*	$2–3 \times 10^6$	–	0.6 f.w.	–	–	–	Banage (1963)
Spiders	Moor House, England	*Calluneto-Eriophoretum*	150	–	–	–	–	–	Cherrett (1964)
Spiders	Moor House, England	*Juncus squarrosus*	450	–	–	–	–	–	Cherrett (1964)
SWAMPS AND FENS									
Phragmatoecia castaneae	Austria	*Phragmites* canopy	7.5	–	–	12.5	–	–	Pruscha (1973)
Lipara lucens	Austria	*Phragmites* canopy	–	–	–	6.5	–	–	Pruscha (1973)
Asellus aquaticus	Austria	*Phragmites* benthos	–	–	–	41.8	–	–	Imhof (1973)
Araneus cornutus	Austria	*Phragmites* benthos	–	0.13	–	0.67	–	5.2:1	Imhof and Burian (1972)
Chironomidae	Austria	*Phragmites* benthos	7	–	–	41.8	–	–	Imhof (1973)
Lymnaea ovata	Austria	*Phragmites* benthos	–	4.6	–	20.1	–	4.4:1	Imhof and Burian (1972)
Lymnaea stagnalis	Austria	*Phragmites* benthos	–	11.3	–	25.1	–		Imhof and Burian (1972)
Planorbarius corneus	Austria	*Phragmites* benthos	–	13.0	–	27.6	–	2.1:1	Imhof and Burian (1972)
Planorbis planorbis	Austria	*Phragmites* benthos	–	8.36	–	32.6	–	3.9:1	Imhof and Burian (1972)
Oligochaeta	California	minerotrophic fen	17 000	–	6.3 f.w.	–	44.0 f.w.	7.0:1	Erman and Erman (1975)
Ceratopogonidae	California	minerotrophic fen	–	–	0.19 f.w.	–	1.0 f.w.	5.3:1	Erman and Erman (1975)
Chironomidae	California	minerotrophic fen	–	–	0.16 f.w.	–	0.96 f.w.	6.0:1	Erman and Erman (1975)
Nematoda	California	minerotrophic fen	–	–	0.01 f.w.	–	0.1 f.w.	10.0:1	Erman and Erman (1975)

[1] Dry weight, except where fresh weight indicated (f.w.).

die the litter may be particularly poor in nutrients, and this may prevent many species becoming established, unless they are adapted to pass on nutrients from one generation to the next — as is the case with *Cognettia sphagnetorum* by its method of reproducing by fragmentation.

With only a limited amount of data on secondary production available, it does not seem useful to generalize about the rates of production which may characterize animals in bogs. It is not easy to use the ratio of annual production: average biomass because in annual and synchronous species the calculation of average biomass ($\bar{B}$) is extremely difficult, particularly where growth is discontinuous. Thus, calculation of the average biomass of a species with, for instance, a prolonged egg stage, or a species with a prolonged late-larval or pupal or adult stage, can lead to errors in the calculation of $\bar{B}$, unless some correction is made for the non-growing period. Any error in $\bar{B}$ would lead to a large error in $P{:}\bar{B}$ which would make comparison between species meaningless. If long-lived and continuously growing animals only are considered, then annual $P{:}\bar{B}$ [population turnover, analogous to the efficiency index r of Blackman (1919)] is still determined by both individual growth rates and the type of mortality suffered by the population, and, unless these parameters are known, trends in the value of annual $P{:}\bar{B}$ ratios are difficult to interpret. The use of $P{:}B$ ratios in production studies is discussed by Standen (1980).

Table 11.2 gives examples which show that annual production by animals in *Calluneto–Eriophoretum* bog at Moor House is estimated to be approximately 100 kJ m^{-2}, while that of *Juncus squarrosus* bog is more than twice this but similar to that of *Phragmites* swamp.

REFERENCES

Abrahamsen, G., 1972. An ecological study of enchytraeids of Norwegian forest soils. *Pedobiologia*, 12: 26–82.

Banage, W.B., 1963. The ecological importance of free living soil nematodes with special reference to those of moorland soil. *J. Anim. Ecol.*, 32: 133–140.

Banoub, M.W., 1975. The effects of reeds on the water chemistry of Gnadensee (Bodensee). *Arch. Hydrobiol.*, 75: 500–521.

Batzli, G.O., 1975. The role of small mammals in arctic ecosystems. In: F.B. Golley, K. Petrusewicz and L. Ryszkowski (Editors), *Small Mammals: Their Productivity and Population Dynamics. I.B.P. Publ.*, No. 5: 243–269.

Beadle, L.C., 1974. *The Inland Waters of Tropical Africa. An Introduction to Tropical Limnology*. Longman, London, 265 pp.

Bibby, C.J., Green, R.E., Peplar,G.R.M. and Peplar, P.A., 1976. Sedge warbler migration and reed aphids. *Br. Birds*, 69: 384–399.

Blackith, R.E., 1974. The ecology of Collembola in Irish blanket bogs. *Proc. R. Ir. Acad.*, 74: 203–227.

Blackith, R.E. and Speight, M.C.D., 1974. Food and feeding habits of *Rana temporaria* in bogland habitats in the west of Ireland. *J. Zool. Lond.*, 174: 67–69.

Blackman, V.H., 1919. The compound interest law and plant growth. *Ann. Bot., Lond.*, 33: 353–360.

Block, W.W., 1966. Seasonal fluctuation and distribution of mite populations in moorland soils with a note on biomass. *J. Anim. Ecol.*, 35: 489–504.

Butterfield, J.E.L. and Coulson, J.C., 1975. Insect food of adult red grouse *Lagopus lagopus scoticus* (Lath.). *J. Anim. Ecol.*, 44: 601–608.

Catchpole, C.K., 1974. Habitat selection and breeding success in the reed warbler (*Acrocephalus scirpaceus*). *J. Anim. Ecol.*, 43: 363–380.

Chambers, M., 1971. The dominance, production and utilization of *Gammarus tigrinus* (section) in the exposed *Phragmites* beds of the Tjeukemeer (Holland). *Hydrobiologia*, 12: 297–303.

Cherrett, J.M., 1964. The distribution of spiders on the Moor House National Nature Reserve, Westmorland. *J. Anim. Ecol.*, 33: 27–48.

Coulson, J.C., 1962. The biology of *Tipula subnodicornis* Setterstedt with comparative observations in *Tipula paludosa* Meigen. *J. Anim. Ecol.*, 31: 1–21.

Coulson, J.C. and Whittaker, J.B., 1978. The fauna of moorland soils. In: O.W. Heal and D.F. Perkins (Editors), *The Ecology of Some British Moors and Montane Grasslands*. Springer-Verlag, Berlin, pp. 52–94.

Cragg, J.B., 1961. Some aspects of the ecology of moorland animals. *J. Anim. Ecol.*, 30: 205–223.

Dvořák, J., 1970a. A quantitative study on the macrofauna of stands of emergent vegetation in a carp pond of south-west Bohemia. *Rozpr. Cesk. Akad. Ved. Rada Mat. Prir. Ved*, 80: 63–108.

Dvořák, J., 1970b. Horizontal zonation of macrovegetation, water properties and macrofauna in a littoral stand of *Glyceria aquatica* (L) Wahlb. in a pond in south Bohemia. *Hydrobiologia*, 35: 17–30.

Ellis, E.A., 1963. Some effects of selective feeding by the coypu (*Myocastor coypus*) on the vegetation of Broadland. *Trans. Norf. Nor. Nat. Soc.*, 20: 32–35.

Erman, D.C. and Erman, N.A., 1975. Macroinvertebrate composition and production in some Sierra Nevada minerotrophic peatlands. *Ecology*, 56: 591–603.

Forrest, G.I., 1971. The structure and production of North Pennine blanket bog vegetation. *J. Ecol.*, 59: 453–479.

Gosling, L.M., 1972. The coypu in East Anglia. *Trans. Norf. Nor. Nat. Soc.*, 23: 49–59.

Green, R.E. and Bibby, C.J., 1973. Sedge warblers and aphids. *Wicken Fen Group Rep.*, 5: 7–11.

Grimshaw, P.H., 1911. The heather beetle. In: Lord Lovat (Editor), *The Grouse in Health and Disease*. Smith and Elder, London, pp. 414–430.

Hadley, M., 1971. Aspects of the larval ecology and population dynamics of *Molophilus ater* Meigen (Tipulidae) on Pennine moorland. *J. Anim. Ecol.*, 40: 445–466.

Hale, W.G., 1966. A population study of moorland Collembola. *Pedobiologia*, 6: 65–99.

Haslam, S.M., 1972. *Phragmites communis* Trin. Biological flora of the British Isles No. 128. *J. Ecol.*, 60: 585–610.

Heal, O.W., 1962. The abundance and microdistribution of some testate amoebae (Rhizopoda: Testaceae) in *Sphagnum*. *Oikos*, 13: 35–47.

Hedberg, O., 1954. A pollen analytical reconnaissance in Tropical East Africa. *Oikos*, 5: 137–166.

Hirai, K., 1970. A quantitative relationship between zooplankters and attached animals on aquatic plants in the littoral zone of Lake Biwa. *Jap. J. Limnol.*, 31: 1–14.

Hodkinson, I.D., 1973. Population dynamics and host plant interaction of *Strophingia ericae* (Curt). *J. Anim. Ecol.*, 42: 565–585.

Houston, W.W.K., 1971. Carabidae (Col.) from two areas of the north Pennines. *Entomol. Mon. Mag.*, 107: 1–4.

Houston, W.W.K., 1973. The food of the common frog *Rana temporaria* on high moorland in north England. *J. Zool. Lond.*, 171: 153–163.

Howard-Williams, C. and Lenton, G.M., 1975. The role of the littoral zone in the functioning of a shallow tropical lake ecosystem. *Freshwater Biol.*, 5: 445–459.

Imhof, G., 1973. Aspects of energy flow by different food chains in a reedbed — a review. *Pol. Arch. Hydrobiol.*, 20: 165–168.

Imhof, G. and Burian, K., 1972. *Energy-Flow Studies in a Wetland Ecosystem (the Reed Belt of Lake Neusiedlersee)*. Austrian Acad. Sci., Spec. Publ. for I.B.P., Vienna, 15 pp.

Jenkins, D., Watson, A. and Miller, G.R., 1963. Population studies of the red grouse *Lagopus lagopus scoticus* (L.) in north-east Scotland. *J. Anim. Ecol.*, 32: 317–376.

Lambert, J.M., 1965. The vegetation of Broadland. In: E.A. Ellis (Editor), *The Broads*. Collins, London, pp. 69–92.

Leopold, A. and Ball, J.N., 1931. British and American Grouse management. *Am. Game*, July–October: 1–32.

Longhurst, W.H., 1960. Big game and rodent relationships to forest and grassland in north America. In: *Ecology and Management of Wild Grazing Animals in Temperate Zones*. I.U.C.N., Warsaw, pp. 305–326.

Lowe-McConnel, R.M., 1975. *Fish Communities in Tropical Freshwaters*. Longman, London, 337 pp.

Macfadyen, A. 1963., The contribution of the microfauna to total soil metabolism. In: J. Doeksen and J. van der Drift (Editors), *Soil Organisms*. North-Holland, Amsterdam, pp. 3–16.

Macfadyen, A. 1963. The contribution of the microfauna to recovery of the benthic fauna of a tropical lake after a dry phase. *Limnol. Oceanogr.*, 20: 54–63.

Mason, C.F., 1976a. Broadland. In: R. Washbourn (Editor), *The Natural History of Norfolk*. Jarrolds, Norfolk, pp.78–89.

Mason, C.F., 1976b. Relative importance of fungi and bacteria in the decomposition of *Phragmites* leaves. *Hydrobiologia*, 51: 65–69.

Mason, C.F. and Bryant, R.J., 1974. The structure and diversity of the animal communities in a Broadland reedswamp. *J. Zool. Lond.*, 172: 289–302.

Mason, C.F. and Bryant, R.J., 1975a. Production, nutrient content and decomposition of *Phragmites communis* (Trin) and *Typha angustifolia* (L). *J. Ecol.*, 63: 71–95.

Mason, C.F. and Bryant, R.J., 1975b. Periphyton, production and grazing by chironomids in Alderfen broad, Norfolk. *Freshwater Biol.*, 5: 271–277.

Mason, C.F. and Bryant, R.J., 1975c. Changes in the ecology of the Norfolk Broads. *Freshwater Biol.*, 5: 257–270.

Meschkat, A., 1934. Der Bewuchs in den Rohrichten des Plattensees. *Arch. Hydrobiol.*, 27: 436–517.

Moss, R., 1972. Food selection by the red grouse (*Lagopus lagopus scoticus*) (Lath.) in relation to chemical composition. *J. Anim. Ecol.*, 41: 411–428.

Newsome, A.E., 1969. A population study of house-mice permanently inhabiting a reed-bed in South Australia. *J. Anim. Ecol.*, 38: 341–360.

Nielsen, C.O., 1961. Respiratory metabolism of some populations of enchytraeid worms and free-living nematodes. *Oikos*, 12: 17–35.

Nixon, D., Okely, E.F. and Blackith, R.M., 1975. The distribution and morphometrics of spittlebugs on Irish blanket bog. *Proc. R. Ir. Acad.*, 75: 305–315.

Norgaard, E., 1951. On the ecology of two Lycosid spiders (*Pirata piraticus* and *Lycosa pullata*) from a Danish *Sphagnum* bog. *Oikos*, 3: 1–21.

Obrtel, R., 1973. Surface-active Coleoptera in the terrestrial reed-belt of Nesyt fishpond. In: J. Květ (Editor), *Littoral of the Nesyt Fishpond*. Akademia, Prague, pp. 133–134.

Osborne, P.L. and Moss, B., 1977. Paleolimnology and trends in the phosphorus and iron budgets of an old man-made lake, Barton Broad, Norfolk. *Freshwater Biol.*, 7: 213–233.

Pelikan, J., Svoboda, J. and Květ, J., 1970. On some relations between the production of *Typha latifolia* and a muskrat population. *Zool. Listy*, 19: 303–320.

Pimlott, D.H., 1960. Ecology and management of moose in North America. In: *Ecology and Management of Wild Grazing Animals in Temperate Zones*. I.U.C.N., Warsaw, pp. 246–265.

Pruscha, H., 1973. Biologie und Produktionsbiologie des Rohrbohrers *Phragmataecia castaneae* Hb (Lepidoptera, Cossidae). *Sitzungsber. Öst. Akad. Wiss., Mat-Nat. Kl., 1*, 182: 1–49.

Rawes, M. and Welch, D., 1969. Upland productivity of vegetation and sheep at Moor House National Nature Reserve, Westmorland, England. *Oikos (Suppl.)*, 11: 1–72.

Rudescu, L. and Popescu-Marinescu, V., 1970. Vergleichende Untersuchungen über benthische und phytophile Biocönosen einiger emerser Makrophyten des Donaudeltas mit besonderer Berücksichtigung von *Phragmites communis* Trin. *Arch. Hydrobiol. Suppl.*, 36: 279–292.

Rychnovsky, B., 1973. Occurrence of *Lipara* spp. at the Nesyt fishpond and their influence on the productivity of reed. In: J. Květ (Editor), *Littoral of the Nesyt Fishpond*. Academia, Prague, pp. 129–131.

Schröder, R., 1975. Release of plant nutrients from reed-borders and their transport into the open waters of the Bodensee-Untersee. *Symp. Biol. Hung.*, 15: 21–27.

Smith, G.R.J., 1973. *Some Aspects of the Biology of* Molophilus ater *Meigen*. Thesis, Durham University, Durham, 96 pp. (unpublished).

Standen, V., 1973. Production and respiration of an enchytraeid

population in blanket bog. *J. Anim. Ecol.*, 42: 219–245.
Standen, V., 1978. The influence of soil fauna on the decomposition of blanket bog vegetation. *J. Anim. Ecol.*, 47: 25–38.
Standen, V., 1980. The production biology of terrestrial Enchytraeidae. In: R.O. Brinkhurst and D.G. Cook (Editors), *Aquatic Oligochaete Biology*. Plenum Press, New York, N.Y., pp. 212–223.
Svendsen, J.A., 1957. The distribution of Lumbricidae in an area of Pennine moorland. *J. Anim. Ecol.*, 26: 411–421.
Waitzbauer, W., 1971. Produktionsbiologische Aspekte Schilffressender Insekten. *Verh. Zool. Ges.*, 65: 116–119.
Whittaker, J.B., 1967. Estimation of production in grassland leafhoppers (Homoptera: Insecta). In: K. Petrusewicz (Editor), *Secondary Productivity of Terrestrial Ecosystems*. Państwowe Wydawnictwo Naukowe, Warsaw, pp. 779–789.
Wortley, J.S., 1974. *The Role of Macrophytes in the Ecology of Gastropods and Other Invertebrates in the Norfolk Broads*. University of East Anglia, Norwich, 185 pp.
Zinevici, V., 1971. Comparative data regarding the bottom fauna living in the reed zones from Carasuhac, Maliuc and Fortuna Lakes (Danube Delta). *Bul. Inst. Cercet. Project Piscic.*, 30: 63–76 (in Rumanian).

SYSTEMATIC LIST OF GENERA[1]

MONERA (PROKARYOTA)

ACTINOMYCETE BACTERIA

Frankiaceae
Frankia

CYANOBACTERIA (CYANOPHYTA)

MYXOPHYCEAE
Chroococcaceae
Gloeocapsa
Enterophysalidaceae
Chlorogloea
Nostocaceae
Anabaena
Nostoc
Rivulariaceae
Calothrix
Scytonemataceae
Scytonema
Tolypothrix
Stigonemataceae
Fischerella
Hapalosiphon

EUBACTERIA

Azotobacter
Bacillus
Clostridium
Desulfotomaculum
Desulfovibrio
Escherichia
Klebsiella
Rhizobium
Rhodospirillum
Thiobacillus

PLANTS

BACILLARIOPHYTA

CHLOROPHYTA

CHLOROPHYCEAE
Chlorococcaceae
Trebouxia

CHAROPHYCEAE
Characeae
Chara

CONJUGATAE
Mougeotiaceae
Zygogonium

RHODOPHYTA

RHODOPHYCEAE
Batrachospermaceae
Batrachospermum

FUNGI

ZYGOMYCOTINA
Endogonaceae
Endogone
Mortierellaceae
Mortierella

ASCOMYCOTINA
Helotiaceae
Pezizella
Nectriaceae
Nectria

BASIDIOMYCOTINA
Agaricaceae
Amanita
Boletaceae
Boletus
Leccinum
Ganodermataceae
Ganoderma

DEUTEROMYCOTINA (FUNGI IMPERFECTI)
Agonomycetaceae
Rhizoctonia
Tuberculariaceae
Fusarium

LICHENES

ASCOLICHENES
Cladoniaceae
Cladonia
Collemaceae
Collema
Leptogium
Lecideaceae
Bacidia
Nephromaceae
Nephroma
Parmeliaceae
Cetraria
Hypogymnia
Parmelia
Peltigeraceae
Peltigera
Solorina
Stereocaulaceae
Stereocaulon
Stictaceae
Lobaria
Sticta

BRYOPHYTA

HEPATICAE
Calypogeiaceae
Calypogeja
Cephaloziaceae
Cephalozia
Cephaloziellaceae
Cephaloziella
Harpanthaceae
Mylia
Odontoschismaceae
Odontoschisma
Pleuroziaceae
Pleurozium

MUSCI
Amblystegiaceae
Acrocladium
Calliergidium
Calliergon
Campylium
Cratoneuron
Drepanocladus
Aulacomniaceae
Aulacomnium
Brachytheciaceae
Homalothecium
Scorpidium
Tomenthypnum
Dicranaceae
Chorisodontium
Dicranum
Leucobryum
Fontinalaceae
Fontinalis
Grimmiaceae
Racomitrium
Hypnaceae
Hylocomium
Hypnum
Rhytidiadelphus
Meesiaceae
Paludella
Polytrichaceae
Polytrichum
Pottiaceae
Tortula
Sphagnaceae
Sphagnum

PTERIDOPHYTA

LYCOPSIDA
Lycopodiaceae
Lepidotis (Lycopodium)
Selaginellaceae
Selaginella

SPHENOPSIDA
Equisetaceae
Equisetum

FILICOPSIDA
Dennstaedtiaceae
Pteridium
Osmundaceae
Osmunda
Schizaeaceae
Schizaea

PINOPHYTA (GYMNOSPERMAE)

PINOPSIDA (CONIFERAE)
Cupressaceae
Chamaecyparis
Juniperus
Thuja
Pinaceae
Abies
Larix
Picea
Pinus
Pseudotsuga
Tsuga
Taxodiaceae
Taxodium

MAGNOLIOPHYTA (ANGIOSPERMAE)

LILIOPSIDA (MONOCOTYLEDONES)
Alismataceae
Alisma
Araceae
Acorus
Callitrichaceae
Callitriche
Cyperaceae
Carex
Cladium
Cyperus
Eleocharis
Eriophorum
Fimbristylis
Gymnoschoenus
Kobresia
Kyllinga
Oreobolus
Rhynchospora
Schoenus
Scirpus
Scleria
Hydrocharitaceae
Hydrilla
Stratiotes
Iridaceae
Iris
Juncaceae
Juncus
Rostkovia
Juncaginaceae
Lemnaceae
Spirodela
Liliaceae

---

[1] The taxonomic relationships of all genera of plants and animals mentioned in this Part A of *Mires: Swamp, Bog, Fen and Moor* are shown. Genera are listed alphabetically under families, arranged into phyla and subphyla or classes; orders are included in the animal kingdom.

LILIOPSIDA (*continued*)

Narthecium
Mayacaceae
Orchidaceae
Dactylorhiza
Epipactis
Gymnadenia
Liparis
Listera
Malaxis
Microstylis
Orchis
Platanthera
Spiranthes
Pandanaceae
Pandanus
Poaceae (Gramineae)
Agrostis
Aira
Ammophila
Arctophila
Arundo
Calamagrostis
Dactylis
Deschampsia
Diplachne
Dupontia
Festuca
Glyceria
Lolium
Molinia
Moliniopsis
Nardus
Oryza
Panicum
Paspalum
Phalaris
Phleum
Phragmites
Poa
Scolochloa
Spartina
Zea
Zizania
Pontederiaceae
Eichhornia
Monochoria
Pontederia
Potamogetonaceae
Potamogeton
Restionaceae
Empodisma
Hypolaena
Sporadanthus
Scheuchzeriaceae
Scheuchzeria
Sparganiaceae
Spharganium
Typhaceae
Typha
Zosteraceae

MAGNOLIOPSIDA (DICOTYLEDONES)
Acanthaceae
Hygrophila
Aceraceae
Acer
Anacardiaceae
Rhus
Annonaceae
Xylopia
Apiaceae (part Umbelliferae)
Cicuta
Aquifoliaceae
Nemopanthus
Araliaceae
Hedera
Asteraceae (Compositae)
Aster
Hieracium
Lactuca
Senecio
Solidago
Avicenniaceae
Avicennia
Betulaceae
Alnus
Betula
Boraginaceae
Myosotis
Brassicaceae (Cruciferae)
Raphanus
Caesalpiniaceae
Gleditsia
Caprifoliaceae
Viburnum
Chenopodiaceae
Atriplex
Suaeda
Clethraceae
Clethra
Convolvulaceae
Ipomoea
Cornaceae
Cornus
Corylaceae
Corylus
Droseraceae
Drosera
Ebenaceae
Diospyros
Empetraceae
Empetrum
Epacridaceae
Styphelia
Ericaceae
Andromeda
Arbutus
Arctostaphylos
Calluna
Chamaedaphne
Erica
Gaylussacia
Kalmia
Ledum
Rhododendron
Vaccinium
Euphorbiaceae
Ricinus
Fabaceae (Leguminosae: Papilionaceae; *see also* Caesalpiniaceae, Mimosaceae)
Aeschynomene
Genista
Glycine
Lathyrus
Lotus
Lupinus
Pisum
Trifolium
Vicia
Fagaceae
Fagus
Quercus
Gentianaceae
Gentiana
Haloragaceae
Myriophyllum
Hippocastanaceae
Aesculus
Hydrangeaceae
Hydrangea
Juglandaceae
Carya
Lamiaceae (Labiatae)
Lycopus
Mentha
Lythraceae
Decodon
Lythrum
Magnoliaceae
Liriodendron
Menyanthaceae
Menyanthes
Myricaceae
Myrica
Myrtaceae
Eucalyptus
Metrosideros
Naucleaceae
Cephalanthus
Nymphaeaceae
Nuphar
Nymphaea
Nyssaceae
Nyssa
Oleaceae
Fraxinus
Onagraceae
Chamaenerion
Epilobium
Polygonaceae
Rumex
Ruprechtia
Symmeria
Potaliaceae
Anthocleista
Ranunculaceae
Caltha
Ranunculus
Rhamnaceae
Frangula
Rhamnus
Rhizophoraceae
Rhizophora
Rosaceae
Acaena
Aronia
Crataegus
Filipendula
Malus
Potentilla
Prunus
Rubus
Sanguisorba
Sorbus
Rubiaceae (*see also* Naucleaceae)
Coprosma
Galium
Salicaceae
Populus
Salix
Sambucaceae
Sambucus
Sarraceniaceae
Sarracenia
Saururaceae
Saururus
Scrophulariaceae
Veronica
Solanaceae
Lycopersicum
Solanum
Tiliaceae
Tilia
Ulmaceae
Celtis
Ulmus
Urticaceae
Urtica

ANIMALS

PROTOZOA

RHIZOPODA (SARCODINA)

PLATYHELMINTHES

CESTODA (CESTOIDEA)

TREMATODA
Digenea
Schistosomatidae
Schistosoma

ASCHELMINTHES

GASTROTRICHA

ROTIFERA

NEMATODA
Chromadorina
Chromadoridae
Chromadora

ANNELIDA

OLIGOCHAETA
Plesiopora Prosotheca
Enchytraeidae
Cognettia
Mesenchytraeus
Prosopora
Lumbriculidae
Kincaidiana
Lumbriculus
Rhynchelmis
Opisthopora
Glossoscolecidae
Alma
Lumbricidae
Allolobophora
Dendrobaena
Scherotheca

HIRUDINEA

MOLLUSCA

GASTROPODA
Aspidobranchia
Lymnaeidae
Lymnaea
Neritidae
Aplexa
Basommatophora
Planorbidae
Biomphalaria
Planorbarius
Planorbis
Segmentina
Stylommatophora

MOLLUSCA (*continued*)

Cochlicopidae
Cochlicopa
Limacidae
Arion
Deroceras
Vertiginidae
Vertigo
Zonitidae
Omphalina

BIVALVIA [= PELECYPODA]
Heterodonta
Sphaeriidae
Pisidium

CEPHALOPODA

ARTHROPODA

ARACHNIDA
Acarina
Araneae
Araneidae
Araneus
Liniphyidae
Lycosidae
Lycosa
Pisauridae
Dolomedes

CRUSTACEA
Amphipoda
Gammaridae
Gammarus
Isopoda
Asellidae
Asellus
Notostraca

COLLEMBOLA
Achorutidae
Friesia
Isotomidae
Tetracanthella
Onychiuridae
Onychiurus
Tullbergia

INSECTA
Coleoptera
Carabidae
Carabus
Chlaenius
Diachila
Oodes
Chrysomelidae
Lochmaea
Dytiscidae
Agabus
Hydroporus
Rantus
Helodidae
Cyphon
Hydrophylidae
Hydrobius
Diptera
Cecidomyidae
Perrisia
Ceratopogonidae
Chironomidae
Chironomus
Cricotopus
Glyptotendipes
Chloropidae
Lipara
Culicidae
Chaoborus
Culex
Theobaldia
Psychodidae
Pericoma
Ptychopteridae
Ptychoptera
Sciomyzidae
Stratiomyiidae
Tabanidae
Tipulidae
Molophilus
Prionocera
Tipula
Ephemeroptera
Caenidae
Caenis
Hemiptera
Aphididae
Colopha
Hyalopterus
Cercopidae
Neophilaenus
Philaenus
Coccidae
Chaetococcus
Coryxidae
Sigara
Notonectidae
Psyllidae
Strophingia
Saldidae
Hymenoptera
Tenthredinidae
Pristiphora
Tenthredo
Lepidoptera
Cossidae
Phragmatoecia
Geometridae
Costaconvexa
Lasiocampidae
Lasiocampa
Macrothylacia
Lycaenidae
Lycaena
Lymantriidae
Laelia
Noctuidae
Leucania
Trachea
Papilionidae
Papilio
Pyralidae
Schoenobius
Saturniidae
Saturnia
Orthoptera
Acrididae
Locusta
Nomadacris
Stethophyma
Tetrigidae
Tetrix
Plecoptera
Porilla
Trichoptera
Brachycentridae
Agrypnia
Polycentropidae
Cyrnus

CHORDATA

PISCES
Lepidosireniformes
Protopteridae
Protopterus
Microcyprini
Cyprinodontidae
Nothobranchius
Osteoglossiformes
Notopteridae
Papyrocranus
Polypteriformes
Polypteridae
Polypterus

AMPHIBIA
Caudata
Ambystomidae
Ambystoma
Salientia
Ranidae
Rana

REPTILIA
Chelonia
Emydidae
Emys
Crocodylia
Squamata
Colubridae
Natrix
Lacertidae
Lacerta

AVES
Anseriformes
Anatidae
Anas
Anser
Ardeiformes
Ardeidae
Plataleidae
Platalea
Charadriiformes
Jacanidae
Jacana
Laridae
Rhynchops
Scolopacidae
Limosa
Philomachus
Ciconiiformes
Ciconiidae
Threskiornithidae
Ibis
Galliformes
Tetraonidae
Lagopus
Gruiformes
Rallidae
Fulica
Passeriformes
Motacillidae
Anthus
Sylviidae
Acrocephalus
Pelecaniformes
Pelecanidae
Pelecanus
Phoenicopteriformes
Phoenicopteridae
Phoenicopterus
Strigiformes
Strigidae
Bubo

MAMMALIA
Artiodactyla
Bovidae
Bubalus
Ovis
Tragelaphus
Cervidae
Alces
Rangifer
Hippopotamidae
Hippopotamus
Carnivora
Mustelidae
Mustela
Insectivora
Soricidae
Neomys
Sorex
Lagomorpha
Leporidae
Lepus
Perissodactyla
Rhinocerotidae
Coelodonta
Primates
Hominidae
Homo
Proboscidea
Elephantidae
Elephas
Mammuthus
Rodentia
Capromyidae
Myocastor
Castoridae
Castoroides
Castor
Cricetidae
Lemmus
Microtus
Ondatra
Synaptomys
Hydrochoeridae
Hydrochoerus
Muridae
Mus

AUTHOR INDEX[1]

Aaby, B., 36, 38, *59*, 147, *150*, 170, 193, 199, 200, *218*
Aario, L., 46, 51, *59*
Aartolahti, T., 54, 56, 58, *59*, 135, 148, *150*
Abrahamsen, G., 377, *380*
Achuff, P., *347*
Adams, M.S., 82, *155*
Adamson, J.A., 250, *255*
Ahti, E., 100, 111, 112, *150*
Albrecht, F.O., 359, *363*
Alekseevsky, V.E., 72, *150*
Aletsee, L., 36, 40, 54, 55, *59*, 136, *150*, *346*
Alexander, M., *255*
Alexander, V., 226, 228, *243*
Allen, J.F., *281*
Allen, K.R., 286, *306*
Allen, L.H., 304, *306*
Allen, S.E., 163, 181, 182, 188, *218*, *220*, 290, *306*
Andersen, S.Th., 38, *59*
Anderson, F.O., 288, 289, 293, *306*
Anderson, J.A.R., 12, 136, *150*, 159, 161, 180, *218*, 325
Anderson, J.H., 290, *306*
Anderson, J.W., 228, 230, *243*
Andreyashkina, N.I., 290, 291, 293, *307*
Andrienko, T.L., 171, *219*
Andriesse, J.P., 136, *150*
Andruss, R.E., *347*
Anonymous, 25, *30*, 75, 77, 91, 105, 106, 110, 114, 123, 129, 133, 143, *150*, 202, *218*
Antie, D.A., *255*
Appleby, P.G., *222*
Arber, A., 265, *280*
Ardakani, M.S., 249, *256*
Aref'eva, A.I., 104, *150*
Armentrout, D., 351, *364*
Armstrong, D.E., *256*
Armstrong, W., 254, *255*, 267, 270, 271, *280*, 341, *344*
Ashmore, M.R., 84, *150*, 300, *306*
Aspiras, R., *255*
Atkinson, D.E., 266, *280*
Aubendok, L.I., *152*
Auclair, A.W.D., 288, *306*
Auer, V., 51, 40, *59*
Averdieck, F.R., *346*
Avnimelech, Y., 251, *255*

Backéus, I., 339, *344*
Baden, W., 67, 71, 76, 89, 91, 104, 107, 108, 134, 141, 143, *151*
Bahnson, H., 161, *218*
Baillie, M.G.K., *222*
Baines, M., 266, *281*
Baker, D.R., 75, *155*, *156*
Baker, G., 275, 277, 278, *282*
Baker, J.H., 196, 198, 206, 207, *218*, *243*
Balátová-Tuláčková, E., 111, 136, *151*
Balboni, A., 20, *32*
Ball, J.N., 376, *381*
Banage, W.B., 378, 379, *380*
Bannister, P., 83, 84, 102, *151*
Bannister, T.T., *282*
Banoub, M.W., 374, *380*
Baradziej, E., 288, *306*
Barber, D.A., 267, *283*
Barber, K.E., *222*
Barèl, D., *219*, *243*
Barry, R.G., 75, *151*
Barry, T.A., 167, 185–188, *224*
Barsdate, R.J., *219*, *243*
Bartels, R., 71, *151*, *156*
Bartholemew, B., *309*
Basilier, K., 181, *218*
Battarbee, R.W., *222*
Batzli, G.O., 376, *380*
Baumgartner, A., 304, *306*
Bavina, L.G., 70, 91, 92, 94, 95, 110, 142–144, *151*
Bay, R.R., 92, 93, 110, 115, 129, 140, 142, 143, *151*
Bayly, I.A.E., 354, *363*
Bazilevich, N.I., 292, *309*
Beadle, L.C., 368, *380*
Beaulieu, J.L. de, *see* De Beaulieu, J.L.
Bear, J., 120, *151*
Beardsell, D.V., 260, *282*
Becker, B., 36, *64*
Behre, K.E., 46, *59*
Belak, S., 23, *30*
Belkevich, P.I., 165, *218*
Bell, D.T., 262, *280*
Bell, J.N., 111, *151*, 261, *280*
Bell, J.P., 101, *151*
Bell, R.G., 248, *255*
Bellamy, D.J., 6–8, 26, *32*, 71, 111, 118, 142, *151*, *155*, 167, 169, 171, 187, 192, *218*, *221*, 227, *244*, 251, *255*, 290, 291, 298, *306*, *310*, 338, 341, *344*, *346*
Bellamy, K., 40, 55, *59*
Bellamy, R., 40, 55, *59*
Belotserkovskaya, O.A., 93, 94, *151*
Benda, I., 165, *218*
Bendixen, L.E., 264, 269, *280*
Benson, R.B., 361, *363*
Berg, A., 302, *306*
Berghen, A. vanden, *see* Vanden Berghen, A.
Berglund, E.R., 94, *151*
Bernard, J.M., 286, 288–290, 293, *306*
Bernatowicz, S., 22, *30*, 97, *151*
Berry, J.A., *309*
Berry, P.L., 167, 210, 213, *218*
Bevege, D.I., *243*
Bharucha, F.R., 137, *153*, 335, *345*
Bibby, C.J., 372, *380*
Bick, W., 4, *30*
Billings, W.D., *223*, 234, *245*, 270, *281*, 287, *309*
Bingham, F.T., *31*, *221*
Birks, H.H., 48, *59*
Birks, H.J.B., 136, *151*, 262, *282*
Bjarnasson, O., 19, *30*, 159, *218*
Björkbäck, F., 339, *344*
Black, C.A., 249, 250, *255*
Blackith, R.E., 354, 356, *363*, 378, *380*
Blackith, R.M., 358, *363*, 377, *380*, *381*
Blackman, V.H., *380*
Blake, G.R., 103, *152*
Blakemore, E.S.A., 234, *243*
Bland, D.E., 176, *218*
Bliss, J.C., 290, *306*
Bliss, L.C., 287–289, 296, *307*, *308*, *310*
Block, W.W., 378, 379, *380*
Blumentals, A., *223*
Blytt, A., 41, *59*
Boate, G., 4
Boatman, D.J., 40, 52, *59*, 71, 110, 138, *151*, 163, 180, 181, *218*, 226, *243*, 261, *281*, 339–341, *344*
Boch, M.S., *see* Botch, M.S.
Bochkov, A.P., 76, *151*
Bodarenko, N.F., 126, *151*

[1]Page references to text are in roman type, to bibliographic entries in italics.

Boekel, P., 168, 169, *223*
Boelter, D.H., 67, 71, 100, 101, 103, 112, 113, 117, 124–126, 131, 134, 140, 142–144, *151*, *152*, 168, *218*, 332, 340, *344*
Boggie, R., 83, 85, *152*, 183, 190, *218*, *219*, *221*, 234, *243*
Bohn, H.L., 18, *30*, 247, *255*
Bond, G., 229, *243*
Bonner, J., 175, *219*
Bonnet, L., 362, *363*
Bonny, A.P., *63*, *346*
Borisov, A.A., 75, 92, *152*
Bormann, F.H., 131, *152*
Borówko-Dużakowa, Z., 45, *59*
Bostock, J., 325
Botch, M.S., 6, 14, 23, 36, 67, 111, 133
Botkin, D.B., *224*
Bouchard, A., *306*
Bouché, M.B., 350, 357, *363*
Bowen, G.D., 240, *243*
Bower, M.M., 40, 52, *59*, 132, 136, *152*
Boyd, C.E., 286, 288–290, 293, *306*, 335, *344*
Brackke, F.H., 181, *219*
Bradbury, I.K., 82, 196, 286, *307*
Bradis, E.M., 171, *219*
Bradstreet, E.D., *309*
Bragg, O.M., 78, 100, 111, *152*
Brand, T. von, *see* Von Brand, T.
Brandt, A., 44, *59*
Bratt, G.T., *256*
Bray, J.R., 97, *152*, 289, *307*
Brelie, G. von der, *see* Von der Brelie, G.
Bremner, J.M., 252, *255*
Bridgewater, P., *306*
Brindley, H.H., 356, *363*
Bristowe, W.S., 350, *363*
Broadfoot, W.M., 261, 262, *281*
Broady, P.A., 228, *243*
Broihan, F., 48, *60*
Broué, P., *282*
Brown, C.L., *282*
Brown, H.W., 330, *345*
Brown, J., 290, *307*
Brown, J.M.B., 125
Brown, R.J.E., 118, *152*
Brown, W.M., *220*
Broyer, T.C., 261, *281*, 300, *307*
Bruzau, F., *282*
Bryant, C., 278, *281*
Bryant, R.J., 231, *244*, 288, 289, *308*, 368, 369, 371–374, *381*
Bryson, R.A., 38, 39, *65*
Buckney, R.T., 354, *363*
Budyko, M.I., 74, 75, 110, *152*
Buell, H.F., 111, *152*, 315, *344*
Buell, M.F., 111, *152*, 315, 330, 336, *344*, *345*
Bülow, K. von, *see* Von Bülow, K.
Bunce, R.G., *31*
Bunnell, F.L., 199, 200, 201, 216, *219*, *220*
Buol, S.W., 217, *219*
Burford, J.R., 252, *255*
Burgeff, H., 189, *219*, 239, 241, *243*
Burges, A., 51, *59*
Burghardt, W., 56, *62*
Burian, K., 22, *31*, 356, 357, *364*, 373, 379, *381*
Burke, W., 141–143, *152*
Busby, J.R., 291, *307*
Butterfield, J.E.L., 171, 173, 198, 201, 203–206, *219*, 231, 232, *243*, 376, *380*
Buttery, B.R., 258, *281*, 288, *307*, 335, *344*

Cain, R.L., 336, *344*
Cajander, A.K., 4, 13, *30*, 35, 40, 46, 47, 51, *59*
Caldwell, M.M., 302, *307*
Cambray, R.S., *222*
Campbell, E.O., 12, 77, 102, *152*, 336
Candolle, A.P. de, *see* De Candolle, A.P.
Cannell, R.Q., 257, *281*
Carlson, C.W., *158*
Casparie, W.A., 44, 55, 56, *60*, 186, 188, *219*, 315, 339, 340, 342, *344*
Catchpole, C.K., 372, *380*
Čatský, J., *157*
Cawse, P.A., 181, 195, *219*, *222*
Chabot, B.F., 300, *309*
Chambers, C., *64*, *347*
Chambers, K., 332, *347*
Chambers, M., 370, *380*
Chamie, J.P.M., *156*
Chandler, G.E., 228, 230, *243*
Chang, J.-H., 75, *152*
Chapin III, F.S., 183, 187, *219*, *223*, 230, *243*
Chapin, M.C., *219*
Chapman, S.B., 71, 111, 143, *152*, 163, 168, 171, 185–187, 190, *219*, 291, *307*
Chapman, T.G., 70, 77, *157*
Chapman, V.J., 286, *307*
Chebotarev, N.P., 120, *152*
Cherrett, J.M., 376, 379, *380*
Chesire, M.V., *255*
Childs, E.C., 101, 112, 120, 149, *152*
Chirkova, T.V., 263, 264, 270, 273, 277, *281*
Chistova, L.R., 165, *218*
Chouard, P., 337, *344*
Chow, T.J., *221*
Churaev, N.V., 101, 125, *158*, 169, *223*
Chvala, M., 358, *364*
Clapham, A.R., 315, 339, *344*
Clarholm, M., 227, *243*
Clausen, H.B., *60*
Clayden, B., 217, *222*
Clements, F.E., 311, 312, 321, *344*
Cleve, K. van, *see* Van Cleve, K.
Clifford, M.H., 315, *345*
Clowes, D.R., *345*
Clymo, R.S., 4, 17, 18, 28, *30*, 52, 55, 101, 117, 146, *152*, 159, 164–167, 169, 170, 172, 175, 183–186, 189, 190, 192–203, 205, 207, 210, 213–215, *219*, 225, 231, 236, 238, *243*, 247, 262, 268, 286, 287, 291, *307*, 332, 337, 341, *344*
Cockayne, L., 102, *152*
Codarcea, F., 134, *152*
Cohen, A.D., 234, *243*
Coleman, N.T., 249, *255*
Collins, N.J., 160, *219*, 296, 297, *308*
Collins, V.G., 165, 170, *219*, 227, 235, 238, 242, *243*
Connell, W.E., 248, *255*
Conner, W.H., 144, *152*
Conway, B.E., 189, *219*
Conway, V.M., 14, 17, *30*, 132, 136, 141–143, *152*, 181–183, 219, 323, 326, 330, 336, 337, 339, *344*, *345*
Coope, G.R., 37, 38, *60*, 360, 361, *364*
Cooper, J., 353, *364*
Cosgrove, D.J., 252, *255*
Coulson, J.C., 171, 173, 198, 201, 203–206, *219*, 231, 232, *243*, 374–379, *380*
Coulter, J.K., 159, *219*
Coutts, M.P., 258, 264, 265, 269, *281*, *282*
Cowan, T.R., 96, 97, *152*
Coyne, P.I., 287, *307*
Cragg, J.B., 235, *244*, 374, *380*
Crawford, R.M.M., 8, 110, 118, *152*, 257, 262, 264, 266–269, 272–278, 280, *282*, 300, 335
Crawford-Sidebotham, T.J., 261, *281*
Crisp, D.T., 298, *307*
Cronquist, A., 279, *281*
Crooke, W.M., *221*
Cubley, S., 301, *310*

Dachnowski, A.P., 323, *344*
Dachnowski-Stokes, A.P., 323, *344*
Dahl, E., 16, 18, *30*, 148, 149, *152*
Dai, T.S., 126, *152*
Dalton, J., 198, *219*
Dam, L. van, *see* Van Dam, L.
Dambska, I., 22, *30*
Damman, A.W.H., 170, 185–187, 192–194, *219*
Danchenko, O.I., *151*

Daniels, R.E., 9–12, *30*, 129, 133, *152*, 334, *344*
Dansereau, P., 326, *344*
Dansgaard, W., 37, *60*
Darbyshire, J.E., 268, *281*
Das, V.S.R., 295, *309*
Dau, J.H.C., 12, *30*, 40, *60*
Dauphin, C., 362, *364*
David, J.H., *256*
Davies, D.D., 274, 277, 278, *281*
Davies, R.I., 249, *255*
Davies, W.E., *244*
Davis, J.F., 249, 250, 253, *255*
Davis, M.B., 323, 333, *344*
Dawson, N., 358, 364
Day, J.H., 169, *219*
Day Jr., J.W., 145, *152*
De Beaulieu, J.L., 48, *60*
De Candolle, A.P., 278, *281*
Deevey, E.S., 42, *60*
Delwiche, C.C., *224*
Demaree, D., 263, *282*
Dempster, J.P., 352, *364*
Dennis, V.G., 290, 293, *307*
Devitt, A.C., *281*
De Vries, D.A., 75, *152*
De Wit, C.T., 302, *307*
De Zwaan, A., 278, *281*
Dickey, B., 356
Dickinson, C.H., 134, *153*, 159, 165, 171, 176, 181, 183, 195, 206, *220*, 235, 236, 238, 241, 242, *243*, *244*, 248–250, *255*, 271
Dickinson, W., 52, *60*, 318, *344*
Dickson, R.E., 261, *281*, 300, *307*
Dieck, 202
Dierschke, H., 111, *152*
Dierssen, B., 18, *30*
Dierssen, K., 18, *30*
Dijk, H. van, *see* Van Dijk, H.
Dilley, A.C., 75, *152*
Dimbleby, G.W., 51, *60*, 239, *243*
Dittrich, J., 127, *152*
Dobson, A.T., 25, *30*
Dokturovskiĭ, W.S., 40, 43, *60*
Donnan, F.G., 175, *219*
Dooge, J., 67, *152*
Dooley, F., 227, *243*
Dorst, J., 352, 355, *364*
Dowding, E.S., *345*
Dowding, P., 230, 231, *243*
Downing, R.A., *156*
Downton, W.J.S., 295, *307*
Doyle, G.J., 290, 291, 307
D'Sylva, B.T., *219*, *243*
Duane, J., 169, *219*
Dubinina, I.M., 277, 278, *281*
Duffey, E., 352, *364*
Duncan, D.P., 335, 336, *344*
Duncan, U.K., 229, *244*
Duncan, W.G., 302, *307*
Dunham, R.J., 299, *307*
Dunnican, L.K., 227, 242, *244*
Du Rietz, G.E., 10, 13, *30*, 35, 36, 40, 46, 52, *60*
Duxbury, J.M., *255*
Dvořák, J., 368, 369, *380*
Dykyjová, D., 293, 294, 302, 303, *307*
Dziedzic, J., 22, *32*

Eakins, J.D., *222*
Eggelsmann, R., 67, 71, 72, 75, 76, 86, 88–91, 104, 106–108, 111, 117, 124, 126, 127, 132–135, 141, 143, 144, *151*, *153*
Ehleringer, J.R., *155*
Einarsson, T., 19, *30*
Eisenlohr, W.S., 98, *153*
Elberling, J.A., *256*
Elgee, F. 127, *153*
Elina, G.A., 52, *60*, 171, *219*
Ellis, B.G., 249, *255*
Ellis, E.A., 373, *380*
Engler, A., 102, *153*
Engler, R.M., 251, 253, *255*
Erickson, A.E., *281*
Eriksson, E., 181, *219*
Erman, D.C., 370, 373, 379, *380*
Erman, N.A., 370, 373, 379, *380*
Ernst, O., 315, *344*
Errington, P.L., 354, *364*
Eurola, S., 20, *30*, 40, 52, *60*, 315, *344*
Evans, F.C., 286, *310*
Evans, J.G., 361, *364*

Fairbridge, R.W., 46, *60*
Fareed, M., *307*
Farnham, R.S., 134, *153*, 163, 169, 171, *220*
Fenton, K.H.C., 160, 166, 168, 171, 196, 198, 213–215, *220*
Ferda, J., 22, *30*, 117, 141, 143, *153*
Fiala, K., *31*, 293, *307*, 354, 355, *364*
Fields, H.H.A., *281*
Fieser, L.F., 173, *220*
Fieser, M., 173, *220*
Finney, H.R., 134, *153*, 163, 171, *220*
Firbas, F., 48, *60*, 78, 82–86, 127, *153*, *157*, 315, 339, *346*
Firsov, L.V., *62*
Fisher, E.M., *222*
Flanagan, P.W., 199–201, *220*
Flint, P.S., 242, *244*
Florschütz, F., 159, *220*
Flower-Ellis, J.G.K., 191, *220*
Forova, V.S., *62*
Forrest, G.I., 196, 209, *220*, *223*, 241, *244*, 286, 287, 290, 291, 293, 302, *307*, 328, 329, 341, *344*, 375, 377, *380*
Forrester, J.W., 216, *220*
Fosberg, F.R., 26, *30*
Fourt, D.F., 337, *346*
Fox, W.E., *155*
Francis, C.M., 273, *281*
Francis, M.J., *255*
Frankland, J.C., 198, *220*, 231, *244*
Freistedt, E., 167, *220*
French, D.D., *220*, 230, 234, *244*
Frenzel, B., 13, 14, 16, 17, 19, 29, 36, 38, 39, 48, 51, 56, *60*, 173, 313, 316, 318, 342
Fries, M., 46, *60*, 333, *344*
Frolik, A.L., 322, 327, *344*
Froment, A., 20, *30*
Früh, J., 22, *30*, 36, 40, 43, 44, *60*
Fryer, G., 353, *364*
Fuchsman, C.H., *224*
Fulkem, L.W., 295, *308*
Fulton, J.M., 278, *281*
Furuya, C., 117, *158*

Gaastra, P., 294, 295, *307*
Gäde, G., 278, *281*
Galvin, L.F., 112, 113, 125, 134, 135, *153*
Gamayunov, N.I., *158*
Gardiner, J.J., 251, *256*
Gardiner, J.S., 360, *364*
Gardner, W.R., 96, *153*, 299, *307*
Gates, F.C., 330, 336, *344*
Geel, B. van, *see* Van Geel, B.
Gehl, O., 53, *60*
Geilser, G., 241, *244*
Gel'bukh, T.M., 97, *153*
Geraghty, J.J., 75, *153*
Gersper, P.L., 242, *244*
Gessel, S.P., *31*, *221*, 248, *256*
Gessner, F., 264, *281*
Giesel, W., *156*
Giles, B.R., 181, 183, 184, *220*
Gilg, E., 102, *153*
Gill, A.M., 331, *346*
Gill, C.J., 261, 263, 264, *281*
Gilliland, M.R., 173, *220*
Gilman, K., 127, *153*
Gilmour, D., 276, 278, *281*
Gimingham, C.H., 82, 83, *153*, 322, *344*
Given, P.H., 134, *153*, 159, 165, 176, *220*, 227, 235, 236, 238, 242, *243*, *244*, 248–250, *255*
Glob, P.V., 202, *220*
Glosser, J., 295, *307*
Godfrey, P.J., 270, *281*
Godwin, H., 16, 17, *30*, 36, 40, 42, 44, 46, 47, 55, 56, *60*, 106, 136, 137, *153*, 171, *220*, 320, 322, 325, 331, 335, 336, 339, *344*, *345*
Goeschl, J.D., 278, *282*
Goit, J.B., 101, 125, *154*

Golley, F.B., 292, *307*
Gonzalez, E., 37, *64*
Goodall, D.W., 362, *364*
Goode, D.A., 67, 92, 106, 142, 144, *153*
Goodman, G.T., 229, *244*
Goodwillie, R., 2, 20–22, *31*
Gopal, B., 25, *31*
Gorchakovsky, P.L., 290, 291, 293, *307*
Gore, A.J.P., 17, 18, 25, *31*, *33*, 43, 44, 54, 67, 71, 72, 145, *153*, 163, 165, 167, 181, 182, 189–191, 199, 209, 212, 213, 215, *220*, 221, *221*, *223*, 226, 235, 241, 242, *244*, *245*, 247, 248, *256*, 263, 291, *307*, 332, 357
Gorham, E., 1, 4, 9, *31*, 71, *153*, 169, 171, 173, 180, 188, 190, *220*, 249, *255*, 288, 289, *307*, *309*, 334, 335, *345*
Gosling, L.M., 372, *380*
Gosselink, J.G., 71, 145, 148, *153*
Göttlich, K., 21, *31*, 132, *154*, 159, *220*
Gould, H.R., 333, *346*
Grace, J., 82, 84, *153*, 196, *220*, 287, 296, 297, 301, 303, *307*, *308*
Gradmann, R., 41, *60*
Graetz, D.A., 248, *255*
Graham-Bryce, I.S., *255*
Granat, L., 181, 193, *220*
Granhall, U., 227, 228, *244*, 252, *255*, 298, *308*
Granhall, V., *218*
Granlund, E., 4, 5, 8, 13–16, 24, 29, *31*, *33*, 44, 55, *60*, 71, 148, 149, *153*, 161, 162, 209, *220*, *223*, 316–319, *345*
Grass, L.B., *256*
Grauze, G., *155*
Greaves, M.P., *224*, 268, *281*
Grebenshchikova, A.A., 117, *153*
Green, B.H., 185, *220*, 320, 321, 336, *345*
Green, R.E., 372, *380*
Greenway, H., 273, *282*
Greenwood, D.J., 241, *244*, 251, *255*
Grego, S., *281*
Grichuk, V.P., 36, 39, 59, *60*
Griffin, K.O., 339, *345*
Grimshaw, H.M., *218*
Grimshaw, P.H., 376, 379, *380*
Gross, H., 43, 44, 56, *60*
Grosse-Brauckmann, G., 36, 40, 45, 48, 52, 55, *60*, *61*
Grundy, S.D. van, *see* Van Grundy, S.D.
Gutman, T.C., 264, 270, *281*

Haag, R., 290, 292, *308*
Haas, H.J., 140, *158*
Haavisto, V.F., 247, 248, *255*
Hadley, E.B., 287, 296, *308*
Hadley, G., 240, *244*
Hadley, M., 377, *381*
Hafsten, U., 48, *61*
Hale, W.G., 378, 379, *381*
Hall, M.J., 142, *153*
Hall, T.F., 262, *281*
Hallberg, R.O., 253, *255*
Hamer, D., 173, *221*
Hamilton, A.C., 12
Hamilton, C.D., *307*
Hammel, H.T., *309*
Hammen, T. van der, *see* Van der Hammen, T.
Hammond, P.M., 361, *364*
Hanau, R., *307*
Hanrahan, E.T., 125, 135, *153*, 213, *220*
Hansen, B., 339, *345*
Hansen, H.P., 42, *61*
Happach, H., 78, 81, *156*, 291, *309*
Harder, W., 354, *364*
Harding, R.J., 72, *154*
Hare, F.K., 75, *158*
Harkness, D.D., *222*
Harley, J.L., 238–240, *244*
Harper, J.L., 286, *308*
Harrar, E.S., 335, *346*
Harris, R.F., *256*
Harrison, J.G., 127, 137, 138, *155*, 274, *282*
Hart, M.G.R., 190, *220*
Hartline, P.H., *281*
Hartman, R.T., 333, *347*
Haslam, S.M., 334, 335, *345*, 372, *381*
Haude, W., 75, 88, 90, *154*
Hawkinson, C.F., 71, *154*
Haworth, E.Y., *63*, *346*
Hayaishi, O., 272, *281*
Hayen, H., 54, 56, *61*
Hayman, D., 240
Hayward, P.M., 101, *154*
Heal, O.W., 191, 198, 202, 205–207, *220*, 230–232, 234, 236, *244*, 351, *364*, 378, *381*
Heath, G.H., 83, 85, *154*, 234, *244*
Heath, J., 361, *364*
Hedberg, O., 377, *381*
Heikurainen, L., 4, *31*, 93, 94, 111, 113, 115, 116, *154*, *221*, 239, *244*
Heilman, P.E., 327, 337, *345*
Heinonen, S., 296, *308*
Heinselman, M.L., 2, 7, 14, *31*, 51, *61*, 319, 320, 323, 333, 337, 341, 342, *345*, *347*
Hejný, S., 23, *31*
Helfferich, F., 181, *221*
Helmond, I., 75, *152*
Henske, G., 176, *224*
Herbert, R.A., *255*
Herbichowa, M., 22, *31*
Herrmann, R., *63*
Heslop-Harrison, Y., 228, *244*
Hess, A.D., *281*
Hess, L.W., 289, 290, *306*, 335, *344*
Heusser, C.J., 38, 43, *61*
Hewetson, V.P., *64*, *347*
Hewlett, J.D., 128, 130, 138, *154*
Hibbert, A.R., 128, *154*
Hibbert, F.A., 36, 38, *61*, *64*, *347*
Hicks, B.B., *155*
Hicks, S.P., 332, *345*
Hill, M.O., 10, *31*, *179*, *221*
Hillam, J., *222*
Hillell, D., 101, *120*, *154*
Hinshiri, H.M., 83, 84, *154*
Hinton, H.E., 350, *364*
Hirai, K., 370, *381*
Hochachka, P.W. 278, 280, *281*
Hodkinson, I.D., 376, 379, *381*
Hofstetter, R.H., 14
Hofstra, G., 286, *307*
Holdgate, M.W., *218*
Holding, A.J., 226, 228, 230, 240–242, *244*, *255*
Hole, F.D., *219*
Holland, D.J., 141, *154*
Holland, P.J., 290, *306*
Holmes, J.W., 101, 112, *155*
Hölzer, A., 20, *31*
Hooghoudt, S.B., 149, *154*
Hook, D.D., 257, 264, 269, 273, *282*
Hornburg, P., 18, *31*
Horton, R.E., 128, 138, *154*
Houghton, J.A., 227, *243*
Houston, W.W.K., 377, 378, *381*
Howard, A.J., 173, *220*, *221*
Howard-Williams, C., 374, *381*
Howson, G., *220*, *244*
Hsiao, T.C., 299, *308*
Hughes, R., *156*, *244*
Huikari, O., 86, 96, *154*
Hulme, P.D., *151*, *218*, *243*
Humboldt, A. von, *see* Von Humboldt, A.
Hunter, R.F., *152*, *219*, *243*
Huntley, B., *345*
Hupkens van der Elst, F.C.C., 102, *154*
Hustich, I., 56, 59, *61*
Hutchinson, J.N., 159, 210, *221*

Ilnicki, P., 21, *30*, *31*
Imhof, G., 22, *31*, 356, 357, *364*, 372, 373, 379, *381*
Incoll, L.D., 84, *154*
Ingham, G., 181, *221*
Ingram, H.A.P., 1, 8, 28, *31*, *33*, 47, 48, 67, 78, 103, 104, 111, 124, 126, 127, 133, 136, 137, 144, 146, 149, *154*, *157*, 168, 171, 189, 200, 210, *221*, *222*, 263, *282*
Inkson, R.H.E., *221*
Irmay, S., *151*
Irwin, R.W., 101, *154*, 169, *221*
Isaak, D., 330, *345*
Ishibashi, A., 304, *308*
Ishizuka, Y., 251, *255*

Isirimah, N.O., 248, 249, 251–253, *255*
Ivanov, K.E., 67, 68, 70, 72, 80, 103, 110, 115, 119, 120, 123, 124, 130, 131, 133–135, 138–141, 143, 146, 147, *154*
Ivanov, L.A., 86
Ivanov, N.N., 75, *154*
Iversen, J., 38, 51, *61*
Ivitskiĭ, A.I., 87, 88, *154*

Jackson, T.A., 253, *255*
Jacobsen, J., 193, *218*
Jacobsen, O.S., *218*
Jahren, A., *308*
James, G.B., *308*
James, W.O., 275, *282*
Jankovská, V., 48, *61*
Janota-Bassalik, L., 200, *221*
Janssen, C.R., 323, 333, *345*
Jarvis, P.G., *157*, 304, *308*
Jasnowski, M., 21, 22, *31*
Jeffers, J.N.R., *220*
Jelgersma, S., 39, 46, *61*
Jenkins, D., 376, *381*
Jenny, H., 17, *31*, 206, 209, *221*
Jensen, C.R., 267, 268, *282*
Jensen, U., 20, *31*, 52, 55, *61*
Jervis, R.A., 288, 289, *308*
Jessen, K., 42, *61*
Jewell, M.E., 330, *345*
Johansson, I., 73, 92, 141, 143, *154*
Johansson, L.G., 296, *308*
John, C.D., 273, *282*
Johnsen, S.J., *60*
Johnson, F.L., 262, *280*
Jonas, F., 46, 54, *61*
Jones, D.G., 228, 229, *244*
Jones, H.E., 17, 18, *31*, 167, 189, 199, 209, 213, 215, *221*, 235, *244*, 264, 271, *282*
Jones, J.A.A., 127, 138, *154*
Jones, M.B., 296, *308*
Junk, W.J., 111, 264, 354

Kahn, S.U., 161, 175, *223*
Kallio, P., 226, *244*, 296, 298, *308*
Kallio, S., 298, *308*
Kamula, A., 169, *221*
Kamura, T., 248, 251, 254, *256*
Kamyshev, N.S., 43, *61*
Kantrud, H.A., 98, *157*
Kärenlampi, L., *157*
Karesniemi, K., 161, 162, 168, *221*
Kastanek, F., 96, *154*
Kats, N.Ya., 4, 26, *31*, 36, 38, 40–42, 52, *61*, 72, 145, *154*
Kats, S.V., 36, *61*
Kaufman, W.J., 133, *154*
Kaule, G., 20, 21, *31*, 132, *154*
Kawano, K., *309*
Kay, B.D., 101, 125, *154*
Keefe, C.W., 285, 308
Keeney, D.R., 164, 180, 182, 225, 242, 248, 249, 251–253, *255*
Keitel, A., 55, *61*
Kelley, J.J., 287, *307*
Kenworthy, P., *281*
Kepczyński, K., 22, *31*
Kerney, M.P., 356, *364*
Kershaw, A.P., 48, *64*
Kessler, J., 75, *155*
Khotinskiĭ, N.A., 37–39, 43, 44, 46, 47, 52, 54, 58, *61*
Kind, N.V., *62*
King, M.L., *364*
Kirkby, M.J., 120, *155*
Kirkham, D., 124, 135, *155*
Kiryushkin, V.N., 44, 46, *61*
Kivekas, V., 251, *256*
Kivinen, E., 24, 26, *31*, 160, 171, 217, *221*, 251, *256*, 339, *345*
Kjelvik, S., 296, *306*, *307*, *308*
Kladec, J.A., *156*
Klopatek, J.M., 288, 289, 293, *308*
Kluytmans, J.H., *281*
Knezek, B.D., 249, *255*
Knight, A.H., *152*, 164, 181, *219*, *243*
Kohler, M.A., 75, *155*
Konstantinov, A.R., 74, 75, 77, 92, *155*
Kormanik, P.P., *282*
Korpijaakko, M., 168, *221*
Koshcheev, A.L., 45, 47, *61*
Koutler-Andersson, E., 162, 167, 173, 174, 176, 179, 185–188, *221*
Kovács, M., 23, *31*
Kovalenko, N.P., *151*
Kramer, P.J., 263, *282*
Krames, K., 39, *62*
Krebs, H., 276, *282*
Kroigaard, –, 20, *31*
Kubitzki, K., 36, 54, *62*
Kukla, St., 47, *62*
Kulczyński, S., 4–6, 8, 14, *31*, 51, 55, 56, *62*, 110, 111, 131, 135, 137, 140, *155*, 161, *221*, 311, 313, 315, 318–321, 325, 332–334, 336, 339, 340, 343, *345*
Kulikova, G.G., 40, *62*
Kuntze, H., 20, *31*, 56, *62*, 71, 100, 114, *151*, *155*
Kurz, H., 263, *282*, 337, *345*
Küster, E., 165, *221*, 251, *256*
Kuzmin, P.P., 75, 92, *155*
Květ, J., 23, *31*, 176, *221*, 288, 289, 293, *307*, *308*, 354, 355, *364*, *381*

Laine, J., 116, *154*
Lake, P.S., *363*
Lakhani, K.H., *364*
Lamb, H.H., 36, 39, *62*
Lambers, H., 277, *282*
Lambert, J.M., 258, *281*, 288, *307*, 326, 333–335, *344*, *345*, 367, *381*
Lance, A.N., 355, *364*
Landsberg, J.J., *308*
Lange, E., 48, *62*
Langway, C.C., *60*
Lasalle, P., 37, *62*
Latter, P.M., *219*, *220*, 227, 231, 235, *243*, *244*
Law, F.M., *156*
Lawrence, D.B., *307*
Lawson, J., 291
LeBarron, R.K., 331, *345*
Lechowicz, M.J., 82, *155*
Lee, G.B., *255*
Lee, J.A., 170, 194, *221*
Leeden, F. van der, *see* Van der Leeden, F.
Lees, J.C., 254, *256*
Leffef, W.J.B., *365*
Legg, B.J., 84, *156*
Leick, E., 81, *155*
Leiser, A.T., 83, *155*
Lemée, M.G., 20, *31*
Lenton, G.M., 374, *381*
Leopold, A., 376, *381*
Levin, I., 26, *33*
Levins, R., 357, *364*
Levkovskaya, G.M., 36, *62*
Lewis, D.H., 238, 240, *244*
Lewis, F.J., 327, *345*
Lhote, H., 353, *364*
Lid-Torsvik, V., 227, 228, *243*, *244*, 298, *308*
Likens, G.E., 131, *152*, *224*
Linacre, E., 67, 77, 78, 92, 97, 98, *155*
Linde, A.F., 355, *364*
Linder, S., 296, *308*
Lindroth, C.H., 361, *364*
Linhart, Y.B., 275, 277, 278, *282*
Lishman, J.P., 45, 48, *63*, *346*
Lishtvan, I.I., *158*
Liss, O.L., *62*
Livett, E.A., 170, 193, *221*
Livingston, R.B., 335, *345*
Liwski, S., *256*
Locci, R., 165, *221*
Lockie, J.D., 136, *155*
Løddesøl, Aa., 18, *31*
Logan, A., *218*
Lohammar, G., 334, 335, *345*
Longhurst, W.H., 375, *381*
Longton, R.E., 291, *308*
Loomis, R.S., *307*
Loon, H. van, *see* Van Loon, H.
Lowe, J.F., 228, *244*
Lowe-McConnel, R.M., 368, *381*
Lowry, K.H., *64*, *347*
Lucas, R.E., 249, 250, 253, *255*
Luck, K.E., 183, *221*

Luckwill, L.C., 83, 85, *154*, *244*
Lund, J.W.G., 321, *345*
Lundqvist, G., 14, *32*, 53, 56, *62*
Luthin, J.N., 101, 120, 124, 135, *155*
Lutz, J.L., 40, *62*
Luxmoore, R.J., *282*
Lyneborg, L., *364*

McAlpine, R.G., 261, *282*
Macarthur, R.H., 362, *364*
McCrackern, R.J., *219*
McCreath, –, 272
MacCrimmon, H.R., 252, *256*
MacDonald, J.G., 286, 288, 290, 293, *306*
Mace, A.C., 94, *151*
Macfadyen, A., 378, *381*
MacFarlane, I.C., 169, *221*
McGuire, J., 48, *64*
Maciak, F., 250, *256*
McIlroy, I.C., 77, *157*
McIntyre, A., 38, *63*
Mackay, –, 170, 196
Mackenzie, A.J., *256*
Mackereth, F.J.H., 45, *62*
McLachlan, A.J., 370, *381*
McManmon, M., 272–274, 277, *281*, *282*
McNaughton, S.J., 289, 293, 295, *308*
McVean, D.N., 9–12, *32*, 40, *62*, 132, 136, *155*, 229, *244*, 291, *308*, 335, *345*
Maggs, G.H., 206, *219*, 235, 236, 241, *243*
Mahon, G., 355, *364*
Makisimow, A., *256*
Malmer, N., 36, 52, 55, *62*, 71, 111, *155*, 163, 164, 187, 188, *221*
Malmström, C., 14, 16, *32*, 101, *155*
Marek, S., 22, *32*
Marino, M.A., 149, *155*
Marks, T.C., 196, *220*, 296, 297, 301, *307*
Marr, J.W., 59, *62*
Marsan, A.A., *153*
Marshall, C., *306*
Marshall, D.R., 272, 273, *282*
Marshall, T.J., 101, 112, *155*
Marshall, W.H., *345*
Martin, N.J., 226, 230, 240–242, *244*
Martin, P.S., 37, *62*, 362, *364*
Masing, V., 4, 6, 14, 23, *32*, 36, 67, 111, 133
Mason, C.F., 225, 231, *244*, 288, 289, *308*, 367–369, 371–374, *381*
Masterson, C.L., 228, 229, *244*
Mathewes, R.W., 333, *345*
Mathews, C.P., 286, *308*
Mattson, S., 162, 167, 173, 174, 176, 177, 179, 185–188, *221*
Mazelis, M., 277, *282*
Meek, B.D., 248, *256*
Mejstřik, V., 22, *32*
Menage, A., 278, *282*
Menéndez Amor, J., *220*
Menke, B., 46, 47, 51, *62*
Menshun, M., *218*
Mentz, A., 20, *32*
Mercer, J.H., 37, *62*
Meschkat, A., 370, *381*
Metsävainio, K., 83, *155*
Michaud, J.-R., *153*
Michna, I., 22, *32*
Milburn, T.R., 296, *308*
Miles, J., 261, *282*
Millar, A., 132, 136, 141–143, *152*
Millbank, J.W., 229, *244*
Miller, D.H., 69, 74, *155*
Miller, D.W., *153*
Miller, G.R., *381*
Miller, P.C., 85, *155*, 300, *309*
Millers, R., *223*
Millington, R.J., 315, *345*
Minderman, G., 211, *221*
Misra, R., 25, *32*
Mitchell, G.F., 315, 339, *345*
Mitchell, J.K., 125, *155*
Moen, A., 18, *32*
Mohrmann, J.C.J., 75, *155*
Moizuk, G.A., 335, *345*
Molchanov, A.A., 93, *155*
Monteith, J.L., 74, *155*, 201, *221*, 292, 294, 295, *308*
Moore, J.J., 210, 214, *221*
Moore, J.M., 275, *282*
Moore, P.D., 6–8, 26, *32*, 111, 118, 139, 142, *155*, 171, *221*, 226, *244*, 325, 337, 339, 341, *345*, *346*
Moravec, J., 47, *62*
Moreau, R.E., 352, *364*
Morel, G., 278, *282*
Moretti, A., 20, *32*
Morgan, J.J., 247, *256*
Morita, H., 176, *221*, 250, *256*
Mörner, N.A., 37, 46, *62*
Mörnsjö, T., 160, 184, 186, 187, 189, 190, *221*, 288, 293, *308*, 338, *346*
Morris, M.G., 363, *364*
Mortimer, C.H., 189–191, *221*
Morton, A.J., 230, *244*
Moss, B., 373, *381*
Moss, C.E., 127, 132, *155*
Moss, E.H., 327, 334, *345*, *346*
Moss, R., 376, *381*
Mosse, B., 240, *243*
Mott, P.J., 105, 106, 109, 110, *155*
Moucha, J., *364*
Muc, M., 290, 293, *308*
Müller, K., 36, *62*
Münnich, K.O., *346*
Munro, J.M.M., *244*
Murozumi, M., 180, 194, *221*
Muskat, M., 120, *155*
Mustonen, S.E., 75, *155*

Næss, T., 18, *32*
Nannfeldt, J.A., 40, 46, *60*
Nascimiento, K.H., *281*
Nature Conservancy [U.K.], 354, *364*, 374
Nauke, W. von, *see* Von Nauke, W.
Neetzel, J.R., 331, *345*
Neill, W.E., 357, *364*
Neĭshtadt, M.I., 36, 40, 42, 43, 51, 53, 54, *62*
Nekrasova, I.V., 110, *155*
Nelson, R.E., *158*
Netherlands Soil Survey, 19, *32*
Neuhäusl, R., 22, *32*, 78, 82, 86, 87, *155*
Newbould, P.J., 186, *222*, 290, *309*
Newman, E.I., 298, 299, *308*
Newsome, A.E., 368, *381*
Newson, M.D., 127, 137, 138, *153*, *155*
Nicol, A.T., 127, 155
Nichol, W.E., 249, *256*
Nicholls, K.H., 252, *256*
Nichols, H., 37, 39, 59, *62*, *63*, 333, *346*
Nicholson, I.A., *156*
Nielsen, C.O., 377, *381*
Nilsson, T., 36, 38, *63*, 318, *346*
Nixon, D., 376, 379, *381*
Nixon, S.W., 363, *364*
Nordenson, T.J., 75, *155*, *156*
Nordhagen, R., 18, *32*
Norgaard, E., 376, *381*
Novikov, S.M., 109, *156*
Novikova, Y.V., 97, *156*
Nųritdinov, N., 266, *282*
Nye, P.H., 111, *156*, 181, *222*, 299, *307*
Nys, L., 103, 117, 133, *156*

Oberdorfer, E., 82, *156*
O'Brien, J.C., *219*
Obrtel, R., 369, *381*
Oechel, W.C., 296, 297, *308*
Ogata, S., 251, 255, *256*
Ohira, Y., 169, *222*
Ohtaki, E., 304, *308*
Okely, E.F., *381*
Okruszko, H., 113, *156*
Olaczek, R., 22, *32*
Olausson, E., 40, 44, 54, 56, *63*, 71, *156*
Oldfield, F., 46, *63*, 170, *222*
Oldroyd, H., 358, *364*
Olenin, A.S., 171, *222*
Olkowski, M., 22, *32*
Olsen, O.B., 168, *222*
Olson, J.S., 17, 18, *31*, 212, 213, *220*
Oltmanns, F., 81, *156*
Omer-Cooper, J., 360, *364*
Ondok, J.P., 302, 303, *307*, *308*, *309*
Oram, R.N., *282*

Orlob, G.T., 133, *154*
Orlova, L.A., *62*
Ornes, W.H., 287, *309*
Osborne, P. L., 373, *381*
O'Sullivan, P.E., 48, *63*
Osvald, H., 4, 6, 13, 14, 18, 29, *32*, 35, 40, 44, 52, 55, *63*, 92, 127, *156*, 160, *222*, 314, 323, 339, *346*
Oświt, J., 21, 22, *32*
O'Toole, M.A., 229, *244*
Ouryvaev, V., 105, *158*
Overbeck, F., 4, 19, 21, *32*, 35–37, 40, 42–44, 52–54, 56, *63*, 78, 81, 127, 132, *156*, 161, *222*, 291, *309*, 318, 339–341, *346*
Oviatt, C.A., 363, *364*
Øvig, J.K., 20
Owen, T.G., *281*

Paarlahti, K., 242, *244*
Paasio, I., 51, *63*
Pacowski, R., 22, *32*
Päivänen, J., 81, 100, 112, 114–116, 126, 134, 135, *154*, *156*, 162, 167, *222*
Pajaczkowski, J., *306*
Pakarinen, P., 24, 26, *31*, 192, 194, *221*, *222*, 291, *309*
Palczyński, A., 21, 22, *31*, *32*
Pallis, M., 23, *32*, 333, *346*
Panychev, V.A., *62*
Parihar, N.S., 80, *156*
Parker, G.R., 288, *309*
Parkin, J., 279, *282*
Parkinson, H.L., *158*
Parkinson, K.J., 84, *156*
Parr, J.F., 248, *256*
Pasák, V., 117, 141, 143, *153*
Passer, M., 250, *256*
Paterson, M.P., 181, *222*
Patil, K.D., *281*
Patrick Jr., W.H., 248, 251, 253, *255*, *256*
Patterson, C., *221*
Patzelt, G., 37, 39, 56, *63*
Paul, H., 252, *256*
Pavlova, K.K., 140, *156*
Peacock, J.M., 304, *309*
Pearcy, R.W., 295, 297, 300, *309*
Pearsall, W.H., 127, *156*, 163, *222*, 288–290, *306*, *307*, *309*, 319, 322, 325, 327–329, 340, *346*
Pearson, G.R., *222*
Pearson, G.W., 170, *222*
Pearson, L.C., 289, *307*, *309*
Pearson, M.C., 16, *152*, 185, *220*, 320, 321, 336, 340, *345*, *346*
Pearson, R.G., 361, *364*
Peirson, D.H., 180, 186, *222*
Pelikan, J., 372, *381*
Penfound, W.T., *281*, 289, *309*
Penman, H.L., 68, 74, 75, 92, 93, 95, 110, *156*
Pennak, R.W., 333, *346*
Pennington, W., 37, 45, 48, *63*, 184, 222, 332, *346*
Perez, H., 300, *309*
Perkins, D.F., 191, *220*, 229, *244*
Perkins, G.L., *364*
Perry, D.A., 274, *282*
Perry, K.M., *223*
Peschke, P., 48, *63*
Peterson, M.L., 264, 269, *280*
Petty, J.A., 269, *282*
Philipson, J.J., 258, 264, 265, *281*, *282*
Pieczyńska, E., *30*
Pierce, R.S., *347*
Pigott, C.D., 161, *222*, 328–330, *346*
Pigott, M.E., 161, *222*
Pilcher, J.R., *222*
Pimlott, D.H., 375, *381*
Piret, E.L., *256*
Pisano (Valdes), E., 11, 12, 25, 72, *156*
Planchais, N., 48, *63*
Polakowski, B., 22, *32*
Polisini, J.M., 288, 289, *309*
Pollett, F.C., 2, 7
Polubarinova-Kochina, P.Ya., 120, *156*
Ponnamperuma, F.N., 120, *222*, 247–254, *256*
Poore, M.E.D., 321, *346*
Pop, E., 23, *32*
Popescu-Marinescu, V., 370, *381*
Poskitt, T.J., 167, 210, 213, *218*
Post, L. von, *see* Von Post, L.
Potonié, H., 44, *63*
Potonié, R., 171, 172, *222*
Powell, C.L., 240, *244*
Pradet, A., *282*
Prat, H., 337, *344*
Pratt, H.K., 278, *282*
Predtechenskiĭ, A.V., *62*
Premazzi, G., 184, *222*
Přibáň, K., *307*
Priddle, J., 197, *222*
Pringle, J.S., *364*
Proceedings 1st International Peat Congress, Dublin, *224*
Proceedings 2nd International Peat Congress, Leningrad, *224*
Proceedings 3rd International Peat Congress, Quebec, *224*
Proceedings 4th International Peat Congress, Otaniemi, *224*
Proceedings 5th International Peat Congress, Poznán, *224*
Proceedings 6th International Peat Congress, Duluth, *224*
Prus-Chacinski, T.M., 142, *153*
Pruscha, H., 372, 379, *381*
Pryor, A.J., 272, 273, *282*
Prytz, K., 87, 89, 104, 114, *156*
Pullen, O.J., *244*
Purves, S., 240, *244*
Purvis, E.R., 238, *245*, 250, 251, *256*
Puustjärvi, V., 163, 165–168, 175, *222*
Pyatetsky, G.Y., 114, *156*
Pyatt, D.G., 100
P'yavchenko, N.I., 51, *63*, 93, *156*

Radforth, N.W., 7, 10, 11, *32*, 168, 169, *221*
Radke, G.J., 48, *63*
Radziej, J., *30*
Ragg, J.M., 217, *222*
Raghavendra, A.S., 295, *309*
Raikes, R., 147, *156*
Rakhmanina, A.T., 82, *156*
Rampton, V., 59, *63*
Ratcliffe, D.A., 9–12, *32*, 40, *62*, 77, 111, 132, 136, *155*, *156*
Raup, H.M., 262, *282*
Ravera, O., 184, *222*
Rawes, M., 375, *381*
Raymond, P., 266, 267, *282*
Read, D.J., 239, *245*
Reader, R.J., 290, 291, 293, *309*
Reddaway, E.J.F., 101, 146, *152*, 165, 197, 198, 203, *219*, 231, 236, *243*, 287, 291, *307*, 332, *344*
Redman, R.E., 301, *309*
Reiners, R.H., *224*
Reiners, W.A., 288, 289, *309*, *344*
Renger, M., 71, 100, *156*
Rennie, R., 251, *256*
Richards, P.W., 188, *222*
Richardson, C.J., 145, *156*
Richardson, C.T., 323, *346*
Rieley, J., 167, 169, 182, 192, *218*, 291, *306*, 338, *344*
Rietz, G.E. du, *see* Du Rietz, G.E.
Rigg, G.B., 248, *256*, 323, 326, 333, 335, *346*
Riley, W.S., *282*
Ripley, E.A., 301, *309*
Ritchie, J.C., 39, 59, *63*, 333, *346*
Roberts, J., 163, *218*
Robertson, A., *30*
Robertson, R.A., 73, 127, 141–143, *155*, *156*, 159, 187, 188, *222*, *223*
Robinson, A.C., 76, *156*
Robinson, W.O., 254, *256*
Rodda, J.C., 76, 97, *156*
Rodewald, M., 39, *63*
Rodewald-Rudescu, L., 23, *32*, 97, *156*, 334, *346*
Rodin, L.E., 292, *309*
Romanov, V.V., 68, 76, 77, 80, 81, 91, 92, 96, 97, 101–104, 114, 117, 123, 132 – 134, 136, 144, 147, *156*, *157*, 180, *222*

Romanova, E.A., 52, *63*, 111, 133, *157*
Rooth, J., *365*
Rose, C.W., 74, 101, 120, *157*
Rose, F.L., 351, *364*
Rosswall, T., 198, 200, 201, *222*, 227, 242, *244*
Røstad, A., 39, *63*
Roux, F., 352, *364*
Rowe, R.N., 260, *282*
Ruddiman, W.F., 38, *63*
Rudescu, L., 370, *381*
Rudolph, K., 41, 52, *63*, 127, *157*, 315, 339, *346*
Rühling, A., 180, 181, 194, *222*
Ruinen, J., 227, *245*
Rump, H.H., 55, *63*
Rusinska, A., 22, *32*
Russell, D.A., 125, *157*
Russell, J.D., *224*
Russell, R. Scott, *see* Scott Russell, R.
Rutter, A.J., 86, 104, *157*, 337, *346*
Ruuhijärvi, R., 2, 4, 6, 15, 29, 40, 43, 46, 52, 53, 56, 58, *60*, *63*, 73, 172
Rybníček, K., 22, *33*, 36, 52, *63*, 78, 111, 136, *157*, 316, *346*
Rybničková, E., 48, *62*, *63*, 78
Rychnovsky, B., 372, *381*
Rycroft, D.W., 15, 16, *33*, 67, 105, 123–127, 133–135, *154*, *157*, 168, 181, 216, *221*, *222*
Rydén, B.E., *152*, *157*

Sæbø, S., 182, 183, 187, *222*, 291, *309*
Sagher, A., 252, *256*
Sainty, G.R., *155*
Sakaguchi, Y., 23, *33*
Salmi, M., 170, *222*
Salmon, L., *222*
Sandberg, G., *221*
Sander, J.E., 131, *157*
Sanger, J., 169, *220*
Sarasto, J., *154*
Sawyer, J.S., 36, *64*
Schallinger, K.M., 26, *33*
Scheidegger, A.E., 120, *157*
Schlesinger, W.H., 288, 300, *309*
Schlüter, H., 48, *62*, *64*, 72, 73, *157*
Schmeidl, H., 55, 57, *64*, 87, 88, 107, 108, 117, 141, 143, *157*
Schmitz, H., 40, *64*
Schneekloth, H., 21, 29, *33*, 35, 46, 47, 53, 54, 56, *64*
Schneider, G., 288, *309*
Schneider, R., *30*
Schneider, S., *30*, 35, 54, *64*, 161, 185, *222*
Schnitzer, M., 161, 175, *223*
Scholander, P.F., 263, 264, *282*, 299, 300, *309*
Scholander, S.I., *282*
Scholtens, J.R., 264, 273, *282*
Schothorst, C.J., 18, 20, *33*, 238, *245*
Schröder, R., 374, *381*
Schroevers, P.J., 354, *364*
Schröter, C., 22, *30*, 36, 40, 43, 44, *60*
Schuch, M., 67, 72, 108, 132, 135, 141, 144, *153*, *157*
Schwaar, J., 25, *33*, 40, 43, 54, *64*
Schwintzer, C.R., 334, *346*
Scorer, R.S., 181, *222*
Scott, D., 287, *309*
Scott, R., *223*
Scott Russell, R., 257, 271, *282*
Scoullar, K.A., 216, *219*
Sculthorpe, C.S., 265, *282*
Segadas-Vianna, F., 326, *344*
Sekiya, S., *255*
Selander, H., 252, *255*
Selling, O.H., 25, 26, *33*
Seo, T., 304, *308*
Sernander, R., 4, 14, 29, *33*, 314–316, 339, *346*, *347*
Šesták, Z., 91, *157*
Shaver, G.R., 183, *223*, 234, *245*
Shaw, M.W., *31*
Shepherd, H., *221*
Shewry, P.R., *309*
Shoham, D., 26, *33*
Sifton, H.B., 271, *282*
Sikora, L.J., 164, 180, 182, 225, 242
Sillanpää, M., 184, 192, *223*
Sinker, C.A., 325, *346*
Sirén, G., 51, *64*
Sjeflo, J.B., 98, *157*
Sjörs, H., 2, 4, 6–8, 13, *33*, 40, 52, 56, *64*, 71, 77, 110, 111, *155*, *157*, 171, 172, 187, 188, *221*, *223*, 338, *346*
Skartveit, A., 72, *157*
Skaven-Haug, S., 166, *223*
Skene, M., 164, *223*, 337, *346*
Skobeeva, E.I., *62*
Skre, O., 296, *309*
Slack, N.G., 111, *158*, 334, *347*
Slatyer, R.O., 74, 77, 83, 91, 101, *157*, 300, *309*
Slavík, B., 75, 83–85, *157*
Smakman, G., 277, *282*
Small, E., 83, *157*, 230, *245*, 271, *282*, 296–298, 300, *309*
Šmid, P., 98, *157*
Smirnoff, –, 267–269
Smith, A.G., 39, 45, *64*
Smith, G.R.J., 377, 379, *381*
Smith, L.P., 75
Smith, R.A.H., 181, 182, 188, 190, 196, 197, *223*, 290, 291, *307*
Smith, R.I.L., 25, *33*
Smith, S.E., 240, *245*
Sobotka, D., 22, *33*
Sokolov, A.A., 70, 77, *157*
Solantie, R., 73, *157*
Solem, Th., 48, *61*
Solsky, B.A., 288, *306*
Soma, K., *158*
Somers, M.G., 288, *307*
Sommers, L.E., 252, *256*
Sonesson, M., 73, *157*, 180, 181, 185–187, 191, 192, 197, *223*
Soó, R., 23, *33*
Sparling, G.P., 183, *223*, 240, *245*
Sparling, J.H., 126, *152*, 261, *282*
Spearing, A.M., 175, *223*
Speight, M.C.D., 110, 358, *363*, 377, *380*
Spence, D.H.N., 312, 326, 330, 334, 335, *346*
Sprent, J.I., 181, *223*
Standen, V., 225, 231, 377–380, *381*, *382*
Stanek, W., *33*
Stanhill, G., 75, *157*
Starichenkov, I.P., *61*
Stavset, K., 183, 196, *223*
Stearns, F.W., 288, 289, 293, *308*
Steckhan, H.U., 54, *64*
Steele, A., *157*
Steele, P., *281*
Steffen, H., 51, *64*
Stehlik, A., 22, *33*
Steindórsson, S., 19, *33*
Steingrover, E., 277, *282*
Stenström, T.-A., *218*
Sternhell, S., *218*
Steubing, L., 75, 87, *157*
Stevens, K.R., 165, 173, 176, 177, 179, 188, *223*, *224*, 250, 251, *256*
Stevenson, F.S., 249, *256*
Steward, K.K., 287, *309*
Stewart, J.M., 187, 188, *223*, 290, 291, 293, *309*
Stewart, R.A., 98, *157*
Stewart, W.D.P., 227, 228, *245*
Stocker, O., 102, *157*
Stoekeler, J.H., 140, *157*
Stolzy, L.H., *282*
Stoner, W.A., *155*, 300, *309*
Storey, K.B., 278, 280, *281*
Stout, J.D., 227, 238, 242, *245*, 250, *256*
Straka, H., 40, *64*
Strebel, O., *156*
Streeter, J.G., 278, *282*
Stribley, D.P., 239, *245*
Strokina, L.A., *152*
Struzer, L.R., 76, *151*
Strygin, N.N., 166, 203, *223*
Stumm, W., 247, *256*
Suess, H., 36, *64*
Summerfield, R.J., 72, *157*, 290, *309*
Summers, R.W., *364*
Suzuki, S., 23, *33*
Svendsen, J.A., 375, *382*
Svensson, B.H., 198, 203, *223*

Svoboda, J., *31*, *307*, *381*
Swain, F.M., 170, *223*
Swain, R., *363*
Swan, J.M.A., 331, *346*
Swartzendruber, D., 122, 125, *157*
Swinbank, W.C., 76, *157*
Switsur, V.R., *61*, 167, *223*, 332, *346*
Syers, J.K., 252, *256*
Szafer, W., 21, *33*
Szczepanek, K., 48, *64*
Szczepański, A., 21, 22, *33*
Szuniewicz, J., 113, *156*

Tait, D.E.N., 199, *219*
Takai, Y., 248, 251, 254, *256*
Takasu, K., *308*
Takhtajan, A., 278, *282*
Takijuma, Y., 254, *256*
Tallis, J.H., 13, 14, 16, 40, 44, 45, 48, 51, *64*, 97, 110, 111, 117, 132, 138, 145, *151*, *157*, 161, 167, 170, 194, *221*, *223*, 261, 262, 280, 282, 319, 320, 325, 327, 331, 332, 336, 338, 340, *346*, 367
Tamm, C.O., 180, *223*, 291, *309*
Tamoshaïtis, Yu.S., 44, *64*
Tanaka, A., 288, 294–296, 302, 304, *309*
Tansley, A.G., 6–13, *33*, 35, 40, 44, 52, *64*, 161, 172, *223*, 315, 326, 328, 329, 339, *346*
Tauber, H., 36, 38, *59*, 161, 199, 200, *218*
Taylor, J.A., 4, 6, 13, 17, 23, 25, 28, 36, 39, 41, 48, 73, 127, 158, 226, 238, 261, 320, 325, 331
Taylor, –, 275
Tennant, G., 328, 329
Tennant, I., 78
Tenney, F.G., 165, 203, *224*
Tereshchenko, K.P., 72, *150*
Terning, R.E., *221*
Tervet, D.J., 72, *158*
Theander, O., 164, 173–175, *223*
Thom, A.S., 304, *309*
Thomas, G.W., 249, *255*
Thompson, J.F., 278, *282*
Thompson, K., 12, 287, *309*
Thomson, P.W., 43, 46, *64*
Thornthwaite, C.W., 75, 87, 88, 93, *158*
Tibbetts, T.E., 197, *223*
Tickle, W.M., *306*
Tieszen, L.L., *219*, 290, 292–294, 297, 302, *307*, *309*
Tikhomirov, L.I., *61*
Tilton, D.L., *156*
Tinbergen, L., 86, 87, *158*
Tinker, P.B., 111, *156*, 181, 183, *222*, *223*, 240, *245*
Tinklin, R., 299, *309*
Tjuremnov, S.N., *62*
Toebes, C., 105, *158*
Tolbert, N.E., *281*
Tolonen, K., *33*, 192, 194, *222*, 315, 324, 333, 339, 340, *345*, *346*
Tomlinson, R.W., 110, 138, *151*, *218*, *243*, 339–341, *344*
Tooley, M.J., 46, *64*
Tottenham, C.E., *364*
Transeau, E.N., 330, *346*
Trautmann, W., 48, *64*
Treacy, K., *219*
Troedsson, T., 180, *223*
Troels-Smith, J., 218, *223*
Troise, F.L., *153*
Tsubota, G., 252, *256*
Tsukada, M., 37, *64*
Tucker, D.S., 354, *364*
Tucker, R.B., 127, *158*
Turcek, F.J., 349, 361, *365*
Turner, F.T., 248, *256*
Turner, J., 48, *64*, 202, *223*, 332, *347*
Turner, J.S., 325, *345*
Turner, R.C., 249, *256*
Turner, R.E., 71, 145, 148, *153*
Turrill, W.B., 274, 276, *282*
Tuschl, P., 97, *158*
Tüxen, J., 2, *33*
Tyler, G., 180, 181, 192, 194, *222*, *223*
Tyler, P., 275, 277, 278, *281*
Tyler, P.A., 354, *363*
Tyrtikov, A.P., 51, *64*

Uchijima, Z., 301, 302, 304, 305, *309*
Uhden, O., 104, 141, *158*
Uhlig, S., 75, *158*
Ullmann, H., 22, *33*
UNESCO, 11, *33*
Unsworth, M.H., 287, *310*
Urquhart, C., 18, *33*, 165, 189–191, *223*, 241, 242, *245*, 247, 248, *256*

Val'k, U.A., 46, 54, *64*
Valk, A.G. van der, *see* Van der Valk, A.G.
Vallane, N., 296, *308*
Van Cleve, K., *219*
Van Dam, L., *282*
Vanden Berghen, C., 20, *33*
Van der Hammen, T., 37, *64*
Van der Leeden, F., *153*
Van der Valk, A.G., 288, 289, *310*
Van Dijk, H., 168, 169, *223*
Van Geel, B., 38, *64*
Van Grundy, S.D., *282*
Van Loon, H., 38, 39, *65*
Van 't Woudt, B.D., *158*
Van Werden, K., *63*
Van Zeist, W., 54, 56, *65*
Van Zinderen Bakker, E.M., 37, *65*, 72, *158*
Vartapetian, B.B., 266, *282*
Vartiovaara, U., 242, *244*
Vasari, A., 44, *65*
Vasari, Y., 44, *65*
Venema, H.J., 75, *152*
Vennesland, B., 277, *282*
Ven Te Chow, 106, 120, 129, *158*
Verry, E.S., 67, 71, 117, 131, 140, 142–144, *152*, *154*
Vester, G., 262, *282*
Veum, A.K., 200, *220*, 226, *244*
Vickers, D.H., 288, *306*
Virta, J., 94–96, 103, 104, 106, *158*
Visser, S.A., 241, 242, *245*
Vitt, D.H., 111, *158*, 291, *309*, 334, 341, 342, *347*
Vogl, R.J., 322, 336, *347*
Voigt, G.K., *347*
Volarovich, M.P., 101, 125, *158*, 169, *223*
Von Brand, T., 278, *283*
Von Bülow, K., 26, 29, *33*, 40, 51, *65*
Von der Brelie, G., 46, *65*
Von Humboldt, A., 117
Von Nauke, W., 161, *223*
Von Post, L., 4, 8, 14, 16, 29, *33*, 40, 42–44, 52, *65*, 113, 115, 134, 161, 162, 168, 174, 188, *223*, 314, 315, 339, *347*
Vorob'ev, P.K., 114–116, *158*
Vorren, K.D., 18, *33*
Vries, D.A. de, *see* De Vries, D.A.

Wadsworth, R.M., 106, *158*
Wager, H.G., 278, *283*
Waid, J.S., 232, *245*
Waine, J., 125, 126, *158*
Waitzbauer, W., 372, *382*
Waksman, S.A., 28, 30, *33*, 165, 173, 176, 177, 179, 188, 203, *223*, *224*, 238, *245*, 250, 251, *256*
Waldheim, St., 35, 40, 51, 52, *65*
Walker, D., 16, *34*, 55, 56, *65*, 77, 111, *156*, 161, 188, *224*, 315, 322, 323, 331, 332, 339, 340, *347*
Walker, J.M., 295, *310*
Walker, P.C., 333, *347*
Walker, P.M., 16, *34*, 55, 56, *65*, 161, 188, *224*, 315, 339, 340, *347*
Wallace, B., *243*
Walley, W.W., 288, 290, 293, *306*
Wallin, B., 37, *62*
Walsh, T., 167, 185–188, *224*
Walter, H., 4, 6, 7, *34*, 75, *158*, 171, *224*, 264, *283*
Wanke, R., *157*
Ward, R.C., 69, 72, 74, 75, 120, 128, 129, *158*
Warming, E., 325, *347*
Warrick, A.W., 146, *158*
Was, St., 45, *65*
Watson, A., *381*

Watt, A.S., 161, *224*
Watt, R.F., 337, *347*
Watts, W.A., 38, *65*
Watts, W.R., 304, *310*
Waughman, G.J., 177, 179, 180, 187, *224*, 298, *310*
Waygood, E.R., 295, *310*
Weast, R.C., 191, *224*
Weatherley, P.E., 298, 299, *309*, *310*
Weaver, J.E., 286, *310*
Webber, P.J., 290, *310*
Weber, C.A., 4, 8, 12, 14, 19, 29, *34*, 40, 44, 52, 53, *65*, 145, 148, *158*, 171, 172, *224*, 311, 313, 340, 342, *347*
Weber, K.Yu., 52, 54, *65*
Webster, J.R., 191, *224*
Webster, R., 100, *158*
Wechmann, A., 135, *158*
Weimarck, H., 40, *65*
Wein, R.W., 85, 86, *158*, 234, *245*, 275, *282*
Weitzman, S., 140, *157*
Welch, D., 234, *245*, 375, *381*
Weller, G., 301, *310*
Wendelberger, G., 22, *34*
Wendland, W.H., 38, 39, *65*
Wentz, W.A., *156*
Werden, K. van,
see Van Werden, K.
Werenskjold, W., 14, *34*, 149, *158*
Werger, M.J.A., 72, *158*
Wertz, G., 133, *158*
West, G., 330, *347*
West, G.C., 290, *307*
West, R.G., 38, *61*, *65*, 166, 218, *224*, 343, *347*, 360, *365*
Westlake, D.F., 286, 288, 289, 293, *308*, *310*, 332, *347*
Wetmore, R.H., *282*
Wetzel, R.G., 333, *347*
Wheatley, R.E., 173, *224*
White, J., 286, *308*
Whitmore, T.C., 159, *224*
Whittaker, J.B., 203, *219*, 375–379, *380*, *382*
Whittaker, R.H., *224*, 289, *310*
Whittow, G.G., *281*
Wickman, F.E., 14–16, 24, *34*, 68, 71, 148, 149, *158*, 209, *224*
Wiegert, R.G., 286, *310*
Wielgolaski, F.E., 290, 296, *306*, *308*, *310*
Wijmstra, T.A., *220*
Wijngaarden, A.V., 361, *365*
Wijsman, T.C.M., 278, *281*
Wilde, S.A., 248, *256*, 338, *347*
Wildehain, W., 176, *224*
Wildi, O., 146, *158*, 216, *224*
Wildung, R.E., 249, *255*
Williams, B.L., 183, *224*
Williams, D.J.A., *33*, *154*, *157*, *221*, *222*
Williams, G.P., 94, *158*
Williams, J., 38, 39, *65*
Williams, J.M., *222*
Williams, W.A., *307*
Williams, W.T., 267, *283*
Willis, W.O., 140, *158*
Williston, H.L., 261, 262, *281*
Wilps, H., *281*
Wilson, C.M., 232, 234, 235, 238, 241, *245*
Wilson, E.M., 129, *158*
Wilson, E.O., 362, *364*
Wilson, J.F., 328–330, *346*
Wilsson, L., 352, 354, *365*
Wind, G.P., 96, *158*
Wit, C.T. de,
see De Wit, C.T.
Witting, M., 55, *65*, 71, *158*
Wolfenbarger, D.O., 359, *365*
Woodwell, G.M., 197, 210, *224*
Woolhouse, H.W., 287, 296, 303, *307*, *308*, *309*
World Meteorological Organization, 36, *65*, 75
Wortley, J.S., 374, *382*
Woudt, B.D. van 't,
see Van 't Woudt, B.D.
Wright, H.E., 38, *65*

Yakovlev, V.I., 106, *158*
Yamaguchi, J., *309*
Yamamoto, S., 126, *158*
Yamanaka, M., 24, *34*
Yamazaki, F., 117, *158*
Yanishevskiĭ, Y.D., 91, *158*
Yefimova, N.A., *152*
Yoshida, T., 248, 250, *256*
Yoshioka, K., 17, 23, 24, *34*
Young, E.C., 353, *365*
Youngberg, C.T., *256*
Younger, J.S., 125, *155*

Zarzycki, K., 21, *33*
Zaslavsky, S., *151*
Zavel'skiĭ, F.S., *62*
Zebe, E., 278, *283*
Zeist, W. van,
see Van Zeist, W.
Zelikzon, E.M., 54, *62*
Zimenko, T.G., 227, 242, *245*
Zinderen Bakker, E.M. van,
see Van Zinderen Bakker, E.M.
Zinevici, V., 370, *382*
Zink, E., 286, *310*
Zoller, H., 39, *65*
Zoltai, S.C., 2, 7
Zwaan, A. de,
see De Zwaan, A.

SYSTEMATIC INDEX[1]

Abies (fir), 39, 331, 383
A. balsamea (L.) Mill., 326, 331
Acaena, 384
A. exigua A. Gray, 26
Acanthaceae, 384
Acarina (mites), 353, 357, 378, 379, 385
Acer (maple), 331, 384
A. negundo L., 262
A. rubrum L., 262, 331, 335
A. saccharum Marsh., 326
A. saccharinum L., 262
Aceraceae, 384
Achorutidae, 385
Acorus, 383
A. calamus L., 23, 293
Acrididae, 385
Acrocephalus, 385
A. schoenobaenus L. (sedge warbler), 372
A. scirpaceus (Hermann) (reed warbler), 372
Acrocladium, 383
A. stramineum (Brid.) Richards & Wallace, 324
Actinomycetes, 250, 290, 383
Aeschynomene, 384
A. pfundii Taub, 371
Aesculus, 384
A. turbinata Blume, 23
Agabus, 385
A. striolatus Gyll., 361
A. vittiger Gyll., 361
Agaricaceae, 383
Agonomycetaceae, 383
Agrostis, 384
A. tenuis Sibth. 274
Agrypnia, 385
A. pagetana Curtis, 369
Aira, 26, 384
Alces, 385
A. alces L. (moose), 355, 375
Algae, *see* Bacillariophyta, Chlorophyta, Cyanobacteria, Rhodophyta; *and in General Index*
Alisma, 383
A. plantago-aquatica L., 288
Alismataceae, 383
Allolobophora, 384
A. chlorotica (Savigny), 357
Alma, 350, 384
Alnus (alder), 12, 47, 161, 184, 229, 269, 276–278, 280, 313, 324, 326, 331, 335, 384
A. crispa (Ait.) Pursh, 327
A. glutinosa (L.) Gaertn., 229, 264, 270, 325, 328, 331, 335, 336, 355
A.incana (L.) Moench, 269, 270, 331
A. japonica Sieb. & Zucc., 23
A. rugosa (Du Roi) Spreng., 288, 290, 326
Amanita, 383
A. muscaria Fr. ex L., 239
Amblystegiaceae, 383
Ambystoma, 385
A. tigrinum Green (tiger salamander), 351
Ambystomidae, 385
Ammophila, 384
A. arundinaria (L.) Link, 272, 273
Amphibia, 351, 385
Amphipoda, 385
Anabaena, 227, 383
Anacardiaceae, 384
Anas, 355, 385
A. platyrhynchos L. (mallard) 355
A. querquedula (garganey), 352
Anatidae (*see also* ducks; swans *in General Index*), 385
Andromeda, 326, 384
A. polifolia L., 12, 73, 82, 89, 239
Angiospermae [Magnoliophyta], 383
Annelida, 278, 280, 350, 384
Annonaceae, 384
Anser, 385
A. anser L. (grey-lag goose), 355
Anseriformes, 385
Anthocleista, 384
A. nobilis G. Don, 263
Anthus, 385
A. pratensis L. (meadow pipit), 376, 377
Anura, *see* Salientia
Aphididae, 173, 355, 372, 385
Apiaceae [part Umbelliferae], 384
Aplexa, 384
A. hypnorum (L.), 368
Aquifoliaceae, 384
Araceae, 383
Arachnida, 385
Araliaceae, 384
Araneae (spiders), 350, 353, 376, 378, 379, 385
Araneus, 385
A. cornutus (L.), 379
Arbutus, 239, 384

[1] In this index, no attempt has been made, for larger taxonomic entities, to list all the pages where subordinate taxa are mentioned. These may found by use of the Systematic List of Genera (pp. 383–385).

Arctophila, 384
A. fulva (Trin.) Rupr., 300
Arctostaphylos, 239 384
A. uva-ursi (L.) Sprengel, 239
Ardeidae, 358, 385
Ardeiformes, 385
Arion, 385
A. ater (L.), 356, 358
Aronia, 384
A. melanocarpa (Michx.) Ell., 296
Arthropoda, 358, 385
Artiodactyla, 385
Arundo, 97, 384
Aschelminthes, 384
Ascolichenes, 383
Ascomycotina, 383
Asellidae, 385
Asellus, 357, 385
A. aquaticus (L.), 369, 373, 379
A. meridianus Rac, 369
Aspidobranchia, 384
Aster, 322, 384
Asteraceae [=Compositae], 384
Atriplex, 295, 384
Aulacomniaceae, 383
Aulacomnium, 337, 383
A. palustre (Hedw.) Schwaegr., 337
Aves (birds) (*see also General Index*), 385
Avicennia, 264, 384
A. nitida Jacq., 264, 265
Avicenniaceae, 384
Azotobacter, 227, 383
Azotobacteriaceae, 227

Bacidia, 229, 383
Bacillariophyta (diatoms), 356, 371, 383
Bacillus, 227, 383
B. polymyxa (Prazmowski) Macé, 227
Bacteria (*see also General Index*), 383
Baeothryon, Section of *Scirpus*, q.v.
B. caespitosum (L.) A. Dietr., *see Scirpus cespitosus*
Basidiomycotina, 383
Basommatophora, 384
Batrachospermaceae, 383
Batrachospermum, 128, 383
Betula (birch), 38, 47, 58, 93, 161, 218, 239, 313, 321, 324, 326, 331, 335, 384
B. ermani Cham., 23
B. glandulosa Michx., 327
B. glandulosa var. *glandulifera* [=*B. pumila* var. *glandulifera*], 326
B. michauxii Spach. [=*B. nana*], 161, 296
B. nana L., *see B. michauxii*
B. odorata Bechst., *see B. pubescens*
B. papyrifera Marsh, 288, 327
B. populifolia Marsh, 296
B. pubescens Ehrh. [=*B. odorata*], 86, 239, 328, 336
B. pumila L., 327
B. pumila var. *glandulifera*, *see B. glandulosa* var. *glandulifera*
Betulaceae, 384
Bilharzia, see Schistosoma
Biomphalaria, 384
B. sudanica (Martens), 350
Bivalvia [=Pelecypoda], 276, 278, 353, 385
Boletaceae, 383
Boletus, 383
B. scaber Fr., 239
Boraginaceae, 384
Bovidae, 385
Brachycentridae, 385
Brachytheciaceae, 383
Brassicaceae [=Cruciferae], 384
Bryales [=order of Musci], 324
Bryophyta, 290, 293, 383
Bubalus, 385
B. bubalis H. Smith (water buffalo), 353
Bubo, 385
B. virginianus (Gmelin) (great horned owl), 354

Caenidae, 385
Caenis, 385
C. horaria (L.), 369
Caesalpiniaceae, 384
Calamagrostis, 97, 183, 327, 384
C. epigejos (L.) Roth, 321, 328
Calliergidium, 228, 383
Calliergon, 228, 338, 383
C. stramineum (Brid.) Kindb., 95
Callitrichaceae, 383
Callitriche, 128, 383
Calluna, 384
C. vulgaris (L.) Hull *see* heather *in General Index*
Calothrix, 227, 383
Caltha, 384
C. palustris L., 273, 274
Calypogeiaceae, 383
Calypogeja, 82, 383
Campylium, 383
C. stellatum (Hedw.) Lange & C. Jens., 338
Caprifoliaceae, 384
Capromyidae, 385
Carabidae (ground beetles), 38, 358, 378, 385
Carabus, 385
C. granulatus L., 349
Carex (*see also* peat, types of *in General Index*), 13, 97, 165, 183, 234, 287, 288, 290, 293, 313, 325–328, 333–335, 383
C. acutiformis Ehrh., 336
C. aquatilis Wahlenb., 199, 200, 234, 288, 297, 300
C. arenaria L., 273
C. atherodes Spreng., 98
C. chordorrhiza Ehrh., 324, 325, 327
C. curta Good., 268
C. elata All., 288, 293, 328, 372
C. flacca Schreber, 275
C. lacustris Willd., 286, 288, 290, 293
C. lasiocarpa Ehrh., 313, 324–327, 330, 333, 334
C. limosa L., 95, 324
C. middendorfii F. Schmidt, 23, 24
C. nigra All., 11

Carex (*continued*)
C. panicea L., 11
C. paniculata L., 325, 326, 335
C. paupercula Michx., 327
C. prairea Dew., 326
C. rostrata J. Stokes, 87, 124, 180, 288, 293, 324, 326–328
C. runssoroensis K. Schum., 377
C. stans Drejer, 239
C. utriculata Boott, 327
C. vesicaria L., 328
Carnivora, 385
Carya (hickory), 384
C. aquatica (Michx. f.) Nutt., 262
Castoroides (giant beaver), 362, 385
Castor (beaver), 45, 258, 354–356, 385
C. fiber L., 352
Castoridae, 385
Caudata [= Urodela], 351, 385
Cecidomyiidae, 372
Celtis, 384
C. laevigata Willd., 262
Cephalanthus, 384
C. occidentalis L., 262, 264
Cephalopoda, 276, 278, 385
Cephalozia, 82, 383
Cephaloziaceae, 383
Cephaloziella, 82, 383
Cephaloziellaceae, 383
Ceratopogonidae, 370, 379, 385
Cercopidae, 385
Cervidae, 385
Cestoda, 278, 384
Cetraria, 229, 383
Chaetococcus, 385
C. phragmitidis (Marchal), 372
Chamaecyparis, 383
C. thyoides (L.) B.S.P., 335, 336
Chamaedaphne, 326, 331, 336, 384
C. calyculata (L.) Moench, 296
Chamaenerion, 384
C. angustifolium (L.) Scop., 322
Chaoborus, 385
C. crystallinus (Oegeer), 369
Chara, 325, 355, 383
Characeae, 383
Charadriiformes, 385
Charophyceae, 383
Chelonia, 385
Chenopodiaceae, 384
Chironomidae, 385, 369, 371–373, 379, 385
Chironomus, 385
C. brevipalpus Kieff., 371
C. tentans Fabricius, 368
Chlaenius, 385
C. costulatus Motschulsky, 361
C. tristis (Tristis), 361
Chlorococcaceae, 383
Chlorogloea, 227, 383
Chlorophyceae, 229, 383
Chlorophyta (green algae), 371, 383
Chloropidae, 372, 385
Chordata, 385
Chorisodontium, 228, 230, 383
C. aciphyllum (Hook. f. & Wils.) Broth., 160, 196, 207
Chromadora, 384
C. bioculata Schultze, 370
C. ratzeburgensis Linst, 370
Chromadoridae, 384
Chromadorina, 384
Chroococcaceae, 383
Chrysomelidae, 385
Ciconiidae, 358, 385
Ciconiiformes, 385
Cicuta, 384
C. virosa L., 373
Cladium (sawgrass), 97, 161, 171, 177, 218, 331, 383
C. jamaicense Grantz, 287
C. mariscus (L.) Pohl, 183, 239, 321, 335, 336, 367
Cladonia, 82, 229, 313, 314, 356, 383
Cladoniaceae, 383
Clethra, 384
C. alnifolia L., 300
Clethraceae, 384
Clostridium, 227, 250, 383
C. nigricans, 253
Coccidae, 372, 385
Cochlicopa, 385
C. lubrica (Mull.), 350
Coelodonta, 385
C. antiquitatis Blumb. (woolly rhinoceros), 361
Cognettia, 384
C. sphagnetorum Vedj., 203, 377, 378, 380
Coleoptera (beetles), 38, 336, 350, 358, 360, 361, 368, 369, 378, 385
Collema, 229, 383
Collemaceae, 383
Collembola (springtails), 354, 356, 358, 378, 379, 385
Colopha, 385
C. compressa Koch, 173
Colubridae, 385
Commelinidae [= subclass of Liliopsida (= Monocotyledones)], 279
Compositae, *see* Asteraceae
Coniferae, *see* Pinopsida
Conjugatae, 383
Convolvulaceae, 384
Coprosma, 26, 384
Cornaceae, 384
Cornus, 384
C. stolonifera Michx., 327
Corylaceae, 384
Corylus, 384
C. avellana L., 328
Coryxidae, 353, 385
Cossidae, 372, 385
Costaconvexa, 385
C. polygrammata Borkh., 361
Crataegus, 384
C. mollis (Torrey & H. Gray) Scheele, 262
C. monogyna Jacq., 321

Cratoneuron, 383
C. commutatum var. *falcatum* (Brid.) Mönk [= *C. falcatum*], 338
C. falcatum, *see C. commutatum* var. *falcatum*
C. filicinum (Hedw.) Spruce, 370
Cricetidae, 385
Cricotopus, 385
C. dizonius Meig., 371
Crocodylia (crocodiles), 278, 353, 385
Cruciferae, *see* Brassicaceae
Crustacea, 351, 353, 357, 368, 370, 385
Cryptostigmata [= Sarcoptiformes, suborder of Acarina], 378
Culex, 385
C. pipiens L., 368
Culicidae, 385
Culicinae [subfamily of Culicidae], 369
Cupressaceae, 383
Cyanobacteria [= Cyanophyta] (blue-green "algae"), 181, 227–229, 252, 383
Cyperaceae (sedges) (*see also* peat, types of *in General Index*) 58, 77, 78, 85, 94, 114, 239, 287, 290, 293, 295, 342, 383
Cyperus, 97, 295, 306, 383
C. exaltatus Retz, 25
C. papyrus L. (papyrus), 28, 242, 287, 295, 355
C. rotundus L., 25
Cyphon, 369, 385
Cyprinodontidae, 385
Cyrnus, 385
C. flavidus (McLachlan), 368, 369

Dactylis, 384
D. glomerata L., 274
Dactylorchis, *see Dactylorhiza*
Dactylorhiza [= *Dactylorchis*], 240, 384
D. lapponica (Laest ex Hartm.) Soō, 240
D. maculata (L.) Soō [= *Orchis maculata* L.], 240
Decodon, 384
D. verticillatus (L.) Ell., 326
Dendrobaena, 357, 384
Dennstaedtiaceae, 383
Deroceras, 385
D. agreste (L.), 356
D. laeve (Müll.), 356
Deschampsia, 384
D. cespitosa (L.) Beauv., 267, 272, 326
D. flexuosa (L.) Trin., 290
Desulfotomaculum, 242, 383
Desulfovibrio, 227, 242, 250, 383
D. desulfuricans (Beijerinck), 253
Deuteromycotina [= Fungi Imperfecti], 383
Diachila, 361, 385
D. arctica Gyll., 361
D. polita Fald., 361
Dicotyledones, *see* Magnoliopsida
Dicranaceae, 383
Dicranum, 327, 338, 383
Digenea, 384
Diospyros, 384
D. virginiana L., 262
Diplachne, 384
D. fusca (L.) Beaur., 371
Diptera (flies), 350, 351, 358, 368, 370, 377, 385
Dolomedes, 351, 385
Drepanocladus, 95, 228, 231, 324, 327, 333, 337, 383
D. aduncus (Hedw.) Warnst., 327, 370
D. fluitans (Hedw.) Warnst., 325
D. intermedius (Hedw.) Warnst., *see D. revolvens*
D. revolvens (Sm.) Warnst. [= *D. intermedius*], 325, 327
D. sendtneri (Schp.) Warnst., 327
D. vernicosus (Lindb.) Warnst., 327
Drosera (sundew), 228, 330, 384
D. anglica Huds., 23
D. longifolia L., 26
D. rotundifolia L., 11, 23, 86, 327
Droseraceae, 384
Dupontia, 384
D. fischeri R.Br., 234, 300
Dytiscidae, 385

Ebenaceae, 384
Eichhornia, 384
E. crassipes (Mart.) Solms, 25
Eleocharis (spike-rush), 383
E. palustris (L.) J.J. Roem & Schult., 288
E. plantaginea R.Br., 25
E. quadrangulata (Michx.) R. & S., 288
Elephantidae (elephants), 349, 385
Elephas, 353, 385
Empetraceae, 239, 384
Empetrum, 82, 93, 384
E. hermaphroditum (Lange) Hagerup, *see E. nigrum* subsp. *hermaphroditum*
E. nigrum L., 11, 12, 82, 83, 239, 261, 275
E. nigrum subsp. *hermaphroditum* (Hagerup) Böcher [= *E. hermaphroditum*], 12, 73
Empodisma, 384
E. lateriflora [= *Hypolaena lateriflora*], 102
Emydidae, 385
Emys, 385
E. orbicularis L. (European pond tortoise), 358
Enchytraeidae, 203, 357, 370, 377, 384
Endogonaceae, 383
Endogone, 240, 383
Enterophysalidaceae, 383
Epacridaceae, 77, 384
Ephemeroptera (may-flies), 358, 368, 385
Epilobium, 384
E. angustifolium L., 272
E. montanum L., 331
Epipactis, 240, 384
Equisetaceae, 383
Equisetum, 383
E. fluviatile L. [= *E. limosum*], 13, 180, 288, 324
E. limosum, *see E. fluviatile*
Erica, 232, 234, 235, 241, 384
E. cinerea L., 239, 271
E. mackaiana Bab., 239
E. tetralix L., 10–13, 20, 82, 83, 87, 89, 232–234, 239, 241, 264, 271, 275, 327
Ericaceae, 77, 83, 184, 218, 239, 290, 293, 323, 326, 384

Eriophorum (*see also General Index*), 383
E. angustifolium Honckeny, 11, 12, 85, 133, 138, 184, 199, 201, 205, 234, 267, 268, 300, 327
E. scheuchzeri Hoppe, 262, 263
E. vaginatum L., *see General Index*
Escherichia, 383
E. coli (Migula), 253
Eubacteria, 383
Eucalyptus, 239, 384
Euphorbiaceae, 384

Fabaceae [= part of Leguminosae: (Papilionaceae)], 384
Fagaceae, 384
Fagus, 47, 384
Festuca, 384
F. ovina L., 204
F. rubra L., 274
Filicopsida (ferns), 24, 25, 383
Filipendula, 384
F. ulmaria (L.) Maxim., 265, 267, 268, 272, 275, 321, 326, 328
Fimbristylis, 295, 383
Fischerella, 227, 383
Folsomia, 354, 384
Fontinalaceae, 383
Fontinalis, 128, 383
Frangula, 384
F. alnus Mill., 331, 335, 336
Frankia, 229, 383
Frankiaceae, 331, 383
Fraxinus, 384
F. excelsior L., 328, 336
F. mandshurica Rupr., 23
F. nigra H. Marsh., 288
F. pennsylvanica Marshall, 262
Friesia, 385
F. mirabilis Tullberg, 378
Fulica, 385
F. atra L. (coot), 355
Fungi (*see also General Index*), 383
Fungi Imperfecti, *see* Deuteromycotina
Fusarium, 336, 383

Galium, 384
G. palustre L., 275, 321
Galliformes, 385
Gammaridae, 385
Gammarus, 385
G. pulex (L.), 369, 373
G. tigrinus Sexton, 370
Ganoderma, 383
G. applanatum (Pers.) Pat., 336
Ganodermataceae, 383
Gastropoda, 384
Gastrotricha, 370, 384
Gaylussacia, 384
G. baccata (Wang.) K. Koch, 296
Genista, 384
G. tinctoria L., 228
Gentiana, 384
G. thunbergii (G. Don) Griseb. var. *minor* (Maxim.) Nakai, 24
Gentianaceae, 384
Gleditsia, 384
G. triacanthos L., 262
Gloeocapsa, 227, 383
Glossoscolecidae, 384
Glyceria, 287, 384
G. aquatica (L.) Wahlberg non Presl, *see G. maxima*
G. maxima (Hartm.) Holmberg [= *G. aquatica*] (reedgrass), 267, 268, 272, 273, 286, 288, 293, 334, 354, 368
Glycine, 384
G. max (L.) Merr. (soybean), 294
Glyptotendipes, 385
G. glaucus Meig., 371
Gramineae, *see* Poaceae
Grimmiaceae, 383
Gruiformes, 385
Gymnadenia, 240, 384
Gymnoschoenus, 42, 383
Gymnospermae, *see* Pinophyta

Habenaria,
H. hyperborea R.Br., *see Platanthera hyperborea*
Haloragaceae, 384
Hammarbya, *see Malaxis*
H. paludosa (L.) O. Kuntze, *see Malaxis paludosa*
Hapalosiphon, 227, 228, 383
Harpanthaceae, 383
Hedera, 384
H. helix L. (ivy), 47
Helodidae, 385
Helotiaceae, 383
Hemiptera (bugs), 350, 368, 369, 376, 385
Hepaticae (liverworts), 314, 383
Heterodonta, 385
Hieracium, 384
H. pilosella L., 272–274
Hippocastanaceae, 384
Hippopotamidae, 385
Hippopotamus, 355, 361, 385
Hirudinea (leeches), 368, 369, 384
Homalothecium, 383
H. nitens (Hedw.) Robins. [= *Tomenthypnum nitens*], 291
Hominidae, 385
Homo, 385
H. sapiens, *see* human bodies; man, influence of *in General Index*
Hyalopterus, 385
H. pruni (Geoff), 372
Hydaticus,
H. stagnalis Fab., 361
Hydrangea, 384
H. paniculata Sieb., 24
Hydrangeaceae, 384
Hydrilla, 383
H. verticillata Presl., 25
Hydrobius, 385
H. arcticus Kuw., 361
Hydrocharitaceae, 383

Hydrochoeridae, 385
Hydrochoerus, 385
  H. hydrochaeris L. (capybara), 354
Hydrophilidae, 385
Hydroporus, 385
  H. angustatus Sturm, 369
  H. lineatus (Fab.), 369
Hygrophila, 384
  H. auriculata (Schumach.) Heine, 25
Hylocomium, 327, 338, 383
Hypnaceae, 58, 383
Hypnum, 94, 114, 133, 177, 338, 383
  H. cupressiforme Hedw., 11, 12
  H. pratense Koch ex Spruce, 291
Hypogymnia, 383
  H. physodes (L.) Nyl. [– *Parmelia physodes*], 82
Hypolaena, 102, 384
  H. lateriflora Benth., *see Empodisma lateriflora*

Ibis, 353, 385
Insecta (insects), 276–278, 336, 385
Insectivora, 385
Ipomoea, 384
  I. aquatica Forsk., 25
Iridaceae, 383
Iris, 278, 325, 383
  I. pseudacorus L., 273, 288, 333
Isopoda, 236, 357, 385
Isotomidae, 385

Jacana (lilytrotter, lotus bird), 351, 385
Jacanidae, 385
Juglandaceae, 384
Juncaceae, 239, 287, 383
Juncaginaceae, 279, 383
Juncus, 97, 287, 288, 383
  J. effusus L., 202, 232, 267, 268, 272, 273, 275, 286, 288, 325
  J. planifolius R.Br., 240
  J. squarrosus L., 206, 231, 234, 235, 239, 325, 376, 377, 379, 380
  J. subnodulosus Schrank, 180, 367
Jungermannineae [suborder of Hepaticae], 77
Juniperus (juniper), 38, 383

Kalmia, 326, 331, 384
  K. angustifolia L., 296
Kincaidiana, 384
  K. freidris (Cook), 370
Klebsiella, 227, 383
Kobresia, 239, 383
Kyllinga, 295, 383

Labiatae, *see* Lamiaceae
Lacerta, 385
  L. vivipara Jacquin, 349
Lacertidae, 385
Lactuca, 384
  L. sativa L. (lettuce), 274
Laelia, 385
  L. coenosa (Hübn.), 361
Lagomorpha, 385
Lagopus, 385
  L. lagopus L. (willow grouse), 355
  L. lagopus var. *scoticus* (Lath.) (red grouse), 355, 356, 376
Lamiaceae [= Labiatae], 384
Laridae, 385
Larix (larch), 39, 324, 383
  L. decidua Miller, 257
  L. laricina (DuRoi) K. Koch, 326, 327, 330, 335, 336
Lasiocampa, 385
  L. quercus (L.) var. *callunae* (northern eggar), 376
Lasiocampidae, 385
Lathyrus, 384
  L. palustris L., 228
Leccinum, 383
  L. scabrum (Fries) S.F. Gray, 239
Lecideaceae, 383
Ledum, 326, 327, 331, 384
  L. groenlandicum Oeder, 271, 296
  L. palustre L., 13, 82
Leguminosae, *see* Caesalpiniaceae; Fabaceae; Mimosaceae
Lemmus (lemming), 376, 385
Lemnaceae, 383
Lepidoptera, 352, 372, 385
Lepidosireniformes, 385
Lepidotis, 383
  L. inundata (L.) C. Börner [= *Lycopodium inundatum*], 23
Leporidae, 385
Leptogium, 229, 383
Lepus, 385
  L. timidus L. (blue hare), 255
Leucania, 352, 385
Leucobryum, 241, 383
  L. glaucum (Hedw.) Schimp., 236
Lichenes (*see also* lichens *in General Index*), 383
Liliaceae, 239, 383
Liliopsida [= Monocotyledones], 262, 267, 278, 279, 287, 383
Limacidae, 385
Limnotragus, subgenus of *Tragelaphus*, q.v.
Limosa, 385
  L. limosa L. (black-tailed godwit), 352
Liniphyidae, 376, 385
Lipara, 385
  L. lucens (Meig.), 372, 379
Liparis, 384
  L. loeselii (L.) L.C.M. Richard, 239
Liriodendron, 384
  L. tulipifera L. (tulip tree), 261
Listera, 240, 384
  L. cordata (L.) R.Br., 239, 240
Lobaria, 229, 383
Lochmaea, 385
  L. suturalis Thoms. (heather beetle), 376, 379
Locusta, 385
  L. migratoria (R. & F.) (migratory locust), 359
Lolium, 384
  L. perenne L., 274
Lotus, 384
  L. corniculatus L., 261

Lotus (*continued*)
  L. pedunculatus Cav., 228
  L. uliginosus Schkuhr, 261
Lumbricidae, 350, 356, 357, 377, 384
Lumbriculidae, 370, 384
Lumbriculus, 384
  L. variegatus (Müller), 369
Lupinus, 384
  L. angustifolius L., 273
Lycaena, 385
  L. dispar L. (large copper butterfly), 352, 361
Lycaenidae, 385
Lycopersicum, 384
  L. esculentum Mill.(tomato), 278
Lycopodiaceae, 383
Lycopodium, 383
  L. inundatum L., *see Lepidotis inundata*
Lycopsida, 383
Lycopus, 384
  L. europaeus L., 268
Lycosa, 385
  L. pullata (Clerck), 376
Lycosidae, 376, 378, 385
Lymantriidae, 385
Lymnaea, 384
  L. ovata (Drap.), 379
  L. peregra L., 373
  L. stagnalis (L.), 379
  L. truncatula (Müll.), 350
Lymnaeidae, 384
Lythraceae, 384
Lythrum, 384
  L. salicaria L., 267, 268, 272

Macrothylacia, 385
  M. rubi (L.) (fox moth), 356
Magnoliaceae, 384
Magnoliophyta [= Angiospermae], 383
Magnoliopsida [= Dicotyledones], 267, 384
Malaxis, 384
  M. monophyllos (L.) Swarz, *see Microstylis monophyllos*
  M. paludosa (L.) Sw. [= *Hammarbya paludosa*], 239, 240
Malus, 24, 384
Mammalia (mammals) (*see also General Index*), 280, 385
Mammuthus, 385
  M. primigenius (Blumenbach) (hairy mammoth), 361, 362
Mayacaceae, 279, 384
Meesiaceae, 383
Mentha, 384
  M. aquatica L., 264, 267–269, 273, 274
Menyanthaceae, 239, 384
Menyanthes, 325, 326, 333, 384
  M. trifoliata L., 95, 163, 180, 239, 324
Mesenchytraeus, 370, 384
Metrosideros, 384
  M. polymorpha Gaudich., 26
Microcyprini, 385
Microstylis, 384
  M. monophyllos (L.) Lindley [= *Malaxis monophyllos*], 240
Microtus, 376, 385
  M. subgenus *Pedomys*, 376
  M. oeconomus (Pallas), 361
Molinia, 10, 136, 137, 230, 234, 328, 330, 331, 356, 363, 378, 384
  M. caerulea (L.) Moench. (purple moorgrass), 11, 12, 20, 23, 184, 229, 239, 288, 291, 293, 321, 328, 356
Moliniopsis, 384
  M. japonica Hayata, 23, 24
Mollusca, 276, 278, 280, 350, 355, 369, 384
Molophilus, 385
  M. ater Meigen, 203, 377, 379
Monera [= Prokaryota], 383
Monochoria, 384
  M. vaginalis Presll, 25
Monocotyledones, *see* Liliopsida
Mortierella, 241, 383
Mortierellaceae, 383
Motacillidae, 385
Mougeotiaceae, 383
Muridae, 385
Mus, 385
  M. musculus L. (house mouse), 368
Musci (*see also* mosses *in General Index*), 383
Mustela, 385
  M. vison Schreber (mink), 354
Mustelidae, 385
Mylia, 82, 383
Myocastor, 385
  M. coypus (Molina) (coypu), 354, 372, 373
Myosotis, 384
  M. palustris (L.) Hill, *see M. scorpioides*
  M. scorpioides L. [= *M. palustris*], 267, 273
Myrica, 136, 137, 218, 229, 326, 384
  M. gale L., 11, 181, 229, 321, 336
  M. gale var. *tomentosa* C.DC., 24
Myricaceae, 384
Myriophyllum, 325, 355, 384
Myrtaceae, 384
Myxophyceae (blue-green "algae"), 181, 227–229, 252, 383

Nardus, 384
  N. stricta L., 239, 265, 267–269
Narthecium, 10, 11, 230, 384
  N. asiaticum Maxim., 23
  N. ossifragum (L.) Hudson (bog asphodel), 11, 12, 72, 85, 164, 184, 205, 206, 234, 239, 261, 267, 268
Natrix, 385
  N. natrix (L.) (grass snake), 358
Naucleaceae, 384
Nectria, 383
  N. cinnabarina (Tede ex Fries) Fries, 336
Nectriaceae, 383
Nematoda, 278, 370, 378, 379, 384
Nemopanthus, 326, 384
Neomys, 385
  N. fodiens (Pennant) (water shrew), 358
Neophilaenus, 385
  N. lineatus (L.), 356, 376, 379
Nephroma, 229, 383
Nephromaceae, 383

Neritidae, 384
Noctuidae, 385
Nomadacris, 385
  N. septemfasciata (Serville) (red locust), 359
Nostoc, 228, 383
  N. muscorum Kuetz, 228
Nostocaceae, 383
Nothobranchius, 351, 385
Notonectidae, 350, 385
Notopteridae, 385
Notostraca, 351, 385
Nuphar, 325, 384
  N. lutea (L.) Sibth. & Sm., 239, 373
  N. nelumbo Gaertn., 274, 276
Nymphaea, 239, 384
Nymphaeaceae, 239, 384
Nymphaeales [order of Magnoliopsida], 278
Nyssa (tupelo gum), 41, 384
  N. aquatica L., 262, 264, 300
  N. sylvatica Marsh., 261
  N. sylvatica var. *biflora* (Watt.) Sarg., 273
Nyssaceae, 384

Odontoschisma, 82, 383
Odontoschismaceae, 383
Oleaceae, 384
Oligochaeta, 370, 373, 377, 379, 384
Omphalina, 241, 385
Onagraceae, 384
Ondatra (muskrat), 355, 372, 385
  O. zibethicus (L.), 354
Onychiuridae, 385
Onychiurus, 354, 385
Oodes, 385
  O. gracilis (Villa), 361
Opisthopora, 384
Orchidaceae, 239, 384
Orchis, 240, 384
  O. ericetorum E.F. Linton, 239
  O. fuchsii Druce, 239
Oreobolus, 383
  O. furcatus H. Mann, 26
Orthoptera, 355, 385
Oryza, 384
  O. sativa L. *see* rice *in General Index*
Osmunda, 383
  O. cinnamomea L., 23, 24
Osmundaceae, 383
Ostracoda [subclass of Crustacea], 368, 369
Osteoglossiformes, 385
Ovis, 385
  O. ammon aries L. (sheep), 374, 375
Oxycoccus, subgenus of *Vaccinium*, q.v.
  O. microcarpus Turcz. ex Rupr., *see Vaccinium microcarpum*
  O. palustris Pers., *see Vaccinium oxycoccos*
  O. quadripetalus Br.-Bl., *see Vaccinium oxycoccos*

Paludella, 383
  P. squarrosa (Hedw.) Brid., 324
Pandanaceae, 384
Pandanus, 384
  P. helicopus (Kurz), 97
Panicum, 26, 384
  P. hemitomon Schultes, 288
Papilio, 385
  P. machaon L. (swallow-tail butterfly), 352, 362
Papilionaceae, *see* Fabaceae
Papilionidae, 385
Papyrocranus, 368, 385
Parmelia, 383
  P. physodes (L.) Ach., *see Hypogymnia physodes*
Parmeliaceae, 383
Paspalum, 384
  P. scrobiculatum L., 25
Passeriformes, 385
Pedomys, subgenus of *Microtus*, q.v.
Pelecanidae, 385
Pelecaniformes, 385
Pelecanus (pelican), 353, 385
Pelecypoda, *see* Bivalvia
Peltigera, 229, 383
Peltigeraceae, 383
Pericoma, 385
  P. nubila (Meigen), 369
Perissodactyla, 385
Perrisia, 385
  P. inclusa (Frauengeld), 372
Pezizella, 383
  P. ericae Read, 239
Phalaris, 97, 384
  P. arundinacea L., 265, 267, 272, 273, 288, 328, 331, 355
Philaenus, 385
  P. spumarius (L.), 356, 376
Philomachus, 385
  P. pugnax L. (ruff), 352
Phleum, 231, 384
  P. pratense L., 88, 204
Phoenicopteriformes, 385
Phoenicopteridae, 385
Phoenicopterus (flamingo), 353, 385
Phragmatoecia, 385
  P. castaneae (Hb),
Phragmites (*see also General Index*), 384
  P. australis (Cav.) Trin. ex. Steud. [= *P. communis*], *see General Index*
  P. communis Trin., *see P. australis*
Picea (spruce), 39, 93, 324, 327, 331, 383
  P. abies (L.) Karts. [= *P. excelsa*], 87
  P. excelsa, *see P. abies*
  P. mariana (P. Mill.) B.S.P., 93, 140, 258, 326, 327, 331, 335–337
  P. sitchensis (Bong.) Carr., 254, 258
Pinaceae, 239, 383
Pinophyta [= Gymnospermae], 383
Pinopsida [Coniferae], 290, 293, 383
Pinus (pine), 58, 93, 114, 134, 161, 203, 218, 239, 240, 269, 313, 324, 325, 327, 342, 383
  P. contorta Dougl. ex Loud., 258, 259, 263, 265, 269, 335
  P. mugo Turra, 132
  P. mugo subsp. *rotundata* (Link) Hoopes, 87

Pinus (*continued*)
P. strobus L., 177
P. sylvestris L., 86, 87, 96, 114, 327, 335–337
Pisauridae, 385
Pisces (fishes), 181, 351, 353, 358, 368, 370, 385
Pisidium, 353, 385
Pisum, 384
P. sativum L. (pea), 273, 274
Planorbarius, 384
P. corneus (L.), 379
Planorbidae, 384
Planorbis, 385
P. albus Müll., 369
P. planorbis (L.) 379
Platalea, 385
P. leucorodia L., 358
Plataleidae, 358, 385
Platanthera, 384
P. hyperborea Lind. [= *Habenaria hyperborea* R.Br.], 240
Platyhelminthes, 384
Plecoptera (stoneflies), 358, 385
Plesiopora Prosotheca, 384
Pleuroziaceae, 383
Pleurozium, 327, 383
P. schreberi (Brd.) Mitt., 291
Poa, 384
P. colensoi Hook. f., 240
Poaceae [= Gramineae] (grasses), 11, 239, 265, 287, 384
Polycentropidae, 385
Polygonaceae, 384
Polypteridae, 385
Polypteriformes, 385
Polypterus, 368, 385
Polytrichaceae, 383
Polytrichum, 96, 228, 327, 383
P. alpestre Hoppe, 160, 196, 215, 291
P. alpinum Hedw., 297
P. commune L. ex Hedw., 196, 325, 327
P. juniperinum Willd. ex Hedw., 291
P. strictum Sm., 57
Pontederia, 384
P. cordata L., 288
Pontederiaceae, 384
Populus, 384
P. deltoides Marsh., 262
P. tremula L., 331, 355
P. tremuloides Michx., 327, 336
Potaliaceae, 384
Potamogeton, 128, 325, 355, 384
Potamogetonaceae, 279, 384
Potentilla, 384
P. anserina L. 272
P. erecta (L.) Räuschel, 11, 12
P. palustris (L.) Scop., 267, 268, 326
Pottiaceae, 383
Povilla, 385
P. adusta Navas, 368
Primates, 385
Prionocera, 385
P. turcica (Fabr.), 369
Pristiphora, 385
P. erichsonii (Hartig), 336
Proboscidea, 385
Prokaryota, *see* Monera
Prosopora, 384
Protopteridae, 385
Protopterus, 368, 385
Protozoa, 250, 278, 351, 353, 370, 384
Prunus, 24, 384
P. armenaiaca L. (apricot), 260
P. domestica L. (plum), 260
P. grayana Maxim., 24
P. persica (L.) Batsch (peach), 260
Pseudotsuga (Douglas fir), 383
P. menziessii (Mirb.) Franco, 327
Psychodidae, 385
Psyllidae, 385
Pteridium, 231, 383
P. aquilinum (L.) Kuhn, 198
P. aquilinum var.*japonicum* Nakai, *see P. aquilinum* var. *latiusculum*
P. aquilinum var. *latiusculum* (Desv.) Underw. [= *P. aquilinum* var. *japonicum*], 24
Pteridophyta, 383
Ptychoptera, 385
P. albimana (Fabr.), 369
Ptychopteridae, 385
Pulmonata [subclass of Gastropoda], 350
Pyralidae, 372, 385
Pyrus, see Aronia

Quercus (oak), 47, 177, 384
Q. falcata, Michx., 262
Q. lyrata Walt., 261, 262
Q. macrocarpa Michx., 262
Q. nigra L., 262
Q. nuttalli E.J. Palmer, 261, 262
Q. palustris Muench L., 262
Q. petraea (Matt.), Liebl., 328
Q. phellos L., 262
Q. robur L., 336

Racomitrium, 160, 325, 383
R. lanuginosum (Hedw.) Brid., 26
Rallidae, 385
Rana, 385
R. temporaria L., 351, 358
Rangifer (caribou), 375, 385
Ranidae, 353, 385
Rantus, 385
R. aberratus (Gemm. & Har.), 361
Ranunculaceae, 384
Ranunculus, 384
R. flammula L., 272–274
R. lingua L., 267, 268, 333
Raphanus, 384
R. sativus L. (radish), 278
Reptilia, 385
Restionaceae, 102, 384
Rhacomitrium, see Racomitrium

Rhamnaceae, 384
Rhamnus, 384
  R. catharticus L., 331, 335
Rhinocerotidae, 385
Rhizobium, 228, 383
Rhizoctonia, 240, 383
Rhizophora, 264, 384
  R. mangle L., 263, 264
Rhizophoraceae, 384
Rhizopoda [=Sarcodina], 340, 384
Rhododendron, 24, 384
Rhodophyceae, 128, 383
Rhodophyta (red algae), 383
Rhodospirillum, 227, 383
Rhus, 24, 384
Rhynchelmis, 384
  R. limosella Hoffmeister, 369
Rhynchops (skimmers), 358, 385
Rhynchospora, 26, 383
  R. alba (L.) Vahl., 23, 24, 57, 85, 95, 184
Rhynchota, *see* Hemiptera
Rhytidiadelphus, 383
  R. squarrosus (Hedw.) Warnst., 325
Ricinus, 299, 384
Rivulariaceae, 383
Rodentia, 354, 385
Rosaceae, 239, 384
Rostkovia, 383
  R. magellanica (Lam.) Hook., 202
Rotifera, 370, 384
Rubiaceae (*see also* Naucleaceae), 384
Rubus, 230, 384
  R. chamaemorus L., 12, 183, 198, 200, 201, 204–207, 211, 231, 239, 296
Rumex, 384
  R. hydrolapathum Huds., 137, 373
Ruprechtia, 300, 384

Saldidae, 350, 385
Salicaceae, 384
Salientia [=Anura] (*see also* frogs *in General Index*), 351, 385
Salix (willow), 47, 161, 264, 270, 278, 290, 293, 324, 325, 327, 355, 372, 384
  S. atrocinerea Brot., 321, 336
  S. cinerea L., 326, 328, 331
  S. interior Rowlee, 262
  S. planifolia Pursh, 290
  S. purpurea L., 328
  S. repens L., 275, 336
Sambucaceae, 384
Sambucus, 384
  S. nigra L., 331
Sanguisorba, 384
  S. officinalis L., 24
Sarcodina, *see* Rhizopoda
Sarcoptiformes [=Cryptostigmata; suborder of Acarina], 378
Sarracenia, 228, 384
  S. purpurea L. 296
Sarraceniaceae, 384
Saturnia, 385
  S. pavonia (emperor moth), 356
Saturniidae, 385
Sauria, *see* Lacertilia
Saururaceae, 384
Saururus, 384
  S. cernuus L., 288, 290, 293
Scherotheca, 384
  S. savignyi, 350
Scheuchzeria, 133, 313, 384
  S. palustris L., 19, 24, 137, 262, 313, 324
Scheuchzeriaceae, 279, 384
Schistosoma [=*Bilharzia*], 350, 384
Schistosomatidae, 384
Schizaea, 383
  S. robusta Bak., 26
Schizaeaceae, 383
Schoenobius, 385
  S. gigantellus (Schiff), 372
Schoenoplectus, section of *Scirpus*, q.v.,
  S. lacustris (L.) Palla, *see Scirpus lacustris*
Schoenus, 378, 379, 383
  S. ferrugineus L., 180
  S. nigricans L. (black bog-rush), 40, 239, 261, 335, 356
Sciomyzidae, 358, 385
Scirpus [=*Trichophorum*; includes Sections *Baeothryon* and *Schoenoplectus*], 97, 218, 287, 288, 295, 333, 334, 378, 383
  Section *Schoenoplectus*, 97
  S. acutus Muhl., 98, 327
  S. americanus Pers. 288
  S. cespitosus L. [=*Baeothryon caespitosum*], 10–12, 20, 78, 95, 136, 206, 218, 234, 267, 268
  S. fluviatilis (Torr.) Gray, 288
  S. lacustris L. [=*Schoenoplectus lacustris*], 23, 288, 293, 325, 328, 367, 370
  S. maritimus L., 23
  S. validus Vahl., 288, 327
Scleria, 295, 383
Scolochloa, 384
  S. festucaceae (Willd.) Link, 98
Scorpidium, 383
  S. scorpioides (Hedw.) Limpr., 180, 324, 325
Scrophulariaceae, 384
Scytonema, 227, 383
Scytonemataceae, 383
Segmentina, 384
  S. nitida (Müll.), 368
Selaginella, 383
  S. deflexa, Brack., 26
Selaginellaceae, 383
Senecio, 384
  S. aquaticus Hill, 272–274, 277
  S. erucifolius L., 272
  S. jacobaea L. 273, 274, 277
  S. squalidus L., 272, 273
  S. sylvaticus L., 273
  S. viscosus L., 272
  S. vulgaris L., 273
Sigara, 368, 385
  S. dorsalia Leach, 369

Sigara (*continued*)
S. falleni (Fieb.), 369
Solanaceae, 384
Solanum, 384
S. dulcamara L., 288
S. nigrum L., 331
Solidago, 322, 384
Solorina, 229, 383
Sorbus, 384
S. commixta Hedl., 24
Sorex, 385
S. minutus L. (pigmy shrew), 358
Soricidae, 385
Sparganiaceae, 279, 384
Sparganium 384
S. emersum Rehmann 293
S. erectum L., 23
Spartina, 295, 384
S. townsendii H. & J. Groves, 267, 268
Sphaeriidae, 385
Sphagnaceae, 383
Sphagnum (bog-moss) (*see also in General Index*), 383
Section *Acutifolia*, 81, 89, 160, 167, 168
Section *Cuspidata*, 137, 160, 194, 324
Section *Cymbifolia*, 89
Section *Palustria*, 81
Section *Sphagnum*, 160, 167, 168
S. acutifolium Ehrh., *see S. capillifolium*
S. auriculatum Schimp. var. *auriculatum* [= *S. obesum*], 78
S. balticum (Russ.) C. Jens., 13, 94, 95
S. capillaceum (Weiss) Schrank, *see S. capillifolium*
S. capillifolium (Ehrh.) Hedw. [= *S acutifolium* = *S. capillaceum* = *S. rubellum*], *see General Index*
S. compactum DC., 23, 95
S. cuspidatum Ehrh. ex Hoffm. emend., *see General Index*
S. dusenii Warnst., *see S. majus*
S. fuscum (Schimp.) Klinggr., *see General Index*
S. imbricatum Hornsch. ex Russ., 13, 40, 161, 174, 175, 340
S. magellanicum Brid., *see General Index*
S. majus (Russ.) C. Jens [= *S. dusenii*], 94
S. molle Sull., 78
S. obesum, *see S. auriculatum* var. *auriculatum*
S. palustre L., 23, 81, 236, 237, 328
S. papillosum Lindb., *see General Index*
S. parvifolium Warnst., *see S. recurvum* var. *tenue*
S. recurvum P. Beauv., *see General Index*
S. recurvum var.*tenue* Klinggr. [= *S. parvifolium*], 13, 95
S. rubellum Wils., *see S. capillifolium*
S. subsecundum Nees, 197, 324
S. tenellum (Brid.) Brid. 174
S. teres (Schp.) Ångstr. 58, 324, 325
Sphenopsida, 383
Spiranthes, 240, 384
S. romanzoffiana Cham., 239
Spirodela, 25, 383
Sporadanthus, 384
S. traversii (F. Muell.) Kirk, 102
Squamata, 385
Stereocaulaceae, 383
Stereocaulon, 229, 383
Stethophyma, 385
S. grassum (L.), 356
Sticta, 229, 383
Stictaceae, 383
Stigonemataceae, 383
Stratiomyiidae, 351, 385
Stratiotes, 383
S. aloides L., 367
Strigidae, 385
Strigiformes, 385
Strophingia, 378, 385
S. ericae (Curt.), 376, 379
Stylommatophora, 384
Styphelia, 26, 384
Suaeda 295, 384
Sylviidae, 385
Symmeria, 384
S. paniculata Benth., 300
Synaptomys (bog lemming), 376, 385

Tabanidae, 358, 385
Taxodiaceae, 383
Taxodium (bald cypress), 41, 383
T. distichum (L.) Richards 261–264, 288, 300
Tenthredinidae, 385
Tenthredo,
T. rossi Panz., 361
Tetracanthella, 385
T. britannica Cassagnau, 356
T. wahlgreni Axelson, 356, 378
Tetraonidae, 385
Tetrigidae (ground hoppers), 355, 385
Tetrix, 385
T. subulata (L.), 355
Theobaldia, 385
T. annulata (Schrank), 368
Thiobacillus, 242, 383
T. ferrooxidans (Temple & Colmer), 253
T. thiooxidans (Waksman & Joff), 253
T. thioparus (Beijerinck), 253
Threskiornithidae, 385
Thuja (cedar) 324, 331, 383
T. occidentalis L., 288, 331, 335, 336
T. plicata Donn ex. D. Don, 264, 327
Tilia (basswood), 384
T. japonica (Miq.) Simonk., 23
Tiliaceae, 384
Tipula, 385
T. subnodicornis Zetterstedt, 203, 376–378
Tipulidae (craneflies), 203, 375–378, 385
Tolypothrix, 227, 383
Tomenthypnum, 383
T. nitens (Hedw.) Crome, *see Homalothecium nitens*
Tortula, 383
T. robusta Hook & Grev., 291
Trachea, *385*
T. atriplicis L., 361
Tragelaphus, 385
T. spekii Sclater (sitatunga), 350
Trebouxia, 229, 383

Trematoda, 384
Trichophorum, *see Scirpus*
 T. caespitosum (L.) Hartman, *see Scirpus cespitosus*
Trichoptera (caddis flies), 357, 358, 368, 385
Trifolium, 384
 T. fragiferum L., 269
 T. subterraneum L., 264, 273
Tsuga (hemlock), 383
 T. heterophylla (Raf.) Sarg., 264, 327
Tuberculariaceae, 383
Tullbergia, 354, 385
Typha (reed mace), 384
 T. angustata Bory & Chaub., 25
 T. angustifolia L., 23, 97, 289, 293, 326, 354, 370, 371, 373
 T. australis Schumach. [= *T. domingensis*], 288, 370
 T. domingensis Pers., *see T. australis*
 T. latifolia L., 23, 288, 290, 293, 295, 326–328, 335, 354, 367
Typhaceae, 279, 287, 384

Ulmaceae, 384
Ulmus, 384
 U. americana L., 264
 U. davidiana Planch, 23
Umbelliferae, *see* Apiaceae
Urodela, *see* Caudata
Urtica, 384
 U. dioica L., 321, 322, 331
Urticaceae, 384

Vaccinium [including subgenus *Oxycoccus*], 26, 82, 83, 384
 V. canadense Kalm., 331
 V. microcarpum (Turcz. & Rupr.) Schmalh. [= *Oxycoccus microcarpus*], 12
 V. myrtilloides Michx., 296
 V. myrtillus L., 12, 73, 82, 83, 87, 239
 V. oxycoccos L. [= *Oxycoccus quadripetalus* – *O. palustris*], 12, 23, 82, 83, 89, 239, 327
 V. uliginosum L., 13, 78, 82, 83, 86
 V. vitis-idaea L., 82, 239
Veronica, 384
 V. peregrina L., 275, 277
Vertiginidae, 385
Vertigo, 385
 V. genesii (Gredder), 361
Viburnum, 384
 V. opulus L., 336
Vicia, 384
 V. faba L. (broad bean), 273, 274

Xylopia, 384
 X. staudtii Engl. & Diels, 263

Zea, 384
 Z. mays L. (maize, corn), 203, 266, 267, 272–274, 289, 293, 302, 304
Zizania, 97, 384
Zonitidae, 385
Zosteraceae (eel grasses), 279, 384
Zygogonium, 314, 383
Zygomycotina, 383

GENERAL INDEX[1]

Aalborg (Denmark), 20
aapa fens, 6, 7, 14, 19, 28, 41, 43, 73, 99, 138, 339
– –, definition of, 27
– –: transition to raised bogs in Finland, 52, 53
Abisko (Sweden), 71, 164, 183, 190, 197, 202, 203, 214
abscission as source of error in productivity estimation, 286, 287
Acadia Forest (N.B., Canada), 275
accumulation of peat, *see* peat accumulation
accumulative wetlands (*see also* peat-forming systems), 2
acetaldehyde, 263, 272–274, 280
– and stem ventilation, 270
– in peat, 250
acetic acid, 250, 278
acetoin, 278
acetone in peat, 250
acetyl methyl carbinol in peat, 250
acetylene formation in mires, 250
– reduction as index of nitrogen fixation activity, 252
acid rain, 181
acidification by mosses, 337, 338
acidity, *see* bogs, pH in; fens, pH of; mires, pH in; peat pH; soil pH; water, pH of
acidophilic plants, 47
acrotelm (*see also* active layer), 17, 103, 104, 134, 138, 142, 144, 146, 171, 177, 189, 210–214, 340
– and rising water-table, 124, 137
–, bulk density of, 133
–, definition of, 28, 82, 104, 189
–, hydraulic conductivity of, 143
–, mosses in, 134
–, permeability in, 126–131, 145
–, seepage in, 119, 123
–, water flow in, 148, 149
–, – storage in, 111, 115, 117, 145, 147
activated carbon from peat, 2, 19
active layer (*see also* acrotelm), 103, 180, 189
adaptation of aquatic plants, 279
– – mire animals, 349–353
– – – plants, 271–280
adaptive enzymes, 277
adenosine phosphates [ADP, AMP, ATP], 266, 277
ADH [alcohol dehydrogenase], 272, 273
Adriatic Coast (Italy), 20

[1] Where the scientific name of a group has been directly anglicized — e.g. "Carabidae"→"carabids" — entries will usually be found in the Systematic rather than the General Index.

advection of heat, 77, 98
aeration of plant roots, 8
–, soil, *see* soil, aeration of
aerenchyma in roots, 265, 267
aerial photography, 6, 7, 138, 139, 328, 329, 342
aerobic conditions in decomposition processes, 1, 18, 182, 183, 185, 200–207, 225
– organisms in peat, 235
– peat, *see* acrotelm
aerosols: absorption by plant communities, 54
Africa (*see also* Central Africa; East Africa; West Africa; *and particular territories and localities*), 2, 276, 287
–, southern, 72, 102, 353
Agassiz Lake Peatlands (U.S.A.), 319, 320, 341
Ageröds Mosse (Sweden), 318
agriculture affecting mire development, 59, 261, 343
–, clearing for, 47
–: flooding effects, 257, 271
– on former peatlands, 2, 18, 19, 251
air, patterns seen from (*see also* aerial photography), 11
Akaiyachi Moor (Japan), 23
Åkhult Mire (Sweden), 163
aktivnyĭ sloĭ, 104
alanine, 276–279
Alaska (U.S.A.) (*see also particular localities*)
–, climatic change in, 59
–, mires in, 7, 28, 338
–, – –, leaf decay in, 201
–, – –, nutrient relations of, 187, 230, 242
–, – –, primary production in, 290, 294, 295, 297, 301, 302
–, – –, standing crop of, 293
–, – –, water relations in, 85, 86, 300
albedo, 74, 79, 94, 98
Alberta (Canada) (*see also particular localities*), 288, 289, 291, 327
alcohol (*see also* ethanol)
– dehydrogenase [ADH], 272–274
– –, inductive responses of, 272, 273
– –, Michaelis constant of, 273, 274
– formation in soil, 250
alder (*see also Alnus in Systematic Index*)
– swamps, 12
Alderfen Broad (England), 368, 371, 374
Alfisols, 217
algae (*see also* blue-green algae)
– as symbionts in lichens, 229
–: consumption by animals, 356, 370, 371
– forming gelatinous surface layer of low permeability, 136
–: growth affected by peat extracts, 176
– in Antarctic moss communities, 228
– in bog hollows and pools, 17, 77, 314, 341

algae (*continued*)

- in lakes, 374
- in periphyton, 370, 371
algal cover on flarks, 27
- hollows in bogs, 17
Algeria, 353
Allerød Interstadial, 38
allochthonous peat, 28
allogenic change, 28, 44, 312, 313, 319, 340, 341, 343
Alnion glutinosae, 22
Alno-Padion, 22
alpine meadows, 290
Alps, European, 21, 22, 39, 109, 132, 141, 143, 360
Alston (England), 84
aluminium, 170, 180, 185–187, 192–195, 249, 252, 261
Amazon Basin, 2
- River, 264, 354
Amazonia, 300
America (*see also* Central America; North America; South America; *and particular territories and localities*), 354
amides in peat, 173
amino acid(s), 280, 371, 376
-- excretion by roots, 241
γ-amino butyric acid, 278
ammonia liberation from peat soils, 250, 251
ammonification, 226, 242
ammonium form of nitrogen in soil, 181, 187, 188, 191, 238, 241, 242
anaerobic conditions in decomposition processes, 1, 103, 182, 185, 189–191, 201, 225, 247–254
--, tolerance of (*see also* anoxia), 260
- organisms in peat, 235
anaerobic peat, *see* catotelm
anaerobiosis (*see also* anoxia), 272–280, 311, 356
Andøya (Norway), 160, 167
animal(s) (*see also particular groups and species*), 1, 349–350
-, anaerobic metabolism in, 276–278
-, densities of, 373, 375–379
-: effects on peat profiles, 185
- in mires, 349–380
- production, *see* productivity, secondary
-: role in decomposition, 198, 203–205, 231, 232
anoxia (*see also* anaerobiosis; oxygen), 258, 260, 276, 277, 280, 368
-, adaptation to, 276–280
- and flooding tolerance, 271–280
Antarctic, 215, 228, 291
- moss-banks, 196, 202, 207, 213
- peat, 168, 188, 199
- Peninsula, 25
- snow, lead in, 180
anthropogenic disturbances, *see* man, influence of
antibiotics to distinguish decomposition by fungi and bacteria, 373
anti-transpirants increasing resistance to flooding, 271
Antrim (Northern Ireland), 48
apatite, 252
apricot (*Prunus armeniaca*), 260
aquatic plant communities, 44
aquiclude, 71, 72
aquifer yield, *see* specific yield
arabinose in peat, 173, 175
arbuscular mycorrhizae, 239, 240
arbutoid mycorrhizae, 239, 240
Arctic bogs, 6, 7
- snow, lead in, 180
- Zone, 2, 28, 45
Ardennes (Belgium), 20, 87
Argentina (*see also particular localities*), 25
Aridisols, 217
aridity index, 109, 110
Arisaig, Sound of (Scotland), 139
Arizona (U.S.A.), 37
artesian water, 72
ash content of peat, *see* peat, ash content of
- from fires affecting mire development, 54, 322
Asia, 136
aspartic acid, 278, 279
aspect, *see* slope aspect
asphalt, 173
asphodel, bog, see *Narthecium ossifragum in Systematic Index*
Athlone (Ireland), 10, 13
Atlantic flora, 22
- Ocean (*see also particular islands*), 38
- period, 52, 54, 58
- raised bogs, *see* raised bogs, types of: Atlantic
Atlantic/Sub-boreal transition, 38
Atlas Mountains (Morocco), 360
atmometers (*see also* evaporation), 75, 78, 82
atmospheric circulation, 39
- physics, 99
ATP [adenosine triphosphate], 266, 277
auger hole method for sseepage estimation, 124, 126, 127
auks, 278
Australasia (*see also particular territories and localities*), 2
Australia (*see also particular regions and localities*), 42, 102
Austria, 22, 372, 379
autochthonous peat, 28
autogenic change, 28, 44, 312–315, 319, 340–342
autoregulation of mires, 1
auxotrophic conditions, 29

Bacteria (*see also* decomposition; micro-organisms; nitrogen fixation)
-, activity of, 165, 235, 250
-, --: affected by peat extracts, 176
-, biomass of, 165, 235
-, consumption of, by animals, 356
- in blanket bog, 165
- in bog pools, 355
- in decomposition, 356, 373
- in peat, 165, 235, 250
- in periphyton, 370
- in rhizosphere, 241
-, nitrogen fixation by, 227
-, populations of, 235, 238
-, sulphur-transforming, 250
Bakchar (Siberia, U.S.S.R.), 51
Balaton Lake (Hungary), 23

bald cypress (*Taxodium*), 41, 261–264, 288, 300
Baltic coast of Poland, 21
– raised bogs, *see* raised bogs, types of: Baltic
– region, 6, 54
Bangladesh, 25
Barabinsk (U.S.S.R.), 114–116
Barnesbury pressure bulb recorder, 105
Barrow (Alaska, U.S.A.), 85, 201, 242, 294, 297, 300, 301
basalt as source of run-off water, 9
baseflow, 129
basins, flat-bottomed, as sites of paludification, 45, 46
basswood (*Tilia* spp.), 23
Bayern [Bavaria] (Germany), 6, 55, 57, 87, 88, 107, 132, 134, 135, 141, 143
bean, broad (*Vicia faba*), 273, 274
–, soy (*Glycine max*), 294
beaver(s) (*Castor*), 258, 352, 354, 356
– as possible cause of paludification, 45
–, giant (*Castoroides*), 362
beech (*Fagus*), 39, 47
beetle(s) (Coleoptera), 38, 336, 350, 358, 360, 361, 368, 369, 378, 385
–, heather (*Lochmaea suturalis*), 376
Belgium, 19, 20, 86
Belorussiya [White Russia] (U.S.S.R.), 14, 23, 134, 227
below-ground component of biomass and productivity, 286, 288, 289, 291–293, 306, 377
below-ground/above-ground biomass ratio (*see also* root/shoot ratio), 292, 293
Beloye More [White Sea] (U.S.S.R.), 6
Berendeevo bog (U.S.S.R.), 52
Bernau (Germany), 87
Berwyn Mountains (Wales), 325
Bethel (Minn., U.S.A.), 97
Betula–Vaccinium forest, 93
bicarbonate, 9, 249, 251
Biebrza River (Poland), 21
Bigfork River (Minn., U.S.A.), 320
biomass of animals, 368, 370, 371, 373, 376–380
– of plants, 184, 371
– of vegetation, 287, 375
birch, *see Betula in Systematic Index*
– carr, 47
bird(s) (Aves) (*see also* waterfowl *and particular groups and species*), 352, 353, 378, 385
–, feeding rates of, 372
– nests as hibernation sites for invertebrates, 351
– –, sites for, 358, 372, 377
bitumen, 173, 250
bivalve molluscs, 276, 278
Black Beck (England), 328
Black Forest (Germany), *see* Schwarzwald
Black Lake (England), 327
Blacklaw Moss (Scotland), 141, 143
blandmyr, 27
Blairgowrie (Scotland), 109
blanket bog(s), animals in, 356, 358, 363, 374, 375, 378
– – as climax, 342
– –, decomposition in, 204, 230, 232, 234, 236
– –, definition of, 4, 27
– –, erosion of, 127, 140, 147
– –, formation and development of, 13, 39, 48, 148
– –, geographical distribution of, 6, 7, 41
– –, hydrograph analysis of, 142, 145
– – in Antarctic, 25
– – in British Isles (*see also* Moor House), 10, 72, 230
– – in Hawaii, 26
– – in Japan, 23, 24
– – in Norway, 19, 72
– –, microorganisms in, 165, 229
– –, models of, 17, 146, 213
– –, nutrients in, 226, 231, 242, 251, 298
– –, peat accumulation in, 209
– –, productivity of, 197, 236
– –, vegetation of, 10
– –, water relations of, 84, 110, 130, 131, 136, 138, 141–143, 145
– peat, *see* peat, blanket
blue-green algae (Cyanobacteria), 181, 227–229, 252
Bodensee (Germany and Switzerland), 374
bog(s) (*see also* blanket bogs; hummocks in bogs; raised bogs)
–, animals in, 367, 374–378
– asphodel, *see Narthecium ossifragum in Systematic Index*
– burst, 138, 170, 342
–, classification of (*see also* –, types of; mires, classification of), 9, 10
–, climatic, 27
–, dating of (*see also* pollen analysis; radiocarbon), 17
–, definition of, 2, 4, 8, 27
–, evaporation from, 92, 94, 98
– expanse, 13, 29
– –, definition of, 27
– flow, see – burst
– forest, 323, 326, 327, 335–337, 342
–, forested, 6, 93
–, formation of, 45
–, geographical distribution of, 6, 7
–, head-water, 28
–, hydrology of, 14–17
– in Germany, 46
– in Hawaii, 25, 26
– in Romania, 23
– lakes, dystrophic, 22
– lemming (*Synaptomys*), 376
–, marginal slopes of, *see* rand
–, nutrients in, 177
–, oligotrophic vegetation of, 144
–, orchids in, 240
–, peat composition of (*see also* peat, composition of), 177–179
–, pH in, 180
–, ponds in, *see* bog-pools
–, primary productivity in, 196, 285, 290, 291, 306
–, secondary productivity in, 374–378
– seepage, 219
– sere, 326, 327
–, standing crop in, 293
– stratigraphy, 24
–, successional relations of, 40, 326, 327, 330, 331
–, tephra layers in, 17, 19

bog(s) (*continued*)

–, types of (*see also* mires, types of; moss(es); raised bogs; *and also names of vegetation Alliances and Associations*)
–, – –: *Calluna–Eriophorum vaginatum* nodum, 10, 12
–, – –: *Calluna–Narthecium* nodum, 10, 11
–, – –: *Erica tetralix–Scirpus* nodum, 10–12
–, – –: mixed *Calluna* nodum, 10
–, – –: *Molinia–Erica tetralix* nodum, 10
–, – –: pine bogs, 29
–, – –: *Sphagnum* bogs, 103
–, – –: *Sphagnum*–sedge–chamaephyte bogs, 105
–, – –: transition bogs [*Übergangsmoore*], 29, 172
–, – –: upland *Calluna–Eriophorum* nodum, 10, 12
–, – –: valley bogs, 6, 7, 28, 130, 184
–, volcanic ash in, *see* –, tephra layers in
–, water supply to, 71
–, weakly soligenous, 29
–, wooded, 6, 13
bog-moss, *see* Sphagnum
bog-pools, 23, 29
bog-rush (*Schoenus nigricans*), 40, 239, 261, 335, 356
Bohemia (Czechoslovakia), 9, 22, 78, 86
Böhmer Wald (Czechoslovakia), 22
Bølling Interstadial, 38
boloto, 27
Boreal period, 58
– Zone, 2, 7, 11, 27, 28, 56
Borneo (*see also* Sarawak), 188
Borofibrists, 217
Borohemists, 217
Borosaprists, 218
bottom fauna, 368–370
bottom-lands, 357
Bourtanger Moor (Netherlands), 339, 340, 342
Bowen ratio, 77, 98
bradycardia, 280
Brahmaputra Delta (Bangladesh), 25
Braunmoor, 27
break-down of organic material, *see* decomposition
Breck fens (England), 335
Bremen (Germany), 88
Brest-Litovsk (U.S.S.R.), 137
Bretagne (France), 20
brine tests for waterfowl, 137
Britain, *see* Great Britain
British Columbia (Canada), 323, 333
British Guiana, *see* Guyana
British Isles (*see also* England; Great Britain; Ireland; Scotland; Wales; *and particular regions and localities*)
– –, blanket bog in, 10
– –, climatic change in, 38
– –, – observations in, 75
– –, hydroseral changes in, 328–331
– –, peat stratigraphy in, 322, 323, 326
– –, productivity in, 287
– –, recurrence surfaces in, 318
– –, raised bogs in, 6
– –, wetland animals in, 355, 360, 362, 374
Brittany (France), 20
broad bean (*Vicia faba*), 273, 274
Broadlands (England) (*see also* Norfolk Broads), 354
Bronze Age, 41, 54
Bruchmoor, 27
Bryants Bog (Mich., U.S.A.), 330
bud dormancy broken by flooding, 257, 258, 269–271
Bug River (U.S.S.R.), 137
bugs (Hemiptera), 350, 368, 369, 376
bulk density, *see* peat, bulk density of
buntings, 358
burrows of beavers, 352
butanediol in peat, 250
butanol in peat, 250
butterfly, large copper (*Lycaena dispar*), 352, 361
–, swallow-tail (*Papilio machaon*), 352, 362
buttress roots, 263, 264
butyric acid, 250, 278
Byelorussia, *see* Belorussiya

^{14}C, *see* radiocarbon
C_4 and C_3 plants, 295, 296
caddis flies (Trichoptera), 357, 358, 368
cadmium, 193, 194
caecotrophy, 354, 355
caesium-137 as marker, 184, 185, 194, 195
Caithness (Scotland), 73
calcium as limiting factor in bogs, 216
– carbonate, 53
– – precipitated in Indian marshlands, 25
– –, solubility of, 191
– – sorbing phosphorus, 252
– fertilizer effects, 291
– in peat, 25, 160, 163, 164, 178, 180, 181, 183, 185–187, 248, 249, 251
– – –, profile of, 185, 192–195
– in rainfall and run-off water, 9
– in wetland water, 71, 334
– leached from litter, 230
– sulphate, solubility of, 191
calcium/magnesium ratio in peat, 71, 185, 186, 338
– – – – recharge water, 71
California (U.S.A.), 275, 295–297, 300, 302, 370, 373, 379
callows, 357
Calluna, see heather
Calluneto-Eriophoretum, 10, 136, 375–377, 380
Callunetum, 10, 107, 108
calorific content of peat, 160, 169
Calthion, 21
Calthorpe Broad (England), 325
cambium, 279
– resisting oxygen diffusion, 269
Cambridge (England), 106, 238
Cambridgeshire (England), 331, 335
Camp Century (Greenland), 37
Canada (*see also particular regions and localities*), 2, 19, 99, 334
–, Arctic mires in, 7
–, climatic changes in, 37, 39, 43, 59
–, muskeg in, 27
–, peat classification in, 171

Canada (*continued*)

–, vegetational changes in, 43
–, wetlands in, 2, 19, 334
–, – –, photosynthesis in, 295, 296
–, – –, productivity of, 288–293
–, – –, seed viability in, 275
–, – –, types of, 11
–, – –: water composition, 252, 253
–, – –: – relations of mire plants, 83, 300, 301
–, – –, water-level changes in, 126
canopy, *see* leaf canopy
capillarimeter, 114
capillary fringe, 96, 100, 101, 111, 115
– rise, 71, 94, 96, 317
capitulum [of *Sphagnum*], 78–81
caproic acid in peat, 250
capybara (*Hydrochoerus hydrochaeris*), 354
carbohydrate(s), energy content of, 292
– exchange with mycorrhizae, 239, 240
– in peat, 173–177
– – –, degradation of, 250
carbon content of litter, 231
– dioxide efflux from soil, 183, 198
– – exchange by vegetation (*see also* photosynthesis), 287, 303
– – flux, 304, 305
– fixation pathways, 295
carbon/nitrogen ratio and decomposition, 251
Carboniferous rocks as source of run-off water, 9
carboxyl groups: formation in soil, 250
Cardamino-Montion, 22
caribou (*Rangifer*), 375, 385
Caricetum dioicae, 86
– *fuscae*, 86
Carici-Festucetum rubrae, 87
Caricion canescentis-fuscae, 21
Caricion davallianae, 21
carnivores (*see also* predators), 369, 373, 376–378
carnivorous plants, 228
Carpathian Mountains, 6, 21, 23
Carpatii Meridionali (Romania), 23
carr (*see also* fen; peat, types of)
–, birch, 47
–, branching habit of trees in, 269
–, definition of, 27, 40
–, fen, 161, 322, 326, 336
–, geographical distribution of, 41, 46
– in England, 328, 329, 331
– in Germany, 46, 47
– in Japan, 23
– in Ireland, 13
– in riverine lowlands, 45
–, instability of surface in, 336
–, plant species in, 324, 331
–: semi-swamp carr, 336
–, successional relations in, 325, 326, 328, 329, 331, 335, 337
–: swamp carr, 326, 336
carrion-feeders, 369
Casparian strip, 300
catchment area of mires, delineation of, 130
– behaviour, 76, 110
– heterogeneity, 130
– hydrology, 76, 90, 142
cation exchange, 163–165, 175, 178–180
– – by *Sphagnum*, 337
catotelm, 17, 82, 103, 117, 137, 145, 147–149, 189
– and ecosystem stability, 139
–, definition of, 28
–, permeability in, 117, 138, 144
–, seepage in, 119, 147
–, water storage in, 147
Cecil Bog (Mich., U.S.A.), 330
cedar (*Thuja*), 264, 288, 324, 327, 331, 335, 336
Cedar Creek Bog (Minnesota, U.S.A.), 330
cellulose decomposition, 250
– in peat, 174, 175, 177, 179, 250
– – –, composition of, 175
Cenozoic Era, 360
Central Africa (*see also particular territories and localities*), 25
Central America (*see also particular territories and localities*), 26
Central Committee for the Survey and Study of British Vegetation, 9
Central Europe (*see also particular territories and localities*), 36–42, 47, 52, 141, 316
cephalodia in lichens, 229
cereal pollen in peat, 49, 50, 170
cesium, *see* caesium,
Ceylon, *see* Sri Lanka
chamaephytes (*see also* dwarf-shrubs; life-forms), 77, 113, 114, 118, 134
charcoal in peat, 49, 50, 325
channels in mires, 70, 71, 132
– – –, classification of, 129
– – –, flow in, 70, 120, 140–145
– – –, – –: base-flow and quick-flow distinguished, 128, 129
–, ontogeny of, 131, 132
Charion fragilis, 21
chelation in peat, 164, 253
chemistry as basis for mire classification, 4
Chemutovka River (Czechoslovakia), 143
Cherangani Hills (Kenya), 37
chernozem, 51
Cherryfield (Maine, U.S.A.), 177
Cheshire (England), 321, 327, 331, 336
Chiemsee (Bayern, Germany), 87, 107, 108
Chiemseemoore (Bayern, Germany), 87, 88, 107, 141, 143
Chile, 2
Chilton (England), 195
Chilwa, Lake (Malawi), 370
China, 24
chionophilous vegetation, 74
chloride, 8, 53
chlorine in precipitation, 180
chlorosis, 270, 271
Cladietum, 336
Claish Moss (Scotland), 139, 339
classification, *see* mires, classification of; peat, classification of; soil, classification of

clearing around lakes affecting trophic conditions, 45
- of forests, 47, 54, 325
clegs, 358
climate (*see also* air masses; bioclimate; cloudiness; cyclones; evaporation; microclimate; precipitation; radiation; temperature; wind)
- and humus accumulation, 51
-, changes in, 36–40, 44, 53, 313, 316–318, 360
- fluctuations in mire development, 14, 16, 55, 330
-, history of, 36–40
-: relation to mires, 1, 35–59
-, trends in, 14
climax (*see also* fire-climax; succession), 342
Cloncreen (Ireland), 187, 188
clover (*Trifolium*), 228
"clothes-line effect" on evaporation, 87, 305, 306
C/N ratio in plant tissue and peat, 251
coal compared with peat, 169, 170, 238
Colchis [Kolkhida] (U.S.S.R.), 36
Colombia, 37
colonization in development of mires, 262, 321, 331
- by animals, 359
colour: effect on evapotranspiration, 94
coma [of *Sphagnum*], *see* capitulum
community architecture and evapotranspiration, 77
compaction effects on bog growth, 17
compartment model of mire hydrology, 69
competition among bog animals, 357
complex formation in organic soils, 249, 253
compression of peat, *see* peat, compressibility of
computer simulation of mire growth (*see also* models), 17, 18
concentration time [in hydrology], 131, 138, 142
concentric mires, 6, 338
conductivity, *see* electrical conductivity; hydraulic conductivity; thermal conductivity
conifers (*see also* forests, coniferous; *and particular groups and species*), 41, 269, 290, 293, 335
Connecticut (U.S.A.), 323, 333, 335
conservation (*see also* nature reserves), 128
- of matter as principle of mire hydrology, 69
- of mires (*see also* mires, protection of), 22, 118
- of peatlands, 20
- of wildlife, 145
Constance, Lake of, *see* Bodensee
consumption by herbivores, 372
continental raised bogs, *see* raised bogs, types of: continental
continuity, principle of, 69, 76, 122, 147
contributing area [in hydrology], 140
Coom Rigg Moss (England), 187
coot (*Fulica atra*), 355
copper, 182, 190, 193, 194, 249
coprophages, 375
cormorants, 278
corn (*Zea mays*), *see* maize
Coteau de Missouri (N.D., U.S.A.), 98
cotton (*Gossypium* spp.), 267
cotton-grass, *see Eriophorum*
counter-erosion: role of peat accumulation, 148
coypu (*Myocastor coypus*), 354, 372, 373
craneflies (Tipulidae), 203, 375–378
Cranesmoor (England), 169, 184
Cratoneurion commutati, 22
creep of peat, 159, 213
Cretaceous rocks as source of run-off water, 9
crocodiles (Crocodylia), 278, 353
Crossens (England), 142
crown-gall tissue, 278
Cruden Moss (Scotland), 187
Cryofibrists, 217
Cryohemists, 217
Cryosaprists, 218
cryogenic factors, *see* frost; ice; permafrost
cryoturbation, 132
cryptodysaptic fens, 137
Cuba, 26
cuckoo-spit bugs, 356
Cumbria (England), 374
cupola of raised bogs, 13, 14, 26, 317, 338
cuticular wax in mire plants, 298
cyanogenetic glucosides, 260
--: hydrolysis on flooding, 260
cycles of population in muskrats, 354
cyclical change(s), 339
- evolution of bogs, *see* hummock-and-hollow regeneration hypothesis
cypress, bald (*Taxodium*), 41, 261–264, 288, 300
cytochrome oxidase, 266, 271, 272
Czechoslovakia (*see also particular regions and localities*)
-, wetland in, 22, 23
-, --, animals in, 368, 369, 372
-, --, canopy structure in, 302, 303
-, --, evaporation from, 86
-, --, microclimatology of, 78
-, --, photosynthesis in, 294, 295
-, --, primary productivity in, 288, 289
-, --, root standing crop in, 293
-, --, water discharge from, 143

dairy grassland on peat deposits in The Netherlands, 20
Dalton's equation, 74, 77, 98
damming by beavers, 352
damping depth, 201
Danube Delta (Romania), 23, 370
- River, 23, 349, 360, 361
Darcy's Law, 15, 71, 121–125, 133, 136, 147, 149, 168, 216
Dářko (Czechoslovakia), 78, 86
Dartmoor (England), 8
datum reference plane, 99, 101, 104, 109
Death Valley (Calif., U.S.A.), 297, 300
decay, *see* decomposition
deciduous and evergreen habit compared, 230
- forest, 41
Deckenmoore (*see also* blanket bogs), 7
decomposition (*see also* humification)
-, models of, 17, 18, 207–217
-, products of, 173
-, rate of (*see also* peat accumulation), 230–238

decomposition (*continued*)

–, rate of, and peat formation, 35, 159–161, 197–216, 225, 311, 332
–, – –, factors affecting, 35, 199–207, 232
–, – –, – –: climate, 16
–, – –, – –: nutrients, 134, 204, 205
–, – –, – –: oxygen, 315
–, – –, – –: pH, 249
–, – –, – –: temperature, 18, 25, 200, 201, 232
–, – –, – –: water content, 25, 200–202, 232
–, – –, – –: water-table, 285
–, – –, in acrotelm and catotelm, 211, 212
–, – –: methods of estimation, 198–200, 204, 206
–, role of invertebrates in, 203–205, 373, 377, 378
–, – – micro-organisms in, 230–238
–, selective, in peat, 17, 18, 54, 211, 212
Deregö Stormyr (Sweden), 320
Delamere Forest (England), 331
denitrification, 226, 230, 241, 242, 248, 250, 251
Denmark (*see also particular localities*), 19, 20, 38, 199, 200, 203, 288, 289, 293, 376, 377
dephosphorylation, 252
Derbyshire (England), 48
deserts and wetlands compared, 362
detergents as source of phosphate eutrophication, 343
detached-organ method for transpiration, 86
detoxification mechanisms (*see also* toxins), 268
detritivores, 198, 349, 356, 357, 368, 369, 373, 377–379
detritus, *see* litter
Devon Island (Canada), 290, 291, 293
Devonian rocks, 47
dew, 70
deyatel'nyĭ sloĭ, 104
diapause, 276, 280
diatoms (Bacillariophyta), 356, 371
Dicrano-Pinion, 22
dictionaries of mire terms, 4
"die-back" of *Frangula*, 336
diffuse surface flow, 70, 71, 119
diffusion within peat, 181, 183, 184, 195
dip wells, 103–105, 116, 144
diplotelmic mires, 137, 145, 146, 149
discharge from mires, 70, 118–145
dispersal of wetland animals, 358, 359
– – – –, cost of, 359
diversity, specific, 287, 360, 367, 369
diving animals, 280
DO [dissolved oxygen], 247
dolines, 357
domed bogs, 4, 14
dormancy (*see also* aestivation; bud dormancy; encystment; hibernation; seed dormancy), 229, 230, 351, 352
– increasing resistance to flooding, 261
Douglas fir (*Pseudotsuga*), 327
dråg, 28
drain(s), open, 126
– spacing and water table, 96, 124
drainage as anthropogenic disturbance, 23, 26, 114, 124
– – – – causing extinctions, 361, 362
– – – –: effects on animals, 359–362
– – – –: – – evapotranspiration, 123, 126, 141
– – – –: – – plants, 257, 261, 312
– – – –: – – productivity, 291
– – – –: – – water table, 96
–: dictionary of drainage terms, 4
– ditches, 124
–, impeded, leading to bog development, 7, 14, 24
– patterns determining mire type, 7, 319
Draved Mose (Denmark), 199, 200, 203, 208–211, 214, 215
drought effects in wetlands, 132, 149
Dryas period, 47
dubh lochan, 138
ducks, 278, 349, 355
duckboarding for observations on mires, 106
Dun Moss (Scotland), 105, 109, 111, 124–126, 133, 135, 149
dunes damming run-off, 45
Dupuit–Forchheimer approximation, 122, 123, 133, 135, 147, 149
Dutch words for mire concepts, 27, 28
dwarf-shrubs (*see also* chamaephytes), 29, 77, 93, 94, 102, 111, 140
–, water relations of, 102
dy, 28, 162
dyes used ot test pipe flow, 137
dysaptic fens, 137
dystrophic bog lakes, 22
– conditions, definition of, 29

East Africa (*see also particular territories and localities*), 37, 359, 377
East Anglia (England), 8, 325, 335
Easter Island [Isla de Pascua], 25
eccentric mires, 6, 170, 338
ecophysiology of plant water relations, 77
ectomycorrhizae, 238, 239
eddy correlation technique, 76
– diffusivity, 304
Edenderry (Ireland), 13
eel grasses (Zosteraceae), 279, 384
efficiency index, 380
efflux of water from mires, 70
eggs of flies adapted for underwater development, 350
– as overwintering stage, 352
–, drought resistance of, 351
eggar, northern (*Lasiocampa quercus* var. *calunae*), 376
egret, 358
Eh (redox potential), 247, 248, 252, 254
Eifel (Germany), 20
Eire, *see* Ireland
electrical analogue models in hydrology, 131
– conductivity of peat, 163
– – – mire recharge water, 71
electron micrography, 232–234, 236, 237
electro-osmotic forces in peat, 121, 122
elephants (Elephantidae), 349
Elephant Island (Antarctica), 25
elution of ions in peat, 195, 196

embolism in roots, 269
emersive plants, 332, 333, 343
Emmen (Netherlands), 19
emperor moth (*Saturnia pavonia*), 356
Empetro-Eriophoretum, 136
Ems River (Germany), 107
encystment, 351
endemism, 359
endotelmic: defined, 146
energy (*see also* radiation)
– balance, 77, 91–94, 98, 266, 300
– – equation, 77
– capture by crops and vegetation (*see also* leaf area index; leaf canopy; photosynthesis), 294, 295, 306
– charge [in nucleotides], 266, 267
– content of peat (*see also* peat, calorific content of), 160, 169
– – – vegetation, 292
– flow, 362, 363, 378
– partition between evaporation and sensible heat, 301
–, peat as source of, 2
– resources distribution within ecosystem, 291, 292
engineering interest in peat properties, 159
England (*see also* British Isles; Great Britain; *and particular regions and localities*)
–, blanket bogs in, 11, 12, 17, 18, 25, 48, 143, 197, 230, 234
–, climatic change in, 38
–, fens in, 2, 23, 231
–: peat accumulation, 183–185, 214
–: – chemistry, 169, 181–183, 187, 195, 196, 338
–: – erosion, 127, 132
–: – stratigraphy, 161
–, raised bogs in, 236
–, wet heaths in, 104
–, wetlands in (*see also* –, blanket bogs in; –, fens in; –, raised bogs in),
–, – –, animals in, 38, 354, 359, 362, 367, 372, 376, 379
–, – – decomposition in, 200, 204, 207, 209, 230, 232, 235, 237
–, – –, drainage of, 2, 321
–, – –, dynamics of, 45, 315, 320, 336
–, – –, microbiology of, 242
–, – –, nutrient relations in, 227
–, – –, productivity of, 197, 288–291, 296, 302
–, – –, root development in, 243, 293
–, – –, water relations of plants in, 300
Entisols, 217
environmental manipulation by beavers, 352
enzyme(s), adaptive, 277
– deletion, 277
epinasty, 271
epiphytic fauna, 368
epizootics, 354
Ericion tetralicis, 22
ericoid mycorrhizae, 239
Eriophoretum, 10, 12
Eriophorum (cotton-grass) (*see also E. vaginatum*; *and Systematic Index*)
–, animals associated with, 173, 377, 378
– as pioneer, 326, 334
–, decay of, 195, 232, 234, 235, 377
– in peat (*see also* peat, types of), 164, 168, 169, 174, 188, 325
– in vegetation (*see also* bogs, types of), 58, 78, 113, 133, 134, 136, 138, 172, 278, 290, 326, 334
–, roots of, 235, 236, 241
–, stomatal resistance of, 85
Eriophorum vaginatum, anatomy of, 85, 232, 233
– –, cation exchange capacity of, 164
– –, decomposition of, 17, 198, 201, 202, 205, 207, 232–234
– – in peat, 55, 126, 161, 163, 167, 194, 218, 340
– – in vegetation (*see also* bogs, types of), 10–12, 17, 22, 24, 78, 89, 95, 138, 141, 180, 275, 290, 293, 314, 338
– –, nutrient relations of, 229, 231, 291
– –, roots of, 234, 239, 241, 267, 268
– –, transpiration of, 86
erosion (*see also* peat, erosion of), 127, 140, 321
– in mire formation, 320
– of mires, 131, 147
– of peat, 136
–, subcutaneous, 127
Erosionskomplex, 44
errors in productivity estimates, 285, 286
Ersterweger Dose (Germany), 104
Erzgebirge, *see* Krušné Hory
establishment of seedlings, 334–336
ester formation in soil, 250
Esthwaite North Fen (England), 328–330
Estonia (U.S.S.R.), 52, 54, 91
Estonian words for mire concepts, 4, 20, 27, 28
ethane evolution from bogs, 198
ethanol (*see also* alcohol), 262, 263, 270, 278–280
– and membrane solubilization, 275
– in peat, 250
– production on flooding, 262, 266, 272–275, 277
– toxicity, 275
ethylene production under hypoxia, 263, 270, 278
Eufuscion, 13
Eurasia (*see also* Asia; Europe; *and particular territories and localities*), 37, 38, 40, 352
Europe (*see also* Central Europe; *and particular territories and localities*)
–, animal distribution in, 349, 356
–, bird migration in, 252
–, conifers in wetlands of, 335, 336
–, evapotranspiration in, 75
–, introduced animals in, 354, 372, 373
–, peatland research in, 4
–, wetland distribution in, 6, 7, 56
eustatic earth movements, *see* continental uplift; sea-level, changes in
eutrophic conditions, definition of, 29
– influences in mire development, 52
– waters associated with immersive plants, 333
eutrophication, 330, 343, 344, 373, 374
– through lumbering, 321, 322
eutrophy in mires, 8, 41, 171, 172, 325
evaporation, 72–99, 111, 116, 122, 144, 301, 303
–, direct, 73
– flux, 144, 304
– from different mire types, 77, 78, 90
– – snow, 92
–, measurement of (*see also* –, pan; atmometers), 74–78

evaporation (*continued*)

–, pan, 75, 92, 93, 95, 98
–, potential, 74, 75
evapotranspiration, 70, 73, 75–99, 124, 142, 145, 147, 336, 337
–, annual, 148
– changed by lumbering, 321
–, estimation of, 76, 88, 90, 95
– from different vegetation types, 90, 93, 94
–, maps of, 92
–, potential, 43, 72, 74, 87–90, 92, 93, 98
–, water table and, 88, 95, 96
evenness of specific composition, 369
Everglades (Fla., U.S.A.), 177, 227, 287
evergreen and deciduous habit compared, 230, 271, 298
evolution of flooding tolerance, 277–280
– – mire insects, 360, 362
– – – plants, 298
exchangeable cations, *see* cation exchange
extinction of wetland animals, 360–362

F layer in litter, 231
faeces in nutrient cycling, 225, 230, 375
fairy shrimps, 351
Falkland Islands, 25
Fallahogy (Ireland), 55
"false bottom" in lakes, 326, 330
Fastebo mosse (Sweden), 4, 5
fastmatsamhällen, 27
fatty acids in peat soils, 250, 251
faunal assemblages in bogs, 357, 363
fen(s) (*see also* aapa fens; carr; lagg fens; *Niedermoore*)
–, animals in, 352, 354, 356, 357, 359, 360, 363, 373, 379
– antecedent to bogs, 13, 40, 52, 53
–, calcareous, 370
–, cations in, 177
–, classification of (*see also* –, types of; mires, classification of), 9
–, decomposition in, 231
–, definition of, 2, 7, 8, 27, 172
–, evaporation from, 91, 92, 94, 98
–, fluvial, 140
–, formation of, 45, 47, 71, 73
–, geographical distribution of, 6, 7, 21, 41, 46
– in Czechoslovakia, 22
– in Germany, 20, 21, 46–48
– in Japan, 23
– in The Netherlands, 19
– in Poland, 21
– in Romania, 23
– in the U.S.S.R., 72, 91
–: marginal to raised bogs, 111, 123
–, micro-organisms in, 229, 238, 242
–, nutrient relations in, 229
–, oligotrophic, 8, 11
–, orchids in, 240
–, peat in (*see also* peat, types of), 135, 178, 179
–, – –, permeability of, 137
–, pH of, 8, 180
–, productivity of, primary, 196, 285
–, – –, secondary, 367–374
-, successional relations of, 325, 328–330, 334–336, 338
–, types of: fen carr, 161, 322, 326, 336
–, – –: fen woodland, 325, 359
–, – –: poor fen, 27, 29, 172, 236, 316
–, – – : rich fen, 27, 180
–, – – : *rimpi* fen, 95
–, – – : sedge fen, 23, 97, 137
–, – – : spring fen, 73
–, – – : string fen, 95
–, water discharge from, 115, 130, 133, 143, 145
– watercourses, 129
–, water-table in, 110, 137, 312
Fennoscandia (*see also* Denmark; Finland; Norway; Scandinavia; Sweden), 22, 336
Fenno-Scandian school of mire ecology, 4, 13
fermentation in soil under anaerobic conditions, 248
ferns (Filicopsida), 24, 25
ferric and ferrous ions and compounds, *see* iron
fertilizer dust affecting rainwater composition, 343
– effects on bogs, 205
fibre in peat, 169
fibric material in soils, 217
Fibrists, 217
Fick's Law, 84, 181
field capacity of soil, 112, 115
Fife (Scotland), 275
Filipendulo-Petasition, 21
filtration flume, 123, 124
Finland (*see also particular localities*)
–, climate of, 43, 73, 75
–, mires in, 13, 52, 53
–, – –, classification of, 9
–, – –, geographical distribution of, 43
–, – –, hydrology of, 95, 96, 112–116, 135
–, – –, microbiology of, 242
–, peat classification in, 171
–, – stratigraphy in, 315, 324, 325, 345
Finnish words for mire concepts, 4, 27, 28
fir (*Abies*), 39, 326, 331
–, Douglas (*Pseudotsuga*), 327
fire (*see also* muirburn), 71, 141, 322
–, effects of, 261
– : role in mire dynamics, 325
fishes (Pisces) (*see also particular groups and species*), 181, 351, 353, 358, 368, 370
Fjällmossen (Sweden), 184, 186, 187, 192
Flachhochmoore (*see also* raised bogs, types of: Atlantic; – –, – –: plateau), 4
Flachmoor, 27, 172
flamingo (*Phaenocopterus*), 353
flark, 27, 104, 138, 139, 147
flies (Diptera), 350, 351, 358, 368, 370, 377
flight muscles of corixids, 353
Floating Bog (Minn., U.S.A.), 330
floating mats or rafts of peat (*see also* peat, floating), 315, 326
– – – – of vegetation (*see also Schwingmoore*), 28, 330, 331, 333, 334

floating mats (*continued*)

-- or rafts of vegetation, as pioneer stage, 326, 327
------ colonized by woody plants, 326, 330, 335, 336
------ encroaching on open water, 326, 327, 330, 331, 334
------, stability of, 335, 336
------, successional status of, 323, 324, 326
------, vertical fluctuations in, 315
floi mires [in Iceland], 19
floods as source of mire water, 70
flooding, 257–280
- effects on crop plants, 257
-, prolonged, 261, 262
- tolerance, 257–280
--, evolution of, 277–280
-- of trees, 257, 261, 262
flood-plain(s), mire formation in, 13
Florida [state] (U.S.A.), 227, 234, 287
floristic basis for mire classification, 4
flow, channel (*see also* channels), 70
- lines (*see also* streamlines), 15, 123
-, pipe, 70, 71, 120
-, sheet, 70, 136, 138, 140
-, surface, 71, 76, 120
-, unsaturated, 71
flumes, 129
-, filtration, 123
-, stream-gauging, 142
flushes, 22
-, vegetation of, 22
flushing of soil by moving water as factor in mire ecology, 263, 289, 312
------ removing toxins, 273
fog, 72
Folists, 217
food sources in tropical marshes, 25
forbs [non-woody broad-leaf plants] defining cover classes in mires, 11
forest(s) (*see also* pine forests; woodlands), 48, 359
- as hydroseral climax, 325–327
-, broad-leaved, *see* -, deciduous
-, clearing of, 47, 48, 325
-, coniferous, 56
-, deciduous, 56, 325
-, gallery, 23
-, mesophytic, as climax, 342
- mires, 29, 117
--, swamp, 29
- soils, paludification of, 13
forestry, 2, 18, 20
formic acid, 250, 278
försumpning (*see also* paludification), 29
fox moth (*Macrothylacia rubi*), 356
Fozy Moss (England), 235–237
France, 19, 20
Frankfurt am Main (Germany), 78
freezing, *see* frost
French words for mire concepts, 27, 28
Freshwater Biological Association [United Kingdom], 330
frog(s), 358, 377
frost, 140, 301, 335
frost effects (*see also* permafrost), 7, 72
frost-heaving, 132
fructose in peat, 173
fruits, ethylene production in, 278
-, formation of, 335
Fuhrberg (Germany), 107
Fukuda-Hygen method, 84
fulvic acid, 175, 249, 252
fumaric acid in peat, 250
Fungi (*see also* decomposition; microorganisms; mycorrhizae)
-, activity of, 165, 356
- as parasites of wetland trees, 336
-, biomass of, 165
- consumed by animals, 356, 378
-, decomposition by, 241, 373
-: growth affected by peat extracts, 176
- in leaf litter, 235
- in peat, 235, 250
- in *Sphagnum*, 236
-, thermophilic, in peat, 165
Fyerte Lake (Hungary), 23

galactose in peat, 173, 175
gallery forests, 23
gall-forming invertebrates, 356, 372
game, land management for, 374
game reserves, *see* conservation; nature reserves
gamma ray attenuation, 101
Ganges Delta (India), 25
garganey (*Anas querquedula*), 352
gas transport in plants, 265
---- : function in root survival, 265
Gas san, Mount (Japan), 24
geese, *see* goose
geochemistry, 71, 72
geogenous mires, 6, 8, 19, 21, 29
geographical distribution of mires, 3, 6, 7
----, maps of, 3, 6, 20, 21, 41, 46
geohydrology, 73
geology, *see particular geological periods and rock types*
geomorphology (*see also* erosion; glaciation; landforms; stream capture), 52
-, changes in, 319
-: features related to mire developement, 6, 13, 28, 45, 46, 48, 323, 357, 360, 367
Georgia (U.S.A.), 288
German school of mire ecology, 4
- words for mire concepts, 4, 7, 27–29, 132
Germany (*see also particular regions and localities*)
-, climatic changes in, 316
-, evapotranspiration in, 87, 88
-: fens formed on river terraces, 48
-, peat in, 159
-, --, accumulation of, 313
-, --, classification of, 171
-, --, composition of, 55, 178, 179, 180
-, --, permeability of, 134
-, --, stratigraphy of, 14, 47, 49, 50, 54, 55, 313, 315, 316, 318, 342

Germany (*continued*)

–, peat science in, 159
–, raised bogs in, 25, 127, 132
–, water-table changes in, 107
–, wetlands in, 19, 22
–, – –, age of, 47
–, – –, geographical distribution of, 7, 20, 46
–, – –, hydrology of, 87, 88, 107, 117, 127, 132, 141, 143, 144
–, – –, productivity of, 291
–, – –, protection of, 21
–, – –, succession in, 313
germination, 260
gills, 350
Glacial Period, 45
glaciation (*see also* Ice Age; moraines), 13, 46
gleization, 342
Glenamoy (Ireland), 141, 142, 214, 227, 242, 376, 379
glossary, 2, 4, 8, 26–30
glucose in peat, 173, 175
glucosides, cyanogenetic, 260
glutamic acid, 277, 278
Glycerio-Sparganion, 21
glycerol, 278–280
– and anoxia in animals, 276
– and flooding in plants, 276, 277
– in peat, 250
glycerophosphate, 278
glycolic acid, 278
glycolysis, 280
–: accumulation of products, 272
– and flooding tolerance, 275
–, end-products of, 277
–, rate of, 273
glycolytic activity, 273
– pathway of metabolism, 272
godwit, black-tailed (*Limosa limosa*), 352
goose, 355
–, grey-lag (*Anser anser*), 355
Gough Island, 25, 43, 54
graminoid(s) microlandscapes, 114
–, water relations of, 85, 86
Grampian Mountains (Scotland), 109
Grand Rapids (Minn., U.S.A.), 92, 93, 113
Grande, Isla (Argentina and Chile), *see* Tierra del Fuego
granite as source of run-off water, 9
grass(es) (Poaceae) [*see also* graminoid(s) *and particular groups and species*], 239, 265, 287, 384
– defining cover classes for mires, 11
grasshoppers, 356
grasslands, 376
grass snake (*Natrix natrix*), 358
grazing, 322, 357, 371, 372, 378
– as source of error in productivity estimates, 286
– effects on plants, 261, 372
–, seasonal patterns of, 372, 373
Great Britain (*see also* England; Scotland; Wales; *and particular regions and localities*),
– –, blanket bogs in, 82, 136, 325
– –, bog development in, 319
– –, management of upland vegetation in, 132, 136
– –, meaning of "bog" in, 172
– –, peat classification in, 171
– –, raised bogs in, 6
– –, valley bogs in, 28
– –, wetlands in, animals of, 349, 361
– –, – –, hydrology of, 71, 72
– –, – –, productivity of, 288, 289
– –, – –, succession in, 325–327
Great Lakes (North America), 38
grebes, 278
green algae (Chlorophyta), 371
Greenland, 37, 263
Grenzhorizont, 14, 19, 29, 40–44, 53–56, 313
grey-lag goose (*Anser anser*), 355
Groningen (Netherlands), 19
Grosses Moor (Germany), 132
groughs, 129, 132, 143
ground beetles (Carabidae), 38, 358, 361, 378
ground heat flux, 74, 91
– hoppers (Tetrigidae), 355
ground-water, 31
–, chemical influence of, 52
–, coefficient of, 68, 116
–, definition of, 101
– discharge (*see also* seepage), 76, 96, 112, 311
– –, lateral flow of, 311
– level as factor in mire formation, 14
– mounds, 149
– recharge, 93, 96, 311
– table, 52, 138, 332
grouse, red (*Lagopus lagopus* var. *scoticus*), 355, 356, 376
–, willow (*Lagopus lagopus*), 355
growth substance production in plants, 271
gungfly, 28
Guyana, 9
gypsum, solubility of, 191
gyttja, 28

H layer in litter, 231
H-scale for humification, *see* Von Post scale
Hachimanti, Mount (Japan), 24
haemorrhagic disease in muskrat, 354
"hags", peat, 132
Hakkoda, Mount (Japan), 24
halamyri mires [in Iceland], 19
Halland (Sweden), 92
halophytes, 295
Halsmose (Denmark), 20
Hamburg (Germany), 88
Häme (Finland), 58, 95, 96, 114
Hamilton (New Zealand), 102
Hannover (Germany), 71, 107
haplotelmic mires, 136, 143, 200
Hardangervidda (Norway), 227, 231, 234
hard-pans (*see also* soil pans), 51
hare, blue (*Lepus timidus*), 255
Harvard Pond (Mass., U.S.A.), 331
harvesting, sequential, for productivity estimation, 285, 286

Harz Mountains (Germany), 20, 52, 132
Haude's method, 75, 90
Hautes Fagnes (Belgium), 20, 86
Hawaii (U.S.A.), 25, 26
hay cut from wetlands, 322
headwater bogs, *see* valley bogs
heat, advection of, *see* advection
–, conductivity of, *see* thermal conductivity
– fluxes affected by peat, 56
–, latent, 77
–, sensible, 77
heath, 8, 38
–, wet grass, 26
heather (*Calluna vulgaris*)
–, animals associated with, 356, 363, 378
–, – – –, selective feeding by, 376
– beetle (*Lochmaea suturalis*), 376, 379
–, decay of, 17, 198, 202, 204–207, 230–232, 234, 235, 377
–, grazing of, 82, 376
– in peat (*see also* peat, types of), 126, 182, 188, 325, 340
– in vegetation (*see also* bogs, types of), 10–12, 57, 82, 87, 89, 95, 141, 228, 287, 290, 293, 301, 314, 321
–, microclimate around, 77, 78
– moor, 379
–, mycorrhizae in, 239
–, photosynthetic relations of, 296, 302, 303
– pollen in peat, 49, 50
–, roots of, 83, 300
–, water relations of, 83, 86, 300
heavy metals in organic soils, 249
Hebrides (Scotland), 136
hedgehog, 276
heliophytes, 52
helminths, 278
helophytes (*see also* life forms), 97, 99, 102, 142
Helsinki (Finland), 93
hemic material in soils, 217
hemicellulose in peat, 175, 177, 179, 250
hemicryptophytes, *see* life forms
Hemist(s), 217
herbivores, 349, 354–356, 368, 369, 375–378
–, unpalatability of mire plants to, 378
herbivory (*see also* grazing), 376
–, defence against, 261
heron, 358
Herving (Denmark), 20
heterocysts in Cyanobacteria, 228
heterogeneity in system modelling, 362
– – wetland vegetation (*see also* pattern), 303, 321, 338–341, 363, 374
heterotrophy in orchid seedlings, 240
Hewlett model of storm discharge, 130, 138, 140
hibernation, 351
hickory (*Carya*), 262
high-moors (*see also* bogs; *Hochmoore* raised bogs), 27
hillocks, *see* hummocks
hill-slope hydrology, 146
Hilsbruch (Germany), 49, 50
Himalaya Mountains, 360
Histosols, 217, 218
Hochmoore (*see also* bogs; raised bogs), 4, 6–8, 25, 27, 28, 172, 317, 323
Hochmoorkultur, 89
Hochmoorrand (*see also* rand), 27
högmosse, 27
Hohe Röhn (Germany), 20
Hohe Venn (Belgium), *see* Hautes Fagnes
Hokkaido (Japan), 23
hollows in mires (*see also* hummock-and-hollow pattern in bogs, 29, 57, 77, 81, 104, 109, 111, 128, 143, 147
– – –, development of, 17
Holocene, 39, 51, 53, 58
homoserine, 277
Honshu (Japan), 23, 24
Hooke's Law applied to peat, 213
horseflies, 359
horticulture, exploitation of peat for, 219
–: flooding effects, 257
Horton's model, 138
house mouse (*Mus musculus*), 368
Hubbard Brook (N.H., U.S.A.), 131
Hudson Bay Lowlands (Canada), 6, 7
human bodies: decay in peat, 202
– influence, *see* man, influence of
humic acids, 249, 250, 252, 254
humidity, 52, 74, 91, 305
–, changes in, during Postglacial period, 36, 37
humification of peat (*see also* decomposition; Von Post scale), 159, 161, 174, 338
– – –, assessment of, 161, 162
– – – in acrotelm and catotelm, 145, 149, 318
– – –, profile of, 174, 188, 339
– – – related to bulk density, 167
– – – – – hydraulic conductivity, 16, 134, 168
– – – – – proportion of fibre, 169
– – – – – water yield, 112, 114
– – –, selective, 54
humins, 250
Hummel Knowe Moss (England), 232, 234, 235, 241
hummock(s) in bogs, 26, 29, 57, 58, 72, 73, 77, 81, 139, 143, 147, 336
– – – and water-table, 111
–, *Sphagnum*, 86
hummock-and-hollow pattern in bogs, 13, 22–24, 28, 160, 161, 314, 338–341
– regeneration hypothesis, 14, 29, 55
humus, accumulation of (*see also* peat accumulation), 51
Hungary, 22, 23
Hunsrück Mountains (Germany), 47, 49, 50
Huntingdonshire (England), 321
Hunze Depression (Netherlands), 342
hyaline cells in *Sphagnum*, 78–81, 236, 237
hydraulic conductivity (*see also* permeability), 71, 96, 99, 111, 114, 117, 124, 125, 298, 300
– – of acrotelm, 133, 138, 143, 340, 341
– – of mires, 15
– gradient, 111
– isopotentials, 120, 123, 124, 138
– resistance, *see* – conductivity

hydrogen accumulating under anaerobiosis, 278
- formation in mires, 248, 250
- ion concentration, see pH
- sulphide in bogs, 253
hydrogen-oxidizing organisms, 250
hydrograph analysis, 130, 140, 145
- of blanket bogs, 142, 145
- of drainage, 145
- of mires, 141
- of raised bogs, 138, 141
- recession, 130, 131
-, storm, 138
-, warm-season, 143
hydrological cycle, 69, 145
hydrology, hill-slope, 144
- of mires (*see also* bogs, hydrology of), 7, 12, 18, 22, 67–150
- - -, mathematical notation for, 68
hydrophytes, 128, 140, 142, 254, 280, 367
hydroseral change, 97, 136, 316, 322, 325–332
- stages, 322, 323, 335
- succession (*see also* succession), 311, 321, 325–330
hydroxides, solubility of, 191
hydrothermic quotient, 43
hydrotrolite in soils, 254
hymatomelanic acid, 175, 249
Hypnum microtopes, 94
Hypnum–sedge microlandscapes, 94, 114
hypodermal plates in *Eriophorum*, 85
hypoxia (*see also* oxygen), 258, 271, 275, 276
-: effects on energy charge, 266
- in roots, 266, 271
- in seeds, 274
Hyytiälä (Finland), 113

ibis, 358
ice, 101, 141
Ice Age, 36
ice budget in Greenland, 37
- effects (*see also* frost; permafrost), 360
Iceland, 19, 28, 159
Icelandic words for mire concepts, 19, 27
Idaho (U.S.A.), 289
igapo lagoons in Amazon Basin, 264
igenväxning (*see also* terrestrialization), 29
ignis fatuus, 166
Ila Delta (Kazakhstan, U.S.S.R.), 97
illumination, *see* radiation
immersive plants, 332, 333, 343
immigration of plants affecting pollen analysis, 38
Inceptisols, 217
India, 25
Indian water lotus (*Nuphar nelumbo*), 274, 276
Indiana (U.S.A.), 333
inert layer, *see* catotelm
inertnyĭ sloĭ, 104
infilling of lakes, *see* terrestrialization
infiltration of water into soils, 71, 115, 122, 128, 140
- capacity, 138
- pipes (*see also* seepage tubes), 126
influx of water to mires, 70
insectivorous plants, 228
insects (Insecta) (*see also particular groups and species*), 276–278, 336
Institute of Hydrology [U.K.], 127
interception of rain, dew and fog, 99, 111, 115, 142, 336, 337
- losses, 70, 73, 77, 93, 94, 96, 115
interflow of water in mires, 132
interfluves as location of raised bogs, 6
interglacial periods, 36, 360, 361
International Biological Programme [I.B.P.], 22, 25, 191, 226, 230, 231, 243
- Geophysical Year, 75
- Hydrological Decade, 92
- Peat Congresses, 159, 224
- - Society, 171
inundation, *see* flooding
invasion of mires by woody plants, 335, 336
ion deficiency, *see* nutrient deficiency
- exchange in organic soils, 249
ionic concentration in water from different sources, 9
- requirements of different wetland communities (*see also* minerotrophy; ombrotrophy), 312
- supply, *see* nutrients
- - to groundwater, 312
Ireland (*see also* British Isles; *and particular localities*)
-, blanket bogs in, 10, 48, 72, 113, 141
-, - - -, hydrology of, 113, 141, 143
-, - peat in, 48, 125, 170
-, peat in, accumulation rate of, 214, 348
-, - -, composition of, 185, 187, 188
-, - -, humification of, 188
-, - -, stratigraphy of, 55, 315, 339, 340
-, raised bogs in, 13
-, wetlands in, animals of, 349, 357, 358, 376–379
-, - -, early classification of, 4
-, - -, microorganisms in, 227, 242
-, - -, productivity of, 290, 291
-, - -, vegetation of, 10, 40
-, Northern, 48
Irish use of word "turf", 28
iron absorption by roots, 269
-: bog iron, 187
- in mire plants, 180
- in peat, 178, 180, 181, 185–187, 193
-: influence on plants, 268, 342
-: precipitation of iron compounds, 51, 182, 191, 192, 252, 253, 268
-: redox relations, 191, 247–249, 251, 254, 268, 271
Iron Age, 54
Isla Grande (Argentina and Chile), *see* Tierra del Fuego
"island effect" in evolution of wetland biota, 362
isopotentials, hydraulic, 120, 123, 124, 138
isostatic land movements, 343
Israel, 26, 75, 251
Italy, 19, 20
Ivanov's equation, 133
ivy (*Hedera helix*), 47

Jack o'Lantern, 166
Jamaica, 26

Japan, 23, 24, 28, 36, 126, 301, 370
Jasiolda [Yaselda] River (Belorussiya, U.S.S.R.), 320
Jiffy-Pots, 18
Jönköping (Sweden), 92
juniper (*Juniperus*), 38
Jylland [Jutland] (Denmark), 20, 89, 104, 114

K_m, *see* Michaelis constant
Kainuu (Finland), 43
Kalina bog (U.S.S.R.), 52
Kaltenbronn (Germany), 55
Kamchatka (U.S.S.R.), 28
Karelia (Finland), 43, 115, 116
Karelia (U.S.S.R.), 58, 115
kärr, 27
karst limestone, 357
Kauai (Hawaii, U.S.A.), 25, 26
Kazakhstan (U.S.S.R.), 97
Kemeri (Latvia, U.S.S.R.), 91
Kengirdam Reservoir (Kazakhstan, U.S.S.R.), 97
kermis [ridges], 72
kermi bogs, 28, 56, 58
ketone formation in soils, 250
kettle holes as locations for floating mats, 28
– – in mire development, 323
Kilmacshane (Ireland), 186
Kilohaua (Hawaii, U.S.A.), 26
Klaukkalan Isosuo (Finland), 340
knee roots, 263, 264
Knudsmose (Denmark), 20
Kolkhida [Colchis] (Georgia, U.S.S.R.), 36
Komosse (Sweden), 73, 92, 141, 143, 314
– type of bog vegetation, 13
Kongsmose (Denmark), 20
Königsmoor (Germany), 88–90, 104, 107, 108, 117, 141, 143
Kopparås (Sweden), 164
Körös River (Hungary), 23
korpi, 27
Korvanen (Finland), 95
Krantz anatomy (*see also* C_3 and C_4 plants), 296
Krestunovo (Belorussiya, U.S.S.R.), 91, 94
Krkonoše [Riesengebirge] (Czechoslovakia), 22, 132
Krušné Hory [Erzgebirge] (Czechoslovakia), 22, 143
Krzna River (Poland), 21
Kuusamo (Finland), 43

L-layer in litter, 231
lactates in plants, 274, 275
lactic acid, 278–280
– – in peat, 250
lacustrine peat deposits, 23
– wetlands, 22
lagg, 26, 123, 135, 136
–, definition of, 27
– fen (*see also* fen, marginal), 123
– –, nutrients in, 148
– streams, 120, 129, 131, 132
LAI [leaf area index], 294, 295, 301–303, 306
lake(s) as origin of mires (*see also* terrestrialization), 13, 21, 44, 45, 131
– basins as sites for mire development, 323
–, endotelmic, 147
–, evaporation from, 97, 98
– sediments, 45, 322
– shores as locations for floating mats, 28
Lake Agassiz, *see* Agassiz
Lake District (England), 195, 325, 328
Lake States (U.S.A.), 323
Lammin-Suo (Karelia, U.S.S.R.), 80, 94, 115, 116, 120, 144
Lanarkshire (Scotland), 141, 143
Lancashire (England), 48
land subsidence leading to paludification, 320
– uplift as a factor in mire formation, 320
landforms subject to paludification, 45
Laplace's equation, 122
Lappland (Finland, Sweden), 95, 96
larch, *see Larix in Systematic Index*
larvae of wetland animals, 351
latent heat, 301
Latvia (U.S.S.R.), 91
lawns [in mires], 77, 88, 193
–, definition of, 27
leaching from leaves by rain, 195, 229
– from litter, 230, 231
– within peat, 195, 196
lead in peat, 170, 182, 192–194
leaf angle, 302, 303, 306
– area index [LAI], 294, 295, 301–303, 306
– canopy (*see also* leaf area index), 303, 306
– – development, 301–303
– fall (*see also* abscission), 195
– suckers, 372
– water stress, 298, 299
leaves, structure of, 232, 233, 271
–, rolling of, 298
Łeba River (Poland), 21
Ledo-Parvifolion, 13
leeches (Hirudinea), 368
legumes (*see also* Caesalpiniaceae; Fabaceae; Mimosaceae *in Systematic Index*), 228
Lemnion, 21
lemming (*Lemmus*), 376
–, bog (*Synaptomys*), 376
Leningrad (U.S.S.R.), 58, 91, 94, 101, 146
lenticels, 264, 265, 270
lenticular structure of peat deposits, 314, 315, 339
Leosaare bog (U.S.S.R.), 52
Lesotho, 22
lettuce (*Lactuca sativa*), 274
Lewis, Island of (Scotland), 136
lichens (Lichenes), 26, 29, 227, 229
– defining cover classes in mires, 11
– in muddy depressions, 314
–, water relations of, 82
Liège (Belgium), 87
Liesneva (Finland), 96
life cycles of wetland animals timed in accordance with primary production, 350
– forms of plants, 9, 11

light (*see also* radiation)
- interception (*see also* leaf canopy), 302, 303
- saturation of photosynthesis, 294, 295, 302
lignin in peat, 174–177, 179, 251
Likhvin Interglacial, 36
Lille Vildmose (Denmark), 20
lily-trotter (*Jacana*), 351
liming of peat soils causing ammonia production, 250
limiting height of bog cupola, 317
- horizon, 319
limnetic communities, 313, 316
- peat, 292, 313, 322–324
limnic peat, 29
limnogenous mires, 8, 13
limnology, 8
ling, *see* heather
lipids in peat, 251
Litorellion, 21
litter, 17, 378, 379
–, accumulation of, 17
- bags for measuring decay rate, 199, 202, 204–207, 209, 230, 238, 373, 377
–, decay of, 17, 229–232
–: horizons within litter layer, 231
Littlefork River (Minn., U.S.A.), 320
liverworts (Hepaticae), 26, 82
–, water relations of, 82
locust(s), 359
–, migratory (*Locusta migratoria*), 359
–, red (*Nomadacris septemfasciata*), 359
Løgumkloster (Denmark), 20
Lombardia (Italy), 20
longevity of trees in mires, 336
Loppi (Finland), 95
lotus, Indian water (*Nuphar nelumbo*), 274, 276
lotus-bird (*Jacana*), 351
lövkärr, 27
Lower Saxony, *see* Niedersachsen
low-moors (*see also* fens; *Niedermoore*), 27, 177
lumbering (*see also* forests, clearing of), 321
lung(s), 350
lung-fish, 351, 368
Luutasuo (Finland), 95
Luvifibrists, 217
Luvihemists, 217
Luvisaprists, 218
Luxembourg, 19, 20
lysigeny of root air spaces, 265
lysimeters, 68, 70, 87–90, 92–95, 99, 116
–, faults in, 76

Macquarie Island, 25
macrotopes (*see also* mire complexes), 139, 146
Madison (Wis., U.S.A.), 253
magnesium (*see also* calcium/magnesium ratio), 9, 25, 163, 164, 178, 180, 181, 183, 185, 191, 193–195, 230, 231, 248, 334
magnetite in soils, 254
Magnocaricion elatae, 21
Maine (U.S.A.), 177, 323
Mak bog, Poles'ye (U.S.S.R.), 321
maize (*Zea mays*), 203, 266, 267, 272–274, 289, 293, 302, 304
makrolandshaft (*see also* macrotope), 146
Malawi, 370
Malaya, 159
Malesia, 2, 97, 99
Malham Tarn (England), 161, 376
malic acid, 275, 277–279
- enzyme, 277
malonic acid in peat, 250
mallard (*Anas platyrhynchos*), 355
Malta, 335
mammals (Mammalia) (*see also particular groups and species*), 280
mammoth, hairy (*Mammuthus primigenius*), 361, 362
man, influence of (*see also* agriculture; drainage; fire; forests, clearing of; peat cutting), 40, 47, 48, 320–322
–, ––, affecting relationship between vegetation and environment, 12
–, ––: causing eutrophication of lakes, 45
–, ––: changes in lake levels, 343
–, ––: fires, 54
–, ––, on animal distribution, 362
Manchuria, 274
mangals (*see also* mangroves), 264
manganese, 180, 182, 185–187, 191–195, 247–249, 251, 254
manganous ions, 268
mangrove(s), 41, 263, 264
Manitoba (Canada), 290, 291, 293, 333
mannose in peat, 173, 175
mantle cavity in pulmonate molluscs, 350
maple (*Acer*), 262, 326, 331, 335
mapping of mires, 3, 6, 9–11, 19–21, 23, 41–43, 46
marginal slope of raised bog, *see* rand
Marion Island, 25
Marlee, Loch (Scotland), 326
marsh(es) (*see also* mires; swamps; wetlands), 2, 25, 27, 306
–, valley, 22
–, productivity of, 285, 288, 289, 305
–, standing crop in, 293
Massachusetts (U.S.A.), 323, 331
matric forces in water movement, 71, 100, 105, 111, 113
Maui (Hawaii, U.S.A.), 25, 26
may-flies (Ephemeroptera), 358, 368
Mayo, County (Ireland), 141, 142
Mazurian Lakes (Poland), 21
meadow(s), alpine, 290
- pipit (*Anthus pratensis*), 376, 377
–, seasonally flooded, 25
–, transpiration from, 87
–, wet, 285
Mediaeval period, 4
Medifibrists, 217
Medihemists, 217
Medisaprists, 218
meiotrophic conditions, 29
meltwater role in mire formation, 28
Mer Bleue (Ont., Canada), 94
mercaptans in bogs, 253
mesolandshaft (*see also* mesolandscapes; mesotopes), 146

mesophytes, 102, 111
mesotopes, 139
mesotrophic conditions, 29
Mesters Vig (Greenland), 263
metamorphism, 351, 352
meteoric recharge of mires, 70, 71
meteorites as source of inorganic materials in peat, 180
meteorology, *see* climate; fronts; frost; precipitation; radiation; temperature
methane, 198, 203, 248–251
– in mires, 165, 166, 251
methyl butyrate, 278
– valerate, 278
Mexico, 26
mica schist, as source of run-off water, 9
Michaelis constant [K_m] of ADH, 273
– – of cytochrome oxidase, 271
Michigan (U.S.A.), 288, 323, 330, 331, 334
microbiology of mires, *see* micro-organisms in mires
microclimate, 78, 303–305
microlandscapes (*see also* microtopes), 110, 114, 133, 146
–, graminoid, 114
–, *Hypnum*-sedge, 114
–, sedge, 114
micrometeorological methods for photosynthesis measurements, 287
micrometeorology, *see* microclimate
micro-organism(s) (*see also* algae; Bacteria; Fungi)
– activities in peat, 104, 163, 165, 173, 191, 225–243, 247, 250, 356
– and redox conditions, 247, 248, 252, 253
–: consumption by animals, 356, 373, 378
– in decomposition, 173, 203, 230–238, 250, 251, 349, 373
– in nutrient cycling, 203, 226–230, 238–243, 251–255
– in mires, 225–255
microrelief (*see also* hummock-and-hollow pattern; ridge-and-hollow pattern), 111
microtopes (*see also* microlandscapes), 77, 133, 147
–, *Hypnum*, 94
–, ridge–flark, 147
–, sedge, 94
–, sedge–*Hypnum*, 94
–, *Sphagnum*–chamaephyte, 94
–, *Sphagnum cuspidatum*, 78
–, *Sphagnum–Eriophorum*, 136
–, *Sphagnum–Scheuchzeria*, 133
Middle Ages, 4
midges, 358, 368
Midlands (England), 320
migrations of wetland animals, 352, 353, 359, 372
– – – – timed with production, 350
mikrolandshaft (*see also* microlandscapes; microtopes), 146
millet planted as "duck food", 355
Mineralbodenwasserzeigergrenze, 40
mineral content of peat, 55
– nutrition, *see* nutrients
– soil: infiltration when frozen, 140
– – water limit, 40, 172
– –, – storage in, 117
– –, – table in, 106
minerotrophic mires (*see also* fens), 7, 8, 19, 27, 29, 72, 313, 324
minerotrophy, 171, 172, 226, 316
mink (*Mustela vison*), 354
Minnesota (U.S.A.) (*see also particular localities*)
–, evapotranspiration in, 92–94, 97
–: peat accumulation in, 333, 341
–: – stratigraphy in, 323
–, rain-water composition in, 9
–, sedge mats in, 315, 334, 336
–, upslope development of peatlands in, 14, 342
–, wetlands in, geochemistry of, 71
–, – –, geomorphological changes affecting, 320
–, – –, hydrology of, 110, 112, 113, 140, 143, 144, 319
–, – –, nutrient relations in, 337
–, – –, productivity of, 288, 289
–, – –, root biomass in, 293
–, – –, shape of, 6, 7
–, – –, succession in, 323, 326, 330, 336, 341, 342
–, – –, surface patterns in, 339, 342
mire(s) [*see also* bog(s); fen(s); swamp(s)]
–, chemistry of, 12
–, classification of (*see also* –, types of), 4, 6, 7, 9–11, 12
– complexes (*see also* macrotopes), 73
–, conservation of, *see* conservation
–, definition of, 27
–, development of, 12–17
–, distribution of, 2–4, 6
–, eutrophic, 23
– expanse, 71, 88, 101, 102, 109–111, 119, 126, 130, 132, 140, 146
– –, definition of, 71
–, forested (*see also* forest mires), 2, 9
–, hydrograph analysis of, 141
–, mapping of, 2, 3
– margin, 71, 109, 110, 123, 140
–, micro-organisms in, *see* micro-organisms
–, minerotrophic, 7
–, mixed, definition of, 27
–, montane, in Hawaii, 25
–, ombrotrophic, 7
–, productivity of, 285–307
–, protection of, 21, 22
–, reclamation of, 2, 132
–, restoration of, 2
–, shape of, 12
–: slope of surface, 14
–, tertiary valley, 6
–, transformation of (*see also* man, influence of), 2
–, transition, 22, 110
–, trophic status of, 8
–, types of (*see also* bogs, types of; carr; fen; mires, classification of; moorlands; swamps, types of),
–, – –: *aapa* mires, 338
–, – –: blanket mires, *see* blanket bogs
–, – –: boreal mires, 338
–, – –: concentric mires, 28
–, – –: eccentric mires, 28
–, – –: oceanic mires, 11, 18
–, – –: *palsa* mires, *see palsas*

mire(s) (*continued*)

–, classification of: valley mires, 6, 22
–, utilization of, 25
–, vegetation of, 12
–, water relations in, *see* hydrology
–, wooded, 2
mites (Acarina), 353, 357, 378, 379
models, 362, 363
–, dynamic, 1, 18, 207–216, 362, 363
–, –, feedback in, 216
–, –, parameter estimation by, 199, 200, 208–216
–, –, prediction by, 199, 209, 215
–, –, testing of, 208, 213
–, –, uses of, 198, 208
– of bog development, 17, 18
–, simulation, 146
moder soils, 231
moisture characteristic, 100, 101
– profile, 76, 95, 115
Molinio–Myrica nodum, 136, 137
Molinieto–Callunetum, 10, 136
Molinietum, 10
Molinion coeruleae, 21
Mollisols, 217
molluscs, 276–278
Molokai (Hawaii, U.S.A.), 25
monsoon rains influencing marsh development, 25
moor(s), 1, 7, 8, 379
–, definition of, 8
Moor [German], 27
Moor House (England): animals, 374–380
– –: *Calluna* transpiration, 84
– –: decomposition, 202–207, 209, 238
– –: micro-organisms, 227, 231, 235, 236
– –: modelling, 17, 199, 209, 214, 215
– –: nutrients, 25, 182, 242
– –: peat accumulation, 214
– –: productivity, 197
Mooratmung, 104, 111, 117
moorgrass, purple, *see* Molinia caerulea *in Systematic Index*
moose (*Alces alces*), 355, 375
mor and mull humus, 51, 231
moraines (*see also* glaciation), 46
–, depressions in, as locations for paludification, 13
–, – –, – – – floating mat development, 28
Moravia (Czechoslovakia), 22, 23, 78, 86, 98
mortality as source of error in productivity estimates, 286, 287
mosaic pattern, *see* pattern
Mösky (Finland), 95
moss(es) (Musci)
– carpets, composition of, 287
–, epiphytes in, 228
– communities (*see also individual community types, and dominant species in the Systematic Index*)
– –: ash and nutrient content, 55
– defining cover classes in mire, 11
–, hypnoid (*see also particular taxa in Systematic Index*), 134, 160, 196, 323, 327, 334, 337, 338, 343
–: influence on soil, 338
– peat (*see also* peat, types of), 323
–, productivity of, 286, 287
mosses [type of bog], 6, 8
moss [undecomposed peat], 28
mosse, 27
moth, emperor, 356
–, fox (*Macrothylacia rubi*), 356
mouse, house (*Mus musculus*), 368
mowing, 322
mucilaginous root sheath, 83
muck, 28
Muckle Moss (England), 340
mud, 28
Mud Lake Bog (Mich., U.S.A.), 330
mud-bottom plant communities, 27
muirburn, 132, 136
mull and mor humus, 231
muscle tissue, metabolic products in, 278
muskeg, 1, 7, 27, 327, 337, 342, 352, 375
–, spruce, 29
muskrat (*Ondatra*), 354, 355, 372
mycorrhizae, 183, 225, 226, 234, 238–240, 243
myr, 27
Myras (Finland), 193
myri, 27

Naarsaapa (Finland), 95
NAD (nicotineamide adenine dinucleotide), 276
Nardo-Festucetum capillatae, 87
Narew River (Poland), 21
Narvik (Norway), 18
Nåthult (Sweden), 164
Nature Conservancy [U.K.], 374
nature reserves (*see also* conservation), 18
Nebraska (U.S.A.), 38, 289
necrophages, 369
Negro, Rio (Brazil), 264, 300
nematodes, 198, 278
Neolithic period, 41, 54
Nesyt fishpond (Czechoslovakia), 23
Netherlands, 18–20, 38, 171, 238, 294, 335, 342, 361, 370
Neusiedler See (Austria), 22, 372, 373
neutron scattering technique, 76, 101
New Brunswick (Canada), 275
New England (U.S.A.) (*see also particular states and localities*), 323, 363
New Hampshire (U.S.A.), 131
New Jersey (U.S.A.), 177, 288, 289
New York [state] (U.S.A.), 288, 290, 293
New Zealand (*see also particular localities*), 25, 99, 102
Newfoundland (Canada), 9, 170, 376
Newton (N.J., U.S.A.), 177
niche(s), 363
– differentiation, 369
nicotineamide adenine dinucleotide (NAD), 276
Niedermoore (*see also* fens; low-moore), 4, 7, 20, 172
Niedersachsen [Lower Saxony] (Germany), 21, 89–91, 104, 107, 108, 124, 132, 135, 136
Niger Basin, 359

Nile, River, 359, 360
nitrate in soil, 241, 242, 248, 250–252, 254
nitrification, 242, 250, 251
nitrite in soil, 241, 242
nitrogen (*see also* ammonium; nitrate), 177, 230, 239, 335, 337, 371, 374, 376
– affecting decomposition rate, 202, 251
– and insectivorous plants, 228
– budget, 298
– compounds in peat, 187, 188
– content of leaves, 290
– – – peat, 25, 176, 179–183, 188, 192, 194
– fertilizer effects, 205, 232, 292
– fixation, 181, 225–230, 248, 250, 252, 290, 298
–, reduced forms, 247, 251, 254
– transformations in peat, 241, 242, 247, 251, 252
– translocation, 229, 298
nitrogenase, 227
nitrophilous colonizers following fire, 322
Nordmjele (Norway), 167
Norfolk (England), 335, 368, 371, 373
– Broads, 23, 231, 326, 335, 367, 373
North America (*see also* Canada; Mexico; United States; *and particular regions and localities*)
– –, climatic changes in, 37, 38
– –, floating mats of vegetation in, 334–336
– –, *kermi* mires in, 56
– –, Pacific Coastal Region of, 323, 326, 327
– –, peat accumulation in, 332, 333
– –, – stratigraphy in, 323, 324
– –, raised bogs in, 6
– –: swamp forests antecedent to bogs, 14
– –, wetlands in: animals, 352, 354
– –, – –: geographical distribution of different types, 42
– –, – –: invasion by trees, 335, 336
– –, – –: lumbering effects, 321
– –, – –: succession, 323, 326, 327
—, —: – following fire, 322
North Dakota (U.S.A.), 98, 289
North Island (New Zealand), 102
North Sea area, 6, 54
North Temperate Zone, 2, 7, 11, 45, 56
Northumberland (England), 232, 234–237, 340
Northwest Territories (Canada), 333
Norway (*see also particular localities*), 18, 19, 41, 70, 80, 82, 160, 167, 171, 227, 231, 290, 302, 377
Norwegian words for mire concepts, 27, 28
Notec River (Poland), 21
Nova Scotia (Canada), 9
nucleotides, 266
Nushiri River (Japan), 24
nutrient(s) (*see also* ionic concentration; minerotrophy; oligotrophy; ombrotrophy; *and particular elements*),
– budgets, 298
– composition of mire water, local variations in, 72, 128, 137
– conservation, 230
– cycling, 195, 225–230, 234, 235, 243, 292
– deficiencies, 325
– in plants and peat compared, 178–180
– ratios in litter, 231
– release from litter, 230
– supply, 163, 243, 335
– – affected by through-flow of water, 146
– – – – water-table movements, 111
– – affecting animal populations, 353, 354, 362, 378
– – determining productivity, 289, 290
– – in relation to mycorrhizal development, 239
– – – – – photosynthesis, 297, 298
– – – – – xeromorphy, 271
– turnover around lakes, 45
– uptake, 238–241

Oahu (Hawaii, U.S.A.), 25
oaks (*Quercus*), 47, 177, 261, 262, 328, 336
oasis effect in hydrology, 94, 305
Obol' (Belorussiya, U.S.S.R.), 242
Obra River (Poland), 21
octopine, 278
Odra River (Poland), 21
Okefenokee Swamp (Georgia, U.S.A.), 300
Oklahoma (U.S.A.), 289
old-field productivity, 289
oligocentral and oligoperipheral development of raised bogs, 119
oligotrophic conditions, definition of, 29
– fens, 8
– mires, secondary productivity in, 374–378
– waters associated with emersive plants, 333
oligotrophy in mires, 8, 13, 144, 160, 169, 172
ombrogenous mires (*see also* raised bogs), 11, 15, 20, 21, 26, 29, 40, 41
ombrophilous vegetation, 29
ombrotrophic conditions, development of, 313–315, 320, 338, 342
– mires, 7, 27, 29, 71, 73, 172, 193, 313, 316, 323, 324
– waters, chemical character of, 71
ombrotrophy, 8, 13, 171, 172, 226, 312
Ontario (Canada), 253, 338
Opatvicky fishpond (Czechoslovakia), 23
open water as habitat requirement for plants, 263
orchidaceous mycorrhizae, 239, 240
Oregon (U.S.A.), 289
Os (Sweden), 192, 194
Ottawa (Canada), 83, 94, 296, 300
Ouse River (England), 106
overgrazing affecting mire development, 59
overland flow (*see also* diffuse surface flow), 119, 130, 137, 140
owl, great horned (*Bubo virginianus*), 354
oxaloacetic acid in peat, 250
oxbow lakes, 360
– –, succession in, 289
oxidase pathways, 277
Oxisols, 217
oxygen (*see also* aeration; anoxia; hypoxia; redox potential; waterlogging), 225
– concentrations affecting animals, 368, 370
– consumption in peat, 201
– – – plants, 272
– debt, 276, 280

oxygen (*continued*)

– debt and glycolytic end-products, 273, 274
– – in animals, 276
– – in mire plants, 261, 262
– demand, 370
–, diffusion of, 191, 201, 267
–, dissolved [DO], 247
– in soil, 247
oxygenation of swamps, 368
Ozegahara Moor (Japan), 23

Pacific Coastal Region [of North America], 323, 326, 327
paddy soils (*see also* rice soils), 248, 254
Padul (Spain), 159
paedogenesis, 351
Pakistan, 25
palaeoclimatic changes, 35–59
palaeoecology, 35–59
palatability of plants to herbivores, 367
palsa(s), 7, 18, 19
–, definition of, 28
– bogs or mires, 6, 7, 14, 43, 99, 140
– – – –, water storage in, 118
paludification, 13, 44–52, 313, 319, 324, 327, 333, 342
–, definition of, 29
–: self-accelerating, 51
papyrus (*Cyperus papyrus*), 28, 242, 287, 295, 355
PAR [photosynthetically active radiation], 292, 294, 296
parasites (animal), anoxia in, 276
Parvifolion, 13
Pascal [unit of pressure] defined, 168
Pascua, Isla de [Easter Island], 25
Pasteur effect and flooding tolerance, 272, 274, 275
Pasteur point, 272
Patagonia (South America), 28
pattern in mires (*see also* concentric mires; eccentric mires; hummock-and-hollow pattern; ridge-and-hollow pattern; string bogs), 2, 6, 7, 17, 314, 338–341
– due to frost, 7
patterned mire (*see also* string bog), 28
P/B ratio, *see* productivity/biomass ratio
pea (*Pisum sativum*), 273, 274
peach (*Prunus persica*), 260
peat(s) (*see also under various botanical constituents*),
– accumulation, 13, 35, 51, 52, 56, 167, 196–216, 285, 315, 316, 349, 356
– – determining wetland succession, 313–315
– –, effects of grazing on, 357
– – initiated by man, 47
– –, models of, 207–217
– –, rate of, 24, 54, 55, 199, 210, 331–333, 343
– –, – –: dependence on aerobic conditions, 201, 211–213
– –, – –: – – decay and productivity, 209
– –, – –: methods of calculation, 199, 210
– –: role of detritivores, 356
–, acidification of, 190
–, acidity of, *see* –, pH of
–: aerobic and anaerobic layers, *see* acrotelm; catotelm
–, age of, 19, 170, 173, 175, 177, 193, 199, 200, 209–216
–, – –: methods of estimation, 19, 170, 173, 175, 177, 185, 193
–, amides in, 173
–, amorphous, 127, 188
– as a historical record, 35–59, 195
–, ash content of, 19, 53, 55, 159, 161, 163, 174, 177, 179
–, blanket, 48, 113, 125, 135, 170, 185, 194, 197, 204, 209, 240, 325, 327, 331, 332
–, bog-type, 178–180
– borers, *see* – sampling
–, botanical composition of (*see also* –, types of), 160
–, buffering in, 249, 373
–, bulk density of, 17, 18, 133, 134, 166–168, 194, 332, 333
–, – – –, profile of, 167
–, calorific content of, 160, 169, 170
–, cation exchange in, 163–165, 181
–, charcoal in, 49, 50, 325
–, chemical properties of, 163–165, 172–195
–, classification of (*see also* -, types of), 4, 160, 171, 172, 217, 218
–, coastal, 159
–, collapse of structure in, 167, 215
–, colour of, 54, 160, 162, 170, 173, 235
–, compaction of, 17, 18, 168, 332
– compared with coal, 169, 170
–, composition of, 25, 172–195
–, – –: inorganic constituents, 163, 177–196
–, – –: – –, relocation of, 180–194
–, – –: – –, sources of, 159, 160, 180
–, – –: organic constituents, 173–177
–, compressibility of, 117, 140, 167, 213–215
–, consistency of, 162
–, consolidation of, 159, 167, 213–215
– creep (*see also* –, mass movement of), 159, 213
– cutting, 2, 321
–, decomposition in (*see also* decomposition), 197–207
– definition of, 28, 159
–, deformation of, 167, 213–215
– deposition, discontinuities in, 188
– deposits, riverine, 23
– depth of, 159, 199, 209
–, development of, vegetation change accompanying, 4
–, diffusion within, 181, 183, 194
–, elastic deformation of, 213
–, electrical conductivity of, 163
–, erosion of, 238, 318, 321, 342, 374
– exploitation, 26
–, fen-type, 5, 18, 135, 178–180
–, fibre content of, 134, 169
–, fire effects on, 322
–, floating (*see also* floating mats), 23, 170, 185
– formation, 311
– –, primary, 13
– – under tropical swamp forest, 25
–, freezing and thawing effects in, 140, 169, 182
–, fuel value of, 169
–, fungi in, 235, 250
–, gas content of, 166–168
– "hags", 132
– harvesting, *see* – cutting
–, heat of combustion of, 169, 170

peat(s) (*continued*)

–, hemicellulose in, 175–177, 179
–, high-altitude, 159
–, humification of, 56, 161–163, 168
–, hydraulic conductance of, 16, 127, 167–169, 195
–: impermeable surface developed following drainage or fire, 321, 322
–, irreversible drying of, 149
– islands, 23
–, lacustrine, 23
–, lenticular structure of, 314, 315
–, macro remains in, 36
–, marketing of, 18
–, mass movement of (*see also* – creep), 170, 340
–, minerogenous, 179, 180
– mining, *see*–cutting
–, ombrogenous, 54, 179, 180, 193
–, oxidation of, 20, 179, 201
–, permeability of, 71, 104, 110, 136, 137
–, – –, determining height of raised bogs, 16
–, pH of, 25, 163, 165, 179, 181, 187, 227, 228, 242, 248, 249, 337
–, plastic deformation of, *see* – creep
–, pore size in, 79, 168
–, porosity of, 80, 81, 103, 111, 134
–, primary, 6
– profiles, prediction of, by simulation, 17, 18
– –, historical value of, 35–39, 195
–: rate of deposition, *see* – accumulation
–, redox conditions in, 163, 191, 194
–, relocation of constituents in, 181–196
–, root effects in, 234, 235
–: sampling technique, 166, 167
–: – error, 193
–, secondary, 6
–, shrinkage of, 169, 238, 321
–, soligenous, 54
–, spontaneous ignition of, 165, 166
–, stratification of, 247
– stratigraphy, 35–44, 53, 55, 147, 174, 186, 311, 316, 322–327, 331, 332, 339, 340, 343
–, structure of, 169, 170
–, subsidence of, 20
–, sugars in, 173
–, sulphides in, 187, 191
–, tertiary, 6
–, topogenous, 54
–, tundra, 226, 242
–, types of (*see also* –, classification of),
–, – –: *Acutifolia* peat, 168
–, – –: bog peat (*see also* –, bog-type), 178–180, 248, 253, 322, 323, 326, 357
–, – –: – shrub peat, 218
–, – –: – wood peat, 218
–, – –: brushwood peat, 135
–, – –: *Calluna* peat, 114, 126
–, – –: *Calluna–Sphagnum* peat, 55
–, – –: *Carex* peat, 135, 161
–, – –: *Carex–Hypnum* peat, 133
–, – –: *Carex*–moss peat, 181, 184, 186
–, – –: *Carex–Sphagnum* peat, 313
–, – –: carr peat, 49, 50
–, – –: *Cladium* peat, 177, 218
–, – –: *Eriophorum* peat, 113, 114, 124, 125, 193
–, – –: *Eriophorum–Sphagnum* peat, 165
–, – –: *Eriophorum vaginatum* peat, 167, 218
–, – –: fen peat (*see also* –, fen type), 5, 18, 135, 178–180, 231, 248, 250, 253, 322, 323, 326, 357
–, – –: – carr peat, 49, 50, 186
–, – –: – moss peat, 218
–, – –: – wood peat, 218
–, – –: fibrous peat, 188
–, – –: forest peat, 323, 324
–, – –: graminoid peat, 53, 115
–, – –: graminoid–moss fen peat, 53
–, – –: graminoid–moss transition peat, 53
–, – –: herbaceous peat, 160, 161, 217
–, – –: Hypnaceae fen peat, 53
–, – –: hypnoid moss peat, 115, 116
–, – –: lake *Scirpus* peat, 218
–, – –: limnic peat, 218
–, – –: Magnocaricetum peat, 218
–, – –: mangrove peat, 234
–, – –: moss peat, 160, 161, 217, 248, 323, 324
–, – –: Parvocaricetum peat, 218
–, – –: *Phragmites* peat, 161, 218, 313, 323, 324
–, – –: rush peat, 115
–, – –: *Scirpus cespitosus* peat, 218
–, – –: sedge peat, 53, 113, 115, 116, 160, 177, 242, 249, 252, 323, 324
–, – –: sedge–Hypnaceae fen peat, 53
–, – –: sedge–*Sphagnum* peat, 113, 114
–, – –: sedge–*Sphagnum* transition peat, 53
–, – –: sedimentary peat, 253, 323, 324
–, – –: *Sphagnum* peat, 5, 49, 50, 53–56, 116, 125, 135, 160, 162, 167, 168, 174, 177, 179, 184, 188, 218, 250, 252, 313, 314, 323, 324, 333, 339, 342
–, – –: *Sphagnum* hollow and pool peat, 53
–, – –: *S. cuspidatum* peat, 193, 218
–, – –: *S. fuscum* peat, 53, 167, 187, 192, 193, 214, 340
–, – –: *S. fuscum–Calluna* peat, 55, 218
–, – –: *S. magellanicum* peat, 192–195
–, – –: *Sphagnum–Eriophorum* peat, 101, 124, 125, 163, 168, 176
–, – –: *Sphagnum–Eriophorum–Calluna* peat, 125, 126, 135, 174, 182
–, – –: swamp peat, 323, 326
–, – –: telmatic peat, 218
–, – –: *Typha* peat, 218
–, – –: wood peat, 160, 161, 186, 217, 313, 318, 322, 323, 326
–, uronic acids in, 164, 175
–, utilization of, 2, 18, 19, 170, 182, 321
–, voids in, 166–168
–, – –: in relation to hydraulic conductivity, 167
–, volume changes in, 315
– water, inorganic solutes in, 164, 178, 179, 187
–, water content of, 166–168
–, – potential in, 168
–, wax in, 173

peat-forming systems, 2, 171
– –, hydrology of, 170, 171
– –, morphology of, 170
peatland(s), 1
–, area of, 159
–, conservation of, 20
–, draining of, 20, 267
–, exploitation of, 18–21, 23, 26
– geology, 23
–, reclamation of, 26
–, regeneration of, 19
–, tundra, 231
–, worldwide area of, 26
pedogenesis, 51
pelican (*Pelecanus*), 353, 385
Penman's method, 74, 93, 95
Pennines (England), 10, 12, 48, 132, 138, 141, 142, 181, 195, 230, 232, 325
Peräpohjola (Finland), 43
Perche (France), 20
perched mires, 71, 132
– water-tables, 315
perennation by underground organs, 25, 279
periphyton, 368, 370, 371
permafrost (*see also palsas*), 7, 25, 27, 40, 51, 103, 118, 214, 337, 361
permeability, *see* hydraulic conductivity; peat, permeability of
permeameters, 121, 125
Perthshire (Scotland), 124–126
pH (*see also* acidification; bogs, pH in; mires, pH in; soil pH)
– affecting animal communities, 367, 368, 378
– changed by fungal action, 239
– influenced by phosphate sorption, 252
– influencing decomposition rate, 249
– in *Sphagnum*, 337
–, measurement of, 163
– of mire recharge water, 71, 334
– of peat, *see* peat
– – – water, 163
– of water, 25
phenolic compounds in *Sphagnum*, 235
Philippines, 288, 294, 296, 302, 304
phoresy, 353
phosphine as possible product of phosphate reduction, 252
phosphorus content of litter, 204, 231, 232
– – of peat, 25, 160, 179–183, 187, 188, 192
– – – –, profile of, 194
– – of precipitation and dust, 182
– – of plants, 180, 291
– deficiency, 337
– determining plant dominance, 335
–, diffusion of, 181
– effects on animals, 371
–, eutrophication due to, 330, 343
–, extractants for, 177
– fertilizer effects, 205, 232, 291, 292
– influencing decomposition rate, 203
– interchange between reed-swamp and lake water, 374
–, mineralization of, 252
–: ^{32}P as tracer in root studies, 183
–, recycling of, 338
– related to productivity, 290, 298
– released during decomposition, 238
– – during thawing, 182
– – when soils are flooded, 252
–, sorption of, 252
– storage in plant organs, 238
–, translocation of, 229, 230
–, uptake of, 83, 238, 240
photogrammetry, 110, 138, 139
photoperiod increasing production in Arctic wetlands, 291, 297
photosynthesis, 183, 230, 287, 292–298, 301–306
–, maximum rates of, 295, 296
–, pathways of, 295
– rate as function of wind speed, 304
– related to nutrition, 297, 298
– unaffected by herbivory, 376
phototropism in roots, 263, 264
– – stems, 269
Phragmites (*see also P. australis*; reed(s); *and in Systematic Index*)
–, anatomy of, 270
–, animals associated with, 368–370, 372, 373, 379, 380
– as an immersive plant, 333
Phragmites australis, aeration requirements of, 262, 270
– –, animals in relation to, 354, 355, 367, 368, 372
– –, canopy structure of, 302, 303
– –, decomposition of, 373
– –, growth of, 286, 291
– – in peat (*see also* peat, types of), 13, 324
– – in vegetation, 22, 23, 325, 327, 328, 333, 334
– –, nutrient relations of, 137, 290
– –, photosynthesis by, 294, 295, 297
– –, productivity of, 288, 293, 294, 367–370
– –, roots of, 239, 262
– –, water relations of, 97, 98, 290, 300
–, canopy structure of, 302, 303
–, decomposition of, 231, 273
–, growth of, 286, 289, 290
–, herbivore relations of, 354, 355, 357
– in peat (*see also* peat, types of), 134, 171, 174, 324
– in vegetation (*see also* swamps, types of), 22, 23, 25, 97, 140, 242, 287, 334, 367
– in vegetational succession, 13, 325–329, 336
–, nutrient relations of, 137, 239, 290, 334
– on lake margins, 334, 374
–, photosynthesis of, 270, 271, 294, 295, 297, 302, 304, 306
–, productivity of, 288, 294, 295, 335
–, roots of, 262
–, standing crop of, 290, 293
–, water relations of, 97, 98, 300
Phragmition, 21
phyllite as source of run-off water, 9
physiognomy of vegetation as basis for classification, 10
phytocoenoses as classification units, 11
phytogeography, 40
phytoplankton, 354
phytosociology, 9
Piceo-Vaccinietum, 73

Piche evaporimeter, 75, 78, 86
Piemonte [Piedmont] (Italy), 20
piezometer (*see also* seepage tube), 99, 104, 124
pigmy shrew (*Sorex minutus*), 358
pine (*see also Pinus in Systematic Index*)
- litter, decomposition of, 211
- forests or woods, 13, 72, 77, 93, 316
Pino-rotundatae–Sphagnetum, 78, 87
Pinsk (Belorussiya, U.S.S.R.), 91, 137, 140
Pinus–Sphagnum forest, 93, 94
pioneer species (*see also* colonization), 325, 326
- trees and shrubs, 335
pipe flow, 70, 71, 120, 137, 144, 147
pipit, meadow (*Anthus pratensis*), 376, 377
Pisa River (Poland), 21
Planhochmoore (*see also* raised bogs, types of: plateau), 4
plankton, 368, 370
plant communitites (*see also* phytocoenoses; vegetation; *for particular communities named after component species, see those species in the Systematic Index, or the Association or Alliance name*)
– –, aquatic, 44
– – for classifying mire types, 4
- density changes used in estimating productivity, 286
- growth after flooding, 257
– – stimulated by grazing, 355
- habit modified by flooding, 269, 270
- production, *see* productivity, primary
plant-suckering insects, 376
plastic deformation of peat, *see* peat creep
plateau raised bogs, *see* raised bogs, types of: plateau
plaur, 28
Pleistocene, 47, 360
Pliocene, 360
plum (*Prunus domestica*), 260
Plynlimon (Wales), 127, 137, 138
pneumatophores, 263–265
pneumostome, 350
Po Valley (Italy), 20
podzols, 19, 51, 159
–, paludification on, 13
podzolization, 315, 342
Pohjoisneva (Finland), 95
Poiseuille's Law, 81
Pohjanmaa (Finland), 43
Poland (*see also particular localities*), 6, 9, 21, 22, 113, 171, 288, 315, 325, 336, 361
Poles'ye (Ukraina and Belorussiya, U.S.S.R.), 4, 6, 8, 14, 72, 91, 93, 110, 137, 318–320
Polish words for mire concepts, 4, 28
pollen analysis, 14, 19, 35, 36, 38, 44, 47, 49, 50, 59, 120, 331, 339, 340
pollution (*see also* eutrophication), 261
polymorphism in flight muscles of corixids, 353
- in salamanders, 351, 352
Polytricheto-Salicetum, 87
polyuronic acids in *Sphagnum*, 337
Pomerania (Poland), 21
pool(s), 17, 72, 77, 133, 138, 139, 147, 313, 314, 341
- and tussock structure in carrs, 336
- complex, 338
- distribution in mires (*see also* hummock-and-hollow pattern; ridge-and-pool pattern), 14, 314
–, fauna of, 353, 354
–, permanent (*see also* dubh lochan), 138
poor fen, 27, 29, 172, 236, 316
population dynamics in plants: relevance to productivity estimations, 286
- cycles in animals, 354
- of animals, *see* animal densities
- turnover rates, 380
Postglacial period, 36–40, 45, 47, 51, 56, 59, 316
Potamion eurosibiricum, 21
potassium, 9, 25, 164, 178, 180, 181, 183, 185–189, 193–195, 229–231, 334, 337
potato (*Solanum tuberosum*), 275
potential water surplus, 73
potholes, 98
prairie, 301
–, wet, 227
Preboreal period, 39, 58
precipitation (*see also* ombrotrophy; rainfall; snow), 70–72, 96, 108, 109, 114, 115, 128, 132, 136, 140–142, 145, 148
–, composition of, 180, 182, 188, 193, 195, 312, 343
- determining equilibrium height of raised bogs, 14–17, 316, 317
- gauges, 99
–, increase in, controlling *Sphagnum* growth rate, 55
–, measurement of, *see* rain gauges
–, nitrogen in, 226, 242
–, occult, 72
–: relation to height of raised bogs, 14–17
–, variation in, affecting bog development, 56
predators (*see also* carnivores), 357, 358, 368, 372
predator/prey interaction, 376, 377
pressure bomb technique, 84, 85
- bulb, Barnesbury, 105
–, hydrostatic, 100
–, water vapour, 74, 77
Pripyat' [Pripet] Marshes (Belorussiya and Ukraina, U.S.S.R.), *see* Poles'ye
- River (Belorussiya, U.S.S.R.) 72, 137
productivity of littoral vegetation, 22
- of wetlands, 25, 285
–, primary, 1, 184, 196, 197, 236, 285–306, 332, 335, 375, 378
–, –, below-ground, 286, 288, 289, 291–293, 306, 377
– –: methods of measurement, 285–287
–, secondary, 1, 367–380
productivity/biomass ratio, 291, 373, 376, 377, 379, 380
proline as end-product of glycolysis, 277, 278
propane evolution from bogs, 198
propanol in peat, 250
propionic acid, 278
protected areas, *see* conservation; nature reserves
protein food from *Eichhornia* cultivation, 25
- in peat, 177, 179, 251
proteolysis, 226, 239
Protozoa, 198
psychrometers, 91
psychrometric constant, 77

Ptich' (Belorussiya, U.S.S.R.), 227, 242
pupae as resting stage of wetland insects, 352
pyranometers, 91
Pyrénées (France), 20
pyrite in soils, 249, 254
pyruvate kinase, 277
pyruvic acid, 278

$Q_{CO_2}{}^{N_2}$, *see* glycolysis, rate of
Q_{O_2}, *see* oxygen demand
quantum requirement of photosynthesis, 292
quartzite as source of run-off water, 9
– as substrate for acidophilic vegetation, 47
Quaternary, 26
Quebec (Canada), 288
quick-flow, 128

radiation, 73, 98, 109, 142, 292
– balance, 301
– interception (*see also* leaf area index), 302
–, net, 74, 91
–, photosynthetically active [PAR], 292, 294, 296
–, short-wave, 74, 94
radiocarbon content in atmosphere, 38
– dating, 18, 19, 35, 36, 52, 200, 203, 318, 331, 332, 342
– residence times in soils, 51
radiometers, 93
radish, 278
rail(s), 358
rain (*see also* precipitation)
– as source of nutrients for mires, *see* ombrotrophy
– gauges, 72, 76, 99, 116, 142
– storms, 115, 128, 130, 143, 149
rainfall (*see also* precipitation)
–, changes in, during Postglacial period, 36
–, effective, 14–17, 44, 72
rain-water, ionic composition of, 8, 9
raised bog(s) (*see also Hochmoore*)
– –, animals in, 356
– –, classification of, 4, 7
– –, convexity of, 14, 16, 317
– –, cupola of, 317
– –, decomposition in, 231
– –, definition of, 27, 172
– –, development of, 14, 27, 28, 56
– –, evapotranspiration from, 88–91
– –, geographical distribution of, 6, 20, 21
– –, height of, 14–17, 316, 317
– –, hydrograph of, 138, 141
– –, hydrology of (*see also* – –, water), 67, 71, 73, 107, 145, 146, 148
– – in Austria, 22
– – in Belgium, 20
– – in Canada, 19
– – in Czechoslovakia, 22
– – in Denmark, 20
– – in Germany, 20, 21
– – in Great Britain, 6, 236
– – in Hawaii, 26
– – in Ireland, 13
– – in New Zealand, 102
– – in Poland, 21, 22
– – in Romania, 23
– – in Sweden, 4–6, 14, 24, 148, 316–318
– –, limiting height of, 317
– –, marginal slope of, *see* rand
– –, microrelief in (*see also* hummock-and-hollow pattern), 57
– –, oligocentral and oligoperipheral development of, 119
– –, ombrotrophy of, 71, 73
– –, peat composition in, 177
– –, shape of, 6, 15–17
– –, size of, 15, 148, 149
– –, stability of, 136
– –, surface gradients in, 136
– –: transition to *aapa* fens in Finland, 52, 53
– –, types of, 4, 6, 7
– –, – –: Atlantic raised bogs, 4
– –, – –: Baltic or concentric raised bogs, 4, 6, 7, 338
– –, – –: continental raised bogs, 4, 67
– –, – –: eccentric raised bogs, 6, 7, 120, 170, 338
– –, – –: oceanic raised bogs, 141
– –, – –: plateau raised bogs, 6, 7
– –, – –: ridge raised bogs, 6
– –, – –: saddle raised bogs, 119, 120, 130, 132, 139
– –, – –: upland raised bogs, 4
– –, vegetation of, 102
– –, water discharge from, 120, 123, 126, 131, 133, 136, 138, 141, 143, 145
– –, water storage in, 100, 114–116
– –, water-table changes in, 107
Ramna (Sweden), 174, 176, 177, 187, 188
rand of raised bogs, 4, 6, 13, 15, 24, 72, 109, 110, 123, 135
– – – –, definition of, 27
Rannoch Moor (Scotland), 127, 266
recharge of mires, 70, 119
reclamation, *see* peatlands, reclamation of
recurrence surfaces, 14, 16, 24, 29, 44, 56, 174, 188, 313, 318, 319, 339
red algae (Rhodophyta), 128, 383
Red Lake Peatland (Minn., U.S.A.), 339
redox potential [Eh], 18, 189–191, 242, 247, 248, 252, 253, 357
– –, factors affecting, 18
reed(s) (*see also* Phragmites), 22, 304, 305, 351, 354, 355, 370
– warbler (*Acrocephalus scirpaeus*), 372
reed-bed(s) (*see also* reed-swamps), 349, 357, 358, 360, 370, 372
– harvesting affecting animal populations, 370
reed-grass, *see Calamagrostis*, *Glyceria maxima in Systematic Index*
reed-mace, *see Typha in Systematic Index*
reed-swamps, 12
–, animals in, 367, 369, 370, 372–374
–, evapotranspiration in, 98
–, floating, 28
– in Czechoslovakia, 22, 23
– in Japan, 23
–, peat accumulation in, 103, 315
–: prehistory in Europe, 316
– requiring alternation of rich and poor nutrient supply, 334

reed-swamps (*continued*)

–, successional relations of, 322–331
–, zonation of, 325–327, 334
refugia, 360, 361
regeneration complexes in bogs, 13, 29, 336, 339
– cycles, 314, 315, 330
Regenerationskomplex, 314
Reisermoor, 27
rejuvenation of bogs, 318
– cycles, 339, 340
rekurrensyta [RY] (*see also* recurrence surface), 29, 44
relocation of mineral elements in mires, 183–196
reserves and protected areas, *see* conservation; nature reserves
reservoirs, survival of trees in, 261
resins, 173
respiration, 295, 373
– of wetland animals, 350
retardation layers, 44, 55, 56
Reykjavik (Iceland), 19
rhamnose in peat, 175
rheophilous mires, 8, 29, 137
Rhein-Graben area (Germany), 54
rhizomes, 279
rhinoceros, woolly (*Coelodonta antiquitatis*), 361
rhizosphere, 227, 232, 238, 241, 243, 356
Rhön hills (Germany), 78
Rhynchosporion albae, 21
rice (*Oryza sativa*), 25, 260, 266, 267, 274, 288, 293–295, 304, 305
– cultivation, 23
–, energy capture by, 294
–: ethanol effects, 273
– germination, 260
–: growth under flooding, 262
–, hypoxia effects on, 266
–, leaf area index of, 302
–, microclimate of, 304, 305
–, photosynthesis in, 296
–, productivity of, 288
–, radiation balance in, 301
– seedlings, 260
– soils (*see also* paddy soils), 251, 253, 254
rich fens, 27, 172, 316
ridges in mires, 28, 56, 95, 96, 101, 104, 134, 137, 147
ridge–flark microtopes, 147
ridge raised bogs, *see* raised bogs, types of: ridge raised bogs
ridge-and-hollow patterns, 27, 28, 338, 339, 342
Riesengebirge, *see* Krkonoše,
Riga (Latvia, U.S.S.R.), 91
rills, 120, 128, 129, 131, 132, 148
rimpi, 27
– fens, 95, 170
Ringinglow (England), 181
Riss glaciation, 46
river terraces as locations of fen and carr, 46, 48
riverine peat deposits, 23
– wetlands, 2, 45
road dust as source of nutrient inputs, 180
rock surfaces, paludification on, 13
Roer River (Belgium), 87
Romania, 22, 23
Roman period, 47
Romanov's evapotranspiration equation, 92, 97
root(s) (*see also* mycorrhizae; rhizosphere)
–, activity of, 183
–, adventitious, 262–264, 270, 279, 280
–: adaptation to lack of oxygen, 258
–, aeration of, 111, 157, 263, 264
–, aerenchyma in, 265, 267–269, 300
–, air spaces in, 267
–, alanine in, 276
–, anaerobiosis in, 265, 277
–, branching of, 264
–, buttress, 263
–: cortex sloughed off, 268
–, cyanogenetic glucosides in, 260
–, decay of, 230, 232–234
– depth, 72, 82, 94, 96, 234, 240
–, distribution of, 183, 262
– disturbance, 76
– endodermis, 300
– exudates, 229, 241
– growth, 269
– – affected by flooding, 265
– hairs, 234, 238, 240
– in peat formation, 232, 234, 235, 238
–, knee, 263
– meristems, anoxia in, 272
– morphology affected by flooding, 265
–, mucilaginous sheaths in, 83
– nodules (*see also* nitrogen fixation), 229, 290
–, nutrient uptake by, 183
–, oxygen reserves in, 267, 268
–, – transport in, 258, 264
–, – uptake by, 272
–, penetration by, 18, 235, 240, 241
– production, *see* below-ground component of biomass and productivity
– regeneration restricted by flooding, 262
– response to aeration, 257, 335
–, seasonal dimorphism of, 262
–, stilt, 264
–, structure of, 232, 233, 240
–, – –, affected by waterlogging, 258, 259
–: suberization in, 264
–: substances accumulating under anaerobiosis, 278
–, surface area of, 268
–: surface roots permitting diffusion of ethanol, 262
– systems, 183
– – of *Empodisma*, 102
root/shoot ratio (*see also* below-ground/above-ground ratio), 300
roughness, surface, 82, 303
Roydon Common (England), 335
Rubellion, 13
Rubello-Fuscion, 13
rubidium as tracer, 83
ruff (*Philomachus pugnax*), 352

Rukwa Valley (Tanzania), 359
Rüllen, see rills
run-off, 96, 108, 115, 118, 142, 144, 147
– causing erosion, 321
–, changes in, during Postglacial period, 36
– from bogs, 16, 17, 318
– water, composition of, 8, 9
run-on, 71
Rur River, *see* Roer
rushes, 114, 261, 262, 270
Russia (*see also* U.S.S.R., *and particular regions and localities*), 36, 37, 44
Russian words for mire concepts, 4, 20, 27, 28
rustir, 19
RY [*rekurrensyta*] (*see also* recurrence surfaces), 44
ryam, 109, 114, 116
rye, 203
ryggmosse type of bog vegetation, 13

Sabana de Bogotá (Colombia), 37
Sahara, 352
Sahel, 352
salamanders, 351
salt marshes, 285, 361, 363
sampling methods affecting mire classification, 16
sapric material in soils, 217
Saprists, 218
saprovores, 375, 377, 378
Sarawak (*see also* Borneo), 159, 161
Saskatchewan (Canada), 301
Satakunta (Finland), 58
sawfly, 336
sawgrass, *see Cladium in Systematic Index*
Sázava River (Czechoslovakia), 86
Scandinavia (*see also* Denmark; Fennoscandia; Norway; Sweden; *and particular regions and localities*), 6, 171, 172, 181
scavengers, 369
schist, 9
schizogeny of root air-spaces, 265
Schleswig-Holstein (Germany), 21
Schoenetum, 10
Schoonebeek (Netherlands), 19
Schwarzes Moor (Germany), 78
Schwarzwald (Germany), 20, 54, 55
Schwingmoore (*see also* floating mats), 28, 111, 170, 185
Scirpeto-Callunetum, see Trichophoreto-Callunetum
Scirpeto-Eriophoretum, see Trichophoreto-Eriophoretum
Scirpetum, 10, 11
sclerenchyma in mire plants, 298
Scotland (*see also* British Isles; Great Britain; *and particular localities*),
–, blanket bogs in, 10, 48
–, climatic changes in, 37, 38
–, fens in, 11
–, Highlands of, 11, 12
–, midland valley of, 7
–, peat composition in, 187
–, peatland area of, 159
–, raised bogs in, 149
–, wetlands in, animals in, 38, 377
–, – –: botanical composition of vegetation, 10–12, 266
–, – –: hydrology, 109, 111, 124, 125, 127, 136, 138, 139, 143, 145, 147
–, – –: nutrient relations, 335
–, – –: precipitation needed for ombrotrophic development, 73
–, – –, seed population in, 275
–, – –, succession in, 326, 330
–, – –, surface patterning of, 339, 340
–, – –: water table related to vegetation type, 312
–, – –, zonation of, 326
screes in mire formation, 47
sea-spray as source of nutrient inputs, 180
sea-bird droppings supplying nutrients to island mires, 25
seals, 258
sea-level changes, 47
– – causing paludification, 46
seasonal changes in environment for wetland animals, 350
– fresh swamp(s), 2, 25
– swamp savannah, 353
seasonally flooded meadows, 25
sedge(s) (Cyperaceae) (*see also* graminoids; peat, types of; *and particular groups and species*)
–, anatomy of, 265, 270
– as colonizers (*see also* floating mats or rafts of vegetation), 13
– as habitat for animals, 358
– communities (*see also* bogs, types of), 328
– –, floating, *see* floating mats or rafts of vegetation
– –: grass–sedge meadows, 290
– –: sedge fens, 23, 97, 137
– –: sedge–*Hypnum* microtopes, 94
– –: sedge meadows, 292, 293, 327
– –: – microlandscapes, 114
– –: – mires, 27
– defining cover classes in mires, 11
– peat, *see* peat, types of
– removal following coypu invasion, 354
–, tussock, 322, 325
– warbler (*Acrocephalus schoenobaenus*), 372
sediment in rivers initiating mire formation, 48
sedimentary peat, 28
sedimentation, 373
seed(s), 272–274, 278
–, anoxia effects on, 274
– burial, effect on viability of, 274–276
– dormancy, 274
– germination, 260
– injured by soaking, 274
–, temperature effects on, 274
seedling establishment, 334–336
– tolerance of flooding, 335
– – – shading, 336
seepage (*see also* groundwater discharge; water through-flow), 70, 71, 120–127, 129, 133, 135–137, 143
– in acrotelm, 120, 123
– – catotelm, 120, 147
–, theory of, 119–123
– tubes (*see also* infiltration pipes), 117, 124, 126, 127, 135

semi-terrestric communities, 313, 316, 322, 343
– peat, 313, 322, 323
senescence of plant tissues, 229
sensible heat flux, 74, 77, 98, 142, 301, 303, 304
seral development, *see* succession
serine, 278
sewage as source of eutrophication, 373
shading by mountains affecting mire development, 22
– – parent trees restricting seedling establishment, 336
Shapwick Heath (England), 171
sheep (*Ovis ammon aries*), 374, 375
Shiel, Loch (Scotland), 139
shikimic acid, 278, 279
Shirinskoye Bog (U.S.S.R.), 80, 101
shoot emergence from subterranean buds, 335
– longevity, 286
shrew, pigmy (*Sorex minutus*), 358
–, water (*Neomys fodiens*), 358
Shropshire (England), 325
shrub(s) in mires defining cover class, 11
Shuvalovo Bog (U.S.S.R.), 52, 58
Siberia (*see also* U.S.S.R. *and particular regions and localities*), 36, 37, 39, 44, 51–53, 114, 116
siderite, 187, 249
Signy Island (Antarctica), 196, 228, 230
silica, 185–187, 254
silkworm, 276, 280
silting in wetland development, 328, 330, 334
silver blackening as evidence of sulphide, 189, 190, 241, 242, 248
similarity principle in micrometeorology, 304
simulation, *see* computer simulation; models
sink holes as locations for floating mats, 28
– strength, 74
sitatunga (*Limnotragus spekii*), 350
Skagershultsmosse, 339
– type of bog vegetation, 13
Skastmose (Denmark), 20
skimmers (*Rhynchops*), 358
slate as source of run-off water, 9
slope aspect of mire surface, 111, 123, 135, 136, 139
Slovakia (*see also* Czechoslovakia), 22
slugs, 356
Småland (Sweden), 317
smoke as source of inputs, 180
snails, 261, 350, 357, 358, 368, 373
snake, grass (*Natrix natrix*), 358
Snöromsmosse (Sweden), 318
snow, 73, 140
– drift, 99
–, evaporation from, 92, 99
– melt, 70, 110, 128–130, 132, 136, 141, 142
soak, 28
sod, 28
sodium, 9, 164, 178, 180, 181, 183, 185–187, 193–195, 231
soil(s) (*see also* peat *and particular soil types*)
–, aeration of, 146
– animals, 377, 378
–, calcareous, 249
–, classification of, 171
–, – –: Seventh Approximation, 171
– conditioners, 2
– dust as source of inorganic constituents in peat, 180
– mesofauna, 377
– moisture, 100
–, organic, *see* peat
– pans, 51, 319
– permeability as factor in mire formation, 14, 51
– pH, 4, 378
– pipes, 127, 128, 132, 136
– profiles, 104, 115
– resistance to water flow (*see also* hydraulic conductivity), 299
–, sodic, 249
– temperature, 247
– terms defined, 28
soligenous water supplies in mires, 8, 11, 26, 29, 40, 47
solar energy conversion efficiency, 294
– –, use by vegetation, 292, 293
Solling Mountains (Germany), 132
Somerset (England), 171, 339
Sonnenburger Moor (Germany), 52
soo, 27
soot as atmospheric contaminant, 343
– in peat, 170, 181
sorbitol, 278
South America (*see also particular territories and localities*), 37, 99
South Australia, 368
South Carolina (U.S.A.), 286, 288, 289
South Dakota (U.S.A.), 289
South Georgia, 25
South Orkney Islands, 25, 228, 230
South Shetland Islands, 25
Southern Hemisphere, 2, 77, 102, 287
Southport (England), 142
Soviet Union, *see* Union of Soviet Socialist Republics (U.S.S.R.) *and particular regions and localities*
soybean (*Glycine max*), 294
Spain, 159
spatial heterogeneity, *see* pattern
specific conductance of mire recharge water, 71
– diversity, *see* diversity
– surface of peat, 101
– yield of water, 68, 76, 89, 112, 113, 115, 116
spectral reflection coefficient, *see* albedo
spectrophotometry of peat colour, 54
Sphagnetum, 10, 78, 79
– *magellanici*, 78
– *medii*, 57
Sphagnofibrists, 217
Sphagnion fuscae, 22
sphagnol [phenolic compound in *Sphagnum*], 235
Sphagnum (bog-moss) (*see also particular Sphagnum species, and also in Systematic Index*)
–, acidification by, 164, 337
–, anatomy of, 78–81, 87, 102, 103, 161
–, animals associated with, 351, 376–378
– as pioneer, 334
–, blue-green algae associated with, 181

Sphagnum (continued)

- capitulum, 78–81
–, cation exchange capacity of, 164
–, composition of, 173–175, 177, 180, 235, 250
–, decay of, 17, 201, 203, 205, 231, 235–237, 251
–: disturbance by erosion, 140
–, geographical distribution of, 25, 26
– in peat (*see also* peat, types of), 28, 41, 43, 49, 50, 160
– in vegetation (*see also* bogs, types of; mires, types of), 20, 42, 58, 77, 78, 86, 88, 93, 94, 133, 134, 136, 138, 141, 143, 180, 228, 241, 242, 275, 287, 290, 300, 301, 303, 313, 316, 334, 336, 338
– in vegetational succession, 326–328, 330, 331, 334, 336, 338
–, nutrient relations of, 184, 185, 196, 197, 251, 337
–, photosynthesis in, 296
–, productivity of, 197
–, temperatures in, 301
–, translocation in, 183
–, water relations of, 80–84, 86–88, 91, 93, 94, 96, 102, 103, 111, 113, 138, 142
Sphagnum capillifolium, decay of, 17, 205, 236
– – in peat, 13, 169, 325
– – in vegetation, 11–13
– –, pH relations of, 337
– –, productivity of, 291
– –, structure of, 78, 79, 81, 82
– –, trophic relations of, 13
– –, water relations of, 78, 81, 82, 87, 111, 197
Sphagnum cuspidatum, decay of, 205, 236
– – in peat (*see also* peat, types of), 19, 55, 124, 169, 325
– – in vegetation, 13, 57, 313, 327
– – microtopes, 78
– –, structure of, 78, 79
– –, trophic relations of, 341
– –, water relations of, 82, 86, 94, 95, 133, 333, 341
Sphagnum fuscum, decay of, 230
– – in peat (*see also* peat, types of), 13, 165, 167, 169, 170, 174, 175, 325
– – in vegetation, 13, 23, 58, 183, 338
– –, nutrients in, 338
– –, structure of, 81
– –, trophic relations of, 13
– –, water relations of, 78, 81, 95, 102, 103, 111
Sphagnum magellanicum in peat (*see also* peat, types of), 19, 123, 124, 170, 195
– – in vegetation, 13, 23, 314
– –, nutrients in, 192
– –, pH relations of, 337
– –, productivity of, 192
– –, structure of, 78, 79, 81, 82
– –, water relations of, 78, 81, 82, 86, 87, 111, 197
Sphagnum papillosum, decay of, 202, 205, 236
– – in peat, 19, 169, 324
– – in vegetation, 11–13, 23, 24
– –, productivity of, 291
– –, structure of, 81, 236, 237
– –, translocation in, 183
– –, water relations of, 81, 95, 197
Sphagnum recurvum, animals in, 378
– –, decay of, 204, 205
– – in peat, 19, 169, 174
– – in vegetation, 13, 327
– –, pH relations of, 337
– –, productivity of, 291
– –, structure of, 78, 79, 81
– –, water relations of, 82, 86
Sphagnum–shrub microlandscapes, 94, 115
spiders (Araneae), 350, 353, 376, 378, 379
spike-rush (*Eleocharis*), 25, 288
Spodosols, 217
spoonbill(s), 358
sporophores of mycorrhizal fungi, 238, 239
springs, 22
springtails (Collembola), 354, 356, 358, 378, 379
spruce, *see* Picea *in Systematic Index*
spruce muskeg, 29
Sri Lanka [Ceylon], 349, 353
stability of mire ecosystems, 136, 138
standing crop, *see* biomass
standing-dead litter, 17, 302
Steiermark (Austria), 22
Steinhuder Meer (Germany), 71
stem borers among herbivores, 356, 372
– structure, 270, 280
– suckers among herbivores, 356, 372
– ventilation in flooding, 269, 270
steppe vegetation during Postglacial period in Europe, 38, 41
Steppenheide theory, 41
Stillstandkomplex, 44, 314
Stockholm (Sweden), 318
stomata, 85, 93, 97, 98, 300
stone-flies (Plecoptera), 358
storage of nutrients in perennial plants, 229, 230
Stordalen (Sweden), 71, 164, 227, 230
Store Mosse (Sweden), 194
Store Vildmose (Denmark), 20, 104
storks, 358
storms, *see* rain storms
Strangmoore (*see also* string bogs), 28
Stränge (*see also* ridges), 28
strängmyr, 28
stratigraphy, 13
stream(s), 69, 128, 129, 142, 148
– capture, 319, 320
– discharge increased by deforestation, 337
–, downcutting by, 319
–, headward erosion of, 319
stream-gauging structures, 131
streamlines (*see also* flow lines), 146, 147
strings, *see* ridges
string bogs or mires (*see also* pattern), 6, 28, 170–172
– fen, 95
Styria, *see* Steiermark
subantarctic peats, 160, 171
Subarctic zone, 45
Subatlantic period, 47, 54, 58
Subboreal period, 38, 39, 47, 58
suberization in roots, 264
subsidence, 320, 343

substance volume, *see* peat, porosity of
succession (*see also* climax; hummock-and-hollow regeneration hypothesis), 13, 40, 44, 45, 289, 311–345
– affecting conclusions from pollen analysis, 38
– deduced from vegetation zonation, 325
–, hydroseral, 311, 321, 325–330
– reflected in peat profiles, 36
succinic acid, 250, 278
Sudan, 28
sudd, 28
Sudety Mountains (Czechoslovakia and Poland), 21
suffosion, 127
sugar beet, 294
sugars in peat, 173
Sulfihemists, 217
Sulfohemists, 217
sulphates, 53, 190, 242
sulphides, 268
–, detection of, 189, 190
– in mires, 165, 166, 182, 187, 189–192
– – –, formation of, 192, 241, 242, 248, 250, 253
– in rice paddy soils, 254
–, solubility of, 191, 192, 194
sulphur cycle, 253
– dioxide as cause of acid rain, 181, 343
–, forms of (*see also* sulphates, sulphides), 247, 254
– in plant nutrition, 228, 230, 242, 243
–, oxidation states of, 191, 241, 242, 253
sulphuric acid accumulation in organic soils, 249
Šumava (Czechoslovakia), 22
Sumpf, 27
Sumpfwald, 27
Sunart (Scotland), 139
sundew, *see Drosera in Systematic Index*
suo, 27
Suomenselka (Finland), 43
superphosphate fertilizer effects, 205
surface flow, diffuse, 68
– roughness, 82, 303
– tension of water, 79, 94, 96, 101
surveying of wetlands, 343
survival of flooding by plants, 257–280
Susaká Interstadial, 37
suspended animation as adaptation of wetland life, 351, 362
swallow-tail butterfly (*Papilio machaon*), 352, 362
swamp(s) (*see also* mangal; mire(s); reed-swamps; wetlands)
–, animals of, 354
– as nutrient regulators, 374
– carr, 336
–, classification of (*see also* mires, classification of), 9
–: conversion to ombrogenous bogs, 52
–, definition of, 27
–, eutrophic, 22, 41
– forest, 23, 25, 159
– – mires, 29
– in river deltas, 367
–, minerotrophy of, 40
– mud, 322
–, productivity of, 285, 288, 289, 305, 306
–, riverine, 2, 45, 67, 354, 367
–, standing crop in, 293
–: tolerance to water-table variation, 312
–, tropical forest, 29
–, types of: alder swamps, 12
–, – –: papyrus swamps, 355
swan(s), 355
Swan River (Minn., U.S.A.), 113
Sweat Mere (England), 325, 326
Sweden (*see also particular regions and localities*)
–, aerial photography in, 6
–, climatic change in, 38
–, evapotranspiration in, 92
–, mires in, classification of, 10, 13
–, – –, decomposition in, 202, 230
–, – –, formation of, 14, 16
–, – –, frost effects in, 335
–, – –, geochemistry of, 71
–, – –, nitrogen fixation in, 227
–, – –, patterning in, 314
–, – –, productivity in, 197, 288, 290
–, – –, roots in, 183, 293
–, – –, water discharge from, 143
–, peat in, bulk density of, 214
–, – –, classification of, 171
–, – –, composition of, 163, 164, 174, 176, 177
–, – –, – –: profiles, 184–188, 190, 192, 194
–, – –, relocation of constituents in, 183
–, – –, stratigraphy of, 174, 176, 184–187, 314, 320
–, rainwater composition in, 9
–, raised bogs in, 4–6, 14, 24, 148, 316–318
Swedish words for mire concepts, 4, 27–29
Switzerland, 22, 146
symbiosis (*see also* mycorrhizae), 228, 229
symbols for hydrology, 68
synchro-devices for telemetry of mire observations, 106
systems ecology, 362, 363

tadpoles, 351
taiga, 375
Taisetsu, Mount (Japan), 24
Takadayachi Moor (Japan), 24
tallmosse , 29
tannin decomposition by mycorrhizae, 239
Tarmanskoye Fen (Siberia, U.S.S.R.), 116
Tartu (Estonia, U.S.S.R.), 91
Tasmania (Australia), 25, 42, 336
tectonic influence in paludification, 46, 320
Tees, River (England), 141
Teichkomplex, 314
telemetry of mire variables, 106
telluric recharge of mires, 70, 71, 119
telmatic communities, 313, 316, 322, 332–334
– peat, 29, 313, 322, 323
telmatology, 68
–, theories of, 145–149
temperature(s) above raised bogs, 78
– adaptations, 196, 197
– affecting flooding survival, 274, 275
– – plant growth rate, 304

temperature(s) (*continued*)

– affecting soil processes, 251
– – water uptake, 300
– and redox potential, 247
– changes during Postglacial period, 36–39, 56
– dependence of respiration, 295
– effects on developing muscles of corixids, 353
– limiting production in tundra, 291, 297
– of telluric recharge water, 71
– – water in rice fields, 301
–, soil, 301
tensiometers, 99, 100, 112, 114
tephra, definition of, 28
– layers in bogs, 17, 19
terrainbedeckende Moore (*see also* blanket bogs; raised bogs, types of: upland), 4, 7, 27
terrestrialization, 12–14, 19, 44, 45, 148, 311, 322–324, 332–334, 342
–, definition of, 29
terrestric communities, 313, 316, 343
– peat, 29, 313
territory size of grouse, 356
Tertiary Era, *see* Cenozoic
tertiary valley mires, 6, 7
Texas (U.S.A.), 289
thatching material from wetlands, 322, 331, 336
thawing in Late Glacial causing mire development, 45
thermal conductivity of soil, 51
Thornthwaite's method, 75, 88, 93
throughflow, 119, 130, 135, 143–145
Thüringer Wald (Germany), 72, 73
Thursley Common (England), 202
tides influencing mire development, 46
Tierra del Fuego (Argentina and Chile), 25, 72
tiger salamander (*Ambystoma tigrinum*), 351
till soils, paludification on, 14
timber harvesting, *see* forests, clearing of
– line, *see* tree line
time series analysis in hydrology, 109
Tisza River (Hungary), 23
titanium, 194
Tjeukemeer (Netherlands), 370
Tjörnarp (Sweden), 164
toadstools, 238, 239
tolerance of anaerobic conditions, 260
tomato (*Lycopersicum esculentum*), 278
Tomichevo (U.S.S.R.), 93, 94
Tomsk (Siberia, U.S.S.R.), 51
Tooma (Estonia, U.S.S.R.), 91, 94, 95
topogenous water supplies and mires, 8, 19, 29, 40, 41, 52
topography affecting mire development, 311
Torf, 28
Torneträsk, Lake (Sweden), 73
tortoise, European pond (*Emys orbicularis*), 358
torv, 28
tourbe, 28
tourbière, 27
toxin(s) affecting decomposition rates, 232, 236, 241
– formation under anaerobic conditions, 247, 255, 258, 262, 268, 271
– removal by flowing water, 8, 262–264
trace elements affecting *Rhizobium* activity, 229
Traneröds Mosse (Sweden), 187
translocation of nutrients, 229, 230
– of organic materials, 294
transpiration, 73, 83, 93, 97, 271
–: control by stomata, 98
–: diurnal pattern, 106
–, stand, 86
transport of wetland animals, 353
Transylvanian Alps (Romania), 23
travel time, 130, 142
Třeboň (Czechoslovakia), 23
tree(s), 98, 118
– colonization as part of mire development, 13
–, death of, through waterlogging, 336
– in mires determining cover classes, 11
– line, changes in, 39, 56
Tregaron Bog (Wales), 17, 202
Trichophoreto-Callunetum, 136
Trichophoreto-Eriophoretum caricetosum, 10, 136
Trichophoreto-Eriophoretum typicum, 10, 136, 137
Trifolio-Festucetum rubrae, 86
Tristan de Cunha, 25
Trondheim Fjord (Norway), 18
tropical forest swamp, 29
– peats, 159, 161
Tropofibrists, 217
Tropohemists, 217
Troposaprists, 218
trypanosomes, 276
tubers, 278
tularaemia, 354
tulip tree (*Liriodendron tulipifera*), 261
tundra, 229–231, 242, 285, 290–294, 297, 301, 302, 306, 352, 358, 360, 361
turbulent diffusion, coefficient of, 287
– transfer, 303, 304, 306
– –, coefficient of, 304, 305
turf, 28
turloughs, 357
turtle, 278, 280
turvas, 28
turve, 28

Übergangsmoore [transition bogs], 29, 172
Uganda, 242
Ukraina (U.S.S.R.), 14
Ulfisols, 217
Ummeln (Germany), 107
Union of Soviet Socialist Republics [U.S.S.R.] (*see also particular regions and localities*)
– – – – –, evapotranspiration in, 75, 77, 93, 96
– – – – –, mires in: *aapa* mires, 6
– – – – –, – –: Arctic mires, 7
– – – – –, – –: formation, 14
– – – – –, – –: geographical distribution, 6, 72
– – – – –, – –: hydrology, 110

Union of Soviet Socialist Republics (*continued*)

–––––, mires in: productivity, 290, 291
–––––, ––: recurrence surfaces, 43, 318
–––––, ––: research, 26, 67
–––––, ––: root biomass, 293
–––––, peat classification in, 171
–––––, – profiles in, 80
–––––, – stratigraphy in, 58
–––––, raised bogs in, 4, 6, 27
–––––: State Hydrological Institute, 91
–––––, vegetation surveys in, 11
–––––, water-table recording methods in, 105
United States of America [U.S.A.] (*see also* Alaska; Minnesota; *and particular regions and localities*)
––––, evapotranspiration estimates in, 75
––––, organic soil classification in, 217, 218
––––, peat classification in, 171, 217, 218
––––, – composition in, 177
––––, – pH in, 248
––––, wetlands in, hydrology of, 115, 129
––––, ––, productivity of, 290, 335
Untersee (Germany and Switzerland), 374
up-slope mire formation, 51
Ural Mountains (U.S.S.R.), 6
uronic acids in peat, 164, 175
––––: relation to cation exchange, 164, 175
Utrecht (Netherlands), 19, 20, 238

Vaasa (Finland), 95, 96
Vaccinio-uliginosi–Pinetum, 78
valeric acid in peat, 250
valley bogs (*see also* headwater bogs), 146, 172
Van der Casteel diagram, 105, 106
vapour diffusion resistance, 83–85, 99
Vassijaure (Sweden), 186
Vasyugan bog (Siberia, U.S.S.R.), 53
vattenbana, 28
veen, 27
vegetation, classification of (*see also* bogs, classification of; mires, classification of),
–, ––: UNESCO system, 11
– cycles in peat bogs, 314
– for classification of mires, 9, 10
– mosaics, *see* pattern
– ordination, 180
–, structure of, 9, 10
Venezia [Venice] (Italy), 20
Verlandung (*see also* terrestrialization), 29, 311, 313
Verlandungshypothese, 313, 315, 341, 342
Vermont (U.S.A.), 323
Versumpfung (*see also* paludification), 29
vertebrates in mires, 349–356, 358, 361, 368, 372, 375, 376
–, metabolic products in, 278
vertical shafts in mires, 137
Vertisols, 217
Vienna (Austria), 22
vivianite, 187
volcanic ash in mires, 17, 19, 24, 51, 159, 180
volcanoes, 24
voles, 361
Von Post scale of humification, 16, 161, 162
Vosges (France), 20

Wachstumskomplex, 44
waders, 353
Wahiawa bog (Hawaii, U.S.A.), 26
Waldhochmoore (*see also* raised bogs, types of: continental raised bogs), 4, 7, 29
Wales (*see also* British Isles; Great Britain; *and particular regions and localities*), 17, 48, 73, 127, 137, 202, 290, 325
warbler(s), 358
–, reed (*Acrocephalus scirpaceus*), 372
–, sedge (*Acrocephalus schoenobaenus*), 372
Warta River (Poland), 21
Washington [state] (U.S.A.), 248, 323, 333
Wasserbahn, 28
waste disposal in mires, 249, 250, 255
water (*see also* aridity; flooding; flow; hydrology; meltwater; precipitation; run-off)
– availability in wetlands, 298
– balance, 69, 70, 149, 150
–– affected by trees, 337
–– changes associated with water-table movement, 108, 111
–– equations, 69, 70, 76, 90, 99, 119, 145
–– for separate microtopes used in hydrological modelling of whole mires, 147
–– of raised bogs, 27
–– used to estimate evapotranspiration, 96
–, bound, 101, 135
– buffalo (*Bubalus bubalis*), 353
– chemistry, 8, 9, 25, 334, 335
– content of peat, 100, 103
– detention, 115
– discharge, 69, 70, 76, 105, 109, 130–145
––, climatic effects on, 140–145
– flow lines in raised bogs, 8, 315
– infiltration, 115
– isopotentials, 121–123
–: latent heat of vaporization, 74, 77
–, lateral flow of, 311, 315, 316, 338, 340
– leakage, 147
– level affecting animal communities, 367, 368
– meadows, 357
–, meteoric, 70–73, 148
–, moving, as factor in mire ecology, 263
–, pH of, 25
–, ponding of, 99
– potential, 83, 85, 97, 99, 111–114, 122, 125, 128, 138, 298–300, 306
– recharge, 69–72, 126, 145
–: relation to mires, 72, 73
– relations of mire plants, 298, 299
– reservoirs, survival of trees in, 261
– shrew (*Neomys fodiens*), 358
– stage, 103, 104, 106, 109, 142
– storage in mires, 69, 76, 95, 101–103, 143, 145, 147
––, anomalous (isopinal), 117, 126
–– in *Sphagnum*, 80, 81,
– stress in leaves, 298, 306

water (*continued*)

– stress in wetlands, 300
– supply for mire plants, 298
– – to mires as distinguishing main types, 72, 73, 172, 180, 185
– – – –, changes in, determining their history, 311
– – – –, chemical character of, 71
– surface, 106
–, surface tension of, 81, 94, 96, 101
–, telluric, 70–73, 140, 145
– tracks, 28, 120, 136
– transfer in *Sphagnum*, 80, 81
– upwelling in mires 71, 72
–, viscosity of, 121
– yield coefficient, 134
water-bodies, vegetation of, 21
waterfowl, 354, 355
– migrations, 352
water-level changes, 367, 368, 370
water-lilies, 278
waterlogging affecting animals, 356
– – plants (*see also* flooding), 8, 103
– as criterion for mires, 67
watersheds, *see* catchments; interfluves
water-table (*see also* drainage)
– affected by gauging structures, 129
– and soil aeration, 18
– and tree establishment, 335
–, capillary fringe and, 100
–, catotelm and, 311
–, changes in, 106–110, 138, 340, 341, 343
–, – –, autogenic, 314, 315
–, – –, caused by topographic change, 318–320
–, – –, diurnal, 106
–, – –, during storms, 115, 116
–, – –, related to surface patterning, 340
–, – –, seasonal, 106, 108, 111, 312
–, decomposition related to, 341
–, drainage effects on, 96, 107, 137, 144
–, evapotranspiration affected by, 93, 95–97
–, gradients in, 123, 135, 311
–, irregularity in surface of, 111, 149, 311
–, measurement and recording of, 103–106
–, oscillation of, 111, 149, 311, 316, 333
–, peat development and, 317, 341
–, perched, 315
–, plant growth related to, 261
–, prediction of, 109, 110
– recession, 93, 96, 99
–, specific yield and, 112–116
–, surface gradient and, 135
–: tolerance range for different wetland communities, 312
–, vegetation and, 110, 111, 261, 312, 343
–, water discharge related to, 143, 147
–, – potential related to, 114
–, – storage related to, 111–116
wave action: effect on animals, 372
wax, cuticular, 298
– in peat, 173, 178, 251
– produced by aphids, 173
webbed feet, adaptive value of, 349
weirs, stream-gauging, 76, 129, 141
Weissmoor, 20
West Africa (*see also particular territories and localities*), 352
West Malesia, 2
wet meadows, productivity of, 285
wetland(s) (*see also* bogs; marshes; mires; swamps; raised bogs), 27
–, cyclical charges in, 313–315
–, drainage of, 20
–, eutrophic, 21
–, lacustrine, 22
–, mesotrophic, 21
–, productivity of, 305, 306
–, riverine, 2, 45
–, utilization of (*see also* agriculture; grazing; peat, utilization of; peatlands, exploitation of; waste disposal), 2, 18, 20, 25, 321, 322, 331, 336, 374
whales, 258
White Sea [Beloye More] (U.S.S.R.), 6
Wicken Fen (England), 106, 110, 117, 227, 238, 242, 331, 335, 359
Wickman's equation, 68, 148
Wildseemoor (Germany), 55
Will o' the Wisp, 166
Willis's Piece (England), 339
willow, *see Salix in Systematic Index*
Wilpatta (Sri Lanka), 353
wilting following flooding, 270, 271
wind(s) (*see also* trade-winds), 74
– effects, 303–306
Windermere, Lake (England), 191
Wisconsin (U.S.A.), 9, 248, 252, 253, 288, 289, 323, 327
Witham River (England), 106
woodlands (*see also* forests), clearing of, 325
– as climax of hydrosere, 325–328, 337
–, deciduous, 325
Woodwalton Fen (England), 321
worms (*see also* Annelida; Nematoda *in Systematic Index*), 350, 355
World Meteorological Organization [WMO], 75
Wraymires (England), 195
Wümme River (Germany), 88
Würm glaciation, 46
Wybunbury Moss (England), 321
Wye River (Wales), 127
Wyoming (U.S.A.), 290

xeromorphy, 82, 83, 264, 271, 298, 300
xerophytes, 102
xylem, gas movement in, 269
– water potentials, 299, 300
xylose in peat, 173, 175

Yasel'da [Jasiolda] River (Belorussiya, U.S.S.R.), 320
yield coefficient, *see* water yield coefficient
–, specific, *see* specific yield
Yoppi River (Japan), 24

Yorkshire (England), 127, 376
Yorkshire Moors (England), 8
Younger Dryas period, 38

Zaïre, River, 368
Žd'ár (Czechoslovakia), 86, 87
Žd'árská Vrchy hills (Czechoslovakia), 86
Zehulanka River (Belorussiya, U.S.S.R.), 319, 320
Zelengorsk (U.S.S.R.), 91
zinc, 192–194, 249
zonation of vegetation around open water, 325–327, 330, 337
zones, climatic, 72
zoogeography of wetland animals, 359–362
Zwischenmoore, 172

PELLETIER LIBRARY, ALLEGHENY COLLEGE
3 3768 00665 4266

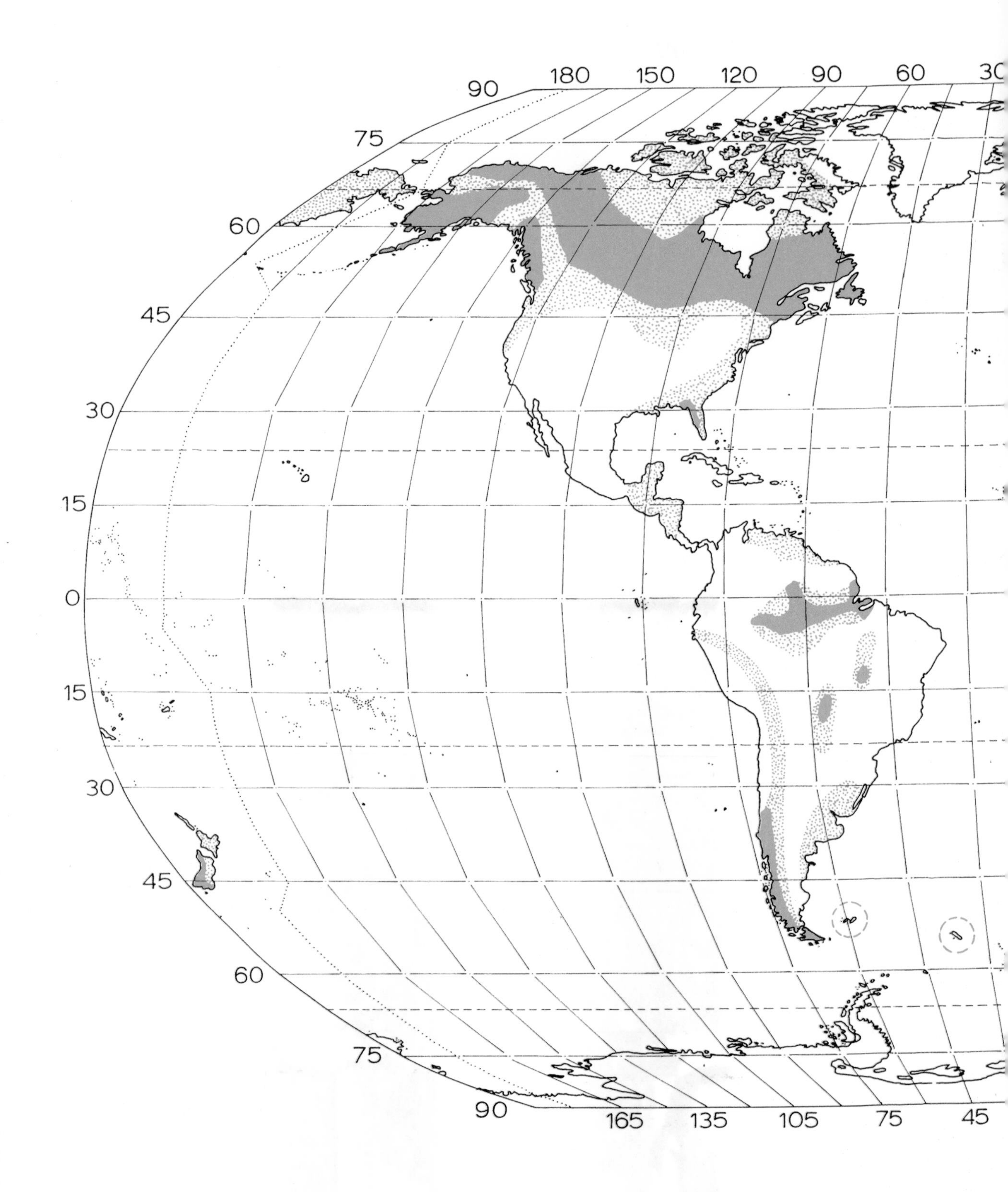
90
180
150
120
90
60
75
60
45
30
15
0
15
30
45
60
75
90
165
135
105
75
45